PERGAMON INTERNATIONAL LIBRARY
of Science, Technology, Engineering and Social Studies
The 1000-volume original paperback library in aid of education, industrial training and the enjoyment of leisure
Publisher: Robert Maxwell, M.C.

INTERNATIONAL SERIES ON
MATERIALS SCIENCE AND TECHNOLOGY
Editor: G. V. RAYNOR, M.A., D.PHIL., D.SC., F.R.S.

VOLUME 24

METALLURGICAL THERMOCHEMISTRY

FIFTH EDITION

Revised and Enlarged

International Series on
MATERIALS SCIENCE AND TECHNOLOGY
Volume 24—*Editor*: G. V. RAYNOR

METALLURGICAL THERMOCHEMISTRY

BY

O. KUBASCHEWSKI
DR. PHIL. HABIL. DRS. H.C.
Technische Hochschule, Aachen

AND

C. B. ALCOCK
Ph.D. A.R.C.S. D.Sc.
University of Toronto, Canada

FIFTH EDITION
Revised and Enlarged

PERGAMON PRESS
OXFORD · NEW YORK · TORONTO · SYDNEY · PARIS · FRANKFURT

U.K.	Pergamon Press Ltd., Headington Hill Hall, Oxford OX3 0BW, England
U.S.A.	Pergamon Press Inc., Maxwell House, Fairview Park, Elmsford, New York 10523, U.S.A.
CANADA	Pergamon of Canada, Suite 104, 150 Consumers Road, Willowdale, Ontario M2J 1P9, Canada
AUSTRALIA	Pergamon Press (Aust.) Pty. Ltd., P.O. Box 544, Potts Point, N.S.W. 2011, Australia
FRANCE	Pergamon Press SARL, 24 rue des Ecoles, 75240 Paris, Cedex 05, France
FEDERAL REPUBLIC OF GERMANY	Pergamon Press GmbH, Hammerweg 6, D-6242 Kronberg-Taunus, Federal Republic of Germany

1st Edition 1951
2nd Edition 1955
3rd Edition 1958
4th Edition 1967
Reprinted 1974
5th Edition 1979
Reprinted (with corrections) 1983

Library of Congress Cataloging in Publication Data

Kubaschewski, Oswald.
Metallurgical thermochemistry.
(International series on materials science and technology; v. 24)
Includes bibliographies and indexes.
1. Thermochemistry. 2. Metallurgy. I. Alcock, C. B., joint author. II. Title
QD511.K8 1977 541'.36 78-17832
ISBN 0-08-020897-5 (Hardcover)
ISBN 0-08-022107-6 (Flexicover)

Printed in Great Britain by A. Wheaton & Co., Ltd., Exeter

Contents

NOTATION ix

I. THE THEORETICAL BASIS 1

1. The mass action law 1
2. Thermodynamic functions 3
   A. The equilibrium constant and thermodynamic functions 3
   B. Heat content and heat of formation 4
   C. Temperature dependence of the heat of reaction 8
   D. Entropy 13
   E. Gibbs energy 17
3. Solutions 27
   A. Partial molar Gibbs energy 27
   B. Vapour pressures and partial Gibbs energies 30
   C. The calculation of integral values from partial molar thermochemical data 34
   D. Dilute solutions 45
      1. Raoult's law 45
      2. Henry's law 46
   E. Atomistics and solution thermodynamics 48
   F. Regular solutions 50
      1. Calculation of activities using equilibrium diagrams 50
      2. Spinodal decomposition 52
      3. Order–disorder transformation 53
   G. Non-regular solutions 54
   H. Ternary solutions 57

II. EXPERIMENTAL METHODS 65

1. Calorimetric methods 66
   A. Measurement of temperature 67
      1. Mercury-in-glass thermometers 67
      2. Platinum resistance thermometers 68
      3. Thermocouples 69
      4. Thermistors 69
      5. Optical pyrometers 70
   B. Determination of water equivalent 70
   C. Isoperibol calorimetry 71
      1. Principle of operation and evaluation of results 71
      2. The bomb calorimeter 74

3. Heats of solution 80
4. Heat contents 88
5. Other methods 92
6. Isoperibol high-temperature calorimetry 98
D. Isothermal calorimetry 101
1. The ice calorimeter 101
2. The diphenyl ether calorimeter 102
E. Adiabatic calorimetry 103
1. Heat capacities 103
2. Heats of reaction 105
F. Other calorimeters 108
1. High-speed methods 108
2. Cooling method 109
3. Tian–Calvet calorimeters 109
G. Heats of fusion and of transformation 111
H. Comparison of methods 113
2. Equilibria with a gaseous phase 114
A. Static methods for the measurement of vapour pressure 116
1. Static methods using radiation 117
B. Gas-condensed phase equilibria in a closed system 119
C. Dynamic vapour-pressure methods 125
1. The transportation method 127
D. Other heterogeneous equilibria 131
1. H_2–CH_4 131
2. H_2–NH_3 132
3. H_2–H_2O 132
4. H_2–H_2S 137
5. H_2–HCl 137
6. Other equilibria 138
E. Methods based on rates of evaporation: effusion methods 139
1. The Knudsen cell–mass spectrometer combination 143
2. Vapour transpiration 153
3. Torsion effusion 155
3. Electromotive forces 157
A. General 157
B. Solid electrolytes 159
1. Solid electrodes in oxygen concentration cells 160
2. Gas electrodes in oxygen concentration cells 165
3. Liquid electrodes 167
C. Other cells 168

III. THE ESTIMATION OF THERMOCHEMICAL DATA 179

1. Heat capacities 179
1. Monatomic gases 179
2. Diatomic gases 180
3. Polyatomic gases 181

4. Solids 181
5. Liquids 184
6. Some average values 185
2. Heats and entropies of transformation, fusion and evaporation 185
1. Evaporation 185
2. Fusion 186
3. Entropy and entropy changes 188
1. Solids 188
2. Gases 191
4. Heats of formation 192
1. General 193
2. Homologous series 195
3. Electronegativities 198
4. Volume change and heat of formation 199
5. Heats of solution 200
6. The packing effect 202
7. Temperature increase during compound formation 206
8. Examples of application 207

IV. EXAMPLES OF THERMOCHEMICAL TREATMENT OF METALLURGICAL PROBLEMS 210

1. Chill factors 211
2. Aluminothermic type reactions 212
A. The production of uranium by reduction of its fluoride with calcium and magnesium 212
B. The production of manganese and chromium by the aluminothermic process 213
3. The chlorination of metal oxides 214
4. The free evaporation of an oxide ceramic *in vacuo* 215
5. The reduction of magnesium oxide with silicon 217
6. Vapour phase transport of silicon carbide 218
7(a). Assessment of standard values 221
Silicon monoxide 221–226
7(b). Assessment of thermochemical values for a system exhibiting wide solution ranges 226
8. Calculation of metallurgical equilibrium diagrams 232
9. Ternary diagrams 235
10. Deoxidation of iron 236
A. Deoxidation with silicon 237
B. Deoxidation with aluminium 238
11. Refining of iron 239
12. Refining of lead 242
A. The removal of zinc from lead 244
13. The decarburisation of iron–chromium–carbon and iron–silicon–carbon liquid alloys 244
14. Metal–refractory interaction 247

15. The oxidation of alloys of platinum and of gold 252
16. The oxidation of iron–chromium alloys 254

V. THERMOCHEMICAL DATA. THE TABULATION OF DATA 258

1. Heats of formation, entropies and heats of fusion 258
2. Vapour pressures 259
3. Heat capacities 260
4. General 260
5. Gibbs energy calculations 262
6. Data for metallic solutions 264

TABLES 267

A. Heats of formation, standard entropies and structures 268
Aa. Structures 324
B. Heats and temperatures of transformation 326
C1. Heat capacities of the elements 336
C2. Heat capacities of the compounds 340
D. Vapour pressures and evaporation constants 358
E. Standard Gibbs energies of reaction 378
F. Thermochemical data of binary metallic systems 386

REFERENCES TO CHAPTERS I, II, III AND IV 414

REFERENCES TO CHAPTER V 423

SUBJECT INDEX 447

Notation

The system of notation used in this monograph is in general accordance with that recommended by the Symbols Committee of the Royal Society in 1951. The adherence is not rigid, however, and there has been no hesitation is using new symbols or elaborating those commonly accepted where it has been found necessary to clarify a theoretical discussion or to avoid confusion in computation. All the symbols used are listed in the following table for rapid reference. The fourth column in the table cites the page in which the symbol is introduced or its use explained.

Confusion often arises in thermochemical calculation with the sign used for the various thermochemical quantities. The convention used in this monograph is the prevalent one nowadays of regarding the heat of any reaction, ΔH, as the increase in heat content of the system, i.e. the heat absorbed. Thus the heat of an *exothermic* reaction is a *negative* quantity. The signs of the Gibbs energy and entropy are most conveniently deduced and remembered from the Gibbs–Helmholtz equation in the form $\Delta G = \Delta H - T\Delta S$.

Symbol	Meaning	Common Units or Value	Page
a	activity (Raoult)	—	31
c	concentration	—	2
c_p	specific heat at constant pressure	cal/g deg	
C_p	molar heat capacity	cal/mole deg	9
E	electromotive force (e.m.f.)	V	18
f	activity coefficient (Henry)	—	
$\mathbf{F}$	Faraday's constant	23,066 cal/V	8
G	free energy (Gibbs' function or thermodynamic potential)	cal	17
ΔG	free energy of formation or reaction	cal	3, 17
$\Delta \bar{G}$	partial molar free energy of solution	cal	29
γ	activity coefficient (Raoult)	—	31
H	heat content, enthalpy	cal	4
ΔH	heat of formation or reaction	cal	4
ΔH_{298}	heat of formation or reaction at 25°C	cal	4
$\Delta' H$	change in heat content due to change in temperature	cal	30
$\Delta \bar{H}$	partial molar heat of solution	cal/mole	2
$K_{p,c,a}$	equilibrium constant expressed in terms of pressures, concentrations, or activities (reaction products in numerator)		
κ	electrical conductivity	$ohm^{-1}\ cm^{-1}$	
ln	natural logarithm, $\log_e$	—	
log	$\log_{10}$	—	
$L_{t,f,e,s}$	latent heat of transformation, fusion, evaporation, sublimation	cal/mole	45
m	mass	g	
M	molecular weight	g	
μ	chemical potential	cal	29
n	number of moles	—	
N_1, N_2	mole fractions	—	

Symbol	Meaning	Common units or value	Page
N	Avogadro's number	$6{\cdot}02 \times 10^{23}$	
p	pressure	mmHg, or atm	
r	radius, also atomic or ionic radius	cm, or Å	
$\boldsymbol{R}$	gas constant	1·986 cal/deg	
$\boldsymbol{R} \ln 10$	—	4·574 cal/deg	
ϱ	density	g/cm^3	
S	entropy	cal/deg	14
S_{298}	standard entropy	cal/deg	15
ΔS	entropy of reaction	cal/deg	14
$\Delta \bar{S}_1$	partial molar entropy of solution	cal/deg mole	30
t	time	sec (min, hr)	
T	absolute temperature	deg°K	
θ	temperature	deg°C	
V	volume	cm^3, or litres	
W	water equivalent of a calorimeter	cal/K	
z	electrochemical valency		8

Brackets around Symbols Denoting Molar Quantities

( )	gaseous
{ }	liquid
⟨ ⟩	solid
[ ]	dissolved (suffix denoting medium)

The Unit of Heat

The unit of heat (enthalpy) used in this book is the calorie, that is the heat required to raise the temperature of one gramme of pure water from 14·5 to 15·5°C.

In place of the calorie, the Joule (= 0·239 cal) has recently become the internationally accepted unit. As experience shows, about one human generation is needed to get accustomed to such a profound change of unit. The authors do therefore *not* apologise for retaining the calorie in this book. In fact, nearly all important tabulations of thermodynamic data of chemical substances retain this unit to-date. Thus, the critical assessors of such data are the ones on whose shoulders the responsibility for eventually changing the physical units is loaded. Since they are dealing with thousands of values, in contrast to authors of original papers who are dealing with only a few, the change-over is likely to take some considerable time.

On the other hand, the present authors reading through the page proofs wish to apologise for the survival of faulty remnants from earlier editions. Apart from those persisting errors *gegen die kein Kraut gewachsen ist*, there remain inconsistencies which matter less, such as the dimensions of entropy being occasionally quoted as "Cl", "e.u.", "cal/°" and "cal/degree" instead of the desirable "cal/K". In the place of "free energy", "Gibbs energy" should now be used, but the authors may not always have succeeded in making the change.

If all these inconsistencies have not been removed—dear reader, please forgive.

CHAPTER I

The Theoretical Basis

1. The Mass Action Law

The problem which most commonly confronts the production metallurgist is that of the preparation of a pure metal by the reduction of one of its compounds; for example

$$Fe_2O_3 + 3C = 2Fe + 3CO$$

or to put the equation in a more general form,

$$M_aX_b + cR = aM + R_cX_b \tag{1}$$

where M represents the metal in question, R the reducing agent and X the radical to be removed. The yield of such a reaction depends firstly on the position of equilibrium between the reacting substances and secondly on the rate at which equilibrium is attained. This latter may be so slow that the yield is still far below the equilibrium after a lapse of time of hours, days, or even years. Simple thermodynamic considerations cannot be applied to such incomplete reactions, though they govern to a certain extent the force driving a given reaction toward equilibrium.

At the high temperatures often encountered in production metallurgy, reaction rates are usually so high that it may be assumed that equilibrium is attained during the process.

A numerical value for the equilibrium state of a reaction is given by the mass action law, first derived by Guldberg and Waage (1867) from kinetic considerations. Its thermodynamic derivation, by van't Hoff's well-known hypothetical equilibrium-box operation (1883) entails that the only components of the reaction which need be taken into consideration are those in the gaseous state or in solution. A pure condensed phase has of course a certain vapour pressure, but this is constant at any given temperature and need not be considered in the mass action relationship.

It is convenient therefore to use symbols to denote the state of aggregation of the components of any reaction. In this monograph we shall make use of various types of brackets. Thus diamond-shaped (angular) brackets will denote the solid state, $\langle NaCl \rangle$; braces the liquid state, $\{H_2O\}$; light parentheses the gaseous state, (O_2); while the state of dilute solution will be indicated by square brackets, $[NaCl]_{aq}$, the suffix denoting the solvent.

In making mass action calculations a distinction must be drawn between heterogeneous and homogeneous reactions. A homogeneous reaction is one which takes place completely in one phase, such as

$$(H_2) + (I_2) = 2(HI)$$

or

$$[CuSO_4] = [Cu^{++}] + [SO_4'']$$

For these examples the mass action law is written

$$K_p = \frac{p_{HI}^2}{p_{H_2}\, p_{I_2}} \tag{2a}$$

$$K_c = \frac{c_{Cu^{++}}\, c_{SO_4''}}{c_{CuSO_4}} \tag{2b}$$

K_p and K_c are the mass action constants at a certain temperature, p denoting pressure in atmospheres, and c the molal concentration.*

For thermochemical calculations the unit of quantity should generally be the gram-molecule or mole, the symbol (H_2) then representing one mole of gaseous hydrogen. The number of moles of a substance which are considered to take part in a reaction appears in the mass action equation as the power of the partial pressure or concentration.

In heterogeneous reactions one or more condensed phases appear in addition to the gaseous or solution phase. The mass action relationship is then expressed as in the following three examples:

$$[FeO]_{slag} + [C]_{Fe} = \{Fe\} + (CO) \tag{α}$$

$$K_{p,c} = p_{CO}/c_C\, c_{FeO}$$

$$\langle Cr_2O_3\rangle + 3(H_2) = 2\langle Cr\rangle + 3(H_2O) \tag{β}$$

$$K_p = \frac{p_{H_2O}^3}{p_{H_2}^3}$$

$$\langle SiO_2\rangle = [Si]_{Fe} + 2[O]_{Fe} \tag{γ}$$

$$K_c = c_{Si}\, c_O^2$$

This should of course be familiar to all chemists, and it is not intended to give fuller details either of the exact derivation or of the experimental proof of the mass action law.

In this monograph we shall follow the usual custom of writing the mass action constant with the active masses of the substances to the right of the chemical equation (the products) in the numerator, and those of the substances to the left (the reactants) in the denominator. The equation of any reaction may of course be written in either direction, but in practice the metal or other substance to be produced will be written on the right. Thus the most favourable conditions for reaction are indicated by a high value of K_p or K_c, while a value far below unity shows that conditions are unfavourable for producing the desired substance. In this latter case the yield can still be made large by the familiar device of removing one of the products, particularly a gas, as it is formed, thereby displacing the system continuously from equilibrium. Thus

* Concentrations will be expressed in this monograph as molal concentrations c_i = number of moles (g-atoms, g-ions, etc.) of solute i per 1000 g solution; mole fractions N_i = number of moles of solute i per total number of moles $n_i/(n_i + n_2 + \cdots)$; atomic per cent = mole fraction × 100; and weight per cent = number of grams per 100 g total solution: $c_1 = [m_1/(m_1 + m_2 + \cdots)] \times 100$.

in reaction (β) described above, a continuous flow of hydrogen would effect this by removing the water formed.

It follows from Le Chatelier's principle (1885) that the magnitude of the total pressure of a system will not affect an equilibrium if the number of moles of gas on either side of the chemical equation are equal. If the numbers differ, an increase in total pressure will displace the equilibrium in favour of the side with the smaller number of moles, and vice versa.

It should be emphasised here that the mass action law in the form given above will apply only in ideal cases where no interaction occurs between the molecular particles of the gas or the solute. This assumption is usually justified for gaseous reactions if the total pressure is not considerably higher than 1 atm. Generally it does not apply to reactions in the solution phase except with very dilute solutions. In more concentrated solutions the interaction of the dissolved components and the solvent affects the equilibrium conditions and K_c is no longer constant. A constant can be obtained, however, by replacing the concentrations c by activities a. The concept of activity will be discussed later.

2. Thermodynamic Functions

A. THE EQUILIBRIUM CONSTANT AND THERMODYNAMIC FUNCTIONS

Generally the aim of any thermochemical calculation in production metallurgy is to obtain values for the equilibrium constant under consideration. The direct method is to measure the relative concentrations of the substances themselves at some selected temperatures, and to derive values of these concentrations or of the constant at other temperatures by interpolation. As a rule, however, this direct determination of the constant either cannot be carried out or is precisely what the chemist is trying to avoid. The number of reactions which can require investigation is astronomical, and a simpler approach is necessary. The advantage of thermochemical calculations is that the vast number of equilibria encountered in metallurgical processes may be computed from a much smaller number of functions characteristic of the separate compounds or elements which take part. These functions can often be determined much more easily than the equilibrium constants.

To illustrate the simple connection between these functions and the equilibrium constant, we may anticipate later discussion by stating that the function usually tabulated instead of the equilibrium constant is the change in standard Gibbs energy for any given process, $\Delta G°$. Its relationship to K_p is expressed by

$$\Delta G° = -RT \ln K_p \tag{3}$$

or

$$\Delta G° = -4{\cdot}574T \log K_p \tag{3a}$$

The change in Gibbs energy is under certain conditions an additive function, which the equilibrium constant is not, and is related to the heat and entropy change of reaction by the simplified Gibbs–Helmholtz equation

$$\Delta G° = \Delta H° - T\Delta S° \tag{4}$$

The following sections deal with the determination of heat changes (ΔH) and entropy changes (ΔS) of reaction at various temperatures, and with the derivation and determination of Gibbs energy values.

The heat change in a reaction will be defined as the sum of the heat contents of the products of the reaction minus the sum of the heat contents of the reactants, and similarly for the entropy change.

B. HEAT CONTENT AND HEAT OF FORMATION

It may be deduced from the first and second laws of thermodynamics that at a given temperature every substance in a given state has a fixed, characteristic value of H and S. If H had not this constancy, then a cyclic process might be envisaged in which heat is produced from nothing, and this has not yet been experienced. The heat content exists in the substance as kinetic and potential energy of its atoms and molecules (energy of translation, rotation, and oscillation). Whatever may be the relative amounts of these constituent energies, their sum remains constant for a given amount of the substance in a given state.

Heat content is so defined that the heat evolved or absorbed during a change taking place in a system* at constant pressure is equal to the change in its heat content. If in the general reaction given in eqn. (1) the heat contents of the participants are respectively $H_{\mathrm{M_aX_b}}$, H_{R}, H_{M} and $H_{\mathrm{R_cX_b}}$, the summation

$$(\mathrm{a}H_{\mathrm{M}} + H_{\mathrm{R_cX_b}}) - (H_{\mathrm{M_aX_b}} + \mathrm{c}H_{\mathrm{R}}) = \Delta H_T \tag{5}$$

is equal to the amount of heat produced if the reaction takes place completely to form the substances on the right of the chemical equation from the reactants on the left. This difference ΔH_T is termed the heat or enthalpy of reaction and can be measured calorimetrically. In the more simple case of the formation of a single compound from its elements

$$\mathrm{aM} + \mathrm{bX} = \mathrm{M_aX_b}$$

(e.g. $\langle \mathrm{Pb} \rangle + (\mathrm{O_2}) = \langle \mathrm{PbO_2} \rangle$) the difference in heat contents

$$H_{\mathrm{M_aX_b}} - (\mathrm{a}H_{\mathrm{M}} + \mathrm{b}H_{\mathrm{X}}) = \Delta H_T \tag{5a}$$

is called the heat of formation. The heats of formation of chemical substances are usually recorded and tabulated at a standard temperature of 25°C (= 298·15°K) and are written as ΔH_{298}.

The numerical values of the heats of formation or reaction will generally be negative, corresponding to an evolution of heat during reaction. This is illustrated diagrammatically in Fig. 1, which is self-explanatory.

If a substance is cooled from a high temperature T to room temperature the heat content will also decrease, the heat being lost by the system to its surroundings. The change represented by $\mathrm{M}_T \rightarrow \mathrm{M}_{298}$ (e.g. $\langle \mathrm{Fe} \rangle_{773} \rightarrow \langle \mathrm{Fe} \rangle_{298}$), is accompanied by the heat *evolution*

$$H_T - H_{298} = \Delta' H_{298}^T \tag{6}$$

* A system may be defined as any number and quantity of chemical substances considered separately from their surroundings.

Alternatively, it may be stated that the change (i.e. increase) in heat content of the system is $-\Delta' H^{T}_{298}$. (The use of the notation Δ' is explained in the footnote on p. 9.)

If the total heat contents of all substances were known they could be tabulated and used additively for the evaluation of the heat of any reaction. These total heat contents would include the heats of formation of atoms from the constituent elementary particles, protons, neutrons, electrons, etc. Since in any chemical reaction the nuclear energy and the energy of the inner electron structure of the atoms is in no way altered, the values of these quantities remain constant and can be omitted from thermochemical consideration. Furthermore, as we are principally concerned with heat

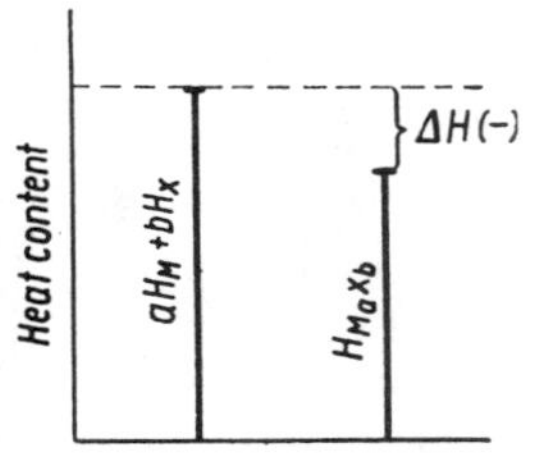

FIG. 1. The heat of compound formation.

changes be they as a result of chemical reaction or changes in temperature of a substance we can, without loss of rigour, select an arbitrary reference state where the substance has zero heat content, *by definition*, and make all of our calculations with respect to this reference or standard state.

Thus we know that the molar heat content of a substance at absolute zero is related to that at, say, 25°C by the relationship

$$H_{298} = H_0 + \int_0^{T=298} C_p \, dT$$

where C_p is the molar specific heat of the substance at constant pressure, and we can define H_{298} as having the value zero for an elementary substance at 25°C. H_0 then becomes a negative quantity, since C_p must be positive, with reference to the element at 25°C as the standard state.

The heat content of compounds at 25°C by the convention which has been laid down is therefore equal to their heat of formation from their elements at this temperature. Thus the heat of formation of liquid water from gaseous oxygen and hydrogen at 25°C

$$(H_2) + \tfrac{1}{2}(O_2) = \{H_2O\}$$

is found to be $\Delta H_{298} = -68{\cdot}31$ kcal, which becomes the heat content of water, since those of hydrogen and oxygen are zero at this temperature. For this reason the heats of formation of compounds are tabulated for this same temperature of 25°C.

One example may be given of the use of these values for calculating heats of reaction:

$$\langle CoO \rangle + (CO) = \langle Co \rangle + (CO_2)$$

$$-57{\cdot}1 \quad -26{\cdot}40 \qquad 0 \quad -94{\cdot}05 \text{ kcal}$$

The respective heats of formation (Table A) are written below each participant. The difference

$$\Delta H_{298} = -94{\cdot}05 - (-57{\cdot}1 - 26{\cdot}40) = -10{\cdot}55 \text{ kcal}$$

is the heat of reaction. This additive property of heats of reaction has long been known as Hess' principle (1840). Further use is made of it in evaluating heats of formation of compounds by combining the heats of various associated reactions. This is often applied in organic chemistry and to numerous inorganic substances. One example will suffice as illustration.

Roth measured the heat of combustion of $\langle Al_4C_3 \rangle$ according to the equation

$$\langle Al_4C_3 \rangle + 6(O_2) = 2\langle Al_2O_3 \rangle + 3(CO_2)$$

and found it to be $-1035{\cdot}4$ kcal/mole. (Probable error $\pm 7{\cdot}5$ kcal.) The heat of formation of $\langle Al_2O_3 \rangle$ is $-400{\cdot}0 \pm 1{\cdot}5$ kcal/mole for constant volume, and that of (CO_2) is $-94{\cdot}05 \pm 0{\cdot}1$ kcal/mole. Then the heat of forming $\langle Al_4C_3 \rangle$ may be calculated:

$$\Delta H^\circ_{298} = 3\Delta H^\circ_{CO_2} + 2\Delta H^\circ_{Al_2O_3} - 6\Delta H^\circ_{O_2} - \Delta H^\circ_{Al_4C_3}$$

$$-1035{\cdot}4 = -282{\cdot}2 - 800{\cdot}0 - 0 - \Delta H^\circ_{Al_4C_3}$$

$$\Delta H^\circ_{Al_4C_3} = -1082{\cdot}2 + 1035{\cdot}4 = -46{\cdot}8 \text{ kcal}$$

One disadvantage of applying Hess' law is demonstrated by this evaluation. The errors of the separate ΔH determinations are additive, giving an overall error of $7{\cdot}5 + 3 \times 0{\cdot}1 + 2 \times 1{\cdot}5 = 10{\cdot}8$ kcal. The calculated heat of formation $\Delta H_{Al_4C_3} = -46{\cdot}8 \pm 10{\cdot}8$ kcal is therefore uncertain to the extent of 24%. This shows how the evaluation of a heat of formation as a small difference of large values makes it subject to a relatively great percentage error even if the original errors are quite small (a maximum of 0·7% in this example). For comparison, the accepted value of the heat of formation of Al_4C_3 is $-49{\cdot}54$ kcal/mole.

The determination of heats of formation will be described in the next chapter, but the principal methods are summarised below:

1. Measurements of the change in temperature resulting from the reaction taking place in a system of known heat capacity (a calorimeter). Hess' law may be applied in connection with this method. For instance, the heat of formation of an alloy may be obtained by measuring the heats of solution of the alloy and of its constituents in a given solvent.

2. Determination of the temperature coefficient of the equilibrium constant of reactions involving gases. The heat of formation or reaction is calculated with van't Hoff's isobar*

$$\Delta H^\circ_T = \boldsymbol{R}T^2 \frac{\mathrm{d} \ln K_p}{\mathrm{d}T} = -\boldsymbol{R} \frac{\mathrm{d} \ln K_p}{\mathrm{d}\,(1/T)} \tag{7}$$

* The term *isochore* is perhaps more familiar, but is a misnomer, so the relationship is referred to as an *isobar* throughout this monograph.

In practice K_p is determined at two or more temperatures ($T_1, T_2, \ldots$) or two values K_{p1} and K_{p2} are taken from a curve of log K_p plotted against $1/T$. Then by integration, eqn. (7) becomes

$$\Delta H_T^\circ = 4{\cdot}574 T_1 T_2 \frac{(\log K_{p1} - \log K_{p2})}{T_1 - T_2} \tag{7a}$$

or

$$\Delta H_T^\circ = 4{\cdot}574 \frac{(\log K_{p1} - \log K_{p2})}{1/T_2 - 1/T_1}$$

which gives a mean value for the heat of formation in the range T_1 to T_2.

If only one of the compounds taking part in the reaction is gaseous, then eqn. (7) can be written

$$\Delta H_T^\circ = \pm RT^2 \frac{\mathrm{d} \ln p}{\mathrm{d}T} \tag{8}$$

or in the integrated form

$$\Delta H_T^\circ = \pm 4{\cdot}574 T_1 T_2 \frac{\log p_1 - \log p_2}{T_1 - T_2} \tag{8a}$$

(which has a positive numerical value if the gaseous component is on the right side of the chemical equation).

Equation (8) is a general relationship, which is perhaps better known in its more particular form as the Clausius-Clapeyron equation applied to the vapour pressure of a pure substance, when eqns. (8) and (8a) may be re-written

$$L_e = RT^2 \frac{\mathrm{d} \ln p}{\mathrm{d}T} = -R\frac{\mathrm{d} \ln p}{\mathrm{d}(1/T)} \tag{9}$$

$$L_e = 4{\cdot}574 T_1 T_2 \frac{\log p_1 - \log p_2}{T_1 - T_2} \tag{9a}$$

Here L_e is the average latent heat of evaporation over the temperature range T_1 to T_2.

3. Equation (8) is also applicable to solutions provided that the dissociation pressures p are measured at various temperatures with the same solution composition. In this case the so-called partial heat of solution of the volatile constituent is obtained (see pp. 27–33). It can be shown that the heat of solution is related to the solubility of the substance (c_{sat}) in the solvent by

$$\Delta H_{\mathrm{soln}} = RT^2 \frac{\mathrm{d} \ln c_{\mathrm{sat}}}{\mathrm{d}T} \tag{10}$$

or

$$\Delta H_{\mathrm{soln}} = 4{\cdot}574 T_1 T_2 \frac{\log c_{\mathrm{sat}(1)} - \log c_{\mathrm{sat}(2)}}{T_1 - T_2} \tag{10a}$$

where ΔH_{soln} is the constant, partial heat of saturated solution over the temperature range T_1 to T_2 and the concentration range $c_{\mathrm{sat}(1)}$ to $c_{\mathrm{sat}(2)}$. It is, of course, essential that the saturated solution should be in equilibrium with the same phase at all temperatures T_1 to T_2, i.e. either with the volatile solute of constant pressure (e.g. 1 atm) or with a compound formed by solute and solvent. Thus, in the special case of eqn. (10) under the restriction mentioned, heats of solution can be calculated from equilibrium data obtained at varying concentration of solute.

Equation (10) is further restricted to the so-called regular solutions the properties of which will be discussed later (pp. 50–54). This restriction implies as a rule that the equation may only be used for rather dilute solutions in which the degree of interaction between the solute atoms or ions is not likely to be significant. It follows that eqn. (10) is useful mainly for low-solubility systems, to which it has frequently been applied, for instance to the metallurgically important solutions of hydrogen in various metals.

As an example, the low solubility of oxygen in liquid bismuth may be mentioned. From the solubility data [1] obtained at various temperatures, a heat of +30,800 cal/mole of oxygen is calculated by means of eqn. (10a). This solution is in equilibrium with solid bismuth oxide, and the corresponding chemical reaction is to be written as follows:

$$\tfrac{2}{3}\langle Bi_2O_3\rangle = 2[O]_{\{Bi\}} + \tfrac{4}{3}\{Bi\}; \quad \Delta H_{soln} = 30{,}800$$

If (O_2) of 1 atm pressure were to be taken as the standard state, the standard heat of formation of $\tfrac{2}{3}\langle Bi_2O_3\rangle$ from liquid bismuth and gaseous oxygen must be added to ΔH_{soln}.

4. The measurement of the electromotive force of a cell in which the chemical reaction under consideration is the agent which produces the current. This provides information about certain thermochemical values, including the heat of formation or of reaction, which is calculated from the Gibbs–Helmholtz equation (pp. 18–19)

$$\Delta H_T = z\boldsymbol{F}\left(T\frac{dE}{dT} - E\right) \tag{11}$$

where $\boldsymbol{F}$ is Faraday's constant (23,066 cal/V), E is the e.m.f. in volts, and z is the number of coulombs which pass if the electrolytic reaction of which ΔH_T is the enthalpy change proceeds to completion.* ΔH_T is expressed in calories.

Over temperature ranges of less than 100°C, the temperature coefficient of the e.m.f. is usually practically constant. This allows eqn. (11) to be used in the form

$$\Delta H_T = z23{,}066\left(T\frac{E_1 - E_2}{T_1 - T_2} - \frac{E_1 + E_2}{2}\right) \tag{11a}$$

C. TEMPERATURE DEPENDENCE OF THE HEAT OF REACTION

For exact calculation consideration must be given to the variation of heats of reaction with temperature. We shall derive the relationship by discussing as an example the reaction

$$\langle Fe_2O_3\rangle + 3\langle C\rangle = 2\langle Fe\rangle + 3(CO)$$

Given the heat of reaction at room temperature, we shall derive its value at 727°C

* Thus for the reaction $ZnCl_2 + Cu = CuCl_2 + Zn$, $z = 2$.

(ΔH_{1000}). This is given, as is seen from Fig. 2, by the equation

$$\Delta H_{1000} = \Delta H_{298} + \Delta' H_R - \Delta' H_L \tag{12}$$

where $\Delta' H_L$ and $\Delta' H_R$ are the changes in heat contents of the reactants and products respectively over the temperature range 25–727°C.* These quantities can be obtained by multiplying together the temperature change and the mean heat capacity values:

$$\Delta' H = (T_2 - T_1)\bar{C}_p \tag{13}$$

In general, however, heat capacities are not tabulated as mean values for various temperature ranges, but as specific heat or atomic heat equations which include temperature factors.

The atomic or molar heat of a substance is the amount of heat required to raise the temperature of one g-atom or one mole respectively by one degree. This equals

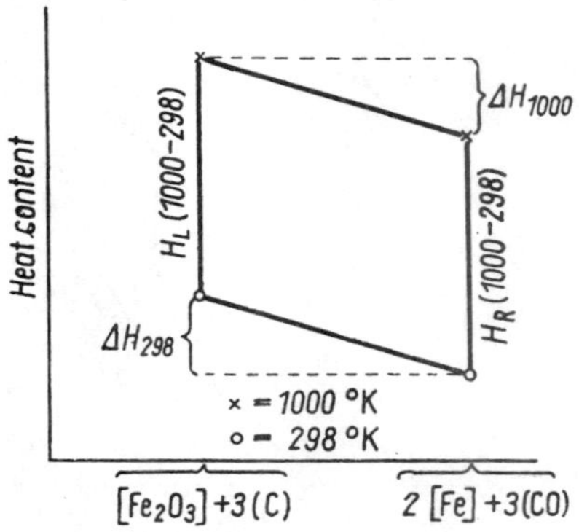

FIG. 2. Illustration of Kirchhoff's law.

the corresponding change in heat content, which we may express as $d(\Delta H)/dT$. In our example

$$\frac{d(\Delta H)}{dT} = 2C_{p(\mathrm{Fe})} + 3C_{p(\mathrm{CO})} - C_{p(\mathrm{Fe_2O_3})} - 3C_{p(\mathrm{C})} = \Delta C_p$$

Integrating between temperatures T_1 and T_2,

$$\int_{\Delta H_1}^{\Delta H_2} d(\Delta H) = \Delta H_2 - \Delta H_1 = \int_{T_1}^{T_2} \Delta C_p \, dT \tag{14}$$

This is simply one way of expressing Kirchhoff's equation (1858). For its evaluation the heat capacities of each participant must be known over the whole temperature range under consideration. Heat capacities are usually expressed in equations of the form

$$C_p = a + bT + cT^2 + dT^{-2} + eT^{-1/2}$$

* In thermochemical calculation, changes in heat content, entropy and free energy fall into two categories, and confusion can arise between the two. The first includes those taking place in an isothermal reaction, and these are denoted by the ordinary Δ symbol. The other describes the change in value of the thermodynamic functions of a substance due to change in temperature, and will be denoted by the symbol Δ'. Equation (12) illustrates the use of the two symbols and the necessity for keeping their identity separate though they represent similar physical quantities.

(of which not all the terms need to be present in any single formula), so no difficulty in integration should be experienced:

$$\int C_p \, dT = aT + \tfrac{1}{2}bT^2 + \tfrac{1}{3}cT^3 - dT^{-1} + 2eT^{1/2} + k$$

where k is a constant.

To determine the change in heat content due to a change in temperature from T_1 to T_2, the integral becomes

$$\int_{T_1}^{T_2} C_p \, dT = a(T_2 - T_1) + \tfrac{1}{2}b(T_2^2 - T_1^2) + \tfrac{1}{3}c(T_2^3 - T_1^3)$$
$$- d(T_2^{-1} - T_1^{-1}) + 2e(T_2^{1/2} - T_1^{1/2})$$

in which the constant k disappears.

In making our calculation, we must first add up the total heat capacities of products and reactants.* Taking the above reaction, over the range 298–1000°K:

$$\left.\begin{aligned} 2 \times C_p(\alpha\text{-Fe}) &= 8{\cdot}36 + 11{\cdot}84 \times 10^{-3}T \\ 3 \times C_p(\text{CO}) &= 20{\cdot}37 + 2{\cdot}94 \times 10^{-3}T - 0{\cdot}33 \times 10^5 T^{-2} \end{aligned}\right\} +$$
$$28{\cdot}73 + 14{\cdot}78 \times 10^{-3}T - 0{\cdot}33 \times 10^5 T^{-2}$$

$$\left.\begin{aligned} C_p(\text{Fe}_2\text{O}_3) &= 23{\cdot}49 + 18{\cdot}6 \times 10^{-3}T - 3{\cdot}55 \times 10^5 T^{-2} \\ 3 \times C_p(\text{C}) &= 12{\cdot}30 + 3{\cdot}06 \times 10^{-3}T - 6{\cdot}30 \times 10^5 T^{-2} \end{aligned}\right\} +$$
$$35{\cdot}79 + 21{\cdot}66 \times 10^{-3}T - 9{\cdot}85 \times 10^5 T^{-2}$$

The difference of the two sums gives

$$\Delta C_p = -7{\cdot}06 - 6{\cdot}88 \times 10^{-3}T + 9{\cdot}52 \times 10^5 T^{-2}$$

On integration between the temperatures 298° and 1000°K we obtain

$$\int \Delta C_p \, dT = -7{\cdot}06(1000 - 298) - 3{\cdot}44 \times 10^{-3}(1000^2 - 298^2) - 9{\cdot}52 \times 10^5(1000^{-1} - 298^{-1})$$
$$= -5848$$

Now ΔH at 298°K is given by*

$$\langle \text{Fe}_2\text{O}_3 \rangle + 3\langle \text{C} \rangle = 2\langle \text{Fe} \rangle + 3(\text{CO})$$
$$-197{,}000 \qquad 0 \qquad\qquad 0 \qquad -79{,}200$$
$$\Delta H_{298} = -79{,}200 + 197{,}000 = +117{,}800 \text{ cal}$$

whence

$$\Delta H_{1000} = \Delta H_{298} + \int_{298}^{1000} \Delta C_p \, dT = 117{,}800 - 5848 = 111{,}950$$

If any of the components taking part in a chemical reaction undergoes change of state of order in the temperature range under consideration (it may be a transformation, fusion or evaporation) the heat effect must be taken into account. The heat of transformation of a reactant must be subtracted from the total while that of a product must be added. In the above reaction, iron undergoes magnetic (order–disorder) transformation in the neighbourhood of 1000°K accompanied by a heat of transformation of 660 cal/g-atom. The corrected heat of formation, taking iron in its normal state at 1000°K, is then

$$\Delta H_{1000} = 111{,}950 + 1320 = 113{,}270$$

* Since it is the principle of this calculation that matters, the numerical values have not been altered compared with those in earlier editions although more recent thermochemical values have partly been adopted in the tables attached to this monograph.

In calculating the change of enthalpy for higher temperatures, the data of the transformed material must be included. Kirchhoff's equation is modified to the form

$$\Delta H_2 - \Delta H_1 = \int_{T_1}^{T_t} \Delta C_p \, \mathrm{d}T \pm L_t + \int_{T_t}^{T_2} \Delta C_p \, \mathrm{d}T \tag{15}$$

L_t is the heat of transformation which takes place at the temperature T_t. It is subtracted for the transformation of a reactant, and added for that of the product. Any further transformation, including melting and evaporation, must be allowed for in the same way.

Table C (p. 336) gives the best values known to the authors of the heat capacities of substances of metallurgical importance. Even if this list were complete as far as the compounds are concerned, the research worker will often require to derive new formulae from more recent and accurate measurements. To illustrate the method used, the curve for the true heat capacity (atomic heat) of nickel over the range 0–900°K is given in Fig. 3. We see that the heat capacity increases from zero at 0°K to about 7 cal/g-atom at 450°K. Between 450° and 650°K there occurs a discontinuity in the C_p curve with a sharp maximum at 630°K corresponding to the magnetic transformation of nickel. These atomic heats are obtained by direct experimental methods.

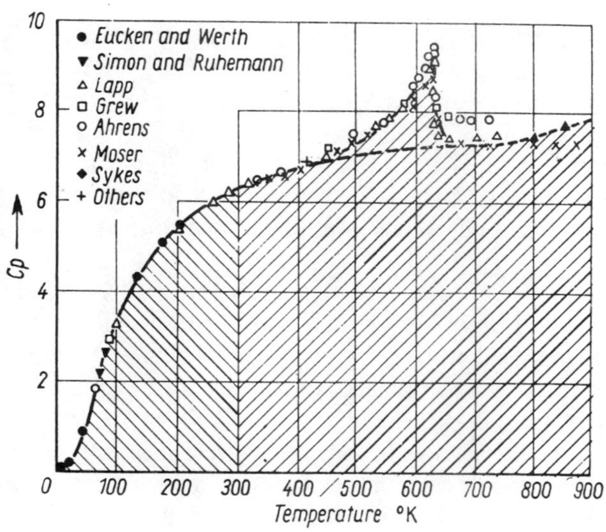

FIG. 3. Atomic heat of nickel from 0° to 900°K.

From such a curve the change in heat content between two temperatures is evaluated by graphical integration. The increase in heat content between 298° and 900°K for instance is given by the right-hand area shaded diagonally. If the drawing is made on graph paper the number of squares in the shaded area may be counted, the total number being multiplied by the caloric equivalent of the unit square.

One may also derive formulae for the atomic heat from smoothed curves of experimental results. Straight lines are expressed by the general equation

$$C_p = \mathrm{a} + \mathrm{b}T \tag{16}$$

where a and b are constants whose values are obtained from two points of the curve. For bent curves these types of formulae are in general use:

$$C_p = \text{a} + \text{b}T + \text{c}T^2 \tag{16a}$$

$$C_p = \text{a} + \text{b}T + \text{e}T^{-1/2} \tag{16b}$$

$$C_p = \text{a} + \text{b}T + \text{d}T^{-2} \tag{16c}$$

The most commonly used four-term expression is:

$$C_p = \text{a} + \text{b}T + \text{c}T^2 + \text{d}T^{-2} \tag{16d}$$

Only in exceptional cases will a T^3 term be needed.

The choice of formulae (16a), (16b) or (16c) depends on the temperature range in which the slope of the curve varies most. If the curvature is more pronounced at lower temperatures, the curve becoming almost linear at high temperatures, use should be made of formula (16c), proposed by K. K. Kelley, or as occasion arises, formula (16b), first used by M. de Kay-Thompson. If the curvature is greater at high temperatures, (16a) should be used. After choosing the type of formula, numerical values at three points on the curve are used to evaluate the constants a, b, and c, d or e respectively. Using (16a) for nickel in the temperature range 273–615°K we obtain

$$C_p = 6{\cdot}10 - 2{\cdot}49 \times 10^{-3}T + 10{\cdot}44 \times 10^{-6}T^2$$

The discontinuity in the curve in Fig. 3 near 630°K is characteristic of magnetic or order–disorder transformation which occurs over a relatively broad range of temperature. In such cases the heat of transformation can be obtained only by drawing the probable continuation of the C_p curve if there were no transformation, as is done in Fig. 3, and evaluating graphically the area between this curve and the one actually observed. The heat of the magnetic transformation in α-nickel is thus found to be 140 ± 20 cal/g-atom. Heat effects of order–disorder transformations are generally not very large.

Phase transformations on the other hand occur normally at definite temperatures. They entail a total rearrangement of the atoms or molecules in a new structure, or a breakdown of the lattice (melting or evaporation). Their heat effects are not evaluated from C_p/T curves in the way described above but are determined by other methods to be described later (pp. 111–13).

True heat capacities are seldom obtained by direct measurements so much as from measurements of change in heat content between a series of high temperatures and room temperature. We shall consider one example of such an evaluation.

Values of several observers for the gram-atomic heat content of cadmium at various temperatures are plotted in Fig. 4. A smooth curve must be drawn through the most reliable experimental points. The height of the discontinuity at the melting point (321°C) is equivalent to the heat of fusion.

Since the form of the C_p curve can be expected to conform to Kelley's empirical formula (16c), we must use the corresponding curve for the heat content

$$\Delta' H_{298}^{T} = \text{A} + \text{B}T + \text{C}T^2 + \text{D}T^3 + \text{E}T^{-1} \tag{17}$$

the values of the five constants A, B, C, D and E being obtained from five points on the heat content curve. Formula (16d) is then obtained by differentiation:

$$\text{a} = \text{B}, \quad \text{b} = 2\text{C}, \quad \text{c} = 3\text{D}, \quad \text{d} = -\text{E}$$

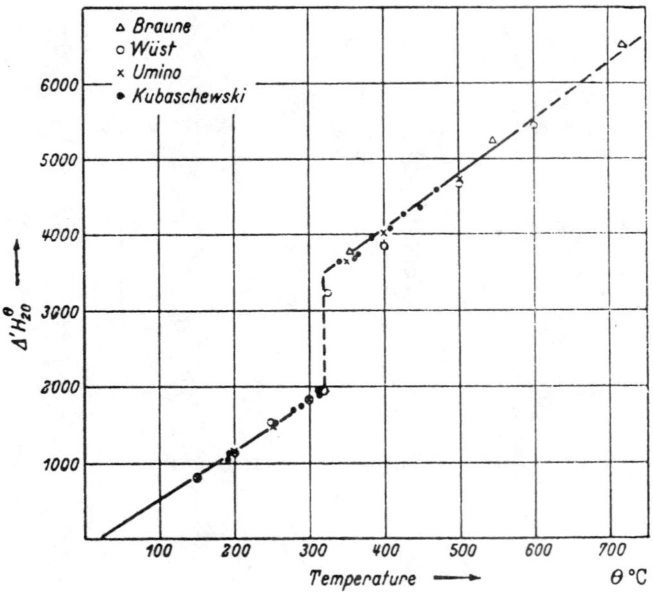

FIG. 4. Change of heat content of cadmium from 20° to θ°C.

For liquid cadmium, the values are not very accurate. Since they do not appear to fall on a definite curve, a straight line must be drawn through the points and a mean value obtained for C_p for that temperature range.

$$\frac{H_2 - H_1}{T_2 - T_1} = \bar{C}_p = C_p \tag{18}$$

D. ENTROPY

The concept and nature of entropy follows from the Second Law of Thermodynamics. Its derivation, usually through the consideration of a reversible Carnot cycle, is a common feature of any textbook dealing with classical thermodynamics, so it will not be repeated here. It is not easily defined directly, but for the purpose of this monograph it may be interpreted in terms of *increase* in entropy, which is equivalent to the decrease in atomic order of a system undergoing any process; quantitatively, this is equal to the heat taken up isothermally and reversibly,* divided by the absolute temperature at which the process takes place (Boltzmann, 1866).

The true heat capacity (molar heat) of a substance is such a reversible heat effect, and the quantity C_p/T for any temperature is an entropy expression. Figure 5 shows the C_p/T curve derived from the C_p values for nickel in Fig. 3. Its form is characteristic of all such curves.

* A reversible process may be defined as one which is carried out under conditions such that the system at all times differs only infinitesimally from a state of equilibrium. A process produces maximum work, i.e. it works most efficiently, when done reversibly. Melting or evaporation at the melting or boiling point are examples of reversible processes. Since the true specific heat is measured in small steps of one degree it represents energy which is to all intents and purposes introduced reversibly.

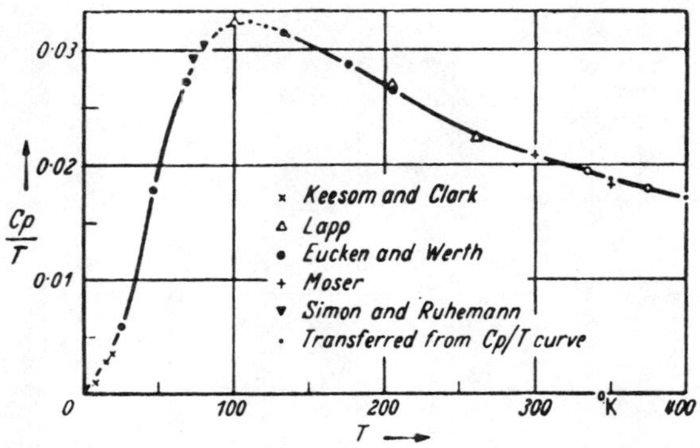

FIG. 5. C_p/T values for nickel between 0° and 400°K.

The equation for the entropy of a substance at any temperature T, corresponding to that for the heat content (p. 5), is

$$S_T = S_0 + \int_0^T \frac{C_p}{T} \, dT \tag{19}$$

It also follows from the Second Law of Thermodynamics that every substance in a given state at a given temperature has a certain unequivocal entropy as well as a heat content. Thus if a substance undergoes a cyclic process involving several intermediate changes, it will at the end have the same entropy as it had initially. We can therefore define an entropy of reaction ΔS_T, corresponding to the heat of reaction, replacing eq. (5) by the relationship

$$\Delta S_T = (aS_M + S_{R_cX_b}) - (S_{M_aX_b} + cS_R) \tag{20}$$

Combining this with eqn. (19) we obtain an expression for the entropy of reaction at any temperature T,

$$\Delta S_T = \Delta S_0 + \int_0^T \frac{\Delta C_p}{T} \, dT \tag{21}$$

It was W. Nernst who first postulated (1906) that the entropy of reaction at absolute zero, ΔS_0, is nil, though at first this was not expressed in so simple a form. He and his co-workers established the validity of this assumption for reactions involving completely ordered crystallised substances only by measuring a number of heat capacities and reaction entropies down to very low temperatures.

Planck extended that deduction, and concluded (1912) that the entropy of any ordered crystalline substance at absolute zero S_0 should be zero. Applying this, eqn. (19) becomes

$$S_T = \int_0^T \frac{C_p}{T} \, dT \tag{22}$$

This conclusion has found general acceptance, and in this form it is the simplest formulation of the Third Law of Thermodynamics. Its significance to us is that, unlike the values of the heat contents (see p. 5), which are related to arbitrary standards, those of entropies can be determined absolutely. They are obtained either from heat capacity measurements down to low temperatures and applying eqn. (22), or from reaction entropies as will be described later. For convenience in calculation, entropy values are tabulated for 25°C (= 298·15°K). Termed S°_{298}, they are known as *standard entropies.*

The metallurgist will rarely be called upon to evaluate standard entropies from C_p measurements, since this is usually done by those who measure the heat capacities. The measurement of C_p in the neighbourhood of absolute zero is obviously rather difficult. If measurements are not extended much below 50°K, the Debye equation for the molar heat of a solid at constant volume, C_v, in terms of its absolute temperature in this low range: $C_v = 465T^3/\Theta^3$, may be applied. In this equation Θ is a characteristic constant for each solid, and can be generally derived from the slope of the C_p curve above 50°K. This enables the C_p curve to be extrapolated to absolute zero, the relationships between C_p and C_v being known.

This equation is based upon the assumption that a crystalline solid has the properties of a perfectly elastic body. This applies to temperatures near zero and the agreement is good for temperatures up to a value of $T = \Theta/12$ (Θ for the elements varies between 50 [calcium] and 1840 [carbon]). Debye's equation cannot be applied at higher temperatures.

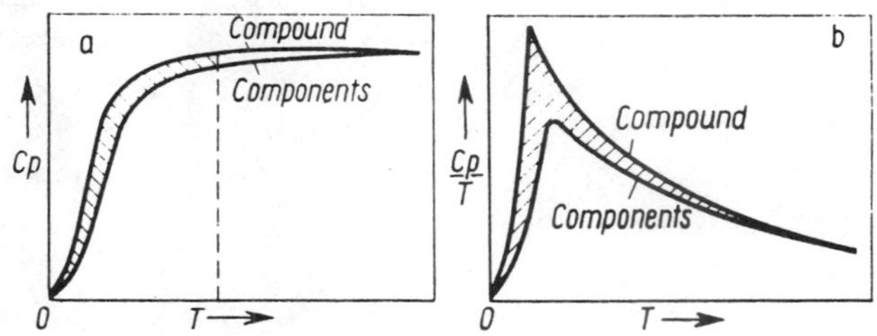

FIG. 6. Molar heat content and entropy of a compound and its component elements.

To clarify further the evaluation of ΔS_T from heat capacity curves, reference may be made to Fig. 6. The molar heat of a compound and the sum total atomic heats of its constituent elements are plotted against temperature in Fig. 6a. The shaded area represents the term $\int_0^T \Delta C_p \, dT$. Plotting the same values as C_p/T against T, we obtain Fig. 6b, in which the shaded area represents the entropy of formation of the compound at the temperature T:

$$\Delta S_T = \int_0^T \frac{\Delta C_p}{T} \, dT \tag{23}$$

If any phase change occurs in the temperature range considered, a transformation (t), fusion (f) or evaporation (e), its effect on the entropy must be taken into account. Such phase changes are thermodynamically reversible, so their associated entropy changes are obtained by dividing the respective latent heats by the temperatures at which they occur.

$$\Delta S_t = \frac{L_t}{T_t}; \quad \Delta S_f = \frac{L_f}{T_f}; \quad \Delta S_e = \frac{L_e}{T_e} \tag{24a–c}$$

Equation (22) may then be expanded to the form:

$$S_T = \int_0^{T_t} \frac{C_p}{T} \, dT + \frac{L_t}{T_t} + \int_{T_t}^{T_f} \frac{C_p}{T} \, dT + \frac{L_f}{T_f} + \int_{T_f}^{T_e} \frac{C_p}{T} \, dT + \frac{L_e}{T_e} + \int_{T_e}^{T} \frac{C_p}{T} \, dT \tag{22b}$$

Nitrogen, for instance, has a transformation at −237·6°C (35·5°K), its melting point at −209·9°C (63·2°K) and boiling point at −195·8°C (77·3°K). Its standard entropy is therefore

$$S_{298}(N_2) = \int_0^{35\cdot5} \frac{C_p}{T}\,dT + \frac{L_t}{35\cdot5} + \int_{35\cdot5}^{63\cdot2} \frac{C_p}{T}\,dT + \frac{L_f}{63\cdot2}$$
$$+ \int_{63\cdot2}^{77\cdot3} \frac{C_p}{T}\,dT + \frac{L_e}{77\cdot3} + \int_{77\cdot3}^{298\cdot1} \frac{C_p}{T}\,dT$$

and each heat effect must be determined in order to evaluate the standard entropy.

We have already mentioned that such evaluations are usually carried out by the observers of these heat effects, and do not require calculation by the metallurgist himself. Equation (22b) is, however, equally applicable to the determination of entropy values above room temperature when phase changes occur within the temperature range covered.

It was emphasised that eqn. (22) applies only to completely ordered, crystalline substances, and eqn. (23) to reactions involving such substances. Alloys are sometimes disordered at room temperature and may even remain so down to absolute zero. In such cases S_0 (eqn. (19)) has a non-zero value and must be taken into consideration in eqns. (22) and (23). The zero-point entropy of completely disordered binary mixtures (e.g. alloys) is given by

$$S_0 = -\boldsymbol{R}(N_1 \ln N_1 + N_2 \ln N_2) \tag{24}$$

N_1 and N_2 being the atomic fractions of the two components of a binary alloy. S_0 amounts to 1·38 cal/deg for the atomic fraction $N_1 = N_2 = 0{\cdot}5$. Thus the formation of a disordered binary alloy phase is accompanied by an entropy change of

$$\Delta S_T = \int_0^T \frac{\Delta C_p}{T}\,dT - \boldsymbol{R}(N_1 \ln N_1 + N_2 \ln N_2) \tag{25}$$

The Third Law of Thermodynamics in addition does not apply to substances in a metastable state such as glass. It also leads logically to the postulate that every substance should be in an ordered crystalline state at absolute zero, if in equilibrium. In so many cases, however (e.g. the gold–silver alloys), this state is unattainable, since the disorder–order transformation temperature is too low to permit any observable atomic rearrangement within any practical length of time.

Entropies and entropy changes can be determined by several methods.

1. The determination of entropies from low-temperature heat capacities, and of reaction entropies from equilibrium measurements has already been outlined.

2. The reaction entropy of a chemical process which can be harnessed in a galvanic cell is obtained from the temperature coefficient of the resultant e.m.f. by the relationship

$$\Delta S = z\boldsymbol{F}\frac{dE}{dT} \tag{26}$$

or, for limited temperature ranges,

$$\Delta S = z23{,}066\,\frac{E_1 - E_2}{T_1 - T_2}\ \text{cal} \tag{26a}$$

3. From measurements of the equilibrium constant at different temperatures, reaction entropy is given by the equation

$$\Delta S^\circ_T = \boldsymbol{R}\left(T\frac{\mathrm{d}\ln K_p}{\mathrm{d}T} - \ln K_p\right) \tag{27}$$

or, for a limited temperature range,

$$\Delta S^\circ_T = 4{\cdot}574\left(T\,\frac{\log K_{p(1)} - \log K_{p(2)}}{T_1 - T_2} - \log K_p\right)\mathrm{cal} \tag{27a}$$

E. GIBBS ENERGY

While Berthelot and Thomsen were, over a century ago, carrying out their work on heats of reaction in the belief that they were identical with the "driving force" of a reaction, it became clear from the work of Carnot, van't Hoff, Gibbs and Helmholtz that there was another thermodynamic function which conformed more exactly to the conception of a driving force. This quantity was the Gibbs energy* G which for any substance under given conditions is the maximum portion of its energy which can be converted into mechanical work. Any process which takes place is accompanied by a change in Gibbs energy of the system, and this change is equal to the net reversible work done by or absorbed during the process. It is a maximum when the process takes place reversibly. Thus the expression

$$\Delta G = G_{\text{products}} - G_{\text{reactants}} \tag{28}$$

represents the maximum mechanical work that can be carried out by a reaction at constant pressure.

The Gibbs energy change during a reaction is related to the various other functions associated with the reaction and these will be dealt with in some detail.

In van't Hoff's equilibrium-box experiment, Gibbs energy is derived as the maximum work ΔG done in a system when the reactants in a chemical process pass reversibly from their initial state to the one in which they are in equilibrium with and are transformed to the products, which are brought in turn to their final state. The Gibbs energy is then given by the relationship

$$\Delta G = -\boldsymbol{R}T\ln K_p + \boldsymbol{R}T\sum n\ln p \tag{29}$$

in which the term $\sum n \ln p$ defines the initial and final states of the system, n and p representing respectively the number of moles and the pressure of the participants in the reaction. (The n values for the reactants are negative, for the products positive.) It is sometimes written $\ln K'_q$, but K'_q is not an equilibrium constant. At any temperature the term $\boldsymbol{R}T \ln K_p$ is constant for any given reaction but $\boldsymbol{R}T\sum n \ln p$ is not, and the Gibbs energy change therefore varies according to the pressures of the gases in the

* The term Gibbs energy (G) used in this monograph is also known as the thermodynamic potential and is related to the heat content H by the equation $G = H - TS$. There is another term (F) which is also described as Gibbs energy, but more precisely as Helmholtz function or the maximum work function which is related to the internal energy U by the equation $F = U - TS$. Thus $G = F + pV$. It should be noted that the symbol $\boldsymbol{F}$ is often used, particularly in American literature, to denote Gibbs energy.

initial and final states of the system. If the pressures are all fixed at unity, the term $\boldsymbol{R}T\sum n \ln p$ vanishes. Thus defined, the term

$$\Delta G^\circ = -\boldsymbol{R}T \ln K_p \quad (29a)$$

becomes the standard Gibbs energy change of the reaction under consideration.

In a reaction involving only one gaseous component, the equilibrium constant consists only of its partial pressure p. This applies similarly to the Gibbs energy change: hence if the gas is a *product* of the reaction,

$$\Delta G^\circ = -\boldsymbol{R}T \ln p \quad \text{or} \quad \Delta G^\circ = -4{\cdot}574T \log p \quad (30)$$

The same equation applies to the Gibbs energy of evaporation of a single substance. At the boiling point or sublimation point p is unity and ΔG° is nil: there is no driving force to alter the equilibrium since it involves the participants in their standard states.

It should be noted that standard Gibbs energy changes are conventionally denoted as ΔG°.

There is another method, in addition to van't Hoff's equilibrium-box operation, by which a reaction can be made to take place reversibly and the associated mechanical work measured. This is to harness the reaction in an electrolytic cell and to measure the reversible electromotive force it produces. In practice this is measured by a null method; i.e. one in which there is no passage of current. Under these conditions, the reaction is in effect tending reversibly toward equilibrium, and its Gibbs energy is given by

$$\Delta G = -z\boldsymbol{F}E \quad (31)$$

where $\boldsymbol{F}$ is the Faraday, $z\boldsymbol{F}$ the number of coulombs which pass if the reaction proceeds to completion, and E is the e.m.f. With ΔG expressed in calories and E in volts, eqn. (31) may be re-written

$$\Delta G = -z23{,}066E \text{ (cal)} \quad (31a)$$

This is the most direct method of measuring Gibbs energies of reaction, which can then be used of course for calculating equilibrium constants. Gases involved in e.m.f. measurements must have a pressure of 1 atm to fulfil the condition discussed above. Electromotive force measurements do not only provide the values of the Gibbs energy of formation of equilibria, but also of complete condensed reactions.

For instance, the e.m.f. of the common Daniell cell gives the Gibbs energy of the complete reaction

$$\langle Zn \rangle + [CuSO_4]_{aq} \rightarrow \langle Cu \rangle + [ZnSO_4]_{aq}$$

or, more precisely,

$$\langle Zn \rangle + [Cu^{++}]_{aq} \rightarrow \langle Cu \rangle + [Zn^{++}]_{aq}$$

The connection between the heat and entropy of reaction and the maximum work was first derived by Gibbs (1875) and independently by Helmholtz (1882). The relationship is nowadays written in the form

$$\Delta G = \Delta H - T\Delta S \quad (32)$$

or, according to its initial derivation from Carnot's cycle,

$$\Delta G = \Delta H + T\frac{d\Delta G}{dT} \tag{33}$$

Equation (32) is valid for values of ΔG, ΔH and ΔS at the one temperature T. Since values for ΔH and ΔS are usually known and tabulated for room temperature, the values of ΔG may be calculated directly for this temperature. For values at higher temperatures heat capacities must be considered, and eqns. (15) and (23) included, to obtain

$$\Delta G_T = \Delta H_{298} + \int_{298}^{T} \Delta C_p \, dT - T\Delta S_{298} - T\int_{298}^{T} \frac{\Delta C_p}{T} dT \tag{34}$$

If any phase transition should occur in the range 298–T°K its entropy and heat have also to be considered, as in eqns. (15) and (22b). It may be permissible, as a first approximation, to neglect the heat capacity terms, thus:

$$\Delta G_T = \Delta H_{298} - T\Delta S_{298} \tag{32a}$$

(see p. 262), but heats and entropies of formation, fusion or evaporation are usually too large to be omitted.

If the values of both ΔG and the equilibrium constant at a given temperature are required, eqn. (34) must be applied. An example will be given of such an evaluation. For this purpose eqn. (34) is expanded into the form

$$\Delta G_T = \left(\Delta H_{298} - \int_0^{298} \Delta C_p \, dT\right) + \int_0^{T} \Delta C_p \, dT$$
$$- T\left(\Delta S_{298} - \int_0^{298} \frac{\Delta C_p}{T} dT\right) - T\int_0^{T} \frac{\Delta C_p}{T} dT \tag{34a}$$

In practice, the numerical value of the term

$$\Delta H_{298} - \int_0^{298} \Delta C_p \, dT = \Delta H_{(0)}$$

does not usually represent the heat of reaction at absolute zero, since the C_p formulae used for these evaluations are generally valid only for temperatures above 298°K. Thus $\Delta H_{(0)}$ is a value used for calculation and has no fundamental or effective meaning. The same also applies to the term

$$\Delta S_{298} - \int_0^{298} \frac{\Delta C_p}{T} dT = \Delta S_{(0)}$$

The fictitious nature of these quantities is denoted by the parenthesised zero in the suffix.

An example will be given of such an evaluation,* that of the reaction

$$2\langle Al\rangle + 1\tfrac{1}{2}(O_2) = \langle Al_2O_3\rangle \tag{α}$$

* The C_p equations used in this example may not exactly agree with those in Table C. However, since the evaluation is simply an exercise, it has been deemed unnecessary to revise these data.

The heat of formation of Al_2O_3 is $\Delta H^\circ_{298} = -400{,}000$ cal/mole and its entropy of formation is

$$\Delta S^\circ_{298} = S^\circ_{Al_2O_3} - 2S^\circ_{Al} - 1\tfrac{1}{2}S^\circ_{O_2} = 12{\cdot}20 - 13{\cdot}54 - 73{\cdot}53 = -74{\cdot}87$$

The sum of the heat capacities of the reactants is given by

$$\begin{aligned} C^\circ_p(1\tfrac{1}{2}O_2) &= 10{\cdot}74 + 1{\cdot}50 \times 10^{-3}T - 0{\cdot}60 \times 10^5 T^{-2} \\ C^\circ_p(2Al) &= \;\;9{\cdot}88 + 5{\cdot}92 \times 10^{-3}T \\ \hline \sum C^\circ_p &= 20{\cdot}62 + 7{\cdot}42 \times 10^{-3}T - 0{\cdot}60 \times 10^5 T^{-2} \end{aligned}$$

which is subtracted from the molar heat of $\langle Al_2O_3 \rangle$

$$C^\circ_p(Al_2O_3) = 27{\cdot}43 + 3{\cdot}06 \times 10^{-3}T - 8{\cdot}47 \times 10^5 T^{-2}$$

to give

$$\Delta C^\circ_p = 6{\cdot}81 - 4{\cdot}36 \times 10^{-3}T - 7{\cdot}87 \times 10^5 T^{-2}$$

Integration of this equation gives

$$\int \Delta C^\circ_p \, dT = 6{\cdot}81T - 2{\cdot}18 \times 10^{-3}T^2 + 7{\cdot}87 \times 10^5 T^{-1} + k_1$$

For $T = 298$,

$$\begin{aligned} \int \Delta C^\circ_p \, dT &= 2029{\cdot}5 - 193{\cdot}5 + 2641 + k_1 \\ &= 4477 + k_1 \end{aligned}$$

Whence

$$\int_{298}^{T} \Delta C^\circ_p \, dT = 6{\cdot}81T - 2{\cdot}18 \times 10^{-3}T^2 + 7{\cdot}87 \times 10^5 T^{-1} - 4477$$

The ΔC°_p equation divided by T and integrated gives in the same manner

$$\int \frac{\Delta C^\circ_p}{T} dT = 6{\cdot}81 \ln T - 4{\cdot}36 \times 10^{-3}T + 3{\cdot}94 \times 10^5 T^{-2} + k_2$$

$$\int^{T=298} \frac{\Delta C^\circ_p}{T} dT = 38{\cdot}80 - 1{\cdot}30 + 4{\cdot}44 + k_2 = 41{\cdot}94 + k_2$$

whence

$$\int_{298}^{T} \frac{\Delta C^\circ_p}{T} dT = 6{\cdot}81 \ln T - 4{\cdot}36 \times 10^{-3}T + 3{\cdot}94 \times 10^5 T^{-2} - 41{\cdot}94$$

Reverting to eqn. (34),

$$\begin{aligned} \Delta G^\circ_T &= \Delta H^\circ_{298} + \int_{298}^{T} \Delta C^\circ_p \, dT - T\,\Delta S^\circ_{298} - T \int_{298}^{T} \frac{\Delta C^\circ_p}{T} dT \\ &= (-400{,}000) + (6{\cdot}81T - 2{\cdot}18 \times 10^{-3}T^2 + 7{\cdot}87 \\ &\quad \times 10^5 T^{-1} - 4477) - (-74{\cdot}87T) - (6{\cdot}81T \ln T - 4{\cdot}36 \end{aligned}$$

$$\times 10^{-3}T^2 + 3{\cdot}94 \times 10^5 T^{-1} - 41{\cdot}94T)$$

$$= -404{,}477 - 15{\cdot}68T \log T + 2{\cdot}18 \times 10^{-3}T^2 + 3{\cdot}93 \times 10^5 T^{-1} + 123{\cdot}62T \qquad (34\alpha)$$

This is typical of the formulae usually obtained for change in Gibbs energy with temperature.

The temperature-independent term is equivalent to

$$\Delta H^\circ_{298} - \int_0^{298} \Delta C^\circ_p \, \mathrm{d}T,$$

which is the fictitious standard change in heat content for the reaction at absolute zero, i.e. $\Delta H_{(0)}$. We may then write the general formula for the standard Gibbs energy change of a reaction as

$$\Delta G^\circ_T = \Delta H^\circ_{(0)} + \mathrm{a}T \log T + \mathrm{b}T^2 + \mathrm{c}T^{-1} + \mathrm{I}T \qquad (35)$$

Continuing our evaluation of the Gibbs energy of formation of alumina, we note that the heat capacity values we have used are valid in the following ranges: $\langle Al_2O_3\rangle$, 273–3000°K; (O_2), 298–1500°K; and $\langle Al\rangle$, 273–932°K (m.p.). Formula (34a) is therefore applicable at temperatures between 298° and 932°K. To evaluate the Gibbs energy of reaction above this temperature range, we must introduce the Gibbs energy of melting of aluminium

$$2\{Al\} = 2\langle Al\rangle \qquad (\beta)$$

The reaction is written in this way so that its simple addition to eqn. (α) will give

$$2\{Al\} + 1\tfrac{1}{2}(O_2) = \langle Al_2O_3\rangle \qquad (\gamma)$$

The Gibbs energy formulae will likewise be additive.

The heat of reaction (β) is double the heat of fusion of aluminium and opposite in sign

$$\Delta H^\circ_{932}(\beta) = -2 \times 2500 = -5000$$

Further,

$$\Delta S^\circ_{932}(\beta) = -5000 \div 932 = -5{\cdot}35$$

The further calculation of $\Delta G^\circ_T(\beta)$ is similar to that of $\Delta G^\circ_T(\alpha)$

$$2C^\circ_p\ (\text{Al, solid}) = 9{\cdot}88 + 5.92 \times 10^{-3}T$$

$$2C^\circ_p\ (\text{Al, liquid}) = 14{\cdot}0 \qquad (932\text{–}1273°\text{K})$$

$$\Delta C^\circ_p = -4{\cdot}12 + 5{\cdot}92 \times 10^{-3}T$$

$$\int \Delta C^\circ_p \, \mathrm{d}T = -4{\cdot}12T + 2{\cdot}96 \times 10^{-3}T^2 + k_3$$

$$\int \frac{\Delta C^\circ_p}{T} \, \mathrm{d}T = -4{\cdot}12 \ln T + 5{\cdot}92 \times 10^{-3}T + k_4$$

$$\int_{932}^{T} \Delta C^\circ_p \, \mathrm{d}T = -4{\cdot}12T + 2{\cdot}96 \times 10^{-3}T^2 + 1269$$

$$\int_{932}^{T} \frac{\Delta C_p^\circ}{T} \mathrm{d}T = -9{\cdot}49 \log T + 5{\cdot}92 \times 10^{-3}T + 22{\cdot}66$$

$$\Delta G_T^\circ(\beta) = \Delta H_{932}^\circ(\beta) + \int_{932}^{T} \Delta C_p^\circ \,\mathrm{d}T - T\,\Delta S_{932}^\circ(\beta) - T\int_{932}^{T} \frac{\Delta C_p^\circ}{T}\mathrm{d}T$$

$$= (-5000) + (-4{\cdot}12T + 2{\cdot}96 \times 10^{-3}T^2 + 1269)$$

$$- (-5{\cdot}35T) - (-9{\cdot}49T \log T + 5{\cdot}92 \times 10^{-3}T^2 + 22{\cdot}66T)$$

$$= -3{,}731 + 9{\cdot}49T \log T - 2{\cdot}96 \times 10^{-3}T^2 - 21{\cdot}43T \qquad (34\beta)$$

This is the equation for the change in Gibbs energy of melting of 2 moles aluminium. As with all equilibrium phase changes, it is zero at the melting point itself, but it has a finite value which is additive in computing free energies above the melting point from values which are valid below that temperature. To obtain the Gibbs energy of reaction (γ), $\Delta G_T(\alpha)$ and $\Delta G_T(\beta)$ are added, giving

$$\Delta G_T^\circ(\gamma) = -408{,}208 - 6{\cdot}19T \log T - 0{\cdot}78 \times 10^{-3}T^2$$

$$+ 3{\cdot}93 \times 10^5 T^{-1} + 102{\cdot}19T \qquad (34\gamma)$$

Since the C_p formula for liquid aluminium is valid up to 1273°K, eqn. (34γ) is reliable from the melting point of aluminium up to this temperature, i.e. from 932° to 1273°K.

The heat capacity equation which has been adopted for these calculations is of a standard form which successfully represents the behaviour of most substances from 298°K and higher temperatures. The term in T^{-2} dominates the curvature of the heat capacity in most circumstances only up to 600–700°K. Above this temperature range, the contribution to the heat capacity from this term is normally negligible, and the term which is linear in the temperature accounts for the variation of the heat capacity of a given phase in the given state. For thermochemical calculations we are more frequently interested in the Gibbs energy change of a reaction than in the heat capacities of the individual reactants and products. An approximate value for the standard Gibbs energy change of a reaction at temperatures a few hundred degrees above 298°K can be obtained using calorimetric data for ΔH_{298}° and low-temperature data for ΔS_{298}°. Thus, corresponding to eqn. (32a):

$$\Delta G_T^\circ\,(\text{approx.}) = \Delta H_{298}^\circ - T\Delta S_{298}^\circ \qquad (36)$$

the difference between the real Gibbs energy change ΔG_T° and this approximate value is given by

$$\Delta G_T^\circ - \Delta G_T^\circ\,(\text{approx.}) = (\Delta H_T^\circ - \Delta H_{298}^\circ) - T(\Delta S_T^\circ - \Delta S_{298}^\circ) \qquad (37)$$

This deviation can be expressed in terms of the heat capacity change equation

$$\Delta C_p = \Delta\alpha + \Delta\beta T + \Delta\gamma T^{-2}$$

by means of the equation

$$(\Delta H_T^\circ - \Delta H_{298}^\circ) - T(\Delta S_T^\circ - \Delta S_{298}^\circ) = \int_{298}^{T} \Delta C_p \,\mathrm{d}T - T\int_{298}^{T} \frac{\Delta C_p}{T}\mathrm{d}T \qquad (38)$$

We define the function, the heat capacity deviation (h.c.d.) by the equation

$$\text{h.c.d.} = \frac{1}{T}\int_{298}^{T} C_p\,\mathrm{d}T - \int_{298}^{T} \frac{C_p}{T}\,\mathrm{d}T \tag{39}$$

for each reactant and product and the change in the h.c.d. function can be related to the heat capacity change thus:

$$\Delta\text{h.c.d.} = \frac{1}{T}\int_{298}^{T} \Delta C_p\,\mathrm{d}T - \int_{298}^{T} \frac{\Delta C_p}{T}\,\mathrm{d}T \tag{39a}$$

and substituting for ΔC_p and using ϕ for the fraction $298/T$, we find

$$\Delta\text{h.c.d.} = \Delta\alpha(1 - \phi + \ln\phi) + \Delta\beta T\left(\phi - \frac{\phi^2}{2} - \frac{1}{2}\right) + \Delta\gamma T^{-2}\left(\frac{1}{\phi} - \frac{1}{2\phi^2} - \frac{1}{2}\right) \tag{40}$$

Each of the bracketed functions will appear in the calculation of the Gibbs energy deviation for any reaction in which the recommended form of the heat capacity equation is employed, and the additional appropriate terms can readily be calculated for other heat capacity equations. If each term in Δh.c.d. is multiplied by the temperature under consideration for a given reaction, then the sum of these terms when added to the ΔG_T° (approx.) equation yields the correct standard Gibbs energy change at the temperature T.

Where a computer is not available, it is useful to tabulate the Gibbs energy deviation contributions since this table is of general applicability. Table I has been prepared for 200°K intervals from 400° to 2000°K, and if each term is multiplied by the appropriate value of $\Delta\alpha$, $\Delta\beta$ and $\Delta\gamma$ for a reaction of interest, then the deviation may be calculated. The reader can bring the values in the heat capacity table and those for the heats of formation and entropy contents up to date as new results appear in the literature and, together with a suitable computer program for the Gibbs energy deviation or a tabulation as presented here, have a flexible calculation procedure to obtain standard Gibbs energy changes of as high a precision as the current "data bank" will allow.

Table I. Values of the Gibbs energy deviation contributions as a function of temperature [2]

T (°K)	$-T(1 - \phi + \ln\phi)$ cal	$-T^2\left(\phi - \frac{\phi^2}{2} - \frac{1}{2}\right)$ cal	$-\frac{1}{T}\left(\frac{1}{\phi} - \frac{1}{2\phi^2} - \frac{1}{2}\right)$ cal
298	0	0	0
400	16	5202	$0{\cdot}15 \times 10^{-3}$
600	118	$0{\cdot}456 \times 10^5$	$0{\cdot}86 \times 10^{-3}$
800	288	$1{\cdot}260 \times 10^5$	$1{\cdot}77 \times 10^{-3}$
1000	509	$2{\cdot}464 \times 10^5$	$2{\cdot}77 \times 10^{-3}$
1200	770	$4{\cdot}068 \times 10^5$	$3{\cdot}82 \times 10^{-3}$
1400	1064	$6{\cdot}071 \times 10^5$	$4{\cdot}88 \times 10^{-3}$
1600	1387	$8{\cdot}476 \times 10^5$	$5{\cdot}97 \times 10^{-3}$
1800	1735	$11{\cdot}280 \times 10^5$	$7{\cdot}06 \times 10^{-3}$
2000	2106	$14{\cdot}480 \times 10^5$	$8{\cdot}16 \times 10^{-3}$

With reference to the values in Table I, it should be observed that the temperature-dependent parts of the heat capacities of substances have coefficients which are typically of the order 1–10×10^{-3} for the coefficient β and 1–10×10^5 for γ, so these deviation terms only amount to a few kilocalories under normal circumstances.

The errors in formulae such as (34γ) (p. 22) vary within wide limits, due mainly to the error in the experimentally determined value of the heats of formation at room temperature. A figure accurate to ± 500 cal may be deemed a good one; errors of 1000–3000 cal are more the rule. The first consequence of this is that figures in equations such as (34α) and (34γ) may be rounded off without loss in accuracy. The next consequence may be demonstrated by drawing the $\Delta G/T$ curves of a number of substances, when it is seen that they do not differ much from straight lines. In most cases, therefore, no loss in accuracy is involved by drawing the ΔG vs. T curve corresponding to an equation such as (34α), and using it to derive a simple equation of the type

$$\Delta G_T^\circ = \Delta H_x^\circ - T\Delta S_x^\circ \tag{36a}$$

ΔH_x° and ΔS_x° are two numerical values determining the $\Delta G^\circ/T$ relationship, and are not necessarily close in value to ΔH_{298}° and ΔS_{298}° for the reaction.

It will be demonstrated later (p. 185) that the change in molar heat contents in a reaction ΔC_p depends to the greatest extent on the change in number of moles of gaseous components when the reaction proceeds from one side to another. Thus in such a reaction as

$$\langle \text{Ta} \rangle + 2\tfrac{1}{2}(\text{Cl}_2) = \langle \text{TaCl}_5 \rangle$$

ΔC_p will have a relatively large value and there will be an appreciable curvature in the plot of ΔG_T against T over a range of a few hundred degrees. If, in addition, all the data used in evaluating ΔG_T are known with good accuracy, then one is justified in introducing a third term to express the $\Delta G_T/T$ relationship:

$$\Delta G_T^\circ = \Delta H_y^\circ + mT\log T + T\Delta S_y^\circ \tag{35a}$$

This equation may be derived by drawing the curve and selecting three points on it, or by direct calculation from eqn. (35a) using three temperatures. If the curves are to coincide at 300°, 700° and 1200°K, the relationship between the constants of eqns. (35) and (35a) are as follows:

$$\Delta H_y^\circ = \Delta H_{(0)}^\circ + 410 \times 10^3 b + 731 \times 10^{-5} c + 6{\cdot}25 d$$

$$m = a + 3{\cdot}125 \times 10^3 b + 1{\cdot}241 \times 10^{-5} c - 0{\cdot}0224 d$$

$$\Delta S_y^\circ = I - 8{\cdot}78 \times 10^3 b - 4{\cdot}37 \times 10^{-5} c + 0{\cdot}0927 d$$

Simplified ΔG° formulae for certain reactions are given in Table E.

In fact, the present authors use this method *only* when they wish to simplify cumbersome Gibbs energy equations published in the literature. On the other hand, when deriving a Gibbs energy equation from the heats and entropies at a given temperature and heat capacities, they rarely take the trouble of going through the elaborate procedure described for alumina on the preceding pages. Normally all the ΔC_p equations up to the temperature of the first boiling point of one of the reactants or products are derived and then the ΔC_p values are calculated at several temperatures, say, 400°, 700°, 1000°, 1300° and 1600°K. These values vary, of course, with temperature but not as much as the individual heat capacities provided that no transition from the condensed to the gaseous state occurs. The ΔC_p values are then averaged, and the operation described above is performed with the average, i.e. constant, ΔC_p value.

From what has been said above about the general accuracy of the other thermochemical properties, this method which is more economical in time should be fully adequate. It is only when all the data incorporated in the evaluation are really reliable and accurate, that the present authors would use a two-term expression for ΔC_p.

Turning back to the Gibbs energy of dissociation of alumina, ΔG° values have been calculated from eqns. (34α) and (34γ) and are given in Table II. Since the reactions involve only one gaseous component oxygen, its equilibrium pressure can be calculated from eqn. (30) which must be used in the form

$$\frac{2}{3}\Delta G_T^\circ = -4{\cdot}574T \log \frac{1}{p_{O_2}} = 4{\cdot}574T \log p_{O_2} \qquad (30\alpha, \gamma)$$

for formulae α and γ and for *one* mole oxygen. Pressures calculated from this equation are also tabulated in Table II.

Table II. Oxygen pressure of alumina and its reduction with hydrogen

°C	°K	ΔG_γ	p_{O_2} (atm)	ΔG_δ	p_{H_2O}/p_{H_2}
727	1000	−324,790	4×10^{-48}	+186,730	$2{\cdot}4 \times 10^{-14}$
927	1200	−308,950	3×10^{-38}	+179,000	$1{\cdot}3 \times 10^{-11}$
1127	1400	−293,200	3×10^{-31}	+171,400	$1{\cdot}2 \times 10^{-9}$

To illustrate the use of these data we can consider whether the reduction of alumina with hydrogen is a practical proposition.

The relevant equation

$$\langle Al_2O_3\rangle + 3(H_2) = 3(H_2O) + \{Al\} \qquad (\delta)$$

may be obtained by combining eqn. (γ) with

$$3(H_2) + 1\tfrac{1}{2}(O_2) - 3(H_2O) \qquad (\varepsilon)$$

so that

$$\Delta G_T^\circ(\delta) = \Delta G_T^\circ(\varepsilon) - \Delta G_T^\circ(\gamma) \qquad (34\delta)$$

We may now turn to the simplified ΔG formulae obtained in the manner described above. These are

$$\Delta G_T^\circ(\varepsilon) = -171{,}750 + 13{\cdot}44T \log T - 6{\cdot}63T$$

$$\Delta G_T^\circ(\gamma) = -405{,}760 - 3{\cdot}75T \log T + 92{\cdot}22T$$

which lead by subtraction to

$$\Delta G_T^\circ(\delta) = +234{,}010 + 17{\cdot}19T \log T - 98{\cdot}85T$$

With this the equilibrium constant of reaction (δ) can be calculated by means of eqn. (29)

$$\frac{1}{3}\Delta G_T^\circ(\delta) = -4{\cdot}574T \log \frac{p_{H_2O}}{p_{H_2}}$$

Table II also includes the values of this Gibbs energy change at 1000° to 1400°K. It is evident that this equilibrium is far towards the left of eqn. (δ) and the reduction could be carried out only in a stream of extremely dry hydrogen, containing less than 3×10^{-12} per cent of water vapour, at 1000°K. Since this purity is hardly attainable in practice it must be concluded that alumina cannot be reduced by hydrogen at these temperatures (a fact which is already well known to the metallurgist). Even if this dryness of the gas were attainable, the reaction would be impractical owing to the enormous volume of circulation of hydrogen which would be required to remove the oxygen at its minute equilibrium pressure.

We have discussed means by which equilibrium constants may be derived from Gibbs energies and other thermal data. Conversely, equilibrium pressure measurements can be used to determine Gibbs energies of reaction or, by suitable combinations, Gibbs energies of formation of compounds. Such reactions as the following are typical:

$$\langle FeO\rangle + (H_2) = \langle Fe\rangle + (H_2O) \qquad (\eta)$$

$$\langle V_2O_5\rangle + (CO) = 2\langle VO_2\rangle + (CO_2) \qquad (\theta)$$

$$\langle CrCl_2\rangle + (H_2) = \langle Cr\rangle + 2(HCl) \qquad (\iota)$$

$$\langle WC\rangle + 2(H_2) = \langle W\rangle + (CH_4) \qquad (\kappa)$$

The Gibbs energy of formation of the metal compounds may be evaluated by measuring the equilibrium of the reaction and combining it with the Gibbs energy data of the complementary reaction,

$$(H_2) + \tfrac{1}{2}(O_2) = (H_2O) \qquad (\eta')$$

$$(CO) + \tfrac{1}{2}(O_2) = (CO_2) \qquad (\theta')$$

$$(H_2) + (Cl_2) = 2(HCl) \qquad (\iota)$$

$$\langle C\rangle + 2(H_2) = (CH_4) \qquad (\kappa')$$

For instance,

$$\Delta G^\circ_T(FeO) = \Delta G^\circ_T(H_2O) - \Delta G^\circ_T(\eta) \qquad (34\eta)$$

Methods of measuring Gibbs energies may be summarised as follows:

1. Gibbs energies of substances or of reactions may be computed from heats of formation, entropies and specific heats, using the relationships expressed in eqns. (34) and (32a).
2. Gibbs energies of reaction or formation are obtained directly from measurements of the equilibrium constants of the reaction, applying the relationships in eqns. (29) and (30).
3. Gibbs energies of reaction are also directly obtainable from e.m.f. measurements if the reaction under consideration can be made to produce current in a galvanic cell, applying the relationship in eqn. (31).
4. Gibbs energies of reaction or formation may finally be obtained by suitable combination of other Gibbs energy equations, as was shown in two examples: eqns. (34δ) and (34η).

3. Solutions

A. PARTIAL MOLAR GIBBS ENERGY

In the foregoing sections we have been dealing with elements or substances which are stable only at a fixed composition of simple atomic proportions, i.e. those to which Dalton's rule of simple and multiple proportions applies to a close approximation. This inflexibility of composition characterises organic and inorganic compounds in which the atoms are held together by the major valency bonds (homopolar and heteropolar). These bonds arise from the sharing or donation of certain of the electrons of the various atoms. Such compounds generally have an ordered structure in the stable solid state, and consequently they have no significant entropy at absolute zero.

The "metallic" binding forces which are found generally in alloy systems do not arise from such inflexible anisotropic exchanges of electrons. Rather, the constituent atoms are (to express it simply) kept in an ordered array by the so-called "electron gas". This binding does not require the constituent atomic species to be present in a simple stoichiometric proportion, and so intermetallic phases are often found to be stable within quite wide composition limits. These are defined by the relative ionic sizes and by the relative strengths of the metallic, homopolar and heteropolar components of the overall binding forces. These forces are rarely present singly in intermetallic phases.

Mixed binding forces also account for the relatively large ranges of homogeneity present in certain systems of oxygen, sulphur and nitrogen with transition metals, such as the titanium–oxygen, zirconium–oxygen, titanium–nitrogen, cobalt–sulphur and nickel–sulphur systems. It is therefore not surprising that "compounds" such as TiO, TiN, VN and MoS_2 exhibit metallic conductivity.

The thermochemical treatment of these phases of variable composition differs from that of pure, fixed compounds. Similar considerations are applicable to liquid mixtures, such as those of metals or of inorganic salts, which are often completely soluble. To illustrate the modified treatment we shall discuss an example of each of the extreme types of system: the system silver–gold (Fig. 7b), which consists at all proportions of a homogeneous, disordered solid solution; and the system tellurium–tin (Fig. 7a) which contains a compound of definite composition and no significant range of solid solution.

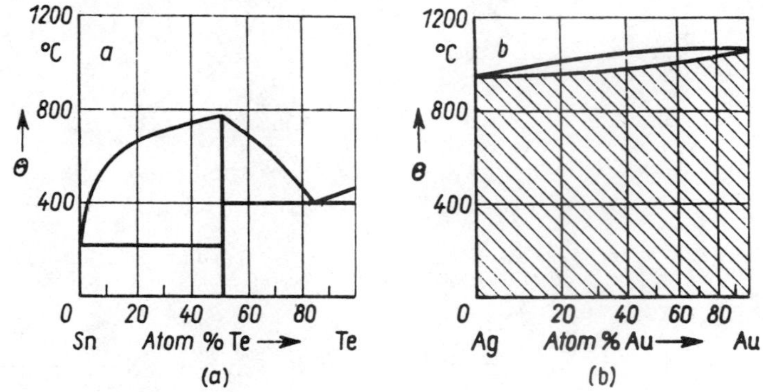

FIG. 7. Phase diagrams of the systems: (a) tin–tellurium and (b) silver–gold.

To compare these systems we must devise an experiment which shows the variation in Gibbs energy of formation over the complete range of alloys.

This is most conveniently done by constructing suitable galvanic cells, the e.m.f. of which is a measure of the Gibbs energies concerned. The negative electrode of such a cell is formed of the more electronegative of the two metals, and its positive electrode consists of the alloy itself. The cation of the electrolyte is the ionised form of the anodic metal. The cell can be represented schematically by

$$\ominus\ \mathrm{M^I}\,|\,\mathrm{M^I} - (\text{cation})\,|\,\mathrm{M^I_x M^{II}_y}\ \oplus$$

When the cell is in operation, the metal M^I is transferred into the alloy $M^I_x M^{II}_y$. The pontentiometric method used in measuring the e.m.f. involves no transfer of M^I, so in practice there should be no change in composition of the alloy. Actually, however, there is always a transportation of a small amount of material which must be considered in assessing the accuracy of the method (see p. 157).

The two alloy systems we are comparing have actually been investigated experimentally by this method.

The tin–tellurium cell [3] for all the alloy compositions between SnTe and Te may be represented by

$$\ominus\ \{\mathrm{Sn}\}\,|\,\{\mathrm{SnCl_2}\}\,|\,\langle\mathrm{Te}\rangle + \langle\mathrm{SnTe}\rangle\ \oplus$$

The e.m.f. is constant over the whole of this range (Fig. 8a), being due to the formation of ⟨SnTe⟩ from liquid tin and solid tellurium. The Gibbs energy change

$$\Delta G^\circ = -2 \times 23{,}066E$$

is that arising from the combination of one mole {Sn} with one mole ⟨Te⟩. Virtually no e.m.f. would be produced by an alloy of composition between SnTe and pure tin, since there would be no free tellurium present to combine with the tin.

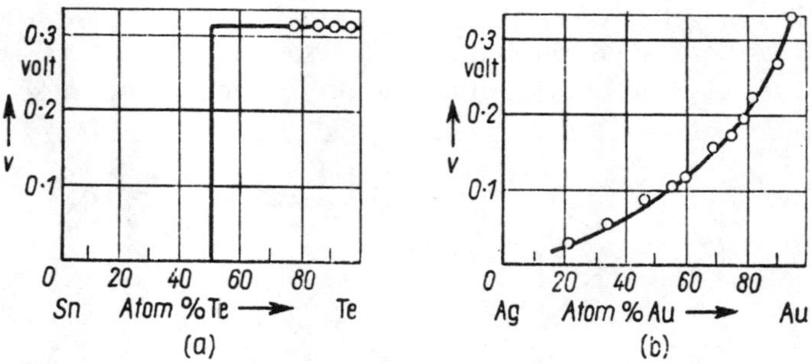

FIG. 8. E.m.f.s of cells with alloy concentration electrodes: (a) tin and tin–tellurium alloys, (b) silver and silver–gold alloys.

The gold–silver cell which was studied [4] may be represented schematically by

$$\ominus\ \langle\mathrm{Ag}\rangle\,|\,[\mathrm{Ag^+}]\text{ in glass}\,|\,\langle\mathrm{Ag_xAu_y}\rangle\ \oplus$$

Its e.m.f., arising from the transfer of ⟨Ag⟩ into the $\langle Ag_xAu_y\rangle$ solid solution, is found to vary over the whole alloy composition range (Fig. 8b). The Gibbs energy change calculated from each composition, from the formula

$$\Delta\bar{G}_{Ag} = -1 \times 23{,}066E$$

is that of the assimilation of 1 mole ⟨Ag⟩ by an infinite amount of the alloy $\langle Ag_xAu_y\rangle$—(i.e. the composition of the solid solution is unchanged by the solution of the silver). It is to be expected that this Gibbs energy change will increase as the gold content increases, since the silver atoms liberated at the electropositive electrode will find a greater concentration of atoms of the other species for which they have an affinity in the gold-rich alloys. The Gibbs energy change is described more fully as the partial molar Gibbs energy of solution of ⟨Ag⟩ in $\langle Ag_xAu_y\rangle$. There are corresponding enthalpies and entropies and these partial molar thermodynamic functions are distinguished by a bar over the symbol, e.g. $\Delta\bar{G}$.

We may enlarge upon the comparison of integral and molar thermodynamic functions by considering a binary alloy to which both types of functions apply.

Integral Gibbs energies are those which are involved in the formation of an alloy from its constituent elements:

$$xA + yB = A_xB_y; \qquad \Delta G(\Delta H, \Delta S) \qquad \text{[integral values]}$$

The quantities x and y can either be in atomic fractions, N_A and N_B (e.g. $Mg_{0 \cdot 6}/Bi_{0 \cdot 4}$), when ΔG, ΔH and ΔS will be expressed in gram-atomic quantities, or denote mole quantities, n_A and n_B (e.g. Mg_3Bi_2), when ΔG, ΔH and ΔS will be expressed in terms of moles.

The partial molar Gibbs energies, etc., are related to the reaction

$$1A + A_mB_n = A_{m+1}B_n; \qquad \Delta\bar{G}(\Delta\bar{H}, \Delta\bar{S}) \qquad \text{[partial values]}$$

where the ratio m/n is equal to the ratio x/y in the previous equation, but the magnitude of m and n is such that $(m+1)$ practically equals m. The partial molar thermodynamic functions are generally expressed in terms of 1 g-atom of the constituent A.

Taking the example of the silver–gold alloys, we can regard the partial Gibbs energy change as the difference between the Gibbs energy of the silver in the pure state and its Gibbs energy in solution in the alloy. These two quantities are often used and referred to (after Gibbs) as *chemical potentials*, denoted by μ. Using the term μ_{Ag} for the chemical potential of the pure metal and $\mu_{[Ag]}$ for that of the dissolved metal, we have

$$\Delta\bar{G}_{Ag} = \mu_{[Ag]} - \mu_{Ag}$$

Chemical potentials are used often in theoretical discussions but rarely in the numerical calculations of the type with which we are concerned. They must be assigned an arbitrary value for substances in a standard state, so are no more absolute than Gibbs energies. We are only concerned with difference in chemical potential and we can relate this to the Gibbs energy G°_T of the substance concerned in its pure state. The authors favour the use of the symbol $\Delta\bar{G}$ in place of the appropriate term in μ so as to emphasise its connection with the other partial molar functions $\Delta\bar{H}$ and $\Delta\bar{S}$ (cf. ΔG, ΔH and ΔS).

In practice it will be found that e.m.f. measurements on concentration electrodes and vapour pressure measurements are primarily methods of deriving partial molar functions, while calorimetric methods usually provide the integral values.

In this section we have referred to measurements of e.m.f.s largely as an illustration of partial thermodynamic functions. We may finally summarise the relationships as follows:

$$\Delta \bar{G}_A = -z \cdot 23{,}066 E_A \tag{41}$$

$$\Delta \bar{G}_A = \Delta \bar{H}_A - T \Delta \bar{S}_A \tag{42}$$

$$\Delta \bar{H}_A = z \cdot 23{,}066 \left(T \frac{dE_A}{dT} - E_A \right) \tag{43}$$

$$\Delta \bar{H}_A = z \cdot 23{,}066 \left(\frac{T_1 + T_2}{2} \frac{E_2 - E_1}{T_2 - T_1} - \frac{E_1 + E_2}{2} \right) \tag{43a}$$

$$\Delta \bar{S}_A = z \cdot 23{,}066 \frac{dE_A}{dT} \tag{44}$$

$$\Delta \bar{S}_A = z \cdot 23{,}066 \frac{E_2 - E_1}{T_2 - T_1} \tag{44a}$$

These equations refer to the reaction $A + A_m B_n = A_{m+1} B_n$, as defined on p. 29, and E is measured in volts.

The integral values of the thermodynamic functions of a solution are related to the partial values by the equations

$$\Delta G = N_A \Delta \bar{G}_A + N_B \Delta \bar{G}_B \tag{45}$$

$$\Delta H = N_A \Delta \bar{H}_A + N_B \Delta \bar{H}_B \tag{45a}$$

$$\Delta S = N_A \Delta \bar{S}_A + N_B \Delta \bar{S}_B \tag{45b}$$

where the integral values are those for a solution having mole fractions of A and B equal to N_A and N_B respectively.

B. VAPOUR PRESSURES AND PARTIAL GIBBS ENERGIES

The Gibbs energy change when one mole of an ideal gas undergoes a change in pressure, at constant temperature, from p_1 to p_2 is given by

$$\Delta G = \boldsymbol{R} T \ln \frac{p_2}{p_1} \tag{46}$$

This expression may be applied to a parallel series of measurements which could be made with the tin–tellurium and silver–gold alloys which were studied by the electrochemical method above. Over each alloy composition there exists a gaseous phase which contains partial pressures of the vapours of each component of the alloy. The partial pressure of each component vapour changes with temperature and composition of the alloy, maintaining the vapour-condensed phase equilibrium. It therefore follows that the Gibbs energy change when one mole of a pure component is added isothermally to a large volume of the alloy so as not to change composition is equal to the Gibbs energy change when one mole of the vapour of the component undergoes the pressure change from p°, the vapour pressure of the pure component at the given temperature, to p the vapour pressure exerted by the component over the alloy. This process can be visualised as follows:

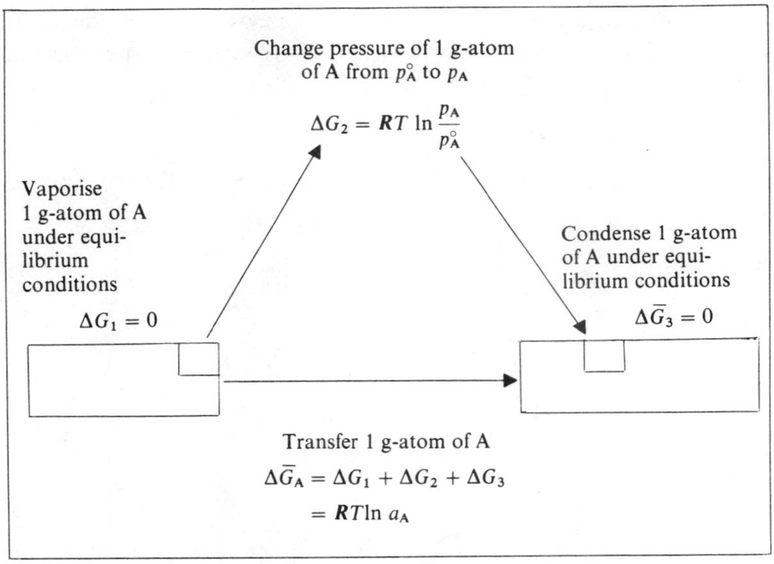

Since these paths constitute a closed cycle the Gibbs energy changes must be the same by either route, hence

$$\Delta \bar{G}_A = \boldsymbol{RT} \ln \frac{p_A}{p_A^\circ} \tag{47}$$

It thus follows that the partial vapour pressure of a component over an alloy will change in exactly the same way as the partial Gibbs energy of solution of the component, its absolute value depending on the vapour pressure of the pure component at the given temperature. Moreover, as the *ratio* of vapour pressures between the pure and solution states is related to our useful thermodynamic function, we shall give this a special name, the *activity* and the symbol a. Thus

$$\Delta \bar{G}_A = \boldsymbol{RT} \ln a_A \tag{47a}$$

This expression has no immediately clear relationship with the composition of the alloy but we may interrelate the two by an expression $a_i = \gamma_i N_i$ where γ_i is called the activity coefficient of the ith component of the alloy and the product $\gamma_i N_i$ must have values between the limits zero and unity. It therefore follows that

$$\Delta \bar{G}_i = \boldsymbol{RT} \ln \gamma_i + \boldsymbol{RT} \ln N_i \tag{48}$$

There are two particularly simple forms of this expression which are sometimes found for the behaviour of the partial Gibbs energy of a component as a function of its mole fraction. The first of these is when γ has the value unity and

$$a_i = N_i \tag{49}$$

This behaviour was first suggested as a general form by Raoult and the equality is a statement of Raoult's law. However, very few systems show this behaviour over wide ranges of composition.

The second simplified form is when γ_i is constant but not necessarily unity over a range of composition. This behaviour which is frequently shown by dilute solutions is called Henry's law and can be stated in the form

$$a_i = k\, N_i \tag{50}$$

where k is a constant for the system, and is represented by the tangents indicated in Fig. 9 as broken lines.

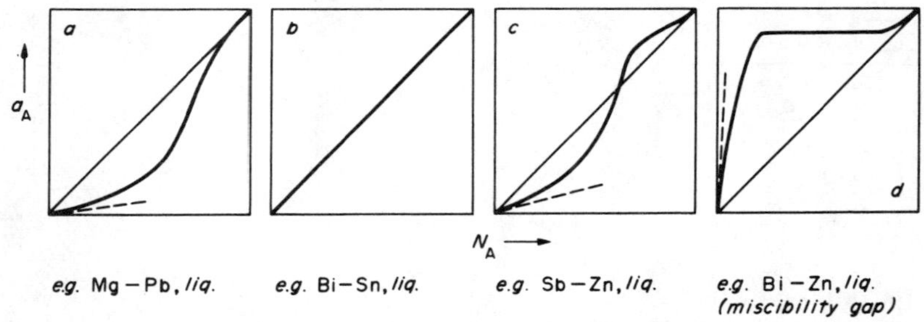

FIG. 9. Activities of single components of binary mixtures (schematic).

There is no simple law to express the vapour pressure of a component of a binary mixture at intermediate concentrations. Both Henry's law and Raoult's law are limiting generalisations, but they are often found to apply over quite large ranges of concentration. In fact they facilitate the construction of activity curves of substances with complete mutual solubility even when only a few measurements are carried out.

The activity of a substance can be derived by measuring the ratio of its vapour pressures over the solution and in the pure state. Its partial molar Gibbs energy of solution is simply related to its activity, as will be seen by considering as an example the zinc–aluminium alloys in the liquid state. We can make the following summation:

$$\begin{array}{lrcll} + & \{Zn\} & = & (Zn); & p_{Zn}, \Delta G_0 \\ - & \{Al_mZn_{(n+1)}\} & = & \{Al_mZn_n\} + (Zn); & p'_{Zn}, \Delta\bar{G}_1 \\ \hline & \{Al_mZn_n\} + \{Zn\} & = & \{Al_mZn_{(n+1)}\}; & \Delta\bar{G}_{Zn} \end{array}$$

Here, ΔG_0 is the Gibbs energy of evaporation of one mole (Zn), while $-\Delta\bar{G}_1$ and $\Delta\bar{G}_{Zn}$ are the partial molar Gibbs energies of solution of gaseous and liquid zinc respectively in the alloy $\{Al_mZn_n\}$.

Since

$$\Delta G_0 = -\boldsymbol{R}T \ln p_{Zn}, \qquad \Delta\bar{G}_1 = -\boldsymbol{R}T \ln p'_{Zn}$$

and

$$\Delta\bar{G}_{Zn} = \Delta G_0 - \Delta\bar{G}_1$$

then

$$\Delta\bar{G}_{Zn} = \boldsymbol{R}T \ln \frac{p'_{Zn}}{p_{Zn}} = \boldsymbol{R}T \ln a_{Zn} = 4{\cdot}574T \log a_{Zn} \tag{51}$$

Activities can be derived from measurements of any equilibria involving a solution and a gaseous phase. Equilibria with hydrogen–steam mixtures, for example, are often used in experimental determinations. This may be illustrated by considering the iron–nickel alloys (which form a continuous series of solid solutions at high temperatures). The activity of the iron is obtained by determining the H_2O/H_2 ratio of the vapour in equilibrium with FeO in contact with the alloys and pure iron:

$$
\begin{array}{llr}
- & \langle FeO\rangle + (H_2) = \langle Fe\rangle + (H_2O); & \dfrac{p_{H_2O}}{p_{H_2}}, \Delta G_0 \\
+ & \langle Fe_mNi_n\rangle + \langle FeO\rangle + (H_2) = \langle Fe_{m+1}Ni_n\rangle + (H_2O); & \dfrac{p'_{H_2O}}{p'_{H_2}}, \Delta\bar{G}_1 \\
\hline
 & \langle Fe_mNi_n\rangle + \langle Fe\rangle = \langle Fe_{m+1}Ni_n\rangle & \Delta\bar{G}_{Fe}
\end{array}
$$

$$a_{Fe} = \frac{p_{H_2O}/p_{H_2}}{p'_{H_2O}/p'_{H_2}}$$

$$\Delta\bar{G}_{Fe} = RT \ln a_{Fe} \tag{51a}$$

Just as partial molar heats and entropies of solution were obtained from e.m.f. values (p. 30), so we may apply the Gibbs–Helmholtz equation to their derivation from activities.

$$\Delta\bar{G}_A = RT \ln a_A = 4{\cdot}574T \log a_A \tag{51b}$$

$$\Delta\bar{H}_A = -4{\cdot}574T^2 \frac{d \log a_A}{dT} \tag{52}$$

$$\Delta\bar{H}_A = -4{\cdot}574T_1T_2 \frac{\log a_2 - \log a_1}{T_2 - T_1} \tag{52b}$$

$$\Delta\bar{S}_A = -4{\cdot}574\left(T \frac{d \log a_A}{dT} + \log a_A\right) \tag{53}$$

$$\Delta\bar{S}_A = -4{\cdot}574\left(\frac{T_1 + T_2}{2}\frac{\log a_2 - \log a_1}{T_2 - T_1} + \frac{\log a_2 + \log a_1}{2}\right) \tag{53a}$$

Since the Gibbs energy of a substance in solution is measurable by its e.m.f. its activity can be determined in the same way.

$$\Delta\bar{G}_A = -zFE$$

$$\Delta\bar{G}_A = RT \ln a_A$$

Hence

$$\ln a_A = -\frac{zF}{RT} E \tag{54}$$

or, introducing the numerical values of the constants.

$$\log a_A = -\frac{5043zE}{T} \tag{54a}$$

We have thus summarised the main sources of activity values. Two further important methods of deriving activities, from equilibrium diagrams and from equilibria in a melt, will be discussed at a later stage.

C. THE CALCULATION OF INTEGRAL VALUES FROM PARTIAL MOLAR THERMOCHEMICAL DATA

Frequently the only experimental methods applicable to the study of a series of solutions are those which lead in the first instance to partial molar data. The integral values of the thermochemical functions are generally derived from these by graphical methods.

For simplicity we shall first consider a system the constituents of which, A and B, form a continuous series of solid solutions. Generally the partial molar values for *both* constituents of a solution will not be available from experimental measurements. Phase rule considerations show that the partial vapour pressures of the two components of a binary system are interdependent. From this we can deduce that their partial molar thermodynamic functions must also be related, and this relationship is given by the development by Margules of a generalisation of Gibbs and Duhem.

The Gibbs–Duhem equation can be used to calculate the change in partial molar Gibbs energy of B over the composition range N'_B to N''_B when the property is known for A over the same composition range. Thus

$$\ln \frac{a'_B}{a''_B} = -\int_{N'_B}^{N''_B} \frac{N_A}{N_B} \, d \ln a_A \tag{55}$$

Obviously the value of the activity of B can be calculated by this means provided that it is known at one of the compositional limits.

For the application of this equation it is important to realise that when no information is available concerning the properties of B, the partial molar functions of the other component A must be known across the whole range of concentration $N_A = 0$–1, and particularly accurately at low concentration of the solute—a condition not easily achieved experimentally. Since the partial molar entropies and Gibbs energies of disordered solutions tend towards infinity when zero concentration is approached, it is convenient to introduce the so-called excess functions $\Delta\bar{S}^E_A$ and $\Delta\bar{G}^E_A$ which are given by

$$\Delta\bar{S}^E_A = \Delta\bar{S}^{xptl}_A + \boldsymbol{R} \ln N_A \tag{56}$$

and consequently

$$\Delta\bar{G}^E_A = \Delta\bar{G}^{xptl}_A - \boldsymbol{R}T \ln N_A = \boldsymbol{R}T \ln \gamma_A \tag{57}$$

and correspondingly for a component B.

The Gibbs–Duhem equation can be used in the analogous form

$$\ln \frac{\gamma'_B}{\gamma''_B} = -\int_{N'_B}^{N''_B} \frac{N_A}{N_B} \, d \ln \gamma_A \tag{58}$$

since $dN_A = -dN_B$ in a binary system. Because the activity coefficient γ_A reaches a constant value for the dilute solution of A in B and has the value of unity close to pure A, the integral involving the excess Gibbs energy can be accurately evaluated whereas the integral which makes use of activity data directly must be estimated because a_A tends to zero as N_A tends to the same limit.

When it is desirable to compute the concentration dependence of the *integral* molar thermodynamic functions from measured values of the partial molar functions of one of the components, the Duhem–Margules equation may be used in the following form, where the symbol Z stands for H, S^E and G^E.

$$\Delta Z_f = N_B \int_0^{N_A} \frac{\Delta \bar{Z}_A}{N_B^2} \, dN_A \tag{59}$$

It is sometimes convenient to use the following formulation:

$$\Delta Z_f = N_B \int_0^{x} \Delta \bar{Z}_A \, dx \tag{60}$$

where $x = N_A/N_B$. The evaluation of the Duhem–Margules equation may be illustrated by one example. The e.m.f. values of Fig. 8b have been converted into excess Gibbs energies and are plotted in Fig. 10a against the atomic ratio N_{Ag}/N_{Au}, which represents the number of g-atoms of silver associated with 1 g-atom of gold at a given concentration. The integral excess Gibbs energy per g-atom of the solution the composition of which is represented by x is obtained by dividing the area under the curve between atomic ratios 0 and x by the quantity $(x + 1)$. These integral values are plotted against the atomic fractions for the silver–gold system in Fig. 10b.

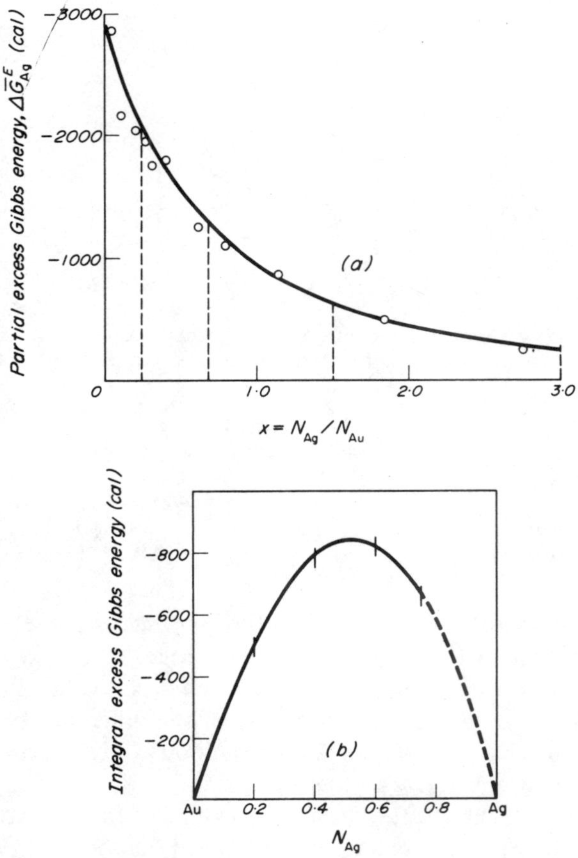

FIG. 10. Partial (a) and integral (b) molar excess Gibbs energies in the system silver–gold.

The integral heats and excess entropies of formation of the solution are obtained in the same manner from $\Delta \bar{H}_A$ as $\Delta \bar{S}_A^E$ respectively.

The integration of partial molar curves is more complicated if the system contains several homogeneous phases separated by heterogeneous fields. To clarify the problem we will first discuss the connection between partial and integral molar values of thermochemical functions in the two limiting types of system. The continuous lines in Fig. 11 show the integral excess Gibbs energies of (a) a binary system containing two compounds but with no significant range of solid solubility, and (b) a binary system forming a continuous series of solid solutions. The tangent to these lines at any point cuts the $N_A = 0$ and $N_A = 1$ ordinates at the values of the partial Gibbs energies of B and A respectively. While these vary continuously through the solid solution phase, they remain constant in heterogeneous phases of two constituents of varying proportion but of fixed composition.

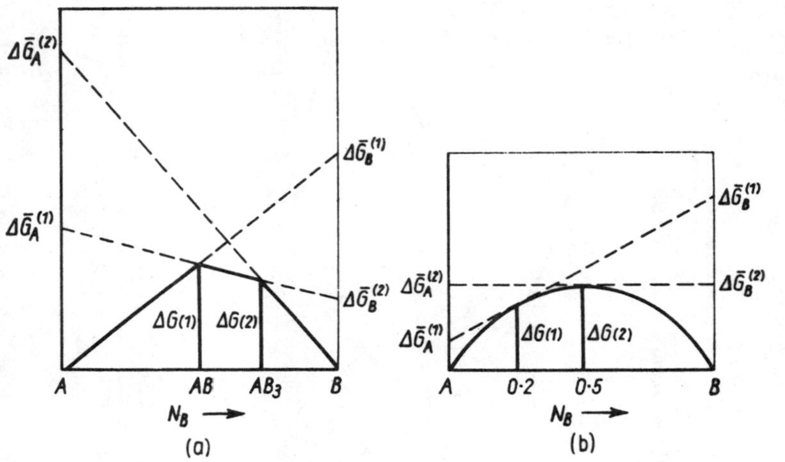

FIG. 11. Relationship between partial and integral excess Gibbs functions in (a) a system forming intermediate compounds, and (b) a system exhibiting complete mutual solubility.

FIG. 11a.

$\Delta G(1)$: $0{\cdot}5A + 0{\cdot}5B = A_{0\cdot5}B_{0\cdot5}$
$\Delta G(2)$: $0{\cdot}25A + 0{\cdot}75B = A_{0\cdot25}B_{0\cdot75}$
$\Delta\bar{G}_B(1)$: $1B + \infty(xAB + yA)$
$= 1AB + \infty(xAB + yA)$
$\Delta\bar{G}_B(2)$: $1B + \infty(xAB_3 + yAB)$
$= 1AB_3 + \infty(xAB_3 + yAB)$

FIG. 11b.

$\Delta G(1)$: $0{\cdot}2B + 0{\cdot}8A = A_{0\cdot8}B_{0\cdot2}$
$\Delta G(2)$: $0{\cdot}5B + 0{\cdot}5A = A_{0\cdot5}B_{0\cdot5}$
$\Delta\bar{G}_B(1)$: $1B + \infty(A_{0\cdot8}B_{0\cdot2})$
$= 1A_4B + \infty(A_{0\cdot8}B_{0\cdot2})$
$\Delta\bar{G}_B(2)$: $B + \infty(A_{0\cdot5}B_{0\cdot5})$
$= 1AB + \infty(A_{0\cdot5}B_{0\cdot5})$

A continuous series of solid solutions gives the continuous curve shown in Fig. 12. In the other type of system the vapour pressure changes in a stepwise manner with the formation of a new compound of higher content of volatile constituent. The magnitude of the pressure increase is a measure of the difference between the Gibbs energy of the phase richer in component A and the total Gibbs energy of component A plus the phase poorer in A (e.g. $G_{(A_3B)} - G_{(AB+2A)}$).

It is often difficult to measure the reaction pressure of the lowest step on account of its very low value. The measurement must be carried out, however, since the integral thermochemical data of the whole system will be incomplete if the values associated with this low vapour pressure are not known. If a pressure measurement cannot be applied, the data must be obtained in some other way.

Consider a two-component system in which A is volatile and for simplicity monatomic, and B and the intermediate compounds AB_3, AB, A_3B are condensed phases.

Let the vapour pressures of AB_3, AB, A_3B and A be respectively p_1, p_2, p_3 and p_4.

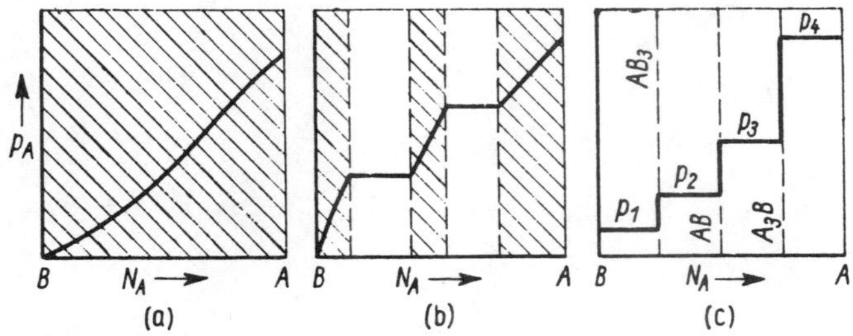

FIG. 12. Schematic vapour pressure curves of (a) a homogeneous system, (b) a three-phase system, and (c) a system containing three compounds.

The Gibbs energy change of each reaction step may be expressed as follows:

$$3B + A = AB_3; \qquad \Delta G_1 = 4{\cdot}574T \log p_1/p_4$$

$$\tfrac{1}{2}AB_3 + A = 1\tfrac{1}{2}AB; \qquad \Delta\overline{G}_2 = 4{\cdot}574T \log p_2/p_4$$

$$\tfrac{1}{2}AB + A = \tfrac{1}{2}A_3B; \qquad \Delta\overline{G}_3 = 4{\cdot}574T \log p_3/p_4$$

Then the integral free energies are given by:

$$3B + A = AB_3; \qquad \Delta G_1$$

$$B + A = AB; \qquad \Delta G_2 = \tfrac{1}{3}\Delta G_1 + \tfrac{2}{3}\Delta\overline{G}_2$$

$$B + 3A = A_3B; \qquad \Delta G_3 = \Delta G_2 + 2\Delta\overline{G}_3$$

Thus all the integral Gibbs energy values depend upon the determination of pressure p_1.

A similar limitation is found in the evaluation of Gibbs energy changes from curves of the type shown in Fig. 12. The accuracy of the integral thermochemical data derived from such curves depends upon the lowest concentration of A at which its vapour pressure can be measured. Below this concentration, an extrapolation must be made, and since its curvature can vary considerably, it is necessary that it should be as short as possible to minimise the error.

Methods for the determination of Gibbs energies (e.g. vapour pressure, e.m.f.) generally lead to *partial* thermodynamic functions. Calorimetric methods, on the other hand, normally produce *integral* heats of formation. The best procedure for assessing the thermodynamic properties of binary systems with homogeneity ranges in such cases is to differentiate the enthalpy vs. concentration curves, to calculate the partial entropies from the partial Gibbs energies and the differentiated heats, and then to integrate the partial excess entropies. Unfortunately, the integral enthalpies have rarely been obtained with the high accuracy necessary for a reliable differentiation.

Differentiation can be carried out graphically as indicated in Fig. 11b. This method is not very precise, and the results must be checked by subsequent graphical integration.

For the differentiation of relatively simple enthalpy (or excess Gibbs energy) vs. concentration curves such as that in Fig. 11b, Guggenheim [5] has suggested the following form of the series expansion for the representation of the integral heats, such as

$$\Delta H_f = N_A N_B[a_0 + a_1(N_A - N_B) + a_2(N_A - N_B)^2 + a_3(N_A - N_B)^3] \tag{61}$$

where a_0, a_1, a_2, and a_3 are coefficients to be derived from the experimental data. For the partial molar heats one obtains:

$$\Delta \bar{H}_A = N_B^2[a_0 + a_1(3N_A - N_B) + a_2(5N_A - N_B)(N_A - N_B) + a_3(7N_A - N_B)(N_A - N_B)^2] \tag{62a}$$

$$\Delta \bar{H}_B = N_A^2[a_0 + a_1(N_A - 3N_B) + a_2(N_A - 5N_B)(N_A - N_B) + a_3(N_A - 7N_B)(N_A - N_B)^2] \tag{62b}$$

Of course, for simple enthalpy curves one can reduce the number of terms. If $a_1 = a_2 = a_3 = 0$, eqns. (61)–(62b) reduce to:

$$\Delta H_f = a_0 N_A N_B \tag{63}$$

$$\Delta \bar{H}_A = a_0 N_B^2 \quad \text{and} \quad \Delta \bar{H}_B = a_0 N_A^2 \tag{64}$$

Truly symmetric heat vs. concentration curves are, however, rarely encountered in practice.

From accurate experimental heats of mixing in the liquid solutions of mercury and lead at 335°C, we have the following values:

ΔH_f	−28	−17	+15	+55	+94	+126	+144	+140	+98 cal/g-atom
N_{Hg}	0·1	0·2	0·3	0·4	0·5	0·6	0·7	0·8	0·9

Applying eqn. (61) to four representative compositions (say, $N_{Hg} = 0{\cdot}1, 0{\cdot}3, 0{\cdot}5$ and $0{\cdot}7$), we find the following coefficients:

$$a_0 = +377, \quad a_1 = +737, \quad a_2 = +16, \quad a_3 = +215$$

There are many instances in the literature where values for the functions $\Delta H_f/N_A N_B$ and $\Delta G_f^E/N_A N_B$ of binary systems are described by the use of a simple power series. Thus

$$\Delta G_f^E/N_A N_B = \sum_{i=0} a_i N_B^i \tag{65}$$

When the results are few for a given system, only a few terms are used in the power series development. However, when more data become available it is frequently found that a more precise description can be made using more terms in the series expansion.

For the simple power series each stage of adjustment usually means a drastic reworking of the expansion with new values for each coefficient a_i. This recalculation is not too time consuming if it is fitted to a flexible computer program, but the changes which occur in the values of a_i at each stage of the "refinement" should caution the theoretician not to make too much "theory" around the values at any given stage of development.

Bale and Pelton have shown that there are advantages in using the Legendre polynomials for the expansion of thermochemical data [6]. The power series discussed previously can now be replaced by the expansion

$$\Delta G_f^E/N_A N_B = a_0(1) + a_1(2N_B - 1) + a_2(6N_B^2 - 6N_B + 1) + a_3(20N_B^3 - 30N_B^2 - 12N_B - 1) + \cdots \tag{66}$$

The quantities in the brackets are the first four Legendre polynomials and the continuation of the series may be made by using the recursion formula for the nth polynomial

$$P_n = \frac{(2n-1)(2N_B - 1)}{n} P_{n-1} - \frac{(n-1)}{n} P_{n-2} \tag{67}$$

It is found, in a number of typical cases, that refinement of the experimental data merely adds further polynomials to an already established expansion without changing the coefficients of the preceding terms very much. Thus the introduction of the fourth polynomial in an expansion which already described the data reasonably well with only three polynomials would not change the values of the coefficients a_1, a_2 and a_3 significantly.

In a complex system which includes heterogeneous and homogeneous regions (e.g. brasses, bronzes and many sulphide systems) the reaction pressure vs. concentration curve assumes the form shown in Fig. 12b. There is a continuous but not necessarily linear change in pressure (or activity) across the homogeneous phase fields. In the heterogeneous fields the reaction pressure of A is again determined by its pressure in the boundary phase richer in component A.

Examples of the two latter types of reaction pressure curves are given in Figs. 13 and 14. The first shows the mercury pressure of the cerium amalgams [7]. There is one compound, $CeHg_4$, in the range considered. The pressure does not increase in perfect rectangular steps, the "corners" on the curve being rounded. This may be due to a small variable concentration range of existence for the phase $CeHg_4$, but more probably results from the experimental method used in its determination. The volatilisation of mercury to fill the relatively large volume of the apparatus resulted in an impoverishment of the condensed phase which should have been of stoichiometric composition.

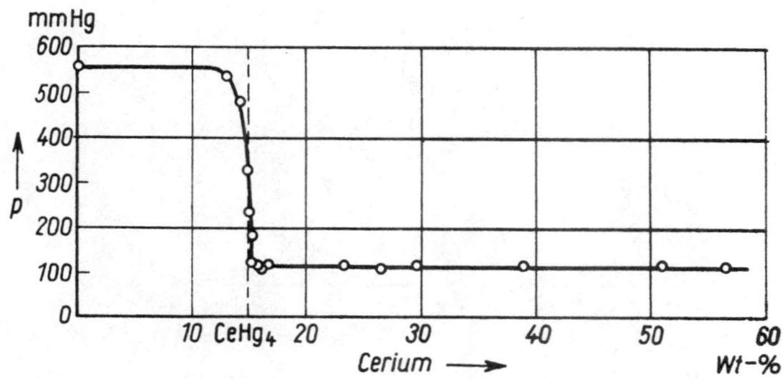

FIG. 13. Mercury pressures of cerium amalgams at 340°C.

Figure 14 shows the partial molar Gibbs energy curve for oxygen in iron, which is described on p. 44.

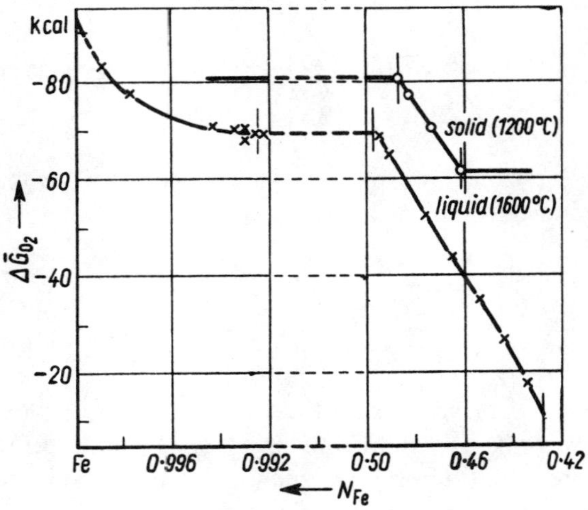

FIG. 14. Partial molar Gibbs energy of oxygen in liquid and solid iron.

The thermodynamic properties of the copper–zinc system are well established [18]. The partial Gibbs energy of solution of liquid zinc in the alloys at 500°C is shown in Fig. 15. As has been discussed above, the Gibbs energies of solution are represented by horizontal lines in the heterogeneous ranges indicated by the vertical lines, while $\Delta\bar{G}_{Zn}$ increases with the zinc content in the homogeneous ranges.

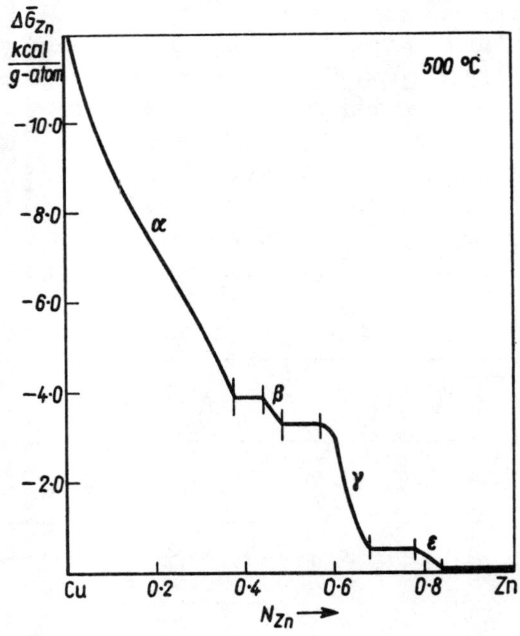

FIG. 15. Partial molar Gibbs energies of zinc in the solid copper–zinc system at 500°C. The liquid is the standard state for zinc.

The partial heats of solution of liquid zinc in the various phases and mixtures of phases is shown in Fig. 16. In the heterogeneous ranges the partial heats are also represented by horizontal lines but no longer form a continuous curve with the individual sections of curve of the homogeneous phases as in the ΔG_{Zn}–N_{Zn} diagram. In fact, the relative positions of the horizontal and curved ΔH_{Zn}–N_{Zn} lines determines the slopes of the phase boundaries with temperature.

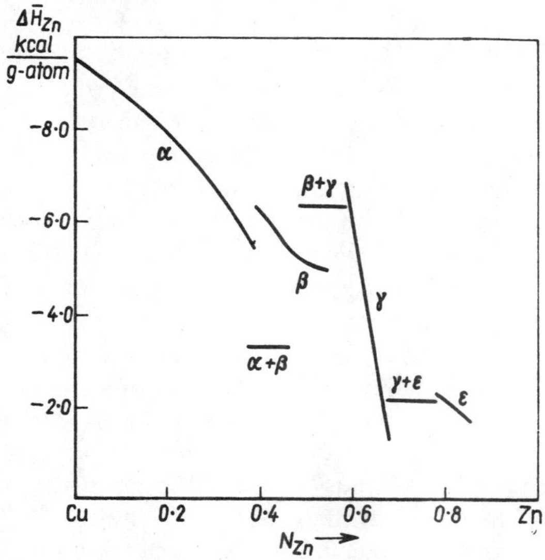

FIG. 16. Partial molar heats of solution of liquid zinc in solid copper–zinc alloys.

The integral Gibbs energies have been derived in the manner described earlier and are shown as a full line in Fig. 17. In the heterogeneous ranges the curve is a straight

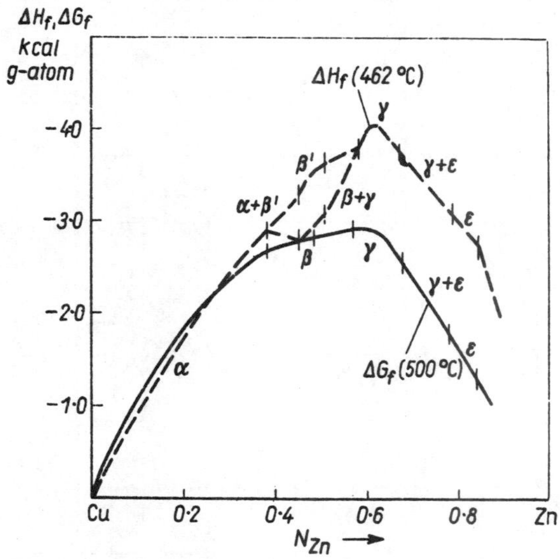

FIG. 17. Integral heats of formation at 462°C and Gibbs energies of formation at 500°C of solid alloys in the copper–zinc system. The reference states are solid copper and liquid zinc.

line. Since there must be no reversal in the dG/dN curve, the integral Gibbs energy curve must not show any indentation; otherwise the phase would be unstable in that range.

This condition does not apply to the heat vs. concentration curve since an indentation can be compensated by a corresponding bulge in the entropy vs. concentration curve.

Also shown in Fig. 17 it is the heat curve integrated for a temperature of about 462°C at which an order–disorder transformation of the b.c.c. β phase occurs. Since the heats of the order–disorder transformation are known, the heat of formation of ordered β' has also been plotted, the Greek letters indicating the phases in the copper–zinc diagram to which the various portions of the thermochemical curves refer.

A more detailed assessment of the thermodynamic of all the phases in the copper–zinc system has been demonstrated by Lumsden [8] who used all the available e.m.f., vapour pressure and calorimetric measurements as well as the established phase boundaries. Lumsden's book may be perused for further studies.

Another system of practical significance is that of the titanium oxides. There are several phases in this system, some of them having quite a wide composition range of homogeneity. The partial molar Gibbs energy of assimilation of oxygen has been obtained for three compositions in the α and β phases by equilibration with Ca/CaO, Mg/MgO and Ba/BaO mixtures at 1000°C. These results were combined with the thermochemical data of the "compounds" (Table A) to construct $\Delta\bar{G}_{O_2}$ curve shown in Fig. 18. The branch of the curve in the concentration range Ti–TiO was obtained in the following manner. The experimental points (circles) within the solid solutions

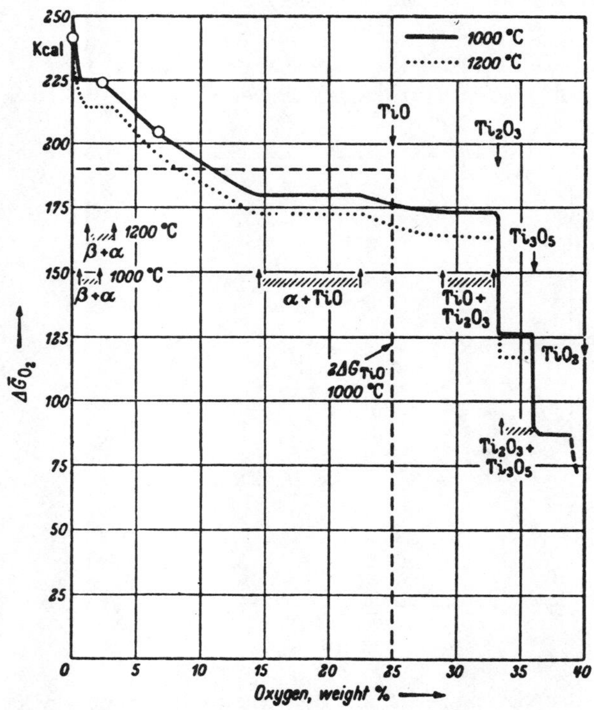

FIG. 18. Partial Gibbs energies of dissociation of 1 mole oxygen in the titanium–oxygen system.

correspond to the Gibbs energies of dissociation of the three alkaline earths (from Table E). From heats of formation, entropies and heat capacities, the value 190,000 cal is obtained for the integral Gibbs energy of the reaction $2\langle TiO\rangle = 2\langle Ti\rangle + (O_2)$ at 1000°C. The connection between the partial and integral values is given by eqn. (60), i.e.

$$\Delta G = \frac{1}{2} N_{Ti} \int_0^x \Delta\bar{G}_{O_2}\, dx$$

Thus, the area under the curve in Fig. 18 (with the abscissa expressed as atomic fraction $x = N_O/N_{Ti}$ instead of at.% oxygen, as shown) must be equal to the integral value given above, and is represented by the broken-line curve in Fig. 18. The experimental part of the curve is extrapolated to the phase boundary $\alpha/(\alpha + TiO)$, thus determining the horizontal part in the $(\alpha + TiO)$ range. To satisfy the condition that the areas beneath the continuous and broken lines should be equal, there can be no more than a slight drop in the value of $\Delta\bar{G}_{O_2}$ within the homogeneity range of TiO. It may be noted that the partial Gibbs energy of dissociation $\Delta\bar{G}_{O_2}$ of 2 "TiO" at 1000°C is 176,000 cal, as compared with its integral Gibbs energy of dissociation of 190,000 cal. Thus the dissociation pressure calculated from the last-named figure would not actually be observed at any but one composition in the range TiO–Ti. The actual dissociation pressures can only be obtained from the partial Gibbs energy curve.

The graph shows that a reducing agent that can overcome the affinity of oxygen to TiO_2, Ti_3O_5, Ti_2O_3 and even TiO may yet fail to produce pure titanium owing to the affinity of oxygen dissolved in the metallic titanium phases. For this reason, simplified thermochemical calculations which do not bring the solution phase into account may sometimes fail, reduction being less complete than would be expected from the calculations. Magnesium or lithium should be capable of reducing titania to titanium containing 2% oxygen by weight, but they will fail to overcome the affinity corresponding to the rapid increase of $\Delta\bar{G}_{O_2}$ in the solution phase towards pure titanium. Only calcium would be effective in this case, the corresponding equilibrium concentration of oxygen in titanium being 0·07% by weight at 1000°C.

Similar problems arise in other systems, especially those of oxygen or sulphur with metals such as titanium, zirconium, vanadium and niobium. Each of these systems contain several compounds, and in addition the oxygen and sulphur have a certain tendency for solution in the metallic lattice. The complete reduction of this metallic phase is essential to the production of the pure metal, and it is usually a very difficult step to overcome.

Figure 19 shows the curve of integral Gibbs energy of formation (of 1 *g-atom* of compound) of the titanium–oxygen system, computed from the partial molar Gibbs energy curve.

The corresponding data for the metallic solution phase in the liquid oxygen–iron system are more accurately known, since the partial molar Gibbs energy of solution of oxygen is much less in iron than in the metals just mentioned. It is more convenient to measure the oxygen pressure of the solution with the reaction

$$\langle FeO\rangle + (H_2) = \{Fe\} + (H_2O)$$

Dastur and Chipman [9] have determined the equilibrium constant of this reaction for various concentrations of oxygen in iron at 1600°C. By subtracting the Gibbs energy of water vapour formation from the partial molar Gibbs energies derived from their equilibrium values of p_{H_2O}/p_{H_2} we obtain the partial molar Gibbs energy of solution of oxygen in iron at various concentrations. These are plotted in Fig. 14. In the miscibility gap between {Fe} + [FeO] and {FeO} + [Fe], the value of $\Delta\bar{G}$, or the oxygen pressure, remains constant until the composition FeO is just exceeded.*

The activity of oxygen in a given solution in iron is given by the ratio of the oxygen pressure of the solution to the oxygen pressure of a saturated solution at the same temperature (i.e. the oxygen pressure of FeO itself).

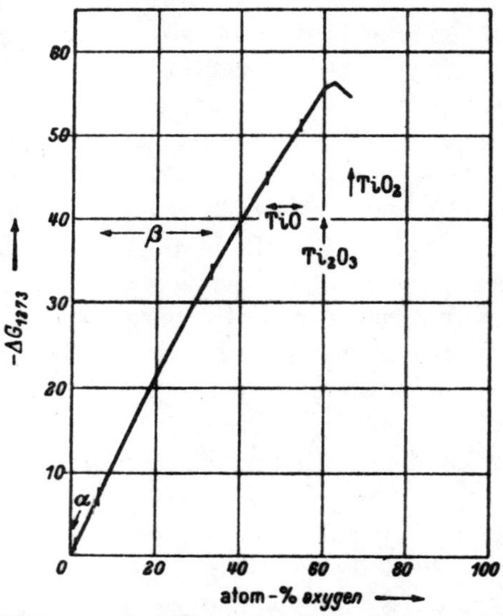

FIG. 19. Integral Gibbs energies of formation of titanium–oxygen compounds per g-atom.

The oxygen pressures calculated from Chipman's figures are given in Table III, the corresponding activities being listed in the fourth column. Then taking the saturation concentration $F_{FeO} = 0{\cdot}775 \times 10^{-3}$ to be unity and evaluating the other concentrations accordingly, we obtain the activity coefficients $f_{FeO} = h_{FeO}/c_{FeO}$ of solutions of oxygen in iron.

Table III. Equilibrium of oxygen with ferrous oxide dissolved in liquid iron

$N_{FeO} \simeq N_O$	$\dfrac{p_{H_2O}}{p_{H_2}}$	$\frac{1}{2}\log p_{O_2}$	f_{FeO}	h_{FeO}
$1{\cdot}05 \times 10^{-3}$	0·127	−4·896	0·135	0·990
$2{\cdot}10 \times 10^{-3}$	0·260	−4·585	0·260	1·015
$2{\cdot}20 \times 10^{-3}$	0·262	−4·582	0·262	0·975
$2{\cdot}35 \times 10^{-3}$	0·263	−4·580	0·263	0·918
$5{\cdot}66 \times 10^{-3}$	0·671	−4·174	0·710	0·968
$6{\cdot}44 \times 10^{-3}$	0·754	−4·123	0·800	0·962
$6{\cdot}65 \times 10^{-3}$	0·755	−4·122	0·800	0·932
$6{\cdot}86 \times 10^{-3}$	0·980	−4·010	1·035	1·168
$6{\cdot}86 \times 10^{-3}$	0·715	−4·146	0·756	0·854
$7{\cdot}00 \times 10^{-3}$	0·983	−4·008	1·041	1·128
$7{\cdot}31 \times 10^{-3}$	0·830	−4·082	0·878	0·928
$7{\cdot}38 \times 10^{-3}$	1·021	−3·991	1·081	1·130
$7{\cdot}75 \times 10^{-3}$	0·848	−4·072	0·898	0·897

* The partial molar Gibbs energies within the $FeO–Fe_3O_4$ range are taken from the work of Darken and Gurry [10].

D. DILUTE SOLUTIONS

1. Raoult's Law

It has been found that Raoult's law, $N_A = a_A$ ($\gamma_A = 1$), holds in a great number of systems for high concentrations of the solvent (see Fig. 9). A simple illustration of why this should be is given by the following schematic consideration. If A represents the solvent and B the solute, then each ultimate particle of B in dilute solution is completely surrounded by A particles (Fig. 20). The attractive A $\leftrightarrow$ B and B $\leftrightarrow$ B forces, being thus hemmed in, as it were, do not affect the escape or addition of A particles from and to the solution. The behaviour of the A particles is therefore simply colligative.

The extent of the concentration range in which Raoult's law applies varies widely from one system to another, but rarely exceeds a lower limit of $N_A = 0{\cdot}85$, i.e. a maximum solute content of 15 molar %.

FIG. 20. Dilute solution of B in A.

Raoult's law has at least one application that is useful to the metallurgist. On adding a substance B to a solvent A there is a lowering of the freezing point of A by an amount given by the Clausius–Clapeyron equation. The exact formulation of this equation, in terms of activities, is

$$L = \boldsymbol{R}T^2(\mathrm{d}\ln a_A/\mathrm{d}T)$$

If we apply this equation to the liquidus line representing the equilibrium between a pure solid A and a liquid solution of B in A, we obtain, on integration,

$$\frac{L_f(T_0 - T)}{\boldsymbol{R}T_0T} = -\ln N_A \qquad (68)$$

in which L_f is the heat of transition of the solid $\langle A \rangle$ into the liquid, i.e. the heat of fusion, T_0 is the melting point of the pure substance A, and T the liquidus temperature at composition N_A.

The overruling proviso that there should be negligible solid solubility of B in A is often found to hold in simple eutectic systems. The relationship is used in all branches of chemistry to determine the molecular weight or degree of dissociation of a soluble substance, provided that the heat of fusion of the solvent is known. Benzene and water are the commonest solvents for this purpose. The metallurgist makes rather more use of its converse application, that of determining a heat of fusion by the freezing-point depression caused by a known molar concentration of solvent. The data required are of course directly available in equilibrium diagrams.

Kelley [11] has evaluated the heats of fusion of a large number of elements by this method, and much still remains to be derived in this manner from the available data.

If there is appreciable solid solubility of the one component in the other, eqn. (57) must be altered to

$$\frac{L_f(T_0 - T)}{RT_0T} = -\ln\frac{N_A}{N'_A} \tag{69}$$

where N'_A is the solid solubility of A in B at temperature T and N_A is the liquidus composition at the same temperature.

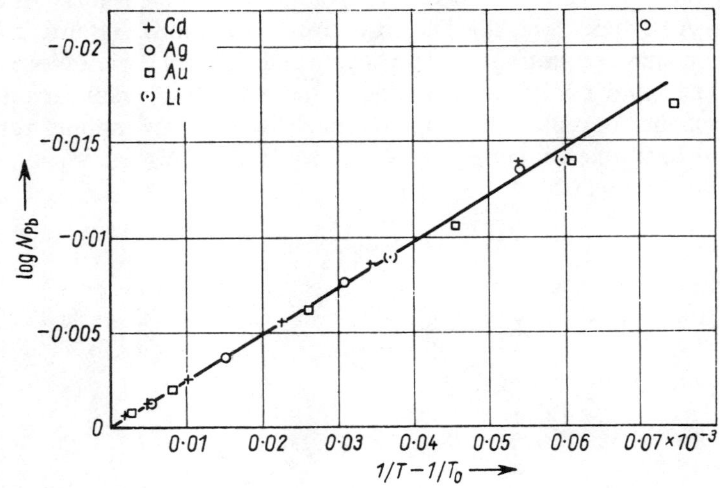

FIG. 21. Evaluation of the heat of fusion of lead from freezing point data.

Figure 21 shows a plot of N_{Pb} against the reciprocal liquidus temperature of alloys of lead with several other metals. The points fall into a straight line of slope -246 and the heat of fusion thus obtained is

$$L_f, (\text{Pb}) = 1130 \pm 60 \text{ cal}$$

which agrees very well with the value found calorimetrically: $L_f = 1140 \pm 30$ cal.

2. Henry's Law

The law to which substances in dilute solution tend to conform is the more generalised one derived from the postulate of Henry that the solubility of a gas in a liquid is proportional to its pressure (provided that its molecular state is the same in solution as in the gas). This will be found to lead to the relationship

$$a_A = \text{constant} \times N_A \tag{50}$$

Like Raoult's law, that of Henry applies within a concentration range the extent of which varies from one system to another, and it is valid only at low concentration.

Henry's law is of greatest practical significance in its application to substances in dilute solution considered separately from the complete system, that is mainly in systems with a limited solubility range. On p. 31, it was expressed in the form

$$\Delta\bar{G}_i = RT \ln \gamma_i + RT \ln N_i \tag{48}$$

Here, the first term on the right-hand side is in fact the excess Gibbs energy of solution

$$\Delta \overline{G}_i^E = \boldsymbol{R}T \ln \gamma_i \tag{70}$$

For more accurate applications, the excess Gibbs energy should be split up into an enthalpy and an entropy term:

$$\Delta \overline{G}_i = \Delta \overline{H}_i + \boldsymbol{R}T \ln N_i - \Delta \overline{S}_i^E T \tag{71}$$

If the solute element i is a diatomic gas i_2, one often finds that the solubility is not directly proportional to the pressure of the gas but to its square root owing to the dissociation of the gaseous species into two atoms upon solution. This observation is known as Sieverts' law, which is a special case of Henry's law, and may be written in the form

$$\Delta \overline{G}_{i_2} = \Delta \overline{H}_{i_2} + 2\boldsymbol{R}T \ln N_i - \Delta \overline{S}_{i_2}^E T \tag{72}$$

In these forms, Henry's law (71) and Sieverts' law (72) respectively are to be used for practical calculations involving dilute solutions (see Chapter IV, examples 11–15). The constant heat and entropy terms of eqn. (76) are given in Table IV for the solution of diatomic gases (H_2, S_2, O_2, N_2) in a number of metals (e.g. ref. [12]) as well as the upper concentration limits for their application.

Table IV. Henrian constants for some solutions of diatomic gases in metals

X = oxygen	$\Delta \overline{H}_{X_2}$ (cal/mole)	$\Delta \overline{S}_{X_2}^E$ (e.u./mole)	X = nitrogen	$\Delta \overline{H}_{X_2}$ (cal/mole)	$\Delta \overline{S}_{X_2}^E$ (e.u./mole)
Solid metals			Solid metals		
Ti-β	−260,000	−43·8	V	+8,500	−3.95
Zr-β	−290,000	−72·0	Ta	−87,800	−28·80
V	−201,900	−48·2	Cr	+16,600	−2·20
Ta	−178,200	−42·2	Mo	+39,800	−12·60
Cr	−106,300	−31·5	Fe, α, δ	+14,200	−22·54
Co	−93,300	−46·3	Fe, γ	−1,360	−28·35
Liquid metals			Liquid metals		
Cu	−41,000	−21·8	Fe	−2,800	−24·00
Ag	−6,760	−20·8	Co	+20,000	−23·20
Ga	−105,360	−31·4			
Ge	−78,000	−25·3			
Sn	−87,700	−30·6	X = hydrogen		
Pb	−56,600	−24·4	Solid metals		
Fe	−55,800	−12·2	V	−13,110	−27·94
Co	−47,000	−18·1	Ta	−13,310	−24·60
Ni	−36,100	−14·7	Cr	+23,400	−17·40
			Fe, α, δ	+13,000	−28·85
			Fe, γ	+10,800	−24·00
X = sulphur			Co	+15,600	−21·92
Liquid metals			Cu	+24,100	−19·80
Cu	−58,400	−13·3	Al	+19,815	−33·95
Ag	−32,800	−17·3			
Sn	−49,900	−28·6	Liquid metals		
Pb	−46,300	−27·5	Fe	+15,000	−17·75
Fe	−60,000	−14·7	Co	+28,400	−20·50
			Cu	+23,590	−14·91
			Al	+25,285	−17·00

Some authors [13] prefer to relate activities to reference states other than mole fraction. They express, for instance, concentrations as atomic percentages (or weight percentages) and take $a = 1$ when $0 = 1$ at.% (or 1 wt.%). Such a definition of activity, which also involves an activity coefficient, is confusing and is not recommended here.

E. ATOMISTICS AND SOLUTION THERMODYNAMICS

One of the objectives of statistical thermodynamics is to make a mathematical bridge between atomic models of systems and their macroscopic thermochemical properties. The highly simplified model of randomly mixed solution of two atomic species forms a convenient basis for discussing the measured values of the partial molar properties of, for example, binary alloys. Let us consider the mixing of n_A atoms of type A and n_B atoms of type B on a three-dimensional lattice which has the crystal structure common to both pure A and B and the whole range of composition of their continuous solid solutions. We will set

$$n_A + n_B = N = 6{\cdot}06 \times 10^{23} \text{ atoms}$$

so that the discussion concerns 1 g-atom of material throughout. To begin with, let the lattice be completely void of atoms which are set apart separately, and randomly choose one atom to place on the first vacant lattice site. We may choose this atom in N ways from the total number. Now we take a second atom and place that on a neighbouring lattice site to the one which was first filled. This choice may be made in $N - 1$ ways since one atom is already located on the lattice. The third atom may be chosen in $N - 2$ ways and so on until the last atom can only be "chosen" in one way.

Clearly we have had a total number of ways of choosing the atoms from the original assembly to fill the lattice in

$$N(N-1)(N-2)\ldots 1$$

i.e. $N!$ ways. Each choice would have placed an atom on a given lattice site, so that n_A sites would have A atoms on them, and n_B sites would be occupied by B atoms. Such an array of sites filled with atoms we will call a configuration. But now if we select two sites occupied by A atoms we could not tell the difference between atom 1 of type A on site 1 and atom 2 of type A on site 2 and the reverse configuration with atom 2 on site 1 and atom 1 on site 2. These two configurations would be indistinguishable. We could only obtain a distinguishable configuration if the atoms were of different types, one of A and one of B. Thus the number of ways of producing distinguishable configurations will be smaller than $N!$ and will be given by

$$\text{Number of distinguishable configurations} = \frac{N!}{n_A!\, n_B!}$$

which is called also the thermodynamic probability (W). It can be deduced from an expression evolved by Boltzmann that the entropy change on mixing two species in this fashion is related to the thermodynamic probability by

$$\Delta S = \boldsymbol{k} \ln W$$

where $\boldsymbol{k}$ is Boltzmann's constant, $\boldsymbol{k} = \boldsymbol{R}/\boldsymbol{N}$. Hence

$$\Delta S = \boldsymbol{k} \ln \frac{N!}{n_A!\, n_B!}$$

According to Stirling's theorem, for a large number x,

$$\ln x! = x \ln x - x$$

Applying this to the equation above, and making use of the facts that $n_A/\boldsymbol{N} = N_A$ (the mole fraction of A) and $n_B/N = N_B$, we find that

$$\Delta S = -\boldsymbol{R}\{N_A \ln N_A + N_B \ln N_B\}$$

Hence the configurational entropy of mixing the components of a binary alloy randomly has the value given above. From the general relationships given earlier between integral and partial entropies one sees that

$$T\,\Delta\bar{S}_i = \boldsymbol{R}T \ln N_i = \boldsymbol{R}T \ln a_i, \quad \text{if} \quad \gamma_i = 1$$

In other words, in a binary alloy where the Gibbs energy of mixing arises simply from the random mixing of the component atoms on a fixed crystal lattice, that alloy would conform to Raoult's law.

Now let us turn to the energy of mixing of a binary alloy. To simplify this calculation, we will assume that each atom only interacts with its nearest neighbours, and that the energy of interaction between two atoms has a constant value depending only on the chemical identity of the atoms and is unchanged by the presence of other atoms. We may then define three interaction energies E_{AA} between the A atoms, E_{BB} between B atoms and E_{AB} between one atom A and one of type B. We will call the number of nearest neighbours which each atom has Z, the co-ordination number.

Now in the binary alloy which we constructed above the *statistical* probability that an A atom occupies any selected site is N_A. The probability that it will be occupied by a B atom is N_B. The probability that any selected neighbouring site has an A or a B atom on it is determined in the same way in a random mixture. Thus the probability that

the two sites are occupied by A atoms is N_A^2

the two sites are occupied by B atoms is N_B^2

the two sites are occupied by an A–B pair is $2N_A N_B$

(The two here is because either site could be occupied by either species to produce an unlike pair.) There will be $\frac{1}{2}NZ$ of such pairs in the mixture as a whole and hence the total interaction energy will be

$$\tfrac{1}{2}NZ(N_A^2 E_{AA} + N_B^2 E_{BB} + 2N_A N_B E_{AB})$$

Now to obtain the energy of formation of the alloy from N_A and N_B g-atoms of A and N_B g-atoms of B we must subtract

$$\tfrac{1}{2}NZ(N_A E_{AA} + N_B E_{BB})$$

from this quantity and thus

$$\Delta E_{\text{mixing}} \cong \Delta H_{\text{mixing}} = \tfrac{1}{2}ZN\{(N_A^2 - N_A)E_{AA} + (N_B^2 - N_B)E_{BB} + 2N_A N_B E_{AB}\}$$

$$= NZN_A N_B\left(E_{AB} - \frac{E_{AA} + E_{BB}}{2}\right) \qquad (72a)$$

It should be noted that the interaction energies are negative quantities and thus when

$$E_{AB} > \frac{E_{AA} + E_{BB}}{2} \qquad \Delta H_{\text{mixing}} \text{ is negative}$$

$$E_{AB} = \frac{E_{AA} + E_{BB}}{2} \qquad \Delta H_{\text{mixing}} \text{ is zero (Raoult's law)}$$

$$E_{AB} < \frac{E_{AA} + E_{BB}}{2} \qquad \Delta H_{\text{mixing}} \text{ is positive}$$

Thus by taking into account the configurational arrangement of the atoms in a binary alloy we can arrive at three ideal laws of mixing.

1. Raoult's law where $\Delta H = 0$; $\Delta \bar{S}_i = -\boldsymbol{R} \ln N_i$.
2. The strictly regular solution where $\Delta H = \alpha N_A N_B$; $\Delta \bar{S}_i = -\boldsymbol{R} \ln N_i$.
3. Henry's law where $\Delta \bar{H}_i \rightarrow \alpha$ as $N_i \rightarrow 0$.

As well as these configurational terms there is also the possibility of thermal contributions arising from deviations from the Neumann–Kopp rule in the alloy. This states that

$$\Delta C_{p_{\text{alloy}}} = N_A C_{p_A} + N_B C_{p_B} \tag{73}$$

It is the presence of these terms as well as possible departures from random mixing, which will be discussed later, which causes real systems to deviate from these ideal laws.

F. REGULAR SOLUTIONS

Thirty years ago it was still assumed that all metallic solutions are approximately regular. In fact, one of the present authors in a monograph written in collaboration with the late Dr. Weibke converted measured Gibbs energies of metallic solutions into heats of solution by means of the equation

$$\Delta \bar{H}_A \equiv \Delta \bar{G}_A^E = 4{\cdot}574T \log \frac{a_A}{N_A} = 4{\cdot}574T \log \gamma_A \tag{74}$$

and presented all the thermochemical data merely in the form of enthalpy curves.

It will be seen in Section G that the underlying assumptions are not borne out by experimental evidence. However, there are a few aspects which have been mentioned in Section D and will be outlined in the present section where the regular solution concept retains some usefulness.

1. *Calculation of Activities Using Equilibrium Diagrams*

We may follow Rey [14] to demonstrate the calculation of chemical activities of the components of liquid solutions from the values represented by the liquidus and solidus lines in the equilibrium diagram. In the equations which follow, L_f denotes the heat of fusion of the pure solvent substance A at its melting point T_0. The activities and corresponding mole fractions of substance A at the liquidus and solidus compositions at a lower temperature T_c are denoted respectively by a, a'; N, N' (Fig. 22). T is an arbitrary temperature at which activities are to be calculated. The suffix A is omitted as all quantities refer to the solvent.

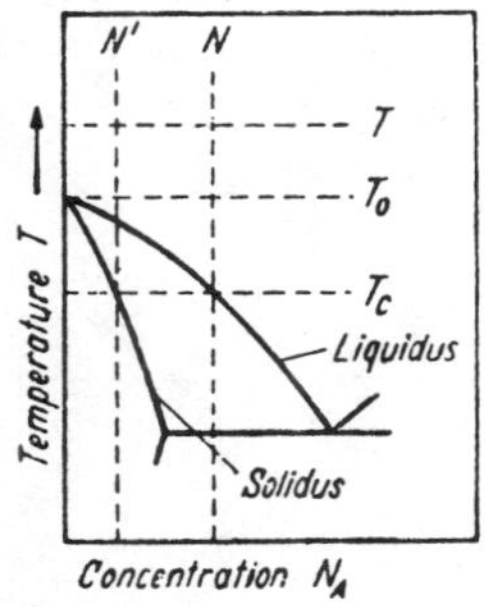

FIG. 22. Calculation of activities from equilibrium diagrams

In systems where there is no solid solubility—that is, the solidus curve coincides with the pure solvent substance—then activities in the liquid mixture can be obtained by the following considerations. Equation (74) may be taken to mean that $\boldsymbol{R}T \log (a/N)$ is independent of temperature. Hence

$$\log a_T = \frac{T_c}{T} \log a_{T_c} + \frac{T - T_c}{T} \log N$$

Combining this with eqn. (72),

$$\log a_T = -\frac{(T_0 - T_c)L_f}{4{\cdot}574T\ T_0} + \frac{T - T_c}{T} \log N$$

or, if there is a range of solid solubility over which Raoult's law is valid (i.e. the solution is ideal: $a' = N'$) the following equations are derived

$$\log a_T = -\frac{(T_0 - T_c)L_f}{4{\cdot}574T\ T_0} + \frac{T_c}{T} \log N' + \frac{T - T_c}{T} \log N \qquad (75)$$

and

$$\log a'_T = \frac{T - T_c}{T}\left(\log N - \frac{L_f}{4{\cdot}574T_0}\right) + \frac{T_c}{T} \log N' \qquad (76)$$

These or similar equations have often been used to derive activities, together with heats of solution (see for example, Fig. 23). Changes in molar heats on solution are occasionally taken into account [13], and the results are sometimes presented in the form of integral heats of solution [15].

There is, however, some objection to the low accuracy of the data so derived. The calculation involves an exponential relationship which requires that the heat of fusion of the solvent should be known accurately which is often not the case. The liquidus and solidus curves should be known with an accuracy well within $\pm 0{\cdot}2°$ to guarantee a fair accuracy of the resulting thermochemical data; thermal measurements have, however, rarely been made with such precision. Deviations from regularity also affect the evaluations considerably; these will be discussed in some detail in Section G. The neglect of the change in heat capacities on melting introduces a further uncertainty, but this is probably of a minor order. All these uncertainties may amount together to considerable errors, and little evidence has so far been forthcoming to show that activities calculated from equilibrium diagrams without independent thermochemical evidence are reliable. Comparisons show quite large discrepancies between these calculated values and those obtained by the more orthodox experimental measurements [15], but their authors claim that the values obtained from the phase diagrams are the more reliable. However, this contention cannot be regarded as proved.

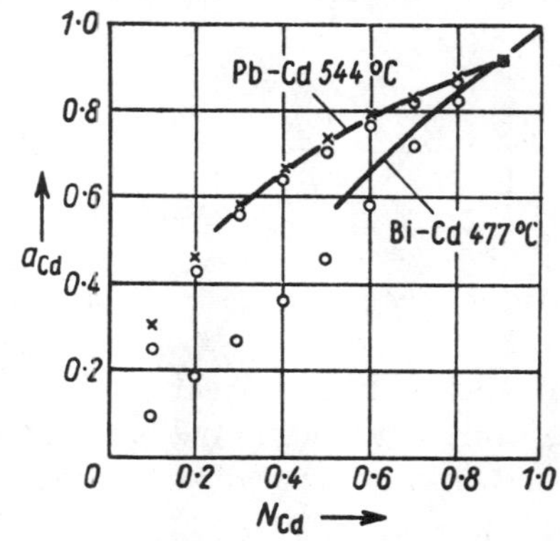

FIG. 23. Activity of cadmium in alloys with bismuth and lead: lines calculated from the equilibrium diagrams (M. Hansen, *Der Aufbau der Zweistofflegierungen*, Berlin, 1936); circles plotted from e.m.f. measurements (N. W. Taylor, *J. Amer. Chem. Soc.*, 1923 **45** 2865); and crosses from vapour pressures (K. Jellinek and G. A. Rosner, *Z. physikal. Chem.*, A 1931 **152** 67)

At present it may be regarded as a useful approximate means of filling gaps in our quantitative knowledge. Calculations of the activity curves of silicate melts by Rey [16] are an example. Owing to the objections raised above, these results are probably of low accuracy but they give quite a good indication of the general shape of the activity curves and illustrate the implications of the shape of the curves.

2. *Spinodal Decomposition*

In a strictly regular solution having a positive integral heat of mixing there must exist a critical temperature above which the components are completely miscible and below which the system consists of a physical mixture of two solutions, one rich in component A, and the other rich in component B. The limiting compositions of these solutions may be calculated by means of the equations obtained above. Consider the Gibbs energy of mixing curve at a temperature below the critical value T_c as shown in Fig. 24.

The common tangent to the integral curves at the composition of the two limiting solutions passes clearly through OQ. These are the two points on the curve where $\partial G^M/\partial N_A = 0$. The two points O', Q', are those where $\partial^2 \Delta G^M/\partial N_A^2 = 0$ and these are called the spinodal points of the system, since it is at these compositions that the two new phases first appear when A–B alloys are cooled from a higher temperature, and not at the equilibrium compositions O and Q. The curve traced out as a function of temperature in the system by the points corresponding to O and Q are those showing the equilibrium compositions on the phase diagram. The spinodal points trace out the spinodal curves between these two limits.

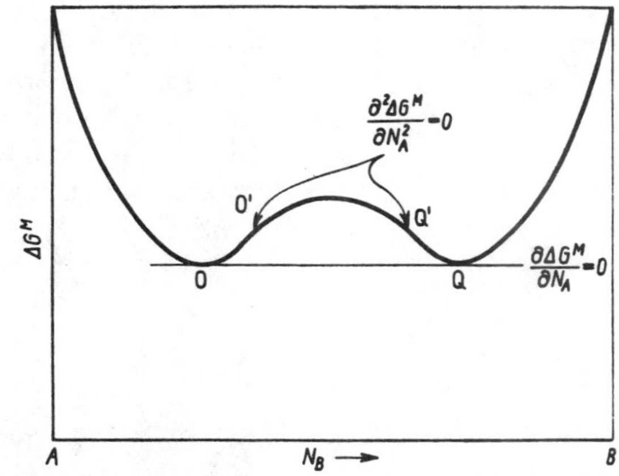

FIG. 24. First and second differentials with respect to mole fraction of A of the integral Gibbs energy of mixing of a binary A–B alloy. The curve is for a temperature below T_c.

From the expression for the Gibbs energy of mixing of a binary strictly regular solution, the phase boundaries are given by

$$\frac{\partial \Delta G}{\partial N_A} = \alpha(1 - 2N_A) + RT\left[\ln \frac{N_A}{1 - N_A}\right] = 0 \tag{77}$$

$$\frac{RT}{\alpha} = (N_B - N_A) \ln \frac{N_A}{N_B} \tag{78}$$

The critical temperature where immiscibility begins must occur at the equimolar composition at the temperature such that

$$RT_c = 2\Delta H_{0\cdot 5}$$

Also,

$$\frac{\partial^2 \Delta G}{\partial N_A^2} = -2\alpha + RT\left[\frac{1}{N_A N_B}\right] = 0$$

at the spinodal compositions.

Hence $N_A N_B = RT/2\alpha$ along the spinodal curves.

3. *Order–disorder Transformation*

Some binary systems having a negative heat of mixing show the order–disorder transformation. This is when, at a given composition, the system undergoes a change in the position of the atoms with increasing temperature. At the lower temperatures all the atoms of each component occupy well defined sublattices, and above a critical temperature the atoms are randomly mixed on one common lattice. If the random mixture is also a strictly regular solution, the algebra of the transformation may be evolved in the following way.

In one g-atom of alloy of equiatomic composition there will be 0·5 g-atoms of A on one sublattice and 0·5 g-atoms of B on the other sublattice in the ordered low-temperature system. In the high temperature, disordered alloy, each sublattice contains 0·25 g-atom of each species. In the low-temperature system, each A atom is immediately surrounded by B atoms and vice versa. At high temperatures the co-ordination of each atom is made of $Z/2$ A atoms and $Z/2$ B atoms. If S represents the fraction of the A sublattice which is occupied by the B atoms, then S takes the value $\frac{1}{2}$ in the random alloy. At some intermediate stage $(0 < S < \frac{1}{2})$ the entropy of the alloy is $\Delta S = -\boldsymbol{R}[(1-S)\ln(1-S) + S\ln S]$ per g-atom and the energy of the lattice is $E = \frac{1}{2}NZ[S(1-S)E_{AA} + S(1-S)E_{BB} + (1-S)(1-S)E_{AB}]$.

The stable configuration at any given temperature can be obtained by minimizing the Gibbs energy with S as the variable

$$\frac{\partial G}{\partial S} = 0 = \tfrac{1}{4}NZ(1-2S)\left[\frac{E_{AA}+E_{BB}}{2} - E_{AB}\right] + \boldsymbol{R}T\left[\ln\frac{S}{(1-S)}\right] \tag{79}$$

As $T \to T_c$ the critical temperature above which the alloy is random, $S \to \frac{1}{2}$, and the equation

$$(1-2S)\frac{T_c}{T} = \ln\left(\frac{1-2S}{S}\right) \tag{80}$$

is obtained for the value of S at any intermediate temperature.

G. NON-REGULAR SOLUTIONS

Experimental evidence has shown that "regular solutions" are rare, but the concept retains its usefulness for didactic purposes. In order to describe deviations from regularity, the "excess Gibbs energy" and "excess entropy" functions have been introduced. By combination of the partial excess functions (p. 34) with those of Lewis (p. 30) one obtains the integral excess functions

$$\Delta S^E = \Delta S^{\text{xptl}} + \boldsymbol{R}(N_A \ln N_A + N_B \ln N_B + \cdots) \tag{81}$$

$$\Delta G^E = \Delta G^{\text{xptl}} - \boldsymbol{R}T(N_A \ln N_A + N_B \ln N_B + \cdots) \tag{82}$$

For a regular solution, ΔS^E is zero across the whole range of solution. In Fig. 25 are plotted the maximum or minimum entropies of mixing against the maximum or minimum heats of mixing for a random selection of systems exhibiting complete mutual solubility, having different types of bonding. It may be seen that the condition for regular behavior, $\Delta S^E = 0$, is rarely fulfilled, strictly only for the "ideal solutions" for which $\Delta H = 0$. Actually, there seems to be a direct relationship between ΔH and ΔS^E as indicated by the curve which is quite empirical. In addition, it may be pointed out that the maxima in the heat and entropy vs. concentration curves is normally not observed at equiatomic composition as required by the regular solution model, but is shifted to higher or lower concentrations for both the enthalpy and entropy—apparently always in the same direction. As might be expected from these observations, a plot of the partial excess entropies against the partial heats at high dilution also produces a direct relationship, as has been shown by Slough [17].

Kleppa [18] has drawn attention to the work by Scatchard [19] on the effect of a net volume change on the entropy of mixing. According to this, one could calculate

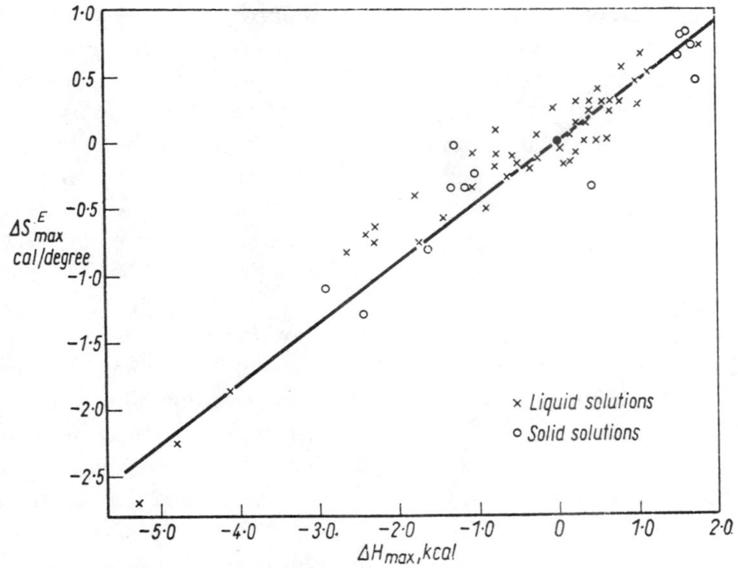

FIG. 25. The linear relationship between the maximum or minimum excess entropies and heats of mixing for a number of binary alloy systems.

the deviations from the ideal entropy of mixing if the volume changes are known, and Kleppa concluded under some simplifying assumptions that they should be approximately proportional, but could not confirm this relationship when comparing the experimental data. However, Table V shows that a rough connection does exist between the entropy and the volume changes on mixing liquid metals. According to Table V there also appears to be an approximate proportionality between the changes in volume and heat of mixing.

Table V. Maximum or minimum values of the integral changes of volume, heat and entropy of mixing of liquid binary alloys

System 1–2	ΔV in %	N_2	ΔS^E in e.u./g-atom	N_2	ΔH in kcal/g-atom	N_2
Fe–Si	−27·0	0·52	−2·5	0·57	−9·0	0·48
K–Hg	−25·4	0·58	−3·9	0·47	−5·3	0·6
Mg–Pb	−7·0	0·33	−0·75	0·38	−2·3	0·38
Mg–Zn	−6·5	0·65	−0·59	0·55	−1·42	0·66
Cu–Zn	−3·1	0·6	(−0·4)	(0·5)	−1·8	0·5
Mg–Al	−3·0	0·4	(−0·5)	(0·5)	(−0·9)	0·5
Na–K	−1·0	0·33	+0·05	0·4	+0·18	0·43
Zn–Al	−0·5	0·45	+0·30	0·51	+0·615	0·48
Pb–Bi	+0·3	0·33	+0·05	0·5	−0·27	0·5
Zn–Cd	+0·6	(0·5)	0·00	—	+0·51	0·47
Cd–In	+0·8	0·43	+0·14	0·45	+0·35	0·45
Cd–Pb	+0·95	0·38	+0·22	0·35	+0·68	0·48
Cd–Tl	+1·0	0·44	+0·23	0·4	0·56	0·44
Sn–Bi	+1·1	0·33	−0·07	0·4	+0·03	0·5
Cu–Ag	+1·3	0·66	+0·27	0·67	+1·0	0·44
Zn–Sn	+1·5	0·32	+0·56	0·45	+0·82	0·43
Cd–Sn	+1·55	0·45	+0·28	0·45	+0·44	0·46
Cd–Bi	+1·8	0·33	+0·38	0·35	+0·22	0·33

Most liquid systems for which the changes on mixing of volume, entropy and heat are known, involve one or two meta- or semi-metals, such as Bi, Sn, Pb, Tl, In, Cd and Zn, for which discussions of structure and other X-ray evidence have amply shown that covalent bonds occur [20, 21]. If a more "metallic" metal is added to the more covalent one, this should result in a partial or complete break-up of the covalent bonds. This process would consume energy and increase disorder. In fact, Table V shows that the alloys concerned are generally formed endothermically and have positive excess entropies of mixing. Since the volume changes are also positive, it follows that the covalent "clusters" are not completely destroyed but are too rigid to allow a suitable close packing.

In Table V, the systems 1–2 are written in such a way that the more covalent metal stands in the second place. It is then seen that the maxima in the entropy and heat curves are nearly always shifted to below equiatomic compositions. This is in line with the hypothesis just outlined: the covalent bonds are obviously more strongly affected when the respective metal is more dilute; in other words, when a covalent metal is dissolved in a less covalent one, more energy and order is consumed than in the reverse process.

We may take metallic bonds to be essentially additive and ascribe the heats of formation of truly metallic phases to packing effects [22] (see p. 202). Then the mutual solutions of true metals should be nearly ideal as long as the component metals differ only little in atomic diameter. Although this is not always observed, regular behaviour is quite often approached in such solutions. As an example, the entropy curve of liquid silver–copper alloys, shown in Fig. 26, is near the ideal one. The entropy curve for zinc–tin, also depicted in this diagram, is higher owing to the destruction of covalent bonds and the resulting increase in vibrational entropy.

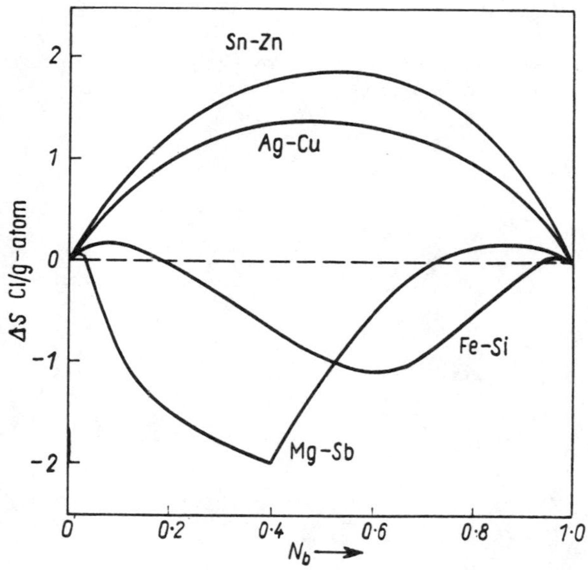

FIG. 26. Integral entropies of mixing in the liquid systems tin–zinc, silver–copper, iron–silicon, and magnesium–antimony.

Alterations in the positional entropy can lead to a minimum ΔS of zero; alterations in the vibrational entropy may make ΔS negative. Suitable systems with which this question may be investigated are the liquid alloys of magnesium with bismuth and antimony. Both systems form a complete series of liquid solutions but solid Mg_3Sb_2 and Mg_3Bi_2 are stabilised partly by polar bonds which conceivably persist to some extent in the liquid state. The entropies of mixing of liquid magnesium and antimony are also shown in Fig. 26. It is seen that the curve is in fact negative over large ranges of composition with a minimum at approximately stoichiometric composition. This behaviour is undoubtedly due to strong polar forces in melts that consist partly of Mg^{2+} and Sb^{3-} or Bi^{3-} ions. These forces exert their influence mainly on the vibrational term of the entropy.

When strong covalent bonds are formed on alloying, the effect on the entropy is similar as with polar bonds. The example of the liquid iron–silicon alloys in Fig. 26 is probably due to this type of bond mechanism.

These are extreme cases, but they indicate the possible magnitude of the effect. Moreover, the alloys have strongly negative heats of mixing with minima at the stoichiometric compositions, namely $-9{\cdot}0$ (Fe–Si), $-8{\cdot}1$ (Mg–Sb) and $-4{\cdot}3$ (Mg–Bi) kcal/g-atom. The corresponding curves of partial heats of solutions are $\sim$ shaped, as was predicted by Wagner on the basis of a similar consideration for the liquid system Ag–Te [23].

These examples illustrate the importance of the behaviour of the valence electrons in considerations of entropy- and heat-changes on mixing. Even when two true metals are mixed together the conduction electrons will shift, provided that the metals are of different valence, and thus bring about bonding between the positive ions. This change in bond will influence the various thermodynamic functions. This is clearly indicated by the observation of Hume-Rothery that solutions for example of germanium, gallium and zinc in copper all become saturated at approximately the same concentration of valence electrons. Further examples of the significance of a change in the concentration of valence electrons are given by Wagner [23]. These changes would affect the heat as well as the entropy terms, but the quantitative elaboration is complicated.

H. TERNARY SOLUTIONS

Three methods for solving the ternary Gibbs–Duhem equation have been given in the literature. The resulting equations make it possible in principle to calculate the activities of two components, say B and C, in an A–B–C ternary alloy, from a knowledge of the partial Gibbs energy of component A over the whole ternary composition range for a given temperature. The consequence of this requirement of such detailed information is that attempts have been made to put workable approximations into the accurate equations in order to obtain the main features of ternary systems and to test the approximations with the limited amount of data which are at present established. Unfortunately, most of the experimental results relate to systems in which the interactions between the alloying components are relatively weak. Such systems do not provide a very rigorous test of the approximations and these are made to look more successful than they really are.

The methods of solving the ternary Gibbs–Duhem which were obtained by Darken and by Wagner have some degree of similarity; the principal difference resides in a sequence of integration and differentiation in Darken's procedure [24] which is reversed in Wagner's analysis [15]. From many points of view, Wagner's approach is more readily applied to experimental data with reasonable accuracy than Darken's, but for the purposes of the present analysis we shall begin our discussion of the ternary systems with an equation which emerges during Darken's analysis. The method of solution of Schuhmann [26], which is also very powerful for experimental data of a limited scope, will not be referred to further, but the interested reader will find a useful discussion of this procedure in Richardson's monograph [27].

Returning to the Darken derivation, we will denote by α_B the function

$$\Delta\bar{G}_B^E/(1 - N_B)^2$$

and then an important equation obtained by Darken may be written as

$$\Delta G^E_{B+A/C} = (1 - N_B) \int_1^{N_B} [\alpha_B \, dN_B]_{N_A/N_C} - N_A \int_1^0 [\alpha_B \, dN_B]_{N_C=0} - N_C \int_1^0 [\alpha_B \, dN_B]_{N_A=0} \tag{83}$$

The first integral on the right-hand side is obtained from values of α_B as a function of N_B and along an integration path with a constant atom fraction ratio of A and C, N_A/N_C (Fig. 27). The excess integral Gibbs free energy on the left is for the ternary system containing component B at the atom fraction N_B, with A and C in the proportion N_A/N_C. The other two integrals on the right-hand side are for α_B in the binary A–B and B–C systems, respectively.

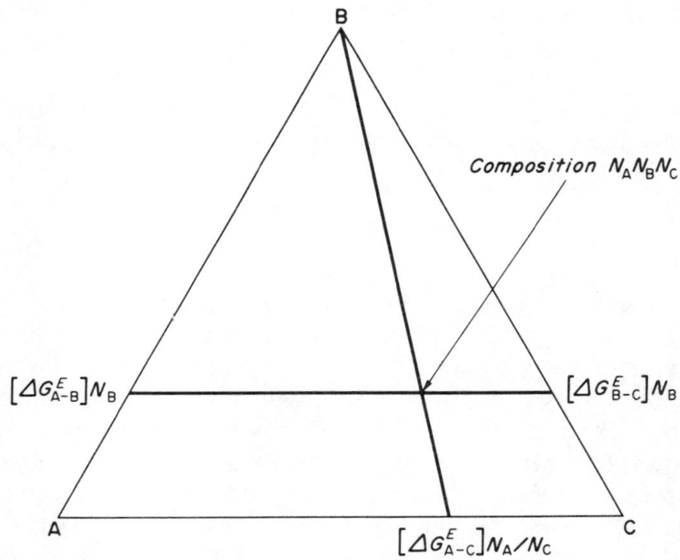

FIG. 27. Illustration of the binary terms used in the application of the Toop and Bonnier equations.

Toop [28] has shown that this equation can be simply integrated for the ternary system in which all solutions are strictly regular, that is to say α_B remains constant along any given integration path with N_A/N_C fixed. The binary A–B and B–C systems must also be regular. Then,

$$\Delta G^E_{B+A/C} = \frac{N_A}{1-N_B}\Delta G^E_{A-B} + \frac{N_C}{1-N_B}\Delta G^E_{B-C} + (1-N_B)^2[\Delta G^E_{A-C}]_{N_A/N_C} \tag{84}$$

Kohler [29] obtained another solution which can also be obtained from this Darken equation by choosing different integration paths (Fig. 28) but again assuming strictly regular behaviour along these. The paths of integration are now along constant ratios N_A/N_B, N_B/N_C and N_C/N_A to yield the equation

$$\Delta G^E_{A-B-C} = (1-N_A)^2[\Delta G^E_{B-C}]_{N_B/N_C} + (1-N_B)^2[\Delta G^E_{A-C}]_{N_A/N_C} + (1-N_C)^2[\Delta G^E_{A-B}]_{N_A/N_B} \tag{85}$$

A further equation has been obtained by Bonnier and Caboz [30]

$$\Delta G^E_{B+A/C} = \left[\frac{N_A}{1-N_B}\Delta G^E_{A-B} + \frac{N_C}{1-N_B}\Delta G^E_{B-C}\right]_{N_B} + (1-N_B)[\Delta G^E_{A-C}]_{N_A/N_C} \tag{86}$$

These equations have been surveyed by Ansara [31] and Spencer *et al.* [32] have compared calculated and experimental results from some ternary systems together with the equation of Margules [33] which contains an empirical ternary constant K_{ABC}, as well as the binary constants, K_{BA}, etc.

$$\Delta G^E_{A-B-C} = N_A N_B[N_A K_{BA} + N_B K_{AB}] + N_A N_C[N_A K_{CA} + N_C K_{AC}] + N_B N_C[N_B K_{CB} + N_C K_{BC}] + N_A N_B N_C K_{ABC} \tag{87}$$

Although these equations have all been written for ΔG^E, they may also be applied in the calculation of ΔH and ΔS^E in the obvious manner.

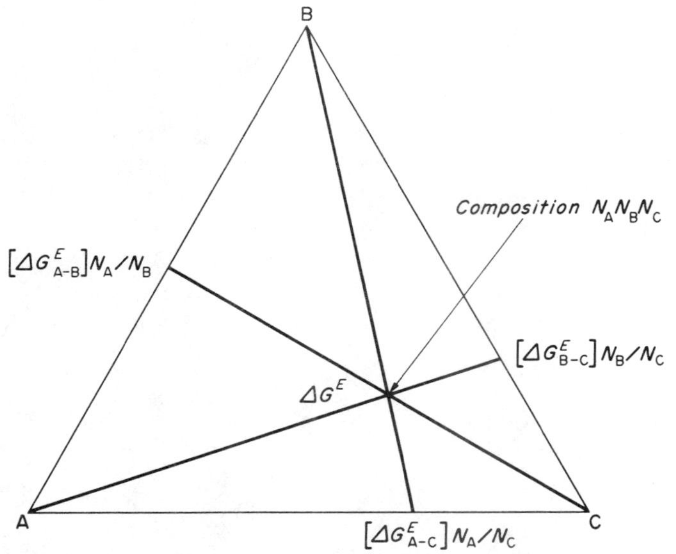

FIG. 28. Illustration of the binary terms used in the application of the Kohler equation.

For low-affinity solutions these methods mostly give similar values and agree quite well with the experimental ones, often within the experimental accuracy. For high-affinity systems, deviations are larger. Figure 29 shows as an example the experimental heats of formation of iron–cobalt–nickel alloys in comparison with the values calculated by means of the Kohler and Bonnier equations [34]. However, it is difficult at the present stage to decide about the relative merits of the various modes of extrapolation, and further experimental values must be awaited for a systematic survey. Ansara [31], who has made a thorough study of the subject, seems to give preference to the Kohler equation. The Margules equation has been employed by Kubaschewski and Counsell to describe the thermochemical properties of the system gold–palladium–platinum [35].

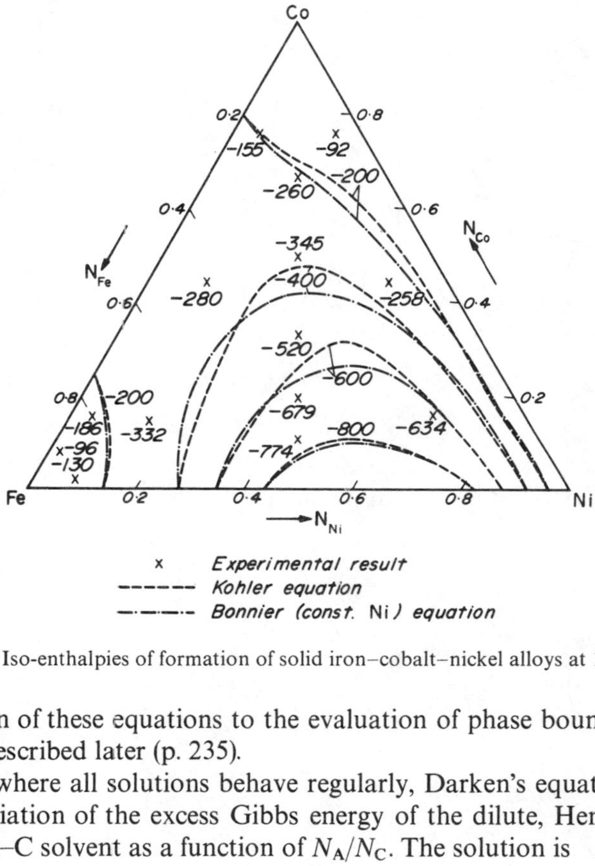

FIG. 29. Iso-enthalpies of formation of solid iron–cobalt–nickel alloys at 1473°K.

The application of these equations to the evaluation of phase boundaries in ternary systems will be described later (p. 235).

In the system where all solutions behave regularly, Darken's equation may be used to obtain the variation of the excess Gibbs energy of the dilute, Henrian, solution of B in the binary A–C solvent as a function of N_A/N_C. The solution is

$$\Delta\overline{G}^E_{B(A-C)} = N_A\Delta\overline{G}^E_{B(A)} + N_C\Delta\overline{G}^E_{B(C)} - \Delta G^E_{A-C} \tag{88}$$

B(A–C) denotes B dissolved in the A–C alloy of mole fractions N_A, N_C and for this solution

$$N_A + N_C \cong 1$$

Alcock and Richardson [36] used an atomistic model in which the B atoms are surrounded randomly by A and C atoms and each A–B and B–C bond is of constant strength at all values of N_A/N_C. This constant pairwise bonding model which together with a fixed co-ordination number Z describes a strictly regular solution also yields equation (88). The model system further suggests, however, that when, say, the A–B bond is much stronger than the B–C bond, then A atoms will tend to cluster around the B atoms to the exclusion of the less strongly bonded C atoms. The population of the Z atoms which surround any B atom is then calculated by the "quasi-chemical" approach to be in the proportion

$$\frac{n_A}{n_C} = \frac{N_A}{N_C} \exp (W/ZRT)$$

where W is the exchange energy for the change of bonds $A - B + C \rightarrow B - C + A$ and therefore

$$W = \Delta \bar{H}_{B(C)} - \Delta \bar{H}_{B(A)} - \Delta \bar{H}_{C(A-C)} + \Delta \bar{H}_{A(A-C)}$$

From these atomistic considerations, the equation

$$\left[\frac{1}{\gamma_{B(A-C)}}\right]^{1/Z} = N_A \left[\frac{\gamma_{A(A-C)}}{\gamma_{B(A)}}\right]^{1/Z} + N_B \left[\frac{\gamma_{C(A-C)}}{\gamma_{B(C)}}\right]^{1/Z} \tag{89}$$

was derived [36]. This equation contains the co-ordination number explicitly but approaches eqn. (88) as Z becomes large. For a typical value in liquid metallic solutions $Z \simeq 10$ the effect of clustering is relatively unimportant in most alloy systems involving three metallic species. Recent results obtained by Jacob and Alcock for the dilute solution of indium in the solid copper–gold alloy where Z is known to be 12 are indicative of the size of the effect [37] (Fig. 30). The more significant effect which is found when the dilute solute is a non-metal such as oxygen has been accounted for by these

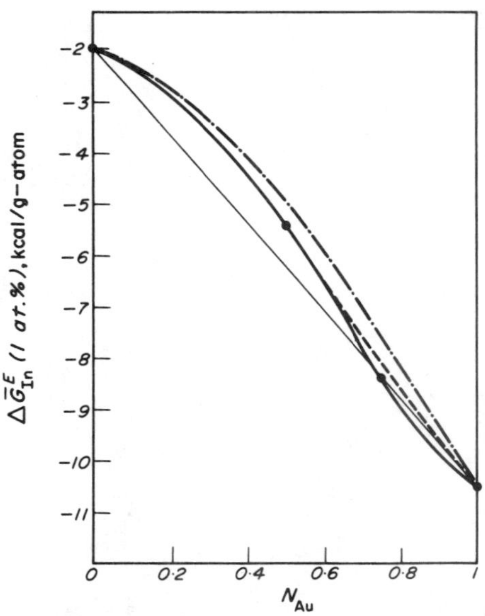

FIG. 30. Variation of the chemical potential of indium at 1 at.% concentration with N_{Cu}/N_{Au} ratio in Cu + Au + In alloys at 900°K: —·—·— regular solution model; ---- quasichemical model $Z = 12$;—●—experimental data.

authors in a further extension of the quasi-chemical model [38]. If an oxygen atom attaches one or two electrons in a metallic alloy solvent, then this negatively charged species will tend to repel the conduction electrons in its immediate environment. In the language of bonds, the oxygen species will tend to cause the breaking of metal–metal bonds in its co-ordination shell. The assumption that oxygen and sulphur form strong bonds with four atoms in the co-ordination shell where half of the metal–metal bonds are broken, accounts for most of the published literature for the dilute solutions of oxygen and sulphur in liquid metallic alloys. The equation for oxygen in dilute solution in the A–C alloys is, then,

$$\left[\frac{1}{\gamma_{\mathrm{O(A-C)}}}\right]^{1/4} = N_{\mathrm{A}}\left[\frac{\gamma_{\mathrm{A(A-C)}}^{1/2}}{\gamma_{\mathrm{O(A)}}^{1/4}}\right] + N_{\mathrm{C}}\left[\frac{\gamma_{\mathrm{C(A-C)}}^{1/2}}{\gamma_{\mathrm{O(C)}}^{1/4}}\right] \tag{90}$$

This equation resembles mathematically that given earlier for dilute metallic alloys, eqn. (89) but differs essentially in the two values of Z which are used. Hydrogen in solution in liquid metallic alloys which might be expected to ionize in the normal "metallic" fashion to H^+, and an electron appears to follow eqn. (89) thus differing from the electro-negative oxygen and sulphur.

The effect of an alloying element on the activity coefficient of a dilute solute in solution in a liquid metal is of considerable importance in the refining of metals. The atomistic models given above demonstrate that a number of factors determine this effect, not the least among these being the metal–metal interactions.

From the many experimental studies that have been made of these effects, collections of empirical data have been made covering many systems of industrial importance. These are usually reported as values of the so-called "interaction" parameter, $\varepsilon_{\mathrm{B}}^{\mathrm{C}}$ which is defined by the equation

$$\frac{\partial \ln \gamma_{\mathrm{B}}}{\partial N_{\mathrm{C}}} = \varepsilon_{\mathrm{B}}^{\mathrm{C}}$$

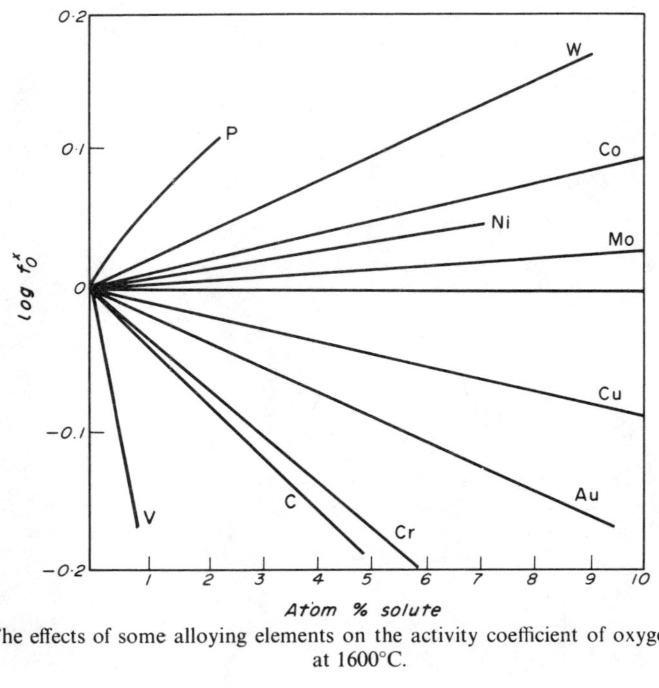

FIG. 31. The effects of some alloying elements on the activity coefficient of oxygen in liquid iron at 1600°C.

for the dilute solutions of B and C in the solvent metal A. This parameter only has a constant value over a limited range of composition, usually less than one atomic percent for each solute. As the concentration of C is increased while that of B is kept small, the change in ε_B^C will largely reflect the competing interactions of B with C and of the solvent A with C. Figures 31 and 32 show some collected data for the Fe–O–X and

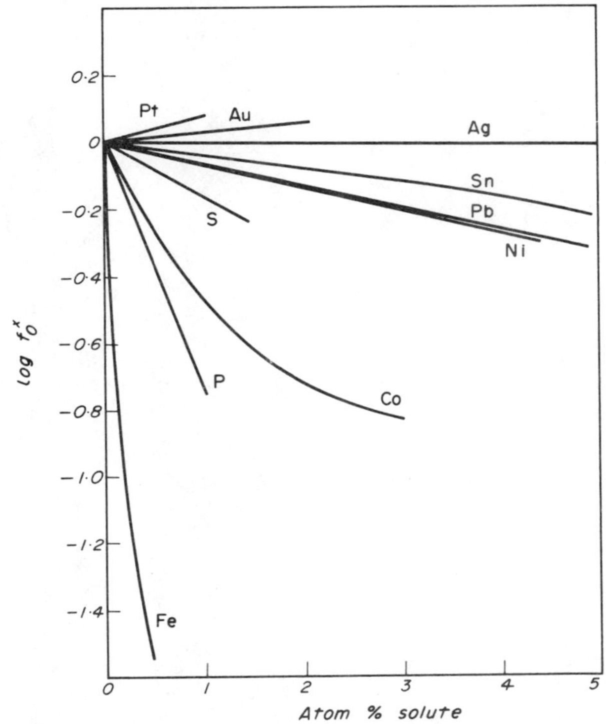

FIG. 32. The effects of some alloying elements on the activity coefficient of oxygen in liquid copper at 1200°C.

Cu–O–X systems, where X is the alloying element and the concentration of oxygen is low. For many uses it is convenient to express the concentration of X in weight percent and to use logarithms to the base ten. Then,

$$\frac{\partial \log \gamma_O}{\partial \text{wt}\%\text{X}} = e_O^X = \frac{M_{Fe}}{M_X}\frac{\varepsilon_O^X}{230{\cdot}3} \tag{91}$$

when the solvent is iron and M_{Fe}, M_X are the atomic weights of iron and of X respectively. This simple relationship only applies for small concentrations of X which are usually of most importance in the industrial applications. The interaction parameter ε_B^C is clearly related to the integral excess Gibbs energy of the ternary system and can be shown to be equal to the conjugate interaction parameter ε_C^B. This results from the fact that the sequence of differentiation of a homogeneous function, in this case ΔG_f^E, with respect to two variables N_B and N_C, does not affect the result. Thus

$$\varepsilon_B^C = \frac{\partial^2 \Delta G_f^E/RT}{\partial N_B\, \partial N_C} = \frac{\partial^2 \Delta G_f^E/RT}{\partial N_C\, \partial N_B} = \varepsilon_C^B \tag{92}$$

Table Va. Interaction coefficients for oxygen in liquid iron and copper

Fe–X–O at 1600°C	ε_O^X	e_O^X	$\gamma^0_{X(Fe-X)}$
Fe–Ni–O	1·4	0·006	0·62
Fe–Mo–O	0·67	0·0035	Raoultian
Fe–Mn–O	−4·7	−0·021	1·33
Fe–Cr–O	−8·5	−0·04	0·87
Fe–Si–O	−15	−1·13	0·0013
Fe–C–O	−22	−0·45	0·57
Fe–V–O	−63	−0·3	0·18
Fe–Ti–O	−118	−0·6	0·01
Fe–Al–O	−433	−3·9	0·058
Cu–X–O at 1100°C			
Cu–Pt–O	+38	0·057	0·05
Cu–Au–O	+6·9	0·007	0·14
Cu–Ag–O	−0·2	0	3·23
Cu–Sn–O	−4·6	−0·0087	0·42
Cu–Ni–O	−7·0	−0·034	2·22
Cu–Pb–O	−7·4	−0·007	5·27
Cu–Co–O	−61	−0·31	1·54
Cu–Fe–O	−553	−2·73	19·5

The standard state for element X is the pure element in the equilibrium state at 1600°C for iron alloys and 1100°C for copper alloys. γ° is the activity coefficient at infinite dilution.

CHAPTER II

Experimental Methods

The quantities conveniently recorded for the thermodynamic description of inorganic and metallic compounds are the heat of formation at 298°K, the standard entropy, the heat capacity (for condensed substances up to the boiling point) and the heats of transformation and fusion where applicable. All these data can in principle be obtained by means of suitable calorimeters. Of course, no single apparatus of this type can be employed to measure all these quantities. Application of each calorimeter is restricted to a certain range of temperature. There are calorimeters operating between 1–50°K and room temperature producing low-temperature heat capacities and standard entropies, provided the substances are in the well-ordered crystalline equilibrium state required by the third law. There are calorimeters specially designed to measure heats of reaction and others for the determination solely of heat capacities and heats of transformation and fusion. Some of the last-mentioned calorimeters can be made to produce not only heat capacities but also heats of reaction or formation. Thus the calorimeter is the most important piece of equipment of the versatile thermochemist.

However, a calorimeter may not always give the desired results. A case in point is the evaluation of low-temperature heat capacities to provide standard entropies. Strongly ionic compounds, it is true, can normally be produced in the highly ordered form required for the application of eq. (22). Intermetallic compounds, the stability of which stems from a dense atomic packing, are also likely to have a near-zero entropy at 0°K, as has been demonstrated within the experimental error for $TaCr_2$ and other Laves phases by Martin *et al.* [39]. Many phases of a predominant covalent–metallic nature, however, are not easily brought into a completely ordered state, and "frozen-in equilibria" prevail—in particular with the preparation of high-melting-point compounds. In such cases it is necessary to find independent thermochemical methods to provide the desired information—and it is here where the usefulness of methods measuring Gibbs energies become apparent. Of such methods there is a great variety, entailing measurements of dissociation pressure, of condensed-state equilibria and of electromotive forces of suitable cells.

When such measurements are made over a range of temperature, heats and entropies of reaction may be evaluated using the relevant equations recorded in Chapter I. However, it cannot be emphasised strongly enough that such evaluations are unreliable when condensed phases take part in the reaction because the temperature coefficients are very sensitive to small experimental errors. The so-called second-law values are not *necessarily* inaccurate, but the trouble is that one never knows. Just unreliable!

This severe restriction has already been pointed out earlier for liquid and solid solutions of alloy systems. Whenever possible, Gibbs energy results should be combined with independently obtained heats in assessing the standard values. It should not be overlooked that in cases where heats of reaction are difficult to obtain by means of the available calorimeters for kinetic reasons (e.g. metal carbides), and Gibbs energy measurements seem to offer the only alternative, their temperature coefficients are even more affected by kinetic checks. In such cases and in alloy thermochemistry, it is sometimes preferable to record heats and entropies not at 298°K but at some convenient higher temperature where the measurements have actually been made—a procedure followed, for instance, by Hultgren *et al.* [40] for the presentation of the thermochemical properties of alloys.

This is not necessarily a severe defect because the data are often required for industrial applications, calculations of equilibrium diagrams, etc., at temperatures which are not far from those of the original measurements. When standard heats and entropies have been derived from the temperature coefficients of Gibbs energy measurements, they are given in the tables of the present monograph with relatively large limits of error. Nevertheless, when such values are incorporated in an actual calculation for practical ends, the result is likely to be more accurate at the temperature of application than the stated limits of error imply.

Naturally, as time goes on, more experimental information becomes available and the standard values can be improved. Unfortunately, there is at present a tendency among experimentalists to employ elaborate equipment that is not always justifiable because simpler thermochemical techniques may produce the same results with similar accuracy. A study of the following pages may therefore be recommended.

1. Calorimetric Methods

The first of the thermodynamic functions to be determined experimentally for a large range of substances was the heat of formation. The foundations of this work were laid with the classic investigations of Berthelot in France and Thomsen in Denmark. At the time, both Berthelot and Thomsen regarded the heat of formation of a compound as the measure of the affinity of its constituent elements for one another, and it was for that reason that their work was carried out on such an extensive scale.

Heats of formation or of reaction are usually determined by calorimetric methods, which are quite simple in principle. The reaction to be investigated is carried out in an apparatus of known heat capacity, the calorimeter, so that the heat effect is measured by the change in temperature during reaction.

The heat capacity of the calorimeter is determined as its water equivalent. Theoretically this is the sum of the heat capacities of all the constituent parts of the calorimeter. In practice, however, various heat interchanges take place which are difficult to account for by calculation, and the water equivalent is always determined experimentally. The water equivalent W multiplied by the change in temperature then gives directly the heat effect of the reaction

$$q_T = W\Delta T \tag{93}$$

The physics of heat conduction, capacity and interchange and their bearing upon the design of calorimeters will not be dealt with in great detail in this monograph. Readers are recommended to study sections on this subject in textbooks of practical physics, and particularly in the second volume of *Experimental Thermochemistry* [41].

The classification of calorimeters will here be based [42] on three main variables: the temperature of the calorimeter T_c, the temperature of its surroundings T_s, and the heat Q produced per unit of time.

If $T_c = T_s =$ constant and only Q varies, we have an *isothermal calorimeter*. Its best-known representative is the ice calorimeter (p. 101) designed in its modern form by Bunsen.

A calorimeter in which $T_c = T_s$ but is not constant as Q varies, is called an *adiabatic calorimeter*. Such calorimeters (p. 103) have been built for the determination of heat capacities as well as heats of reaction, but because the temperature of the enclosure must at all times be the same as that of the calorimeter, its application should be limited to relatively slow reactions, for instance the dissolution of metals, etc., in acids or the formation of endothermic alloys.

Somewhat simpler in design than the adiabatic calorimeters are the *heat-flow calorimeters* for which $T_s - T_c =$ constant. These are suitable for the direct determination of heat capacities and heats of transition but not for the study of heats of reaction because it cannot be "stopped" and held at constant temperature. Heat-flow calorimeters will not be described in this monograph, but the reader may be referred to a paper by Hagel *et al.* [43] in which such an apparatus, originally designed by C. S. Smith, has been described.

The most widely used type of calorimeter is the *isoperibol calorimeter** (usually misnamed "isothermal") in which the enclosure is held at constant temperature ($T_s =$ constant) and T_c is measured before, during and after a reaction.

The quantities to be determined in almost all calorimetric experiments are thus the change in temperature during the experiment, and the water equivalent of the calorimeter, which is found by calibration. In addition it is essential to know exactly what amount of substance takes part in the reaction, and to ensure that no secondary reactions occur. The exact chemical analysis of reaction products is, therefore, of prime importance in calorimetry. Without it, attempts to measure physical quantities, such as temperature change, with great accuracy become meaningless.

A. MEASUREMENT OF TEMPERATURE

Temperature measurements in calorimetry usually involve the accurate determination of changes in temperature of a few degrees only. The importance of accuracy in these measurements is such that all those engaged on this work should study the rudiments of thermometry. Two short monographs by Hall [44] give a concise introduction to the subject with accounts of the main types of precision thermometers.

Mercury-in-glass thermometers, platinum resistance thermometers, and thermocouples have all been used extensively in calorimetry, and the account which follows is intended to indicate the salient practical features and advantages of each type.

1. *Mercury-in-glass Thermometers*

To measure temperatures with the necessary accuracy, these thermometers must have a fine-bore capillary and a large reservoir. The range of mean temperatures to be

* ἡ περιβολή = the surroundings.

covered may be very wide, however, and since there are practical limitations to the length of a thermometer stem two methods have been developed to make calorimetric and similar work more flexible. The first is to use a set of thermometers, each calibrated over 6°C with overlapping ranges from 9° to 33° [45]. These thermometers are very robust, and are ideal for the routine work carried out, for instance, in a fuel control laboratory.

The second method is to use an adjustable zero thermometer of the type introduced by Beckmann, which is also graduated over a 6°C range. Because this instrument is rarely used in metallurgical thermochemistry nowadays and an experienced thermochemist made the point in a review of an earlier edition of the present book that he had never seen a Beckmann thermometer since his student days, the authors considered omission of its description. However, they found themselves very reluctant to do so because the instrument can and should continue to have many uses even in metallurgical thermochemistry, being just as accurate as most electrical thermometers in the hands of a skilled observer and cheaper and simpler in operation. The "Beckmann" is, of course, restricted in temperature of application: the zero-point temperature may be varied from $-10°$ to $+100°$C or more by confining superfluous mercury to a reservoir at the top end of the capillary. It is an instrument for individualists (like the late W. A. Roth) who through continual acquaintance have become thoroughly familiar with its idiosyncrasies and virtues.

Before a Beckmann thermometer is put to use it should be calibrated to determine the correction for any irregularity in the bore. A thread of mercury about one degree in length is detached by tapping and its length in terms of scale graduations is measured at half-degree intervals along the whole scale. This length depends on the average diameter of the portion of capillary occupied. The deviations from an average value are plotted against the mean scale reading and the resulting curve is integrated graphically. In subsequent experiments each reading is corrected from a calibration curve. The actual temperature change represented by one degree on the scale varies also according to the zero temperature of the thermometer since adjustment of the zero changes the mass of mercury in the bulb. The change caused by a small adjustment is very small, however, and the use of a widely different range would generally be accompanied by a new determination of the water equivalent of the calorimeter; this will eliminate corrections to mean scale readings.

Though graduated only to 0·01°C, with good Beckmann thermometers an accuracy in reading of $\pm 0{\cdot}0005$°C is attainable with some practice. The thermometer must be immersed always to the same depth in the calorimeter liquid, and before each reading it should be tapped near the meniscus with the fingers or a rubber-covered rod. Whenever its temperature is changed by, say, 40°C or more, it should preferably be left for 24 hours before other measurements are made.

2. Platinum Resistance Thermometers

The accurate determination of temperature by the measurement of the resistance of a coil of platinum wire was developed initially by H. L. Callendar in the years 1887–92, and the principle of the method may be found in textbooks of heat. If the electrical measuring instruments are reliable and sensitive, these thermometers can be eight times more accurate than the mercury-in-glass type, so they are the obvious choice for work of high precision.

The conventional type of resistance element may prove too cumbersome for calorimetric work. Barber [46] has described a more compact thermometer the bulk of which is no larger than that of a Beckmann thermometer. It has a lag of 5 sec only, and its zero point is reproducible within 0·002° after temperature variations from −180° to 600°C. Much calorimetric work is carried out near room temperature, however, and Maier [47] has described a modified resistance thermometer for this range.

It may often be desirable to dispense with the normal self-contained resistance element and to incorporate the platinum resistance within the construction of the calorimeter itself. This has been done, for example, in an adiabatic calorimeter operated at 650°C [48], and in working with fluoride solutions [49] (which would have attacked glass). In constructing resistance thermometers for these special purposes, steps should be taken to observe as far as possible the principles followed in the design of the standard thermometers. The platinum element, once mounted, should be kept quite free from strain at all temperatures. It must be supported with a completely non-conducting material, and any possibility of condensation occurring on the wire must be rigidly excluded. In precision thermometry nowadays it is usual to measure the current flowing through the element and the potential difference across its ends. For much calorimetric work the earlier Wheatstone bridge method, involving the use of compensating leads, will be found sufficiently accurate and may be found simpler to set up with the equipment available in some laboratories.

3. Thermocouples

The sensitivity of measurements on a single thermocouple is too small for most calorimetric purposes, but may be increased by connecting a number of couples in series. In addition to multiplying the voltage, this can give an average temperature for a large body if the couple junctions are disposed judiciously. As many as 1000 couples have been combined in this way [50] to enable temperature differences as low as 10^{-7}°C to be measured. The number of thermocouples cannot be increased indefinitely, as they may introduce other sources of error, such as the thermal conductance of the connections. The usefulness of the method in giving a direct measure of temperature difference has, however, not yet been fully exploited. The use of ten to fifty thermocouples in series can be made normal practice. With ten copper-constantan couples 1 mV is equivalent to 2·34°C, and with the same number of chromel-constantan couples 1 mV ≡ 1·58°C. Using a mirror galvanometer a temperature change of 10^{-4}°C may be discerned. This high precision may be offset in calorimetric work by the thermometric lag which in most instances, due to the thickness of couple sheaths, will be relatively high.

4. Thermistors

In a thermistor the resistance element is a semiconductor, such as silicon carbide. Its high resistivity enables it to be made much more compactly than the usual metallic resistance winding. Also the temperature coefficient (which is negative) is much larger than that of metals, increasing the sensitivity. However, they are useful only over limited temperature ranges. Thermistors are theoretically capable of measuring temperatures with a precision of a few microdegrees without elaborate precautions. Their chief difficulty thus far has been an instability; after a temperature cycle they do not return to the same reading. This difficulty is probably being overcome.

5. *Optical Pyrometers*

Optical pyrometers may be mentioned because they are being employed to measure the temperature of the specimen before the drop in drop calorimetry combined with levitation melting, but a description is obviously not called for in the present context.

B. DETERMINATION OF WATER EQUIVALENT

Ideally, determinations of water equivalent should always be made under conditions similar to those in which the reaction to be measured takes place. In small calorimeters, particularly, the apparent water equivalent varies with the magnitude of the temperature change. Consequently, the known heat effects used in calibration should be as near as possible in magnitude and in rate of evolution to those to be studied in subsequent experiments.

If the reaction to be investigated is endothermic, calibration is made by dropping into the calorimeter a quantity of metal of known heat capacity cooled initially to a known temperature below 0°C. The metal used may be annealed copper or silver, or mercury enclosed in a glass bulb.

For high-temperature heat content measurements a similar method is employed for the calibration as for the actual determination, using metals of known heat capacity (see "drop method", p. 88). Aluminium oxide in the form of sapphire is recommended as the standard substance. Precise measurements have been made at IRSID as well as NBS [51], and Ferrier and Olette have suggested the following equation to represent the available data:

$$H_T - H_{298} = -11{,}155 + 27{\cdot}43T + 1{\cdot}53 \times 10^{-3}T^2 + 8{\cdot}47 \times 10^5 T^{-1}$$

The authors have usually used silver for heat content measurements, though it has the disadvantage of giving up its heat slowly and so prolonging the duration of measurement. This should be roughly equal for the calibration and the subsequent experiment. A coating of graphite on the silver does not materially affect the heat loss.

For exothermic reactions generally, the calibration is best carried out by electrical heating. A detailed description of the method is given by Roth [52] for instance. An electric current is passed for a given time through a heating coil of known resistance inside the calorimeter, and the resulting temperature increase is determined. The input of energy is

$$q = RI^2t \times 0{\cdot}2389 \text{ cal} \tag{94}$$

where R is the resistance of the coil in ohms, I the current in amperes and t the time in seconds for which it is passed. Reliable electrical instruments must, of course, be used, their precise nature and mode of operation depending upon the accuracy it is intended to attain. The resistance of the heater may be measured at room temperature by a bridge circuit, and any correction due to change in resistance with rise in temperature is calculated from the temperature coefficient of the heater material. Current is best determined by measuring the potential drop across a standard resistance included in the heater circuit using a potentiometer. For accurate work the power being dissipated in the heater during an experiment may be measured by measuring the voltage across its ends as well as that across the standard resistance. The heat input is then given by the equation

$$q = 0{\cdot}2389 \frac{E_S E_H}{R_S} t \tag{95}$$

where E_H and E_S are the potential drops across the heater and the standard resistance (R_S) respectively. In practice it will rarely be possible to carry out continuous observations with potentiometers during experiments since there are so many other observations to be made. The precise application of the electrical calibration method therefore demands a supply of perfectly constant voltage for the duration of the experiment (which should, however, not exceed the very few minutes taken by the thermal reaction subsequently to be studied). For this purpose, van Artsdalen and Anderson [53] have made use of a United States Army radar unit rectifier giving a current of 0·50 A at about 300 V d.c. Voltage fluctuations were in general less than 1 part in 6000 and never more than 3 parts in 6000.

For some calorimetric work sufficient accuracy in electrical energy measurement may be attained by the use of an ordinary voltmeter and ammeter. The water equivalent of the calorimeter and its contents is obtained by dividing the heat input by the corrected temperature increase,

$$W = q/\Delta\theta \tag{96}$$

The design of an electrical heater will depend on the available space and on the voltage, which should in general be kept as low as practicable for reasons of safety. A typical calibration heater, one which was used in connection with the determination of the heat of solution of hydrogen peroxide in water [54], is constructed as follows: a nichrome wire 0·3 mm diameter of resistance 6·7 ohms is wound non-inductively on a glass rod. As protection against solvent attack and to eliminate electrolysis effects, the wire is coated by dipping in paraffin wax. This does not melt in operation, due to the rapid removal of heat by the convection of the water. The ends of the nichrome wire are soldered below the liquid surface level to stout copper terminals of negligible resistance.

It is important that heaters should have a low heat capacitance and a low heat lag, and for this reason the heater is often immersed unprotected in the calorimeter fluid. This method is impracticable in solutions of electrolytes, and with this in mind Murphy [55] has described the fabrication of heater containers of aluminium (and possibly tantalum), electrically insulated by a sealed anodised coating.

C. ISOPERIBOL CALORIMETRY

1. *Principle of Operation and Evaluation of Results*

The essential components of an isoperibol calorimeter are shown in Fig. 33. The outer jacket A must be maintained at a constant temperature during the experiment.

For precision work its temperature should be controlled by some device such as a toluene regulator, with a low thermal lag. The reaction takes place in the inner vessel B, the calorimeter proper. The temperature of the inner vessel will always tend to become the same as that of the outer jacket. The tendency follows Newton's law, which involves an asymptotic approach of temperature θ_B to the temperature θ_A. C is an intermediate shield of metal sheet (first introduced by W. P. White) which should generally be used to reduce exchange of heat by convection.

For an exothermic process, the temperature–time curve observed will be one of the three types shown in Fig. 34. The type of curve depends on the initial temperature of the

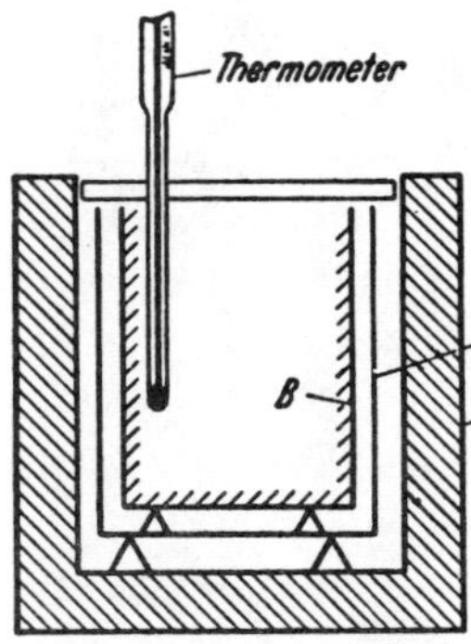

FIG. 33. The principal components of an isoperibol calorimeter.

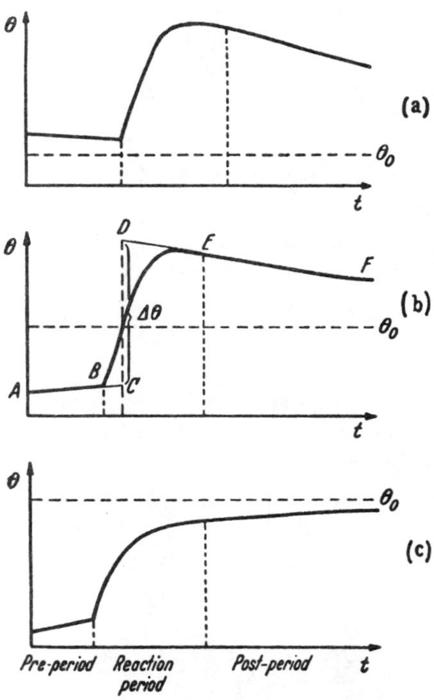

FIG. 34. Temperature–time curves in isoperibol calorimetry: (a) initial temperature higher than the convergence temperature; (b) convergence temperature between initial and final temperatures; (c) final temperature below the convergence temperature.

inner vessel relative to that of the outer jacket, which is shown in the diagrams as a dotted line.

For evaluating the heat evolution (or absorption) in the calorimeter, a correction must be made for the exchange of heat between the calorimeter and its surroundings during the process. For this purpose, the course of the temperature change is divided into three periods. The first period is observed for several minutes prior to the reaction. During reaction and for a short while afterwards, the temperature of the inner vessel increases due to the liberation of heat by the reaction. The third period begins when the rate of cooling of the inner vessel again follows Newton's law.

When the highest accuracy is not required the corrected temperature increase is determined as indicated in Fig. 34b. A vertical line is drawn through the curve at the point corresponding to half the determined total temperature increase. The temperature–time curves in the preliminary and final periods, AB and EF respectively, are extrapolated to intersect this line at C and D. The temperature difference CD is then taken as the corrected temperature increase $\Delta\theta$, which when multiplied by the water equivalent of the system gives the amount of heat evolved in the reaction.

To be more exact one must take into consideration the fact that AB and EF are not straight lines, and that the mean temperature of the calorimetric surrounding θ_0 cannot be determined with accuracy.

The more precise methods of evaluation make direct use of Newton's law, that rate of cooling is proportional to temperature difference. Methods can be used which do not require that θ_0 should be known accurately. To illustrate this more specifically we shall find the corrected temperature change during the experiment summarised in Fig. 35, which shows the temperature–time curve for a calorimeter during the solution of copper in Herschkowitsch's solvent [56] at a mean temperature of 90°C.

The following symbols are used:

t_1 time at commencement of reaction, i.e. at the end of the preliminary period of Newtonian heating or cooling.

θ_1 temperature at commencement of reaction.

t_2 time at beginning of Newtonian cooling period.

θ_2 temperature at beginning of Newtonian cooling period.

θ_0 temperature of surroundings.

θ' mean of consecutive temperature readings in the reaction period.

z number of readings in the reaction period.

k constant in Newton's law of cooling for the apparatus.

v_1 rate of change of temperature in initial period.

v_2 rate of change of temperature in the cooling period. According to Newton's law

$$v_1 = k(\theta_0 - \theta_1) \quad \text{or} \quad \theta_0 = \frac{v_1}{k} + \theta_1$$

$$v_2 = k(\theta_0 - \theta_2) \quad \text{or} \quad \theta_0 = \frac{v_2}{k} + \theta_2$$

Hence

$$k = \frac{v_1 - v_2}{\theta_2 - \theta_1}$$

In order to obtain the corrected temperature change we must add to the observed temperature difference $(\theta_2 - \theta_1)$ the product of Newton's constant and the area difference $(A_2 - A_1)$, shown in the figure.

The area $(A_2 + A_3)$ equals the sum of the products of the successive temperature differences $(\theta' - \theta_1)$ and the unit of time between each observation. For simplicity this is taken as one minute.

$$(A_2 + A_3) = \Sigma(\theta' - \theta_1)1$$

Further,

$$(A_1 + A_3) = (\theta_0 - \theta_1)(t_2 - t_1) = (\theta_0 - \theta_1)z$$

Hence

$$(A_2 - A_1) = \Sigma(\theta' - \theta_1) - (\theta_0 - \theta_1)z$$

But since

$$(\theta_0 - \theta_1) = v_1/k$$

$$(A_2 - A_1)k = k\,\Sigma(\theta' - \theta_1) - v_1 z$$

The readings upon which Fig. 35 is based are given in Table VI. From them we obtain:

$$\begin{array}{rlrlr} \theta_2 = & 3{\cdot}7080 & v_1 = & 0{\cdot}0024 & \\ \theta_1 = & 2{\cdot}3340 & v_2 = & -0{\cdot}0025 & \Sigma(\theta' - \theta_1) = 21{\cdot}45 \\ \hline \theta_2 - \theta_1 = & 1{\cdot}3740 & v_1 - v_2 = & 0{\cdot}0049 & k = 0{\cdot}0036 \end{array}$$

The correction therefore becomes

$$(A_1 - A_2)k = (0{\cdot}0036)(21{\cdot}45) - (0{\cdot}0024)(20) = 0{\cdot}0292$$

This must be added to the observed temperature difference

$$\begin{array}{rl} \theta_2 - \theta_1 = & 1{\cdot}3740^\circ \\ \text{correction} & 0{\cdot}0292 \\ \hline \Delta\theta = & 1{\cdot}4032 \end{array}$$

which is the corrected temperature increase due to the reaction.

Though the method was applied in this example to a heat of solution, it may be used for any heat effect, exothermic or endothermic.

2. *The Bomb Calorimeter*

The bomb calorimeter, introduced by Berthelot, is probably the best-known piece of equipment used in thermochemistry, and a brief description of it should suffice here.

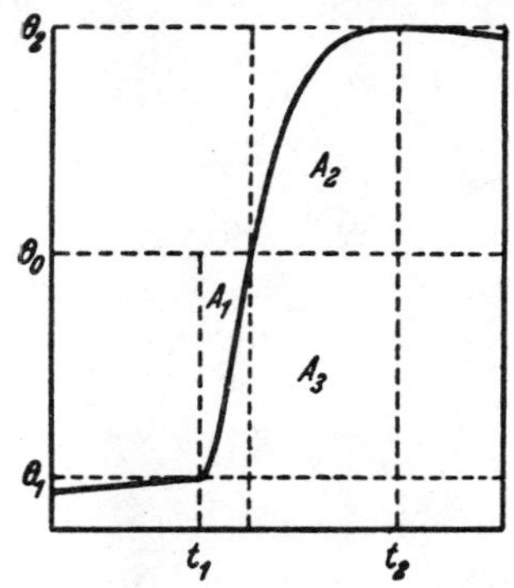

FIG. 35. Correction of temperature change for radiation losses.

It is used to determine heats of combustion, the substance under consideration being burnt in an excess of oxygen, which is usually introduced up to a pressure of 25 atm.

The bomb must remain gas-tight during reaction, which may occur with explosive violence, so it consists of a stout shell of high tensile steel, closed with a heavy screw cap. A tight seal is ensured by the use of a suitable washer of lead, or in some instances rubber. The substance to be oxidised is provided in the form of powder or thin foil and is placed in a crucible supported by thin platinum or iron wires about 0·25 mm in diameter which also serve to carry the electric current used to start the reaction. Oxygen is introduced at pressure and is retained by a needle valve or a spring-loaded

Table VI

Initial period		Reaction period			Final period	
θ	$\Delta\theta$	θ	θ'	$\theta' - \theta_1$	θ	$\Delta\theta$
2·3030		2·3340			3·7080	
	+0·0025		2·39	0·06		−0·0020
2·3055		2·45			3·7060	
	+0·0025		2·56	0·23		−0·0025
2·3080		2·67			3·7035	
	+0·0020		2·79	0·46		−0·0025
2·3100		2·90			3·7010	
	+0·0025		3·00	0·67		−0·0025
2·3125		3·10			3·6985	
	+0·0025		3·18	0·85		−0·0025
2·3150		3·25			3·6960	
	+0·0025		3·31	0·98		−0·0020
2·3175		3·37			3·6940	
	+0·0025		3·42	1·09		−0·0025
2·3200		3·46			3·6915	
	+0·0020		3·49	1·16		−0·0025
2·3220		3·52			3·6890	
	+0·0025		3·55	1·22		−0·0025
2·3245		3·57			3·6865	
	+0·0025		3·59	1·26		−0·0030
2·3270		3·61			3·6835	
	+0·0025		3·63	1·30		−0·0025
2·3295		3·64			3·6810	
	+0·0020		3·65	1·32		−0·0025
2·3315		3·66			3·6785	
	+0·0025		3·67	1·33		−0·0030
2·3340		3·675			3·6755	
	—		3·68	1·34		−0·0025
		3·687				
Mean	+0·0024		3·69	1·35		
		3·6960				
			3·70	1·36		
		3·7005				
			3·70	1·36		
		3·7035				
			3·71	1·37		
		3·7065				
			3·71	1·37		
		3·7075				
			3·71	1·37		
		3·7080		—		
				21·45		

non-return valve. The capacity of the bomb is usually between 350 and 500 cm^3. It is immersed during experiment in a water-filled calorimetric vessel, that is, in the inner vessel of Fig. 33. This is contained in its turn in an outer jacket maintained at constant temperature. For all work other than the combustion of organic materials the bomb should be lined with enamel, platinum, gold or some similar material to protect the steel from attack.

Those intending to carry out work with a combustion bomb are recommended to read the description of the apparatus developed at the United States Bureau of Standards [57], and used in the work cited to determine the heat of combustion of graphite. Points of experimental interest include the following: use of a device to measure the small leak which always occurred during combustion; determination of the trace of carbon monoxide in addition to the dioxide formed; use of a constant quantity of vaseline during each experiment for lubricating the bomb threads; use of a platinum crucible which was shallow to provide large area of contact with oxygen and thin so as not to conduct too much heat from the burning carbon. A gold gasket was used for sealing the bomb.

The combustion bomb is generally calibrated by the heat of combustion of pure benzoic acid, although electrical calibration may be preferable. The heat of combustion of benzoic acid is 6315 cal/g at constant pressure under bomb conditions, but this value may vary slightly with the origin of the sample. When the melting point of the acid was between 121·8 and 122·0°C the variation in the heat of combustion was found to be less than 10 cal/g [58]. Succinic acid, with a heat of combustion of 3003 cal/g, is also recommended for calibration.

Though the bomb calorimeter is used primarily for the combustion of organic compounds, it has also been used to determine the heat of formation of metallic oxides and the heats of formation and combustion of halides, carbides and sulphides.

The determination of the heat of formation of metallic oxides is described in several of Roth's papers [52, 59, 60]. Becker and Roth have also reported experiments with a "micro-bomb" of water equivalent approximately 500 cal/deg on the heats of combustion of rare metals.

The substances investigated must be as pure as possible since contaminations may introduce a disproportionately large error into the results. Following Roth [60], we may list as illustration the heat of formation of alumina, from bomb experiments only, against the year of determination:

Observers	Year	$-\Delta H^\circ_{298}$ (kcal)	Purity of Al (%)
Berthelot	1881	380·2	99·5
von Wartenberg and Witzel	1919	396·0 $\pm$ 0·4	95·0
Moose and Parr	1924	375·8 $\pm$ 0·2	98·4
Roth and Müller	1929	380·8 $\pm$ 0·4	99·83
Meichsner and Roth	1934	393·3 $\pm$ 0·4	99·83
Roth, Wolf and Fritz	1940	402·9 $\pm$ 0·3	99·9
Snyder and Seltz	1945	399·0 $\pm$ 0·24	99·98
Holley and Huber	1951	400·29 $\pm$ 0·31	99·996
Schneider	1954	400·3 $\pm$ 1·6	99·99+

The values tend to increase with improvement in technique and in purity of the substance used. Since we can hardly attempt to extrapolate a value for the year 2000 with any hope of accuracy we must seek out the reasons for these discrepancies. In addition to lack of purity of materials, inadequate analysis and inaccurate calorimetric measurement, sources of error include fluxing between the molten alumina and the crucible material, and inequality between calibration and combustion conditions. In the case under discussion, a further uncertainty arises from the doubt that has been cast on the crystallographic nature of the combustion product. It has previously been found or assumed that α Al_2O_3 is formed. Schneider, however, observed that his product gave mainly the X-ray pattern of a δ modification of alumina which is apparently similar to γ Al_2O_3. This matter will have to be investigated in more detail.* The point we wish to make in this connection is that the significance of the limits of error in the table above should be carefully studied. The observers have considered the precision of their measurements, but have largely disregarded the other errors that can arise. In assessing the reliability of all thermochemical data one must keep this lesson constantly in mind.

The purity of the materials may not always be truly determined by the common methods of chemical analysis, namely by analysis of the solution of the substance. For instance, the heat of combustion of cementite, previously taken to be -2150 cal/g, was found on using a sample in a completely degassed state [62] to be -2027 cal/g.

It is sometimes necessary to start the combustion of rather inert substances such as carbides with an easily inflammable material. Paraffin, vaseline and benzoic acid have been used for this purpose, and Roth especially recommended cellophane. The mass and heat of combustion of this initiator must be known accurately, and subtracted. It is advisable to determine the heat of combustion with each new batch separately as there may be slight differences. Approximate values for some materials are as follows:

Liquid paraffin	11,000 cal/g
Vaseline	11,050 cal/g
Cellophane	3700 cal/g

Whenever the additive heats of combustion or formation of two or more substances are introduced, it must be remembered that the errors are also additive. If the heat of combustion of the substance investigated by this method is given by a small difference in two large quantities, the error will be larger.

A similar problem arises when the heat of formation of a substance such as carbide, silicide or phosphide, is obtained by Hess' law of constant heat summation as the difference between its heat of combustion and that of its constituent elements. As a result of the relatively small adjustment in the heat of *combustion* of cementite just mentioned (2150–2027 cal/g), its accepted heat of *formation* was changed from 26,000 to 3900 cal/mole!

McGraw *et al.* [63] determined the heat of combustion of tungsten carbide and then calculated its heat of formation by means of Hess' law:

$$\begin{array}{llr} -\langle WC\rangle + 2\tfrac{1}{2}(O_2) = \langle WO_3\rangle + (CO_2): & \Delta H^\circ = & -285{,}800 \pm 80 \\ +\ \ \langle C\rangle + \ (O_2) = (CO_2): & \Delta H^\circ = & -94{,}050 \pm 11 \\ +\ \ \langle W\rangle + 1\tfrac{1}{2}(O_2) = \langle WO_3\rangle: & \Delta H^\circ = & -195{,}700 \pm 900 \\ \hline \ \ \ \langle W\rangle + \langle C\rangle \ = \langle WC\rangle: & \Delta H^\circ = & -3950 \pm 1000 \end{array}$$

* Kleppa and Yokokawa [61] have measured the heats of solution of α Al_2O_3 and meta-stable δ Al_2O_3 in a lead–cadmium borate melt at 705°C. From these, $\Delta H_{978} = -2700 \pm 400$ cal for the reaction $\langle Al_2O_3\rangle_\delta = \langle Al_2O_3\rangle_\alpha$.

the heat of combustion of tungsten being taken from earlier work by Moose and Parr. The authors then wisely decided to redetermine this quantity, and found [64]

$$\langle W\rangle + 1\tfrac{1}{2}(O_2) = \langle WO_3\rangle : \Delta H^\circ = -200{,}160 \pm 100$$

whence

$$\langle W\rangle + \langle C\rangle = \langle WC\rangle : \Delta H^\circ = -8410 \pm 190$$

We strongly urge that in such an evaluation all the associated reactions should be investigated always in the same apparatus, except those for which the data are no longer disputable (e.g. CO_2, H_2O). It is seen that this type of work calls for extreme accuracy in chemical analysis and rigorous self-criticism by its authors. The errors stated above, for instance, are only observational errors and do not include any systematic errors inherent to the reaction investigated. More allowance must be made for these. One should bear in mind a statement made by the experienced thermochemist W. P. White 60 years ago [65] that a "glance at almost any comparative table of calorimetric values will suffice to show that calorimetry is one of those subjects in which the systematic errors are usually much larger than the accidental, and deserve, therefore, the more careful consideration".

Systematic errors often result from reactions between one of the participating substances and the crucible material. This may take the form of fluxing between a molten metallic oxide and a (usually) siliceous refractory. This was found, for instance, by Siemonsen [66] as a reaction between manganous oxide and silica. Alternatively, elements of high reducing power, such as titanium, may react with siliceous crucibles with formation of amorphous silicon monoxide [55]. To avoid this trouble it is preferable whenever possible to use crucibles of the oxides of the same or similar metals as those to be investigated. If such crucibles are too friable, they may simply be used as linings of the normal crucibles [67]. The oxide, mixed with starch paste, is smeared over the metal or silica crucible, dried, and fired to dispose of the carbonaceous material. Klemm and Schnick [68] protected silica crucibles with gallium oxide in this way for investigating gallium combustion. V. Wartenberg [69] used aluminium crucibles lined with magnesium fluoride to investigate the reaction between chromic fluoride and magnesium. However, the non-stoichiometric nature of even refractory oxides, which increases with temperature, must be borne in mind. Zirconia, for instance, becomes oxygen deficient at high temperatures even when the oxygen pressure is several atmospheres.

When reaction takes place violently there is a risk that the yield may be incomplete, due to dispersion of the powdered material [67]. The use of a closed crucible has been recommended [70], but this has the disadvantage of prolonging the reaction period, which increases the normal error of the method.

In discussing the combustion of sulphides and halides in the Berthelot bomb, Roth [59] has stated that the reaction products are never of fixed composition. Chlorides yield a mixture of chlorine and hydrogen chloride. Bromides give a mixture of liquid, dissolved and gaseous bromine. Iodides are relatively free from complication, but they have not been investigated thoroughly. To eliminate the complication of a mixed product it has become the practice to take up the chlorine with a reducing agent.

As_2O_3 solution is commonly used for this but certain precautions must be taken to ensure that this reduction goes to completion.

The direct chlorination (see also p. 96) of metals in a bomb under a pressure of 7–8 atm has been performed by Siemonsen [71]. The difficulty in this case is, of course, to find a material that is resistant to chlorine under high pressure. Siemonsen found an enamelled steel bomb to be satisfactory. The various leads were made of silver or platinum; magnesia crucibles were recommended as containers for the metal. For the insulation of the electrical leads a seal made of asbestos string impregnated with polyvinyl chloride was selected after some failures with other materials. The valve could, of course, not be enamelled, but Siemonsen claims that the reaction of the steel with the chlorine was too small to have any effect on the thermochemical results. There was some corrosion of the enamel; it was observed, however, that this was due to the attack not of the chlorine but of hydrogen chloride produced by hydrolysis after opening the bomb.

The successful development of a bomb calorimeter for combustions in an atmosphere of fluorine gas has been reported by investigators at the Argonne National Laboratory [72]. The bomb is made of nickel which is rendered inert to attack during the combustion process by a careful pre-fluorination treatment, so that the inner surfaces become coated with a stable nickel fluoride film. Combustion experiments have been made on several metals and inorganic compounds.

The combustion of nitrides [67] in oxygen involves no great difficulty as the combined nitrogen is completely converted into the gaseous N_2 form. Their heats of formation are quite simply obtained as the difference between their heat of combustion and that of the metal.

The heats of formation of some nitrides have been determined by direct nitriding in the bomb [73]. Since the rate of these reactions is very low at room temperature, the metals were contained in a platinum crucible which could be heated with a small resistance furnace. The electrical energy input was measured and subtracted from the measured amount of heat liberated. Chromium, manganese, thorium, uranium and cerium were almost completely nitrided within 2 minutes at 900°C in nitrogen at 25 atm pressure. The nitrogen was purified by storing it under pressure for at least 6 hours in a cylinder containing an absorption mixture. This was composed of 25 cm^3 concentrated ammonia solution, 25 cm^3 saturated ammonium carbonate, 50 cm^3 water, 10 g ammonium chloride, 300 g copper and some tin.

Some of the nitriding reactions would only take place with the addition of a catalyst; thus 25% magnesium oxide was added to magnesium. Sodium fluoride has also been used for this purpose.

The bomb has also been used in the measurement of heats of decomposition [74], that of copper nitride, Cu_3N, for example. Such reactions may be started by using a combustion to raise the temperature. In this determination, the copper nitride was compressed into a pellet and sealed into a small silica bulb. This was placed on a pellet of benzoic acid which was ignited in the usual way. The heat of combustion was sufficient to cause complete decomposition of the nitride without breaking the silica bulb. The heat of decomposition was obtained by deducting the heat of combustion of the benzoic acid from the observed heat effect. The decomposition of cadmium nitride in this way invariably destroyed the bulb, so the more usual method was adopted of heating the nitride electrically by means of a platinum wire.

3. *Heats of Solution*

The measurement of heats of solution of substances in aqueous solvents at room temperature is relatively simple, and by applying Hess' law of constant heat summation they may be used to evaluate heats of formation or reaction that are difficult to determine directly. To take a simple example, the heat of formation of hydrated copper sulphate $CuSO_4{\cdot}5H_2O$ from the anhydrous sulphate is obtained by measuring the heats of solution of the two salts in water. Many of the heats of formation of the metallic compounds which we know today were obtained by this method, which figured prominently even in Thomsen's work carried out almost a century ago [75].

Strictly speaking, the term "solution" should be applied to the formation of a single homogeneous phase from two or more substances. In thermochemical work the term is used for experimental purposes in a wider sense to include certain processes in which not only solution but reaction and formation of a second phase, gaseous or possibly solid, occur. For example, the heat of formation of an intermetallic compound of magnesium and zinc is determined by measuring the following heats of solution in hydrochloric acid.

$$\langle Mg\rangle + 2[HCl] = [MgCl_2] + (H_2); \Delta H_a \tag{a}$$

$$2\langle Zn\rangle + 4[HCl] = 2[ZnCl_2] + 2(H_2); \Delta H_b \tag{b}$$

$$\langle MgZn_2\rangle + 6[HCl] = [MgCl_2] + 2[ZnCl_2] + 3(H_2); \Delta H_c \tag{c}$$

From these equations we obtain

$$\langle Mg\rangle + 2\langle Zn\rangle = \langle MgZn_2\rangle; \Delta H_x$$

$$\Delta H_x = \Delta H_a + \Delta H_b - \Delta H_c \tag{d}$$

Here, again, the heat of formation of a compound is the difference between its heat of solution and that of its constituent elements in a given dissolving medium.

The heats of formation of ordinary metallic salts are often determined in a similar way, though the various stages are sometimes more complicated than in the last example. We may take as a simple illustration the determination of the heat of formation of magnesium chloride. The actual values of the heats of solution are tabulated against the reaction observed [76].

$$\begin{array}{lll} +\langle Mg\rangle + 2[HCl]_{aq} = [MgCl_2]_{aq} + (H_2); & \Delta H_\alpha = -111{,}322 \pm 41 & (\alpha) \\ -\langle MgCl_2\rangle = [MgCl_2]_{aq}; & \Delta H_\beta = -\ \ 36{,}566 \pm 18 & (\beta) \\ +(H_2) + (Cl_2) = 2[HCl]_{aq}; & \Delta H_\gamma = -\ \ 78{,}460 \pm 100 & (\gamma) \\ \hline \langle Mg\rangle + (Cl_2) = \langle MgCl_2\rangle; & \Delta H_x = -153{,}216 \pm 159 & \end{array}$$

The validity of this experimental method depends upon solution taking place uniformly and upon the final state of the various solutions being the same. In the latter example, for instance, the magnesium chloride must be dissolved (reaction β) in hydrochloric acid solution of the concentration of the residual acid in reaction (α). Similarly the hydrogen chloride formed by gaseous reaction (γ) must be dissolved in an acid solution initially of this same concentration.

No subsidiary reactions in addition to those considered must take place, and the solvent must be carefully chosen with this object in view. If hydrogen is evolved during

Table VII. Aqueous solvents for calorimetric investigations

No.	Composition	Examples of application	Remarks	References
	Hydrochloric acid			
1.	$HCl:H_2O$ = 1: 500 to 1:1000 (~N/20)	Mg, rare earth metals, their chlorides and iodides	Concentration of HCl to be increased progressively during solution	[77]
2.	$HCl:H_2O$ = 1:53·5 (1N)	Mg, MgO, $MgCl_2$, $Mg(NO_3)_2$, $Ca(NO_3)_2$		[76]
3.	$HCl:H_2O$ = 1:18 to 1:25 (~2·5N)	ZnO, $ZnCO_3$, $CaSiO_3$, $CaCO_3$	50°C	[81]
		Ca	partially paraffin coated	
		Ca–Zn, Ce–Mg, Ce_3Al		[80]
		Mg_3N_2, LaN, CeN, CrN		[82]
		MgS, CaS	solns. sat. with H_2S	[83]
4.	$HCl:H_2O$ = 1:8·8 (~5N)	Mg, Ca, Na, Cd, Al, MgCd, CoAl,	accelerated with $PtCl_4$ or $PtCl_3$	[80]
		$CaAl_3$, $CeAl_4$, Na–Hg $FeAl_3$, La–Mg		[84]
		Fe oxides	100°C; Fe mixed with Pt powder	[85]
5.	$HCl:H_2O$ = 1:3·1 to 1:4 (~10N)	Fe oxides, MnO, $MnCO_3$		[86]
6.	20%HCl, mixed with 20% HF	CaO, $CaSiO_3$, SiO_2, FeO, Fe_2SiO_4, slags	77°C	[87]
7.	40% HF solution	Al silicates		[49]
8.	30% $HCl:ICl_3:I_2$ = 10:10:1·5 by weight	AuSn, $AuSb_2$	90°C	[88]
9.	$Br_2:KBr:H_2O$ = 2:1:2 by weight	Cu–Cd, Cu_3Sn, Au–Zn, Au–Sn, $FeBr_2$, FeI_2, Ga, Ga halides		[79, 88]
10.	$Br_2:HBr:H_2O$ = 2:1:2 by weight	AuSn, $AuSb_2$	90°C	[88]
11.	10 g $FeCl_3$ + 20 cm^3 HCl (conc.) +30 cm^3 solution No. 12	Cu_3Sb, Cd–Sb		[84]
12.	25 g KI + 12·5 g I_2 per 1 l H_2O	cementite		[89]
13.	35 g I_2+1 Citre 90% C_2H_5OH, with a few cm^3 HCl	cementite		[89]
	Sulphuric acid			
14.	$H_2SO_5:H_2O$ = 1: 50 (2N) to 1:105 (1N)	Te_2O, Mn, MnO		
15.	$H_2SO_4:H_2O$ = 1:4 (12·5N)	BeS	80°C	[83]
16.	1N–H_2SO_4 + KI	Mn_3O_5		[76]
17.	2N–H_2SO_5 + $Fe(NH_4)(SO_4)_2$	Mn_3O_4, MnO_2		[76]
18.	3/4N–Na_2S	Sb_2S_3		[90]
19.	$KOH:H_2O$ = 1:272 (N/5)	$Al_2(SO_4)_3$, $KAl(SO_4)_2$		[91]

solution, for instance, it must either be completely evolved in gaseous form or else completely oxidised by suitable reagents: otherwise the results will be inconclusive. The other requirement is that solution should take place neither too slowly nor too rapidly.

These experimental requirements can be met by careful attention to certain details. Table VII gives a number of solvents which have been found suitable for the compounds listed and their constituent elements. Hydrochloric acid of various concentrations is used extensively as hydrogen is evolved from it quantitatively. It is applicable to many metals and intermetallic compounds, to chlorides, oxides (MgO), nitrides (CrN) and to sulphides (MgS). For the solution of sulphides the acid must first be saturated with hydrogen sulphide to eliminate a correction for additional thermal effects due to the formation and solution of this gas. Iodides may also be dissolved in hydrochloric acid [77], provided it is at a low concentration (N/20).

At such dilution it may be assumed that the nature of the anion is of no importance, and the heat of formation of the iodides may be simply taken as the difference between their heats of solution and that of the corresponding metal in this solvent.

Since the heats of formation of the halides of hydrogen are frequently involved in the evaluation of results in solution calorimetry, their values are given in Table VIII which is taken from the Selected Values of the National Bureau of Standards (Circular 500).

Metals too electropositive to be dissolved by hydrochloric acid should be investigated by oxidation with suitable agents in a calorimeter. Nitric acid should not be used, as it has been proved to be unsuitable as a calorimetric solvent [78].

Table VIII. Heats of formation in kcal/mol of the hydrogen halides in aqueous solution at 25°C

N	$[HF]_{aq}$	$[HCl]_{aq}$	$[HBr]_{aq}$	$[HI]_{aq}$
10	−76·235	−38·556	−27·953	−12·335
50	−76·316	−39·521	−28·719	−12·956
100	−76·340	−39·657	−28·815	−13·014
200	−76·358	−39·740	−28·874	−13·048
500	−76·423	−39·812	−28·926	−13·082
1000	−76·531	−39·850	−28·958	−13·106
5000	−77·14	−39·905	−29·006	−13·149
∞	−79·50	−39·952	−29·05	−13·19
Gas	−64·8	−22·062	−8·70	+6·33

N = moles H_2O per 1 mole of solute.

Solvents which contain bromine as an oxidising agent, such as Herschkowitsch's solvent [79], (Table VII), have found useful application. They must be used with caution, however, as they have been found [80] to be sensitive to variations in concentration and even in the quantity of solution used, and the rates of solution vary widely from one metal to another.

As a further complication, interaction may possibly take place when two metals (e.g. of an intermetallic compound) are dissolved simultaneously which does not occur when they are dissolved separately. To counteract this, the metallic constituents should also be dissolved simultaneously.

Sulphuric acid has not been used as frequently as hydrochloric acid or bromine- and bromide-containing solvents. It has proved useful for the solution of beryllium sulphide (at 80°C), and for manganese oxides. In the latter case it was mixed with potassium iodide and ferric ammonium sulphate [92], and the mechanism for obtaining the heat of formation of the oxide was quite complex, as may be seen from the following representation. Manganese and mangano-manganic oxide were dissolved in N–H_2SO_4 containing potassium iodide:

$$3\langle Mn\rangle + 6[H^+] = 3[Mn^{++}] + 3(H_2); \qquad \Delta H = -162{,}774 \pm 142$$

$$\langle I_2\rangle + [I^-] = [I_3^-]; \qquad \Delta H = +1136 \pm 16$$

$$\langle Mn_3O_4\rangle + 8[H^+] + 3[I^-] = 3[Mn^{++}] + [I_3^-] + 4\{H_2O\}; \quad \Delta H = -78{,}243 \pm 62$$

$$4(H_2) + 2(O_2) = 4\{H_2O\}; \qquad \Delta H = -273{,}272 \pm 40$$

$$(H_2) + \langle I_2\rangle = 2[H^+] + 2[I^-]; \qquad \Delta H = -25{,}021 \pm 200$$

$$3\langle Mn\rangle + 2(O_2) = \langle Mn_3O_4\rangle; \qquad \Delta H = -331{,}650 \pm 460$$

That solution takes place by this mechanism must, of course, be confirmed analytically. The heat effect of each step must be determined separately, and the errors are naturally additive.

It is impossible to predict with certainty the suitability of any solvent for a given compound. The list in Table VII may be useful as a guide, but preliminary trials must be made to test the solvent chosen. The trial may suggest slight modifications, or it may show that the solvent is quite unsuitable.

In dissolving a compound and its component elements separately in a solvent, it is often found that one of the substances dissolves too slowly or rapidly relatively to the others. This must be adjusted if at all possible.

The rate of solution may be increased by using the powdered material in a finer grain size; by carrying out the solution at higher temperatures; or by using more vigorous stirring. The solution of silicates, for instance, takes place very slowly even in hydrofluoric acid. Neumann [49] therefore ground aluminium silicates for 100 hours in an agate mill, separated the coarse grains by sedimentation and repeated the process until he obtained a sample of grain diameter less than 0·1 μ. It must be remembered, however, that at extremely low particle sizes the surface energy can no longer be neglected. Fricke and co-workers [93] showed this effect to become perceptible at grain diameters of about 10^{-4} cm, and Giauque [94] showed that MgO prepared from $Mg(OH)_2$ at 1000°C and 300°C respectively differed in heat content at room temperature by as much as 888 cal/mole, the 300° powder being fine, almost colloidal.

To accelerate the solution of metals and alloys in hydrochloric acid, chloroplatinous acid (H_2PtCl_4) was adopted as the accelerator [95]. The heat of deposition of platinum by the reaction

$$H_2 + PtCl_4^{2-} = Pt + 2H^+ + 4Cl^-$$

has been found in this application to be $-0{\cdot}21 \pm 0{\cdot}01$ cal/mg of platinum.

The rate of solution is reduced by stirring more slowly or by covering the test pieces with an inert substance that is more or less impermeable to the solvent.

To decrease the violence of solution of magnesium in hydrochloric acid, Biltz and Hohorst [96] painted their metal sample with collodion. Calcium was coated with paraffin wax, leaving a small part of the surface unprotected. Sodium was dipped in liquid paraffin which forms a covering which is displaced sufficiently when the metal is immersed in acid to allow the reaction to proceed slowly.

Solutions other than aqueous have been used on occasion. Kraus and Ridderhof [97] measured the heats of solution of various compounds (e.g. Na_2Te) in liquid ammonia at −33·4°C, i.e. at its boiling point. The ammonia volatilised by exothermic reactions was absorbed in water, weighed, and the heat of reaction calculated from its heat of evaporation.

Mercury is a particularly useful solvent for alloys since the heats of solution of metals in mercury are much less than in aqueous solutions. Consequently the various heats of solution are smaller than usual in relation to their calculated difference and the percentage experimental error is similarly lowered.

The use of mercury as a solvent for alloys was introduced by Tayler [98] who applied it to the alloys of lead with tin, bismuth, and zinc, and of zinc with tin and mercury. The most favourable conditions are obtained when mercury is one of the alloying elements, and the method of mercury solution has frequently been applied to the amalgams [99]. Heats of solution of a number of Group B metals in mercury have been determined calorimetrically at 150°C over fairly wide ranges of composition by Kleppa [100]. Heats of dilution of liquid amalgams have been measured systematically by Richards and co-workers [101].

The apparatus most commonly used for the determination of heats of solution is either an ice calorimeter (p. 101), which is the more useful for slow solution, or one of the simple isoperibol type. This latter consists essentially of a Dewar flask containing the solvent, a stirrer, a thermometer and device used to introduce the substance to be dissolved. For some reasons, for instance to prevent atmospheric oxidation, it is sometimes necessary to enclose this test substance in a sealed bulb. The dead space in such a bulb must be as small as possible or else it must be evacuated; otherwise the air evolved on breaking the bulb may cause the loss of heat by evaporation of the solvent, which is increased if a gas is evolved during solution. The correction to be applied may be determined by a blank experiment in which dry air is bubbled through the solution and the decrease in temperature per unit volume of air is determined.

The heat generated in stirring need not be considered in isoperibol calorimetry if the rate is the same in the actual experiment as in calibration. The same applies to the normal evaporation of the calorimetric liquid.

Stirring should be sufficiently rapid to prevent a "dip" or sharp concavity in the cooling curve. This results from a time lag in the dispersal throughout the calorimeter fluid of the heat generated by the reaction. Whether it is observed depends largely upon the position of the thermometer in the calorimeter, and it may perhaps be desirable in some calorimeters to place the thermometer fairly near the reaction vessel in a test experiment to check the efficiency of the stirring. When this "damming up" of heat occurs, there is a sharp drop in temperature from the maximum, before normal Newtonian cooling sets in, and correction with such a curve is rendered uncertain.

A solution calorimeter of good design which has been used for a number of salts (oxides, sulphates, nitrates and halides), has been described by Southard [102] (US Bureau of Standards). A half-gallon Dewar flask (Fig. 36) is immersed in an oil bath maintained at 25° ± 0·01°C. A stirrer is used which can provide vigorous agitation to accelerate the solution of substances such as gypsum. Temperature changes are measured by a specially constructed resistance thermometer. Metal parts are made acid-resistant with a coating of a Bakelite lacquer. The sample is introduced in a thin-walled glass bulb E, sealed to the end of a glass tube which passes through the hollow stirrer shaft to the outside of the calorimeter. At the appropriate time the operator breaks the bulb by jerking it upward against three prongs which project from the stirrer blade.

In his extensive work on heats of solution of intermetallic compounds, Biltz used Weinhold vessels, which are similar to Dewar vessels except that they are not silver coated. This allows the course of solution to be observed. The inner vessel of the calorimeter [96] (Fig. 37) is 90 mm in diameter and 240 mm in height. The stirrer is of platinum and the sample is introduced in a basket of platinum gauze, closed with a

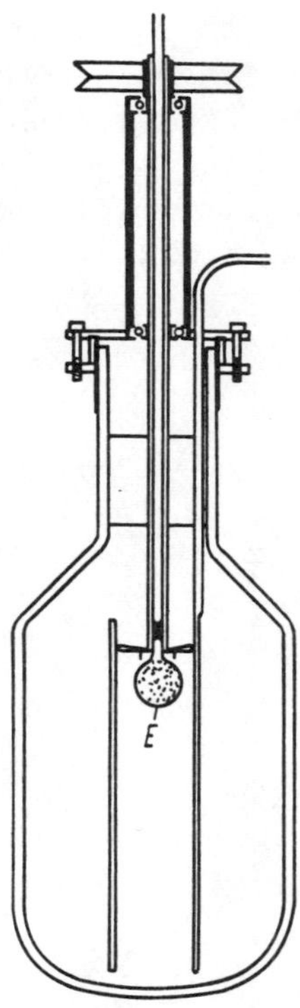

FIG. 36. Calorimeter for the determination of heats of solution.

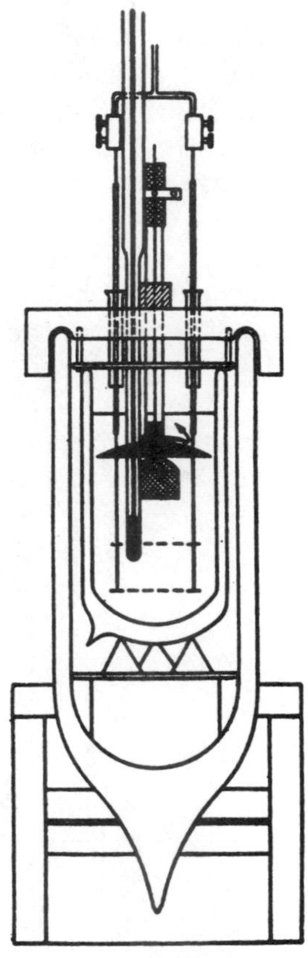

FIG. 37. Calorimeter for heats of solution (Biltz).

cover of platinum sheet. To ensure full heat exchange between the evolved gases and the solution, a double baffle of platinum is mounted in the solution above the basket. The gas bubbles are collected by the lower shallow bell-like portion, pass through a hole in its centre, and are distributed radially by the upper portion as indicated by the arrow. This simple device, first suggested by Richards and Burgess, is very effective. Owing to the low specific heat of gases it is unnecessary to use more elaborate heat exchangers than this.

Heats of formation of alloys may be determined directly by the use of a differential solution calorimeter of the type employed by Buck *et al.* [95] for studying magnesium–cadmium alloys. Samples of alloys and equivalent mixtures of the constituent elements were dissolved simultaneously in the same quantities of hydrochloric acid in two similar vessels. The temperature difference between the two vessels was observed with 25 copper–constantan thermocouples connected in series with a galvanometer. The difference in heats of solution was measured by determining the amount of electrical energy supplied to one of the vessels to bring the temperature difference between the two to zero. The two calorimeter vessels were Dewar flasks, immersed in a thermostat maintained at 25° ± 0·003°C. Each flask was supplied with a stirrer, a heater and a reaction chamber of tantalum. The equipment was sufficiently sensitive to detect a change of 0·005 cal in the differential heat of solution, which varied from 0·9 to 4·2 cal with the quantities of sample used. The total heat evolved in the calorimeter varied from 100 to 200 cal. Possible errors due to the use of an accelerator (see p. 83) and to the loss of heat with the escaping hydrogen were accounted for, and it was found that the greatest errors were systematic, resulting chiefly from the weighing of samples. In the worst cases, an error of 0·1 mg in the mass of magnesium introduced an error of 450 cal in the heat of formation of the alloy.

For the measurement of the heats of solution in mercury, Tayler [98] used an apparatus in the form of two concentric test tubes joined at the top. This vessel, which

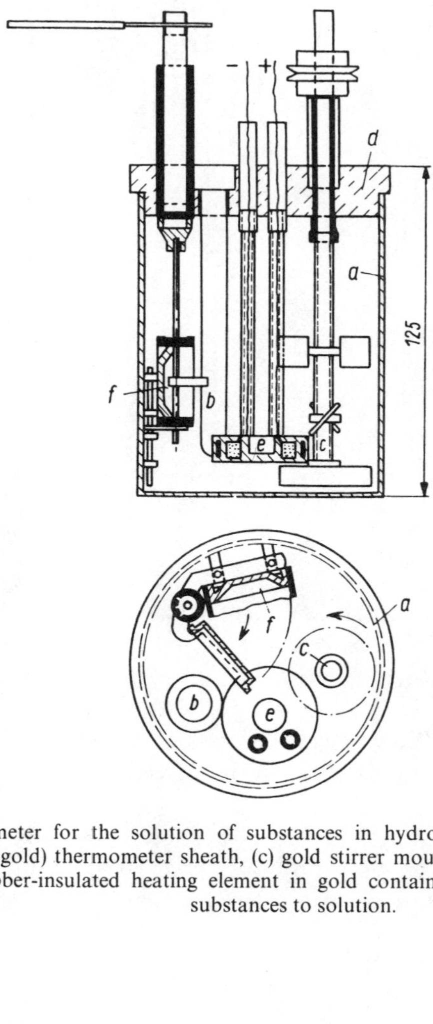

FIG. 38. Calorimeter for the solution of substances in hydrofluoric acid: (a) Silver canister, (b) platinum (or gold) thermometer sheath, (c) gold stirrer mounted on glass rod, (d) amber lid, (e) silk- and rubber-insulated heating element in gold container, (f) device in gold for adding substances to solution.

had a capacity of 300–500 g mercury was silver-plated to reduce radiation and was mounted in the neck of a wider glass bottle. Richards and Forbes [101] built a calorimeter for measuring the heats of dilution of amalgams.

A calorimeter for dissolving silicates and slags was constructed by Roth and Troitzsch [87], who used as a solvent a mixture of hydrochloric and hydrofluoric acid at 77°C. It was found that the purest silver resisted the attack of the acids, and so to avoid the expense of using platinum the apparatus was constructed of pure silver, gold, rubber and amber. Since a calorimeter such as this provides a notable means of investigating these metallurgically important substances, its description is given with some detail in Fig. 38. The Beckmann thermometer is surrounded by a thimble containing mercury to give a quick response to temperature change. Whenever economic considerations permit, platinum should be used for the solution vessel, as it does not require such extreme care in selection and handling. All glass equipment which is likely to be exposed to fluoride fumes should be protected with paraffin wax.

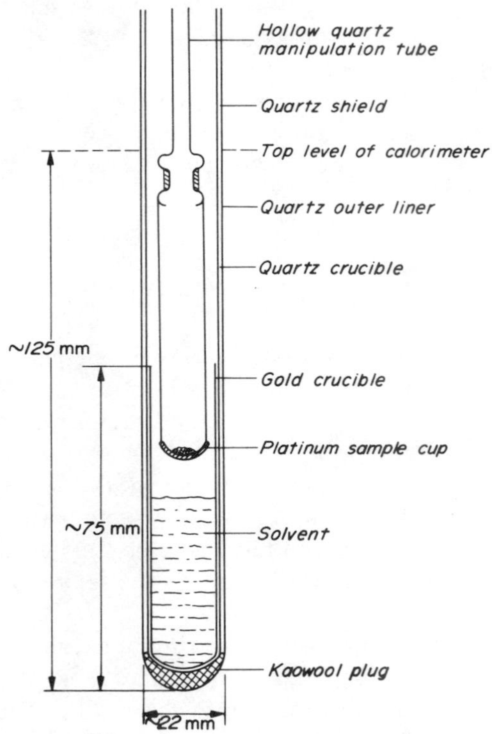

FIG. 39. Schematic diagram of experimental arrangement inside calorimeter proper.

Hydrofluoric acid solution calorimetry of great precision was further developed by Torgeson and Sahama [103]. Although the results appeared to be reliable, particularly those obtained on silicates, inadequacies become apparent, and the search for more suitable solvents began. Kleppa [104], with his associates Yokokawa, Holm and Navrotsky, was the first to replace the aqueous acid of conventional solution calorimetry by an oxide melt. This, of course, required the use of high-temperature calorimeters operating at temperatures not less than 700°C. In view of the slowness of the solution processes of oxides in oxide melts at these temperatures, a sensitive twin microcalorimeter of the Calvet type (p. 110) was employed.

Figure 39 shows the experimental arrangement inside the calorimeter. The sample, in the form of a 150–250 mesh powder, is held in a shallow platinum cup suspended from the end of a silica manipulation tube and positioned a few millimetres above the solvent contained in a cylindrical gold crucible of 17 mm o.d. and 7·5 cm height. To initiate reaction, the cup is dipped into the melt and then moved up and down. After each stirring operation, which involves a standard length of time and number of up-and-down movements, the cup is returned to its original position. The heat effect generated by each stirring operation is determined after completion of the experiment proper and applied as a correction to the observed heat of solution.

The difficulties arising in this manipulation can be largely overcome by selecting the best solvent. Kleppa and his associates have experimented with various liquid solvents, such as vanadium pentoxide, lead borates and sodium molybdates. In the $PbO–B_2O_3$ system, the molar ratio 2:1, and in the $Na_2O–MoO_3$ system, the ratio 3:4 have been selected. Among the substances investigated have been the aluminium silicates [105], many spinels [106] and several tungstates [107].

Because of the very low rates of solution of compounds containing chromic oxide at 700°C, Müller and Kleppa [108] raised the temperature of the calorimeter to 900°C for the investigation of the chromite spinels. Thus the gold of the container had to be replaced by platinum.

4. *Heat Contents*

Specific heats are often determined by measuring the change in heat content between room temperature and a number of higher temperatures. Isoperibol or isothermal calorimeters are used for this purpose. The method entails heating the substance to the desired temperature and dropping it into the calorimeter in which the heat effect is measured—hence the name "drop method" applied to this technique.

In principle the isoperibol calorimeters are the same as those used in bomb and solution calorimetry. In practice certain modifications are desirable, and the literature contains descriptions of a number of calorimeters specially designed for the measurement of heat contents.

It may be necessary to heat the substance in a suitable container, either to prevent it from becoming oxidised or decomposed, or because it is in powdered or liquid form. The hot sample is dropped either directly into the calorimetric fluid (water, aniline, paraffin oil, etc.) or into a metal vessel surrounded by water. In the first method there is a risk of error due to the evaporation or splashing of the fluid. The use of the second method increases the main heat-exchange period of the experiment and so consequently the error in calculating the corrected temperature increase. Occasionally a dual method has been employed: that of dropping the sample into an inner vessel and allowing the liquid to come into direct contact with it after the bulk of the heat has been dissipated. This ensures that the main heat exchange period is not unduly prolonged.

Awbery and Griffiths [109] dealt with the problem by dropping the specimen into a massive inner container which was kept closed while most of the heat of the piece was given up. The calorimeter water was then allowed to enter to accelerate the final distribution of the heat throughout the calorimeter. The calorimeter used in these experiments was exceptionally large, the heat of metal specimens weighing 1·8–4 kg being dissipated in 28 l of water. While the large quantities handled lowered the error, it remained considerable, due mainly to manipulative difficulties.

Roth and Bertram used a eutectic mixture of sodium and potassium nitrates (m.p. 210°C) as the recipient for the specimen. The hot test piece melted its way into the mass of salt while giving up its heat.

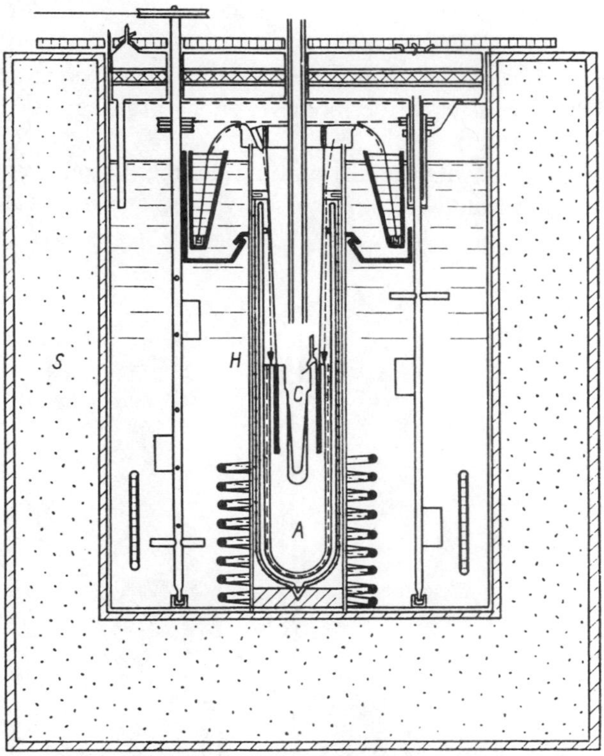

FIG. 40 Aluminium block calorimeter (Jaeger).

For some purposes certain observers favour apparatus in which the calorimeter proper is entirely metallic, i.e. contains no liquid. One of these, which is shown in Fig. 40, was constructed by Jaeger and Rosenbohm [110]. The Dewar flask H is mounted, in a thermostatic bath. Inside it, and separated by a layer of felt, is contained a 6 kg aluminium block, suspended on silk cords. The hot specimen is dropped into the cavity C and the temperature increase is measured by 36 thermocouples in series. This calorimeter was used for a number of metals and intermetallic compounds, and the evaluation of the corrections to the results have been described [111].

The main disadvantage of metal-block calorimeters is that the heat is not dispersed rapidly throughout the calorimetric medium as it is by enforced convection. The lower rate of heat dissipation by conduction may result in a damming-up of heat and in the form of temperature curve described on p. 84. One can, of course, extrapolate the cooling portion of the temperature–time curve in which Newton's law is obeyed [112] to obtain the corrected maximum temperature, but such an extrapolation may be subject to an appreciable error. This trouble, if it arises, can be largely eliminated by the distribution of the thermocouples in series over all parts of the block.

The necessity of moving the furnace is eliminated in an arrangement described by Southard [113] (Fig. 41) and by Olette and Ferrier [114] in which the furnace is permanently fixed over the calorimeter. These two parts of the apparatus are separated by a water-cooled door D except during the instant when the specimen is dropped. Thus there is no heat exchange between furnace and calorimeter even when the former is at 1500°C. The calorimeter C is a gold-plated copper cylinder (12·6 × 20·3 cm) immersed to level H in an oil bath maintained at 25·00°C. The initial temperature of the sample is measured with a Pt/Pt–Rh thermocouple and the temperature changes in the calorimeter with a resistance thermometer. An atmosphere of carbon dioxide is maintained in the central tube.

When there is no reaction with the surrounding gas, the specimen may be heated without protection, but more usually, and of course always in the case of a molten substance,

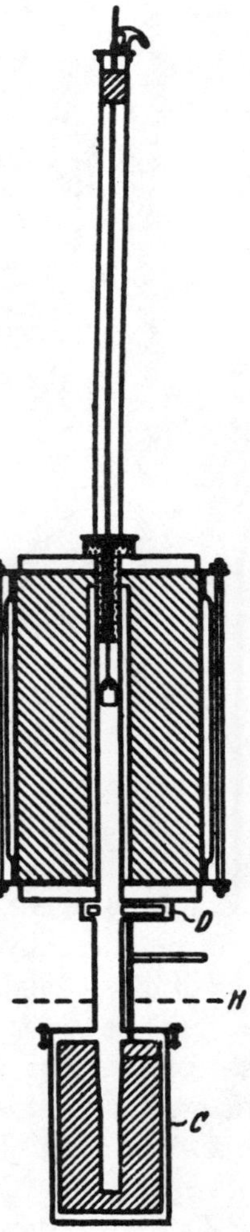

FIG. 41. Southard's calorimeter with superposed heating furnace.

it must be contained in a vessel. Glass and silica are suitable since they enable the specimen to be completely sealed without difficulty. The heat content of each batch of these materials must be determined, however, as it is very sensitive to changes in composition. Values quoted in literature should not be used without verification. An iron vessel may be used at temperatures up to 500°C: its heat content will require checking. Platinum is frequently used and, failing other materials, graphite may sometimes be found useful. For the heat content up to 900°C of highly reducing substances such as magnesium the authors have used an iron crucible enclosed in a quartz tube.

Most of the results obtained by drop calorimetry to date do not exceed 1800°K, that is the maximum temperature of a platinum resistance furnace. At higher temperatures two problems become paramount: the chemical reactivity of the specimens and the attainment and maintenance of high temperatures. To some extent, electromagnetic levitation has been the answer to these problems, its main difficulty being good temperature measurements. Quite a few attempts have been made to employ electromagnetic levitation. Of these, the work of Chekhovskoi *et al.* [115] and Berezin *et al.* [116] on solid and liquid molybdenum up to 3000°K may be presented in some detail.

Their apparatus consists of two main parts: an isoperibol block calorimeter and a furnace chamber in which the sample is levitated and heated. The former need not concern us here. The furnace chamber is made from a seamless drawn copper tube and has a volume of about 13 l. Use is made of various purpose-designed inductors differing both in shape and size. A removable flange made of brass incorporates two copper sleeves in which are mounted a quartz prism and a sight glass; both are shielded from incident radiation and deposition of sublimation products by means of movable metal screens. The prism allows the temperature of the sample to be measured by a disappearing filament pyrometer. The shape of the samples turned on a lathe takes account of the need to stabilize their position in the inductor so that the blackbody model used for temperature measurements faces vertically upwards. In assessing the temperature of the sample, absorption in the prism and imperfection of the blackbody model are taken into account.

Measurement of the enthalpy of liquid molybdenum [116] is complicated because it is not possible to simulate blackbody conditions and there are no reliable values of the spectral emissivity coefficient. Thus, readings are taken of the brightness temperature at the melting point of molybdenum held in a two-phase state for some length of time. The experimental brightness temperature at the melting point, together with the corresponding true temperature of 2890°K, is then used to calculate the spectral emissivity coefficient of liquid molybdenum at its melting point. For solid specimens the temperature distribution could be measured, and its inhomogeneity was found to be insignificant. The molten metal samples were justifiably assumed to be even much closer to being isothermal owing to the intensive electromagnetic stirring of the melt.

Underneath the coil in the furnace chamber, a movable water-cooled stainless steel disc is located. A recess in the disc houses a heavy copper pan designed for catching samples which for some reason drop out of the coil. The disc and the pan also shield the calorimeter from radiation from the sample. The lower flange of the chamber through which the connections pass to the resistance thermometers is shielded by 0·1 mm stainless steel screens.

Prior to the beginning of each experiment, the sample is placed in the inductor and held there with the aid of a manipulator. After the required vacuum or argon pressure has been created in the apparatus, the temperature variation of the calorimeter is recorded for 15–20 min in electrical resistance units. Then the generator is switched on, the sample becomes suspended in the inductor, and the manipulator is removed from under it. The sample temperature is controlled by varying the power fed to the inductor. Thermal equilibrium is achieved in 1–2 min. After the sample temperature has been taken, the gate and the cover are opened with the aid of the manipulator and a linkage inside the chamber, the disc is removed from under the inductor, and the sample drops into the calorimeter. Experimental conditions are always selected so as to minimise the correction for heat transfer between the calorimeter block and the jacket. In order to prevent liquid specimens from spattering as they hit the bottom of the receiver cup, the latter is provided with a helically wound 3 mm copper strip tightly fitting into the cup. The distance between the cup bottom and the lower edge of the insert was 10 mm and the height of the insert about one-third of the depth of the cup.

5. *Other Methods*

Heats of formation of compounds have mostly been determined by the methods already described with the application of Hess' law. Direct methods of measuring heats of formation have, however, been developed both for room temperature and for higher temperatures. Though there is no fundamental difference between the two, consideration of high temperature methods will be left to the next section.

Direct-reaction calorimetry has easily been extended to the determination of heats of formation other than oxides. The reaction of a mixture of iron and sulphur powders is frequently being used in chemical lectures for demonstration purposes and is well known to many students. It obviously lends itself conveniently to the bomb technique and was thus applied, for instance, by Parravano and de Cesaris [89]. Zeumer and Roth [117] used a micro-bomb in an ordinary water calorimeter provided with a Beckmann thermometer. The stoichiometric mixture of iron and sulphur powders was placed in an iron crucible lined with iron powder. The crucible was covered with a thin mica sheet to prevent scattering of the powders and supported by another in order to reduce the rate of heat loss during reaction. The bomb was evacuated before reaction and filled with nitrogen shortly afterwards to assist the dispersion of heat.

This calorimetric experiment might well be remembered at a time when calorimeters of ever-increasing sophistication are being constructed or, often, when thermochemists desist from changing their equipment because of this very reason. There is still plenty of scope for more primitive methods of adequate accuracy. Just to give one example, an experiment by Dench [118] may be reported which demonstrates that there are other metallurgically interesting reactions that may be performed in room temperature calorimeters of simple design.

When the strongly exothermic heats of formation from the component oxides of some of the molybdates and tungstates of the alkaline earth metals were noted, an equimolar mixture of barium and molybdenum oxides was compacted to form a cylinder 14·4 mm in diameter and 11 mm long. A thermocouple was inserted and the compact ignited in air by a small oxy/gas flame. Reaction set in quickly and proceeded without further energy supply forming a transient liquid phase of MoO_3 to produce a rise in temperature to 1073°C recorded on a digital voltmeter. The mean heat capacity of $(BaO + MoO_3)$

between 300° and 1350°K is 37·3 cal/deg mole. When the heat loss of the compact to the surroundings is disregarded, the observed temperature increase thus corresponds to a heat of reaction of about −40 kcal/mole $BaMoO_4$. The second-law value of its heat of formation is −49·8 kcal/mole. Since this type of experiment produces in our experience roughly 80% of the total heat of reaction (see p. 206), the agreement of the resultant estimate of −50 kcal/mole with the experimental value is encouraging, though its closeness must be fortuitous.

When the exothermic heat of reaction is smaller and its rate consequently more sluggish, the device of introducing a little electrical furnace into the calorimeter that heats the powder mixture to near the melting point of one of the components may be helpful, the electrical energy being measured accurately. The calorimeter employed by Zeumer and Roth [117] for the synthesis of silver sulphide is shown in Fig. 42. It was

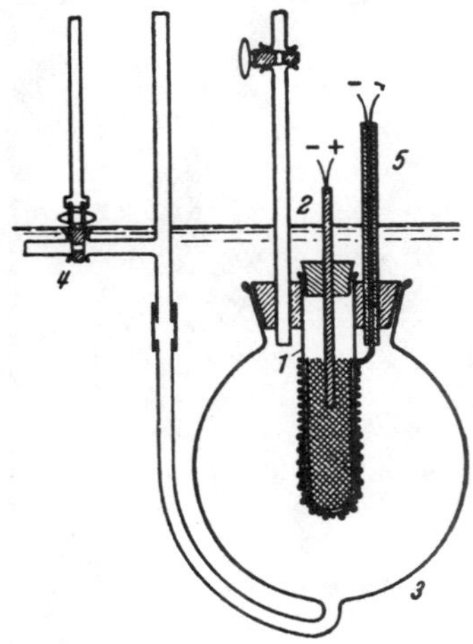

FIG. 42. Small electric furnace for calorimetric work.

operated at 90°C. The completeness of the reaction is influenced by the grain size of the powders, and this was investigated in preliminary trials. The powder mixture in the glass tube (1) was heated by the coil (5) until an accelerated increase in temperature, measured by the thermocouple (2), indicated the onset of reaction. When the temperature began to fall, the stopcock (4) was opened to allow the calorimeter water to enter to provide easier heat dispersion.

When the melting point of the lower melting element is higher than that of sulphur, the temperature of spontaneous reaction is correspondingly higher and the construction must be more elaborate—in particular that of the little furnace inside the calorimeter. Kubaschewski and Dench [119] devised the apparatus shown in Fig. 43 for the investigation of the heats of formation of transition-metal aluminides.

For the concentration of heat, the compact made of metal and aluminium powders (A) was suspended in a small furnace (B) made of molybdenum wire surrounded by ten radiation shields of 0·03 mm nickel foil (E). These were enclosed in an aluminium block (F) supplied with gas outlet holes (H) and suspended (J) within a brass vacuum container (K), which in turn was placed in a thermostat controlled at 25·00°C. Ten copper–Eureka thermocouples in series had their hot junctions (P) clamped to representative positions on the surface of the aluminium block and the cold junctions to the brass block (Q). The brass container could be evacuated through M, but instead of admitting an inert gas for the faster dissipation of heat, a somewhat novel evaluation was used. The furnace (B) was heated electrically until alloying took place rapidly, and the electrical energy supplied was measured by a watt-hour meter (accuracy: $\pm 0{\cdot}2\%$). When the calorimeter was again at 25°C, an amount of electrical energy was put into the furnace so that the calorimeter block was raised to the same maximum temperature as in the reaction run: the difference between the input electrical energy in the reaction-run and calibration-run was then

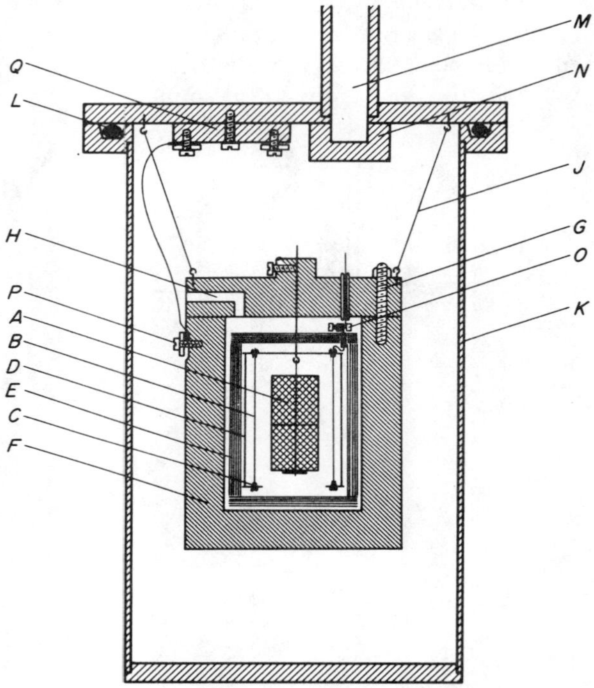

FIG. 43. The "Öfchen"-Calorimeter.

the energy evolved by the reaction. In this way the need to plot the cooling curve for a period of time was avoided, and the physical errors were restricted to the accuracy of the watt-hour meter. The water equivalent of the calorimeter need not be known. When the temperature of the compact rises above the melting point of the alloy formed (e.g. nickel–titanium alloys [120]), the compacts must be encased, in molybdenum crucibles for instance. An attempt to react transition-metal mixtures with silicon failed in this calorimeter. It was found that at least one of the eutectic temperatures must be below 1100°C for reaction to take place.

A method has been developed by Oelsen [121] for the determination of the heats of formation of silicides, aluminides and other alloys of the iron metals. Briefly the method consists in pouring the one molten component on to the other which is contained in a calorimeter at room temperature. If reaction is incomplete under these conditions, then both components are heated above their melting points and poured into the calorimeter simultaneously. The heat content of the molten components must be subtracted from the measured heat effect, but the error involved is much smaller than those in computing the heats of formation from heats of combustion or solution using Hess' law: this is the great advantage of these direct methods. The calorimeter used by Oelsen was large, and consisted of a can containing 6·3 l of water, fitted with a Beckmann thermometer, and insulated from its surroundings by wood fibre. Figure 44 shows the reaction vessel, a thick-walled iron cylinder *f* containing a crucible *d* of sintered sand, the annular space *e* also being filled with sand. The calorimeter vessel containing the cold component is temporarily removed from its jacket for the addition of the other substance which is melted under a protective slag. To ensure complete reaction the vessel is shaken and then replaced in the calorimeter. The whole manipulation requires two or three operators, and the time during which the vessel is out of the water need not exceed 20–40 sec. It was found that the manipulation itself gave rise to a negligible error owing to the large water equivalent of the calorimeter.

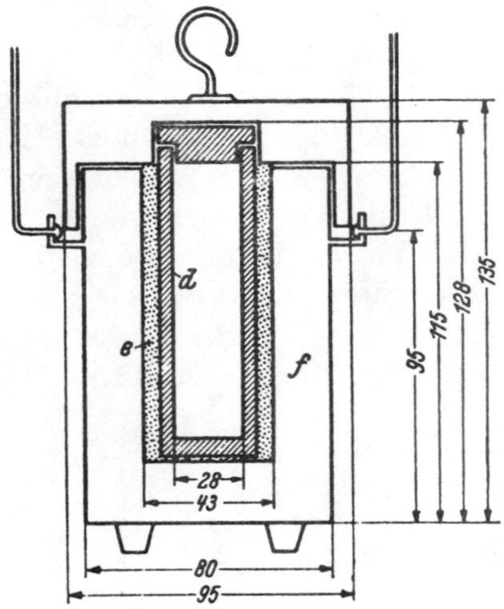

FIG. 44. Reaction vessel, to be enclosed in a calorimeter, for the synthesis of high-melting-point alloys.

A similar method was developed at Stuttgart [122] for the heats of formation of alloys of the alkali metals. The calorimeter is smaller ($W = 1400$ cal/deg) and the whole experiment is carried out under argon. The alkali metal is contained in a graphite crucible, and the other component (lead, bismuth, etc.) is heated in an iron crucible. The operation of removing the graphite crucible from the calorimeter and the iron one from the furnace, pouring the hot metal over the cold one and replacing the graphite crucible in the furnace, need take no more than 15 sec.

Roth estimated a maximum experimental error in his work of ± 100 cal. The work of Oelsen and of Kubaschewski is subject to a 6% error. The main error in this work lies in the risk of incomplete reaction due to the rapid cooling of the system and in oxidation due to incomplete protection by the salt cover or the inert gas. The error due to the actual manipulation is very small.

For the determination of the heat of mixing of liquid sodium-potassium alloys, McKisson and Bromley [123] used an inert-atmosphere "dry box" which allowed handling of large quantities of the molten metals under argon. Since the heat effects associated with the mixing of sodium and potassium are small, total quantities of two to three g-atoms were mixed either in a molybdenum container or a 0·5 pint (280 ml)

Dewar vessel. The method of McKisson and Bromley is obviously quite accurate, but is, of course, restricted to low-melting-point alloys, e.g. those of the alkali metals.

Direct *fluorination* and *chlorination* in bomb calorimeters has already been mentioned on p. 79. Developments seem to point to such operations being carried out at lower pressures. An early attempt was made by v. Wartenberg and Schütte [124], who used an apparatus consisting of a vertical cylinder containing a suspended crucible, made entirely of platinum. At the beginning of the experiment the cylinder was swept out with nitrogen, which was then replaced by fluorine.

Mielenz and v. Wartenberg [125] succeeded in chlorinating beryllium, and Bertram and Roth [126] silver, in a calorimeter. Similar attempts with sodium, zinc and cadmium failed. The difficulties are probably mainly due to the relatively great volume of the chlorine ions which cover the metal surfaces and protect them against further attack. Chlorination should be possible, however, if the chloride formed has a low boiling point, and can be distilled by heating the original metal in Cl_2 gas and condensing the chloride within the calorimeter. This principle has been used by Kubaschewski and v. Goldbeck [127] to determine the heat of formation of vanadium tetrachloride.

The apparatus (Fig. 45) consists of a platinum tube (100 cm long, 0·5 cm in diameter) immersed in a room-temperature calorimeter. The gas inlet portion has a bent section which is covered with a layer of lead glass, obtained simply by buttering the metal with the glass; a 94 cm length of nickel–chromium wire was wound on the section of the tube thus insulated. Metallic vanadium (*ca.* 0·5 g) was placed in this part of the platinum tube which for the actual measurement is heated electrically for 6 min to *ca.* 500°C while

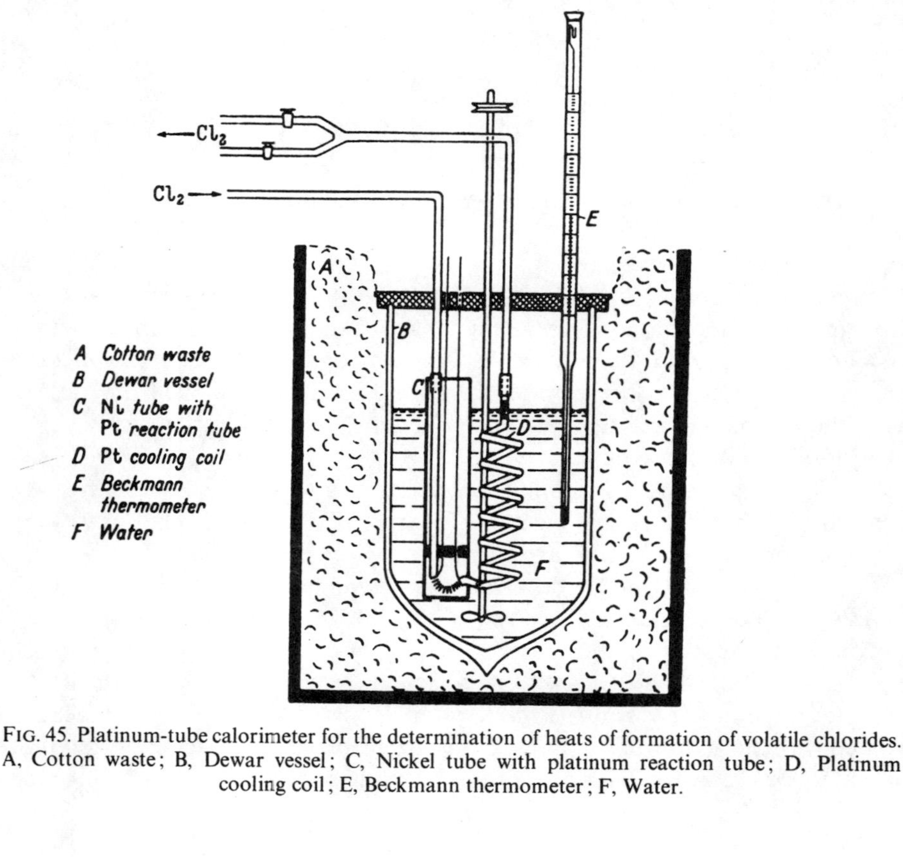

FIG. 45. Platinum-tube calorimeter for the determination of heats of formation of volatile chlorides. A, Cotton waste; B, Dewar vessel; C, Nickel tube with platinum reaction tube; D, Platinum cooling coil; E, Beckmann thermometer; F, Water.

a stream of dry chlorine is passed over the metal. The volatile VCl_4 formed is condensed just beyond the heating section where the platinum tube is coiled into a spiral and is in contact with the water of the calorimeter. Thus formation, evaporation and condensation occur entirely within the calorimeter and the thermal effect observed is due to the formation of the chlorides and to the input of electrical energy which must be measured with accurate instruments. The amount of chloride formed must be determined for each run by chemical analysis and a correction must be applied for the amount of chloride carried away by the gas steam due to its vapour pressure at calorimeter temperature; analysis must involve determinations of both vanadium and chlorine; since the reagents are not entirely pure, the product contains some vanadium oxychloride which must be accounted for in the final evaluation. It was also observed that the blank values with electrical heating but without vanadium present, were higher (by 0·8%) when chlorine instead of argon was passed through the tube, and this was ascribed to a slight reaction between the platinum and the chlorine. It was thus necessary to carry out the actual measurements under exactly the same conditions as the blank runs.

During experiment the heat effect due to VCl_4 formation was only 6–11% of the electrical heat input and future experiments along the same lines should therefore be aimed at increasing the temperature of the heating coil without increasing the energy input, i.e. to improve the thermal insulation.

Gross *et al.* [128] determined the heat of formation of titanium tetrachloride by chlorinating titanium under pressure in a pyrex glass combustion vessel, shown in Fig. 46. Inside the reaction vessel (A) (about 100 ml) was suspended a beryllia crucible

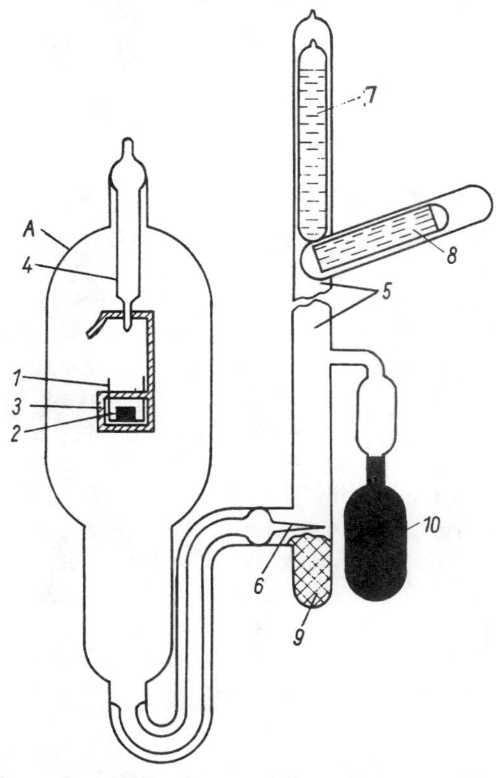

FIG. 46. Pyrex glass chlorination vessel (Gross, Hayman and Levi).

(1) containing titanium pieces. Tube 4 permitted the vessel to be evacuated after introduction of the titanium. A small bulb (10), containing a known amount of liquid chlorine and kept at constant temperature (−75°C) was joined to the slide tube 5. When the vessel had reached equilibrium in a calorimeter the glass barb 6 was broken by allowing a glass tube (7) containing mercury to fall down the slide tube. Reaction began spontaneously after a few minutes and was generally complete within 10–15 min. The titanium tetrachloride, saturated with chlorine, collected on the bottom of the vessel forming a liquid seal at the capillary 5. The chlorine continued to bubble through this liquid thus making the solution homogeneous and bringing it into equilibrium with Cl_2 vapour. The heat of solution of titanium tetrachloride in liquid chlorine was determined separately in the same apparatus.

Another method of determining heats of formation, which may find a wider application in future than it has done hitherto, consists in measuring the heat of reaction between a salt and a metal that has a greater affinity for the anion than has the cationic metal. Gross *et al.* [128], for instance, determined the heat of the reaction: $3PbF_2 + 2Al = 2AlF_3 + 3Pb$, the heat of formation of lead fluoride being known. The reaction was carried out in a small brass bomb fitted with electrical contacts and a needle valve for filling with argon. The reaction mixture, in briquette form, was placed in an alumina crucible. The reaction was ignited electrically; it was complete in a few seconds. In order to ensure completeness of reaction an excess of magnesium was used, and the results corrected for the formation of Mg_2Pb.

The suitability of reactions for determinations of this kind should be established by preliminary experiments. The method should be more suitable for fluorides than for chlorides, as the latter tend to form protective layers owing to the larger size of the chloride ion.

6. *Isoperibol High-temperature Calorimetry*

All the methods described so far are used at initial and final temperatures near room temperature or at slightly elevated temperatures, up to 100°C. For higher temperatures an adiabatic method has been recommended and will be described later (pp. 105, 106).

One of the major difficulties with isoperibol methods is that of holding the calorimeter at uniform temperature. Reaction vessels of low water equivalent should therefore be used. Several such methods have been devised by Kawakami [129] for instance. In each of these calorimeters the temperature measured has been that of a melt, and the reaction vessel was of low water equivalent. Thus the accuracy depended to a large extent on an accurate knowledge of the specific heats of the reactants. In addition, where ceramic reaction vessels were used, the temperature of their outer surface was uncertain, and it is on these temperatures that heat losses to the enclosure depend. It is significant that the only high-temperature heat-of-mixing calorimeters to yield accuracies as good as ±1% had the reaction vessel enclosed in a metal block, and the temperature measured was that of the block. The fact that the temperature rise measured was smaller in these cases was offset by the possibility of using multiple thermocouples, and by the fact that the rate of heat loss to the surroundings was lower.

Of the two more accurate heat-of-mixing calorimeters the one designed by Wittig and Huber [130] is shown in Fig. 47. It gave the same accuracy as the one of Kleppa [131]

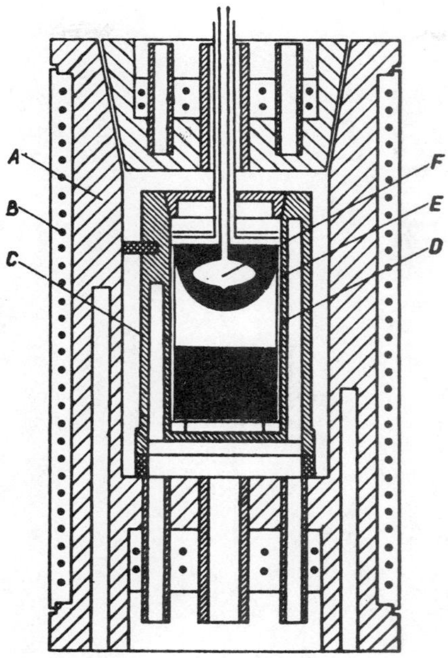

FIG. 47. Calorimeter by Wittig and Huber for mixing molten metals.

but was smaller and handier, and allowed measurements to be carried out in a shorter time. The enclosure A was heated by a non-inductive winding of 0·4 mm tantalum wire B, with additional heaters in the top cover and bottom. Standing on ceramic struts within the enclosure was the calorimeter block C. These parts were made of chromium-plated steel. Four Ni/Ni–Cr differential thermocouples were used to measure the temperature difference between the block and the enclosure. Within the block was a glass reaction vessel D containing one metal, the other being held above it in a thin-walled glass vessel which was broken by the stirrer F to initiate the reaction. The enclosure was supported in an atmosphere of argon in a ceramic tube which was also provided with a heater winding and surrounded by insulation. Care was taken to screen off the thermocouple leads and the compensation apparatus, and to earth the calorimeter block and enclosure, in order to avoid errors due to eddy currents and stray e.m.f.s. The water equivalent was 120 cal/deg. The apparatus was used in the range 450–550°C to measure heat effects up to 200 cal.

Endothermic and small exothermic heats of reactions, in particular of solids, are more difficult to measure directly, since they are not self-heating, and the mixed components must therefore be held at a sufficiently high temperature in order to ensure complete reaction within a reasonable time. Difference methods may therefore be sometimes preferable provided that the heats of solution of reactants and products are comparatively low compared with the heat of reaction to be determined by subtraction.

Such a calorimeter suitable for measurements of heats of solution in low-melting-point metals up to 320°C was devised by Ticknor and Bever [132] and is shown in Fig. 48. It consists of a Dewar flask completely immersed in a constant-temperature bath, a mixture of sodium, lithium and potassium nitrates. A multiple iron–constantan thermocouple consisting of two pairs of junctions in series is used for measuring the metal-bath temperature, and a similar thermocouple for following the temperature of the salt bath.

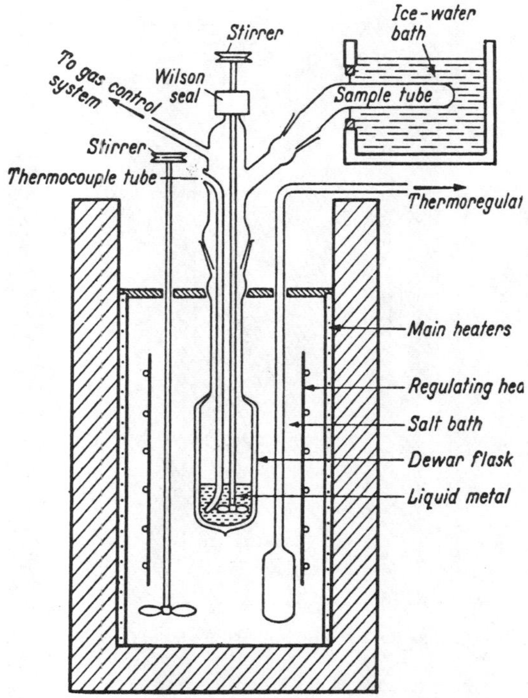

FIG. 48. Liquid-tin calorimeter.

In the experiments described, tin was the liquid metal in the flask. This was inserted into the salt bath and connected to the gas-control system. The air was pumped out, the system swept with hydrogen, and again evacuated. After the tin was melted (a process taking 4 hr) the metal-bath stirrer was started. A sample of solute metal (e.g. silver) was placed in the sample tube which was cooled in an ice-water bath. About 4 hr after melting the temperature of the tin attained a steady state, i.e. the temperature variations were less than $\pm 0{\cdot}01°C$. The sample tube was then rotated into the vertical position and the solute metal dropped into the tin. About 5–10 min were required for the cooling or heating of the metal bath to become nearly exponential and about 3 hr elapsed before the bath returned virtually to its original temperature. The major sources of error lay in the determination of the temperature change in the calorimeter and the evaluation of the heat exchanged between the calorimeter and the salt bath. The uncertainty in the final values of the heats of solution was estimated to be $\pm 5\%$. The authors stated that a slight oxidation of tin could not be prevented. In view of the relatively high heat of formation of the oxide, this could be another major source of error, but one may assume that most of this oxidation took place before the actual mixing stage of the experiment.

McKisson and Bromley [133] have described an isothermal calorimeter for the determination of heats of formation of liquid alloys (e.g. sodium–tin) up to 1500°K. The calorimeter vessel is of graphite and is immersed in a bath of molten tin contained in another graphite vessel. The assembly is heated at constant temperature by a machined graphite coil, and the whole is packed with graphite inside a large copper case. A unit is provided for adding samples, initially at room temperature, to the melt. Much of the heat of mixing of the metals is taken up in heating the solute metal to the temperature of the calorimeter. This increases the overall efficiency of each determination if the specific heat of the solute is known especially precisely. The method of correction for heat loss during experiment is described at length in McKisson and Bromley's paper.

D. ISOTHERMAL CALORIMETRY

1. *The Ice Calorimeter*

The ice calorimeter or Bunsen-type calorimeter is the best-known representative of the truly isothermal calorimeters. The reaction container is surrounded by ice and water, the proportions of which are normally kept constant by an outer vessel also maintained at 0°C. Any heat transfer due to reaction in the inner vessel causes melting of ice or freezing. Thus heats of reaction are translated into latent heat of fusion of ice at 0°C, and the quantity of heat is measured by the change in volume of the water–ice mixture.

The types of ice calorimeter which have been used do not differ greatly one from another. We may consider as a typical example the one constructed by Fischer and Biltz [134], profiting by the experience of Oberhoffer.

The apparatus (Fig. 49) is made throughout of glass. The main bulb, enclosing the reaction tube, is filled with distilled water, while its lower portion and the capillary tube contain mercury. The colorimeter is placed in a beaker containing ice, and the whole apparatus should for preference be stored in a refrigerator when not in operation.

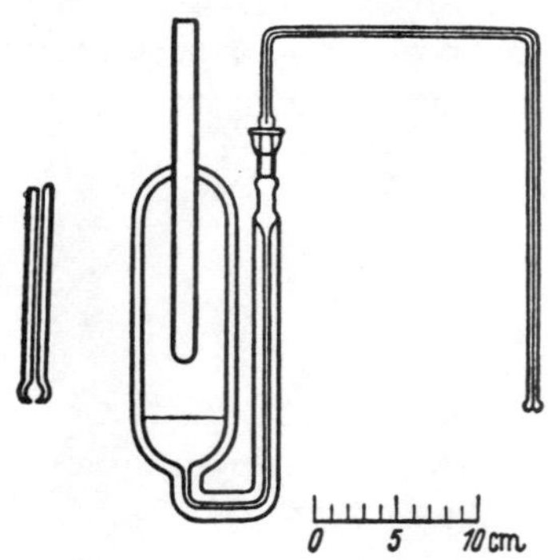

FIG. 49. The ice calorimeter.

The coating of ice required around the reaction tube is produced by successive additions of small amounts of solid carbon dioxide to a quantity of alcohol contained in the tube. The ice layer should initially be 10–15 mm thick, and should not be allowed to become thinner than 3 mm. In addition, the layer of liquid between the ice and the reaction tube formed during an experiment should not exceed 3 mm in thickness. The outer end of the capillary is preferably of the form shown in the lesser figure, the orifice diameter being 0·5–0·8 mm. This open end is immersed in a small cup containing mercury. The heat effect is determined by the change in weight of mercury in the cup: 1 cal = 15·49 mg of mercury. The water equivalent of the apparatus plays no part in the determination as there is no change in temperature. Fischer and Biltz determined heats of solution with an estimated accuracy of ±100 to ±300 cal/mole [134].

The most accurate work done with this apparatus was carried out by Precht [135], who measured the heats of radioactive transformations. He recommended limiting the diameter of the reaction tube to less than 13 mm in an apparatus such as the one described. Condensation of atmospheric moisture in the upper part of the tube is avoided by keeping it closed, except for stirring.

The great advantage of the ice calorimeter is that it permits the study of reactions which are slow or in which the heat exchange is small. It has been used, for instance, to measure the heat of breakdown of the β phase in the aluminium–zinc system [136]. The transformation is arrested by quenching, after which it takes place at 0°C within about 100 min. The maximum heat evolved was no more than 9·5 cal/g. In Schröter's experiment [136] the reaction tube contained alcohol to improve the conduction of heat. The amount of mercury sucked in was measured by an electrical drop-counting device.

Great care has to be taken in forming the ice mantle in the calorimeter that no cavities are formed. Determinations of the density of ice have in the past given scattered results, and since measurements with the calorimeter depend upon the difference in densities of ice and water, a 0·01% variation in the former will give rise to an error of 0·1% in the calibration factor. Careful attention was given to the question of density of ice by Ginnings and Corruccini [137], who described an ice calorimeter with which they determined thermal properties of alumina between 0°C and 900°C. The water was distilled in an apparatus designed to remove practically all the dissolved gases, and they carried out electrical conductivity tests to determine whether the water became contaminated in use. An electrical heater was soldered to the inner vessel and used for calibrating the calorimeter, and other provisions for accurate operation included a steady flow of CO_2 gas at 0°C to prevent atmospheric moisture from freezing in the throat of the apparatus.

2. *The Diphenyl Ether Calorimeter*

Proposals have been made to replace the water in the ice calorimeter by an organic liquid. The suggested substitutes include anethole, *o*-cresol, diphenyl methane and diphenyl ether [138]. Of these substances, diphenyl ether has become rather popular. It has several advantages over ice as a working substance [139]. It is more sensitive, displacing more than three times the amount of mercury per calorie. Its melting point is 26·88°C, close to room temperature, decreasing possible heat leaks and rendering unnecessary the flow of dry carbon dioxide used with the ice calorimeter. Standard enthalpy data are referred to 25°C which is very near the working temperature.

Diphenyl ether appears perfectly stable under working conditions, even after years of operation. Its thermodynamic properties have been thoroughly determined by the National Bureau of Standards [140].

A specific apparatus has been described by Hultgren *et al.* [139] who are using it for the measurement of heat contents of metals and alloys. The design is very similar to the ice calorimeter (Fig. 49). The thermostatically controlled water bath is maintained slightly above the melting point of diphenyl ether, purified by a series of fractional crystallisations, with a fluctuation of not greater than $\pm 0{\cdot}002°C$. The pool of mercury in the bottom communicates with a horizontal calibrated 0·5 mm capillary tube 600 cm long. Readings of the mercury column are taken at one minute intervals before and after a sample is dropped in order to establish the drift rate. Before a run, about 25% of the diphenyl ether is frozen and adheres to the chamber as a solid mantle. During a run, only a small proportion of it is melted. This is refrozen in preparation for the next run by adding a few drops of alcohol to the chamber and evaporating it in a gentle stream of air. Since diphenyl ether is a poor conductor of heat, it is important to have a freshly frozen mantle before each run.

E. ADIABATIC CALORIMETRY

In the adiabatic calorimeter the temperature of the outer jacket is kept the same as that of the calorimeter proper throughout the course of a determination: $T_c = T_s$ (p. 67).

1. Heat Capacities

Heat capacities are often calculated from measurements of heat contents, as described in Section I2C. If a transformation occurs in the temperature range over which the heat content has been measured, especially one with a small heat effect such as magnetic change in nickel, the heat capacity must be measured directly. That is, one must measure the heat required to raise the temperature of the substance by a few degrees at a time.

Adiabatic methods of measuring heat capacities at temperatures above room temperature have been developed by Moser and by Sykes, among others.

With the apparatus shown in Fig. 50 Moser [141] obtained an accuracy of about 0·5% in measurements in the range 50–670°C. A sample piece v is placed in well-fitting silver

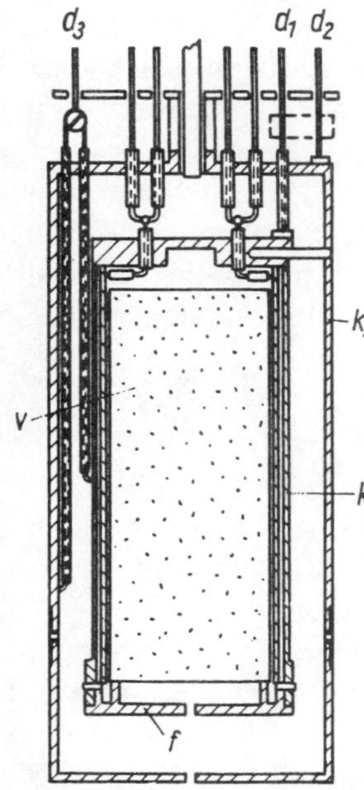

FIG. 50. Moser's calorimeter for the determination of high temperature heat capacities.

cylinder k_1, 23 mm in diameter, which has a removable base. The cylinder is heated by a platinum heater set in a helical groove in a surrounding magnesia sleeve. The rate of heat input is determined from the heating current and voltage, using a calibrated ammeter and voltmeter. The resistance of the heater is 3 ohms at 25°C. The outer cylindrical silver container can be evacuated, and is filled with argon at a pressure of 30 mmHg for each experiment. To eliminate errors that may arise from induced eddy currents, a copper winding is provided outside cylinder k_2, having the same number of turns as the platinum heater and carrying the same current in the opposite direction.

Constantan wires joining the cylinders to a common lead d_3 and silver leads d_1, d_2 constitute thermocouples. These are used to measure the temperature rise of the inner cylinder and to ensure that the outer cylinder is heated to the same temperature so that there is no heat exchange between them. The true specific heat c_p of a sample of mass m is calculated from the relationship

$$c_p = \frac{1}{m}\left[\frac{0{\cdot}239EI - a}{\Delta\theta} - W\right]\frac{1}{1+b}$$

where W is the water equivalent of the empty calorimeter and a and b are corrections for the loss of heat during measurement and for difference between the increase of temperature of the calorimeter and of the sample.

Further particulars of the method are available in the original paper. Moser's apparatus has become the prototype for a number of adiabatic calorimeters for the measurement of heat capacities, heats of transition and heats of reaction. Sykes and Jones [142] applied it to the investigation of various order–disorder transformations and to age-hardening phenomena.

The specimen in the form of a closed hollow cylinder is placed inside and thermally insulated from a closed copper cylinder. This is kept in an inert atmosphere in a silica tube which can be heated by a resistance furnace. The specimen is heated also by a small nichrome coil mounted inside it. Potentiometer leads are provided to measure the heat input by this auxiliary heater. The temperatures of specimen and block being T_S and T_B respectively, the measurements taken are those of T_B and $(T_S - T_B)$. In carrying out a determination, T_B is increased at a steady rate of the order of 1°C per min. Meanwhile the current through the auxiliary heater is adjusted at intervals so that $(T_S - T_B)$ varies between plus and minus 0·1°C. T_S thus varies with time in the manner shown by curve C_{T_S} (Fig. 51). Where this curve intersects the line made by plotting T_B against time, heat interchange between block and specimen is zero. Consequently, at these points

$$Q = mp\left(\frac{dT_S}{dt}\right)_{T_S = T_B}$$

T_S is given by the measurements $[(T_S - T_B) + T_B]$.

Q is the power input of the auxiliary heater, m the mass of the specimen and c_p its specific heat. The heat capacity of the coil and thermocouples are usually negligible compared with that of the specimen.

Backhurst has extended the range of the adiabatic method to 1600°C for solid and liquid metals [143]. The cylindrical specimen and its crucible are separated by three concentric radiation screens from an outer enclosure made of molybdenum. The outer enclosure is heated by tungsten spirals evenly distributed over the inner surface of a box formed by six ceramic blocks. The apparatus stands in an evacuated container. Heat is supplied to the specimen at a constant rate by means of an internal heater. Meanwhile the outer enclosure is kept at the same temperature as the crucible. The temperature of the specimen is recorded on a chart, and the specific heats are calculated using the gradients of this curve and the power supplied to the internal heater.

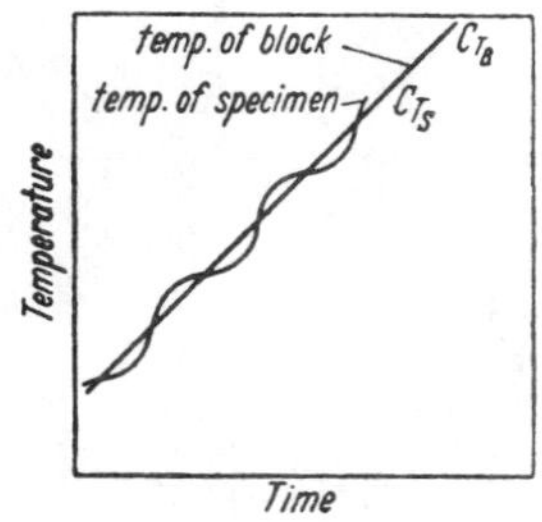

FIG. 51. Temperature–time curves in Sykes and Jones' method.

2. *Heats of Reaction*

Dench [144] has adopted the adiabatic principle and constructed an apparatus somewhat similar to those mentioned which operates in the range 500–1400°C and can be used for the determination of heat capacities and heats of transition as well as of heats of reaction. It has so far been used for the study of metals and alloys.

The specimen A in Fig. 52 is prepared by compacting a mixture of the powdered reactants into a cylindrical form, 19 mm diameter × 32 mm high. This is placed in the adiabatic enclosure B which consists of three concentric cylindrical radiation shields made from tantalum sheet. Around the adiabatic enclosure is the vacuum furnace C, which has a tantalum winding and tantalum and molybdenum radiation shields. Part of the winding is arranged around the sides of the furnace, the remainder being formed into two coils—one near the top, the other near the bottom. The furnace is enclosed in a water-cooled brass vacuum chamber D. Three Pt–Pt/Rh differential thermocouples E in alumina sheaths are arranged round the middle of the adiabatic enclosure, each having one junction against the outside of the enclosure walls. Any temperature difference between the inside and outside of the enclosure is indicated by a deflection in a galvanometer to which the differential thermocouples are connected in series. The current in the surrounding furnace winding is continuously adjusted in order to keep the temperature difference at the small finite value corresponding to zero heat transfer across the wall of the enclosure. Heat may be supplied to the specimen via a tungsten resistance heating coil in an alumina sheath F in a hole down the axis of the specimen. The energy supplied to the heater is measured by means of a precision watt-hour meter. An alumina sheath G in a second hole in the specimen carries a Pt/Pt–Rh thermocouple by means of which the specimen temperature is measured.

In measuring heats of formation, the following procedure is adopted. Preliminary experiments are carried out on small specimens to find the highest temperature at which the mixture of the powders to be reacted may be held without any appreciable proportion of the reaction occurring, the "safe temperature", and also to find the temperature range within which the mixture must be held to ensure complete alloying within an hour or so.

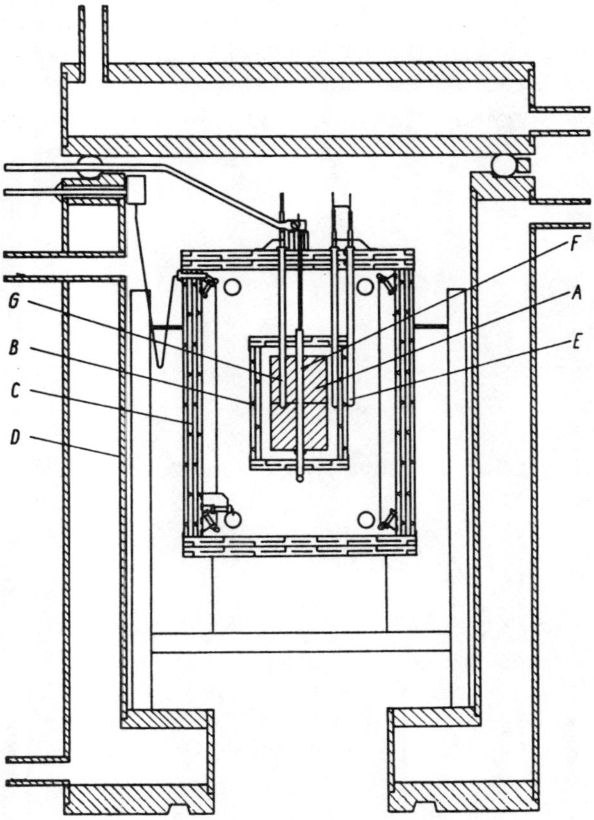

FIG. 52. Adiabatic reaction calorimeter for use in the temperature range 500–1400°C.

The specimen is then heated adiabatically from the safe temperature to the reaction temperature range using the internal heater which is eventually switched off. During the heating, the reaction will have begun, and, after switching off, the temperature of the specimen will continue to rise if the reaction is exothermic, or will begin to fall if it is endothermic. The rate of change of temperature will gradually decrease as the reaction proceeds, a constant temperature being eventually attained when it is complete.

The heat of reaction is obtained by subtracting from the measured input energy the change in heat content of the reactants and empty calorimeter between the safe temperature and the final temperature, these heat contents being measured separately in the same apparatus. An estimated accuracy of ± 100 cal/g-atom is attained.

Mean heat capacities may be obtained with an accuracy of $\pm 2\%$ by heating the specimen adiabatically over temperature intervals of approximately 25°C [145]. Heats of transformation may be measured by heating the specimen in a series of steps arranged so that one of the steps contains the complete transformation, the heat capacities of the phases just above and just below the transformation temperature being obtained by extrapolating the mean heat capacity curve.

The early applications of the adiabatic method to reaction calorimetry were to measurements at room temperature. The method was first used by Person (1849) and was brought into prominence by Richards [146]. In the first adiabatic calorimeter, the outer vessel contained dilute alkali the temperature of which was raised as required by the addition of acid. Subsequently electrical heating has been used almost without exception owing to its simplicity. Uniform heating may be obtained by passing the current through the calorimetric liquid itself, using the outer and inner vessels as electrodes. The temperature control can be made automatic with suitable relays. When doing this, attention should be paid to such details as compensating for the heat input due to the stirrer by maintaining the temperature of the outer jacket slightly lower than that of the reaction vessel [147]. It is generally more convenient to make such adjustments than to attempt to make calculated corrections.

An adiabatic calorimeter operating in a narrow temperature range at relatively high temperatures has been constructed at Stuttgart [48, 148]. The apparatus is shown in Fig. 53. The thermostat A is an electric furnace; the calorimeter C and the cylindrical vessel B (which carries an additional heating element) are made of massive heat-resisting stainless steel. The thermocouples in series are set in holes in both B and C, and are connected differentially to a galvanometer. Any rise in temperature due to reaction in C must be compensated by heating B so that the galvanometer at II always indicates a

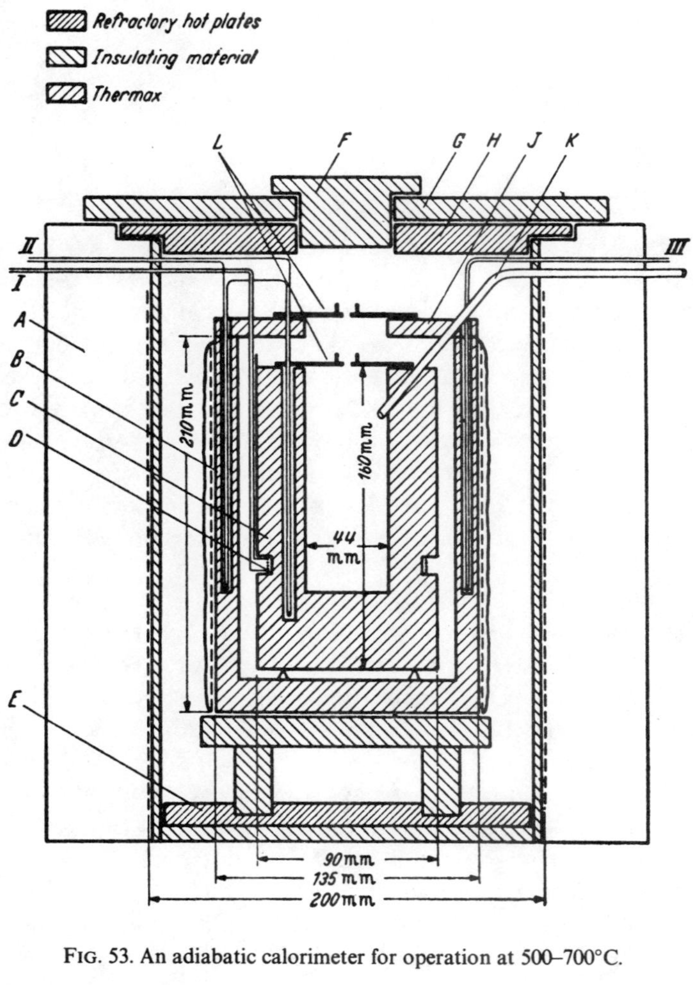

FIG. 53. An adiabatic calorimeter for operation at 500–700°C.

zero e.m.f. The increase in temperature is measured with the platinum resistance thermometer I at D. This is wound non-inductively. The lid and base of the calorimeter contain heater elements, and an overall constant temperature is maintained. The reaction space is kept filled with a slow flow of argon. The water equivalent of this calorimeter was 1060 cal/deg, the operating temperature 600–700°C, and the experimental error was estimated to be $\pm 2{\cdot}5\%$.

The calorimeter was made relatively large in order to have thermal capacity and space for dropping in metal powder compacts which reacted at the experimental temperature. It thus suffered from the same disadvantage as did Backhurst's calorimeter, namely that of maintaining uniform temperature. An additional difficulty with the method was that of having to compensate the sudden changes in temperature, e.g. the initial fall in temperature due to the heating of the cold compact and the subsequent rise in temperature due to the reaction. Since the jacket responded relatively slowly to changes in the heating current, the observer was forced to "anticipate" the behaviour of the reaction mixture. For future applications, the dimensions of the calorimeter should be reduced as far as possible, and it should be used for rather slower than spontaneous reactions.

F. OTHER CALORIMETERS

1. *High-speed Methods*

As the operating temperature increases above the region of 1500°K severe problems are encountered in conventional calorimeters owing to increased rates of evaporation, mechanical creep, diffusion into and reaction with the container and support materials, heat transfer by radiation, etc. The effects can be reduced by shortening the duration of the experiment, and this has led to the development of pulse or high-speed calorimeters for the measurement of heat capacities and enthalpies of transformation (as reviewed by Cezairliyan [149]). They employ heating durations usually shorter than 10 sec; they have been applied so far mainly to electrical conductors and semiconductors, and could be better classified in a non-quantitative manner as calorimeters making use of direct electrical resistive heating for supplying energy to a specimen in the form of a wire, rod or tube.

The energy may be supplied from a battery or a regulated power supply, the power input being determined by measurements of current and potential difference, or it may be supplied as a discrete pulse by discharging a calibrated capacitor-bank through the specimen, the energy being determined by the capacitance and the initial potential. The specimen temperature may be measured by means of a thermocouple attached to the specimen, or it may be derived from the electrical resistance (this requiring calibration in separate steady-state experiments), or it may be measured by optical automatic methods. The potential differences produced by the transducers used in the measurements are recorded either on an oscillograph or by photographing oscilloscope traces or by using high-speed data-acquisition systems.

Several different modes of operation may be distinguished—among them the following:

(i) The specimen is initially in thermal equilibrium with the surroundings which are kept at room temperature. A continuous current is then supplied to the specimen, and measurements of potential difference, current and temperature are continuously recorded vs. time during the heating, giving a heat capacity vs. temperature curve for the whole range covered. For example, Cezairliyan *et al.* [150] measured the heat capacity of molybdenum between 1900° and 2800°K as well as its electrical resistivity and thermal radiation properties.

(ii) The specimen is initially in thermal equilibrium at an elevated temperature in a furnace. An increment in temperature of a few Kelvins is produced by delivering a pulse of electrical energy to the specimen, one heat capacity value being obtained for each pulse. This procedure is then repeated for different starting temperatures. For example, Parker [151] measured the enthalpy of transformation of α- to β-titanium at 1156°K.

(iii) The surroundings are kept at room temperature and the specimen is held initially in thermal equilibrium at an elevated temperature by passing a steady current through it. The input current is then made to undergo a step-change, and measurements are made as in (i) during the subsequent heating. For example, Kollie [152] measured the heat capacity of iron between 300° and 1175°K.

In (ii) the specimen is held for a relatively long time at elevated temperatures and some of the advantages of the pulse method are lost. However, it does allow steady-state measurements of starting temperatures to be made and requires a less elaborate data-recording system than (i). Method (iii) seems to have little advantage over (i).

2. *Cooling Method*

Another method of determining true specific heats is that first introduced by Dulong and Petit (1819) in the course of their well-known investigation of heat capacities. It involves the analysis of cooling curves, which depend on the specific heat of the substance concerned. Though at first sight this may appear a simple matter, in practice both the experimental method and the numerical evaluation require the utmost care to obtain dependable results. A comprehensive discussion of an application of this method is that of Knappwost [153]. For various reasons gases alone should be used as the heat conducting medium, and Knappwost used a 2 mm layer of air or hydrogen between the cylindrical test sample and its enclosing crucible.

Oelsen *et al.* [154] simplified the evaluation by determining the rate-of-heat-loss vs. time curve of the specimen in addition to the cooling curve. The specimen, in a thin-walled container provided with a thermocouple, was heated, then placed on pinpoints inside another container, also thin-walled, immersed in water in a room-temperature calorimeter. The true specific heat at any given temperature was calculated from the gradients of the two curves at the corresponding time. This simple method was used at temperatures between 100° and 1050°C, the errors being surprisingly small.

3. *Tian–Calvet Calorimeters*

Calvet has constructed a number of calorimeters operating on the basis of the thermopile (Calvet and Prat [155]). These apparatus are extremely sensitive and have been used to study, for instance, biochemical reactions and radioactive processes. As long as they were only operated at room temperature, they were of less interest to the metallurgical thermochemist. In recent years, however, quite a number of calorimeters of the Tian–Calvet type have been constructed for higher temperatures. Brief reviews by Laffitte [156], Gerdanian [157] and Kleppa [104] are available.

Two types of apparatus are used at high temperatures—adiabatic and isothermal. The latter is really a heat-flow calorimeter (see p. 67) and is suitable for the investigation of slow chemical processes, whereas the adiabatic type is used for the study of fairly rapid processes.

Figure 54 shows a twin microcalorimeter as constructed and used by Kleppa [104] at temperatures up to 500°C and, when the aluminium jacket was replaced by one made from high-grade nickel, up to 800°C. The temperature change resulting from a reaction in one calorimeter is compared with the temperature change in the other, similar, calorimeter. Kleppa's calorimeters differ from the commonly used Calvet-type apparatus

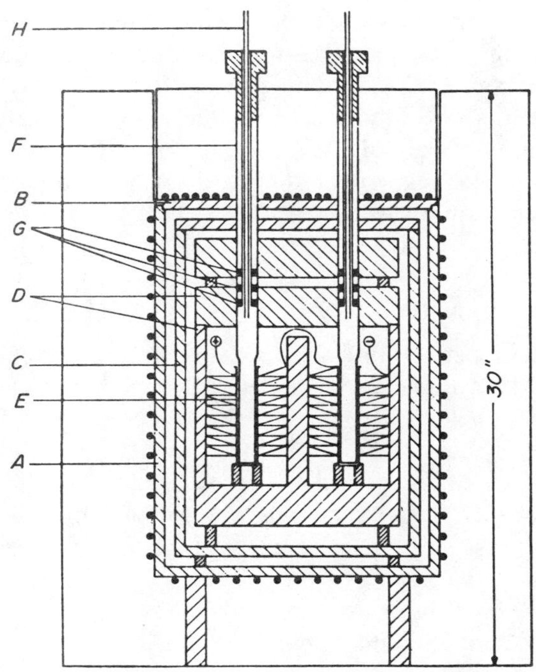

FIG. 54. General layout of Calvet type twin calorimeter. A, main heater; B, top heater; C, heavy shield; D, aluminium jacket; E, calorimeter; F, protection tube; G, radiation shields; H, manipulation tube; ⊕–⊖, thermopile.

in having a working diameter of 20–25 mm, which allows enough space for crucibles and manipulation devices. Whereas their thermopiles consist of 48 thermocouples using a d.c. amplifier, the calorimeters built at Marseille [156] employ a much greater number of couples and permit work at temperatures up to 1200°C, a temperature beyond which the alumina of the thermocouple support becomes weakly conducting.

By eliminating the contact between the couples and the alumina, Dodé, Gerdanian and their associates have been able to work at still higher temperatures. With the latest design [158], an operating temperature of 1500°C has been reached. Because of the high sensitivity of the apparatus, it has been possible to measure thermal effects which are virtually *partial* heats of solution—in their case, of oxygen in oxide phases, by admitting very small amounts of oxygen to the samples in each run. Thus, the Orsay group has determined the partial heats of solution of oxygen in titanium, zirconium and hafnium [158] as well as in non-stoichiometric phases, such as UO_{2+x} and $Fe_{1-x}O$ [157].

Here we will desist from giving detailed instructions on how to build a Calvet-type calorimeter. As Laffitte points out [156]: anyone not already very familiar with the construction of high-temperature calorimeters is well advised to purchase one of the commercially available apparata and not to try to construct one's own.

G. HEATS OF FUSION AND OF TRANSFORMATION

Methods of determining heats of fusion have already been described in connection with specific heats. We shall summarise them again, with added remarks on the special precautions and techniques associated with the measurement of these particular quantities.

1. The heat of fusion of a substance is usually obtained from measurements of change in heat content between room temperature and several temperatures in a range extending above and below the melting point. The evaluation is simple and can be deduced from Fig. 4 and the description on p. 12.

Any errors in determining the effective heat content of the containing crucible are eliminated if they vary continuously with temperature. The accuracy of the determination is improved by increasing the number of measurements of heat content above and below the melting point. Values obtained by a careful application of this method are very reliable.

The method is not particularly suitable for the determination of heats of transformation. It involves rather rapid cooling rates which may prevent or delay complete transformation, or introduce errors due to quenching stresses. Values so obtained must therefore be considered as minimum values. Values of heats of fusion obtained by this method can also be unreliable where slow transformations occur between the melting point and room temperature [159].

2. Heats of fusion may be measured directly (e.g. [160]) by a method similar to the determination of heat capacities. That is, the heat input or evolution is measured at the one temperature, that of melting. A correction must be made, using known values of specific heats, for any small change in temperature. Such corrections will generally be much smaller than those for loss of heat by radiation. At high temperatures the correction for radiation may be of similar order to the heat of fusion and will render the result uncertain.

The method of direct measurement of heat capacity is most suitable for the determination of heats of transformation, especially those which extend over a range of temperature (see pp. 12 and 106).

3. Wittig [161] has devised a method in which a crucible containing the metal is heated and the temperature difference across the wall of the crucible is measured by means of a differential thermocouple. The graph of this temperature difference plotted against time shows a discontinuity at the melting point of the metal. The area A enclosed by this discontinuity is a measure of the latent heat. The proportionality factor is determined

by electrical calibration. A part of the heat of fusion is provided by an immersion heater within the metal; this quantity of heat Q is measured accurately. The area A plotted against Q gives a straight line whose extrapolation intersects the Q axis (i.e. $Q = 0$) at the value of the heat of fusion.

4. The cooling method originally introduced by Dulong and Petit (p. 109) has occasionally been used also to obtain latent heats. Knappwost has, however, shown that the results are rather uncertain. Tammann [162] has suggested that the ratio of the heats of fusion of two metals might be obtained by comparing the times taken by the two metals to crystallise on cooling. Contrary to Tammann's assumption, however, there is no direct proportionality between the arrest periods and latent heats. In its stead, a relationship suggested by Plato

$$L_f = \alpha \frac{\gamma}{gm} \tag{97}$$

[L_f in cal/g; m = mass of metal in grams; g = rate of cooling before, and γ after solidification] has proved useful. The value of the factor α must be found by the determination in the same apparatus of cooling curves of substances of known latent heat. Details of the method were given by Jenckel and Wilsing [163].

The advantage of this method is that a useful indication of latent heats is obtained by simple means, but the results are not very accurate.

5. Heats of fusion may be calculated from measurements of the lowering of the freezing point. This method, the theoretical basis of which is discussed on p. 45, has been extensively applied by Kelley [11]. Conditions which must be observed are that the solution in the range under consideration must be practically ideal, and the constituents should not combine together to form main valency compounds. A linear relationship must hold between $\log(N'_A/N_A)$ or $\log N_A$ and $1/T$. Unless the liquidus and solidus curves are known to an accuracy better than $\pm 0{\cdot}2°$, the accuracy of the values obtained is rather low. It may be improved by investigating several systems with the one common solvent metal and taking the mean value of the heats of fusion determined.

6. Though of very limited application, the lowering of the melting point of a substance by the application of high pressures may be used for the determination of the heat of fusion of a substance which melts without molecular change. The relationship is given by Clapeyron's equation

$$L_f = T(V_{\text{liq}} - V_{\text{sol}}) \frac{dp}{dT} 0{\cdot}02345 \text{ (cal)} \tag{98}$$

For most practical purposes dp/dT may be taken as $(p_1 - p_2)/(T_1 - T_2)$, where T_1 and T_2 are the temperatures of melting at pressures p_1 and p_2 respectively, and $T = (T_1 + T_2)/2$. The factor 0·02345 relates the volume change in cm^3/g to the heat of fusion in cal/g.

Measurements using this principle have for the most part been carried out and evaluated by Bridgmann [164], who used pressures up to 12,000 kg/cm^2 and even higher. The method is the only one which has been applied to heats of fusion at high pressures. Extrapolation of the value of the heat of fusion to atmospheric pressure is somewhat uncertain, and for this pressure Bridgmann related his measurements to calorimetric results.

7. Heats of order–disorder transformations or of precipitation phenomena in alloys have been determined by methods developed at the Royal Institute of Technology, Stockholm. The methods involve thermometric measurements on specimens cooled from annealing temperatures and placed in thermostats developed for this work: either a vapour thermostat in which a liquid is boiled at a controlled pressure, or a large metal block in a regulated electrical resistance furnace (1949: Borelius). In a method used for lead–tin (1949) and gold–copper (1950) alloys [165], it has been possible to measure the heats of reactions taking place within a very few minutes of cooling from the temperature of homogenisation. The sample, in the form of a wire about 0·7 mm in diameter, is mounted along the axis of a hollow copper cylinder. Tappings are soldered near the ends of the wire for voltage measurements (for the determination of energy input and of temperature of heat treatment). The wire is electrically insulated and protected against oxidation with silicone varnish, and a number of thin-gauge thermocouples are connected in series and mounted so as to record the temperature difference between the wire and the cylinder. The assembly is mounted in a suitable thermostat, and a current is passed through the wire to raise its temperature to that at which it becomes homogenised or disordered. When the treatment is judged complete, the current is cut off, the wire cools toward the temperature of its surroundings and the subsequent rise in its temperature due to reaction is recorded and plotted against time. The heat liberated is determined by calibration, known amounts of electrical energy being dissipated per unit length of the wire. It is possible for this type of determination to give quantitative information on the kinetics as well as the thermochemistry of the reactions observed.

H. COMPARISON OF METHODS

The practical determination of heats of formation, heat capacities and other thermal data always begins with the selection of a suitable method. In many instances only one or two of all the methods that have been developed may be applicable.

The most accurate way of determining heats of formation is to measure directly the heat effect of the actual reaction taking place in a calorimeter. This method requires the fewest corrections, and therefore involves a very small error.

The combustion bomb is still the standard apparatus for the heats of formation of oxides. Common as this apparatus has become, the need for skilful handling, careful analysis and considerable practical experience cannot be over-emphasised. It is very disquieting to see values of heats of reaction quoted as having an error of less than $\pm 1\%$ altered by more than this amount by redeterminations after a decade or so.

Direct chlorination in the calorimeter has shown encouraging recent developments (pp. 79, 96 and 97). In particular the method by Gross, Hayman and Levi appears to be of very satisfactory precision.

The direct calorimetric determinations of Oelsen, of Kubaschewski, and of Kawakami do not attain this accuracy. The error reported by these observers for the final values averaged $\pm 6\%$ or more. This was due not to the fundamental method involved but to the difficulty of handling and pouring molten metals, many of them of a high melting point. In addition the reaction products might not have been in complete equilibrium. These aspects of the work are capable of improvement and should be made to yield results of higher accuracy. The most accurate calorimeters to date for the determination of the heat effects of mixing liquid metals are those devised respectively by Wittig and Huber and by Kleppa and in particular, at a price, the Calvet calorimeters built and employed at Marseille. For solid reactions of low enthalpy, the adiabatic calorimeter of Dench is likely to have a wide range of application.

Probably the most important recent development in metallurgical calorimetry is the use, by Dodé and Gerdanian, of the very sensitive Calvet calorimeter for the determination of quasi-partial gas/metal reactions at elevated temperatures (p. 110).

Most recorded heats of formation are, however, based on difference methods. As mentioned already this involves an increase in the error, and the initial measurements must be carried out at a much higher accuracy than in direct calorimetry to obtain the same accuracy in the final result. This applies especially to combustion calorimetry, as discussed on p. 78. Reports of the determination of heats of reaction should be regarded with mistrust if full details and errors of each experimental step are not included. The purity of the materials used must be high and established beyond question.

Though the technique is simpler, the same criticism applies to solution calorimetry in acids etc. Solution calorimetry requires more time than direct calorimetry. The advantages of the method lie in its wide applicability, and the fact that the metals and compounds can be investigated in any state. This makes it suitable for investigating heat effects associated with age hardening and other delayed or arrested transformations. Alloys may be given any heat treatment before solution. Thus this method is, for instance, the only one that has been found suitable for measuring the dependence of the energy of powders on grain size, i.e. the surface energy [166]. Since such heat effects are generally small their measurement requires the greatest accuracy in determining the heats of solution. The differential solution calorimeter devised by Buck *et al.* (p. 85) is the most attractive apparatus for the purpose.

The error in solution calorimetry can be made much smaller if one of the components is itself used as the solvent, e.g. mercury of amalgams or water for hydrates, and more generally if a solvent is selected in which the components have a low heat of solution. This accounts for the popularity of the liquid-tin calorimeter (e.g. Ticknor and Bever).

Heat capacities may be derived from heat-content measurements, obtained by the drop method, if the substance under consideration undergoes no transformation other than melting in the temperature range concerned. Where transformations occur, especially order–disorder transformations, the direct method must be used. This requires more care and experience than the drop method.

2. Equilibria with a Gaseous Phase

As has been mentioned earlier, the standard Gibbs energy change of a chemical reaction can, in principle, be obtained by a combination of calorimetric data for the heat and entropy change of the reaction. It may be seen from the preceding section that a number of calorimetric studies must be combined in order to yield the Gibbs energy change, and therefore in most circumstances it is preferable to obtain the standard Gibbs energy change at a given temperature from the value of the equilibrium constant at that temperature. The determination of the equilibrium constant for the reaction

$$A + B \rightarrow C + D$$

requires a measurement of the activities of the components in the equilibrium state since

$$K = \frac{a_C a_F}{a_A a_B} = \exp - \frac{\Delta G^0}{RT} \tag{99}$$

The simplest forms of chemical reaction are those which involve a gaseous phase and pure condensed phase or phases only, or a gaseous phase and dilute condensed phase solutions when the equilibrium constants also take a simple form.

$$\langle A \rangle \rightarrow (A) \qquad K = p_A$$

$$\langle A \rangle + (B) \rightarrow \langle AB \rangle \qquad K = \frac{1}{p_B}$$

A and AB are both pure solids or liquids.

$$\langle A \rangle \rightarrow [A] \qquad K' = \frac{c_{[A]}}{p_A}$$

The value of the constant K' can only be obtained under conditions where the A atoms in solution do not interact with one another significantly, and this usually requires that the concentration of A in the solution is less than one atom per cent or so.

In high-temperature systems there are considerable experimental difficulties in the measurement of equilibria involving gaseous phases at pressures greater than one atmosphere, and therefore the following considerations will be limited to those situations where the total pressure of the gaseous phases, be they simple or mixtures, is less than or equal to one atmosphere pressure. Even with this limitation, there are a number of variants of each particular experimental technique which are designed to suit a specific reaction, and only typical examples will be referred to here. There now exist a number of more exhaustive reviews of high-temperature experimental techniques than will be attempted in this work, and the reader seeking further examples is encouraged to consult these [167, 168].

The Equilibrium Constant for Vaporisation Reactions

When a pure solid or liquid vaporises, the equilibrium constant will only contain the partial pressure of the vapour species and, providing this is a gas of simple composition, then the equilibrium constant is equal to the saturation vapour pressure of the solid. Many metals vaporise to simple atomic species or to a gas phase which contains a negligible proportion of polyatomic molecules. For example, the vapour phases in equilibrium with copper, silver and gold consist mainly of the monatomic species and contain negligible amounts of Cu_2, Ag_2 and Au_2 gaseous molecules under most experimental circumstances. The measurement of the vapour pressure of these elements can therefore be made with simple apparatus under simple assumptions. Numerous examples are known, however, where the gaseous phase is a complex mixture of species such as MoO_3, which vaporises to give gaseous molecules of Mo_3O_9, Mo_4O_{12} and Mo_5O_{15} in significant proportions in the temperature range 800–1000°K. With a system showing this kind of behaviour, the *saturation vapour pressure* can be measured using techniques which yield the total pressure directly, but the equilibrium constants for the individual vaporisation processes

$$3MoO_3\,(s) \rightarrow Mo_3O_9\,(g)$$

$$4MoO_3\,(s) \rightarrow Mo_4O_{12}\,(g)$$

$$5MoO_3\,(s) \rightarrow Mo_5O_{15}\,(g)$$

cannot be obtained from such a simple measurement, and it is necessary to devise a technique which permits the measurements of each individual gaseous species. This is usually an extremely difficult and complicated task to carry out with a high accuracy which in the present context means $\pm 1\%$ of the correct value.

A. STATIC METHODS FOR THE MEASUREMENT OF VAPOUR PRESSURE

The direct measurement of pressure by means of manometers has played a minor role in high-temperature thermodynamics, largely because of the problems associated with materials of construction.

Jenkins [169] used a static manometer to measure the vapour pressure of liquid zinc. The liquid metal was held in a quartz U-tube, one limb of which was closed and contained only the metal vapour in equilibrium with the liquid metal. The other, open, limb was joined to a "buffer" bulb containing nitrogen, the pressure of which could be measured relative to atmospheric pressure through a mercury manometer, and which could be varied to balance the vapour pressure of the zinc. The pressure was measured over the range 20–1500 mmHg and a temperature range of 900–1250°K.

Fischer *et al.* [170] constructed an apparatus for the direct measurement of vapour pressure in which a cylindrical molybdenum bell dipped into molten tin was contained in a closed silica vessel. The apparatus was used for the study of the vapour pressures of aluminium halides, a sample of which was placed inside the bell and then the apparatus was evacuated. On increasing the temperature, the force exerted by the contained vapour on the molten tin would raise the bell, the movement of which was taken up by a spring and indicated by an attached pointer. The apparatus was then filled with an inert gas until the pointer regained its original position, and the pressure of gas required to restore the pointer was read on a mercury manometer. Pressures from a few millimetres up to atmospheric could be measured by this technique with a reproducibility of 0·1 mmHg.

Finally, the vapour pressure of liquid tellurium was measured directly by the use of a quartz spiral manometer by Brooks [171] and a Bourdon manometer by Machol and Westrum [172]. In the spiral manometer, a flattened quartz tube is formed into a spiral, and one end of the tube is joined to a bulb containing the sample. The other end of the spiral is joined to a pointer which carries a mirror. This end is contained within another vessel in which the pressure can be regulated. As the apparatus is heated, the quartz spiral deforms and a pressure equal to that of the vapour pressure of the sample must be exerted in the second vessel to restore the pointer to the null position. In the Bourdon manometer, a flattened thin-walled bulb replaces the spiral and the bulb containing the sample. Once again, the difference in pressure between the vapour pressure of the contained sample and that of the surrounding vessel must be reduced to zero to keep a pointer at the null position. The results for the vapour pressure of liquid tellurium which were obtained in these two studies agree well and cover a pressure range of 0·5°–200 mmHg, and a temperature range of 750–1125°K. These techniques can be called static methods since there is no significant transfer of the material to be vaporised around or through the apparatus.

1. Static Methods Using Radiation

Measurements of the vapour pressure in a sealed system may be made when a well-defined volume of the vapour can be characterised with respect to the number of atoms or molecules of the vapour species it contains. Any physical property of this mass of vapour which can be observed from outside can be used to determine the vapour pressure providing that some calibration point or an overlap with some other technique can be obtained. Optical absorption or emission have been used up to temperatures of about 1000°C when the condensed sample and its vapour can be contained in a quartz cell.

Hirst and Olson [173] determined the concentration of mercury atoms in atmospheres in equilibrium with thallium amalgams by measuring the absorption of light corresponding to certain of the mercury resonance wavelengths. In this way they were able to measure pressures of the order of 10^{-3} mmHg. An ionisation manometer was used by Poindexter [174] for the measurement of the vapour pressures of alkali metal amalgams in the range 10^{-2} to 10^{-8} mmHg.

Herbenar, *et al.* [175] used an absorption spectrum method for measuring the zinc vapour pressure of six α-brasses over the temperature range 650–970°C. Millings of the various alloys were sealed in an absorption vessel of transparent silica (Fig. 55), which was located in a tubular furnace with transparent ends. Light from a spark source having

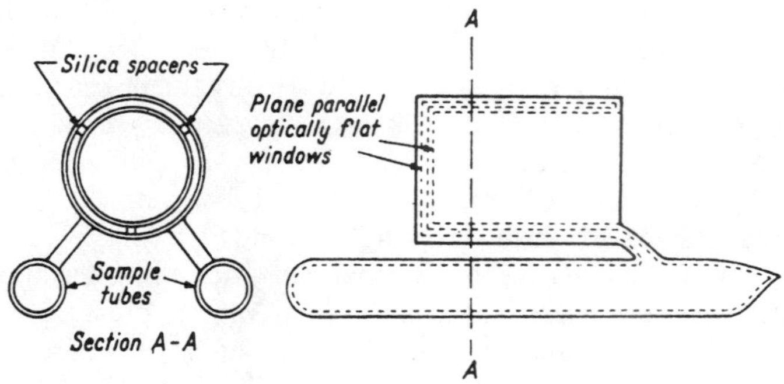

FIG. 55. Light-absorption cell for zinc vapour.

two zinc electrodes passed through the optically flat part of the vessel which confined the zinc vapour within two planes with a separation of 0·8 mm. The transmitted light was analysed by a spectrograph and recorded photographically, a stepped intensity-density pattern being imposed by a tungsten filament source on each photographic plate for calibration purposes. Photometric observations were made of two resonance lines in the zinc spectrum; the 3035 Å line was not absorbed and was therefore used as a standard for measuring the absorption of the 3076 Å radiation. The relationship between absorption, measured by the logarithmic intensity ratio of the two spectral lines, and the vapour pressure p and absolute temperature T is given by the equation

$$-\log \frac{I_{3076}}{I_{3035}} = K \frac{p}{T} d$$

where K is a constant, I the absorption coefficient, and d the thickness of absorbing space. Preliminary measurements were made with liquid zinc over the temperature range 450–745°C to confirm the validity of this law and to determine the value of the absorption coefficient.

The determination of the partial pressure of an atomic species in the vapour above an alloy system by no means exhausts the potential applications of optical absorption studies. Mixtures of atomic and molecular species in the vapour phase above solids can, in principle, be measured separately since the absorption line corresponding to the excitation of the atomic species may be in another part of the spectrum from the band spectrum of the molecular species. Rice and Ragone [176] have made a theoretical and experimental study of the Bi–Bi_2 mixture over liquid bismuth. They included a sample of lead in the system to provide an independent check of the accuracy of the method.

The optical density of atomic bismuth was measured at 3067 Å from 770° to 1280°K, and this corresponded with a pressure range of 2×10^{-8}–10^{-3} atm. The absorption due to lead atoms was measured at 2833 Å in the same temperature range where the vapour pressure of lead changes from 3×10^{-8} to 2×10^{-3} atm. Finally, the Bi_2 band spectrum was observed starting at 2731 Å, and the optical density was measured between 1010° and 1280°K covering a pressure range of 10^{-5}–10^{-3} atm.

The fact that the vapour pressures have the Clapeyron dependence on temperature was invoked to obtain vapour pressures of these species directly from the optical density results for monatomic bismuth gas. This system obeys the simple absorption law only at low pressures, and deviates markedly at high pressures, but a mathematical treatment involving an extrapolation from the Beer's law region using the Clausius–Clapeyron equation enables the measured absorption curve to be interpreted. Alternately it was found that the presence of an inert gas, helium, at a constant pressure, simplified the absorption behaviour of the atomic species and the partial pressure p_{Bi} could be evaluated. The heat of vaporisation of lead which was obtained in the study, $\lambda_e = 47{,}240 \pm 850$ cal was in good agreement with published data.

More elaborate studies have been made by Brebrick and Strauss [177] of the tellurides of lead and tin. These compounds evaporate congruently and thus the vapour pressures of PbTe and SnTe were established by conventional vapour pressure measurements [178, 179]. The experimental technique involved the measurement of the optical densities of the vapour phase due to the presence of Te_2 and PbTe or SnTe molecules over a range of composition in solid alloys ($N_{Pb} = 0{\cdot}05 - 0{\cdot}534$; $N_{Sn} = 0{\cdot}483 - 0{\cdot}504$) at wavelengths of light which were selected by interference filters from the spectrum emitted by a mercury lamp. The pressure–optical density relationship for Te_2 was established in calibrating studies made with pure tellurium and that for the molecules PbTe and SnTe from the Russian studies of conventional vapour pressures. It appears to be the general case, with a few exceptions, that this relationship must be established between optical density and partial pressure at some point in the composition–temperature range in the study of a system by optical absorption, and that absolute measurements of the vapour pressure can only rarely be achieved by the use of this technique.

In the results which have been described above, the absorption cells were usually made of quartz, but this need not necessarily be the case. It is possible to operate a ceramic cell, with two orifices to allow the transmission of a light beam, in a vacuum furnace to function as the container of the equilibrium vapour. Some uncertainty then exists concerning the path length of light through the vapour phase because of the effusion of vapour out of the cell and into the surrounding vacuum. Such effects might yield to a calibration procedure.

B. GAS-CONDENSED PHASE EQUILIBRIA IN A CLOSED SYSTEM

Hargreaves employed an elegant dew-point method [180] (originally proposed for liquids by Lescoeur, 1889) making use of an electric furnace with two independently heated portions. The brass test piece A (Fig. 56) is placed in an evacuated silica tube at the end situated in the hotter part of the furnace. Both ends of the silica tube are provided with small re-entrant tubes which carry thermocouples. The temperature of the whole furnace is raised to the required value, and then the end away from the test piece is slowly cooled, the coolest portion of the silica tube being observed through viewing windows at G. At a certain temperature, measured by the thermocouple, pure zinc condenses in small droplets. By repeated heating and cooling of this part of the tube the "dew-point" of zinc can be determined within $\pm 1°C$. The vapour pressure of zinc at that temperature equals the zinc pressure of the brass at the temperature of A. This is, of course, a relative method. It can also be applied to other than metallic vapours, for example to the titanium tetrachloride pressures in a reaction such as $2\langle TiCl_3\rangle = \langle TiCl_2\rangle + (TiCl_4)$ for the investigation of which the vapour pressures of pure $TiCl_4$ must be known.

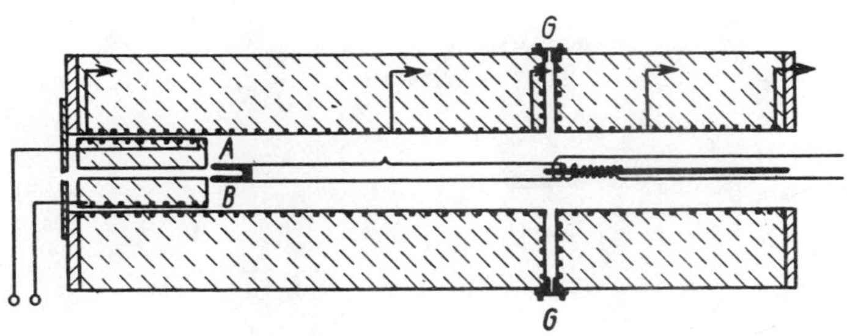

FIG. 56. Apparatus for the dew-point determinations on metals.

Whereas Hargreaves' method has nearly exhausted its usefulness, a similar method devised by Herasymenko [181] continues to find fairly wide application under the name *isopiestic method*. The principle again involves the establishment of an equilibrium vapour pressure between an alloy specimen held at the high temperature end of a sealed tube and the pure volatile component of the alloy held at the low-temperature end. In this technique, the temperatures of the hot and cold ends of the tube are fixed and the equilibrium composition of the alloy determined. The activity of the volatile component in the alloy at the temperature of the hot end can then be calculated provided the vapour pressure/temperature relation for the volatile component is known. As a time-saving device the hot end may be supplied with a row of several alloy specimens held in a temperature gradient rather than at constant temperature. Since the activity of the volatile component in the alloys varies with temperature, subsequent analysis of the alloy specimens produces a number of results per run.

Komarek and Silver [182] have made an elegant use of the method by combining it with a technique employed by Kubaschewski and Dench (see p. 42) for the determination of partial molar Gibbs energies of solution of oxygen in titanium-group metals. For these studies specimens consisting of, for example, titanium and an alkaline earth, MgO, in close physical contact were placed within the temperature gradient of the sealed tube with an excess of the pure alkaline earth metal being held at the cooler end. Reaction was allowed to proceed until equilibrium was reached where the partial pressure of oxygen over the $[O]_{Ti}$ alloy was equal to the partial pressure of oxygen in the reaction $(Mg) + \frac{1}{2}(O_2) = \langle MgO \rangle$. By varying the vapour pressure of the alkaline earth metal from one run to another, the chemical potential of oxygen in the Mg + MgO system was changed and hence, in turn, the chemical potential of oxygen in the $[O]_{Ti}$ alloy. The oxygen content of the alloys was calculated from the mass gain of the specimens after equilibration.

Richardson and Webb [183] used an isopiestic technique to obtain the activities of PbO in liquid PbO–SiO_2 mixtures by equilibrating a slag sample placed in one crucible with a bead of liquid lead contained in a separate crucible. The two condensed phase systems were contained within a larger crucible, and equilibrium was achieved when the PbO pressures of the liquid silicate and PbO dissolved in the metal sample were equal. Since the latter solution is very dilute and hence obeys Henry's law, the activity of PbO in the silicate can be obtained from the ratio of the oxygen content of the metal divided by the saturation oxygen content of lead in equilibrium with pure PbO at the same temperature.

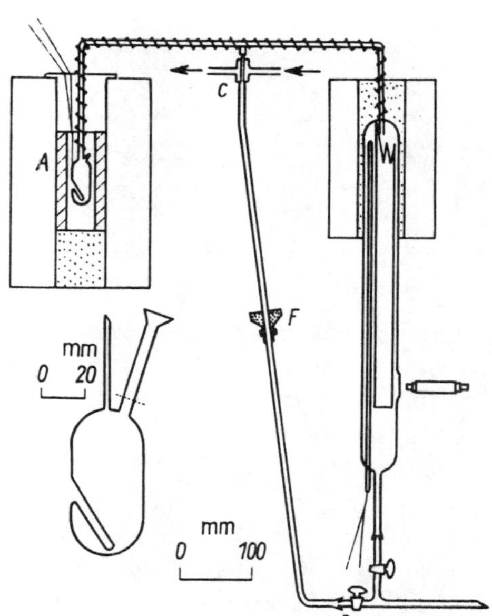

FIG. 57. Tensi-eudiometer for thermal decomposition and measurement of vapour pressure of sulphides.

The sulphur and phosphorus pressures of a large number of sulphides and phosphides were investigated by Biltz and his co-workers. Their long series of experiments began with the determination, by vapour pressure measurements, of the various steps in the progressive decomposition of inorganic ammonia complexes.

The apparatus used to measure the dissociation pressures of sulphides [184] (Fig. 57) has several refinements of experimental interest. The highest valency sulphide of the metal under consideration is enclosed in the bulb A (shown enlarged in the smaller diagram) which is heated to the required temperature. The experiment consists in measuring the vapour pressure of the sulphide after the removal of successive fractions of sulphur which are afterwards determined. The vapour pressure is measured by means of a quartz spiral manometer.* In order to isolate the sulphur vapour at its equilibrium pressure, use is made of a so-called "sulphur valve", invented originally by M. Bodenstein. The side tube leading to the sulphur receiving tube is constructed at C to a diameter of 0·5 mm (Fig. 57a). Passage of cold water through a jacket surrounding C results in the condensation of a plug of sulphur, which is vacuum-tight when 1–3 mm long. When this plug is broken, sulphur distils into the receiving tube CD. In order to measure the

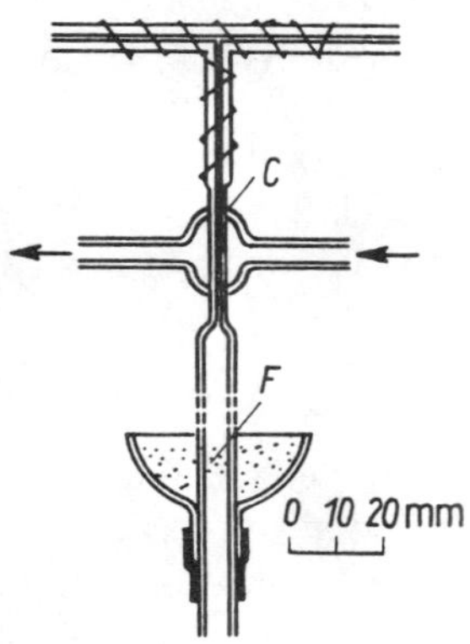

FIG. 57a. "Sulphur valve" in apparatus in Fig. 57.

increments of sulphur withdrawn from A, each one is collected at a different position along the receiver by means of the movable cooling device F, which is filled with a solid carbon dioxide-alcohol cooling mixture. Condensation in other parts of the apparatus is prevented by heating them to a temperature higher than that of the heating vessel A. At the end of each experiment the receiving tube is cut into portions, and the various deposits of sulphur are analysed. It may be noted that the apparatus is made of one piece, and this may necessitate the use of vitreous silica for a refractory sulphide such as that of platinum.

A similar arrangement has been used for phosphides. A full description of the method was given by Haraldsen [187] and a summary, together with the main experimental results, are available in a German paper [188].

Since phosphorus has not the same sealing properties as sulphur, the "phosphorus valve" must be narrower, 0·33 mm diameter, and the cooled portion at least 1·5 cm in

* The quartz spiral manometer was originally developed for such purposes by Bodenstein [185]. A detailed description of this instrument and its application to indium halides was given by Robert [186].

length. For the collection of the phosphorus distillates a straight receiver tube is unsuitable, since one experiment may last several weeks. Instead, a comb form of reservoir is used [189] (Fig. 58). One "tooth" at a time is cooled, and is sealed off when the phosphorus valve is closed.

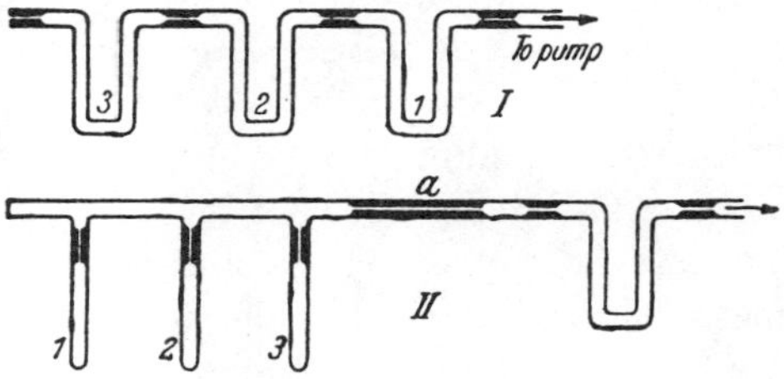

FIG. 58. Devices for condensing phosphorous fractions.

Halogen pressures have rarely been measured directly, due to their low value. Fischer and Biltz [190] measured the halogen pressures of AuCl, AuBr, and the $CuCl_2/Cu_2Cl_2$ system by a static method, using a zero instrument. Syrupy phosphoric acid was used as the manometric liquid. So-called chlorine-proof waxes were found to be unsuitable for experiments of long duration, and all junctions which came in contact with chlorine were homogeneously sealed. Generally, however, dynamic methods should be more suitable for the few halides whose dissociation pressures can be determined directly.

Many reactions of metallurgical importance involve reduction by carbon to form carbon monoxide as, for example,

$$\langle SiO_2 \rangle + 3\langle C \rangle = \langle SiC \rangle + 2(CO)$$

These equilibria can be measured by determining the carbon monoxide pressures. The static method is not a very convenient one owing to the high temperatures which are generally required. It has, however, been used, for instance by Brunner [191], who described an apparatus which was applied to several reactions including the one mentioned, in the range 1300–1700°C. The apparatus was subjected to prolonged vacuum treatment during heating to remove all contaminations and occluded gases before each experiment. A graphite crucible in the furnace was then charged with the finely ground mixture of carbon and oxide in stoichiometrical proportion. Observations were made of the rate of change of pressure at pressures above and below the equilibrium value, which was then determined by interpolation.

Static methods of vapour pressure measurements, if they can be applied, are usually to be preferred to dynamic methods. The fact that equilibrium has been attained, or rather, that the results derived apply to equilibrium conditions, may be ensured by approaching the equilibrium from higher and lower temperatures. Both approaches should lead to the same result at the required temperature. An illustration of this is given in Fig. 59 which shows the variation with time of the sulphur vapour pressures of the osmium sulphides.

In all equilibrium measurements of the kind described in this section account must be taken of the molecular aggregation of the volatile component. If this exists in more than one form (e.g. S_2, S_4, S_6, S_8; P_2, P_4; Bi, Bi_2) its dissociation equilibrium at the temperature of measurement must be known and used in the evaluation of results. In these cases the mass spectrometer is an indispensable apparatus.

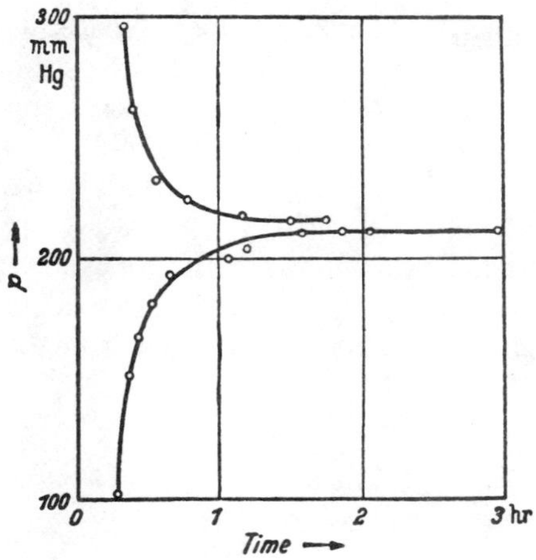

FIG. 59. Approach to equilibrium vapour pressure from both higher and lower pressures.

A considerable amount of work has been carried out by Sieverts *et al.*, Steacie and Johnson, Smithells, Ransley and others on the solubility of gases, more especially hydrogen, nitrogen and oxygen, in metals. In so far as these measurements refer to true equilibria they may be used to evaluate the thermodynamic data of the solution phase and even of the lowest valency compound of the metal with the gaseous element.

A method which is most generally applicable was developed by Sieverts [192]. The metal is heated in a silica or porcelain vessel connected to a burette (Fig. 60) which is also connected via a three-way stopcock to a pump and a gas supply. A known volume of gas is introduced to the reaction space, and the decrease in volume due to solution is observed by the level in the burette. The solubility of the gas at the operating pressure is thus given directly.

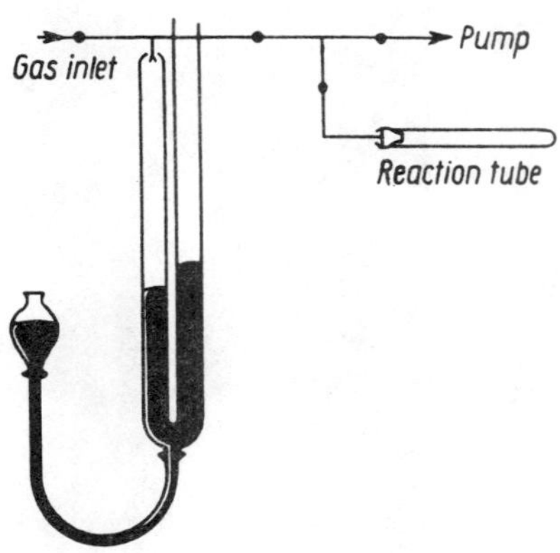

FIG. 60. Sievert's apparatus for determining the solubility of gases in metals.

Considerable advances have been made in the measurement of the solubility of gases in liquid metals and alloys with the advent of induction heating. Weinstein and Elliott [193] have described an apparatus for the measurement of the solubility of hydrogen in liquid iron and its alloys at temperatures around 1600°C (Fig. 61). Because the metal sample, which is both heated and stirred by the induction field, can be contained in a crucible mounted within a water-cooled quartz tube, the permeability of the containing vessel, which was an important source of error in Sievert's apparatus, can be reduced to negligible proportions. The hot volume of the assembly could be maintained at only 40 cm^3 by filling out the "dead" space, and equilibrium between the metal and gas was reached in 5 min.

The temperature of the melt was obtained by optical pyrometric measurement to within $\pm 2°C$, and the solubility in pure iron at 1592°C which was found to be 24·47 ppm was obtained with a standard deviation of $\pm 0·24$ ppm, from the results of fifteen different experiments.

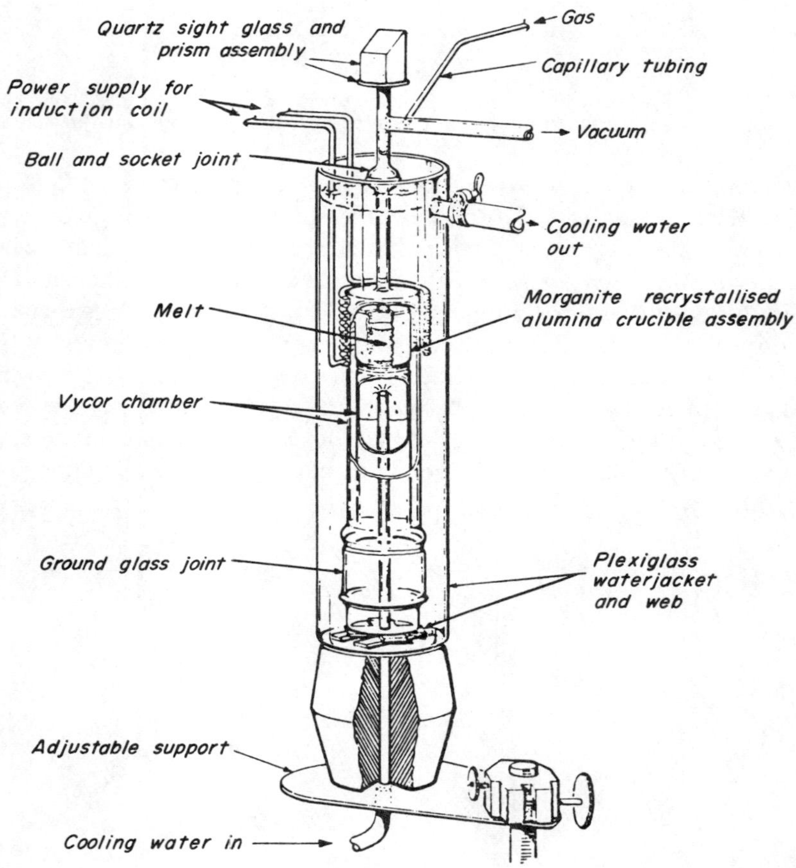

FIG. 61. Apparatus for the measurement of hydrogen solubility in liquid metals using induction heating (Weinstein and Elliott) [193].

In measuring the hot volume of the apparatus, an inert gas is substituted for the gas which is soluble in the liquid metal. In order that the temperature gradients in the apparatus are satisfactorily reproduced, the thermal properties of the calibration gas, the heat capacity and thermal conductivity, are matched as closely as possible to the gas under study. Thus helium should be used for calibration of an apparatus which is to be employed for the study of hydrogen solubilities in liquid metals, and argon is well suited as the calibration gas for studies of oxygen or nitrogen solubilities.

C. DYNAMIC VAPOUR-PRESSURE METHODS

The best-known thermochemical method that is truly dynamic is the determination of the boiling point of a substance. This, of course, gives the temperature at which the vapour pressure of the substance equals the pressure of the atmosphere with which it is in contact.

Early applications of this method to metals were not very successful owing probably to the porosity of the containing vessel, which was commonly of graphite. As a result, the vapour pressure values obtained were too low. Fischer showed [194] that Ruff's method [195] should be improved by using a completely impervious material for the crucible and by removing other lesser sources of error.

The principle of Ruff's method is to observe the temperature at which there is a sudden decrease in weight of a substance as it reaches its boiling point. The substance is enclosed in a crucible suspended from a spring balance inside a furnace. The boiling point is in effect determined from a discontinuity in the weight–temperature curve. The disadvantage of the method is that the change in slope of the curve at boiling is sometimes not sharply defined and, in addition, since the observations are taken at changing temperature, the observation of the actual temperature of the substance is subject to some error.

A modification of Ruff's method was therefore used by W. Fischer and Rahlfs [196], who worked with constant temperature and changing pressure. In this method the boiling point is much more clearly marked, as may be seen in Fig. 62. The apparatus which they used for aluminium halides is shown in Fig. 63. A bulb containing the substance was suspended from a steel spring, and an indication of the weight was given by the level of the mark M_2.

The methods used by Leitgebel [197] and Fischer [194] for observing boiling points at constant pressure are less responsive than Ruff's to porosity of the containing vessel.

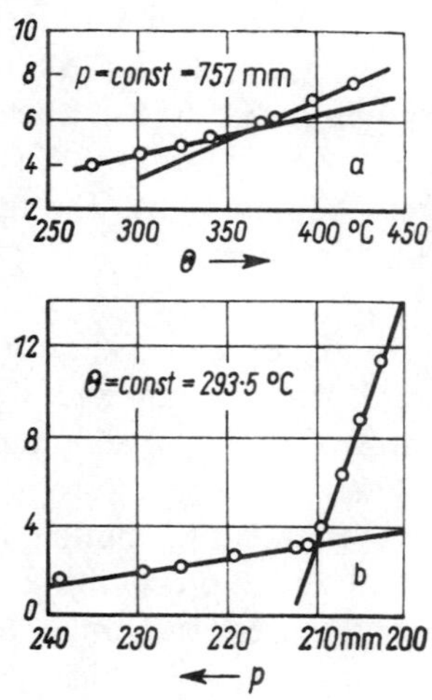

FIG. 62. Boiling point determinations: (a) with varying temperature and (b) with varying pressure.

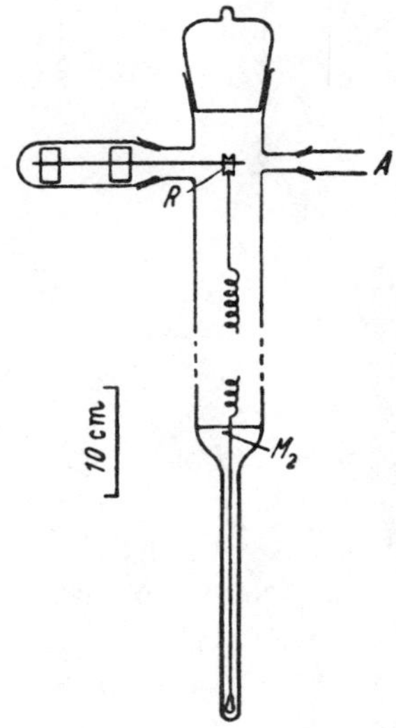

FIG. 63. Ruff type apparatus for evaluating boiling points (Fischer and Rahlfs).

Both observers determined the boiling of metals and alloys in the range 700–1750°C. Leitgebel used a closed graphite crucible, and a thermocouple to indicate the temperature, while Fischer used an alumina crucible and an alumina tube terminating in a hollow graphite "black body" which was completely immersed in the melt for optical determination of temperature.

The boiling point is marked by a well-defined arrest in temperature increase. Its observed value can be high only as a result of superheating, and any such error, as far as metallic systems at least are concerned [194], is small and within the limits of accuracy of the temperature observation.

Even this small possibility of error can be eliminated by using the condensation method in which the vapour temperature is measured at various heights above the boiling liquid. Within a certain distance from the liquid, at which its vapour is condensing, the temperature is constant and is equal to the boiling point of the liquid at the pressure of the buffer gas. This method gives very reliable values but is applicable only at temperatures at which thermocouples can be used (i.e. below 1500°C). The apparatus can be quite simple; the one used by Fischer and Rahlfs [196] for aluminium halides is shown in Fig. 66. Hydrogen was used as the buffer gas. The temperature during boiling was constant within 3–6 cm above the liquid surface.

Gattow and Schneider [198] have studied and emphasised the usefulness of the simple boiling-point method devised by Bauer and Brunner [199] which consists in the determination of the boiling point of a substance at a given pressure. The buffer gas of the evaporation vessel is separated from a large volume of gas by a drop of mercury in a horizontal glass capillary. When the temperature is gradually increased the movement of the drop is accelerated the moment the boiling temperature is attained.

Barton and Bloom [200] took the added precaution of bubbling nitrogen through the melt in their studies of the vapour pressures of a number of metal halides. This procedure helps to remove the possibility of overheating which can occur in liquids of relatively low thermal conductivities. In this apparatus, the boiling point was measured by a thermocouple which was set just below the surface of the liquid attached to a double-bore sillimamite tube which carried nitrogen gas down to the bottom of the melt. The absolute vapour pressures of NaCl and KCl were obtained in this way over a range of temperatures and applied external pressures (15–500 mmHg). The results were combined with data from subsequent transportation studied to yield the monomer/dimer ratio of these vapours as a function of temperature, and hence the heats of dimerisation

1. The Transportation Method

The more general dynamic method of determining vapour pressure is known sometimes as the transportation method. In the application to the measurement of the vapour pressure of a metal, for example, or of the volatile component of a binary metallic alloy, a steady, measured stream of inert gas is passed over the substance under investigation, which is maintained at a constant temperature. The gas removes the vapour or volatile component of the substance at a rate which is dependent upon the relative pressures and upon the rate of gas flow. The vapour is condensed or collected by absorption or chemical combination at a cooler portion of the apparatus. The rate of removal of vapour is measured at different rates of gas flow.

In a system which is completely isothermal, saturation is attained at a zero gas flow rate and the value of the vapour pressure can be determined by extrapolation to zero rate. The value so obtained should be equal to the value observed by the use of static methods. In the typical high-temperature system in which a temperature gradient exists between the sample, which is at the centre of the furnace, and the cold end of the furnace where the vapour is analysed, saturation can be achieved at low, but finite flow rates, and an error is introduced by extrapolation to zero flow rate of the transporting gas.

Alcock and Hooper [201] in a study of the vapour pressure of gold by this technique found that saturation could be achieved over a wider range of flow rates the more the evaporating surface could be made to fill the cross-section of the reaction tube. Their results shown in Fig. 64 indicate this effect.

At low flow rates incorrect results are obtained because the gold atoms diffuse down the temperature gradient at a different mean velocity than the argon atoms, resulting in apparently higher evaporation rates than the equilibrium rate.

At high flow rates the transporting gas does not have a long enough residence time in contact with the sample to reach saturation. Mertens [202] has analysed the transportation system for the case of laminar flow of the transporting gas over the condensed phase which covers the wall of the reaction tube along the length of the even temperature zone of the furnace. The conclusion from the experimental studies that a long "plateau" can be achieved, where the measured vapour pressure is independent of the carrier gas flow rate, when the condensed phase intercepts the flow pattern of the gas as much as possible, suggests that in a typical real system the gas is in turbulent flow around the sample, and such a theoretical analysis is of limited application.

The partial pressures being measured are calculated from the volumes of transported substance V_s and of transporting gas V_g at s.t.p., assuming the validity of Boyle's law, from

$$p_s = \frac{760 V_s}{V_g + V_s} \text{(mmHg)}$$

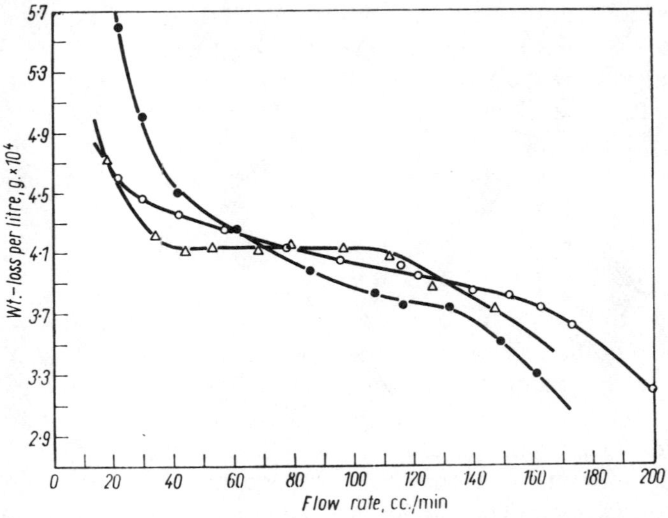

FIG. 64. The apparent vapour pressure of gold obtained by the transportation method, as a function of sample to reaction tube diameter ratio. △, large beads with plugs nearly filling the tube cross-section on either side of the sample; ○, small beads with plugs; ●, small beads without plugs.

This method is very commonly applied to organic and inorganic compounds at relatively low temperatures. It was first applied by v. Wartenberg [203] at the higher temperatures to metals, and his type of apparatus has been used with very little modification in subsequent work on inorganic salts and metallic systems by Jellinek and Rosner [204] and others. The diagram of the apparatus in Fig. 65 is taken from a paper by Schneider and Stoll [205].

Constrictions are made in the furnace tube on both sides of the sample to minimise counterdiffusion of the vapour. This may cause trouble especially when hydrogen is used, and whenever available the heavier gas argon is to be preferred. The transporting gas must be quite pure. The total gas flow and streaming rate are measured with a volume- rather than a flowmeter, since the fundamental quantity being measured is the total volume over a known time, not the rate at any instant.

The furnace is so mounted on wheels that it can be moved away from the test portion of the main tube. Before the experiment the furnace is moved well to the left (Fig. 65) and its temperature raised to about 50° higher than the desired final value. When it is pushed to its normal position the test sample is rapidly heated while the furnace temperature drops to within a few degrees of its desired value.

In the apparatus used by Fischer and Rahlfs [196] for the study of aluminium halides, the reaction and condensation chambers were connected by a heated tube. This allowed experiments to be done with a very slow gas flow which ensured complete saturation without any loss of transported vapour by condensation elsewhere than in the collecting vessel.

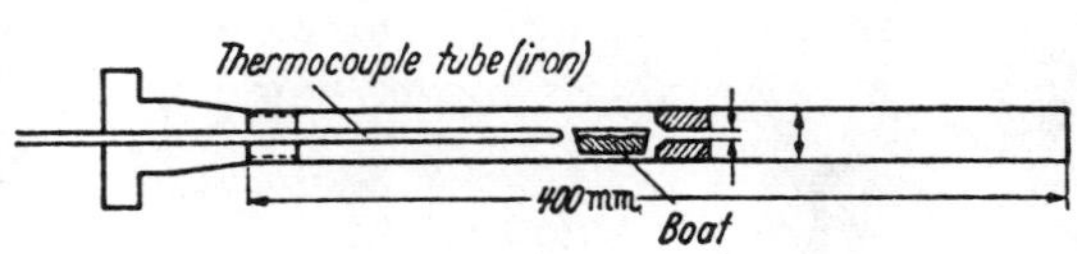

FIG. 65. Reaction tube for the measurement of vapour pressures by the transportation method.

Another method of ensuring that all the vapour in the slowly moving saturated gas is quantitatively transferred into the condensation chamber is found in Fischer and Gewehr's apparatus [206]. It was used to study the formation of aluminium chloride on passing chlorine or hydrogen chloride over heated alumina. The accuracy of the method was tested by measuring the transfer of pure mercury by an inert gas. The basis of the method is the use of a further "carrier" gas to take the saturated gas, diluted from the hot portion of the apparatus to the condensation chamber. The saturated gas leaves the reaction chamber R (Fig. 66) by a capillary tube C. The carrier gas enters at 3, mixes with the saturated gas at the end of C, and transports it to the chamber A. The portion B is heated sufficiently to keep the temperature of C higher than that of R.

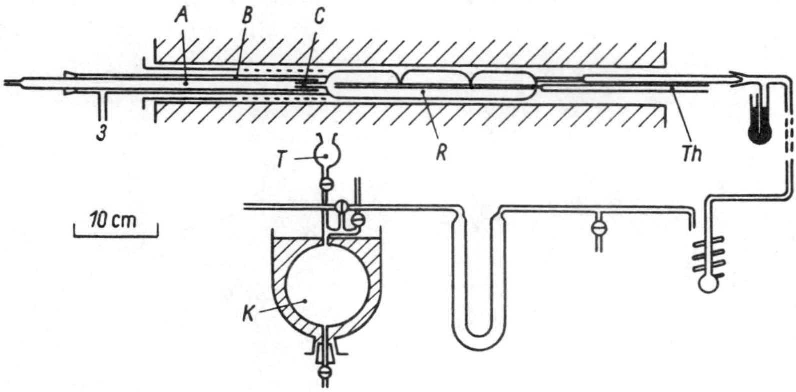

FIG. 66. Apparatus for the transportation method, utilising a carrier gas (vessel K is drawn on a reduced scale).

When the condensed phase is a liquid, saturation of the carrier gas can be rapidly achieved if the gas is bubbled through the liquid phase. Morris and Zellars [207] first used this method in a measurement of the vapour pressure of liquid copper in the temperature range 1450–1600°C using both argon and helium as carrier gases. The mouth of the bubbling tube was placed just in contact with the surface of the liquid metal so as to create ripples on the surface. To sample the equilibrium gas, a sample mixture of known volume was drawn from immediately above the liquid metal through a narrow-bore tube. The quantity of copper in the measured volume of gas was determined from the mass of condensate within this sampling tube.

The transportation technique can be extended to use where the carrier gas undergoes a reaction with the condensed phase to form gaseous products. Again the most advantageous design of the reaction chamber brings the gas and the solid phases into intimate contact, thus establishing equilibrium at a finite flow rate, and the existence of a plateau in the apparent vapour pressure/flow-rate graph.

Typical examples of successful studies of this nature are those by Bookey [208] who studied the reaction

$$\langle Ca_4P_2O_9\rangle + 5(H_2) = 4\langle CaO\rangle + (P_2) + 5(H_2O)$$

$$K = p_{P_2}\left(\frac{p_{H_2O}}{p_{H_2}}\right)^5; \quad p_{P_2} = \tfrac{1}{5}p_{H_2O}$$

by passing hydrogen through a reaction chamber containing pellets of the solid reactants which occupied a large fraction of the cross-section of the reaction tube. Kitchener and Ignatowicz [209] passed hydrogen in nitrogen/hydrogen mixtures through a bed of zinc oxide granules and condensed a sample of the zinc transported as a result of the reaction

$$\langle ZnO \rangle + (H_2) \rightarrow (Zn) + (H_2O)$$

in a known volume of the carrier gas. Since $p_{Zn} = p_{H_2O}$ under these circumstances the equilibrium constant, and hence the Gibbs energy of the reaction was obtained.

Oxygen was used by a number of workers to establish the thermodynamics of the gaseous oxides which are formed by the platinum group metals at high temperatures [201, 210, 211]. The oxide PtO_2, is now well established as a result of the studies in which oxygen was passed over heated platinum samples in a transportation apparatus, and, again, good plateaux could be obtained in the apparent vapour pressure/flow-rate curves (Fig. 67).

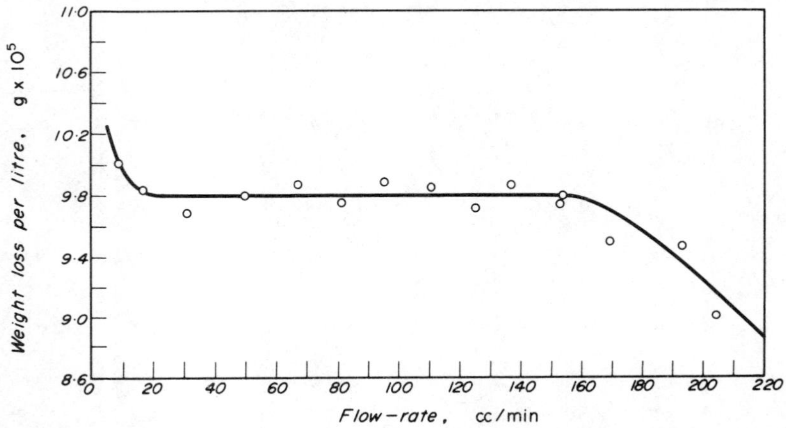

FIG. 67. Weight loss of a platinum sample as a result of (PtO_2) formation in oxygen at 1398°C as a function of the gas flow rate.

It is generally so, in the use of the gas transportation method as a technique for the measurement of vapour pressures, that the material which vaporises is collected in the condensed state. In order to apply the equation given above to the determination of the vapour pressure, it is necessary to know the molecular species in the vapour phase in order to calculate V_s from the mass which is condensed. In many cases of the metals, the monatomic species predominates, but in the case of substances which are evaporated as molecules, or vapours which are formed as the result of reaction between the transporting gas and the condensed phase, it is usually necessary to establish the composition of the vapour species in order to obtain the vapour pressure.

In the general vaporisation reaction

$$m\langle M \rangle + \frac{n}{2}(X_2) \rightarrow (M_mX_n)$$

the equilibrium constant will depend on the partial pressures of the gaseous species and the activity of the metal

$$K = \frac{p_{M_mX_n}}{a_M^m p_{X_2}^{n/2}}$$

The response of the vaporisation process to changes in the partial pressure of X_2, e.g. by dilution with an inert gas or by a reduction of the total gaseous pressure, will indicate the number of X atoms in the gaseous species. A study of the dependence of the partial pressure of M_mX_n upon the activity of M in an alloy with a less-reactive metal will then yield the number of M atoms in the molecule. This procedure is clearly only of any use when the vapour phase is relatively simple. When a number of polymers occur together in significant amounts, e.g. $(MoO_3)_n$ or a number of valency states for the metal e.g. MX, MX_2, MX_3, then another and much more difficult situation arises, and one which cannot usually be resolved by the use of this technique alone.

D. OTHER HETEROGENEOUS EQUILIBRIA

The methods used in the study of systems involving two or more gaseous components are very similar to those described in the preceding section. The transportation method, for instance, may be used to determine the rate of evaporation of two volatile components. Their respective vapour pressures are evaluated and treated individually, and such a system lies completely within the scope of the foregoing section.

There are many factors peculiar to systems including more than one gas, however, and they may be considered separately. A large number of such systems have been investigated, and some of the methods used are best described by a few examples. We shall limit these considerations to reactions between a gas and a condensed phase giving rise to at least one volatile product. The reactions will be classified according to the gas used as the reducing (or oxidising) medium in place of distinction between methods. Most of the equilibria measured hitherto have included hydrogen as the reducing agent.

1. H_2-CH_4

Equilibria between hydrogen, metals and their carbides, and methane have been determined on a large scale by Schenck and his co-workers [212]. Their method is simple and is in principle the one generally used for such equilibria with hydrogen. The partly carburised metal is heated in contact with a mixture of hydrogen and methane. The progress of reaction is observed by the change in pressure. When this has become steady, the gaseous and condensed phases are analysed. Sensitive methods of analysis must be used, and many have been developed by Schenck for this purpose. To investigate each system, experiments are carried out with a number of different mixtures of hydrogen and methane, and equilibrium is approached from the two directions by carburising or decarburising. The gas mixture is slowly circulated, and to ensure that the correct equilibrium is attained and measured, the cooler dead space in the furnace chamber is kept to a minimum with blocks of ceramic material. Even with the best materials, all heated furnace tubes are to some extent pervious to hydrogen at high temperatures, and a correction must be made for this. Since the method is static the results also require correction for thermal diffusion in the gaseous phase.

Systems involving binary iron alloys (with manganese or tungsten, for example) have been studied [213] in addition to those which include only one metal. The kinetics of carbide formation, however, involve a number of incalculable difficulties, and thermochemical data obtained by this method are therefore not entirely satisfactory.

2. H_2–NH_3

Little work has been done on these equilibria, though of this work the investigation of the iron nitriding reaction

$$2\langle Fe_4N\rangle + 3(H_2) = 2(NH_3) + 8\langle Fe\rangle$$

by Emmett *et al.* [214] is of some importance. A weighed amount of ferric oxalate was placed in the reaction tube A (Fig. 68) and reduced in a stream of hydrogen. When the exit gas no longer contained water vapour, a mixture of hydrogen and ammonia of fixed composition was passed into the tube, which was maintained at a constant temperature. The ammonia content of the outgoing gas was periodically analysed, and when it became constant, the gas mixture was replaced by nitrogen. Finally, some of the nitrided iron

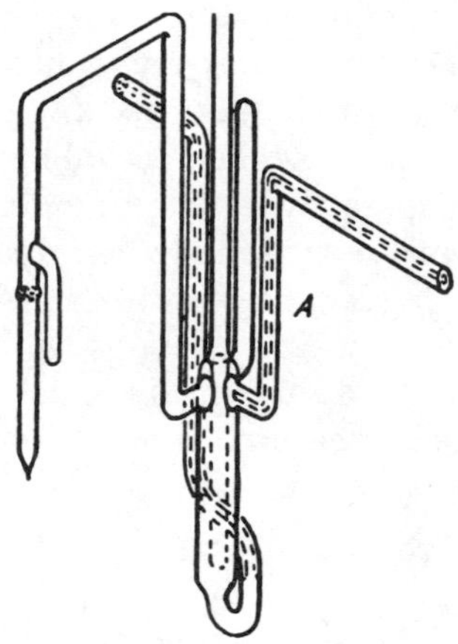

FIG. 68. Glass reaction vessel for the determination of NH_3/H_2 equilibria over iron.

powder was shaken into the thin-walled side tube in which it was sealed for subsequent X-ray analysis.

3. H_2–H_2O

Since all metals are produced from their ores by reduction methods, the thermochemical data of the reduction of metallic oxides are of great fundamental importance, and have been investigated directly by numerous workers.

A static method which is widely applicable was originally devised by Deville (1870) and used subsequently by a number of observers, notably by Eastman and Evans [215] to study the reduction of iron oxides by hydrogen. The apparatus (Fig. 69) contains water vapour at a pressure determined by the fixed temperature of the water in A. This water vapour is allowed to react with iron powder in the boat B which is heated to the required temperature in the silica furnace tube. Measurement of the total pressure after attainment of equilibrium gives on subtracting the water vapour pressure the equilibrium pressure of hydrogen. Variation in the composition of the solid is obtained by successive high-temperature treatments in a stream of water vapour, beginning with the pure metal, or by controlled hydrogen reduction of the oxide. The composition is calculated from the change in weight during the treatment.

A great source of trouble and error in this method is the evolution of occluded gases from the walls of the apparatus on heating. The apparatus must be thoroughly baked and evacuated and blank determinations carried out before each series of measurements.

The rapidity with which equilibrium is attained in Eastman and Evans' experiments decreased with increasing oxygen content of the iron. Over the range 0 to 10% O_2 a time of 30–45 min was taken at 990°C.

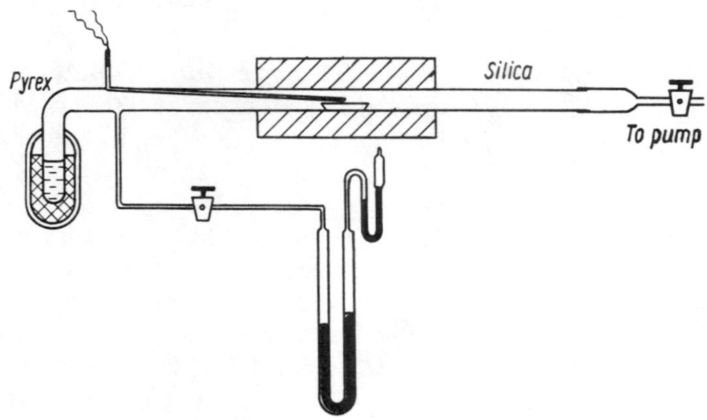

FIG. 69. Static measurement of H_2O/H_2 mixtures in equilibrium with iron and iron oxides.

It was shown by Emmett and Shultz [216] for the equilibrium 'FeO' + H_2 = Fe + H_2O that the results obtained by various observers by static and dynamic methods are discordant and the difference was ascribed to the effect of thermal diffusion in static measurements. Shibata and Kitagawa have determined [217] this effect for H_2O/H_2 mixtures at various temperatures and various compositions. Their results can be expressed approximately by the empirical relationship

$$\log\left(\frac{p_{H_2O}}{p_{H_2}}\right)_{T_2} = \left(1 + 0{\cdot}065 \log\frac{T_2}{T_1}\right)\log\left(\frac{p_{H_2O}}{p_{H_2}}\right)_{T_1} - 0{\cdot}205 \times 10^{-3}(T_2 - T_1), \tag{100}$$

which represents the experimental data (400–1021°C) within $\pm 0{\cdot}005$ on the logarithmic scale. T_2 denotes the higher, T_1 the lower temperature. Whalley [218] pointed out a mistake made by Shibata and Kitagawa in the discussion of their experimental results, and has recalculated certain values of theoretical importance. This does not affect the above equation, however, which for experimental purposes should be quite adequate for calculating H_2O/H_2 ratios in the hot zone of any apparatus from the ratios measured in a cooler part, thus making the static method as useful as the dynamic one.

These observations have been confirmed in an investigation [219] of the oxidation of the iron–nickel alloys at high iron contents between 650° and 950°C. The method of Eastman and Evans was quite suitable for this system, since only the iron is oxidised in the temperature and concentration ranges covered. The simplicity of the method combined with its reliability makes it quite attractive. It must be determined in any series of experiments, however, whether the effect of thermal diffusion necessitates a correction.

Deville's method is not, however, well adapted to equilibria in which the hydrogen pressure is less than a few millimetres, or to those in which the equilibrium vapour pressure is very low. Dynamic methods are needed for such systems.

The reduction of cobalt and nickel oxides, for instance, proceeds with a very low hydrogen pressure. In a study of these equilibria [216, 220], steam and electrolytic hydrogen were mixed in fixed proportions and passed over the heated oxides. The volume of ingoing hydrogen was measured from the electrical power input, the exit steam was condensed and weighed, and the exit hydrogen determined with a volumeter.

At all proportions of water vapour to hydrogen, precautions must be taken to ensure that there is no thermal or back diffusion of the vapour before or after passage through the equilibrium chamber.

Equilibrium measurements at much higher temperatures, those of *molten* iron for example, require more elaborate equipment and refractory materials which must remain practically impervious to gases on heating. The most serious difficulty encountered in practice is that furnace tubes found to be gas-tight on first heating often become pervious after a few more thermal cycles.

The solution to this trouble may lie in the use of induction heating methods. These have been developed to a high degree of refinement by a succession of workers under the direction of Chipman at the Massachusetts Institute of Technology. The present basic form of apparatus was developed by Dastur and Chipman [9], and its elaborated version, which was used by Gokcen and Chipman [221] for the study of the silicon–oxygen equilibrium in liquid iron, is shown in Fig. 70. The iron alloy is melted in an alundum crucible supported on a tube which serves to carry the exit gas and to lower the crucible for rapid cooling when equilibrium has been established. The gas mixtures are passed into the furnace through a refractory tube wound internally with a platinum resistance heater which is maintained as nearly as possible to the temperature of the melt. The design of the furnace, and preliminary experimental work, were intended to eliminate thermal diffusion (see p. 127) and to ensure accuracy in the measurement of temperature. Apparent temperature readings taken by training an optical pyrometer on the surface of the melt were calibrated [222] by the use of crucibles fitted with re-entrant tubes in the base. True temperature readings were made both with the insertion of thermocouples in and with the sighting of a pyrometer into these re-entrant tubes.

As a result of these basic studies in which the equilibrium between oxygen dissolved in liquid iron and H_2–H_2O gas mixtures was established, it has been possible to extend the technique to understand the "equilibration" between these gas mixtures and levitated droplets of molten iron or its alloys. The advantages of this procedure over the conventional technique are, firstly, that there can be no metal-refractory interactions because no crucible is used, and, secondly, the whole droplet can be rapidly quenched for analysis by turning off the radio-frequency heating source and allowing the droplet to fall into a copper mould or "splat-cooling" device.

One disadvantage clearly results from the fact that the gas-levitated droplet is by no means a classical, isothermal, equilibrium system, and steep thermal gradients exist in the gas phase, thus emphasising thermal diffusion effects. Secondly, the temperature of the droplet cannot be measured by thermocouple as is possible when a crucible is present, and observations, usually by two-colour pyrometer, must be relied upon for temperature measurement.

As would be expected from the general pattern of thermal diffusion behaviour in gas

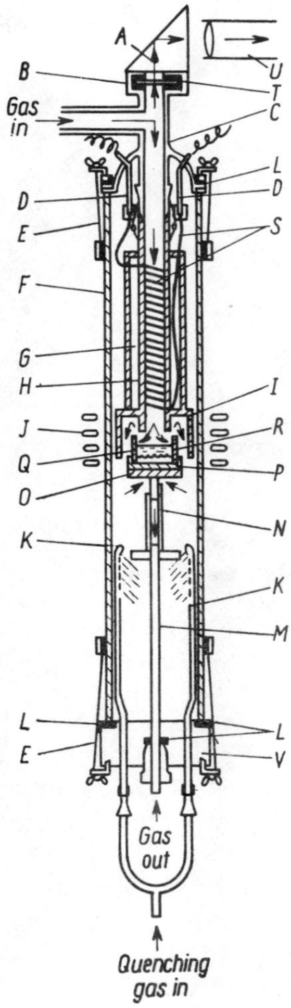

FIG. 70. Gokcen and Chipman's furnace tube for the measurement of H_2O/H_2 equilibria over molten iron.

mixtures, the errors due to this source are greater when H_2–H_2O mixtures are used to fix the oxygen than when CO–CO_2 mixtures are employed. In the latter case, reasonably good agreement was obtained with the results of conventional studies, and the effects of thermal diffusion in H_2–H_2O mixtures can be reduced by introducing argon into the gas mixture. The presence of this gas raises the average molecular weight of the gas, and an extrapolation of the values of log K of the equilibrium constant to the theoretical gas of infinite molecular weight yields a value of log K which is within 0·05 of the value obtained by the conventional technique (Fig. 71).

In an investigation of the reduction of chromic oxide Wartenberg and Aoyama [224] suspended the chromium test piece on thermocouple wires in a furnace supplied with hydrogen whose water content was fixed by cooling by suitable freezing agents. The test piece was heated slowly, and the temperature at which the surface oxidation disappeared was noted. Similarly, the temperature at which oxidation recommenced on cooling was

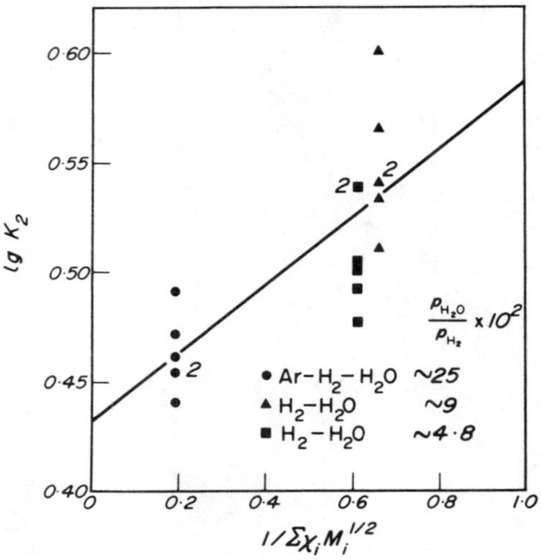

FIG. 71. The effect of thermal diffusion on the apparent equilibrium constant at 1700°C for the reaction $(H_2) + O \rightarrow (H_2O)$. χ_i is the mole fraction of i and M_i is the molecular weight of i. [After A. McLean. High Temp–High Pressure 6, 21 (1974)]

recorded. By repeating these observations several times with slow temperature changes, the equilibrium temperature for the given water–hydrogen mixture was determined.

The visual observation of surface oxidation is not reliable, however, for surface colours can be misleading. The method was therefore improved by Grube and Flad [225], who dispensed with visual observations by substituting determinations of the change in weight of the test sample. The water content of the hydrogen after passing the sample was determined by absorption with phosphorus pentoxide. In these experiments the hydrogen must be very carefully purified.

Grube and Flad [226] extended their equilibrium pressure measurements to determine the thermodynamic data of alloy formation. By determining the water/hydrogen pressure ratios, (1) for pure chromic oxide and (2) for the oxide in intimate admixture with nickel powder, the free energy of chromium–nickel alloy formation is obtained from the relationship

$$\Delta\bar{G}_{Cr} = \tfrac{3}{4}RT\left[\ln\frac{p_{H_2(1)}}{p_{H_2O(1)}} - \ln\frac{p_{H_2(2)}}{p_{H_2O(2)}}\right]$$

The reverse procedure, i.e. the determination of equilibria with alloys containing oxygen at known activities in order to obtain the dissociation pressure of a pure oxide, will rarely be used, presumably. Chiche [227], however, has done so. He first obtained the activities of copper in copper–gold alloys by an e.m.f. method and then made a direct determination of the dissociation pressures of Cu_2O–Cu–Au mixtures. He was thus able to calculate the dissociation pressure of cuprous oxide which is too low to be measured directly with convenience. Since the temperature coefficients of the e.m.f. of copper–gold alloys obtained by Chiche appear to be rather uncertain, it is better to take the assessed activities of copper in copper–gold [40] and to combine these with the dissociation pressures determined by Chiche. The resulting values for the dissociation pressure of pure Cu_2O then appear reasonable.

4. H_2–H_2S

The field of heterogeneous equilibrium measurements discussed in this chapter was first explored by Pélabon who developed a static method, similar to that which Bodenstein used for homogeneous gas reactions, and applied it to sulphides and selenides. This work [228] formed the basis of later static measurements.

Pélabon enclosed the metal sulphide or a mixture of sulphide and metal with hydrogen or a hydrogen/hydrogen sulphide mixture in a tube heated to the required temperature. After allowing a certain time for equilibrium to be established, the tube was quickly removed from the furnace, on the end of a wire, and rapidly aircooled by swinging it around. The gas, its composition retained at its high-temperature equilibrium value, was transferred to a eudiometer under mercury and analysed.

This method was used in an improved form by Keyes and Felsing [229] and also by Watanabe [230]. Schenck and Forst studied H_2/H_2S equilibria by a method similar to that used for H_2/CH_4 equilibria, described on p. 131. In the early investigations of Schenck the effect of thermal diffusion was neglected, but precautions were taken against it in later work [231].

Equilibria between hydrogen/hydrogen sulphide mixtures and molten iron–sulphur alloys have been studied by Sherman *et al.* [232]. The furnace equipment was similar in principle to that described on pp. 134–5, but the melts were sampled periodically during each run to determine the approach to equilibrium by measurement of sulphur content.

Rosenqvist [233] has investigated, apparently very accurately, the equilibrium $Ag_2S + (H_2) = 2Ag + (H_2S)$ for the solid and liquid states in the range 600–1280°C. The experimental system was closed and gas-tight, and the gas mixture was circulated by means of a glass propeller, driven by an external magnet, until equilibrium was established which took about 5 hr. The circulation rate was high to minimise thermal diffusion, the gas being sufficiently preheated before it reached the metal. In its circulation the gas mixture passed through a chamber containing a buoyancy balance, where the density of the gas was determined by magnetic balancing. From the measured density, corrected for pressure and temperature, the H_2S content which varied from 1 to 25% was determined. This method has the advantage that it is more accurate than the usual chemical-analytical methods and that the attainment of equilibrium is being observed in a closed system. After each run, the sulphur content of the alloy was determined by analysis.

Another important feature of Rosenqvist's work is that a correction was applied for the S_2 content of the gas in the hot zone using the known dissociation constants of H_2S. This correction may be as high as 10% and has been overlooked by most previous investigators of equilibria with H_2S/H_2 mixtures.

5. H_2–HCl

The investigation of hydrogen/hydrogen chloride equilibria by static methods has been reported by several workers. Jouniaux studied the equilibrium with silver chloride [234] by a method similar to that of Pélabon for sulphide equilibria.

The apparatus of Ishikawa and Watanabe [235] is shown in Fig. 72. During the reaction, the gas mixture in A was agitated by moving the mercury reservoir F. The gas was sampled by means of the gas burette G.

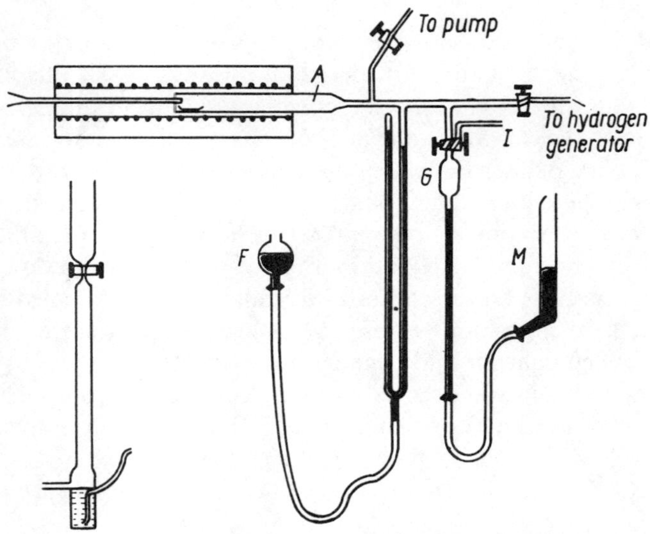

FIG. 72. Static determination of HCl/H_2 equilibria.

Jellinek applied the dynamic method of passing hydrogen into the apparatus and analysing the exit gas to the equilibria with hydrogen of chlorides [236], fluorides [237], iodides and bromides [238].

6. *Other Equilibria*

While those equilibria in which hydrogen is the reducing agent have had the most attention, others, such as those with $CO-CO_2$ mixtures have also been investigated. By and large the static and dynamic methods which have been used are similar to those already described.

The attention of the reader may be drawn to a more detailed survey of the equilibria and transport phenomena involving gas mixtures and condensed phases by Schwerdtfeger and Turkdogan [239].

The heat change which accompanies a chemical reaction was employed in the study of the thermodynamics of sulphate formation from metal oxide, SO_2 and O_2 by Dewing and Richardson [240]. In a particular example, a powdered specimen of MgO was packed into a spiral of Pt–13%Rh wire which had a thermocouple junction placed at its centre. A reference thermocouple junction was exposed to the gas phase, and the whole system was supported in an alumina boat. This was placed in the even zone of a furnace reaction tube down which a metered mixture of SO_2, O_2 and N_2 was passed. The furnace temperature was raised and $MgSO_4$ soon formed over the MgO sample. Heating was continued until a thermal signal from the thermocouple system showed that the temperature of the MgO sample lagged behind that of the gas phase. This indicates decomposition of the sulphate and at this temperature $MgO-MgSO_4$ is in equilibrium with the gas phase, which at the high temperature consists of an equilibrium mixture of SO_2, O_2, N_2 and SO_3. This composition can be calculated from the pertinent thermochemical data, and the constant for the reaction

$$MgO + SO_2 + \tfrac{1}{2}O_2 \rightarrow MgSO_4$$

can be established at the temperature of the thermal arrest. On cooling the furnace, a heat evolution is observed accompanying the re-formation of the sulphate. By varying the input gas mixture, the value of the equilibrium constant can be obtained over a range of temperatures. Results have been obtained by the use of this technique for a number of metal sulphates, and the results for this technique, employing a flowing gas system, agree well with the static dissociation pressures for these sulphates reported by Kellogg [241] which have been obtained by Ingraham *et al.* The agreement indicates that thermal segregation errors in SO_2–O_2–SO_3 mixtures are not significant in this relatively low temperature range. This was further demonstrated by Ingraham and Kellogg [242] who showed that when the gases were circulated in an apparatus which could be used for static measurements, no significant change was observed in the total pressure of the system. The apparatus, used for the study of zinc sulphates, consisted of a furnace tube which was closed at one end by a pyrex bellows capable of transmitting the total pressure of the SO_2, O_2 and SO_3 mixture generated by decomposition of the sulphate to a mercury manometer. When the gases within the furnace tube were recirculated by means of a pyrex centrifugal pump, the total pressure changed for a short time, and then returned to the original pressure exerted by the static gas. This indicates that the total pressure, which would change with a significant thermal diffusion contribution, was unaffected by recirculation and hence significantly free of this source of error.

E. METHODS BASED ON RATES OF EVAPORATION: EFFUSION METHODS

Two principal methods have been developed of determining the vapour pressure of a substance by its rate of evaporation *in vacuo*: they are the methods of Knudsen [243] and Langmuir [244]. The pressure p (atm) is calculated from the formula

$$p = \frac{m}{tA}\sqrt{\frac{2\pi \mathbf{R}T}{M}} = 0{\cdot}02256\frac{m}{tA}\sqrt{\frac{T}{M}} \tag{101}$$

where m (g) is the mass of vapour of molecular weight M which evaporates from an area A (cm^2) in time t (sec) if the Knudsen method is used, but the mass which evaporates when the Langmuir method is used is often *less* than the value given in eqn (101). This is because the evaporation of solids under the conditions required in the Langmuir method does not occur under equilibrium conditions and the fraction of the mass evaporated to that calculated from the corresponding Knudsen experiment is called the vaporisation coefficient.

In the effusion method of Knudsen the substance is enclosed in a gas-tight container into the lid of which a small orifice is drilled. The conditions for the application of eqn. (101) to evaporation out of this container are that the orifice must be knife-edged and have a diameter less than one-tenth of the mean free path of the vapour inside the cell.

A channel may be used instead of a knife-edged orifice, but then it is important that the length-to-diameter ratio of the orifice is known. The correction factor to be applied to the equation above for this method was first deduced theoretically by Clausing, and has been calculated and tabulated by Searcy and Freeman [245].

In the Langmuir method the sample is suspended freely in the vacuum system so that there is no impediment to evaporation from the whole of the specimen surface area.

Although the true surface of a solid is never planar, the projected or apparent surface area is normally used in eqn. (101).

The weight loss of a sample m_L in a Langmuir experiment can be calculated from a knowledge of the equilibrium vapour pressure p_K established by the Knudsen method, through the equation

$$m_L = 44tA\alpha p_K\sqrt{\frac{M}{T}} \tag{102}$$

where the symbols have been defined above, and α is the vaporisation coefficient $(0 < \alpha \leqq 1)$.

The weight loss during a Knudsen or Langmuir experiment was formerly obtained by weighing the cell and its contents at room temperature before and after an experiment. This should now be avoided since the outgassing error is difficult to allow for, and many satisfactory continuous weighing devices have been described in the literature following the early designs of Gulbransen and Andrew [246].

In the latest modification of the Knudsen-effusion weight–loss method by Kubaschewski and Chart [247] a vacuum microbalance with a load-carrying capacity of 25 g and a sensitivity of 1 μg on its most sensitive range was employed. The cell was suspended from the microbalance in a vacuum furnace. By continuous weighing, with the cell at temperature, the progress of de-gassing and the onset of steady-state conditions could be determined. The method was designed to measure the activity of silicon in transition-metal silicides. Since direct dissociation would have required temperatures of well over 2000°K, the silicides were intimately mixed with silica, thus producing silicon monoxide vapour in the cell and reducing the temperature of measurement by 700°K or more compared with the temperature required for direct dissociation. One of the resulting advantages was that silica glass could be used to make the Knudsen cell.

For the highest sensitivity it is preferable to use radioactive isotopes and to condense a known fraction of the vapour on a target. The weight loss can then be obtained by radiochemical assay. Care must be taken in using this method to ensure that a significant fraction of the measured condensate has not been collected after scattering from the plates defining the solid angle of collection. The apparatus of Schadel and Birchenall [248] is a successful design for this method of vapour-pressure measurement.

The combination of effusion and tracer methods has also been applied by Kubaschewski, Heymer and Dench [249] to alloys of chromium. Higher temperatures (*ca.* 1400°C) were required, and resistance heating has been preferred to induction heating (Fig. 73). In addition to the main resistance heating element enclosed in radiation shields, heating elements at the top and bottom could be operated separately in order to establish uniform temperature within the molybdenum cell. The temperature was measured by means of two thermocouples. The effused chromium condensed on the molybdenum tube and on the target disc and was collected by solution in acid. The specimens were made from chromium which had been made radioactive, and the mass effused was determined by comparing the radioactivity of the solution with that of a standard solution of chromium from the same batch. The advantage of using the funnel is that all the escaping chromium could be collected, thus making the method even more sensitive. An improved method of determining the radioactivity has been devised by Feschotte and Kubaschewski [250]: the molybdenum tube and target are burnt, together with the radioactive deposits, in oxygen, the molybdenum trioxide formed compacted into pellets and their radioactivity compared with that of a pellet obtained in the same way from an effusion experiment with pure chromium.

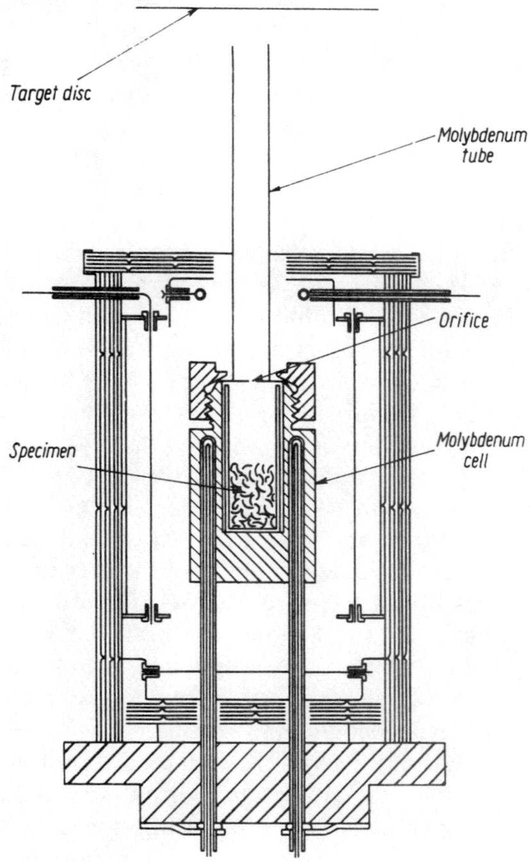

FIG. 73. Effusion cell and furnace for the determination of the vapour pressures of metals between 1200° and 1400°C.

Two main ingredients for a successful measurement of the vapour pressure of a simple substance where the molecular weight of the effusing species is known, are the use of chemically inert materials for the crucible and lid of the assembly and the reduction to a minimum of temperature gradients in the cell. The effects of interaction between the sample and the container material may change the apparent vapour pressure either to a higher or a lower apparent value depending on the nature of the interaction.

Ward and co-workers [251, 252] have published a valuable study of Knudsen effusion at high temperatures (1000–2000°K) which demonstrates the effects of container materials in reducing the results by chemical interaction. The particles effusing through the orifice of a Knudsen cell should follow Lambert's cosine law which in two dimensions is given by

$$I_\theta = I_0 \cos\theta \tag{103}$$

where I_θ is the intensity in a direction at an angle θ to the axis of the orifice where the intensity is I_0. The actual distributions of particles which were directly measured by Ward *et al.* rarely conformed to this simple law, the departures usually being greater the more the sample reacted with the container. Gold and plutonium were used as metal samples, representing extreme cases of chemical reactivity and graphite and tantalum, and Y_2O_3, MgO and ThO_2 were among the container materials. The two metals are both liquid over the major part of the temperature range which was employed, and have approximately the same vapour pressure.

Gold placed in tantalum cells on ThO_2, Y_2O_3 and graphite cups yielded distribution curves relatively close to the ideal, but with some "peaking" of the distribution immediately along the line joining the orifice and a normal to the evaporating gold surface. This latter effect was found to change in magnitude depending on the curvature of the metal sample which changed as the surface tension altered with sample temperature. Some loss of effusate could be detected with the tantalum cell, probably due to alloy formation with the gold atoms, but an all-graphite assembly yielded the results most close to theoretical.

In the case of plutonium very variable results were obtained. With a Y_2O_3 cup placed inside a tantalum cell, the distribution approached the ideal as a number of successive experiments were carried out, but this was never achieved. With a ThO_2 cup, and more so with MgO, the plutonium vapour was severely depleted by interaction with the walls, and the great majority of the atoms leaving the cell came directly from the plutonium sample. In these containers approximately 80% of the anticipated effusate was lost by interaction. These results would therefore lead to an approximately five-fold error in measurements where the total weight loss was measured, or all of the effusate was collected, irrespective of distribution. For a spherical sample, under the conditions of these experiments, the material effusing directly from the sample was to be found within an angle of 10°. It can be concluded that the effusing beam from a Knudsen cell mirrors the relative emission rates from the various surfaces within the cell. Relatively minor losses of material at the cell walls can account for a drastic reduction of effusate from these areas. The importance of finding a chemically inert container, or as close an approximation to this as possible, emerges clearly from this study. It also demonstrates the value of repetition of vapour pressure measurements in a number of container materials where possible. Where container interactions cannot be sufficiently minimised, then a measurement of the total effusate coming *directly* from the sample surface will give the closest approximation to the correct value. The Knudsen experiment has now become virtually a Langmuir experiment since the effects of the vaporisation coefficient now become important.

Storms [253] has carried out a study of the effects of temperature gradients in the cell on the apparent vapour pressures of substances. The probability of there being significant temperature gradients in a Knudsen apparatus will vary very much with the average temperature of the study, and with the method of heating. Under relatively ideal circumstances, at moderate temperatures where resistance-heated furnaces having substantial even zones of temperature can be made, as in the case cited earlier, the effects will be minimal. If a cooled target is brought close to the sample in order to collect a defined beam of effusate, then the inevitable cooling of the Knudsen cell lid by radiative heat loss can cause condensation around the orifice. If vaporisation from these areas contributes significantly to the effusate which should be collected, the effects could be serious. If the effusate is collected over a narrow target angle so that direct effusion from a sample with unit vaporisation coefficient is the objective, this effect could be negligibly small except if the orifice becomes increasingly blocked during an experiment. At much higher temperatures, above 2000°K, an electron bombardment furnace is frequently used; the cell is heated directly by energy transfer from accelerated electrons emitted by surrounding filaments. The assembly is contained in radiation shields to reduce radiative heat loss from the cell. Storms made use of such an arrangement with three separate electron sources allowing for independent irradiation of the top, centre and bottom of a Knudsen cell. The temperature profile of the cell could thus readily be modified and gradients of 200°C could be imposed. The temperature was normally measured either directly through the orifice and on the surface of the evaporating sample or in a "black-body" hole in the bottom of the cell. Using gold in a specially designed graphite crucible (Fig. 74) as the main object of study, it was shown that the error introduced into a measurement of the vapour pressure of gold was equivalent to an error of 5°C when the orifice above the sample was 177°C hotter than the bottom of the cell, but the effusion sample surface was observed for temperature measurement. The error was equivalent to 100°C when the black-body hole in the base of the cell was used for temperature measurement. It was suggested that since the vaporising surface controls the effusion rate, the effects of a significant temperature gradient causing errors in an effusion study should be looked for routinely. This could be done by comparing the apparent vapour pressure when the sample is placed on the bottom of the cell with that obtained when a disc is placed under the sample so that it now occupies a position nearer the top of the cell. Only when these two results agree is the absence of a significant temperature gradient established.

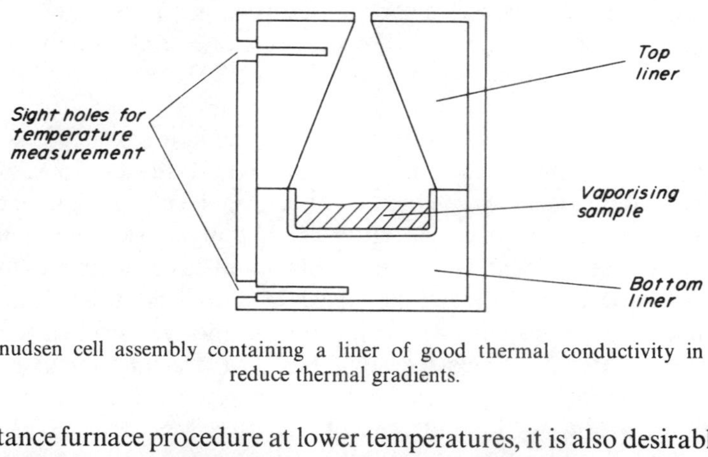

FIG. 74. Knudsen cell assembly containing a liner of good thermal conductivity in order to reduce thermal gradients.

In the resistance furnace procedure at lower temperatures, it is also desirable to mount thermocouples at the top and bottom of the cell to check for the absence of a temperature gradient.

1. *The Knudsen Cell–Mass Spectrometer Combination*

It will have become apparent a number of times during this discussion of the measurement of the vapour pressure that the complexity of the vapour phase must be understood in any situation if a reasonably complete description is to be given of a vaporisation process. The traditional methods can be made to yield an answer in very many simple circumstances and still have a significant role to play, even when this is only to provide, for example, the total weight loss from a Knudsen measurement of a system which is not completely analysed for vapour complexity.

The mass spectrometer is now playing an increasingly dominant role in supplying information about the nature of this complexity because of the capability of separating, for mass determination, the components of complex vapour mixtures.

The separation of the vapour species can only be carried out in the mass spectrometer after a process of ionisation, and this is carried out by bombarding a molecular beam, effusing freely from a Knudsen cell orifice, with a beam of monoenergetic electrons. Since the number of ions produced per unit time is proportional to the flux of molecules through this ion source, and the flux from a Knudsen cell follows the inverse square law of intensity, the ion source must be placed as close as possible to the Knudsen cell. This is a very important aspect of the design of Knudsen cell–ion source systems and in most cases the entry to the ion source is placed directly above the Knudsen cell. This arrangement necessarily limits the length of the furnace, and hence the ease with which an even zone of temperature can be achieved. An alternative system in which the beam effuses from an orifice in the side of the Knudsen crucible, but which permits the use of a longer furnace, is shown in Fig. 75 compared with a "conventional" electron bombardment arrangement.

Having optimised the beam intensity into the ion source and produced the ions by electron bombardment, two methods of separating the ions into the various groups of charged species are mainly used. In the more traditional of these, the ions are extracted continuously from the ion source by means of a fixed accelerating potential and are separated in a magnetic field into groups of constant charge to mass ratio. The alternative procedure is to accelerate the ions with a pulsed accelerating potential and allowing them to separate into groups of constant "time of flight" (TOF) down a straight vacuum chamber.

The immediately obvious advantage of magnetic resolution is that ions are extracted continuously instead of at repeated intervals as in the TOF spectrometer. The sensitivity of magnetic resolution machines can therefore normally be made higher than that of the TOF machine. An advantage of the latter is that high-speed electronic recording techniques may be used to display the arrival of a number of ion groups at the detector end of the "flight tube" apparently simultaneously. Although it is possible to scan through the masses in a magnetic resolution machine, this has not yet reached the speed of the TOF device. The latter has a decided advantage in the simultaneous recording of a number of ionic species. These are the major differences between the two alternatives which are mainly used in high-temperature studies.

These systems for mass analysis have a common requirement of a well-defined source of molecular beam and subsequently of ions, to function in a comprehensible manner. It is desirable to mount a movable slit or shutter between the Knudsen cell and the ion source so that ions which arise unambiguously from molecules leaving the Knudsen cell can be separated from those from background sources. If the shutter can be traversed across the molecular beam, then additional evidence on the beam angular distribution similar to that obtained by Ward *et al.* [251, 252] can be used to reinforce this information. An alternative to a beam-scanning shutter is the capability to rotate the Knudsen cell so that the molecular beam can be removed from alignment with the entry port to the ion source.

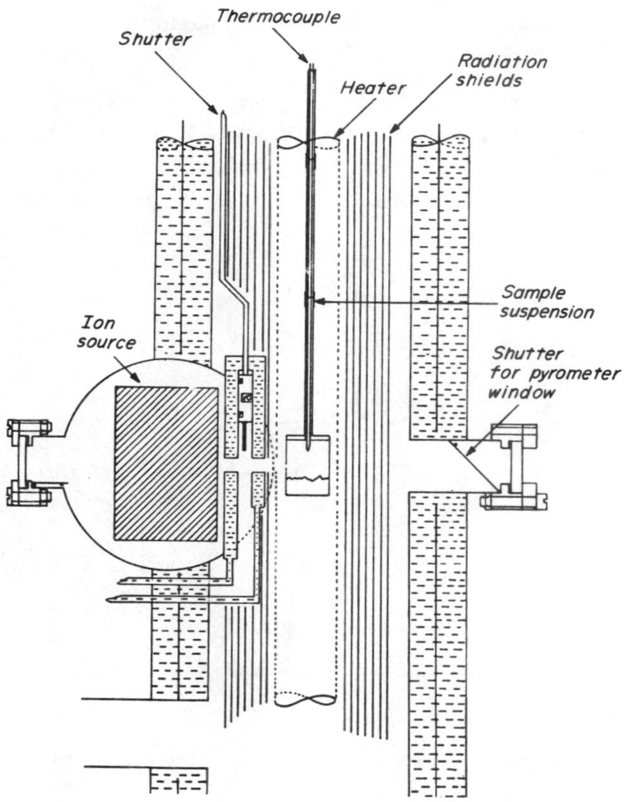

FIG. 75a. Resistance-heated Knudsen cell-mass spectrometer source using a vertical furnace but a horizontal molecular beam from the Knudsen cell.

The ion current which is produced from a given molecular beam intensity into the ion source is markedly dependent on the energy of the bombarding electron beam. A typical ionisation efficiency curve for a simple monatomic species is shown in Fig. 76a. Here virtually no ions are produced until the electron energy passes a critical value, the "appearance" potential of the ion species, and then the ion current is a linear function of the electron energy over a wide range of electron energies until a maximum is reached. The slight curvature in the ionisation efficiency curve just above the appearance potential is the result of lack of precisely uniform conditions for ionisation, such as a small range of energies in the bombarding electron beam. The appearance potential is normally only a few electron volts, and a spectrometer is commonly operated at a little over 10 V for the bombarding electron energy.

In the more complex situations for which there is need of mass spectrometry, a number of molecular and atomic species are to be ionized. In this case non-linear ionisation efficiency curves can be obtained for a given mass number resulting from the fragmentation of complex molecules to smaller masses by electron bombardment (Fig. 76b). This can complicate the interpretation of the resolved ion currents, but variations in the thermodynamic activities of the components in the Knudsen cell can be used to elucidate the true state of affairs. Thus the dimeric form of a metal vapour M_2 could contribute to the monomer ion beam M^+ as a result of the fragmentation process

$$M_2 + e^- \rightarrow M^+ + M + 2e^-$$

increasingly so as the electron beam energy is increased. However, the intensity of the correct M^+ ion beam will vary directly as the activity of M, whereas that of the M_2 species will vary as the square of the activity of M, and hence the two effects can be separated. There are obviously numerous elaborations upon the effects of fragmentation and their recognition.

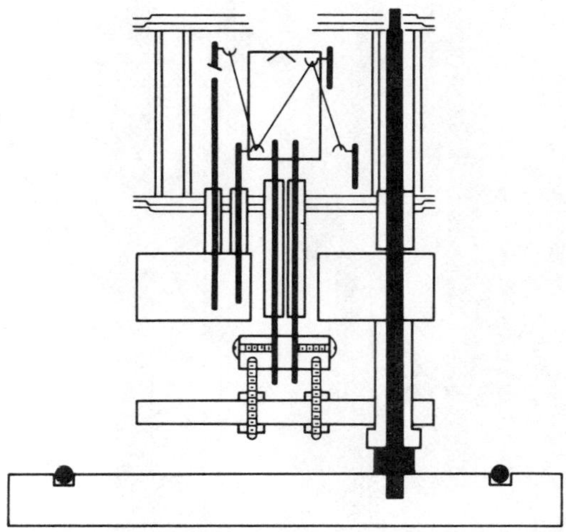

FIG. 75b. Electron bombardment furnace with a coaxial molecular beam from a Knudsen cell.

The ion beam is normally recorded by means of photomultiplier or other electron multiplying devices, and the response of these is a function of the mass of the ion which is detected. This instrumental factor is another contribution to the overall relationship between the intensity of a species in the mass spectrometer and the recorded intensity at the detector end of the spectrometer. The general relationship is

$$p = \frac{KI^+T}{\sigma D \Delta E} \tag{104}$$

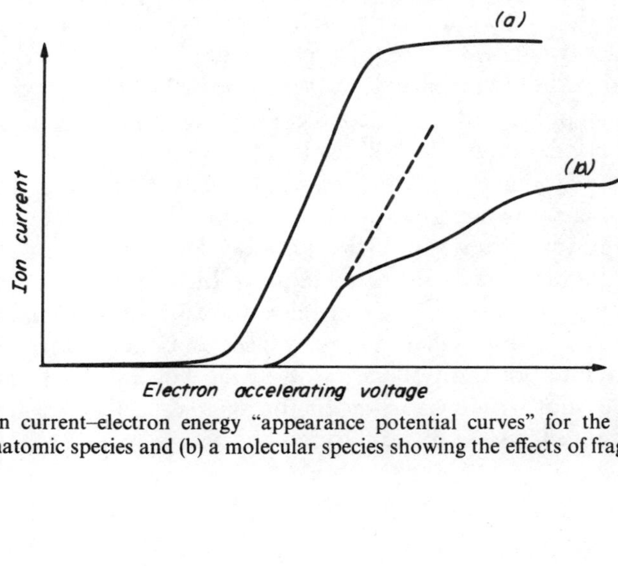

FIG. 76. (a) Ion current–electron energy "appearance potential curves" for the ionisation of a simple monatomic species and (b) a molecular species showing the effects of fragmentation.

where K is a geometric constant, I^+ the recorded ion current, T the absolute temperature of the Knudsen source, σ the ionisation cross-section, D the detector efficiency, and ΔE the electron beam energy minus the appearance potential.

Dimers of metallic elements have been detected in the vapour phase, and these may be used to advantage to reduce the effects of the instrumental factor on mass spectrometric studies. For example, if it is required to obtain the thermodynamics of the gaseous intermetallic species MN, one method is clearly to measure the heat change for the reaction

$$(\mathrm{M}) + (\mathrm{N}) \rightarrow (\mathrm{MN})$$

using the ion currents as a measure of the vapour pressures, and, applying the second law,

$$\frac{\partial}{\partial 1/T} \log \frac{I^+_{\mathrm{MN}} T}{I^+_{\mathrm{M}} I^+_{\mathrm{N}}} = \frac{-\Delta H^0}{R} \tag{105}$$

and to estimate the entropy using the third law.

Alternatively, if the dimeric species M_2 or N_2 can be observed, the constants for the reactions

$$\mathrm{M_2 + N \rightarrow MN + M} \tag{a}$$

and

$$\mathrm{N_2 + M \rightarrow MN + N} \tag{b}$$

can be approximated very closely by the ion current ratios

$$K_a = \frac{I^+_{\mathrm{MN}} I^+_{\mathrm{M}}}{I^+_{\mathrm{M_2}} I^+_{\mathrm{M}}}; \quad K_b = \frac{I^+_{\mathrm{MN}} I^+_{\mathrm{M}}}{I^+_{\mathrm{N_2}} I^+_{\mathrm{M}}} \tag{106}$$

This procedure has the advantage that the instrumental factors relating to the spectrometer, and the geometrical factors relating to the position of the sample and of the orifice of the Knudsen cell in the furnace, do not appear in the equilibrium constant because of cancellation. It follows that a value of the equilibrium constants K_a and K_b can be obtained satisfactorily from measurements of each of the four ion currents in any one experiment for a given temperature, and even though the instrumental and geometrical factors may alter before the experiment at the next temperature, a consistent value of the equilibrium constant will again be obtained. This is not so for the case where only the monomeric species M^+ and N^+ are used.

One disadvantage of the procedure involving the dimers is that the pressure of a dimeric species is usually at least an order of magnitude smaller than that of the corresponding monomer, and hence when the signal for the M_2^+ ion reaches an easily detectable level, the pressure of the monatomic gaseous species is frequently nearing the upper limit for the conditions of Knudsen cell operation.

Clearly ratios of pressures for a given molecule can readily be obtained providing the geometric factor, which reflects the position of the Knudsen orifice with respect to the ion source, can be accurately reproduced.

As this is not generally the case, then an internal calibrating substance of known vapour pressure placed in the Knudsen cell with the unknown material is useful if the equilibrium remains undisturbed within the cell. Silver is very commonly used as an internal calibrating agent because of its well-established vapour pressure and relatively low chemical reactivity. If interaction is unavoidable, a double-cell system can be employed, one cell of which contains the calibrating substance (Fig. 77).

When this has been considered, the product *D* is the remaining source of error in the mass spectrometric determination of absolute vapour pressures. This product can rarely be determined to better than a factor of two except in the trivial cases where the vapour pressure is already known in which case the product *D* is being measured. However, absolute pressure to within $\pm 100\%$ can be obtained in most circumstances when an unambiguous ionisation efficiency curve is at hand.

Finally, the limit of intensity is often set not by the detectability of the ion current but by the resolution of the required ion species from others of neighbouring mass. These latter can arise from a number of sources depending on the system, but one common source in high-temperature mass spectrometry is hydrocarbon fragments from vacuum-pump oil and other extraneous sources. The resolving power of the spectrometer can play an important part in deciding the lower limit of sensitivity, and the magnetic resolution machine normally has a considerably greater resolving power than the TOF machine.

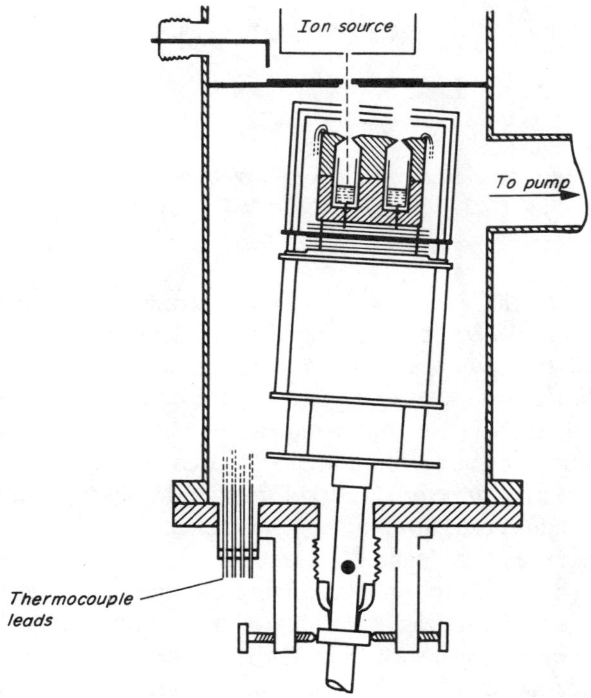

FIG. 77. Double-cell assembly for a Knudsen cell-mass spectrometer measurement with a calibrating standard material in one cell.

Apart from its use in the elucidation of complex evaporation mechanisms and the approximate determination of absolute vapour pressures, one of the increasingly developing applications of the mass spectrometer in metallurgical research is in the study of solution thermodynamics. There are two main routes along which this development is taking place, the common feature being the determination of activities from ion current ratios. The first technique makes use of the partial pressure ratios of gaseous polymers [254]. For the metal M which yields the monatomic and diatomic species and the ions M^+ and M_2^+

$$a_M = \frac{p_M}{p_M^\circ} = \frac{I_M}{I_{M^+}^\circ}; \quad a_M^2 = \frac{P_{M_2}}{P_{M_2}^\circ} = \frac{I_{M_2^+}}{I_{M_2^+}^\circ} \tag{107}$$

Hence

$$a_M = \frac{I_{M_2^+}}{I_{M_2^+}^\circ} \frac{I_M^\circ}{I_{M^+}} \tag{108}$$

The ratio of the ion current I_M/I_{M_2} can normally be obtained independently of all instrumental and geometrical factors and hence if the ratio is measured over the pure element and then over the alloy, the activity is readily obtained. The complexity of the vapour phase is thus employed in the solution of thermodynamic problems instead of making them experimentally intractable.

The other technique which uses ion current ratios for the determination of activities was developed by Belton and Fruehan [256] and, independently, by Neckel and Wagner [257]. The Gibbs–Duhem equation can be manipulated to produce a form which is of particular value in mass spectrometry thus; in an AB mixture

$$\log \gamma_B = -\int_1^{N_B} N_A \, d\left[\log \frac{a_A}{a_B} - \log \frac{N_A}{N_B}\right]$$

which, in the useful form for mass spectrometry, is

$$\log \gamma_B = -\int_1^{N_B} N_A \, d\left[\log \frac{I_{A^+}}{I_{B^+}} - \log \frac{N_A}{N_B}\right] \tag{109}$$

This technique is most valuable for systems where the partial pressures of the two components over an alloy are approximately equal. The ion current ratio will then have a value around unity. Systems showing this kind of behaviour are those which are most difficult to resolve by traditional methods, and hence the mass spectrometer has its greatest potential for quantitative application in those very systems where its use is almost mandatory.

Very few systems have been studied by more than one group at the same temperature using what is ostensibly the same mass spectrometric method, but the liquid copper–germanium alloys have been measured by at least three groups [258, 259, 260] and their results for the ion current ratios at one temperature and the resulting activities are shown in Figs. 78a and b respectively. These graphs should give a representative picture of the state-of-the-art in alloy thermodynamics at the present time.

The copper–gallium system has been studied by this technique [259] and also by Laffitte and co-workers using a double cell assembly [259a]. Figure 79 shows the close agreement between activity measurements which can be achieved by such widely differing approaches. It should be clear from this discussion that mass spectrometric studies of activities in metallurgical solutions will probably become the method of preference in the future despite the many pitfalls to precise measurement which exist at present. An understanding of these in the particular context of high temperature studies is rapidly growing with experience, and a review recently published by Stafford [261] serves as a good guide to the present status.

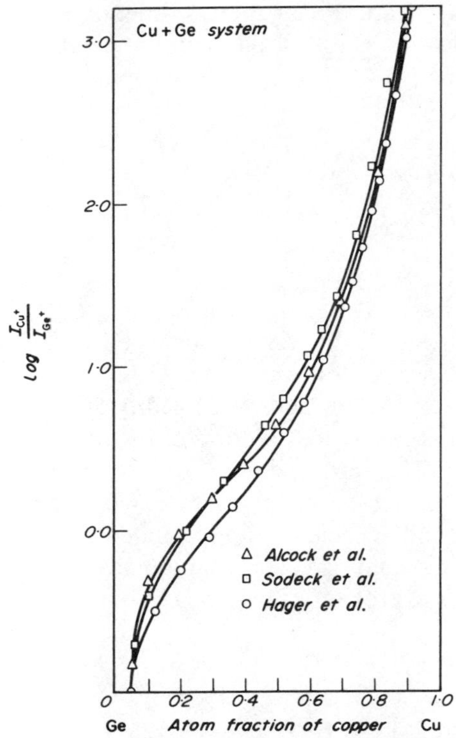

FIG. 78a. Ion current ratios at 1400°C for the liquid system copper–germanium from three separate mass spectrometric studies [258–260].

Whereas the lower pressure limit of application of the Knudsen method is usually about 10^{-4} mmHg, the method of Langmuir enables measurements to be made of pressures considerably lower than this. The Knudsen equation, multiplied by the vaporisation coefficient, can be used with samples with very much larger projected areas than the normal Knudsen orifice when this method is applied, but problems might be anticipated with the measurement of the temperature of the evaporating surface, which must be known. The Langmuir method is more often adopted to enlarge the rate of weight loss to a readily measurable quantity than to avoid difficulties associated with the availability of a suitable container for the Knudsen method. It therefore is most usefully applicable to substances having low vapour pressures at normal temperatures, and hence possessing high energies of sublimation.

Marshall *et al.* [262] obtained some of the first dependable results on metals using the apparatus shown in Fig. 80. They used metal samples in the form of an annular ring 22 mm o.d., 11 mm i.d., thickness 2·5–6 mm. These were heated by induction in an evacuated silica vessel, and the temperature was measured with an optical pyrometer. In measuring the vapour pressure of liquid copper, the metal was contained in a small annular trough of molybdenum.

A number of other studies have been made on pure metals by this technique or variants in which continuous weighing of the sample could be performed [263, 264] and the results have been in agreement with data from Knudsen measurements. This shows that the vaporisation coefficients of metals are unity to within experimental error. With oxide

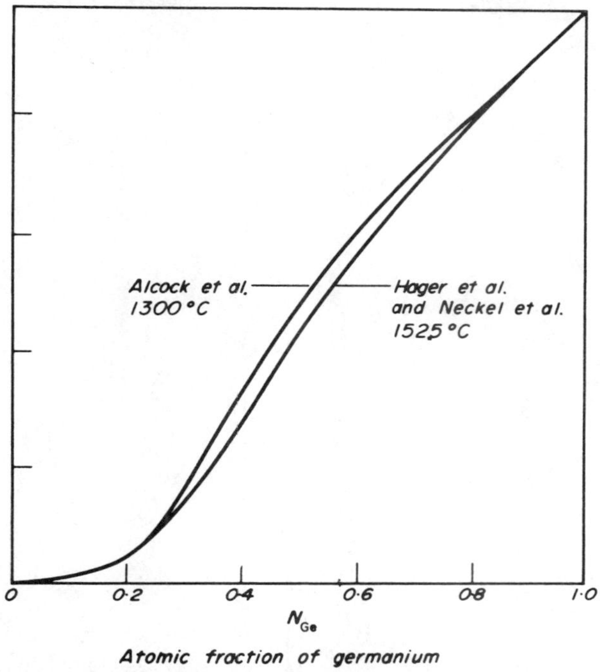

FIG. 78b. Activities for copper–germanium system at 1525°C obtained by Hager *et al.* [260] and Neckel *et al.* [258] and at 1300°C according to Alcock *et al.* [259].

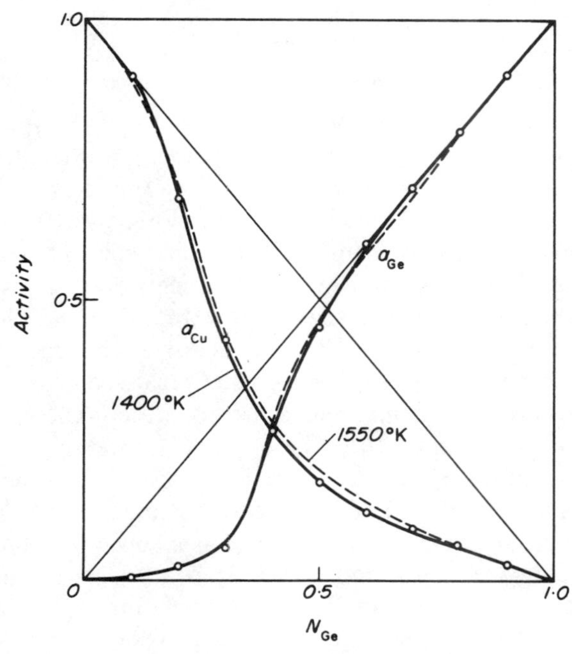

FIG. 79. Activities in the liquid copper–gallium system according to Carbonel *et al.* [*Colloques Internationales du C.N.R.S.* No. 201 (1972), 311] and Alcock *et al.* [259].

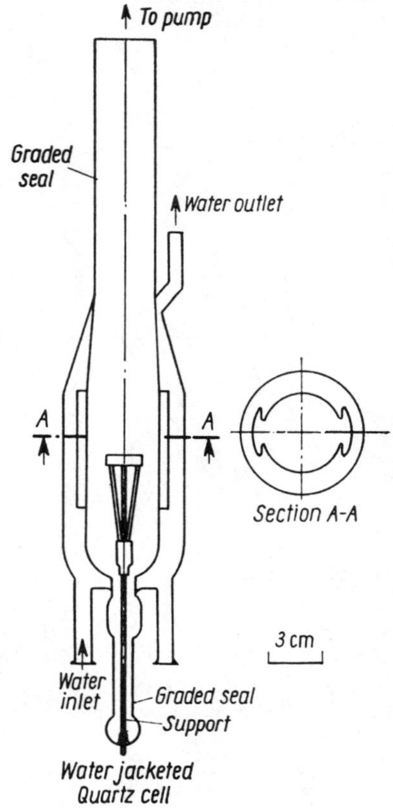

FIG. 80. Langmuir type apparatus for determining reaction pressures.

systems, in most cases a value of about 0·3 seems more appropriate for the solids. Burns [265] made a study of the vaporisation of Al_2O_3, Ga_2O_3 and In_2O_3 in both Langmuir and Knudsen conditions, using a mass spectrometer to analyse the vapours. He found that liquid Al_2O_3 and Ga_2O_3 showed unit vaporisation coefficients whilst the solids were all around 0·3. His analysis of the individual components of Al_2O_3 vaporization, Al, O, O_2, Al_2O, and AlO being the vapour species, showed that all of these had about the same value for the vaporisation coefficients. Because the oxides are relatively poor thermal conductors it is tempting to follow Littlewood and Rideal's conclusions, made with low-temperature systems [266], that since metals are good thermal conductors and have a vaporisation coefficient of unity, then the apparent value of 0·3 for oxides which dissociate on evaporation, and hence have large heats of vaporisation, is due to evaporative cooling of the surface. In order not to rely on temperature measurement on the freely evaporating surface, most workers employ a "black-body hole" in the sample for this measurement. A 10–30° temperature drop accounts for a reduction of the vaporisation rate to a third when the heat of sublimation is 100–200 kcal, and this probably exists near the surface of an evaporating oxide. The difference in thermal conductivities of these solids and metals at high temperatures is probably of the order of a hundredfold.

Wolff and Alcock [267] measured the free evaporation rates of alumina and thoria at temperatures around 2000°C. Their values for the vaporisation coefficient of Al_2O_3 were in agreement with that found by Burns, but the rate of evaporation of ThO_2 was quite in accordance with the Knudsen studies of Ackermann indicating α equal to one. The heats of vaporisation of Al_2O_3 to the predominant atomic species and of ThO_2 to ThO_2 gas, which is the principal species at 2000°C, are 725,000 and 158,000 cal/mole respectively. It is probably this large difference which accounts for the difference of approximately three in the vaporisation coefficients of these two solids. The heat of vaporisation of ThO_2 is not very much larger than those of some of the metals of established unit vaporisation coefficients, i.e. *ca.* 100,000 cal/g-atom.

2. *Vapour Transpiration*

The general limitation of the Knudsen method at the upper range of pressures results from the fact that the orifice diameter, which must be small compared with the mean free path in the vapour, becomes too small to manufacture. This practical limit corresponds in most systems to a vapour pressure of about 10^{-3} atm. The transportation method also has practical limitations at about this pressure, and there therefore exists a gap in the range 10^{-3}–10^{-2} atm between these two dynamic techniques and the direct pressure measurement techniques which were described earlier for use in the pressure range 10^{-2}–1 atm. A transpiration technique which was introduced by Gross *et al.* [268] could probably be very usefully applied in this intermediate range. The apparatus, which is shown in Fig. 81, consists of a closed end reaction tube in which the sample is held

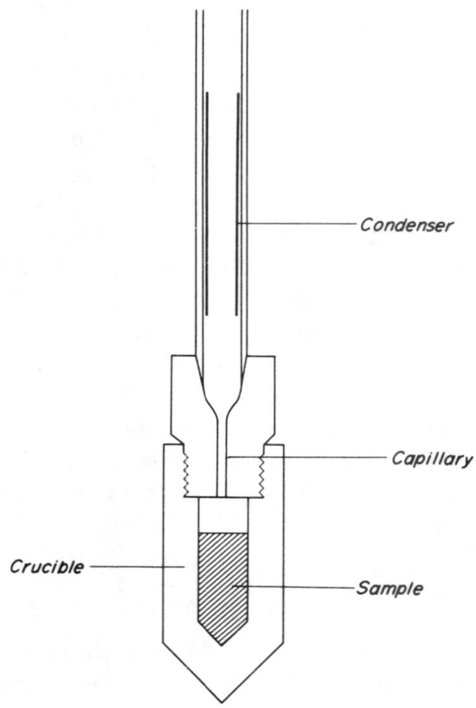

FIG. 81. Vapour transpiration apparatus of Gross *et al.* [268].

in the even zone of a furnace, a length of capillary tube which closes the open end of the reaction tube, and a condenser tube which receives the products escaping through the capillary from the reaction tube. The system is evacuated before being heated, and then the even zone of the reaction tube is brought up to the desired temperature. After a fixed time the tube is cooled by removal of the furnace, and the contents of the condenser tube are analysed for the material which has evaporated from the sample. A number of experiments are made of varying duration, so that the appropriate allowance can be made for the amount of material which is lost by transpiration during the heating-up period. Since the capillary tube has a pressure drop from a few millimetres on one side where normal Poiseuille flow would occur, to a vacuum on the other side, where effusion would be the mode of gas motion, it is impossible to relate by means of a simple equation the mass of condensate which is collected per unit time to the vapour pressure of the substance within the reaction chamber. However, a number of standard substances are now available which have well-established vapour pressures over a wide pressure range. For example, the vapour pressure of lead has been established by the Knudsen and transportation techniques at pressures below 10^{-3} atm [269] and by static and boiling-point methods from 10^{-2} to 1 atm [270, 271]. Gross *et al.* were therefore able to use this element as a calibrating substance in the study of the equilibrium between aluminium and AlF_3 to form AlF. The calibration method can conveniently be applied with the standard substance incorporated with the system of unknown pressure provided no significant chemical interaction can take place.

Vapour transpiration has also been used in conjunction with levitation melting to measure vapour pressures of alloy systems at very high temperatures where the use of a container would be almost impossible because of severe chemical interaction. Mills and Kinoshita [272] have reported activity measurements on liquid titanium–vanadium alloys in the temperature range 1800–2000°C. The system, which showed slight positive departures from ideality, was the second to be studied by this technique, and no comparable results are available to assess the accuracy. The technique was established, however, in an earlier study of the iron–nickel liquid system in the temperature range 1900–2300°C [273]. This system had already received considerable study by a variety of methods; Zellars *et al.* used the gas transportation method [274], Speiser *et al.* used the Knudsen method [275] and Belton and Fruehan [256] used the ion current ratio version of mass spectrometry. The system can therefore be said to be reasonably well documented. In the levitation–transpiration technique a sample of the alloy of known composition is levitated in helium gas contained in a water or air-cooled quartz tube. The temperature of the alloy bead can be controlled by the opposing heating effect of the electromagnetic field, and the cooling effect of the helium, which is a function of the gas flow rate past the sample. The principle of the method is that the flow of metal atoms from the hot sample to the cold walls, where they condense, occurs across a gas–metal boundary layer which is common to both species. Therefore, in the steady state

$$\frac{J_{Fe}}{J_{Ni}} = \frac{D_{Fe}}{D_{Ni}} \frac{P_{Fe}}{P_{Ni}}$$

where J is the flux of atoms, D the diffusion coefficient of the metal atoms in helium and P the equilibrium vapour pressure of the component at the sample temperature.

The flux ratio can be obtained by analysing the condensate on the quartz walls for iron and nickel. Although a number of theoretical treatments are in existence, it is probable that no serious error would be involved in assuming that the diffusivity ratio is equal to one. Then the activity ratio is obtained from the equation

$$\frac{N_{Fe}}{N_{Ni}} = \frac{a_{Fe}\, p^{\circ}_{Fe}}{a_{Ni}\, p^{\circ}_{Ni}}$$

where N_{Fe}/N_{Ni} is the ratio of the atom fractions of iron to nickel in the condensate. Speiser *et al.* produced the form of the Gibbs–Duhem equation which should be used in conjunction with this technique to obtain the single activities in the liquid phase, viz.

$$\ln a_{Fe} = \int_{N_{Fe}=1}^{N_{Fe}} \frac{N^1_{Ni}}{N_{Ni}} \mathrm{d} \ln N_{Fe} = \int_{N_{Fe}=1}^{N_{Fe}} \frac{N^1_{Ni}}{N_{Ni} N_{Fe}} \mathrm{d}\, N_{Fe}$$

where N^1_{Ni} refers to the atom fraction of nickel in the liquid phase and the other atom fractions are for the condensate.

The results which were obtained in this study were quite consistent with the extrapolation of the earlier data to the higher temperature region. The high temperature which was used was the result of the necessity to balance the heating and cooling effects in the system. It is possible that improved design of the levitation system would make it possible to use this system in the more conventional temperature range if so required.

3. *Torsion Effusion*

The use of the Knudsen technique requires a precise knowledge of the vapour species under equilibrium conditions, but the pressure may be obtained directly from a measurement of the recoil momentum which is imparted to a Knudsen cell as a result of the effusion process. This measurement does not require a knowledge of the vapour species, and is valuable as a means for obtaining the total vapour pressure of a system in which a number of different species are present in the equilibrium vapour to a significant extent.

The method was first applied by Volmer [276] who used a two-orifice cell, the orifices pointing in opposite directions on either side of a suspension fibre of known torsion constant. The size of orifice and the vapour-pressure range over which these may be used are subject to the same criteria as for Knudsen cells, and the vapour pressure may be calculated from the equation

$$p = 2D\beta(q_1 A_1 + q_2 A_2) \tag{110}$$

where q_1 and q_2 are the distances of the two orifices of area A_1 and A_2 from the axis of suspension. The torsion suspension fibre from which the double-orifice cell is suspended is turned through an angle β and has a torsion constant D. Useful materials for the suspension are quartz, tungsten and phosphor bronze, the last-named being preferred for strength and reproducibility.

Pratt and Aldred [277] constructed a torsion cell of graphite for the study of solid and liquid alloy systems. The suspension wire was of tungsten, and reliable measurements were obtained in the pressure range 10^{-8}–10^{-5} atm at temperatures up to 1000°C. Hildenbrand and Hall [278] calibrated their apparatus using gold as a vapour pressure standard, and this procedure is to be recommended. Searcy and Freeman [279] built an inductively heated apparatus having a graphite torsion cell which could be used up to

1900°C, but corrections for the torque effect of the radio frequency field on the cell had to be made. Peleg and Alcock [280] used an apparatus having a non-inductive doubly wound tungsten resistance heater for the study of the vaporisation of ceramic oxides up to 2500°C.

The torque which is produced by vaporisation can be measured directly, or in a null position by applying a measured current to solenoids mounted around a permanent magnet which is incorporated in the cold part of the suspension system. In their apparatus, Alcock and Peleg made use of a torsion head which could be smoothly rotated by means of a highly geared electric motor which enabled the whole suspension system to be rotated and returned to the null point. The angle of rotation was obtained by counting the number of revolutions required on the motor shaft with a revolution counter. The entire suspension system could be aligned in the vertical direction through a Wilson seal and horizontally by translation of the torsion head so as to place the orifice of the Knudsen cell in the centre of the furnace. The whole suspension was electromagnetically damped in the cold part of the apparatus (Fig. 82).

As long as the thermal requirements can be met to keep the torsion cell at a constant temperature, it is advantageous to keep the heating element and radiation shields as

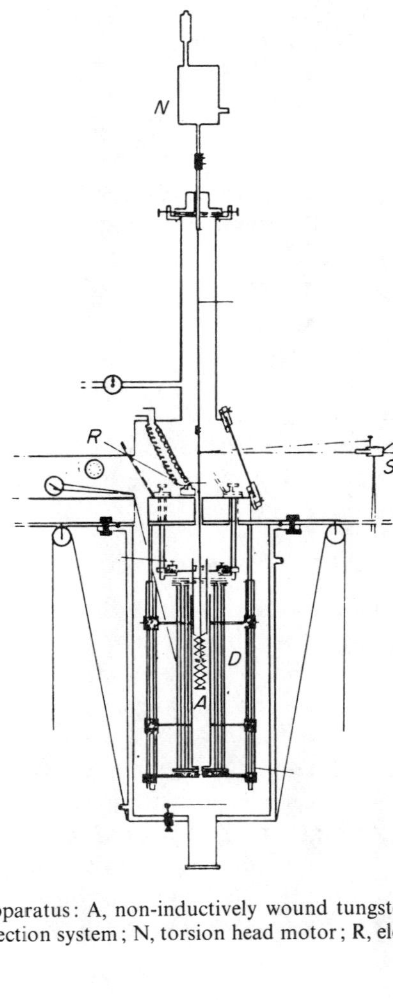

FIG. 82. Torque-effusion apparatus: A, non-inductively wound tungsten heater; D, radiation shields; S, rotation detection system; N, torsion head motor; R, electromagnetic damping.

remote as possible from the Knudsen orifices. This is because atoms and molecules can rebound from the heater and radiation shields to the torsion cell, and depending on the precise geometric disposition of these furnace components with respect to the cell, an apparent torsion can be produced due to asymmetry in this rebound effect. The effect is more noticeable as the vapour pressure increases, and thus assumes its greatest significance at pressures where the technique might be considered in its most satisfactory range.

3. Electromotive Forces

A. GENERAL

The use of e.m.f. measurements to derive thermochemical data has been discussed on pp. 8, 18, 27–30 and 33. The chemical reaction to be investigated must be capable of being harnessed in a galvanic cell in such a way that its energy produces electromotive force. The principal problem which faces the experimentalist who wishes to apply the galvanic cell technique in order to measure thermochemical quantities is to find a suitable electrolyte for the cell which is envisaged. The second most important aspect of these measurements is the identification of a single electrode process which occurs reversibly at each electrode. Consider a cell in which the oxygen dissociation pressure of a metal/metal oxide system is to be measured. This could consist of one electrode made of a platinum sheet attached to a pelletised mixture of the metal and its oxide in powder form, and the other electrode could be oxygen gas at 1 atm pressure in contact with platinum. The cell would be represented diagrammatically thus:

$$\mathrm{Pt}\langle \mathrm{M, MO}\rangle|\text{electrolyte}|(\mathrm{O_2}, 1\ \mathrm{atm})\mathrm{Pt}$$

The electrode reactions will then be at the right-hand electrode

$$\tfrac{1}{2}\mathrm{O_2} + 2\mathrm{e^-} \rightarrow \mathrm{O^{2-}}$$

and at the left-hand electrode

$$\mathrm{M} + \mathrm{O^{2-}} \rightarrow \mathrm{MO} + 2\mathrm{e^-}$$

giving the overall cell reaction

$$\tfrac{1}{2}\mathrm{O_2} + \langle \mathrm{M}\rangle \rightarrow \langle \mathrm{MO}\rangle$$

The cell

$$\langle \mathrm{Ag}\rangle|\langle \mathrm{AgBr}\rangle|(\mathrm{Br_2}), \mathrm{C}$$

has the electrode reactions

$$\mathrm{Ag} \rightarrow \mathrm{Ag^+} + \mathrm{e^-}$$

and

$$\tfrac{1}{2}\mathrm{Br_2} + \mathrm{e^-} \rightarrow \mathrm{Br^-}$$

with the overall cell reaction being

$$\mathrm{Ag} + \tfrac{1}{2}\mathrm{Br_2} \rightarrow \mathrm{AgBr}$$

In these two typical examples there is one electrode at which electrons would be consumed—the cathode—and one electrode at which electrons are produced—the anode—when the cell reaction proceeds. If the electrolyte system conducts electricity by the migration of ions only, then the electrons must move from cathode to anode via an external electronic conductor. It is along this conductor that the electric potential which is generated by the cell reaction can be measured. When the resistance of this conductor is very high, the e.m.f. E is obtained, the so-called "open-circuit" potential. If the electrode components are all present at unit thermodynamic activity, the e.m.f. is the standard e.m.f. which is related to the standard free energy change for the cell reaction by the equation

$$-\Delta G^\circ = z\boldsymbol{F}E^\circ \tag{31}$$

Here, z is the number of electrons which are found in the separate electrode reactions which, added together, make the cell reaction. Thus for those cells shown above, z has the values 2 and 1 respectively.

In most practical high-temperature systems the so-called electrolyte usually has a small component of non-electrolytic conduction. The cell is never then in the open-circuit state because electrons can migrate through the electrolyte and permit the cell reaction to proceed at a speed which is not determined by the resistance of the external electronic conductor. A real electrochemical system therefore has a potential which is given by the equation

$$E = \frac{\boldsymbol{R}T}{z\boldsymbol{F}}\int_{a'_i}^{a''_i} t_i \mathrm{d}\ln a_i \tag{110a}$$

where t_i is the transport number of the ionic species which is conducted through the electrolyte having the thermodynamic activity a'_i at one electrode and a''_i at the other. When the "electrolyte" conducts significantly by electron migration, then the cell reaction can proceed spontaneously and one electrode is depleted of atoms whilst the other receives atoms.

It follows that the electrodes can only be maintained at a constant chemical potential of all components providing that equilibrium can be established more rapidly than the rate of arrival or loss of material at the electrode–electrolyte interface. Consider a cell which has two metal–metal oxide electrodes with a "leaky" electrolyte where t_{ion} is less than one. The electrode with the higher oxygen dissociation pressure can lose oxygen through the electrolyte by the arrival of electrons *through* the electrolyte to form oxygen ions, and hence the oxide is reduced at this electrode. At the other electrode oxygen ions are absorbed and the electrode is oxidised continuously. Providing the electrode reactions can accommodate this "corrosion reaction" as a result of metal–oxygen reactions at the electrodes which restore the original mixture *at the electrode–electrolyte interface*, the cell can be used for thermodynamic studies. Failing this, the e.m.f. will drift continuously and the results are difficult to interpret. At the higher temperature, solid electrolyte systems can be permeable to gases, and interactions between the electrodes and their surrounding gaseous atmospheres can also play an important role in making experimentation difficult.

The porosity of a solid electrolyte depends very much upon the manufacturing method as well as on the temperature of the cell operation, and therefore no general description can be given of the conditions under which this effect could become significant. The effects of atmospheric reactions with electrode systems can be minimised, or eliminated to all practical extents, by providing samples of the electrode to the gas phase for pre-equilibration of the gas approaching the cell electrode and by separating the atmospheres surrounding the electrodes from one another.

The corresponding phenomena in molten salt electrolytes which can lead to drifting e.m.f.s, are reactions which result from the solubility of metals in the molten electrolyte, which confers electron-conducting properties on the electrolyte, or the possibility of cations of more than one valency for a given cation in the melt. Both mechanisms can lead to the transport number of the ions being less than unity and to oxidation/reduction reactions at the electrodes.

In all electrochemical studies for the determination of thermodynamic properties it is therefore important to ascertain the conditions under which the electrolyte shows the closest approximation to the ideal system and to attempt to work in the region of best performance. It is in solid electrolyte systems, especially those in which oxide ions are the rapidly moving species, that the most complete study of the factors contributing to the ionic transport number have been carried out. By comparison, the studies of molten salt systems is more complicated since this involves the possibility of the migration of a number of, usually, cationic species, and the results are specific for a given cation. We shall therefore begin our discussion by consideration of solid oxide electrolytes.

B. SOLID ELECTROLYTES

Kiukkola and Wagner [281] aroused renewed interest in oxygen ion conducting solids with the fluorite structure when they reported results for simple cells of the type

$$\mathrm{Pt}\langle \mathrm{Ni, NiO}\rangle | \mathrm{CaO{-}ZrO_2} | \langle \mathrm{Fe, FeO}\rangle \mathrm{Pt}$$

in which the cell reaction is

$$\langle \mathrm{NiO}\rangle + \langle \mathrm{Fe}\rangle = \langle \mathrm{FeO}\rangle + \langle \mathrm{Ni}\rangle$$

Since these studies a number of other solid oxide electrolytes have been studied in detail, the most important property being the nature and magnitude of the electrical conductivity as a function of both oxygen pressure in the surrounding atmosphere and temperature.

At a given temperature the general features of all systems show a region of semi-conduction by electron migration (n-type $\ominus$) at low oxygen pressures then follows a region of electrolytic conduction ($t_{\mathrm{ion}} \simeq 1$), and finally at high oxygen pressures, another region of semiconduction develops with positive hole (p-type $\oplus$) conduction. All three mechanisms must be considered as operative at a given oxygen pressure, and we can merely discuss the regions of oxygen pressure where one mechanism may predominate with the other mechanisms playing a very minor role. Under any condition, the total conductivity is related to the number n_i, charge $|z_i e|$, and mobility μ_i of each species by the equation

$$\sigma = \sum_i n_i |z_i e| \mu_i = \sum_i \sigma_i$$

where σ_i is the partial conductivity due to species i. It is usually found that the semi-conduction components of the conductivities of oxides follow, approximately, a simpler power law relationship to the oxygen pressure. Thus, after Schmalzried [282].

$$\sigma_{\ominus} = p_{\ominus} p_{O_2}^{-1/n}; \qquad \sigma_{\oplus} = p_{\oplus} p_{O_2}^{-1/m} \tag{110b}$$

where $p_{\ominus}$, $p_{\oplus}$, n and m are constants for a given system. The ionic conductivity is independent of the oxygen pressure, and is related to the diffusion coefficient of oxygen in the solid by the Nernst–Einstein equation in the form

$$\frac{\sigma_{\text{ion}}}{D_{O_{2-}}} = \frac{n_{O^{2-}} 4e}{kT} \tag{110c}$$

It follows that the range of oxygen pressures over which electrolytic behaviour will extend should be higher when the diffusion coefficient of oxygen is high, and when the change of cationic valencies which can introduce *n*-type (reduction) and *p*-type (oxidation) is not readily brought about. Stable cationic valencies in the principal cations is therefore an important contribution to the achievement of satisfactory electrolytes, and the presence of impurities which have variable valencies, e.g. Fe^{2+}, Fe^{3+} should be kept to a minimum. Because of the extreme difficulties associated with the preparation of samples, the control of atmospheres at high temperatures and the measurement of electrical conductivities, the data for the electrolyte systems show some scatter. It is quite well established [283] that the ionic conductivities of zirconia-based electrolytes are about an order of magnitude greater than those of the corresponding thoria-based electrolytes. The ionic transport number remains sensibly unity in the zirconia-based electrolytes up to 1 atm oxygen pressure, but is about 0·6 in the thoria-based electrolytes at 1000°C at 1 atm pressure of oxygen. At the lower limit, the zirconia-based electrolytes begin to show *n*-type semiconduction at p_{O_2} equal to 10^{-20} atm and 1000°C, whereas the thoria-based electrolytes function satisfactorily down to 10^{-30} atm. It is thus quite apparent that zirconia is the superior base material for oxygen potential measurements when the oxygen partial pressure is high in a given electrode, but thoria is to be preferred at the lower oxygen pressures. If a cell is made in which one electrode has a high oxygen partial pressure and the other side has a low oxygen partial pressure, it is possible to work with two electrolytes in series thus:

$$Pt(O_2, 1\,\text{atm})\,|\,ZrO_2\text{–}CaO \,\|\, ThO_2\text{–}CaO\,|\,Cr\text{–}Cr_2O_3\,|\,Pt$$

has been found to be a useful system up to 1200°C.

In these systems the activation energy of the ionic electrical conductivity is typically 25–30 kcal/mole and thus the change in conductivity in the typical temperature range of study, 600–1600°C, is found from the Arrhenius equation to be about 4 orders of magnitude with ZrO_2–CaO electrolytes having a specific conductivity of about 1 ohm^{-1} cm^{-1} at 1000°C.

1. Solid Electrodes in Oxygen Concentration Cells

A number of reference electrode systems have been established for oxygen concentration cells, and data for most of these have been collected in a review article by Rapp and Shores [284]. This shows that a reference system is established at every decade of oxygen pressure from 1 to 10^{-18} atm at 1000°C, with three systems at 10^{-24} atm (Mn–MnO, NbO–NbO_2 and Ta–Ta_2O_5). It is therefore possible with our present fundamental knowledge to carry out a number of electrochemical studies in which oxide electrolyte galvanic cells can be used to obtain thermodynamic data.

The importance of having a number of reliable reference electrode systems available is that the correct one may be chosen which most nearly matches the oxygen potential of the unknown system under investigation. By minimising this difference, and hence the e.m.f. of the cell, the effects of any electronic short circuits within the electrolyte will also be minimised. Furthermore, it would be possible in many circumstances to reverse the polarity of the cell such that the unknown electrode is the cathode with one reference electrode and the anode with another reference. In this way the electrode can be checked for reversible behaviour. In this connection it has been found recently that the imposition of a small a.c. perturbation is useful in equilibrating an oxide-electrolyte interface [285]. This potential must be removed during the measurement of the e.m.f.

The assembly of equipment for making such measurements covers a wide range of possibilities, as might be expected, and only a few typical instances will be given here. The apparatus used by Steele and Alcock [283] to establish a number of reference electrodes, maintained a high vacuum around pellets of electrolytes and electrodes stack sequentially, and held under compression in an alumina assembly (Fig. 83).

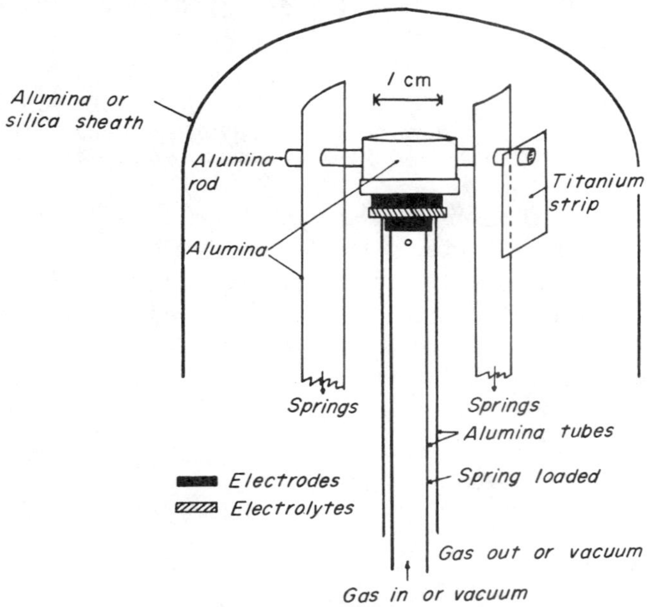

FIG. 83. Vacuum apparatus used by Steele and Alcock [283] for metal–metal oxide electrodes.

Charette and Flengas [286] used an electrolyte in the form of a closed end tube. The tube was evacuated on the inside, which contained a metal–metal oxide pellet pressed against the bottom of the tube and a compacted mixture of metal and metal oxide powders or pure oxygen gas contacted the base on the outside of the tube (Fig. 84). The stated reproducibilities of e.m.f. in both studies were ± 1 mV for nearly all of the cells which were measured.

The pioneering study made by Kiukkola and Wagner, in which the stack of electrode and electrolyte pellets was exposed to a single flow of inert gas, and these results agreed well with the other studies.

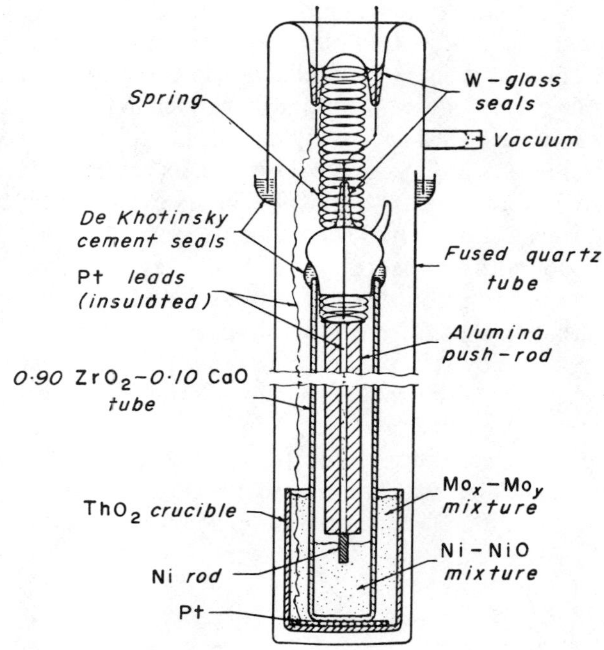

FIG. 84. Evacuated apparatus used by Charette and Flengas [286] incorporating a stabilised zirconia electrolyte tube for study of metal–metal oxide electrodes.

This general state of affairs is by no means typical, and Rapp [287] for example, found in a study of the Mo–MoO_2 system using the "open-stack" system with Ni–NiO and Fe–FeO reference electrodes, that the e.m.f. appeared to be very gas-flow-rate dependent using helium as an "inert" atmosphere, and therefore he redesigned the cell to eliminate this effect. The electrolyte was formed into an H-shape (Fig. 85) so that electrode pellets could be recessed into the electrolyte and protected from parasitic reactions with the gas phase.

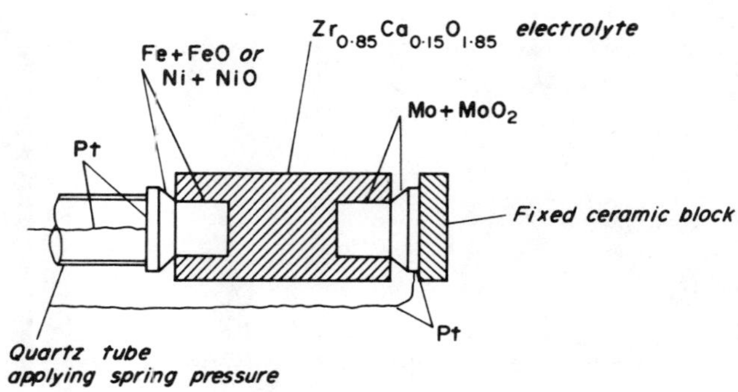

FIG. 85. Cell assembly used by Rapp [287] to minimize gas-electrode interactions.

The most critical test of the satisfactory separation of electrodes from mutual interference is when one electrode is an alloy mixed with an oxide of one of the components, or a single phase of variable composition. Any transfer of oxygen to or from an alloy electrode either via the gas phase or the electrolyte must necessarily involve a change in the composition of the alloy, and hence cause a drift in the e.m.f. of the cell. Dench and Kubaschewski [288] report a successful study of the cobalt–copper solid alloy system in which the first step was to eliminate the effects of thermal gradients by first placing the electrodes within the even zone to yield nearly zero for the e.m.f. of the symmetrical cell

$$\mathrm{Pt}\,|\,\mathrm{Co, CoO}\,|\,\mathrm{ZrO_2{\cdot}CaO}\,|\,\mathrm{Co, CoO}\,|\,\mathrm{Pt}$$

A precision of $\pm 0{\cdot}2$ mV could then be obtained from the cell

$$\mathrm{Pt}\,|\,\mathrm{Co, CoO}\,|\,\mathrm{ZrO_2{-}CaO}\,|\,\mathrm{Co{-}Cu, CoO}\,|\,\mathrm{Pt}$$

with the electrodes and electron leads disposed within the furnace even zone in the light of this knowledge.

The results for the cell

$$\mathrm{Pt}\,|\,\mathrm{Cu, Cu_2O}\,|\,\mathrm{ZrO_2{-}CaO}\,|\,\mathrm{Ni, NiO}\,|\,\mathrm{Pt}$$

agreed very well with those previously reported by Steele and Alcock [283] and by Rizzo *et al.* [289].

The effects of a temperature gradient within an electrochemical cell are to add another source of e.m.f. to that determined by the electrodes. The Seebeck coefficient which measures this effect for unit temperature gradient has been studied mainly with gas electrodes and with symmetrical cells. The indications are that a finite Seebeck coefficient must be anticipated when oxide electrolytes are used and that this coefficient will be a function of the absolute magnitudes of the oxygen potentials of the electrodes. Fischer [290] concluded from his study at relatively high pressures, 10^{-2}–1 atm of oxygen that the Seebeck coefficient for the ZrO_2–CaO electrolyte can be described by the equation

$$\frac{\Delta V}{\Delta T} = 0{\cdot}46 + 0{\cdot}0219 \ln \frac{159}{p}\ \mathrm{mV} \tag{110d}$$

where p is the oxygen pressure in Torr, and 159 is the partial pressure of oxygen in air.

Goto *et al.* [291] concluded that when air or oxygen was the gas in contact with the electrodes in a symmetrical cell, the coefficient α was $0{\cdot}095 \pm 0{\cdot}005$ mV/°C for ZO_2–CaO and $0{\cdot}050 \pm 0{\cdot}005$ mV/°C for ThO_2–CaO if it can be assumed that the e.m.f. of a cell can be expressed as

$$E = \frac{R}{AF}\int_{T_1, p'_{O_2}}^{T_2, p_{O_2}} t_{O^{2-}}\, dT \ln pO_2 + \alpha(T_2 - T_1) \tag{110e}$$

(p'_{O_2} was the same value at each electrode). It should be remarked here that whereas t_{ion} was unity for the zirconia-based electrolyte in this study, the value for the thoria-based electrolyte was less than unity.

Ruka *et al.* [292] studied the e.m.f.s which are generated across a ZrO_2–CaO electrolyte in the temperature range 700–1280°C, and oxygen pressures from 1 to 10^{-18} atm. They analysed the results using irreversible thermodynamics and considered the relationship

$$\frac{\Delta V}{\Delta T} = \frac{1}{F}\left[\frac{1}{2}\Delta\bar{\bar{S}}_{O^{2-}} - \Delta\bar{\bar{S}}_{(e^-,\mathrm{Pt})} - \frac{1}{4}\Delta\bar{S}_{O_2}\right] \tag{110f}$$

in which the first two terms on the right-hand side of the equation are the irreversible entropies of transfer for the oxygen ions and electrons across the temperature gradient and the third term is the reversible entropy change. They showed that the difference in $\Delta V/\Delta T$ between two different gas pressures but the same temperature gradient was constant as indicated by the equation above, and after application of the data of Cusack and Kendall [293] for $\Delta \bar{\bar{S}}_{(e^-,Pt)}$ obtained a constant value of $\Delta \bar{\bar{S}}_{O^{2-}}$ of about 11 cal/deg mole.

These terms were not so clearly separated by Fischer and by Goto *et al.*, but the results cover a much wider range of oxygen pressures. Further studies are needed, however, before a complete analysis is at hand for the effects of thermal gradients in galvanic cells of this type.

A cell can be used to measure the activity of a component of a binary alloy system such as copper–nickel [294], or the Gibbs energy of formation of an intermetallic compound, such as the compound Co_3W which was obtained from the cell

$$Pt\,|\,Fe + FeO\,|\,ThO_2\text{–}La_2O_3\,|\,Co, Co_3W, WO_2\,|\,Pt$$

by Rezukhina and Proshina [295]. The e.m.f. can only be interpreted to yield the activity of a metal in the electrode providing the oxide has a very narrow range of stoichiometry or there exists relatively complete information about the metal and oxygen potentials in the non-stoichiometric oxide. Thus in the cell for the study of copper–nickel alloys, Rapp and Maak could assume that nickel oxide was present at unit activity no matter what the nickel activity in the alloy. Hence the activity of nickel could always be obtained directly from a knowledge of the Gibbs energy of formation of the alloy and the oxygen potential of the electrode. Similarly in the cell for the study of Co_3W the oxide WO_2 could be assumed to have unit activity throughout.

However when the oxide phase is FeO_{1+x} which has a wide range of stoichiometry, the produce $a_{Fe}p_{O_2}^{1/2}$ is not a constant since the *wüstite* activity, with respect to the phase in equilibrium with pure iron, does not remain at unity across the whole phase field. A correction had therefore to be applied in the study of the cell

$$Pt\,|\,Fe, FeO\,|\,ThO_2\text{–}Y_2O_3\,|\,Fe\text{–}Au, FeO\,|\,Pt$$

to take account of this before the correct activities of iron were obtained [296]. It should be remarked that thoria-based electrolytes were used in this study because of the well-known interaction between iron oxides and zirconia-based electrolytes which is observed during prolonged contact at temperatures around 1000°C.

Finally, with solid electrodes which are single phase, it is almost mandatory to seal the electrode in a closed compartment one wall of which is the electrolyte in order to eliminate reactions with the gas phase. Several designs have been used to achieve this end, and two are shown in Figs. 86a and 86b [297, 298]. Because diffusion is slow in the oxide phase it is not advisable to use a sintered pellet of the oxide under investigation. Instead this can be placed loosely as a powder over the surface of the electrolyte, within the sealed chamber, and contacted with a platinum lead. The platinum then probably functions as the electron lead to an oxygen gas electrode in which the gas phase is the main instrument for distributing oxygen within the closed electrode compartment. If a controlled current is passed through the cell, the composition of the oxide can be altered, because the passage of current requires the removal of oxygen from one electrode and its transport to the other. The direction of transport depends upon the polarity of the impressed voltage which must be imposed on the cell to cause current to flow. We shall return to this topic later when discussing liquid electrode systems containing oxygen.

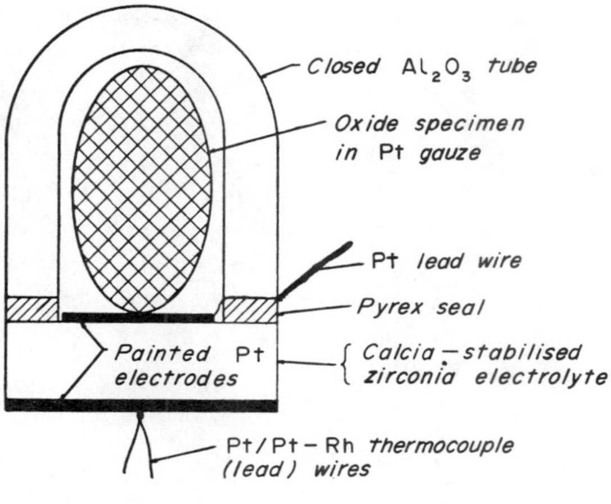

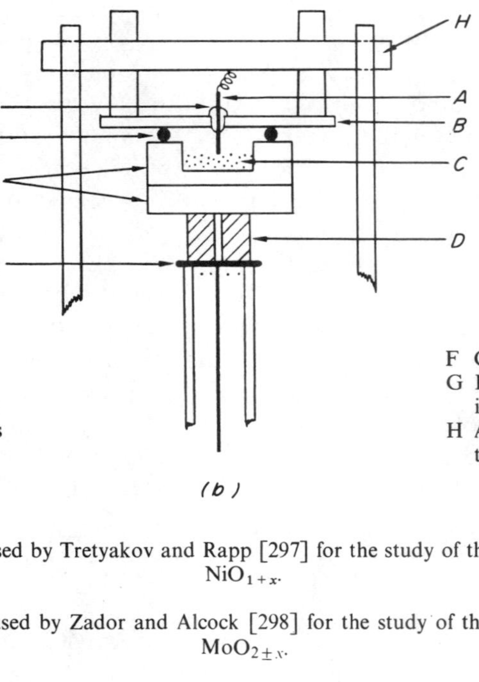

A Platinum leads
B Al_2O_3 lid
C Powdered electrode
D Reference electrodes
E Electrolyte crucible

F Gold seal
G Platinum contact wire in glass seal
H Al_2O_3 rods to compress the cell

FIG. 86a. Apparatus used by Tretyakov and Rapp [297] for the study of the non-stoichiometry of NiO_{1+x}.

FIG. 86b. Apparatus used by Zador and Alcock [298] for the study of the non-stoichiometry of $MoO_{2\pm x}$.

2. *Gas Electrodes in Oxygen Concentration Cells*

It has been found that when a gas of fixed oxygen partial pressure is flowed over a solid oxide electrolyte which has a platinised surface, that the system behaves as a reversible oxygen electrode [299]. This system may be applied for measuring the oxygen potential of O_2-inert gas, H_2–H_2O, CO–CO_2 gas mixtures and others producing a defined oxygen potential over the temperature range 400–1700°C.

The gas electrode provides a very convenient reference electrode and has been used to establish the thermodynamics of metal–metal oxide solid electrodes, to monitor the gas phase in the equilibration of condensed non-stoichiometric phases with gases of fixed oxygen potential, and on the industrial scale to control annealing furnace atmospheres for alloy steels.

The most convenient assembly for the construction of a gas phase monitor involves the use of a closed-end ZrO_2–CaO tube which is platinised on the outside, and a platinum wire contacts this surface and acts as the electron lead in the e.m.f. measuring circuit. Inside the tube a solid or gas reference electrode may be used, but a gas must be kept flowing around the platinised electrode in order to eliminate possible thermal segregation in the mixture of gases which determine the oxygen potential. The high-temperature limit of a gas electrode depends on the oxygen availability of the gas, thus Alcock and Chan [300] found that the e.m.f. of the cell

$$\mathrm{Pt}\,|\,\mathrm{CO{-}CO_2}\,|\,\mathrm{ZrO_2{-}CaO\ tube}\,|\,\mathrm{air}\,|\,\mathrm{Pt}$$

was close to the theoretical value at 1000° and 1200°C but at 1400°C and 1600°C the results were 15 mV and 300 mV lower when the CO–CO_2 ratio was 100/1. The deviations were shown to be due to a flux of oxygen through the tube which increased to a significant value above 1200°C, but when Mo–MoO_2 was used as the reference electrode in place of air, the results were again close to the theoretical value. The gaseous reference electrode should therefore either consist of a gas with a low oxygen dissociation pressure or with a thick electrolyte to minimise this effect. A good indicator of a satisfactory gas electrode is that the measured e.m.f. will be found to be independent of gas flow rate over a wide range of flow rates for a given gas mixture. Any leaks in the furnace tube which allow air to mix with the metered gas flow will, of course, vitiate the flow-rate test.

Smith, Meszaros and Amata have provided results for the permeability of ZrO_2–CaO, HfO_2–CaO and pure ThO_2 tubes [301] and the results show that these have approximately the same oxygen permeability in the temperature range 1100–2050°C. It would appear that the lower diffusion coefficient of oxygen ions in ThO_2- and HfO_2-based tubes is compensated for by the higher concentration of positive holes to provide equal fluxes of oxygen atoms in all of these systems.

The use of gas electrodes to monitor oxygen potentials is not only confined to metal–oxygen systems. Larson and Elliott used these electrodes for the study of sulphide systems [302] in such cells as

$$\mathrm{Pt}\,|\,\langle\mathrm{MnO}\rangle,\langle\mathrm{MnS}\rangle(\mathrm{SO_2},\ 1\ \mathrm{atm}\,|\,\mathrm{ZrO_2{-}CaO}\,|\,(\mathrm{O_2},\ 1\ \mathrm{atm})\,|\,\mathrm{Pt}$$

Since the MnO and MnS have a common manganese activity, the oxygen and sulphur potentials are fixed by the requirement of equilibrium with these phases and with SO_2 at a fixed pressure. The electrode reaction at the left-hand side may be written thus:

$$\mathrm{MnS} + 3\mathrm{O}^{2-} \rightarrow \mathrm{SO_2} + \mathrm{MnO} + 6e^-$$

This is a system with low oxygen and sulphur pressures, but at the other end of the scale in a study of PtS, the left-hand electrode consisted of a mixture of platinum metal and the sulphide. This is because platinum does not form a condensed phase oxide at high temperatures and with oxygen pressures less than 1 atm. Finally, if the oxide is very stable relative to the sulphide, as, for example, Ta_2O_5 compared with TaS_2, then the left-hand electrode is brought to equilibrium with a mixture of sulphur gas and SO_2.

3. Liquid Electrodes

The high-diffusion coefficients which are found in liquid systems, when compared with typical solids, ensure that electrodes in the liquid state reach equilibrium very much more rapidly than is normally found with solids. There are a number of advantages in using liquid reference electrodes, not the least of which is their rapid equilibration. The ease with which contact can be made to a liquid electrode is a definite advantage in comparison with the pressure-assisted contact with a solid electrode. The major disadvantage appears to be the corrosion of metallic contact materials by the liquid electrode, and a great deal of attention should be directed to this aspect in designing cells with liquid electrodes.

Apart from this, the high diffusion coefficients are an advantage in the study of Henry's law behaviour of dilute solutions of oxygen in liquid metals. The possibility of coulometric oxygen transfer from one electrode to the other was first exploited by Alcock and Belford [303] in studies of dilute, unsaturated solutions of oxygen in liquid lead and tin. It was found that conformity to Henry's law could be demonstrated in a cell using, say, Ni/NiO as a reference electrode and titrating known amounts of oxygen in and out of the liquid metal. A known quantity of the metal held in a crucible made of electrolyte material served as the electrode, and the e.m.f. of the cell at various oxygen levels up to saturation was measured. Figure 87 shows a typical titration curve for the cell

$$Pb[O]/ThO_2{-}Y_2O_3/Cu_2O, Cu$$

at 700°C.

When a high temperature of operation is required, such as 1600°C for steelmaking studies, the long-term survival of a simple electrochemical cell requires that the electrolyte and reference electrode system are immersed slowly into the melt so as to avoid thermal shock to the electrolyte. The use of tubes, which are now commercially available, made of stabilized zirconia which contains a suitable reference electrode (Fig. 88), is mainly for laboratory studies of long duration and, again, the principal experimental problem consists of finding good contact materials which will remain in the liquid metal for a significant period without corrosion. Cermets of alumina or zirconia with refractory metals such as molybdenum have been found valuable for studies in liquid copper as well as liquid iron. For control devices which must be rapidly inserted into a liquid metal at temperatures above 1000°C, the "oxygen-probe" can be used for periods of immersion of 30–60 sec. Several commercial forms of this device have been proposed in most of which the tube made of electrolyte material is replaced by a pellet of electrolyte either flame-sealed or cemented into the end of an open quartz tube (Fig. 89). Reference electrodes for such oxygen probes can be made of molybdenum/MoO_2 [304], chromium/Cr_2O_3 [305], or gaseous systems O_2, CO_2 or air in contact with platinum.

Finally, there have been a number of studies in which oxide activities in a liquid slag have been obtained from studies using solid oxide electrolytes. Typical of these are the measurement of PbO activities in liquid lead silicates which were made by a number of groups of workers [306–309] with the cell

$$Pb, PbO{-}SiO_2/\text{electrolyte}/\text{reference electrode}$$

The results show a scatter in the measured activities which is large compared with the best available studies using this method. For example, at a mole fraction of PbO of 0·6,

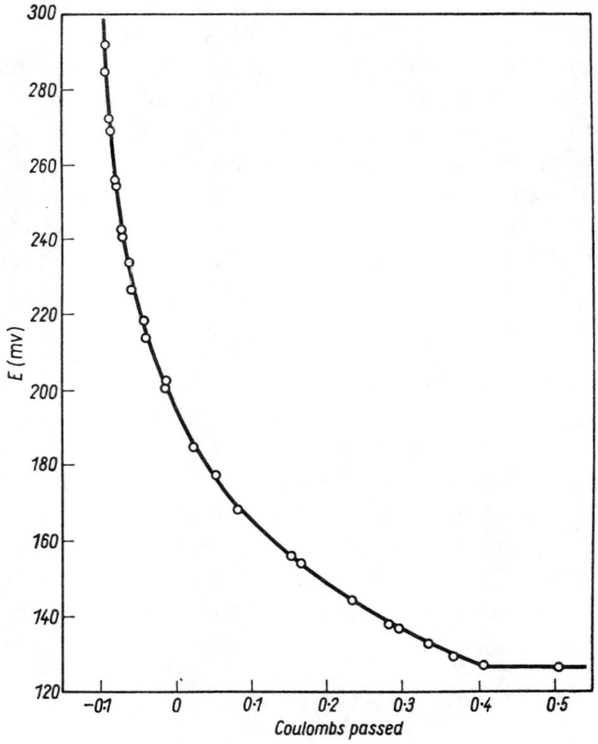

FIG. 87. Variation of the e.m.f. of the cell $Cu|Cu_2O|ThO_2{-}Y_2O_3|[O]_{Pb}$ as a function of the quantity of current passed through the cell. The curved portion corresponds to a solution of oxygen in lead, the flat portion is where PbO separates.

the range of measured activities at 1000°C is from 0·22 to 0·38 if all results are included. This corresponds to a range of partial free energies for PbO of ±700 cal approximately, or a range of e.m.f. values of ±15 mV.

C. OTHER CELLS

One of the main advantages of the zirconia and thoria electrolytes, apart from the valuable aspect of permitting measurement of oxygen activities, is that the solid electrolyte makes an effective barrier between the electrodes no matter what state is involved, and the electrolytes can be shaped into crucibles and tubes for electrode separation. This suggests that further developments of solid electrolytes would be very valuable in the study of high-temperature systems because of this constructional advantage. Unfortunately, very few additional electrolytes for use at high temperature have emerged during more than a decade of studies. Two of the most promising are calcium fluoride (CaF_2) which conducts electricity by fluoride ion migration and β-alumina which can be used for studies of sodium, potassium and silver activities. The original studies of conduction in CaF_2 made by Ure [310] show conductivities of the same order as the zirconia electrolytes, the moving ion being the interstitial fluoride ion. There appears to be little advantage in adding other fluorides in order to enhance the

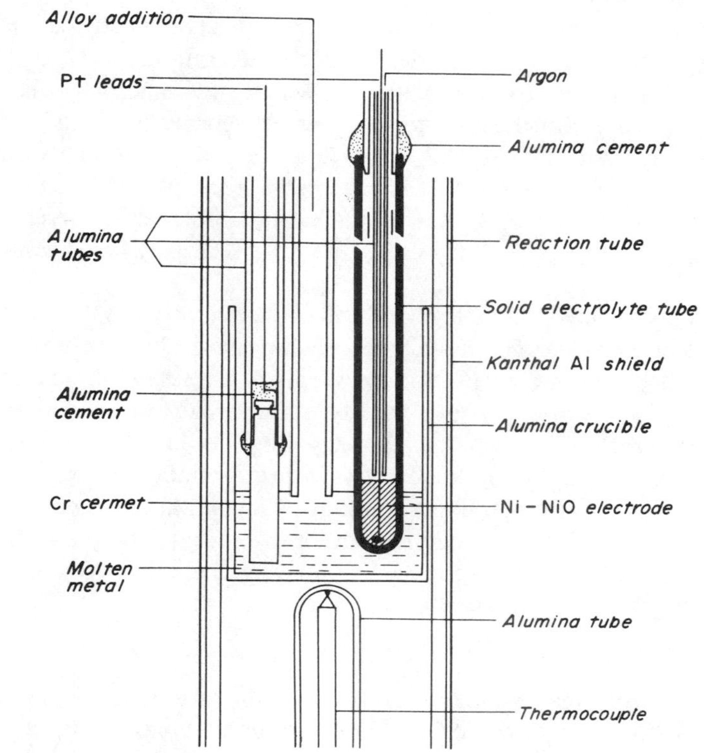

FIG. 88. Apparatus for the measurement of the dilute solution of oxygen in a liquid metal, incorporating a zirconia tube.

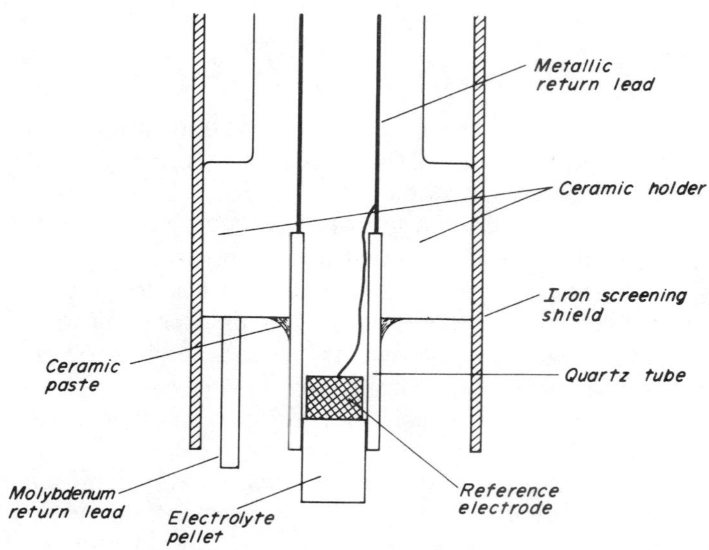

FIG. 89. An oxygen probe for use at steelmaking temperatures *ca.* 1600°C.

conductivity, and a single crystal of CaF_2 serves very well as the electrolyte. The material has been used up to 900°C with varying degrees of success, but it is necessary to pay close attention to the gaseous atmosphere which is allowed in contact with the electrolyte. The principal gas is argon or helium, which must be rigorously de-oxidised and dried before use, and even then it is useful to place some CaF_2 powder upstream of the electrolyte in order to equilibrate the gas phase with a sample of the solid other than the electrode surface. When this is successfully achieved, a number of reference electrodes can be used, e.g. Mg, MgF_2 [311], Cr, CrF_2 [312], U, UF_3 [313] as a result of careful earlier studies.

The immediately apparent advantage of the fluoride electrolyte is the fact that studies may be made using electrodes containing metals which have a large solubility of oxygen, and hence would be unsuitable in oxide electrolyte studies. A disadvantage of the fluorides, in such a comparison, comes mainly from our lack of knowledge of metal–fluorine phase diagrams. The non-stoichiometry ranges of fluorides could well be less than those of oxides, but at present there is very little information one way or the other.

Another field where a considerable advantage is gained through the use of fluoride electrolytes is in the study of metallic carbides and borides. The pioneering work of Egan [314] using this electrolyte produced a value for the free energy of formation of thorium dicarbide (ThC_2) with the cell.

$$Th, ThF_4/CaF_2/ThF_4, ThC_2, C$$

The cell obviously measures the activity of thorium in the carbide which is in equilibrium with carbon. A similar study by Aronson [315] produced values for the free energies of formation of ThC and the terminal carbon saturated α-thorium which is in equilibrium with the $ThC_{0\cdot7}$. The solid solution contains about 10 at.% carbon, and therefore the metal activity is reduced to 0·84 in the α-Th-$ThC_{0\cdot7}$ electrode. The cells which were operated between 800° and 1000°C are represented thus:

$$Th, ThF_4/CaF_2/ThF_4, ThC_2, ThC$$

and

$$Th, ThF_4/CaF_2/ThF_4, ThC_{0\cdot7}, \alpha\text{-}Th$$

Aronson and Auskern [316] used the same technique to study cells with thorium boride electrodes in the temperature range 800–950°C, thus opening another relatively unexplored field of solid state e.m.f. studies.

Because of the problems associated with hydrolysis of the electrolyte, Rezukhina and Pokarev [317] suggest that the cell be evacuated and preheated to 500–700°K before admitting the gas. Thorium, yttrium or lanthanum chips held in a tantalum boat were then put close to the cell in order to keep the argon atmosphere neutral. The cell was placed in double-wall silica vessels inside the furnace, and inert gas was passed continuously between the walls of the silica vessel.

Crucibles can easily be made from calcium fluoride by hot pressing techniques, but no e.m.f. application has been made of these materials to date.

β-alumina is another promising ceramic material which has been used in the measurement of sodium activities in alloy systems. This material, formula $NaAl_{11}O_{17}$ allows preferential, rapid, migration of sodium, potassium and silver ions along widely spaced planes, so that the material can be used satisfactorily from room temperature to about

800°C. Whittingham and Huggins [318] showed that silver β-alumina has an electron transference number less than 10^{-4} over the temperature range 550–790°C, and a wide range of oxygen pressure from 1 to 10^{-24} atm.

The material may be fabricated into pellets or tubes by pressing isostatically at about 20,000 psi and firing to about 1700°C [319]. Joglekar *et al.* [320] succeeded in joining a β-alumina pellet to an alumina tube by the use of a Kovar glass-ethylene glycol paste which slowly heated to 300°C and then fired at 1000°C for a short time. This device could be used as a sodium probe, and was used to measure sodium in solution in liquid tin at 500°C. During the sintering of the β-alumina material there is a tendency for sodium loss to occur at high temperatures. There is therefore a considerable advantage to be obtained by fabricating components by hot pressing, where densification is normally achieved at lower temperatures than at atmospheric pressures. A novel way of minimising the vaporisation loss under atmospheric conditions is the zone-sintering technique described by Wynne-Jones and Miles [321] for the production of long tubes of β-alumina.

Fray and Savory [322] used β-alumina tubes in a study of sodium–lead alloys in the temperature range 350–500°C and $N_{Na} = 0{\cdot}044$–$0{\cdot}78$. The cell can be represented thus:

$$\text{stainless steel, Na}/\beta\text{-alumina}/\text{Na–Pb, stainless steel}$$

Some coulometric titrations were carried out from the sodium to the lead electrode in order to prepare the dilute sodium–lead alloys, and generally e.m.f.s could be reproduced to ± 1 mV. Equilibrium was found to be very rapidly established during heating and cooling cycles. Very good agreement was found with the results of Hauffe and Vierk [323], who used a glass electrolyte, and other electrochemical studies of this system.

The construction of cells used for e.m.f. measurements with aqueous solutions is relatively simple, and many have been described in the literature. A number of these measurements were carried out, for instance, by Japanese workers (Ishikawa, Shibata, Ueda, Watanabe) in the period 1930–5, most of their work being published in scientific reports of the Tôhoku Imperial University. The cell, usually of glass, consists of a simple H-shaped tube or of two beakers connected by a bridge containing the solution. In this work great attention must be given to the purity of the substances used [324, 325]. Mercury should be distilled several times before use, amalgams should be melted under hydrogen, and salts should be recrystallised several times. Dakin and Ewing [324] have observed that equilibrium is attained much more readily when preparing electrodes (such as $\{Hg\}|\langle Hg_2Br_2\rangle|[KBr]_{aq})$ in the absence of air than otherwise. Bates [325] has pointed out the importance of ensuring that compounds are in their most stable state to avoid the occurrence of undefined transitions during experiment.

For the investigation of liquid amalgams with aqueous electrolytes, Richards and Forbes [326] constructed the cell shown in Fig. 90. It was later used by other workers. It can be loaded with several samples simultaneously so as to save time in making a series of measurements. The amalgam is introduced with a special pipette. The conductors are made of platinum wire, and the space above the solution is filled with purified hydrogen.

Liquid ammonia as a solvent was used by Elliot and Yost [327] in the arrangement

$$\langle Sn\rangle|\langle ZnCl_2\cdot 10NH_3\rangle|[NH_4Cl]_{NH_3}|\langle TlCl\rangle|\langle Tl\rangle$$

The H-form cell was fitted with a special device for introducing distilled ammonia.

As an example of the diversity of solvent media that may be employed we may mention an investigation [328] in which the e.m.f.s of a silver–silver nitrate concentration cell using various solvents were compared with values deduced from conductivity measurements. Excellent agreement with the theoretical value was found with ethyl alcohol, pyridine and quinoline; good agreement with acetonitrile, benzonitrile, aniline; and poor agreement with methyl alcohol, isoamyl alcohol and acetone.

The cell used for the investigation of liquid alloys at high temperatures (e.g. [329, 330]) is usually also of the H-form and is made of Pyrex or fused silica. The two sides carry the two electrodes, covered by the solution which also fills the transverse junction. At these temperatures the usual aqueous solutions are replaced by molten salt mixtures. Conductors are usually of platinum, carbon or iron. An iron cage has been used to hold liquid magnesium below the surface of a denser molten salt.

The electrolytes frequently used for high temperatures consist of alkali chlorides in which a small amount of a salt of the transported metal is dissolved. Eutectic mixtures of lithium and potassium chlorides (m.p. 359°C) and lithium and rubidium chlorides (m.p. 312°C) are commonly employed. For still lower temperatures the eutectic mixture of sodium and potassium acetates (m.p. 233°C) has been recommended. This however decomposes with charring at 300°C.

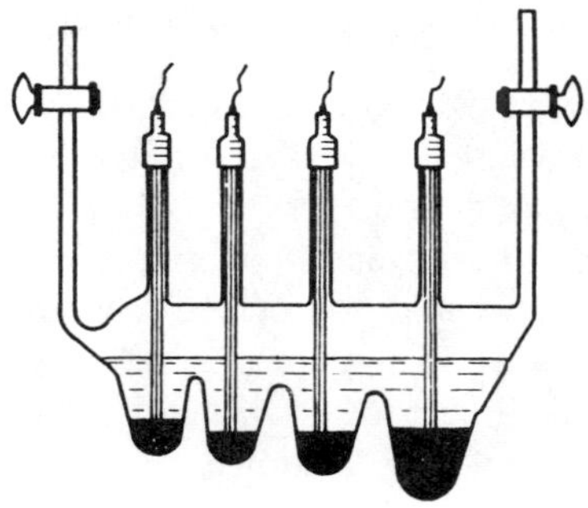

FIG. 90. Cell for e.m.f. measurements on liquid amalgams by T. W. Richards.

Careful attention must be paid to the choice and preparation of these electrolytes. When hygroscopic salts, such as lithium chloride, are used, they must be thoroughly dehydrated to avoid trouble arising from moisture of the formation of oxy-salts (e.g. [331]). Elliott and Chipman [331a] added 0·6% fused potassium hydroxide to a KCl–LiCl electrolyte to neutralise the hydrogen ions which might have been formed from hydrolysis of moisture in the salts. If the metal of the electrolyte is stable in more than one valency state, only one of these must be present. This may be ensured by the choice of a salt mixture which forms suitable complexes.

Certain alkali–metal amalgams cannot be studied with aqueous solutions owing to their reaction with water. Lewis [332] used instead solutions of sodium or potassium iodides in ethylamine. Hauffe [333] overcame the difficulty in an elegant manner, using glass as a solid electrolyte in the electrochemical investigation of the molten Na–Hg and Na–Cd systems at 300–400°C. His apparatus is shown in Fig. 91. According to Warburg [334] sodium ions in glass carry electric current, Faraday's law being obeyed. By using a glass of composition 70·2% SiO_2, 9·5% CaO, 14% K_2O, 6% Tl_2O, 0·3% As_2O_3 as the electrolyte, Vierk [335] was able to apply Hauffe's apparatus to the investigation of liquid thallium alloys, e.g. thallium–tin. In this case, however, equilibrium was attained more slowly, owing obviously to the smaller diffusion rate of

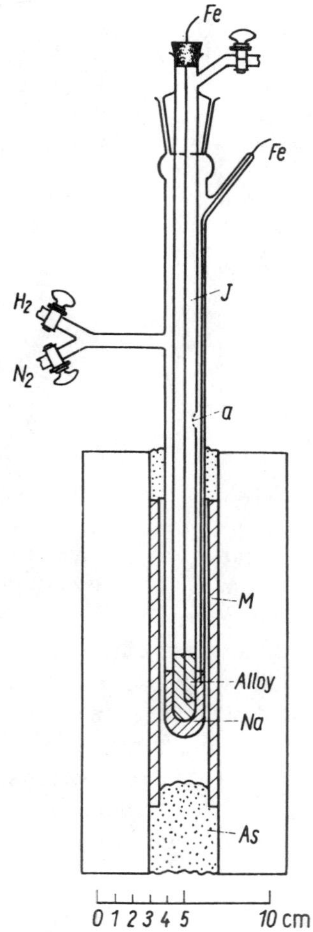

FIG. 91. Cell with glass electrolyte for e.m.f. measurements on liquid alloys. I = inner glass vessels, H to high vacuum, N to nitrogen supply, a = opening to inner vessel, M = silumin tube.

thallium ions in glass relative to that of sodium ions. Nevertheless the Hauffe–Vierk cell is probably the most suitable for the study of alloys containing sodium, potassium, thallium and possibly certain other metals as the less noble constituent.

Richards and Conant [336] also devised a useful technique for investigating liquid alloys, such as sodium amalgams, which oxidise rapidly in contact with an aqueous electrolyte (in their case a sodium hydroxide solution). The two amalgams investigated drop continuously from two capillary tubes, thereby maintaining a practically fresh surface in contact with the electrolyte. (A similar device is used in the Heyrovsky type polarograph so extensively used nowadays for the routine analysis of alloys.)

For the measurement of the e.m.f.s of *solid* alloys against mercury with an aqueous electrolyte, Ölander [337] used the cell shown in Fig. 92. It was 2·5 cm in diameter and was supplied with hydrogen to prevent oxidation. Four electrodes at a time were mounted in the cell.

Weibke and v. Quadt [338] developed the apparatus shown in Fig. 93 to permit a carefully prepared molten salt mixture to be brought into contact with the electrodes with-

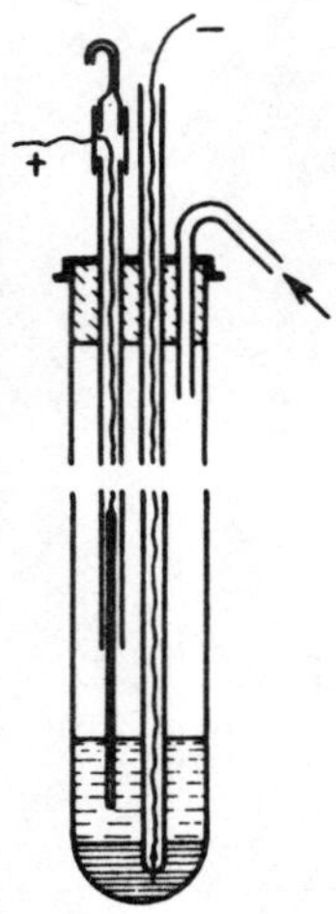

FIG. 92. Simple cell for e.m.f. measurements (Ölander [337]).

out risk of contamination. The apparatus was of resistance glass ("Supremax"). The salts are purified and melted in the reservoir on the right, a vacuum being applied at the same time. The cell is then tilted and the vacuum released in order to drive the liquid into the cell proper, which is then sealed off at the constriction.

The molten salt electrolytes (halide or acetate mixtures) caused considerable trouble in the experiments of Ölander and Weibke, even when they were carefully prepared, i.e. dehydrated and degassed. The alkali chlorides were volatile and condensed on the electrodes and conductors and attacked other parts of the apparatus. Thermo-e.m.f.s may be set up when there is a temperature gradient along the electrodes. These cause localised electrolysis resulting in a change in the composition of the alloy. Weibke and

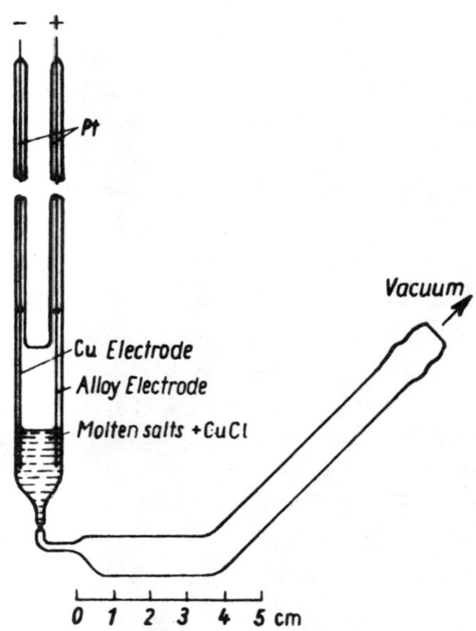

FIG. 93. Cell for e.m.f. measurements on solid alloys. Side tube for preparation of electrolyte.

Matthes even observed in some experiments [339] a tapering of the alloy electrode with a deposition above the taper of a garland of crystals of the more electronegative metal. They obtained a more uniform temperature distribution along the electrodes by enclosing the cell inside their furnace in an iron or copper cylinder. The use of shorter electrodes also diminishes the thermoelectric effect.

In order to eliminate the difficulties that arise in the use of molten salts, attention has been given to the use of solid electrolytes. Results have been encouraging and much further development may be expected in this direction. The earlier work was done on the silver–gold alloys. Wachter [340] used solid silver chloride as the electrolyte. It is known to be a pure cationic conductor. The use of solid salts, however, introduces other complications. Copper halides and chalcogenides, for instance, exhibit metallic as well as electrolytic conductivity, and so do many other salts. Faraday's law does not apply to such substances. Kubaschewski and Huchler [4] used solid glass as the solvent for silver ions in work on the system silver–gold. It was established that silver dissolves in glass in the ionic form [341]. The type of apparatus used for such experiments is influenced, of course, by the low conductivity of the glass. That used by Kubaschewski and Huchler is shown in Fig. 94. A thin silver disc (3) is sealed on to the plate of glass (4) (0·12 mm thick) and is placed in contact with a more massive piece of silver (2) set in a

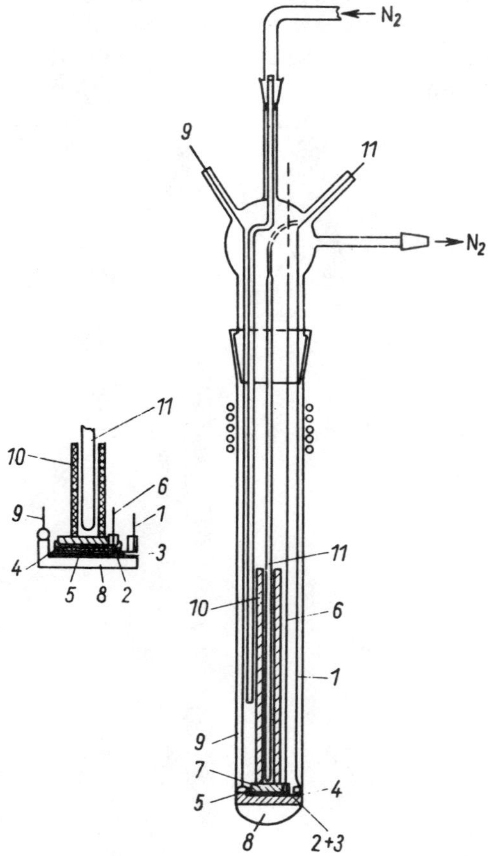

FIG. 94. Cell for e.m.f. measurements on solid alloys, with glass electrolyte.

ceramic cup (8) of 1·9 cm diameter. The gold–silver alloy, also in the form of a disc (5) is pressed on to the upper surface of the glass by an alumina tube (10) which also contains the thermocouple (11). The platinum conductors (1, 6) are welded to the alloy piece and to the lower piece of silver. The resistance of the cell described was 2500 ohms at 600°C. Great care was taken with this apparatus to exclude oxygen and to avoid small extraneous currents. The results were found to be quite reproducible and reliable.

The e.m.f. of halide formation at high temperatures was investigated by Lorenz and Velde [342] in a cell (Fig. 95) based on the scheme

$$\langle C\rangle\{M\}|\{MCl_x\}|(Cl_2)\langle C\rangle$$

The electrolyte consisted of the fused chloride of the metal. The graphite electrodes were sealed into glass to prevent their combustion; one dipped into the molten metal and the other into the molten salt. To prevent diffusion of chlorine through the melt this latter electrode was surrounded by another wider glass tube. A special arrangement was used for magnesium cathodes. In order to saturate the anode with chlorine a polarizing current (0·2–0·3 A, 120 V) was applied, while chlorine gas was passed into the cell. Saturation took as much as 8 days and more. When saturation was complete, the polarising current was switched off and the e.m.f. measured. When all the necessary precautions had been taken the resulting thermodynamical data proved to be very accurate. The values obtained were somewhat different when the anode and cathode compartments were separated by a glass diaphragm, even though the electrolyte on either side was of the same composition.

According to Lorenz no junction potentials arise between the chlorides of different metals, so that the e.m.f.s of cells [343, 344] of the type

$$\{Zn\}|\{ZnCl_2\}|\{PbCl_2\}|\{Pb\}$$

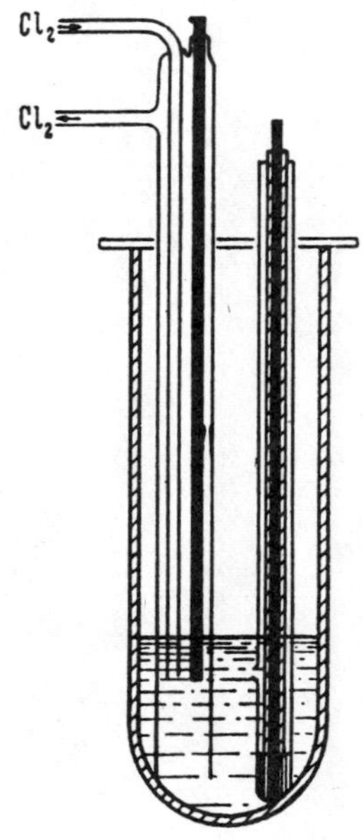

FIG. 95. Cell for e.m.f. measurements on molten chlorides (Lorenz).

can be used additively in calculations. This does not, however, apply to the e.m.f.s of cells containing mixtures of liquid salts, or to cells in which the liquid chlorides are separated by a glass diaphragm [343].

Rose *et al.* [345] reported a series of e.m.f. measurements on the reducibility of chlorides, oxides and sulphides of metallurgical interest. They pointed out that two serious sources of difficulty were hydrolysis of fused salts by atmospheric moisture, and the tendency of molten metals to disperse in a "fog" in the fused electrolyte. Both these troubles appeared to have been obviated to a large extent in the experiments on chlorides by passing a stream of dry hydrogen through the salt. The method of measuring e.m.f.s consisted in taking current/voltage measurements and extrapolating to zero current. They were able to verify Lorenz's results for the cell

$$Ag \,|\, AgCl \,\vdots\, PbCl_2 \,|\, Pb$$

and their results for the cell

$$Sn \,|\, SnCl_2 \,\vdots\, PbCl_2 \,|\, Ag$$

gave a linear e.m.f./temperature relationship whereas Lorenz's results had been scattered.

The oxide cells investigated were:

$$O_2(Ag \text{ or } Pt) \,|\, NaOH, SnO_2 \,|\, Sn$$

$$O_2(Ag) \,|\, NaOH \,\vdots\, NaOH, PbO \,|\, Pb$$

The gas electrode was built to the same pattern as those used for measurements on aqueous solutions, but together with the containing vessel was made of non-siliceous sintered alumina. The oxygen electrode was surrounded only by molten caustic soda in the second cell in view of the tendency of PbO to become oxidised easily. Throughout these experiments allowance was made for thermo-e.m.f.s set up at junctions between the electrode metals and other metals leading to the measuring instruments.

Some of the details to which attention must be paid when carrying out e.m.f. measurements have already been mentioned. They may be summarised as follows:

1. The attainment of equilibrium by alloys after any change in temperature must be verified by observing the course of the e.m.f. over a period of time. The variation of e.m.f. with time after a temperature change with gold–copper alloys is shown in Fig. 96.

2. The transference of ions in the electrolyte must take place without secondary reaction. The various ions must be present in one valency state only. If molten salt mixtures are used, the ions transferred must be the least electronegative in the melt. To

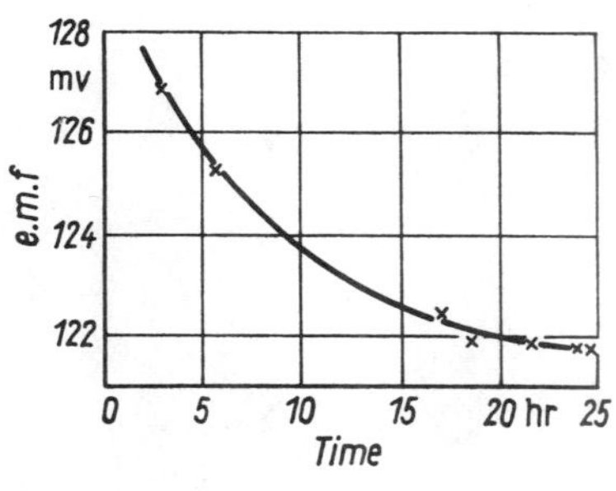

FIG. 96. E.m.f.–time curve for a copper–gold alloy containing 49% copper.

ensure these conditions the pre-treatment (e.g. drying) and preparation of the melt must be undertaken carefully.

3. A protecting atmosphere must be provided.

4. The temperature used must be sufficiently high to ensure a diffusion rate in the electrode great enough to avoid changes in composition at the alloy electrode surface.

5. Thermo-e.m.f.s must be avoided.

6. It should be mentioned that in cells of the type applied to alloys the affinity of the anion of the electrolyte to the more electropositive metal must be much less than its affinity to the other constituent of an alloy. This implies that e.m.f. measurements are likely to be unsatisfactory with binary alloys whose components are close together in the electropotential series. Weibke and Efinger [346] encountered difficulty in measuring the e.m.f. of silver–antimony alloys against antimony with an electrolyte of $SbCl_3$ dissolved in a molten LiCl/KCl mixture. They concluded that the anticipated e.m.f. was overlapped by another due to a concentration cell

$$\langle Sb \rangle \,|\, [SbCl_3]_{conc} \,\vdots\, [SbCl_3]_{dil} \,|\, \langle Sb\text{—}Ag \rangle$$

This cell arises from a reaction of the silver of the alloy electrode with the antimony trichloride of the electrolyte causing an impoverishment of the $SbCl_3$ in the immediate vicinity of the electrode. The reaction is quite probable since the heat of formation of 3 moles AgCl (3 × 30·2 kcal) is nearly equal to that of 1 mole $SbCl_3$ (91·5 kcal).

Wagner and Werner [347] have estimated the magnitude of the error due to the displacement reaction in galvanic cells and tested the basic assumptions for the estimate by measuring the e.m.f. with certain alloy systems.

It is difficult to make a general assertion about the accuracy of e.m.f. measurements, since it varies from one system to another. It is usually expressed as being within 0·1–0·3 mV. Seltz (e.g. [330]) with some systems including bismuth quotes accuracies of ±0·03 mV but such a figure, though it may cover the experimental error, can hardly take into account the possible deviation from equilibrium conditions.

CHAPTER III

The Estimation of Thermochemical Data

No especial skill or experience is required to learn the main thermodynamical formulae and to use them for calculating equilibria from tables of heat capacities, heats of formation and transition, and entropies. Every student of chemistry is expected to carry out some of the calculations in the course of his study, and very little revision is necessary in taking up similar work at later stages in his career. Unfortunately the published data are far from complete, especially in the ever-widening field of metallurgical chemistry, where much of the available information is not very reliable and requires some means of independent checking. This is where ordinary thermochemical knowledge fails, and it is one of the functions of the experienced thermochemist to estimate the missing data with reasonable accuracy. This task is not as difficult as it might first appear, as the most of the principles which are applied have been well established for a considerable time. Nevertheless, it requires some experience to enable the best choice of method to be made in each particular case, and if necessary to develop new methods. Some of the methods used, with examples of their application, are given in this chapter.

1. Heat Capacities

We shall not be concerned with the rules which apply at low temperatures, such as Debye's equation, or with the methods used to calculate electronic heats. These quantities are incorporated in normal entropy values, which will be discussed later. For metallurgical purposes it is sufficient to be able to estimate heat capacities at room temperature and above.

1. Monatomic Gases

Monatomic gases such as metallic vapours have a constant heat capacity of 5 cal/g-atom. This was established by kinetic considerations, and has been verified experimentally for a number of elements. In these gases the three translational degrees of freedom provide the only means of storing up energy. On the basis of Clausius' law of equipartition of energy,

$$C_v = \tfrac{3}{2}\boldsymbol{R} \simeq 3 \text{ cal}$$

$$C_p = C_v + \boldsymbol{R} = \tfrac{5}{2}\boldsymbol{R} \simeq 5 \text{ cal} \qquad (111)$$

2. Diatomic Gases

These gases have two further degrees of freedom, those of rotation. This gives C_p a theoretical value of 7 cal/deg, which generally holds good at room temperature. At higher temperatures an oscillation sets in along the line joining the two atoms of the molecule. Since an oscillation involves both potential and kinetic energy, each vibrational degree of freedom must be counted twice for the purpose of equipartition of energy. Therefore C_p for a diatomic gas may amount to a maximum value of 9 cal/deg. Table IX shows that this maximum value is generally approached at high temperatures, though chlorine and bromine for instance have a value higher than 7 cal/deg even at room temperature.

Table IX. Molar heat capacities of gases

Gas	Molecular weight	C_p in cal/deg mole			
		298°K	700°K	1200°K	1800°K
K	39·1	4·96	4·97	4·98	4·99
Ti	47·9	5·84	5·17	5·13	5·62
Cu	63·5	4·97	4·97	4·97	5·04
Sn	118·7	5·08	7·18	7·86	7·09
H_2	2·0	6·89	7·09	7·46	7·93
CO	28·0	6·96	7·45	7·96	8·55
HCl	36·5	6·96	7·16	7·68	8·33
AlCl	62·4	8·28	8·86	9·02	9·11
S_2	64·1	7·76	8·66	8·90	9·00
I_2	253·8	8·81	9·01	9·09	9·18
H_2O	18·0	8·02	8·98	10·25	11·78
CO_2	46·0	8·90	11·65	13·00	14·38
SO_2	64·1	9·54	12·18	13·33	13·82
Al_2O	70·0	10·92	12·98	13·56	13·75
NH_3	17·0	8·48	11·23	14·28	17·90
SO_3	80·1	12·11	16·82	18·63	19·30
$COCl_2$	98·9	13·73	17·45	18·85	19·42
MoO_3	143·9	14·35	18·18	19·23	19·58
$SiCl_4$	169·9	21·57	24·83	25·48	25·68
$TiCl_4$	189·7	22·84	25·18	25·58	25·74
WCl_4	325·7	23·62	25·37	25·68	25·81
N_2O_5	108·0	23·01	31·89	34·41	35·18
WCl_6	396·6	34·40	37·08	37·52	37·65
Al_2Cl_6	266·7	37·73	42·38	43·25	43·51

Experimental results show that translational and rotational movements are in full operation at room temperature. The excitation of rotation is governed by the moment of inertia of the molecule. This is sufficiently high even in the lightest molecule—hydrogen—to ensure that almost all polyatomic molecules rotate freely at 300°K. On the other hand, only a small fraction of the molecules of a gas are in oscillation at room temperature, though the proportion rises rapidly with increasing temperature. The extent of excitation depends upon the atomic masses and the rigidity of the atomic bond; for a simplified representation one may imagine two massive balls connected by a helical spring with lateral rigidity. Little can be known of the interatomic binding in a molecule of which not even the heat capacity is known, but of course the masses are given by the atomic weight(s). Thus chlorine and gaseous sulphur are heavier than oxygen and nitrogen, and the former elements may be expected to have the greater heat content. This general tendency is shown quite clearly in Table IX. The room temperature heat capacity of diatomic gases in which the two atoms are comparable in mass increases regularly from 7 to 9 cal/deg mole with increasing atomic weight. The regularity breaks down with hydrogen compounds in which one of the atoms is much lighter than the other and the behaviour is that of a substance of a lower molecular weight. For the lighter molecules (up to $M = 40$) the formula

$$C_p = 6{\cdot}7 + 1{\cdot}0 \times 10^{-3}T$$

holds good within fairly narrow limits from 300° to 2300°K, while C_p may be taken as a constant 9·0 within the same temperature range for substances in which the molecular weights exceed 100.

3. *Polyatomic Gases*

It is more difficult to estimate the heat capacities of gaseous molecules with more than two atoms. Values are usually calculated, together with those for the entropies, from molecular constants. Where this information is incomplete, assumptions may be made concerning the molecular configuration and missing data for the fundamental vibrational frequencies, bond distances, bond angles and/or electronic levels estimated. Stull and his colleagues [348] have indicated the mode of evaluation of heat capacities and entropies from the molecular constants, calculated the values for a large number of gaseous molecules at various temperatures and, in doing so, estimated missing information. The *JANAF Thermochemical Tables* are thus recommended for perusal.

4. *Solids*

Generally it will more often be necessary to estimate the heat capacities of solid substances, and here use can be made of Dulong and Petit's rule. In its original form this rule stated that the atomic heats of the elements generally approximate to the value 6·2 cal/deg at room temperature. This rule was also established from considerations of equipartition of energy. Since the atoms of a solid are fixed in a lattice, they are incapable of rotation or translation, but there are three vibrational degrees of freedom (each of which must be counted twice). Hence $C_v = \frac{6}{2}R \approx 6$ cal/°K g-atom when the solid lattice is in a state of complete oscillation.

The relationship between C_p and C_v for the solid state is more complicated than for the gaseous state and is given by a formula which includes the compressibility of the solid and its coefficient of thermal expansion. These quantities are generally not known very exactly, but this is not of such vital consequence since the difference between the two heat capacities amounts only to 0·2–0·5 cal/deg at room temperature, giving a C_p value of about 6·2–6·5 cal/°K g-atom.

Kellogg [349] has suggested an improved method for the estimation of the heat capacity of solid, predominantly ionic, compounds at 298°K. It is analogous to

Latimer's method for the estimation of standard entropies (pp. 188–9) and consists in adding together contributions from the cationic and anionic groups in the compound (here denoted Ξ_{cat} and Ξ_{an}). Working empirically, Kellogg derived from a compilation of experimental data average values of Ξ_{cat} for the metals independent of their valency and of Ξ_{an} for the anions—in the latter case, Latimer style, depending on valency. The estimated heat capacities at 298°K were then obtained by adding up the various contributions, i.e.

$$C_p(298) = \Sigma\Xi$$

For aluminium sulphate, $Al_2(SO_4)_3$, for instance: $C_p(298) = 2\Xi_{Ae^{+++}} + 3\Xi_{SO_4^{=}}$.

Ünal [350] revised the tables of Kellogg with the aid of a computer program using experimental $C_p(298)$ values of 90 compounds. It emerged that, within the scatter of the individual values, the Ξ_{an} values may be represented by a single number for each anion independent of valency. This is in contrast to Kellogg's findings but simplifies the procedure. The values finally arrived at are listed in Tables X and XI. As an example, the heat capacity of one mole of aluminium sulphate at room temperature may be estimated:

$$C_p(298) = 2\Xi_{Al} + 3\Xi_{SO_4} = 9{\cdot}4 + 54{\cdot}9 = 64{\cdot}3$$

which compares with the measured value of 62·0.

Table X. Cationic contribution to the heat capacity at 298°K

Metal	Ξ_{cat} (cal/°K)	Metal	Ξ_{cat} (cal/°K)
Ag	6·15	Mg	4·70
Al	4·7	Mn	5·6
As	6·0	Na	6·2
Ba	6·3	Nb	5·5
Be	(2·3)	Nd	5·8
Bi	6·4	Ni	(6·6)
Ca	5·9	P	3·4
Cd	5·5	Pb	6·4
Ce	5·6	Pr	5·8
Co	6·7	Rb	6·3
Cr	5·5	Sb	5·7
Cs	6·3	Se	5·1
Cu	6·0	Si	—
Fe	6·2	Sm	6·0
Ga	(5·0)	Sn	5·6
Gd	5·6	Sr	6·1
Ge	4·8	Ta	5·5
Hf	6·1	Th	6·1
Hg	6·0	Ti	5·2
Ho	5·5	Tl	6·6
In	5·8	U	6·4
Ir	(5·7)	V	5·3
K	6·2	Y	(6·0)
La	(6·1)	Zn	5·2
Li	4·7	Zr	5·7

The heat capacity of solids increases with temperature and at the melting point is roughly the same per ion or atom for all compounds. Kelley [351] took this value to be 7·0 cal/°K g-atom. Ünal [350], with the more recent experimental evidence,

Table XI. Anionic contribution to the heat capacity at 298°K

Anion	Ξ_{an} (cal/°K)	Anion	Ξ_{an} (cal/°K)
H	2·1	SO_4	18·3
F	5·45	NO_3	15·4
Cl	5·9	P	(5·6)
Br	6·2	CO_3	14·0
J	6·3	Si	(5·9)
O	4·4	CrO_4	21·7
S	5·85	MoO_4	21·6
Se	6·4	WO_4	23·3
Te	6·5	UO_4	25·6
OH	7·4		

increased this average to 7·25 ± 0·5 cal/°K g-atom. Not included in the evaluation of the mean deviation are the data for those substances that undergo a solid–solid transformation below the melting point nor for those that have rather low melting points—say, below 150°C. If one includes the first transformation rather than considering the melting point alone, the average value turns nearer to Kelley's original value—a point that should be noted. These observations* taken together with the values in Tables X and XI may be used to recommend the estimation of the constants in a heat capacity equation of the type

$$C_p = a + b \times 10^{-3}T + c \times 10^5 T^{-2} \tag{112}$$

Kelley [351] estimated quite a number of C_p vs. T equations which have been widely employed in the thermochemical literature. To do so, he used a two-term expression. The introduction of a T^{-2} term, however, reflects the pronounced curvature at the lower temperatures above 298°K. A list of values for c in eqn. (112) evaluated by the investigators themselves or by Ünal [350] shows a considerable scatter due more to the inaccuracy of the differentiation than the actual relations. A mean value of $c/n = -1{\cdot}0$ cal/°K is obtained from the T^{-2} terms of 200 inorganic substances.

It is now possible to derive expressions for the constants in eqn. (112). These are as follows:

$$a = \frac{T_m 10^{-3}(\Sigma\Xi + 1{\cdot}125n) - 0{\cdot}298n10^5 T_m^{-2} - 2{\cdot}16n}{T_m 10^{-3} - 0{\cdot}298} \tag{113a}$$

$$b = \frac{6{\cdot}125n + 10^5 n T_m^{-2} - \Sigma\Xi}{T_m 10^{-3} - 0{\cdot}298} \tag{113b}$$

$$c = -n \tag{113c}$$

Here, T_m is the absolute melting temperature of the compound and n the number of atoms in the molecule.

If $C_p(298)$, the heat capacity at 25°C, is known from low-temperature measurements, the experimental value is to be used instead of $\Sigma\Xi$, the estimated value. If the heat

* In fact, the average value of C_p(m.pt.)/n somewhat increases with molecular weight, as one might expect. However, in view of the scatter, this effect is not great enough to warrant the proposal of a two-term expression in terms of log M.

capacities of compounds similar in mass and chemical nature to the one under consideration, they should be used to adjust the estimated equation.

If nothing better is known about the heat capacity of a compound, then ΔC_p may be assumed to be zero for *condensed reactions* without affecting most calculations too gravely. This postulate of the additivity of the heat capacities of the elements or the reactants to give the heat capacities of the compound or the products in a reaction is known as Neumann and Kopp's rule. It has been found to be valid for a large number of reactions and frequent use has been made of it.

In an investigation of the heat capacities of refractory borides with a drop calorimeter over the range 400–1200°K, Mezaki *et al.* [353] found that the heat contents are approximately equal to the sums of the elementary heat contents or, more accurately, for a compound Me_xB_y:

$$\frac{x(H_T - H_{298})_{Me} + y(H_T - H_{298})_B}{(H_T - H_{298})_{Me_xB_y}} = 0{\cdot}85 \text{ to } 1{\cdot}05 \tag{114}$$

with values around 0·9 being most common. The same ratio may be applied to $(xC_{p_{Me}} + yC_{p_B})/C_{p_{Me_xB_y}}$ in order to obtain reasonably reliable estimates up to 1200°K.

Neumann–Kopp's rule holds good especially for alloy phases as may be seen from Schübel's figures [352] in Table XII. It can also be applied as a first approximation to compounds with a co-ordination lattice. In other words, for reactions involving these substances the Gibbs–Helmholtz equation can be used in the simplified form

$$\Delta G_T = \Delta H_{298} - T\Delta S_{298}$$

for any temperature T. At the transition points the heats of transformation, fusion or evaporation must of course be brought into consideration. Their effect is much greater than that of variation in heat capacity, and we shall discuss their estimation later.

Table XII. Deviation of the specific heats of intermetallic compounds from those calculated additively, in parts per cent (Schübel)

	18–100°C	18–300°C	18–500°C
Cu_2Mg	−1·5	−1·2	+1·1
$CuAl_2$	−0·2	−0·5	−0·6
Cu_3Al	−1·3	−1·0	−0·4
Ag_3Al	0·0	+1·7	+1·1
Cu_3Sb	+4·8	+4·4	—
$MgZn_2$	−1·5	−1·7	—
Ni_2Mg	−5·0	−6·0	−8·0
Co_2Sn	+0·8	—	—

For accurate estimations, the Neumann–Kopp rule should not be applied to reactions in which gases are involved.

5. *Liquids*

The heat capacities of molten inorganic substances do not differ greatly from those of the solid materials. This indicates a close connection between the solid and the liquid state, and although we cannot envisage a simple model which adequately describes the liquid state, we must assume that some form of potential energy is present, analogous in some way to the intermolecular binding in the solid state. However this may be, the heat capacity of an organic liquid amounts to 7 to 8 cal/deg g-atom depending to some degree on atomic weight. The value 7·5 may be used if measurements are not available; alternatively the atomic heats of the liquid constituent elements may be taken additively.

Little is known of the temperature dependence of C_p of inorganic substances in the liquid state. A tendency of the heat capacities of liquid metals to decrease with increasing temperature has been observed [354] and continues to be confirmed by measurements in the laboratory of R. Hultgren. At present it is still customary to use constant C_p values for liquids, but we suspect that substantial errors enter calculations that are extrapolated to very high temperatures.

6. *Some Average Values*

The following list gives estimated *average values* of the change in molecular heat for some types of reaction involving gases. They may be applied, when other information is lacking, over a large temperature range. The values are, however, intended to be no more than a rough guide. Some experience will enable the computer to make minor adjustments to these values for a particular chemical reaction. A and B represent two elements, x and y stoichiometric indices

Type of reaction:	ΔC_p:
$\langle A \rangle = (A)$	$-1{\cdot}8$
$\{A\} = (A)$	$-2{\cdot}3$
$\langle A_x B_y \rangle = (A_x B_y)$	$-2{\cdot}3(x+y)$
$\{A_x B_y\} = (A_x B_y)$	$-2{\cdot}7(x+y)$
$\langle A \rangle + x(B_2) = \langle AB_{2x} \rangle$	$+3{\cdot}0x$
$\langle A \rangle + x(B_2) = \{AB_{2x}\}$	$+3{\cdot}4x$
$\langle A \rangle + (B_2) = (AB_2)$	$-2{\cdot}2$

2. Heats and Entropies of Transformation, Fusion and Evaporation

As already mentioned, a knowledge of the various heats of transition is more important than that of variation in heat capacity since they have a relatively greater effect on the value of the free energy of a substance. Fortunately the methods available for estimating their value, given a knowledge of the molecular structure of the substance, are fairly reliable.

1. *Evaporation*

The heat of evaporation of any substance can be obtained by a rule first put forward by Pictet (1876) and rediscovered and made more widely known by Trouton (1884). It states that the entropy of evaporation (i.e. the heat divided by the absolute temperature of evaporation) at the normal boiling point is approximately the same for all substances.

$$\Delta S_e = L_e/T_e \approx 22 \text{ cal/deg mole} \tag{115}$$

Table XIII shows, however, that Trouton's constant somewhat increases with boiling temperature. It is also affected when there is an association or dissociation of the substance in the gaseous state. Recently more mass spectrometric data have become available. They show that virtually every compound species more or less dissociates and/or associates in the gas. The vapour over molten lead difluoride for instance (Table XIII) consists predominantly of PbF_2 but also contains some PbF, PbF_4 and Pb. This somewhat detracts from the value of Trouton's rule as a means of estimating heats of evaporation, but it is still useful for checking experimental results.

Entropy is a measure of state of order. The entropy of evaporation thus represents the difference between the states of order of a liquid and its vapour. Gases at one atmosphere pressure behave almost as perfect gases, so their "state of order" may be regarded as constant. We may also assume that the "state of order" of any liquid is roughly the same, and it therefore follows that the entropy of evaporation is an almost constant quantity.

Table XIII. Entropies of evaporation at the normal boiling points

Substance	L_e (cal/mole)	T_e (°K)	L_e/T_e (cal/deg)
$SiCl_4$	6,840	330	20·7
PCl_3	7,300	348	20·9
Hg	14,150	630	22·47
Zn	27,270	1180	23·36
$FeCl_2$	30,000	1285	23·35
Mg	30,450	1363	22·34
PbF_2	38,300	1565	24·47
KOH	32,000	1600	20·00
$MgCl_2$	37,340	1691	22·08
Tl	39,220	1746	22·46
ZnF_2	44,200	1778	24·86
AgCl	42,500	1838	23·13
Ag	59,890	2436	24·59
MgF_2	65,300	2605	25·07
CaF_2	73,840	2805	26·32
Sn	70,700	2876	24·58
Fe	83,550	3135	26·65

2. *Fusion*

The entropy of fusion is not as constant as that of evaporation. The change in state of order on melting is not as great as on evaporation, and the variation in the state of order of a solid, due to the variety of the possible binding forces, has therefore a proportionately large effect on the entropy of fusion.

Crompton (1895) and Richards (1897) were the first to point out that the entropy of fusion of metals should be nearly constant. An average value of 2·2 cal/deg was suggested by Tammann [355]. This value, however, only applies to f.c.c. and h.c.p. metals, and closer examination reveals that it somewhat increases with temperature. Available data may approximately be represented by

$$\Delta S_m = 1{\cdot}77 + 0{\cdot}37 \times 10^{-3}\ T_m\ [\text{cal/°K g-atom}] \qquad (116)$$

For b.c.c. metals the values lie on average a little lower, say

$$\Delta S_m = 1{\cdot}62 + 0{\cdot}17 \times 10^{-3}\, T_m \text{ [cal/°K g-atom]} \tag{117}$$

Those for the rare-earth metals except europium are even lower than that.

If the lattice is bound together more or less by covalent bonds (as in Ga, Si, Ge, Sn, Sb, Bi, Se, Te), the entropy of fusion is substantially higher than 2·2. However, since the heats of fusion of metals melting below 2000°K are virtually all known, there is little point in a detailed analysis of the phenomenon from the practical aspect.

Entropies of fusion of alloy phases have been discussed in several published papers [356]. If the alloy is completely disordered, they may be calculated additively from the entropies of fusion of the component metals. Table XIV shows that this rule applies quite satisfactorily. If the alloy (or compound) is completely ordered, a disorder portion σ has to be added to the value calculated additively. This portion depends upon the composition of the alloy phase, which for a binary alloy is given by

$$\sigma = -\boldsymbol{R}(N_1 \ln N_1 + N_2 \ln N_2) \tag{118}$$

where N_1 and N_2 are the atomic fractions of the two components. Values calculated in this way are compared with experimental values in Table XIV. The agreement may be regarded as satisfactory. Where alloys are only partly ordered the quantity to be added is between 0 and σ, and can only be estimated roughly.

Table XIV. Entropies of fusion of ordered and disordered alloy phases

Phase	State	ΔS_f (cal/deg) (obs.)	ΔS_f (cal/deg) (add.)	Δ	σ
$\frac{1}{2}$CdSb	ordered	5·25	3·98	1·27	1·38
$\frac{1}{3}MgZn_2$	ordered	3·70	2·34	1·36	1·23
$\frac{1}{3}NaHg_2$	ordered	3·29	2·22	1·07	1·23
$\frac{1}{3}CuAl_2$	ordered	3·48	2·33	1·15	1·23
$\frac{1}{2}$AuSn	ordered	4·43	2·85	1·58	1·38
$\frac{1}{2}$AuZn	ordered	2·85	2·29	0·56	1·38
$\frac{1}{7}Na_2Pb_5$	partially ordered	2·86	2·13	0·73	(1·19)
$\frac{1}{13}Cu_5Cd_8$	partially ordered	2·84	2·43	0·31	(1·33)
$\frac{1}{3}Bi_2Tl$	partially ordered	3·82	3·82	0·00	(1·23)
$\frac{1}{8}Tl_5Pb_3$	disordered	2·07	1·93	0·14	0
$\frac{1}{5}Bi_3Tl_2$	disordered	3·55	3·62	−0·07	0
$\frac{1}{7}Hg_5Tl_2$	disordered	1·7	2·19	−0·49	0
εAu–Zn	disordered	2·33	2·30	0·03	0
εAg–Cd	disordered	2·34	2·37	−0·03	0

The method outlined above of estimating entropies of fusion of alloy phases makes use of the entropies of fusion of the constituent elements. When even these are unknown, the approximate values $\Delta S_f = 3{\cdot}5$ cal/deg and $\Delta S_f = 2{\cdot}2$ cal/deg may be applied to one g-atom of ordered and disordered binary alloy phases respectively [357].

It is rather more difficult to predict the entropies and, thus, the heats of fusion of inorganic compounds, as the random collection in Table XV shows. The entropies of fusion, that is the difference in entropy of the solid and liquid, depends on the difference in both atomic order and bond mechanism which are interrelated. To a certain extent, crystallographic structure of the solid provides an indication of bond mechanism, but

AgCl and NaCl or $CaCl_2$ and MgF_2 having the same structure have rather different entropies of fusion. That the values for substances with molecular structures are lower than those for co-ordination lattices is understandable. With the former only the inter-molecular bonds are "broken" on melting, but the intra-molecular bonds much less so.

Fortunately for practical applications, heats of fusion are comparatively small, and even an approximate estimate will serve the purpose of calculating equilibria in which liquid species partake. For such ends, entropies and heats of fusion may be estimated by comparison with similar compounds, such as those listed in Table XV.

Table XV. Entropies of fusion of inorganic compounds

Compound	T_m (°K)	ΔS_m (cal/degree) per mole	ΔS_m (cal/degree) per g-atom	Structure
NaF	1269	6·24	3·12	NaCl
NaCl	1074	6·26	3·13	NaCl
KF	1131	5·76	2·88	NaCl
KCl	1044	6·02	3·01	NaCl
MgO	3098	5·98	2·99	NaCl
AgCl	728	4·26	2·13	NaCl
BeO	2850	6·76	3·38	Wurtzite
FeS	1468	5·26	2·63	—
TlCl	702	5·41	2·71	CsCl
TlBr	733	5·35	2·68	CsCl
$MgCl_2$	987	10·44	3·48	$CdCl_2$
$MnCl_2$	923	9·75	3·25	$CdCl_2$
$FeCl_2$	950	10·8	3·6	—
$CaCl_2$	1045	6·51	2·17	Rutile
MgF_2	1536	9·06	3·02	Rutile
TiO_2	2143	7·44	2·48	Rutile
$FeCl_3$	581	15·59	3·9	$FeCl_3$
$ScCl_3$	1240	13·0	3·25	$FeCl_3$
$HoBr_3$	1192	10·0	2·5	$FeCl_3$
$CeCl_3$	1090	11·7	2·92	$Y(OH)_3$
$PrBr_3$	966	11·7	2·92	$Y(OH)_3$
YCl_3	994	7·6	1·9	$AlCl_3$
$HoCl_3$	993	7·3	1·82	$AlCl_3$
MoO_3	1074	10·80	2·70	—
WO_3	1745	10·04	2·51	—
Al_2O_3	2327	12·15	2·43	Cr_2O_3
V_2O_5	943	16·54	2·36	—
PbI_2	680	5·69	1·90	Layer lattice
$PbCl_2$	772	7·13	2·38	Layer lattice
$TeCl_4$	497	9·05	1·81	Molecular l.
$SnCl_4$	239	9·20	1·84	Molecular l.
UF_6	337	13·62	1·95	Molecular l.

3. Entropy and Entropy Changes

1. Solids

The standard entropies of all elements of metallic interest are known more or less accurately so there is no need of methods for estimating them. Very few entropy data are available for alloy phases and intermetallic compounds. No great error is involved in this event if the normal entropies of *ordered* alloys are taken additively from those of their solid components. For binary *disordered* alloys a similar value is used with the addition of the proportion of the amount

$$\sigma = -\boldsymbol{R}(N_1 \ln N_1 + N_2 \ln N_2) \text{ [per g-atom]}$$

dependent upon the degree of order. Since this amount has a maximum value at $N_1 = N_2 = 0{\cdot}5$ of 1·4 cal/deg g-atom it may even be neglected if the state of disorder is unknown.

In special cases, one may improve on this cursory method of estimation. Johnson [358], for instance, has reviewed the thermochemical data for the alloy phases of plutonium, thorium and uranium with other metals and has suggested an empirical relation for predicting the entropy of formation of these compounds. Such a procedure can, of course, be applied to other basis metals provided entropies are available for a sufficient number of intermetallic compounds.

For the estimation of standard entropies of predominantly ionic compounds, Latimer [359] has devised a method by which the entropies are obtained additively from values found empirically for the cationic and anionic constituents. For this he has taken into account the effects of ionic size and charge in addition to the effect of ionic mass. Revised values for the standard entropies of cations and anions are given in Tables XVI and XVII. In order to obtain from these tables the standard entropy of a solid compound, the appropriate value for its cation in Table XVI is multiplied by the number of cations in the molecule and added to the value for the anion. The latter figure is obtained by multiplying the value given in Table XVII according to the charge on the cation by the number of anions in a formula weight. Thus the standard entropy of $Al_2(SO_4)_3$ is deduced as $S_{298} = 2 \times 8{\cdot}0 + 3 \times 13{\cdot}7 = 57{\cdot}1$ cal/deg. The accepted value is $S_{298} = 57{\cdot}2$ cal/deg.

Table XVI. Entropies in cal/degree of the elements in solid ionic compounds at 298°K

Ag	12·8	Eu	14·1	Nb	12·2	Sm	14·1
Al	8·0	Fe	10·4	Nd	13·9	Sn	13·1
As	11·45	Ga	11·2	Ni	10·5	Sr	12·0
Au	15·3	Gd	14·3	Os	15·1	Ta	14·9
B	4·9	Ge	11·3	Pb	15·5	Tb	14·3
Ba	13·7	Hf	14·8	Pd	12·7	Te	(13·4)
Be	4·3	Hg	15·4	Pr	13·8	Th	15·9
Bi	15·6	Ho	14·5	Pt	15·2	Ti	9·8
C	5·2	In	13·0	Ra	15·8	Tl	15·4
Ca	9·3	Ir	15·2	Rb	11·9	U	16·0
Cd	12·9	K	9·2	Re	15·0	V	10·1
Ce	13·8	La	13·8	Rh	12·5	W	15·0
Co	10·6	Li	3·5	Ru	12·5	Y	12·0
Cr	10·2	Lu	14·8	S	(8·5)	Yb	14·7
Cs	13·6	Mg	7·6	Sb	13·2	Zn	10·9
Cu	10·8	Mn	10·3	Sc	9·7	Zr	12·1
Dy	14·4	Mo	12·3	Se	(11·6)		
Er	14·5	Na	7·5	Si	8·1		

The less ionic the nature of the compounds, the larger are the deviations from the entropies estimated by the Latimer method. The strongly covalent sulphides of the transition metals, for instance, tend to have lower entropies than those estimated.

A further contribution due to the mixing of cations can occur in double oxide compounds such as spinels MX_2O_4 or $MO{\cdot}X_2O_3$. In this structure one-third of the cations occupy tetrahedral sites, and two-thirds occupy octahedral sites in the close-packed oxide lattice. If all of the tetrahedral sites which are occupied contain only M^{2+} ions and the octahedral sites contain only X^{3+} ions, this structure, the normal spinel, does not have a cation-mixing contribution. If, however, some of the tetrahedral sites which are occupied contain X^{3+} ions, and some of the octahedral sites contain M^{2+} ions, there is such a contribution. In the general spinel of formula $(M_xX_{1-x})[M_{1-x}X_{1+x}]_2O_4$ where x is equal to zero in the normal spinel, and equal to unity in the *inverse* spinel, the value of x can be calculated from the equilibrium constant for the exchange reaction

$$(M) + [X] \quad [M] + (X) \qquad \begin{array}{l}(\,) = \text{tetrahedral site} \\ [\,] = \text{octahedral site}\end{array}$$

$$\Delta H_{\text{exchange}} = -\boldsymbol{R}T \ln \frac{(1-x)^2}{x(1+x)} \tag{119a}$$

ΔH exchange here is the exchange energy for M and X ions between tetrahedral and octahedral sites as indicated. Apart from considerations of Madelung energy, which normally cause the higher valency cations to seek the more highly co-ordinated octahedral sites, the cations of the transition metals have a contribution arising from the d electron population. This is the reason for Co_3O_4 and Mn_3O_4 having the normal spinel structure, whereas Fe_3O_4 has the inverse structure [285]. The entropy contribution due to cation mixing is given by the equation

$$S = -\boldsymbol{R}\left[x \ln x + (1-x)\ln(1-x) + (1-x)\ln\frac{(1-x)}{2} + (1+x)\ln\frac{(1+x)}{2}\right] \tag{120}$$

In a compound such as Zn_2TiO_4 or $2ZnO{\cdot}TiO_2$, the Ti^{4+} ion occupies half of the occupied octahedral sites because of the high cationic charge, and hence x has the value 0·5. The entropy of mixing can be less than that given by equation above due to ordering of the Zn^{2+} and Ti^{4+} ions on the octahedral sites. This effect cannot be easily predicted without detailed knowledge of the structure of the oxide and its deviations due to distortion from the ideal spinel structure. There will always be a finite contribution from cation mixing, however, which will yield a "residual" entropy for the compound, which cannot be calculated from the usual thermal data.

Table XVII includes proposed values for the $SiO_3^{=}$ ion. The standard entropy of silicates is also approximately equal to the sum of the entropies of the constituent oxides [360]. The same applies to other double oxides, such as some normal spinels [361]. To be more exact, significant negative deviations may occur when the heats of formation from the component oxides are pronouncedly negative.

Entropies of formation of carbides are rather small. It seems that the interstitial carbides are formed from their elements with decrease in entropy, whereas the more complex carbides are formed with increase in entropy [362]. For the entropy of formation of transition metal carbides, Worrell [363] derived the following relationship empirically:

$$\Delta S_{298} = -6{\cdot}0 + 6{\cdot}0(\theta_M/\theta_C) \quad [\text{cal/°K g-atom carbon}] \tag{119}$$

where θ_M and θ_C are the melting points of the metal and the carbide, respectively, in °C.

Grønvold and Westrum [364] have discussed the estimation of the standard entropies of the chalcogenides of the transition elements in some detail.

Table XVII. Entropy contribution in cal/degree of negative ions in solid compounds at 298°K

Negative ion	Charge on positive ion +1	+2	+3	+4
F^-	5·3	4·4	(4·0)	4·6
Cl^-	9·7	8·1	6·9	8·1
Br^-	13·0	10·9	9·9	10·7
I^-	14·6	13·6	12·5	13·0
OH^-	(5·0)	4·5	3·0	
$O^=$	1·3	0·5	0·5	1·0
$S^=$	(8·0)	5·0	5·15	3·65
$SO_4^=$	22	17·2	13·7	(10)
$Se^=$	(10·1)	7·1	8·0	7·8
$Te^=$	(10·9)	10·2	10·3	(8·8)
$CO_3^=$	15·2	11·4	(8·0)	
$SiO_3^=$	14·5	10·5	(7·0)	
$PO_4^=$	(24)	17·0	(12)	

2. Gases

The molecular heat capacity of gases follows certain rules implicit in Clausius' principle of equipartition of energy. The temperature at which the various degrees of freedom become operative in the molecule depends on the masses of and distance between the atoms, and on the rigidity of the inter-atomic bonds. The integral represented by the area beneath the C_p/T curve from absolute zero to 298°K forms one part of the normal entropy of gases. The entropies of fusion and of evaporation constitute further instalments which also obey certain rules. The ease with which a solid lattice disrupts to form a liquid, and with which that liquid evaporates, depends broadly on the atomic size and type of bond. Thus one may expect to find a certain regularity in the normal entropy of gases in relation even to the molecular weight alone. That this is indeed so is shown by Fig. 97, in which the normal entropies are plotted against molecular weight.

The proper way to estimate the standard entropies of gaseous species is via the molecular parameters as has been mentioned earlier (p. 181). However, the present authors sometimes use, as a short-cut, a quicker mode of estimating the standard entropies of gases based on molecular weight alone. Having plotted S°_{298} against the logarithm of molecular weight, the following relationships, depending on the number of atoms in the molecule, have been obtained empirically. (The numbers in brackets indicate the mean deviations from average of the experimental values.)

No. of atoms in molecule	S°_{298} in cal/deg mole	
1	$+26·5 + 7·9 \log M$	(±1·6)
2	$+24·2 + 16·3 \log M$	(±1·4)
3	$+9·0 + 26·7 \log M$	(±1·8)
4	$-1·8 + 35·0 \log M$	(±1·6)
5	$-31·5 + 49·5 \log M$	(±2·7)

If there are more than five atoms in the molecule, the last equation may still be used, but, when more experimental values become available, additional empirical equations may be obtained.

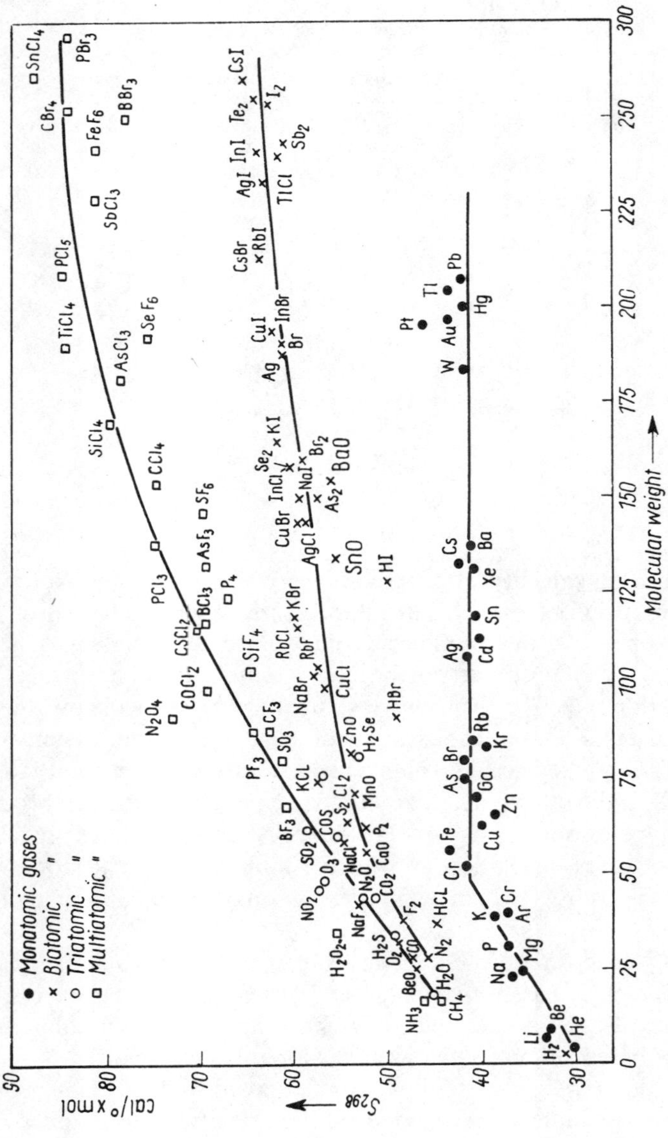

FIG. 97. Standard entropies of gases (at 298·1°K).

4. Heats of Formation

The entropy change of a reaction is an indication of the temperature coefficient of the corresponding equilibrium. The estimation of entropies goes a little way only toward evaluating equilibria or free energies, however, so long as heats of formation or reaction (i.e. enthalpy values) are unknown. Unfortunately, the methods of estimating heats of formation are not very reliable, and in any case cover a relatively small number of compounds. Their uncertainty is such that as many methods as possible should be used to estimate a single value.

The enthalpy of the elements at their standard state at 298·15°K is by agreement taken to be zero. With this as our basis we shall consider the heats of formation of binary compounds only. Since the variation of heats of formation with temperature is generally small, values may be assumed to be approximately constant at all temperatures so long as no change occurs in the state of aggregation.

1. General

If two elements form several compounds of different composition, there are certain fairly general relationships between their heats of formation. In the ensuing discussion, the numerical values of heat of formation will be taken on the g-atomic basis in place of the mole. Thus the heat of formation of ferric oxide, Fe_2O_3, is $-\Delta H_{298} = 197{\cdot}0$ kcal/mole, which will be regarded as $197{\cdot}0 \div 5 = 39{\cdot}4$ kcal/g-atom.

If the melting points of the compounds in a given binary system are known, the one with the highest melting point may be expected to have the (numerically) highest heat of formation per g-atom. If the melting points of the other compounds are considerably lower, their heats of formation may be regarded as additive (i.e. in the relative proportions of the high melting-point compound and one element). If their melting points are not much lower than the highest, their heats of formation will have a value between the additive and the maximum. This may be clarified by considering a few particular systems.

In the iron–chlorine system, $FeCl_2$ has the highest melting point and that of $FeCl_3$ is lower. The former has a maximum in the ΔH vs. N curve (Fig. 98), whereas the latter is only marked by a slight discontinuity.

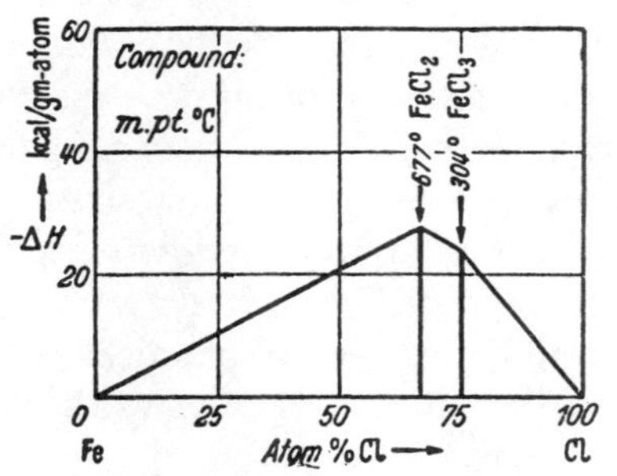

FIG. 98. Heats of formation in the system iron–chlorine.

Figure 99 gives the heat of formation curve of a system containing a number of compounds. Again, the compound with the highest melting point (Li_7Sn_2) has the highest heat of formation. The compositions Li_4Sn and LiSn, both of which show a maximum in the liquidus curve, are marked in the heat of formation curve by pronounced angularities. The other compounds in the system are formed peritectically, and in accordance with this there is practically no indication of them in the $\Delta H/N$ curve. These diagrams are quite typical for systems such as these.

If several compounds in an intermetallic system have nearly the same melting point, the highest heat of formation may be expected from the compound nearest to equiatomic composition, as is known for the sodium–lead alloys, for instance. On the other hand,

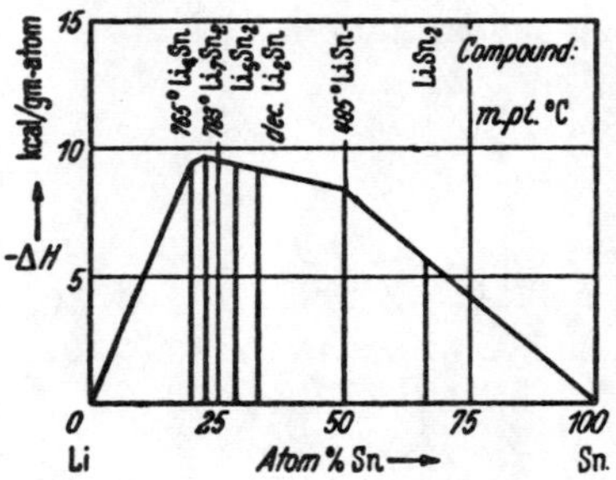

FIG. 99. Heats of formation in the system lithium–tin.

in the manganese–oxygen system (Fig. 100) the heat of formation of Mn_3O_4 is higher than that of MnO. This is probably due to structural differences, the effects of which will be discussed below.

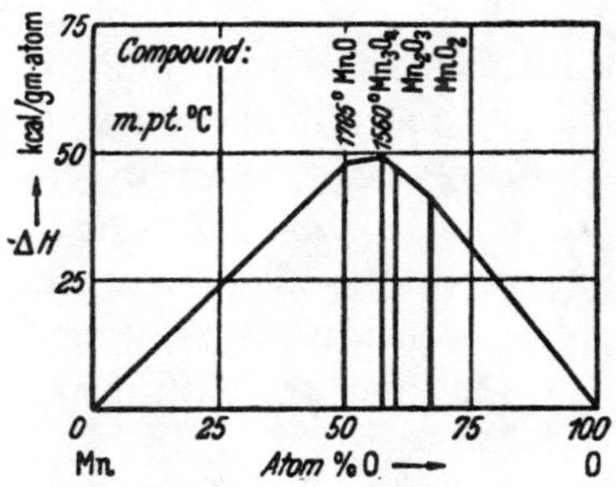

FIG. 100. Heats of formation in the system manganese–oxygen.

It is often necessary to derive the mean values of the data of several observers for the compounds in one system. An indication of how this is done is illustrated in Fig. 101. The crosses show the figures evaluated by Bichowsky and Rossini (1936) in an earlier monograph of the Bureau of Standards. The line connecting these values (shown dotted) has two concave angles corresponding to the compounds VO and VO_2. This means that these compounds should be formed from their two respective neighbouring phases with an absorption of heat. While such behaviour *can* arise even with stable compounds, it is unlikely and is rarely observed in practice. The continuous line shows a much more probable mean of the values. In a later work of the Bureau of Standards these probable errors were corrected by Rossini.

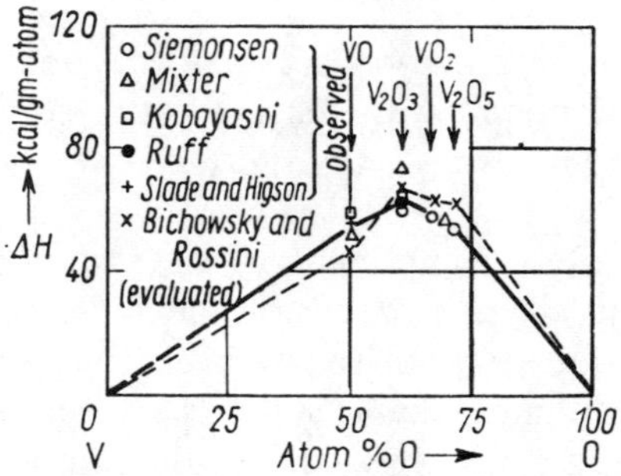

FIG. 101. Heats of formation in the system vanadium–oxygen.

This example illustrates the importance, when evaluating thermochemical data critically, of considering a system as a whole and not the constituent compounds separately.

If the heat of formation of one of the compounds in a binary system is known, the same principle may be applied. The melting points of the other compounds and any statement in the literature about the stability of the compounds (determined empirically) are used to provide a measure of the relative stability of the various phases of the system.

2. *Homologous Series*

There is some evidence that a certain relationship holds between the heat of formation of metal compounds and the atomic number of the metal in compounds of the same stoichiometric proportion and the same common radical. This correlation has been demonstrated, for instance, by Roth [60] with all the available data for the oxides and chlorides. In Figs. 102 and 103 some of his diagrams are reproduced with only minor changes. There is no support for the linear relationship which has sometimes been proposed, but regular curves certainly exist. They may show a maximum or a minimum, and unknown heats of formation can be estimated with some certainty. Roth verified experimentally some of the values he predicted from these curves. Similar regularities have been noted among the metal nitrides [365].

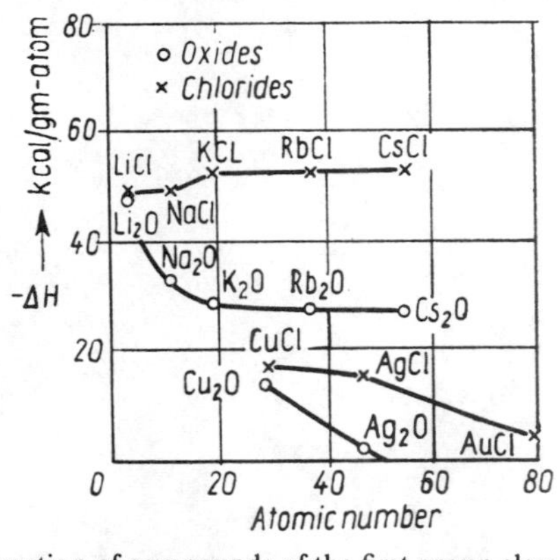

FIG. 102. Heats of formation of compounds of the first group elements with oxygen and chlorine.

The same regularities may be observed in Table XVIII which contains the heats of formation of the compounds of magnesium, calcium and barium with the elements in the fourth to the seventh main group of the Periodic Table. They appear in the ΔH values of homologous compounds of the same electronegative component. Both these interrelationships should be taken into consideration when estimating heats of formation. Much discretion must be used in applying it, dependent upon the substance under consideration.

We may evaluate as an example the probable heat of formation of calcium telluride. From the vertical column CaO–CaTe we extrapolate a value $-\Delta H_{CaTe} = 27$–34 kcal/g-atom. The horizontal Ca_2Sn–CaI_2 gives $-\Delta H_{CaTe} = 35$–42, and, finally, by comparison with the magnesium compounds and the value for MgTe we arrive at $-\Delta H_{CaTe} = 30$–40. From this we deduce that it is highly probable that the heat of

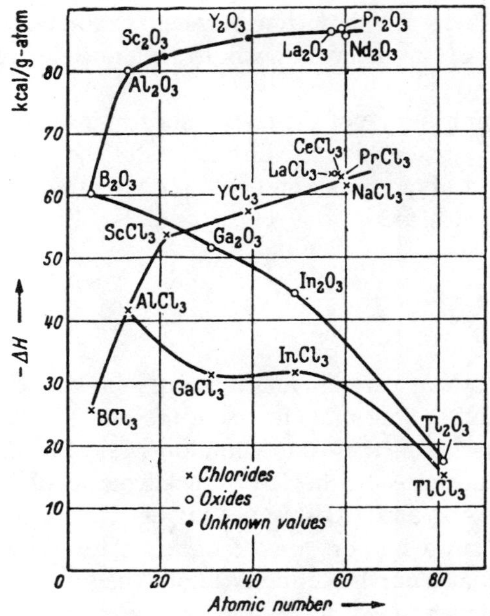

FIG. 103. Heats of formation of compounds of the third group elements with oxygen and chlorine.

Table XVIII. Heats of formation ($-\Delta H_{298}$) in kcal/g-atom of the stoichiometrical compounds of the alkaline earth metals with the elements in the fourth to the seventh group of the Periodic Table

Group							
IV		V		VI		VII	
—		Mg_3N_2	22·1	MgO	71·9	MgF_2	88·7
Mg_2Si	6·3	Mg_3P_2		MgS	41·5	$MgCl_2$	51·1
Mg_2Ge	9·2	Mg_3As_2		MgSe		$MgBr_2$	41·3
Mg_2Sn	6·1	Mg_3Sb_2	15·8	MgTe	25·0	MgI_2	28·7
Mg_2Pb	4·2	Mg_3Bi_2	8·1				
—		Ca_3N_2	21·0	CaO	76·0	CaF_2	97·3
(Ca_2Si)	16·5	Ca_3P_2	24·0	CaS	56·9	$CaCl_2$	63·8
Ca_2Ge		Ca_3As_2		CaSe	37·4	$CaBr_2$	54·4
Ca_2Sn	25·0	Ca_3Sb_2	34·8	CaTe		CaI_2	42·6
Ca_2Pb	17·2	Ca_3Bi_2	25·2				
—		Ba_3N_2	16·3	BaO	69·5	BaF_2	95·9
Ba_2Si		Ba_3P_2		BaS	55·0	$BaCl_2$	68·4
Ba_2Ge		Ba_3As_2		BaSe	37·1	$BaBr_2$	60·1
Ba_2Sn	30·0	Ba_3Sb_2	35·0	BaTe		BaI_2	48·2
Ba_2Pb	23·0	Ba_3Bi_2	32·0				

formation of CaTe is given by $-\Delta H_{298} = 70{,}000 \pm 7000$ cal/mole. The authors would place more reliance upon this value than on a widely different one determined experimentally.

We should be going to excess were we to derive an empirical formula for each class of compounds, but it can be done when necessary to solve a particular problem.

General formulae for the derivation of heats of formation of oxides, sulphides and halides from consideration of the Periodic Table have been given independently by two French authors Süe [366] and Trombe [367], and by the Russian thermochemist Kapustinski [368]. Kapustinski, for instance, suggests the use of the formula

$$\log Z_A = -\Delta H/z + \text{const}$$

where A is any element in a vertical or horizontal column of the Periodic Table combined with a certain other element B, Z_A is its atomic number, z its valency and ΔH is the standard heat of formation of the solid compound. The value of the constant is obtained from known heats of formation for each series. However, a close examination of the tables of the various authors cited reveals that large errors can arise when these methods are used; and the deviations observed are a strong argument in favour of using them in conjunction with all other possible methods of estimation.

From considerations of the theoretical basis of these regularities, we cannot expect them to be more than approximate. By means of the Haber–Born cycle, one can resolve the heat of formation into five terms: the electron affinity of the anion, the ionisation voltage of the metal, the heats of sublimation and of dissociation of the two components, and the lattice energy. These factors arise in a reaction between a divalent metal Me and a halogen X_2, for example, at the stages indicated in the following schematic representation:

H (ionisation)	$(Me^{++}) + 2(X^-)$	Electron affinity	Lattice energy
	$(Me) + 2(X)$	ΔH (dissociation)	
L (sublimation)	$\langle Me \rangle + (X_2)$		
	$\langle MeX_2 \rangle$	ΔH (formation)	

Keeping, for instance, the electro-negative component the same, its electron affinity and heat of dissociation remain constant factors. The ionisation voltage changes more or less regularly with electronic configuration, and so does the heat of sublimation. The lattice energy of substances of the same structural type depends only on the interatomic distance and so again depends on atomic number. This regularity is approximately observed even with compounds of different structural types, as the effect of distance is the more important. From all this we conclude that the heats of reaction of one element with metals of one group should change regularly with atomic number of the variable element, but the regularity can only be expected to be approximate, and is applicable only to ionic compounds.

Since the energies of sublimation and ionisation of the metals, on the one hand, and the electron affinities and heats of dissociation of the halogens, on the other, are mostly known, the more accurate method would appear to be to estimate the lattice energies rather than the heats of formation. An empirical method of doing this has, in fact, been used by Brewer *et al.* [370]. The procedure followed was that of calculating the lattice energy for the halides for which heats of formation were available by going through the Haber–Born cycle, plotting these lattice energies per atom of halogen in each halide against the sum of the cationic and anionic radii, estimating unknown lattice energies from the graphs, and calculating the heats of formation by going backward through the cycle. The points in the graphs obtained for the fluoride, chloride and bromide of a metal lay almost invariably on a straight line of negative slope. In any group of metals such as the alkali metals, the lines connecting the fluoride, chloride and bromide values for each metal were very nearly parallel, and the slight deviation from a common slope varied regularly with increasing atomic weight. In addition, the lines for several groups of metals in the same oxidation state, e.g. the cadmium and lead halides and the alkaline–earth metal halides, were found to fit in very well with one another.

3. *Electronegativities*

A somewhat simpler though less accurate method of estimating heats of formation of predominantly ionic compounds, based on considerations of Born's cycle, has been introduced by Pauling. Since the diameters of the ions are nearly constant, the heat of formation of the compound is really the difference of the virtually constant energy values of the two components which may be computed independently. These characteristic values of the elements are called the "electronegativities".

Relative values for the electronegativities have been derived by various authors (e.g. [369]) from the heats of formation of the metal halides. The empirical electronegativity values used by the present authors are listed in Table XXII. These may be employed to estimate heats of formation of ionic compounds using the following equation:

$$\Delta H_{298} = -23{\cdot}066z(\varepsilon_A - \varepsilon_B)^2 \tag{121}$$

where z is the number of valency links, ε_A and ε_B are the empirical electronegativities and ΔH_{298} is the heat of formation in kcal/mole. In Table XIX the experimental heats of formation of magnesium compounds, reaching from the ionic to the metallic types, are compared with those calculated by means of eqn. (121). The table conveys a representative impression of the accuracy that can be expected when estimating heats of formation of predominantly ionic compounds from electronegativities.

Table XIX. Heats of formation of some compounds of magnesium at 25°C in kcal/mole as observed experimentally and estimated from

$$\Delta H = -23{\cdot}07z(\varepsilon_{Mg} - \varepsilon_B)^2\ [\varepsilon_{Mg} = 1{\cdot}2]$$

Compound	ε_B	$-\Delta H_{298}$ Experimental	Estimated
MgF_2	3·5	266·0	244
$MgCl_2$	3·0	153·4	150
MgI_2	2·55	86·0	84
MgO	2·9	143·7	133
MgS	2·5	83·0	78
$MgTe$	2·1	50·0	51
Mg_3Sb_2	1·85	79·0	59
Mg_3Bi_2	1·8	40·5	50
Mg_2Sn	1·7	18·3	23
Mg_2Pb	1·6	12·5	14·5
$MgTl$	1·5/1·9	12·0	4·1/22·5
$MgAg$	1·8	9·2	16·5
$MgNi_2$	1·8	13·5	16·5

4. *Volume Change and Heat of Formation*

If the heat of formation is required of a compound whose density and structure is known, one can use a method of estimation which was first applied to intermetallic compounds [371]. Its basis is that the deformation or polarisation of the atoms of the two metals which occurs on alloy formation depends upon their affinity. Attention was first brought to this relationship by Richards [372] who compared the heats of formation of the halides with the difference between their molecular volume and the total atomic volumes of their constituents. From this striking connection he was led to the belief that heat of formation was a heat of compression.

In the formation of simple inorganic salt-like compounds the change in density is mainly due to the formation of ions, but surprisingly enough the relationship we have just mentioned still applies. A measure of the change in atomic size is given by the contraction in molecular volume, taking certain rules into observation. The contraction on formation of phases with the NaCl and CsCl structure is calculated in the following way. The atomic volumes of the metals are taken in their state of co-ordination number 12. The atomic volumes of nitrogen, oxygen and sulphur are not known for high co-ordination, but for this purpose they are assumed to be 5·0, 9·0 and 14·5, respectively. The atomic volumes are taken as those of solid state. The molecular volume of the compound should be that of closest packing, in the face-centered cubic form. This requires that the measured molecular volume (molecular weight/density) should be multiplied by a factor e which has the values 0·95 and 0·825 for the CsCl and the NaCl types of lattice respectively. The percentage volume contraction is then given by

$$\Delta V = \frac{100\,(e\,\mathrm{MV} - \Sigma \mathrm{AV})}{\Sigma \mathrm{AV}} \qquad (122)$$

where MV is the measured molecular volume of the compound and ΣAV the sum of the atomic volumes of the two components.

The values of ΔV for a number of compounds are plotted in Fig. 104 against their heats of formation. The plots are seen to lie fairly close to a curve from which, with a few exceptions, they deviate no more than ±6000 cal/g-atom. The wide deviation of lithium thallide is explained by the large difference in the two ionic sizes which permits a much closer packing than is represented by the factor 0·95. To attempt to make an allowance for these ionic differences by introducing an atomic size ratio into the formula would make the relationship much more complicated without even then obtaining complete agreement. A similar relationship holds between compounds of other structures, such as the zinc sulphide or γ-brass type [373]. It needs further detailed consideration, however, and at this stage it can only be taken as an empirical relationship which gives an approximate value for the heat of formation. Its results should be used only in comparison with those of other methods of estimation, since large deviations can occur and since the densities are not always known with sufficient accuracy.

Instead of plotting the data of all binary compounds in one diagram it is preferable to consider the compounds of each structural type separately, using for the volume change term the expression

$$\Delta V = \frac{100(\mathrm{MV} - \mathrm{AV})}{\Sigma \mathrm{AV}} \qquad (123)$$

The unknown heat of formation may then be estimated from the intersection of the curve by the corresponding ΔV value.

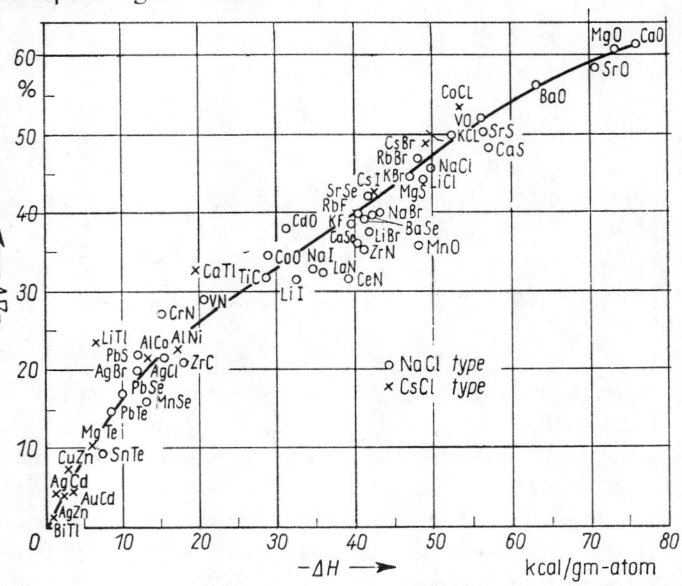

FIG. 104. Heat evolution and contraction on formation of compounds of simple structural type.

5. Heats of Solution

It follows from the considerations in Section II.1.C.3 that the heat of formation of a compound can be calculated from its heat of solution in a solvent and those of the constituent elements in the same solvent provided that the final systems are identical. As a generalisation of the scheme given on p. 80 one may write for the solution of a compound of a divalent metal Me and a gaseous halogen:

$$
\begin{aligned}
\langle \mathrm{Me} \rangle &= [\mathrm{Me}^{++}]_{\mathrm{aq}\infty}\,; && \Delta H_1 \\
(\mathrm{X}_2) &= 2[\mathrm{X}]_{\mathrm{aq}\infty}\,; && \Delta H_2 \\
\langle \mathrm{MeX}_2 \rangle &= [\mathrm{Me}^{++}]_{\mathrm{aq}\infty} + 2[\mathrm{X}^-]_{\mathrm{aq}\infty}\,; && \Delta H_3 \\
\hline
\langle \mathrm{Me} \rangle + (\mathrm{X}_2) &= \langle \mathrm{MeX}_2 \rangle\,; && \Delta H_4 \\
\Delta H_4 &= \Delta H_1 + \Delta H_2 - \Delta H_3
\end{aligned}
$$

It is conventional to take the standard states of ΔH_1 and ΔH_2 at infinite dilution at which any mutual interaction of the cations and anions can be neglected. These values are found by determining the heats of solution at various concentrations and extrapolating to infinite dilution, denoted by the suffix "aq ∞".

The heats of solution of ionic compounds in water (i.e. ΔH_3) are generally relatively small. On the other hand, the heats of formation of a large number of anions and cations in water are known with quite a high accuracy, and it is possible to make use of these two facts for another method of estimating unknown heats of formation. The hydrogen ion in water at infinite dilution is taken as the standard state in that the heat of the reaction

$$\tfrac{1}{2}(\mathrm{H}_2) = [\mathrm{H}^+]_{\mathrm{aq}\infty} \qquad (\Delta H = 0)$$

is taken to be zero and all other heats of formation of ions are related to this value so that, for instance, the heat of formation of hydrogen chloride in water at infinite dilution,

$$\tfrac{1}{2}(H_2) + \tfrac{1}{2}(Cl_2) = [HCl]_{aq\infty} = [H^+]_{aq\infty} + [Cl^-]_{aq\infty},$$

is the same as the standard heat of formation of the chlorine ion:

$$\tfrac{1}{2}(Cl_2) = [Cl^-]_{aq\infty}$$

In Table XX are given the standard heats of formation of cations and anions according to the "selected values" of the Bureau of Standards, supplemented or corrected by some more recent measurements. Further values for more concentrated solution of the hydrogen halides are obtainable from Table VIII.

If the heats of solution are measured or can be estimated, Table XX supplies much information on heats of formation of inorganic salts. Many values in Table A are derived in this manner from known heats of solution. Further values are best estimated

Table XX. Standard heats of formation of ions in aqueous solution at infinite dilution

Ion	ΔH_{298} in $\frac{\text{kcal}}{\text{g-atom}}$	Ion	ΔH_{298} in $\frac{\text{kcal}}{\text{g-atom}}$	Ion	ΔH_{298} in $\frac{\text{kcal}}{\text{g-atom}}$	Ion	ΔH_{298} in $\frac{\text{kcal}}{\text{g-atom}}$
H†	0·0	La^{3+}	−167·0†	Cu^+	+12·4	F^-	−78·6
Li^+	−66·5	Ce^{3+}	−167·2‡	Cu^{++}	+15·4	Cl^-	−40·0
Na^+	−57·7*	Pr^{3+}	−165·3†	Ag^+	−25·3	Br^-	−28·9
K^+	−60·4*	Nd^{3+}	−163·2‡	Au^+	+46·4	I^-	−13·4
Rb^+	−58·9	Th^{4+}	−184·4§	Zn^{++}	−36·4	OH^-	−55·0
Cs^+	−62·6‡‡	U^{3+}	−123·0	Cd^{++}	−17·3	$S^=$	+7·8‖
Mg^{++}	−110·4	U^{4+}	−146·7	Hg^{++}	+41·6	SH^-	−4·1‖
Ca^{++}	−129·7	Mn^{++}	−52·3	Ga^{3+}	−50·4	$SO_4^=$	−216·9
Sr^{++}	−130·4	Fe^{3+}	−11·4	In^+	−12·5**	$Se^=$	(−23)
Ba^{++}	−128·6	Fe^{++}	−21·0	In^{3+}	−43·4**	$CO_3^=$	−161·6
Al^{3+}	−125·4	Co^{++}	−16·1	Tl^+	+1·3	PO_3^-	−235·6
		Co^{3+}	+21·2	Tl^{3+}	+27·7	PO_4^{3-}	−306·9
		Ni^{++}	−15·3	Pb^{++}	+0·5	$CrO_4^=$	−208·0††

* E. E. KETCHEN and W. E. WALLACE, J. Amer. Chem. Soc., 1951 **73** 5810.
† H. R. LOHR and B. B. CUNNINGHAM, J. Amer. Chem. Soc., 1951 **73** 2025.
‡ F. H. SPEDDING and C. F. MILLER, J. Amer. Chem. Soc., 1952 **74** 4195.
§ L. EYRING and E. F. WESTRUM, J. Amer. Chem. Soc., 1950 **72** 5555.
‖ J. W. KURY, A. J. ZIELEN and W. M. LATIMER, J. Electrochem. Soc., 1953 **100** 468.
** W. KANGRO and T. WEINGARTEN, Z. Elektrochem., 1954 **58** 505.
†† C. N. MULDRON and L. G. HEPLER, J. Amer. Chem. Soc., 1957 **79** 4045.
‡‡ H. L. FRIEDMANN and M. KAHLWEIT, *ibid.*, 1956 **78** 4243.

individually by comparison with the known heats of solution of chemically similar compounds. Some values are therefore given in Table XXI, and others may be obtained from the "Technical Notes 270" of the National Bureau of Standards. In order to extrapolate the heats for infinite dilution a series of values at various concentrations must be determined, but if one value only has been measured in fairly dilute solution (molar ratio, H_2O : solute > 400 : 1) this may be taken to represent approximately the heat of solution at infinite dilution, particularly if the heat effect is small.

Table XXI. Heats of solution of some substances in water

Substance X	Dilution, molar ratio H_2O/X	ΔH_{298} kcal/mole	Substance X	Dilution, molar ratio H_2O/X	ΔH_{298} kcal/mole	Substance X	Dilution, molar ratio H_2O/X	ΔH_{298} kcal/mole
NaF	∞	+0·22	$CeCl_3$	∞	−34·4	CsOH	∞	−7·0
KF	∞	−4·2	$ScCl_3$	∞	−48·1	$Ca(OH)_2$	∞	−3·88
RbF	∞	−6·3	$ScBr_3$	∞	−56·1	$Ba(OH)_2$	∞	−12·38
CsF	∞	−9·0	YCl_2	∞	−53·1	TlOH	2000	+2·9
LiCl	∞	−8·9	$AlCl_3$	600	−56·1	Na_2S	800	−15·2
NaCl	∞	+0·93	$MnCl_2$	400	−16·7	K_2S	400	−9·9
KCl	∞	+4·13	$FeCl_2$	∞	−19·5	Li_2SO_4	∞	−7·2
CsCl	∞	+4·3	$NiCl_2$	∞	−19·7	Na_2SO_4	∞	−0·6
NaBr	∞	−0·15	$CdCl_2$	∞	−4·4	K_2SO_4	∞	+5·7
CsBr	∞	+6·2	$ZnCl_2$	∞	−17·08	Cs_2SO_4	∞	+4·05
LiI	∞	−15·15	$PbCl_2$	∞	+6·2	$MgSO_4$	∞	−21·8
NaI	∞	−1·8	$CuCl_2$	2000	−12·2	$SrSO_4$	∞	−2·08
KI	∞	+4·9	$CuBr_2$	400	−8·8	$BeSO_4$	5000	−19·1
RbI	∞	+6·2	$CdBr_2$	∞	−0·67	$MnSO_4$	∞	−15·0
CsI	∞	+7·9	$ZnBr_2$	∞	−16·06	$FeSO_4$	200	−15·5
CaF_2	∞	+3·2	$PbBr_2$	∞	+6·2	$NiSO_4$	∞	−19·2
$MgCl_2$	∞	−37·05	CdI_2	∞	+3·34	$CuSO_4$	∞	−17·5
$CaCl_2$	∞	−19·82	ZnI_2	∞	−13·2	$CdSO_4$	∞	−12·85
$SrCl_2$	∞	−12·43	PbI_2	∞	+15·5	$ZnSO_4$	∞	−19·45
$BaCl_2$	∞	−3·16	TlF	800	−0·5	Tl_2SO_4	∞	+7·55
$MgBr_2$	∞	−44·5	TlCl	∞	+10·35	Na_2CO_3	400	−5·6
$BaBr_2$	∞	−6·07	TlBr	∞	+13·7	K_2CO_3	1000	−7·6
MgI_2	∞	−51·15	TlI	∞	+17·7	Cs_2CO_3	2000	−11·2
CaI_2	∞	−28·7	$GaCl_3$	∞	−45·1	$MnCO_3$	∞	0·0
SrI_2	∞	−21·6	AgF	∞	−4·85	Li_2SiF_6	1500	−2·2
BaI_2	∞	−11·4	LiOH	∞	−5·06	Na_2SiF_6	600	+5·8
$NdCl_3$	∞	−37·5	KOH	∞	−13·2			

6. *The Packing Effect*

Kubaschewski [374] suggested that the stability of many multicomponent, truly metallic, phases stems mainly from a decrease in heat content due to an increase of co-ordination on formation from the component metals. To demonstrate this, it is necessary to re-define the co-ordination numbers of the crystallographers. Only *changes* in co-ordination during the formation of a phase from the component metals need to be considered, and a knowledge of the total energy of a lattice is not required.

Most metals crystallise in the f.c.c., b.c.c. and h.c.p. lattices. In the f.c.c. and h.c.p. structures every atom is surrounded by 12 nearest neighbours at a distance of $2r_A$, the gaps being sealed by 6 atoms at a distance $2r_A\sqrt{2}$. For the present purpose only the changes taking place on alloying within the distance up to $2r_A\sqrt{2}$ are being considered. The simplification is made that the metallic bond energy varies linearly with $1/d$ within the chosen distance d, which was found to be a reasonable approximation. When A forms a new phase with another metal B, the A atoms are surrounded by A and B atoms at various distances d_A and d_{AB}, respectively. If these distances are less than those corresponding to the atomic radii, i.e. $d_A < 2r_A$ and $d_{AB} < (r_A + r_B)$, the bond energy will be increased and the "effective" co-ordination number C_A higher than the actual co-ordination number, C_A, by a factor of either $2r_A/d_A$ or $(r_A + r_B)/d_{AB}$:

$$C_A^* = \sum \frac{2r_A}{d_A} C_A + \sum \frac{(r_A + r_B)}{d_{AB}} C_{AB} \tag{124}$$

A corresponding relationship applies to the B atoms.

For distances $d_A > 2r_A$ and $d_{AB} > (r_A + r_B)$ a limit has been set above, the formal implication of which is that an A atom at the distance $2r_A\sqrt{2}$ just ceases to participate in the mutual attraction, the corresponding limit for A–B bonds being $(r_A + r_B)\sqrt{2}$. The corrections for obtaining the effective co-ordination numbers from the actual ones are then

$$C_A^* = \sum \frac{2(2r_A) - d_A}{(\sqrt{2} - 1)2r_A} C_A + \sum \frac{2(r_A + r_B) - d_{AB}}{(\sqrt{2} - 1)(r_A + r_B)\, C_{AB}} \tag{125}$$

and corresponding ones for the B atoms.

For the pure metals the atomic radii are taken to be equal to half the shortest atomic distances in their normal crystallographic structures. With the interpretation above, the f.c.c. and h.c.p. metals retain their co-ordination number of 12. For the b.c.c. metals the effective co-ordination number becomes 11·75. For hexagonal metals, in particular zinc and cadmium, the c/a ratios of which differ from the ideal one, the co-ordination number is less than 12. The accepted atomic radii and the corresponding effective co-ordination numbers of the metals are listed in Table XXII. The designation of the structural types is that of *Strukturberichte* also used in *Metals Reference* Book [ed. C. J. Smithells] from which the details of most of the structures listed in Table XXII have been obtained.

After having re-defined the co-ordination number, we now make the assumption that the energy of the A–B bonds may be obtained additively from those of the A–A and B–B bonds, and that the energy of attraction between the atoms of the pure metals, represented by the heats of sublimation (L_A and L_B), is equally distributed over all the "links" up to a distance of $2r\sqrt{2}$, the energy being inversely proportional to the distance.

With these assumptions, ΔH_f becomes simply the negative product of the percentage increase in the effective co-ordination number and the heats of sublimation of the components:

$$-\Delta H_f = \frac{N_A L_A[C^*(A)_{\text{alloy}} - C^*(A)_{\text{metal}}]}{C^*(A)_{\text{metal}}} + \frac{N_B L_B[C^*(B)_{\text{alloy}} - C^*(B)_{\text{metal}}]}{C^*(B)_{\text{metal}}} \tag{126}$$

Using this equation, the heats of formation of a number of intermetallic phases have been estimated and are listed in Table XXIII and compared with the experimental values. Since the calculations are rather sensitive to small errors in the lattice dimensions, and since the experimental heats of formation are often accurate to only ± 500 cal/g-atom, agreement between calculated and experimental values to within ± 1 kcal/g-atom may be regarded as an indication that the model is a reasonable one and that the simplifying assumptions are fair approximations.

There are notable exceptions to the quantitative application of the model. For instance, the experimental heats of formation of the higher transition-metal aluminides are rather more exothermic than the estimated ones [375], whereas better agreement

Table XXII. Structure, atomic radius, effective co-ordination number, heat of sublimation and electro-negativity of metals

Metal	Structure	r in Å	Co-ordination number	L_{sb} (298°K) (kcal/g-atom)	ε
Li	A2	1·515	11·75	38·1	0·95
Na	A2	1·855	11·75	25·6	0·9
K	A2	2·31	11·75	21·33	0·80
Cs	A2	2·62	11·75	18·2	0·75
Be	A3	1·11	11·6	77·5	1·5
Mg	A3	1·60	12·0	35·0	1·2
Ca α	A1	1·98	12·0	42·6	1·0
β	A3	1·99	12·0	42·4	
Ba	A2	2·17	11·75	43·5	0·9
Y	A3	1·78	11·65	101·5	1·2
La	A3, A1	1·87	11·95	103·0	1·15
Ce	A3, A1	1·825	12·0	101·0	1·1
Th α	A1	1·80	12·0	137·5	IV: 1, 4
β	A2	1·72	11·75	136·9	
U γ	A2	1·50	11·75	125·5	IV: 1·4
Ti α	A3	1·45	11·8	112·3	1·6
β	A2	1·43	11·75	111·4	
Zr α	A3	1·585	11·7	145·5	1·5
β	A2	1·56	11·75	144·6	
Hf α	A3	1·57	11·67	148·0	1·4
V	A2	1·315	11·75	122·9	III: 1·5
Nb	A2	1·425	11·75	172·5	1·65
Ta	A2	1·43	11·75	186·8	III: 1·3; V: 1·7
Cr	A2	1·25	11·75	95·0	II: 1·4; III: 1·6
Mo	A2	1·36	11·75	157·3	IV: 1·6
W	A2	1·37	11·75	203·0	IV: 1·6
Mn α	A12	1·25	11·5	67·7	II: 1·4
β	A13	1·28	11·85		III: 1·5
γ	A1		12·0		
Re	A3	1·37	11·95	185·2	V: 1·8; VII: 2·2
Fe α	A2	1·25	11·75	99·3	II: 1·7
γ	A1	1·26	12·0	97·9	III: 1·8
Co	A1	1·25	12·0	102·4	1·7
Ni	A1	1·245	12·0	102·8	1·8
Ru	A3	1·325	11·72	155·7	2·0
Rh	A1	1·34	12·0	132·2	2·05
Pd	A1	1·375	12·0	90·0	2·0
Os	A3	1·34	11·7	188·4	2·0
Ir	A1	1·355	12·0	160·0	2·1
Pt	A1	1·38	12·0	135·0	2·1
Cu	A1	1·275	12·0	80·5	I: 1·8; II: 2·0
Ag	A1	1·44	12·0	67·9	1·8
Au	A1	1·435	12·0	88·0	2·25
Zn	A3	1·33	10·65	31·17	1·5
Cd	A3	1·49	10·5	26·72	1·5
Al	A1	1·43	12·0	78·7	1·5
In	A6	1·62	11·25	58·0	1·5
Tl	A3	1·71	11·90	43·25	I: 1·5; III: 1·9
Sn	A5	1·51	5·75	72·0	1·7
Pb	A1	1·745	12·0	46·6	1·6

is found for the lower aluminides (Cr_2Al, Cu_3Al; Table XXIII). Even so, comparisons between the experimental enthalpies and those estimated by means of the model contribute to the discussion of the bond mechanism in alloy phases. The bond mechanism of the B32 type (NaTl), originally discussed by Zintl (see Schneider and Heymer [376]), derives from the interposition of two diamond lattices for the A and B atoms respectively, indicating the presence of covalent and polar as well as metallic bonds. It appears that the formation of the polar–covalent bonds absorbs some energy (see NaTl: Table XXIII).

Table XXIII. Heats of formation of intermetallic phases calculated from the increase in co-ordination

Compound A_xB_y	Structure type	Lattice constant (Å)			Co-ordination number		$-\Delta H_f$ (cal/g-atom)	
		a	*b*	*c*	A	B	Calculated	Experimental
KNa_2	C14	7·48	—	12·27	14·45	11·20	0·82	0·25
LiTl	B2	3·42	—	—	12·85	14·7	6·9	6·4
NaTl	B32	7·47	—	—	14·25	13·85	6·3	4·5
$CaMg_2$	C14	6·22	—	10·10	15·75	11·96	4·35	3·15
$MgNi_2$	C36	4·81	—	15·77	16·45	12·2	5·45	4·4
$MgCu_2$	C15	7·05	—	—	15·65	11·9	3·0	2·8
MgAg	B2	3·32	—	—	13·9	12·3	3·55	4·5
MgCd	B19	3·22	5·27	5·00	12·55	11·4	1·96	1·95
Mg_3Cd	Do19	6·26	—	5·07	12·15	11·25	0·8	1·35
$Mg_{17}Al_{12}$	A12	10·54	—	—	13·4	11·3	0·57	(0·65)
$CaZn_5$	D2d	5·39	—	4·22	19·85	11·45	6·55	(5·5)
$CaZn_{13}$	D23	12·13	—	—	19·8	11·55	4·6	4·1
TiNi	B2	2·98	—	—	13·7	11·4	6·75	8·0
$TiNi_3$	Do24	5·09	—	8·29	15·3	11·85	7·25	8·3
$NbFe_2$	C14	4·835	—	7·88	13·85	11·25	7·4	7·35
Cr_2Al	C11	3·00	—	8·63	13·55	11·50	2·05	2·6
CoAl	B2	2·86	—	—	12·58	14·68	11·1	12·95
Cu_3Al	Orthorhombic	4·52	5·21	4·23	11·85	14·75	3·7	(3·9)
AuCu	L1o	3·955	—	3·66	12·9	11·15	1·12	2·0
$AuCu_3$	L12	3·74	—	—	14·0	11·35	1·6	1·60

Further, where substantial differences between calculated and experimental values are found, the deviations tend to be of the same order and sign in homologous series—as has been pointed out by Alcock *et al.* [377].

The model implies that, for stability, it is essential that the larger atom should be the "stronger binder", i.e. have the higher heat of sublimation, because it is the larger atom that will gain in co-ordination while the smaller atom will lose. As a result one may qualitatively predict the type of equilibrium diagram two metals, A and B, would form. For an empirical test, 350 binary phase diagrams have been ordered in groups and entered into a diagram of $(r_A - r_B)/\frac{1}{2}(r_A + r_B)$ vs. $(L_A - L_B)/\frac{1}{2}(L_A + L_B)$, where A is always the larger atom. An also empirical, mostly small, correction term, $(\varepsilon_A - \varepsilon_B)^2$, has been added to the L-term in order to account for the "electrochemical factor", ε_A and ε_B being the electro-negativities of the components, also listed in Table XXII.

The results are shown in the simplified representation of Fig. 105. Systems forming large ranges of disordered solutions have been included in the survey although these are not covered by the model. A sharp distinction between simple eutectic systems and those showing liquid miscibility gaps cannot be made because it depends on the melting points whether the miscibility gap appears in the phase diagram or not.

Of the 350 systems examined of combinations between the metals of groups IA to VIIIA, IB and IIB (except Hg) less than 20 clearly do not fit into the pattern. The so-called "Hume-Rothery systems" also do not agree well with the present pattern, probably because of the relatively low co-ordination of pure zinc and cadmium.

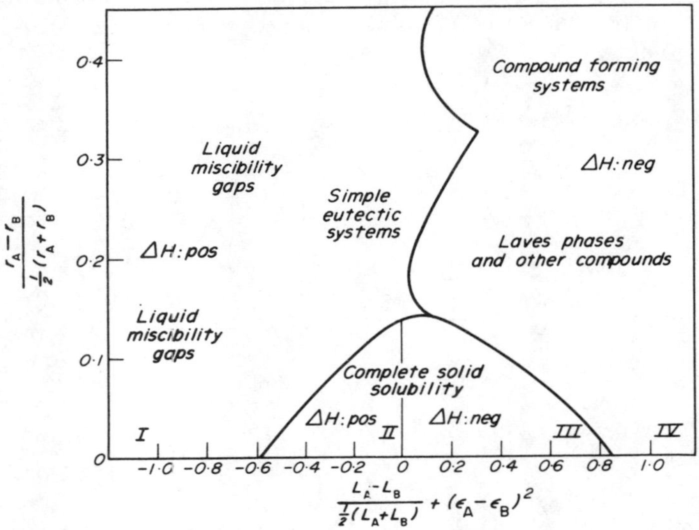

FIG. 105. Diagram indicating, in terms of atomic radius, heat of sublimation, and electro-negativity of the component metals A and B, regions of preference for the formation of certain types of binary equilibrium diagrams.

7. *Temperature Increase During Compound Formation*

Descriptions of the preparation of alloys for metallographical work often contain statements of considerable increase in temperature on mixing the components in powder or liquid form. In suitable cases this information may be found quite useful. If the increase in temperature is rapid, little heat can be lost from the reacting materials to their surroundings, and in this event the increase in temperature multiplied by 7·25 (atomic heat) will give a useful minimum value for the heat of formation per g-atom of alloy or compound. Allowance must of course be made for the heat of any melting (of components or alloy) which occurs during the reaction. If the heat of fusion of the alloy is unknown, the value 3·5 × (m.p., deg K) for ordered or 2·3 × (m.p., deg K) for disordered alloys may be used.

The values of $-\Delta H$ calculated in this manner are low as some of the heat of reaction is imparted to the surroundings from the moment reaction begins. They are only slightly lower when reaction is rapid and when contact with the crucible is poor. If the powder mixture is packed closely in a metallic crucible surrounded by thermal lagging material, the change in heat content of the crucible may be added. A knowledge of the relative weight of charge and crucible is of course necessary.

The figures in Table XXIV give an indication of how the minimum values for heats of formation obtained by this means differ from the actual values. It may again be emphasised that the extent of the departure depends upon the rate of reaction. The measure of agreement found in the table shows that careful assessment of apparently casual information to be found in the literature can be of great value to the desperate thermochemist.

Table XXIV. Minimum heat of formation from temperature increase during compound formation
(all caloric values: per g-atom)

System	Composition	Temperature of reaction (°C)	$\Delta\theta$	State after reaction	Reference	L_f(cal)	$\Delta\theta \times 7{\cdot}25$	Minimum $-\Delta H$ cal	$-\Delta H$ (meas.)
Fe–Si	80% Fe	1250	560	molten	[378]	(4000)	4050	>8050	8200
Pb–Tl	34% Tl	330	52	solid	[379]	—	380	>380	520
Mg–Sn	Mg_2Sn	450	250	molten	[379]	3800	1810	>5610	6250
Mg–Sn	Mg_2Sn	590	220	molten	[380]	3800	1600	>5400	6250
Mg–Bi	Mg_3Bi_2	480	450	molten	[380]	5040	3260	>8300	8140

This method cannot be expected to be reliable for determining exothermic heats of formation much in excess of 15,000 cal/g-atom, and so it can only be applied to intermetallic compounds or to other suitable reactions.

Another useful source of information is found in the readiness with which a metallic compound is reduced by another metal as, for example, occurs in aluminothermal reactions. In such a process the heat of reaction between the reducing element and the radical removed must be greater than the heat of formation of the compound reduced. The results of a systematic investigation [381] of the sulphides on these lines are given in Table XXV. The mixtures of metal and sulphide marked + showed a discontinuity in the heating curve at the temperatures given (deg C). Mixtures marked O showed no thermal evidence of reaction up to 1000°C, while reaction of those marked by an asterisk took place only in the liquid state, probably owing to the coarseness of grain of the materials. In such a series as this a heat of formation may be estimated to be intermediate between those of the two neighbouring compounds. The information required to place the compounds in such a series is usually dispersed throughout the literature, but a search is often well worth while for estimating unknown heats of formation.

8. *Examples of Application*

It is obvious that none of the methods of estimation just described, taken singly, give exact values. Nevertheless, if two or more of these methods can be applied to the determination of one value, and if there is fairly close agreement between the values obtained, then the estimated ΔH values may be expected to be correct within 5–10 kcal/mole. Values with errors even as high as this can be of practical value in dealing with a problem.

To illustrate the agreement to be expected from these estimations we may consider the heat of formation of the compounds of the second group metals with sulphur, selenium and tellurium, comparing the various estimated values as far as possible with those determined experimentally. This comparison is made in Table XXVI. The good agreement of the estimations from homologous series [2(M), the metal being kept the same; 2(X) the electro-negative component being the common one] is somewhat misleading as the experimental values for these compounds were used in constructing the curves. Since all the curves passed smoothly through the values for the compounds in question (including also the heats of formation of the oxides and of beryllium compounds) they would be almost as reliable even if fewer of the ΔH values were known.

Table XXV. The sequence of metals in their affinity to sulphur (after Tammann and von Samson-Himmelstjerna)

Sulphide metal	MnS	ZnS	FeS	CdS	Cu_2S	NiS	CoS	PbS	Bi_2S_3	Ag_2S
Mg	+ 360°	+ 440°								
Mn		+ 680°								
Zn			+ 390°							
Al		○	* 760°	+ 430°						
Cr		○	+ 850°	+ 440°						
Mo		○	+ 880°					+ 880°	+ 740°	
Fe				+ 200°	+ 380°					
Cd					* 350°	* 460°	* 400°			
Sn			○		* 460°	* 410°				
Cu			○			+ 640°	+ 420°	+ 240°	+ 325°	+ 210°
Ni							+ 660°	+ 540°		
Co								+ 560°		
Pb			○		○				+ 275°	
Bi										* 270°
Ag										

The table shows that the homologous series give the best results, whereas estimations from volume changes are the least accurate. This is not necessarily so with other compounds, so it does not follow that estimations from volume changes should be excluded in deriving mean values. The agreement as it stands is quite good and is the best that can be expected from such estimations. The discrepancies in the values for mercury compounds are much more serious, and one cannot feel particularly confident about deriving a mean value from them. In general the metals of atomic number greater than that of lanthanum (i.e. those subject to the lanthanide contraction) deviate from the rules more than the others. This is possibly due to their high polarisation force, and this must be taken into account in making estimations.

Table XXVI. Estimated and observed heats of formation of some compounds

Compound	$-\Delta H$ in kcal/mole, estimated. Method used:						ΔH in kcal/mole observed
	2(M)	2(X)	3	4	Misc.	Mean	
MgS	98·0	83·0	78·0	79·5	(94·4)*	84·5	84·0 ± 0·5
MgSe	62·0	63·0	66·5	61·0	75·0†	65·5	—
MgTe	50·0		37·5	51·0		44·5	50·0 ± 5·0
CaS	115·0	112·0	104·0	104·5	114·0*	110·0	113·8 ± 1·0
CaSe	75·0	75·0	90·5	67·0		77·0	76·5 ± 5·0
CaTe	59·0		55·8	59·0	90·0†	65·5	—
SrS	108·0	108·0	104·0	96·0	114·3*	106·1	108·2 ± 3·0
SrSe	79·0	79·0	90·5	84·5		83·2	95·0 ± 9·0
SrTe	62·0		55·8	60·0		59·5	—
BaS	106·0	106·0	118·0	82·0	112·5*	105·0	106·1 ± 3·0
BaSe	74·0	75·0	104·0	77·7		82·6	94·0 ± 9·0
BaTe	61·0		66·5	46·0	84·0†	64·4	—
ZnS	48·0	48·0	46·0	42·5	49·5*	46·8	49·0 ± 0·5
ZnSe	34·0	36·0	37·5	35·5		35·7	38·0 ± 2·5
ZnTe	29·0	30·0	16·5	29·0		26·1	28·5 ± 0·4
CdS	34·5	34·5	46·0	34·5	22·5‡	34·7	35·7 ± 0·5
CdSe	25·0	28·0	37·5	28·0		29·6	34·6 ± 0·5
CdTe	20·5	22·0	16·5	22·0		20·3	24·1 ± 0·2

* Method 5.
† From Table XVIII.
‡ From Table XXV.

CHAPTER IV

Examples of Thermochemical Treatment of Metallurgical Problems

The real *raison d'être* for the continuation of extensive experimental research in metallurgical thermochemistry is the potential application of its principles and data to practical, in particular industrial, problems. For this purpose the gathering of raw experimental data is obviously not enough. Missing numerical information must be supplemented by estimates to which the preceding Chapter III has therefore been devoted. Raw data must be sifted and critically evaluated to provide for every chemical system a *consistent* set of thermochemical properties. One example of such an evaluation is elaborated in the present chapter, and a large store of critical data is presented in the final Chapter V.

In addition, however, metallurgists of all denominations must increasingly be made aware of the help to be expected from the thermochemists. This is why the present chapter deals with a number of practical applications pertaining to chemical equilibrium. The number of examples is admittedly small compared with the large number of possibilities, but the perceptive reader will soon see that every chosen example can very similarly be employed in many other cases.

In practice, it is true, the knowledge of reaction rates is as important as that of equilibrium, if not more so, but the kinetic problems can only be tackled when the thermodynamic ones have been settled. It is also true that, in practice, metallurgical reactions are more involved than those described below, but then one has to start somewhere, and with some effort it will be found that even complicated chemical processes may be broken up into simpler reactions which are accessible to normal thermodynamic evaluation.

With respect to the accuracy of the results of such calculations, one should distinguish two types of question that may be asked. Often only a cursory answer is needed to the question whether the equilibrium conditions of a particular chemical process are such that the process is feasible. To this type of question an answer is almost always possible, since the accuracy of the pertinent data need not be better than a few kilocalories. In other cases the inquirer may be far more demanding and wish to know if an impurity of, say, a few hundredths of a per cent can still be reduced under given conditions. In such cases the thermochemist may often find that the available data are just not accurate enough, and this is why the improvement of his experimental methods and measurements must continue.

1. Chill Factors

The knowledge of the change in temperature of a bath of liquid iron caused by the addition of a master alloy, is of considerable interest to the steelmaker—especially when the addition is made to the ladle, isolated from an external heat source, because of the necessity to arrive at a predetermined teeming temperature.

The temperature change of a bath of molten metal due to the addition of a master alloy is given by the total heat effect caused by adding this alloy divided by the heat capacity of the alloy produced. The total heat effect is composed of the heat required to raise the master alloy from room temperature to the bath temperature and the heat of solution of the alloy in liquid iron. It is convenient to consider the total heat effect as taking place in three stages: dissociation of the master alloy into its component elements, followed by raising the individual elements from room- to bath-temperature, and their subsequent dissolution in liquid iron, and then to invoke the First Law of Thermodynamics.

The procedure is best illustrated with a practical example, say, the dissolution of a 50 at.% Si–Fe alloy in iron to give a 1 at.% Si solution.

$$\langle Fe_{0\cdot5}Si_{0\cdot5}\rangle_{298°K} = \tfrac{1}{2}\langle Fe\rangle_{298°K} + \tfrac{1}{2}\langle Si\rangle_{298°K} \qquad \Delta H_1 = +9200 \text{ cal}$$

$$\tfrac{1}{2}\langle Fe\rangle_{298°K} = \tfrac{1}{2}\{Fe\}_{1873°K} \qquad \Delta H_2 = +9010 \text{ cal}$$

$$\tfrac{1}{2}\langle Si\rangle_{298°K} = \tfrac{1}{2}\{Si\}_{1873°K} \qquad \Delta H_3 = +10{,}910 \text{ cal}$$

$$\tfrac{1}{2}\{Fe\}_{1873°K} = \tfrac{1}{2}[Fe]_{\{Fe_{0\cdot99}Si_{0\cdot01}\}1873°K} \qquad \Delta H_4 = -0{\cdot}25 \text{ cal}$$

$$\tfrac{1}{2}\{Si\}_{1873°K} = \tfrac{1}{2}[Si]_{\{Fe_{0\cdot99}Si_{0\cdot01}\}1873°K} \qquad \Delta H_5 = -15{,}675 \text{ cal}$$

Summing up,

$$\langle Fe_{0\cdot5}Si_{0\cdot5}\rangle_{298°K} = \tfrac{1}{2}[Fe]_{\{Fe_{0\cdot99}Si_{0\cdot01}\}1873°K} + \tfrac{1}{2}[Si]_{\{Fe_{0\cdot99}Si_{0\cdot01}\}1873°K}$$

$$\Delta H_1 + \Delta H_2 + \Delta H_3 + \Delta H_4 + \Delta H_5 = +13{,}445 \text{ cal}$$

For the heat capacity of the final solution one may take that of pure liquid iron, i.e., $C_p = 11{\cdot}0$ cal/deg mole. Since a positive sign indicates an endothermic heat effect, the temperature of the bath is lowered. Hence

$$\Delta\theta = -0{\cdot}02 \times 13{,}445/11 = -24{\cdot}4°C$$

independent of total mass.

Thus, if the pertinent enthalpy data are known or can be reliably estimated, the calculation of the temperature changes is fairly simple. "Chill factors", defined as the temperature change accompanying the formation of a 1 wt.% master alloy solution from hypothetically pure liquid iron at 1600°C and solid master alloy at 25°C, have occasionally been published, most recently for 53 commercial master alloys by Chart [382] on behalf of the Metallurg Group of Companies. The number of possibilities is, of course, legion.

These chill factors have been calculated to minimise the departure from the desired casting temperatures of steels which occurs in practice and is an important factor affecting the reject rate of steel ingots. A secondary advantage of the use of chill factors is increased furnace life. This is because in many instances the melt will not have to be preheated in the furnace to such an extent, prior to the addition of the ferro-alloys, when the chill factors are applied.

2. Aluminothermic Type Reactions

A. THE PRODUCTION OF URANIUM BY REDUCTION OF ITS FLUORIDE WITH CALCIUM AND MAGNESIUM

A number of metals are produced by reduction of a suitable compound with a very reactive metal. In such a process it is desirable to arrange it so that the heat generated by the chemical reaction will be sufficient to raise at least one of the products of the reaction to the liquid state. This has the practical advantage of allowing a liquid/liquid or liquid/solid separation to occur. The process is used for the reduction of transition metal chlorides in the so-called Kroll process or of their oxides by aluminothermic reaction. All these processes have strongly negative Gibbs energies at all practical temperatures. The example selected here has been taken from the monograph of Rand and Kubaschewski [383].

The reaction of uranium tetrafluoride with calcium,

$$\frac{\langle UF_4\rangle + 2\langle Ca\rangle}{\text{at } 298°\text{K}} \rightarrow \frac{2\{CaF_2\} + \{U\}}{\text{at temp. } T}$$

provides sufficient heat to raise the temperature after ignition above the melting points of calcium fluoride and uranium. The maximum temperature increase may be estimated in the following way. The standard heat of formation of two moles of calcium fluoride is $-584{,}000$ cal, that of 1 mole of uranium tetrafluoride $-453{,}500$ cal, the difference being $-130{,}500$ cal in favour of the formation of calcium fluoride. The change in heat of reaction with temperature may be neglected for such an approximate estimate, but the heats of fusion of the products must be added if these are to be heated above their melting points. The combined heat effect of the transformation and fusion for $2CaF_2$ is 16,500 cal, that for uranium 4400 cal. The heat available for heating the mixture is thus

$$\Delta H = -584{,}000 + 453{,}500 + 16{,}500 + 4400 = -109{,}600 \text{ cal}$$

The reacting mixture may now be considered as a calorimeter, the water equivalent of which may be obtained from the molar heats of the products. If these are not known, it is sufficient to ascribe 7 cal/deg to every g-atom or g-ion present, which in this case amounts to 49 cal/deg since there are 7 g-atoms + g-ions in ($2CaF_2$ + 1U). (From the actual values of the heat capacities, one finds an average value of 47 cal/deg over the temperature range 300–2000°K.) The maximum temperature increase is then obtained by dividing the heat produced by the water equivalent, i.e.

$$\Delta\theta_{max} = 109{,}600/49 = 2337°\text{C}$$

In fact the temperature can never increase by this amount because heat is being conducted away as soon as the reaction starts. To what extent it can be approached depends on the rate of reaction and on the insulation. In the present case, with good insulation, the boiling point of calcium will certainly be attained, and the evaporation of excess calcium would consume more heat. It is clear, however, that the heat of reaction suffices to melt the reaction products so that a dense billet of uranium metal separated cleanly from the slag is likely to be obtained.

This is, however, not true when magnesium is used in a similar manner as the reducing agent. We may repeat the above calculation for the reaction

$$UF_4 + 2Mg = 2MgF_2 + U$$

The heat of formation of magnesium fluoride is $-266{,}000$ cal/mole, its melting point 1263°C and its heat of fusion 13,900 cal/mole. The total heat available to heat and melt the reaction products is therefore $\Delta H = -532{,}000 + 453{,}500 + 27{,}800 + 4400 = -46{,}300$ cal. The average molar heat of the products is 48 cal/deg between 500° and 1750°K, whereas the estimated figure, which we shall use here, is again 49 cal/deg. The maximum temperature increase thus amounts to

$$\Delta\theta_{max} = 46{,}300/49 = 945°C$$

This is clearly not enough to heat the reaction mixture above the melting points of MgF_2 and uranium. Since it is desirable to attain a temperature of, say, 1500°C, a preheating of the reactants to about 600°C is required, the alternative of using a "booster" being ruled out because of the danger of contamination. Harper and Williams [384] have indeed devised a satisfactory method for the reduction of uranium tetrafluoride by magnesium based on a preheating procedure.

In view of the heat losses during reaction owing to radiation and conduction, more elaborate calculations of the temperature increase in this type of reaction are not worth while.

B. THE PRODUCTION OF MANGANESE AND CHROMIUM BY THE ALUMINOTHERMIC PROCESS

In the aluminothermic reduction of manganese, one of the variables under the control of the operator is the initial state of oxidation of the metal. In the ore pyrolusite, manganese is quadrivalent, but MnO, Mn_3O_4 and Mn_2O_3 are also well-defined lower oxides. The heats of formation of the oxides are

$$Mn + \tfrac{1}{2}O_2 = MnO; \qquad -\Delta H° = 92{,}000 \text{ cal/mole}$$

$$3Mn + 2O_2 = Mn_3O_4; \qquad -\Delta H° = 331{,}400 \text{ cal/mole}$$

$$2Mn + 1\tfrac{1}{2}O_2 = Mn_2O_3; \qquad -\Delta H° = 228{,}700 \text{ cal/mole}$$

$$Mn + O_2 = MnO_2; \qquad -\Delta H° = 124{,}300 \text{ cal/mole}$$

Using the heat of formation of alumina,

$$2Al + 1\tfrac{1}{2}O_2 = Al_2O_3; \qquad -\Delta H° = 400{,}000 \text{ cal/mole}$$

we may calculate the heat evolved in the reduction of each oxide by aluminium.

In the process, the products must reach a temperature of about 2000°C, and hence we should calculate the heat required to raise manganese metal and alumina to this temperature from room temperature. The results of this calculation using data in the tables are as follows:

$$Mn, \quad H°_{2300} - H°_{298} = 24{,}000 \text{ cal/g-atom}$$

$$Al_2O_3, \quad H°_{2300} - H°_{298} = 85{,}000 \text{ cal/mole}$$

The alumina is here in the liquid state at the upper temperature. A table is now compiled showing the heat required and that available to heat the products to 2300°K for three of the oxides, for 1 mole of alumina produced. Also shown is the Mn/Al ratio since the lower this ratio becomes, the more expensive is the reduction.

Oxide	*Heat required*	*Heat available/mole Al_2O_3*	*Mn/Al*
MnO	157 kcal	124 kcal	1·5
Mn_3O_4	139 kcal	152 kcal	1·125
MnO_2	121 kcal	214 kcal	0·75

It can thus be seen that the process should become self-sustaining if the Mn/O ratio in the oxide charged into the reactor is between that of MnO and Mn_3O_4. It should be noted that the reduction of MnO_2 would probably be explosively fast once initiated owing to the great excess of heat available.

In practice there will be heat losses from the reduction assembly which have not been accounted for in the calculation so far. An approximate value for these losses may be obtained, purely by chance, from a consideration of the aluminothermic reduction of chromium sesquioxide.

The standard heat change for the reaction

$$Cr_2O_3 + 2Al = 2Cr + Al_2O_3$$

is $\Delta H° = -130$ kcal. The heat required to raise the products of this reaction from room temperature to 2300°K is found to be 129 kcal. Thus the heat available just balances the heat produced, and if there were no heat losses the reduction could be made satisfactorily. However, it is industrial practice to pre-heat the mixture of chromium sesquioxide and aluminium to about 550°C before charging to the aluminothermic crucible. The heat required to do this amounts to 25 kcal per mole of Cr_2O_3, and hence this value must be close to the heat lost by radiation and convection, during the reaction, to the surroundings.

3. The Chlorination of Metal Oxides

An important first step in the extraction of a number of the nuclear metals, such as zirconium, vanadium and titanium, is the chlorination of the oxides which are the naturally occurring form. From the type of the equation for chlorination to form a gaseous chloride,

$$\langle MeO \rangle + (Cl_2) = (MeCl_2) + \tfrac{1}{2}(O_2)$$

and the corresponding equilibrium constant

$$K = \frac{p_{MeCl_2}\, p_{O_2}^{1/2}}{a_{MeO}\, p_{Cl_2}}$$

it can be seen that chlorination efficiency can be increased by lowering the oxygen potential [385]. This is usually achieved in industry by briquetting the oxide with carbon. The oxygen pressure which can be placed in the equilibrium constant above will now be that for the C–CO equilibrium above about 800°C, and close to that for

C–CO_2 below 600°C. This is the oxygen potential suggested by Goodeve [386] as being appropriate to the situation in the iron blast furnace.

Using the available Gibbs energy data it can be shown that the chlorination of all the important metal oxides is close to complete under normal industrial operating conditions where the temperature is between 500° and 1000°C in most cases.

What is a more interesting problem to resolve is that of the selection of an unreactive lining material for the chlorination reactor. Silica can be used successfully in most cases, but in the chlorination of beryllia to form gaseous $BeCl_2$, graphite must be used. The attack of silica by $BeCl_2$ can be demonstrated by the following argument.

For the chlorination of cristobalite,

$$\langle SiO_2\rangle + 2(Cl_2) = (SiCl_4) + (O_2); \qquad \Delta G^\circ = 62{,}000 - 10{\cdot}5T$$

and for beryllia,

$$2\langle BeO\rangle + 2(Cl_2) = 2(BeCl_2) + (O_2); \qquad \Delta G^\circ = 123{,}200 - 47{\cdot}8T$$

On combining these two equations we obtain the one for the attack of silica by $BeCl_2$ vapour:

$$\langle SiO_2\rangle + 2(BeCl_2) = (SiCl_4) + 2\langle BeO\rangle; \qquad \Delta G^\circ = -61{,}200 + 37{\cdot}3T$$

which has a negative standard Gibbs energy change in the temperature range of industrial operation.

On the other hand, for titania,

$$\langle TiO_2\rangle + 2(Cl_2) = (TiCl_4) + (O_2); \qquad \Delta G^\circ = 38{,}500 - 13{\cdot}5T$$

and for the attack of silica by $TiCl_4$ vapour,

$$\langle SiO_2\rangle + (TiCl_4) = (SiCl_4) + \langle TiO_2\rangle; \qquad \Delta G^\circ = 23{,}500 + 2{\cdot}8T$$

which is positive in the same temperature range.

It may thus be concluded that the attack of a silica lining in a TiO_2 chlorination plant will be virtually absent except as a result of the direct access of chlorine to the lining, whereas in the BeO operation, a silica lining cannot be employed.

4. The Free Evaporation of an Oxide Ceramic in Vacuo

When ceramic materials are used as structural components in high-temperature vacuum furnaces it is important to consider the loss in weight and size which will occur as a result of vaporisation at high temperature. The calculation of this rate may be made by assuming that the equilibrium vapour pressure is exerted under conditions of free evaporation. This means that the vaporisation coefficient is assumed to be unity for the material in question. We will consider as an example of this calculation the weight loss of magnesia at a temperature of 2000°K in a vacuum where the residual gases offer no resistance to evaporation of the solid.

The vaporisation mechanism of magnesia is principally

$$\langle MgO\rangle = (Mg) + (O)$$

and the Gibbs energy change for this process may be obtained from data for the

Gibbs energy of formation of MgO at 298°K, thermal data for the oxide up to 2000°K and the Gibbs energy of vaporisation of magnesium and that of dissociation of oxygen. This yields a simplified equation:

$$\Delta G^\circ = 240{,}800 - 67{\cdot}4T \text{ cal/mole of oxide}$$

where

$$K = \frac{p_{Mg}\, p_O}{a_{MgO}}$$

Now concerning ourselves only with pure magnesia ($a_{MgO} = 1$), there is a relation between p_{Mg} and p_O resulting from the fact that equal numbers of magnesium atoms n_{Mg} and oxygen atoms n_O must leave the surface in unit time in order to retain the stoichiometric ratio in the solid. Thus the weights of the two elements leaving the surface must be in proportion to their atomic weights:

$$\frac{n_{Mg}}{n_O} = \frac{m_{Mg}}{M_{Mg}} \frac{M_O}{m_O} = 1$$

The weight of each constituent leaving unit area in unit time is given by Knudsen's equation

$$m_{Mg} = 44{\cdot}4 p_{Mg} \sqrt{\frac{M_{Mg}}{T}}$$

Hence

$$\frac{n_{Mg}}{n_O} = \frac{p_{Mg}}{p_O} \sqrt{\frac{M_O}{M_{Mg}}} = 1; \quad p_O = p_{Mg} \sqrt{\frac{M_O}{M_{Mg}}}$$

Substituting in the equilibrium constant, then,

$$K = 0{\cdot}81 p_{Mg}^2 = 3{\cdot}98 \times 10^{-13} \text{ at } 2000°\text{K}$$

$$p_{Mg} = 7 \times 10^{-7} \text{ atm}$$

The weight loss per unit area of the solid is then

$$\Sigma m = m_{Mg} + m_O = 44{\cdot}4 \frac{p_{Mg}}{\sqrt{T}} [\sqrt{M_{Mg}} + 0{\cdot}81 \sqrt{M_O}]$$

$$= 5{\cdot}67 \times 10^{-6} \text{ g/cm}^2 \text{ sec}$$

The alternative mode of vaporisation would involve the evaporation of oxygen as O_2 molecules. The Gibbs energy change for this process is

$$\langle MgO \rangle \rightarrow (Mg) + \tfrac{1}{2}(O_2)$$

$$\Delta G^\circ = 181{,}600 + 7{\cdot}37T \log T - 75{\cdot}7T \text{ (cal)}$$

which yields at 2000°K

$$\Delta G^\circ = 78{,}850 = -9150 (\log p_{Mg} + \tfrac{1}{2} \log p_{O_2})$$

As can be concluded from the previous discussion, the relationship between p_{Mg} and p_{O_2} is such that the fluxes are in the ratio 2:1 so that the stoichiometry of the oxide remains at 1:1. Hence

$$p_{Mg} = 2p_{O_2}\sqrt{\frac{M_{Mg}}{M_{O_2}}} = 1{\cdot}74p_{O_2}$$

Hence solving for p_{Mg} and p_{O_2}, we find $p_{Mg} = 2{\cdot}14 \times 10^{-6}$ atm; $p_{O_2} = 1{\cdot}23 \times 10^{-6}$ atm at 2000°K.

Under equilibrium conditions the evaporation of MgO to give magnesium vapour and O_2 molecules is slightly favoured over the evolution of oxygen atoms. In a Knudsen experiment the correct ratio p_O/p_{O_2} should be obtained for the equilibrium condition when at 2000°K.

$$\frac{p_O}{p_{O_2}} = \frac{0{\cdot}568}{1{\cdot}23} = 0{\cdot}46$$

but under free evaporation conditions the less-complicated desorption of oxygen atoms would seem to be the more probable event.

5. The Reduction of Magnesium Oxide with Silicon

Magnesium, like aluminium, is generally produced by electrolysis of its fused salts. Since direct heating is cheaper than the use of electrical power in this way, any thermal reduction method which might prove feasible would be of great metallurgical interest.

For this purpose, silicon has been proposed* as a reducing agent for magnesium oxide. The vapour pressures of magnesium over mixtures of ferrosilicon and magnesia or calcined dolomite have been measured by several observers [387, 388, 389], and we may verify their values by calculation. To do this we must first determine the chemical reaction which takes place in the reduction. Schneider and Hesse [387] established by X-ray analysis that in the reduction of magnesia with silicon, magnesium orthosilicate is formed:

$$4\langle MgO\rangle + \langle Si\rangle = 2(Mg) + \langle Mg_2SiO_4\rangle \qquad (\alpha)$$

When calcined dolomite is used in place of magnesia the main product is calcium orthosilicate.

$$2\langle CaO{\cdot}MgO\rangle + \langle Si\rangle = 2(Mg) + \langle Ca_2SiO_4\rangle \qquad (\beta)$$

This equation may be confirmed by comparing its equilibrium constant (which is equal to the magnesium pressure) with the magnesium pressure observed experimentally, as follows.

For the Gibbs energy of formation of calcium orthosilicate from its constituent oxides

$$2\langle CaO\rangle + \langle SiO_2\rangle = \langle Ca_2SiO_4\rangle \qquad (\gamma)$$

* W. F. Bleecker and A. P. Morrison, **1311** 378; **1311** 379 (1919); W. M. Grosvenor, A. P. **1239** 178 (1916); G. Gire and R. Touquet, F. P. **733** 294 (1931); **750** 977 (1932); B. P. **382** 899 (1932). For further literature see Schneider [357a] and Ritter and Jaeckel [357b].

we have the value (Table E)

$$\Delta G^\circ_\gamma = -30{,}200 - 1{\cdot}2T$$

The Gibbs energy of formation of quartz and magnesia are also taken from Table E

$$\langle Si\rangle + (O_2) = \langle SiO_2\rangle \qquad (\delta)$$

$$\Delta G^\circ_\delta = -217{,}400 - 3{\cdot}0T\log T + 52{\cdot}22T$$

$$2(Mg) + (O_2) = 2\langle MgO\rangle \qquad (\varepsilon)$$

$$\Delta G^\circ_\varepsilon = -363{,}200 - 14{\cdot}74T\log T + 151{\cdot}4T$$

We may assume that the Gibbs energy of formation of calcined dolomite from its constituent oxides,

$$2\langle MgO\rangle + 2\langle CaO\rangle = 2\langle MgO{\cdot}CaO\rangle \qquad (\eta)$$

is so small that it may be neglected. Now if we subtract the sum of eqns. (ε) and (η) from the sum of eqns. (γ) and (δ) we obtain eqn. (β). Likewise, the sum

$$\Delta G^\circ_\beta = \Delta G^\circ_\gamma + \Delta G^\circ_\delta - \Delta G^\circ_\varepsilon - \Delta G^\circ_\eta$$

gives the free energy of reaction (β). Thus

$$\Delta G_\beta = 115{,}600 + 11{\cdot}74T\log T - 100{\cdot}38T$$

From this equation the magnesium pressures in mmHg are calculated by the relationship

$$\log p_{Mg} = -\frac{\Delta G^\circ_\beta}{2\times 4{\cdot}574T} + 2{\cdot}88$$

The curve showing the relation between the magnesium pressures and temperature is shown in Fig. 106. The calculated pressures agree well with the results of Schneider and Ebert [388] and reasonably well with those of Pidgeon and King [389] obtained experimentally for the equilibrium of calcined dolomite with silicon, thus indicating that eqn. (β) represents the true chemical mechanism.

The diagram corresponding to Fig. 106 in earlier editions showed substantially lower calculated pressures of magnesium, and the present revision is due to a re-determination of the important heat of formation of silica which was found to be in error by about 7 kcal! Indeed, discrepancies such as the one just indicated, pointed out by Chipman [390], have led to the re-determination of $\Delta H^\circ_f\langle SiO_2\rangle$.

More thermochemical values will continue to be found seriously in error, and the present example demonstrates the need for reliable measurements as well as for the critical attitude the assessor of experimental data has to observe even with "established" values.

6. Vapour Phase Transport of Silicon Carbide

Transport reactions involving volatile halides are of considerable use in the purification of metals such as zirconium by the van Arkel process, and in vapour phase coating and single crystal growth. What these processes have in common is the formation and subsequent decomposition of gaseous halides in a system incorporating a

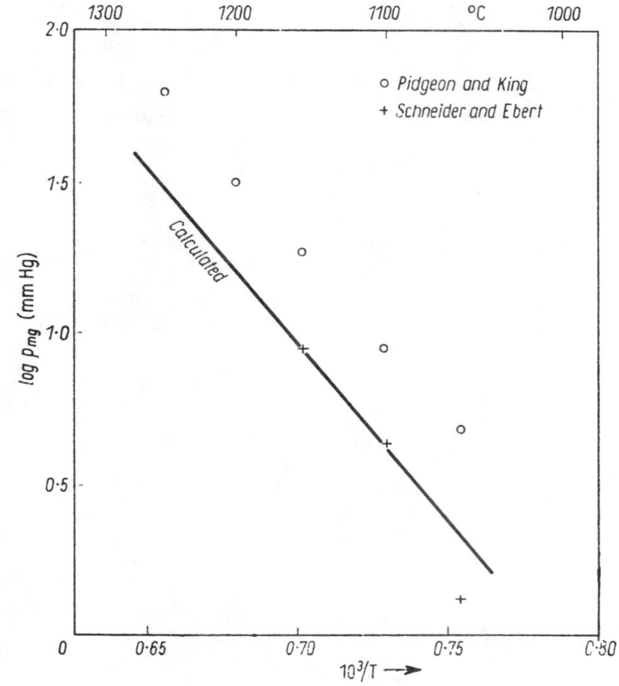

FIG. 106. Vapour pressures of magnesium in the reduction by silicon of calcined dolomite.

temperature gradient. In some instances the gaseous compound decomposes usefully in going to a higher temperature, and in others the decomposition occurs at lower temperatures.

The van Arkel process mentioned above uses the formation of ZrI_4 at low temperatures, around 600°C, between iodine vapour and impure zirconium, and decomposition of the tetraiodide at higher temperatures, around 1400°C, to yield zirconium of a high purity.

In the Gross process an aluminium alloy is reacted with $AlCl_3$ gas to form the monohalide AlCl at a high temperature, around 1200°C, and this then disproportionates at a lower temperature, about 700°C, to produce the pure metal and to regenerate $AlCl_3$ which is used again.

One criterion of a successful vapour transport reaction is obviously that the equilibrium constant should change appreciably over the temperature gradient. Secondly, it is necessary that the partial pressure of the transporting species is reasonably large so that a satisfactory rate of transport may be achieved. It is because the rate of vapour transport of a metal can be increased by the formation of a volatile halide, above the rate of vaporisation of the pure metal, that this process is sometimes referred to as "catalytic distillation".

The change in the value of the equilibrium constant for the formation of a volatile halide with temperature depends mainly on the entropy change for a reaction, and this in turn depends mainly on the change in the number of gaseous molecules in the vapour transport reaction. It follows that the greatest possible valency changes are most desirable in the metal being transported when the gaseous species is formed or is decomposed. The value of the equilibrium constant for the formation of a compound should be relatively close to unity in order that the entropy change can have the largest effect. This is shown in Table XXVII which was prepared by Alcock and Jeffes [391] in a survey of the thermodynamics of vapour phase transport reactions. For a reaction $\langle X\rangle + (Y_2) \rightarrow (XY_2)$ the equilibrium constant will change over a range of temperatures according to the van't Hoff equation

$$\frac{\partial \ln K}{\partial(1/T)} = \frac{-\Delta H^\circ}{R}$$

(clearly a reaction where ΔH° is equal to zero is of no interest in vapour phase transport).

Table XXVII. Pressures of the species XY_2 formed by the reaction $\langle X\rangle + (Y_2) \rightarrow (XY_2)$ as a function of the value of the equilibrium constant K

K	p_{XY_2} (atm)	$p_{XY_2} - p^\circ_{XY_2}$ (atm)
10^{-5}	$9{\cdot}1 \times 10^{-6}$	0·49999
10^{-4}	$9{\cdot}1 \times 10^{-5}$	0·4999
10^{-3}	$9{\cdot}1 \times 10^{-4}$	0·499
10^{-2}	$9{\cdot}1 \times 10^{-3}$	0·49
10^{-1}	$9{\cdot}1 \times 10^{-2}$	0·41
1	0·50 ($p^\circ_{XY_2}$)	0·00
10	0·91	0·41
10^2	0·99	0·49
10^3	0·999	0·499
10^4	0·9999	0·4999
10^5	0·99999	0·49999

The table shows the values of p_{XY_2} for a total pressure of 1 atm ($p_{Y_2} + p_{XY_2}$) when K changes as a result of a change in temperature from 10^{-5} to 10^5, and how each value of p_{XY_2} differs from the value $p^\circ_{XY_2}$ which is the equilibrium pressure when K is equal to unity.

It is apparent from these calculated values that the largest change in p_{XY_2} occurs around $K = 1$, and at the extremes a change of K by one order of magnitude changes p_{XY_2} very little. Thus, although a large value of ΔH° will produce a large temperature coefficient of K, this must be balanced by a sufficiently large entropy change of the same sign, so that, ideally, K should pass through unity in some experimentally useful temperature range.

In the same paper it was also pointed out that the large collection of thermodynamic data on gaseous metal halides indicated that in the general reaction

$$\langle M\rangle + \frac{n}{2}(X_2) \rightarrow (MX_n)$$

the entropy changes are, as would be expected, approximately constant for a given value of n but that the heat change for the formation of a metal bromide was about 23 kcal per mole Br_2 and the iodide about 55 kcal per mole I_2 less negative than the heat change for the corresponding chloride per mole Cl_2. Thus if a given chloride reaction was found to have too large a negative heat change for the equilibrium constant to approach unity in some useful temperature range, the bromide and more so the iodide would lower this temperature range providing ΔH° did not then approach too closely to zero for reasons given above.

Turning to the particular example of the transport not of an element, but of the compound silicon carbide, the same general guiding principles must be applied. Now, however, a vapour phase transport reaction must be used which will transport, effectively, not only silicon, as one of the tetrahalides, but carbon as well. Clearly no pure halogen is of much use in this respect because of the very low affinity of carbon when compared with silicon for the halogens at temperatures where equilibrium is achieved [392].

If HCl is used as the source of chlorine, the partial pressure of $SiCl_4$ which could be generated would be considerably reduced because hydrogen has a comparable affinity for chlorine as has silicon. The hydrogen which would be displaced by $SiCl_4$ formation would now be available to react with carbon forming gaseous methane. Methane is a suitable candidate for the transport of carbon and the low stability of silane SiH_4 in comparison with CH_4 and with silicon halides suggests that a probable transport reaction for SiC would be

$$\langle SiC \rangle + 4(HCl) \rightarrow (SiCl_4) + (CH_4)$$

$$\Delta G^\circ = -72{,}000 + 60{\cdot}1T; \quad \log K = \frac{15{,}738}{T} - 13{\cdot}14$$

The considerable reduction in the chlorine potential which is brought about by the introduction of hydrogen into the system lowers the partial pressure of $SiCl_4$ gas and it is thus made comparable with the partial pressure of the carbon-carrying species CH_4. The ratio of CH_4/H_2 which is achieved at equilibrium is unity at 500°C and 0·03 at 950°C and the equilibrium constant for the vapour transport reaction changes from approximately 10^{-7} to 0·5 in the same temperature range. It follows that this reaction is satisfactory in terms of the criteria for vapour phase transport which were elaborated above. The equilibrium constant achieves the value of unity at 925°C.

7(a). Assessment of Standard Values Silicon Monoxide*

It is often found that several independent measurements involving a particular substance are available which when considered as a whole will permit a more or less accurate assessment of its thermochemical standard values. Unfortunately, individual authors are sometimes inclined to see only their own results and to disregard, apart from an honorable mention, those of their fellow observers. Their evaluations thus require reassessment in the light of the total information available. This is particularly true of equilibrium measurements. Authors usually overlook the fact that the temperature coefficients of such results, on which heat and entropy values depend, are rarely reliable and occasionally quite inaccurate despite the care with which the measurements may have been made. It is generally safer to combine the observed Gibbs energies with estimated entropies rather than to derive the entropies from the temperature coefficients.

* The most recent results on the equilibrium $\langle Si \rangle + \langle SiO_2 \rangle = 2(SiO)$ by Chart [397] have not been incorporated in this evaluation but were found to agree well with the assessed values.

In this section the thermochemical properties of silicon monoxide may be assessed as a demonstration of how available information is to be digested. This information is first summarised, equilibrium pressures being converted initially into Gibbs energy equations by means of eqn. (29a).

The equilibrium

$$\langle SiO_2 \rangle + (H_2) = (SiO) + (H_2O) \tag{a}$$

was studied by means of a transportation method by Tombs and Welch [393] (1228–1653°C), Grube and Speidel [394] (1200–1500°C) and Ramstad and Richardson [395] (1425–1600°C), the results being represented by

$$\text{T \& W (1710°K)}: \quad \Delta G^\circ_a = \;\;78{,}200 - 21{\cdot}0T \tag{α}$$

$$\text{G \& S (1620°K)}: \quad \Delta G^\circ_a = 112{,}000 - 35{\cdot}5T \tag{β}$$

$$\text{R \& R (1800°K)}: \quad \Delta G^\circ_a = 127{,}100 - 45{\cdot}07T \tag{γ}$$

The average absolute temperatures are indicated.

The SiO pressure over mixtures of silicon and silica (presumably in the form of cristobalite) was measured by Tombs and Welch [393] (1300–1647°C) and Ramstad and Richardson [395] (1237–1485°C) who used an entrainment technique, and by Schäfer and Hörnle [396] (1050–1250°C) who used the Knudsen method. Tombs and Welch's results were presented by

$$\Delta G^\circ = 58{,}550 - 25{\cdot}45T \tag{δ}$$

per mole SiO. Since silicon is liquid over most of the temperature range studied by the last mentioned observers, the Gibbs energy of fusion of half a g-atom of silicon, $\Delta G = 6050 - 3{\cdot}6T$, should be added to reaction (δ) in order to get the corresponding Gibbs energy equation for solid silicon, according to

$$\tfrac{1}{2}\langle SiO_2 \rangle_{cr} + \tfrac{1}{2}\langle Si \rangle = (SiO) \tag{b}$$

By using the Gibbs energies of formation of water vapour and cristobalite given in Table E, the experimental Gibbs energies of reaction (a) may be converted to yield the corresponding Gibbs energies of reaction (b). We thus obtain six experimental two-term expressions for the Gibbs energy of reaction (b), as follows:

$$\text{T \& W (1750°K), from } \delta: \quad \Delta G^\circ_b = 64{,}600 - 29{\cdot}05T \tag{ε}$$

$$\text{S \& H (1420°K), directly}: \quad \Delta G^\circ_b = 77{,}000 - 37{\cdot}1T \tag{ζ}$$

$$\text{R \& R (1600°K), directly}: \quad \Delta G^\circ_b = 81{,}460 - 38{\cdot}62T \tag{η}$$

$$\text{T \& W (1710°K), from } \alpha: \quad \Delta G^\circ_b = 29{,}300 - 13{\cdot}35T \tag{θ}$$

$$\text{G \& S (1620°K), from } \beta: \quad \Delta G^\circ_b = 63{,}100 - 27{\cdot}85T \tag{κ}$$

$$\text{R \& R (1800°K), from } \gamma: \quad \Delta G^\circ_b = 78{,}200 - 37{\cdot}42T \tag{λ}$$

Gel'd and Kochnev [398] claim to have prepared amorphous silicon monoxide by heating an intimate mixture of silica and silicon to 1250° to 1350°C. This is somewhat puzzling in view of the observations of the other authors, but for the time being their claim to have measured the vapour pressure of the monoxide in this temperature range may be accepted. They give their results in the form of the equation

$$\text{G \& K (1300°K):} \quad \Delta G_c^\circ = 76{,}020 - 33{\cdot}5T \qquad (\mu)$$

for the reaction

$$\{SiO\}_{am} = (SiO) \qquad (c)$$

There are a few more thermochemical values available for SiO. Wartenberg [399] determined the heats of solution of silicon and amorphous silicon monoxide, made by sublimation of $\langle SiO_2 \rangle + \langle Si \rangle$ mixtures and condensation at 900°C, in hydrofluoric acid. His results have been recalculated, and the value

$$\text{v.W (298°K):} \quad \Delta H_d^\circ = -106{,}500 \pm 5000$$

has been obtained for the heat of formation of amorphous silicon monoxide, according to the equation

$$\langle Si \rangle + \tfrac{1}{2}(O_2) = \{SiO\}_{am} \qquad (d)$$

Spectroscopic data were used by Kelley to calculate the standard entropy of gaseous silicon monoxide as 50·55 cal/°, and by Zintl [400] to calculate the heat of the reaction

$$(Si) + (O) = (SiO) \qquad (e)$$

as

$$\Delta H \simeq -170 \text{ kcal}$$

All this information will now be used for the assessment.

First it is of interest to compare the entropy terms of eqns. (ε) to (λ) by converting them to a common temperature of 25°C and calculating the "standard" entropy of (SiO). For this the change in molar heat ΔC_p must be estimated for reaction (b). The molar heat of $(SiO)_{gas}$ is assumed to be about half-way between that of (CO) and that of (S_2). The molar heats of $\langle Si \rangle$ and $\langle SiO_2 \rangle$ are known from Table C, and an average $\Delta C_p = -2{\cdot}8$ is obtained. Its inaccuracy is not likely to introduce an appreciable error in the present calculations. Then the standard entropy of (SiO) may be obtained from the relationship

$$S_{(SiO)} = \tfrac{1}{2}S_{\langle SiO_2 \rangle} + \tfrac{1}{2}S_{\langle Si \rangle} + \Delta S_{298}$$

where

$$\Delta S_{298} = \Delta S_b - \Delta C_p \ln T + \Delta C_p \ln 298$$

Hence, with $\Delta C_p = -2{\cdot}8$,

$$S_{(SiO)} = \Delta S_b + 6{\cdot}45 \log T - 8{\cdot}6$$

where ΔS_b is the negative factor of T in eqns. (ε) to (λ). The following values for $S_{(SiO)}$ at 25°C are thus derived:

for (ε), 41·4; (ζ), 48·85; (η), 50·7; (θ), 25·55; (κ), 39·85; (λ), 49·8

This comparison clearly demonstrates how little trust can be placed on the evaluation of the temperature coefficients of equilibrium measurements. The most reliable value for the standard entropy is probably the one given by Kelley: 50·55. The estimate by means of the pertinent equation on p. 192 yields $S_{(SiO)} = 51{\cdot}0$ in good agreement with the spectroscopic value. The present authors are quite generally inclined to use estimated entropies rather than those obtained from equilibrium measurements when calorimetric or spectroscopic values are not available, and observers are urged to use estimates, at least as a check on their experimental entropy terms; it will save them and the later assessors much unnecessary work.

It may be noted that the fair agreement between the estimated value and the experimental values of Kelley, Ramstad and Richardson, and Schäfer and Hörnle also indicates that SiO is a molecular gas.

The entropy value of 50·55 may now be used to calculate the entropies of reaction (b) at the various average temperatures of measurement, by reversing the procedure employed above: in fact, this would have been the course taken right in the first place if the discussion had not been extended for the sake of demonstration. In other words, a third-law evaluation should generally be preferred to a second-law evaluation. By combination of the ΔS_b value thus obtained with the experimental ΔG_b value at the same temperature, values for ΔH_b may be calculated, and the standard heat of reaction is derived from the relationship

$$\Delta H_{298} = \Delta H_b - \Delta C_p T = \Delta H_b + 2{\cdot}8T$$

For the individual ΔG_b equations the results are as follows:

	T°K	ΔS_b°	ΔG_b°	ΔH_b°	ΔH_{298}°
T & W (ε)	1750	38·2	13,750	80,550	85,450
S & H (ζ)	1420	38·8	24,300	79,450	83,400
R & R (η)	1600	38·47	19,600	81,150	85,650
T & W (θ)	1710	38·35	6,450	72,000	76,800
G & S (κ)	1620	38·45	17,950	80,300	84,850
R & R (λ)	1800	38·15	10,800	79,500	84,550

By comparing the ΔH_{298} values it is now seen that the disagreement between the various sets of measurements is not so bad as it might have appeared in a comparison of the original equations. The average value, disregarding (θ), is $\Delta H_{298} = 84{,}800$ cal.

With the selected heat of reaction (b) and the heat of formation of cristobalite, the standard heat of formation of gaseous monoxide becomes

$$\langle Si \rangle + \tfrac{1}{2}(O_2) = (SiO) \qquad \text{(f)}$$

$$\Delta H_{298} = 84{,}800 - \frac{216{,}000}{2} = -23{,}200 \pm 2{,}000 \text{ cal}$$

From Zintl's spectroscopic heat of reaction (e), the heat of dissociation of O_2, and the heat of sublimation of silicon, an approximate value of $\Delta H_f = -23{\cdot}5$ kcal is obtained but, since the probable error is rather large, the excellent agreement with the third-law value is obviously fortuitous. However, it tends to support the above assessment to some extent.

We may now turn to the evaporation reaction (c). The heat of evaporation follows from the heats of formation of the gaseous and amorphous monoxide,

$$L_{298}(c) = 106{,}500 - 23{,}200 = 83{,}300 \pm 7000$$

The entropy of amorphous {SiO} must be estimated. For the hypothetically crystalline substance the value $S_{\langle SiO \rangle} = 8{\cdot}6$ follows from Latimer's tables (pp. 189/90); for the entropy of fusion one would expect about 5 cal/deg from Table XV, but empirical experience suggests halving the entropies of fusion for solid-amorphous transformations. Then the entropy of evaporation at 25°C becomes

$$\sigma_{298}(\mathrm{c}) = 50{\cdot}55 - 8{\cdot}6 - 2{\cdot}5 = 39{\cdot}45 \pm 2{\cdot}0$$

Hence

$$\Delta G_{\mathrm{c}}^{\circ} = 83{,}300 - 39{\cdot}45T \quad (298°\mathrm{K})$$

Since

$$\Delta G_{\mathrm{b}}^{\circ} = 84{,}800 - 43{\cdot}2T \quad (298°\mathrm{K})$$

$$\Delta G_{\mathrm{c}}^{\circ} - \Delta G_{\mathrm{b}}^{\circ} = -1{,}500 + 3{\cdot}75T$$

for the reaction

$$\{\mathrm{SiO}\}_{\mathrm{am}} = \tfrac{1}{2}\langle \mathrm{SiO_2} \rangle + \tfrac{1}{2}\langle \mathrm{Si} \rangle \tag{g}$$

From various sources it is found that the transformation of cristobalite to silica glass is accompanied by a Gibbs energy change represented by

$$\tfrac{1}{2}\langle \mathrm{SiO_2} \rangle_{\mathrm{cr}} = \tfrac{1}{2}\{\mathrm{SiO_2}\}_{\mathrm{gl}}; \quad \Delta G_{\mathrm{h}} = 700 - 0{\cdot}55T \tag{h}$$

at room temperature. Then

$$\Delta G_{\mathrm{c}} - \Delta G_{\mathrm{b}} + \Delta G_{\mathrm{h}} = -800 + 3{\cdot}2T = \Delta G_{\mathrm{i}}$$

For this reaction, namely

$$\{\mathrm{SiO}\}_{\mathrm{am}} = \tfrac{1}{2}\{\mathrm{SiO_2}\} + \tfrac{1}{2}\{\langle \mathrm{Si} \rangle\} \tag{i}$$

ΔC_p is probably negligibly small, and the ΔG_{i} equation may be used at any temperature. It follows, if $\Delta G_{\mathrm{i}} = -800 + 3{\cdot}2T$, that amorphous silicon monoxide should be stable in relation to silicon and silica at any temperature above 250°K. This contradicts the observation of Grube and Speidel, Schäfer and Hörnle, and others, that reaction (i) proceeds to the right of the equation at 1000–1150°C. On the other hand, the entropy term of ΔG_{i} appears to have the correct sign and to be of the order to be expected ($\pm 1{\cdot}5$ cal/deg). The heat term should therefore be more negative. It is possible, though it is unsupported by X-ray analysis, that Gel'd and Kochnev's claim is correct that eqn. (i) moves to the left at 1250–1350°C. If ΔC_{i} were zero at about 1150–1200°C, then, with the entropy term left unchanged, ΔG_{i} becomes about

$$\Delta G_{\mathrm{i}} = -4600 + 3{\cdot}2T$$

This may be taken to imply that the heat of formation of $\{\mathrm{SiO}\}_{\mathrm{am}}$ is $-102{,}700$ rather than $-106{,}500$, the alteration being within the limits of error of v. Wartenberg's experimental determination. With this alteration the Gibbs energy of evaporation of $\{\mathrm{SiO}\}_{\mathrm{am}}$ becomes

$$\Delta G_{298}(\mathrm{c}) = 79{,}500 - 39{\cdot}45T$$

The change in molar heat of this reaction may be estimated to be $\Delta C_p = -5{\cdot}5$ on an average, and the complete Gibbs energy equation may then be obtained in the manner described previously in this monograph.

$$\Delta G_c = 81{,}200 + 12{\cdot}65T \log T - 76{\cdot}25T$$

corresponding to the vapour pressure equation

$$\log p_{SiO}\,(\text{mmHg}) = -17{,}750T^{-1} - 2{\cdot}77 \log T + 19{\cdot}55$$

ΔG_c becomes zero at about 2430°K, that is, at the normal boiling point, and the heat of evaporation at this temperature is

$$L_e = 81{,}200 - 5{\cdot}5T = 67{,}800$$

The entropy of evaporation is therefore 27·9 cal/deg which is a little higher than Trouton's constant but may be accepted as reasonable.

At the average temperature of Gel'd and Kochnev's measurements (1300°K) ΔG_c is found to be 33,300 cal whilst the Russian authors quote 32,500 cal. The corresponding value for ΔG_b is 28,650 cal. It thus appears to be probable that Gel'd and Kochnev did in fact have silicon monoxide in the condensed state, but the differences in free energies for the two reactions are too small to allow definite conclusions to be drawn, particularly in view of the observations of Schäfer and Hörnle, Grube and Speidel, and Wartenberg that {SiO} is unstable at about 1100°C.

We have attempted to show that a full assessment of all the available information renders the resulting standard values for a particular substance far more reliable than the evaluation of a single series of measurements is likely to accomplish. It follows necessarily, however, that the evaluator must preserve an open mind and must be prepared to review his assessment whenever new measurements are forthcoming.

From the present example, the values derived from all the results available to date may now be summarised.

Thermochemical properties of silicon monoxide:

Gas:

$$\Delta H^\circ_{298} = -23{,}200 \pm 2{,}000 \text{ cal/mole}$$

$$S_{298} = 50{\cdot}55 \pm 0{\cdot}1 \text{ cal/mole °K}$$

Amorph.:

$$\Delta H_{298} = -102{,}700 \pm 5{,}000 \text{ cal/mole}$$

$$S_{298} = 11{\cdot}1 \pm 1{\cdot}5 \text{ cal/mole °K}$$

b.p. 2160°C:

$$L_e\,(298°\text{K}) = 79{,}500 \pm 6{,}000 \text{ cal/mole}$$

$$L_e(\text{b.p.}) = 67{,}850 \pm 6{,}000 \text{ cal/mole}$$

7(b) Assessment of Thermochemical Values for a System Exhibiting Wide Solution Ranges

In the preceding section the assessment of standard values for a pure compound has been demonstrated with gaseous silicon monoxide. In the present section the consistent evaluation of the thermochemical properties, including the phase boundaries, will be described for a system characterised by substantial solubility ranges. For this, the system chromium–nickel has been selected [401]. The equilibrium diagram as eventually accepted is shown in Fig. 107.

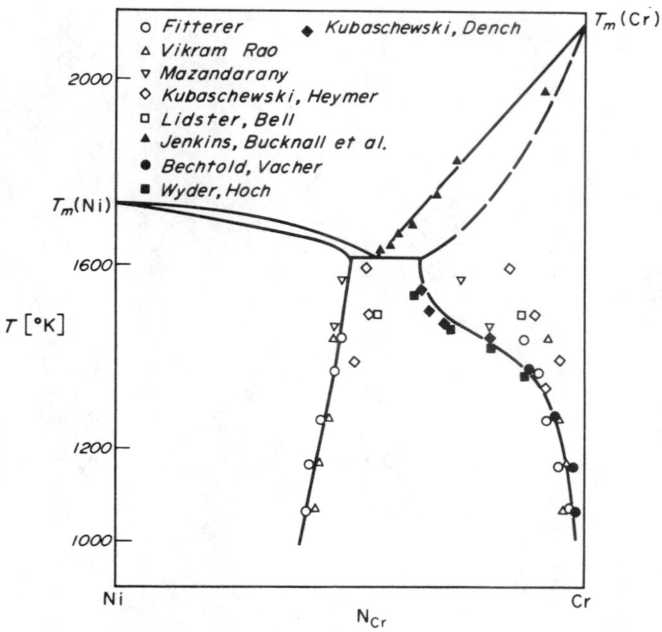

FIG. 107. Experimental and calculated equilibrium diagram of the system chromium–nickel.

Although such an evaluation may vary from system to system, there are certain rules of general significance to be obeyed. Since truly accurate experimental information is hardly ever available, the emphasis of the evaluation should be on consistency rather than ideal accuracy. Whenever possible, the evaluation should begin with good calorimetric heats of mixing in either the solid or the liquid state. In the present case, the heats of mixing of solid nickel–chromium have been determined by adiabatic calorimetry at 1514–1571°K by Dench [144] (Table XXVIII). The results may be taken as independent of temperature, as has been found for some selected compositions [40, 402].

Heats of mixing are usually obtained as integral values whereas experimental Gibbs energies are obtained as partial values. Thus for combination the former must be differentiated or the latter integrated in order to evaluate the entropies of reaction. Differentiation of heat vs. concentration curves may be carried out graphically or by expressing them by convenient polynomials (p. 38) with subsequent conversion. The former course leads at first to inaccurate values which have to be adjusted during a following integration. In Table XXVIII integral heats of Dench and the partial heats of Hack are listed for readers who wish to repeat the procedure(s) as an exercise.

The next step is to combine the partial heats of solution with the measured Gibbs energies of solution. Several investigators who employed different experimental methods have provided such Gibbs energies for the solid solution of chromium in nickel–chromium. Electromotive force measurements have been carried out by Panish *et al.*

Table XXVIII. Thermochemical values for the formation of solid alloys from solid metals

X_{Cr}	ΔH	$\Delta\bar{H}_{Cr}$	$\Delta\bar{H}_{Ni}$	ΔS	$\Delta\bar{S}_{Cr}$	$\Delta\bar{S}_{Ni}$	ΔS^E	$\Delta\bar{S}^E_{Cr}$	$\Delta\bar{S}^E_{Ni}$
0	0	−4200	0	0	∞	0	0	0·675	0
0·05	−158	−2153	−53	0·462	5·99	0·171	0·068	0·038	0·069
0·10	−210	−210	−210	0·733	5·315	0·224	0·087	0·740	0·014
0·15	−172	1096	−396	0·978	5·030	0·263	0·138	1·260	−0·060
0·20	−64	2178	−625	1·213	4·852	0·303	0·219	1·654	−0·139
0·25	115	3053	−864	1·432	4·628	0·367	0·315	1·873	−0·204
0·30	336	3740	−1123	1·643	4·500	0·419	0·429	2·108	−0·291
0·35	595	4257	−1377	1·845	4·300	0·523	0·559	2·214	−0·332
0·40	855	4622	−1656	2·026	4·150	0·610	0·689	2·329	−0·404
0·45	1165	4853	−1853	2·200	3·955	0·764	0·833	2·368	−0·423
0·50†	1520	4969	−1929	2·344	3·815	0·873	0·967	2·438	−0·504
0·55		3537	−497	2·384	2·728	1·964	1·017	1·540	0·378
0·60		3537	−497	2·423	2·728	1·964	1·089	1·713	0·153
0·645‡	2105*	2820*	806*	2·453	2·087	3·118	1·160	1·216	1·058
0·65	2114*	2741*	950*	2·448	2·055	3·118	1·162	1·199	1·093
0·70	2148	2014	2461	2·377	1·715	3·922	1·163	1·006	1·529
0·75	2069	1398	4082	2·223	1·295	5·007	1·106	0·723	2·255
0·80	1879	895	5815	1·999	0·945	6·215	1·005	0·502	3·017
0·85	1577	503	7663	1·687	0·685	7·365	0·847	0·362	3·395
0·90	1163	224	9614	1·308	0·535	8·265	0·662	0·326	3·686
0·95	637	56	11,676	0·817	0·273	11·153	0·423	0·171	5·211
0·98	268	9	12,959	0·428	0·109	16·059	0·233	0·069	8·209
1·00	0	0	13,827	0	0	∞	0	0	—

* Extrapolated.
† Phase boundary Ni f.c.c. at T_{eut}.
‡ Phase boundary Cr b.c.c. at T_{eut}.

The unit for ΔH is $\left[\dfrac{\text{cal}}{\text{g atom}}\right]$ and for ΔS $\left[\dfrac{\text{cal}}{\text{g atom °K}}\right]$.

[403] who used a molten NaCl–RbCl mixture for the electrolyte at 1023° and 1238°K, by Fitterer and Pugliese [404] who employed stabilised zirconia at 1073–1448°K and by Mazandarany and Pehlke [405] who used stabilised thoria at 1273–1573°K. For the assessment the suitability of the experimental method, in this case particularly the nature of the electrolyte, should be considered. Molten chlorides should only be employed below, say, 800°C. At such temperatures, however, diffusional delays in the electrodes may produce errors in systems melting above, say, 1500°C; zirconia electrolytes are on the borderline of usefulness for cells having solid Cr, Cr_2O_3 as the one electrode, because of significant electronic conductivity, so that one would rely most on the results obtained with thoria electrolytes. Rao and Flores-Magon [406] suggested a modified evaluation of the results of Fitterer and Pugliese.

In addition, Dench and Heymer [249] have measured dissociation pressures, that is the partial pressures of chromium, at 1399–1600°K by means of the Knudsen effusion method combined with tracer analysis of ^{51}Cr—a method that has been found to give reasonably good results. The first to investigate the present system, Grube and Flad [407] equilibrated the alloys with H_2O/H_2 mixtures and Cr_2O_3. The method is prone to certain errors (p. 136), but can give quite reasonable results.

All the individual partial Gibbs energies obtained by these investigators have been converted with the accepted partial heats of solution to yield partial excess entropies of solution which are plotted against the atomic fraction of chromium for the solutions in nickel in Fig. 108. The scatter of the data which stem from the various Gibbs energy measurements is not unusual and should make the compilers as well as the investigators think about the precision that can be achieved. Obviously, a consistent set of thermochemical properties cannot be assessed on the basis of these data alone. Additional information must be sought.

The curve shown in Fig. 108 is somewhat arbitrary, but is drawn with due regard to the relative reliability of the experimental methods. The next step is to accept the phase boundaries of the heterogeneous region above 1200°K as presented by Hultgren *et al.* [40] and to calculate the partial Gibbs energies of chromium at the compositions of its saturated solutions in nickel from its established partial heats and entropies of solution. These values are identical with those at the chromium-rich phase boundary for a given temperature. When combined with the partial heats of solution of chromium derived from the results of Dench (Table XXVIII) the excess entropies of solution shown as full curve in Fig. 108 are obtained. Again, it may be seen that the corresponding values derived from the experimental Gibbs energies are rather scattered around the calculated curve.

Finally, for the solid system the partial Gibbs energies are integrated at, say, 1600°K after Duhem–Margules (pp. 34, 35) and combined with the integral heats of mixing to yield integral entropies of mixing. These are tabulated together with the excess functions in Table XXVIII which provides a consistent set of thermochemical data for solid nickel–chromium alloys.

For a check, the phase boundaries of the heterogeneity region are evaluated by means of the common-tangent method (p. 36) from plots of the integral Gibbs energies derived, at various temperatures, from the tabulated heats and entropies of mixing. Figure 108 is an example of such an evaluation at a temperature of 1500°K. During this last stage it is usually found that some slight adjustments to the assessed values well within the experimental accuracy must be made. This is because of the high sensitivity of the position of the calculated phase boundaries to minor deviations in the thermochemical values—as a glance at Fig. 108 shows.

To complete the assessment, consistent data for the liquid solutions must be derived. However, relatively little experimental evidence on liquid nickel–chromium is available. Fruehan [408] measured the e.m.f. of cells of the type $\langle Cr\rangle\langle Cr_2O_3\rangle/\langle ZrO_2–CaO\rangle/\langle Cr_2O_3\rangle\{Cr–Ni\}$ at 1873°K. Here, again, doubts exist as to the suitability of the zirconia electrolyte. Gilby and St. Pierre [409] determined the activities of chromium in {Cr–Ni} solutions with $N_{Cr} = 0{\cdot}01–0{\cdot}5$ at 1873°K via the vapour pressures, using a mass spectrometer.

Since heats of mixing for the liquid are not available nor can be derived, various simplifications may be introduced. One may accept the heats *or* the entropies of mixing of the solid for the liquid solutions, for instance. Most commonly, however, one assumes the liquid solutions to be regular and, hence, takes the excess Gibbs energies as equal to the heats, the entropies being then, of course, the ideal ones.

The liquidus curves, where they have been reliably determined, are used as additional information: in the present case in the range 0–60 at.% Cr [40], including the composition and temperature of the eutectic. There is no easy mathematical way for this incorporation. The integral Gibbs energies of formation of the solid solutions are therefore plotted at the eutectic temperature and other temperatures above; various curves for the integral Gibbs energies of formation of the liquid solutions are also plotted at the respective temperature using the ideal entropies until, by the common-

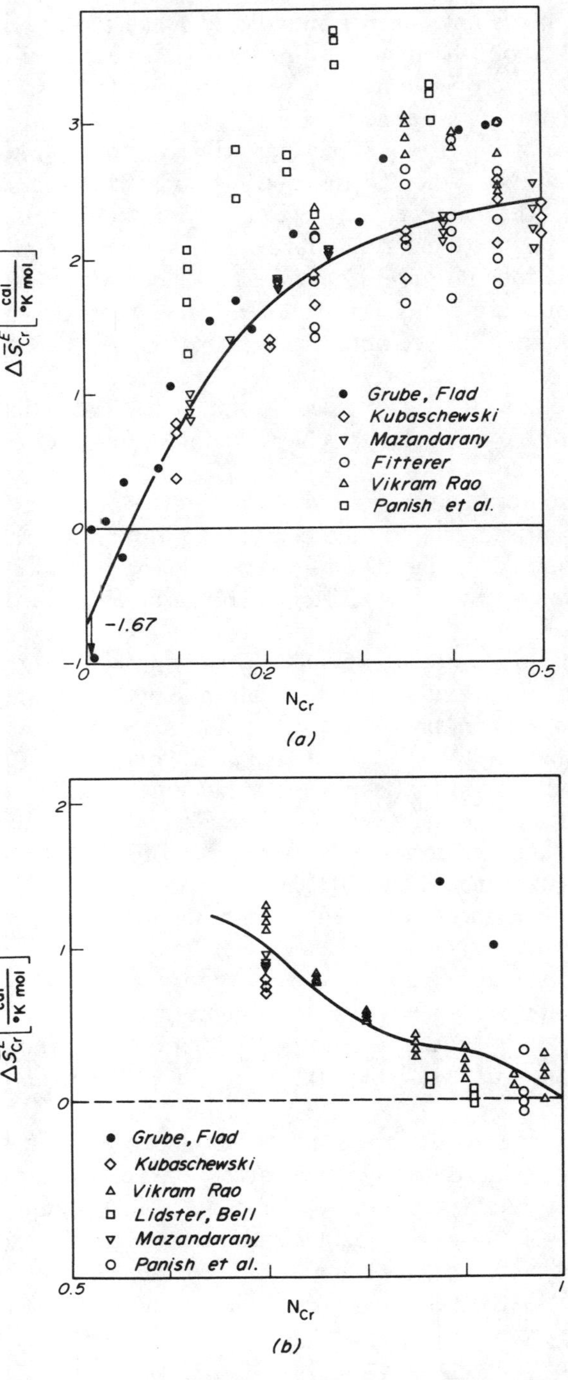

tangent method, the curve is found that reproduces the liquidus curve exactly. (Naturally, the ΔG curves for both the solid and liquid solutions must be related to the pure components in their state of aggregation at that temperature by incorporating the Gibbs energies of fusion for the supercooled or superheated metal, as the case may be.)

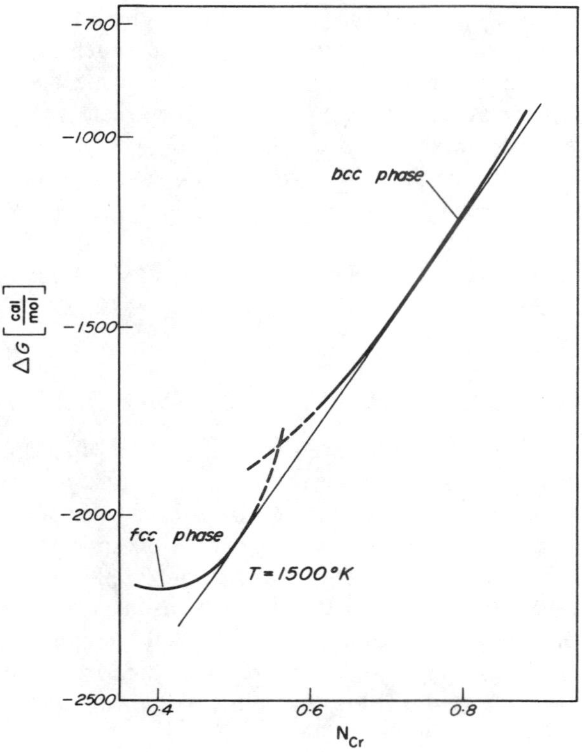

(c)

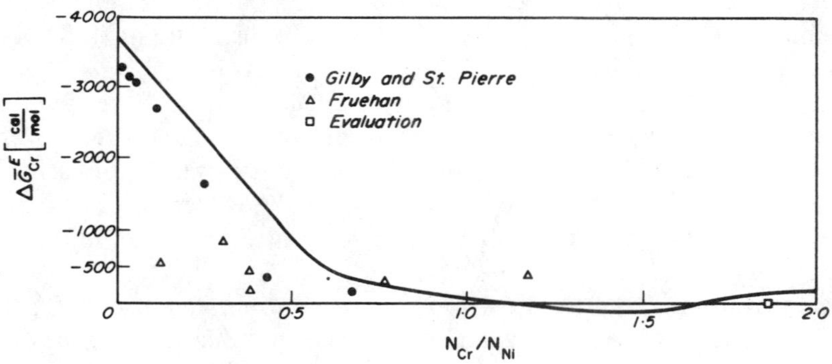

(d)

FIG. 108. Partial excess antropies of chromium: (a) in the nickel solid phase solution; (b) in the chromium solid phase solution; (c) integral Gibbs energies of the f.c.c. and b.c.c. phases in the system chromium–nickel indicating the graphical evaluation of phase boundaries; (d) experimental and assessed partial excess Gibbs energies of chromium in liquid chromium–nickel alloys.

Thus the mode of evaluation is one of trial and error. It must produce a single ΔG^E vs. N curve which, in the present case, may then be extrapolated to concentrations above 60 at.% Cr.

The differentiation of this curve yields the $\Delta \overline{G}^E_{Cr}$ vs. N_{Cr}/N_{Ni} curve in Fig. 108 which also shows the experimental results of the investigators mentioned above. It may be emphasised that, once the curve in Fig. 108 has been selected, the full curves in Figs. 108(b) and 108(c) follow without any reference to the experimental Gibbs energy results. The only additional experimental evidence that has been incorporated are the phase boundaries at temperatures high enough to warrant complete equilibration—provided the investigators were not impatient.

This does not apply in general and in this particular case to the solidus curves which, for reasons pointed out elsewhere, are rarely equilibrium data. Thus the solidus curves in Fig. 107 are to be preferred to those obtained by conventional methods of phase-boundary determination. Here the calculated solidus–liquidus gaps are narrower than those found experimentally—a common observation.

8. Calculation of Metallurgical Equilibrium Diagrams

Applied chemical thermodynamics are essentially phase diagram work, giving the state of equilibrium in terms of concentration, temperature and pressure. It is thus possible to calculate phase boundaries in a T–c or p–c diagram provided the enthalpies and entropies of the phases entering the equilibrium are known.

For the purpose of demonstration, let us take a few examples of binary, metallurgical, equilibrium diagrams [410]. Probably the simplest case is a eutectic system in which the solid solubilities in the component metals are negligibly small. The absolute liquidus temperatures T_{liq} at the composition N_i may be calculated from the heats of fusion of the pure components i and the heats and entropies of solution of i in the liquid mixtures by means of the following equation:

$$T_{\text{liq}} = (\Delta\bar{H}_i + L_i)/(\Delta\bar{S}_i + \sigma_i) \tag{127}$$

The calculation has to be carried out for the two component metals i, the two branches of the liquidus curve intersecting at the eutectic point. If the liquid mixtures form ideal solutions, eqn. (127) simplifies to eqn. (68) (p. 45).

The thermodynamic properties of the system *gallium–zinc* have been assessed (Table F), and the liquidus curve calculated by means of eqn. (127). The resulting curve is shown in Fig. 109. Since this was found to disagree with results of earlier cooling curves near the eutectic temperature, the thermal analyses were repeated employing particularly slow cooling rates [410]. The results, indicated by crosses in Fig. 109, then agreed well with the calculated liquidus curve.

The divergent results are obviously due to the strong tendency of gallium to supercool, equilibration being consequently delayed at the lower temperatures. This example demonstrates the superiority of the thermochemical method where kinetic difficulties obstruct the correct assessment of a phase boundary by conventional thermal analysis, the thermochemical data being obtained at higher temperatures where the equilibration rate is sufficiently rapid.

Slow rates of formation more often delay the equilibration of solid rather than liquid alloys. Solutions that are near-regular but have a positive heat of mixing must form a miscibility gap at some lower temperature, as has been discussed on p. 52 and illustrated in Fig. 24. The underlying principle of evaluation has been applied to the system *chromium–molybdenum* which forms a continuous series of solid solutions at higher temperatures, and the heats and entropies of which have been measured [410]—see Table F. It was calculated that the miscibility gap closes at about 900°C. Annealing

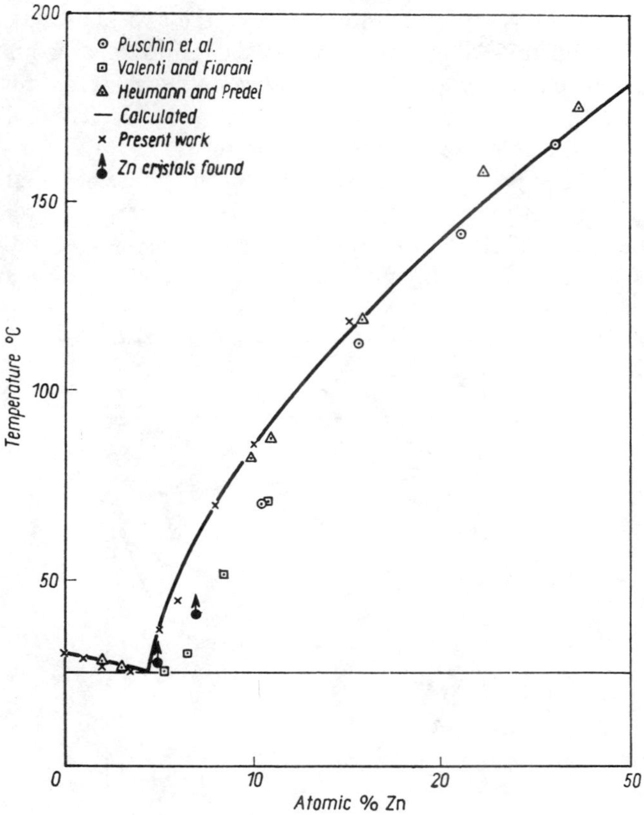

FIG. 109. Equilibrium diagram of the gallium–zinc system near the eutectic composition.

experiments, coupled with X-ray analyses, carried out to check the results of the calculations showed a distinct line separation at 800°C, for instance, only after 6 months, and it was estimated from the activation energy of diffusion that at, say, 650°C, a similar degree of demixing would require 250 years.

This example illustrates a significant case of applying thermochemical information to the determination of phase boundaries. Many solid solutions, given time, disintegrate at relatively low temperatures, and the location of the respective phase boundaries will require annealing times of a length which may be forbidding. Thermochemical measurements can be carried out at higher temperatures where equilibration rates are faster and then be used for the calculation of the extent of the low-temperature gap. Of course, if the equilibrium time runs hundreds of years, the practical significance of such results decreases.

Another system that has been studied thermochemically in some detail is that of *chromium* and *iron*, which is, of course, of considerable practical significance. Since the heat of formation of the α solid solutions is positive (Table F), they are expected to form a miscibility gap at lower temperatures. However, chromium and iron also form an ordered phase having the so-called σ structure. The heats and entropies of formation of the α solutions as well as the heats and entropies of the transformation $\sigma \rightarrow \alpha$ have been determined [144]; the heats and entropies of formation of σ could thus be evaluated.

Figure 110 shows the Gibbs energies of formation of the two phases at three temperatures. The figure also illustrates the graphical method of evaluating the phase boundaries. At a given temperature and composition, it is always the phase or mixture of phases that has the lowest Gibbs energy of formation which is stable. The Gibbs energy of a mixture of two phases is given by the tangent to the ΔG–N curves of the two neighbouring phases. The points of contact represent the phase boundaries at temperature T. The graphical method of evaluating phase boundaries is therefore termed the "common-tangent method". It is more convenient than using calculations which can become very involved even if simplifying assumptions concerning the shape of the ΔS–N and ΔH–N curves are made.

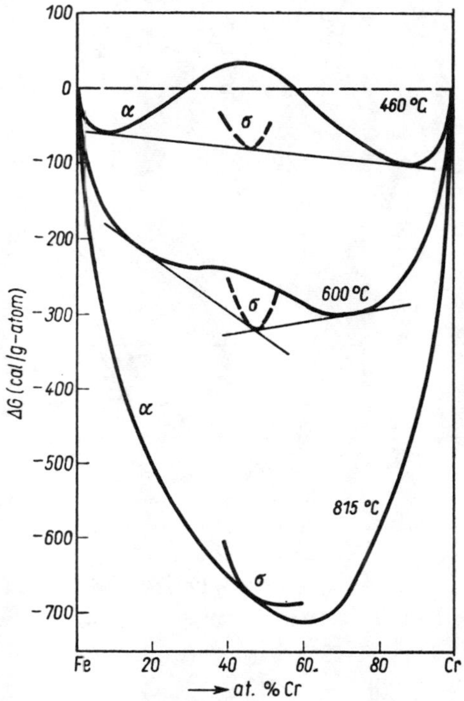

FIG. 110. Gibbs energies of formation of the α and σ phases in the iron–chromium system at three temperatures.

In the chromium–iron system, above 815°C, $\Delta G(\sigma)$ lies above the corresponding $\Delta G(\alpha)$ curve, and α solutions only are formed. At 815°C (Fig. 110), the $\Delta G(\sigma)$ curve just touches the $\Delta G(\alpha)$ curve; between 460° and 815°C (e.g. 600°C, Fig. 110), $\Delta G(\sigma)$ lies below $\Delta G(\alpha)$, the tangents giving the boundaries of the two $(\sigma + \alpha)$ fields. At about 460°C there is a common tangent to $\Delta G(\sigma)$ and the free energy curves of α solutions rich and poor in chromium, i.e. the three phases are in equilibrium. Below this temperature, σ is no longer stable, being replaced by a miscibility gap.

The corresponding equilibrium diagram is shown in Fig. 111. It resembles in most respects the phase diagram based on direct studies and proposed by Williams [411]. At higher temperatures, it differs from the diagram by Cook and Jones [412] with respect to the $\alpha/(\alpha + \sigma)$ boundary, but there is independent evidence that the solubility of σ in iron is less than suggested by Cook and Jones.

Since the practical interest in chromium steels is mainly confined to temperatures below 500°C, the usefulness of the present method for estimating phase boundaries should be obvious.

P. Gilles, at a conference of the International Atomic Energy Agency in Vienna (1965), suggested calculating pressure–composition diagrams in a similar manner to that described above for temperature–composition diagrams. This is, of course, a feasible and rewarding proposition.

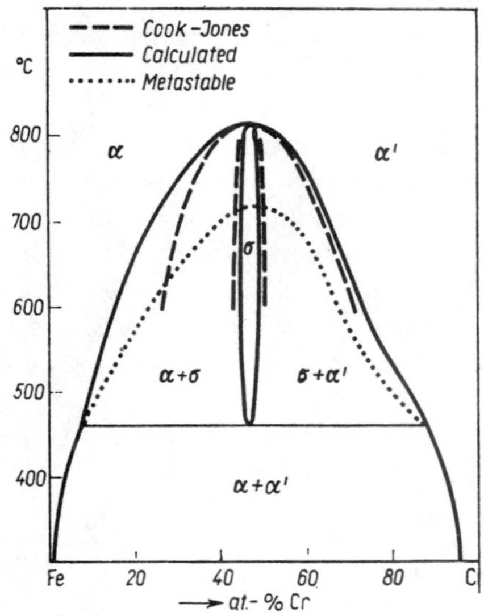

FIG. 111. Part of the equilibrium diagram for the system iron–chromium.

9. Ternary Diagrams

The most important potential use of the thermochemical method of phase boundary determination lies in its application to multicomponent systems. On pp. 58–60 it was shown that relatively simple equations interconnect the thermochemical properties of binary and ternary systems within the experimental errors. The next step is to use the thermochemical data of the ternary systems thus derived to calculate phase boundaries. Progress in this field has been quite rapid in recent years, thanks to the efforts in particular of Kaufman and Bernstein [413] and Ansara [414].

The principle of evaluation is the same as with binary systems, only that the Gibbs energy vs. concentration curves become Gibbs energy vs. concentration surfaces, and the tangents become tangent planes. The graphical evaluation, however, becomes cumbersome and tedious so that genuine progress was only made when the help of the computer could be invoked. For this it is necessary to express the Gibbs energy surfaces by suitable equations.

At high temperatures, solid gold, platinum and palladium are mutually completely soluble. The thermochemical properties of these solutions may be derived from various experimental measurements and expressed by two equations [415], one for the heat and the other for the excess entropy of mixing:

$$\Delta H = N_{Au}N_{Pd}(-11{,}141N_{Au} - 3065N_{Pd}) + N_{Pd}N_{Pt}(-5234N_{Pd} - 3021N_{Pt}) + N_{Pt}N_{Au}(5691N_{Pt} - 634N_{Au}) + 3036N_{Au}N_{Pd}N_{Pt}$$

$$\Delta S^E = -3N_{Au}N_{Pd} - 2{\cdot}9N_{Pd}^2N_{Pt} - 1{\cdot}7N_{Pd}N_{Pt}^2 - 0{\cdot}9N_{Pt}^2N_{Au} - 3{\cdot}5N_{Pt}N_{Au}^2 + 9N_{Au}N_{Pd}N_{Pt}$$

At lower temperatures the system gold–palladium–platinum exhibits a ternary miscibility gap, the boundaries of which may now be calculated by computer from these two equations. The results are compared with the experimental boundaries [416] at two temperatures in Fig. 112. Agreement, it is true, is not exactly impressive, but the calculated phase boundaries are very sensitive to small errors in the thermochemical data, the deviations from the true values being enhanced by the transformation into polynomials which cannot easily reproduce the refinements of the $\Delta G°$ vs. N surfaces. Nevertheless, the speed with which such phase boundaries can be produced, though only approximately, speaks for the usefulness of the method. On the other hand, it should be noted that the deviations are the larger the smaller the differences in Gibbs energy of the partaking phases and that in such cases, owing to the small "driving force", it is correspondingly difficult to bring the alloys into equilibrium, essential to the application of conventional methods of phase-boundary determination.

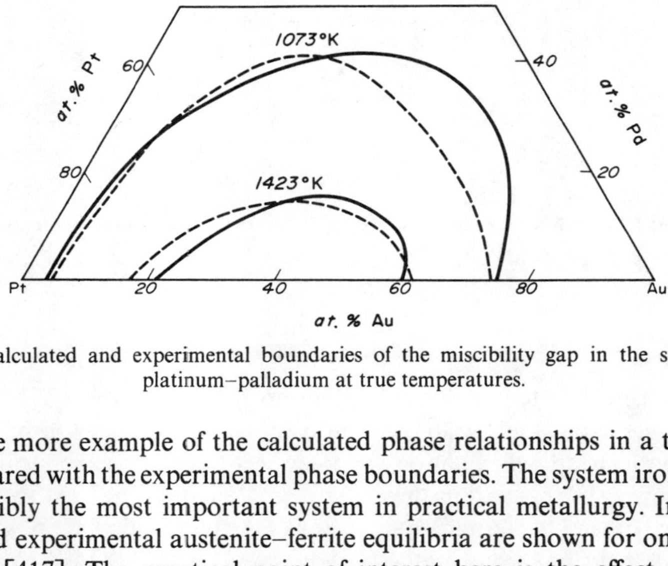

FIG. 112. Calculated and experimental boundaries of the miscibility gap in the system gold–platinum–palladium at true temperatures.

Finally, one more example of the calculated phase relationships in a ternary system may be compared with the experimental phase boundaries. The system iron–chromium–nickel is possibly the most important system in practical metallurgy. In Fig. 113 the calculated and experimental austenite–ferrite equilibria are shown for one temperature out of many [417]. The practical point of interest here is the effect of fourth and more elements on the shift of the phase boundaries: in other words, the calculation of quaternary and higher phase diagrams. Although we are still at the beginning of this development, the problem of calculating the phase boundaries in multicomponent systems has in principle been solved.

10. Deoxidation of Iron

Dissolved oxygen may be removed from a molten steel bath by the addition of an element the oxide of which is more stable than that of iron. The extent of deoxidation

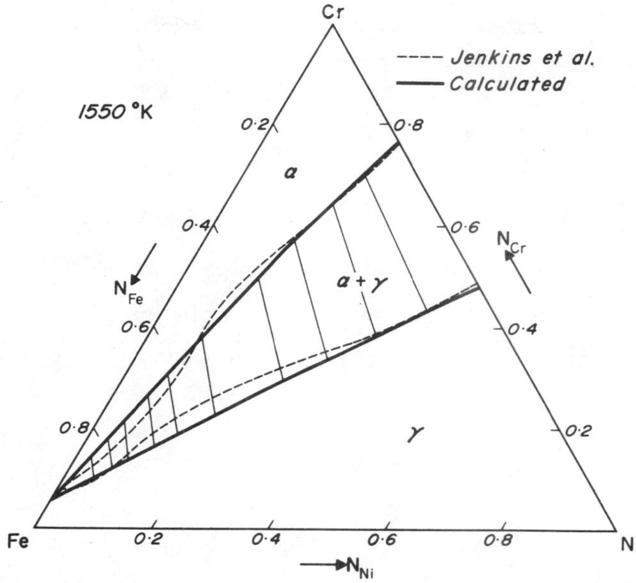

FIG. 113. Calculated and experimental phase boundaries of the ferrite–austenite equilibrium at 1550°K.

depends primarily on the Gibbs energy of formation of the oxide and on the concentration of the deoxidising element remaining in the liquid steel when equilibrium has been attained. The relationship between residual oxygen and oxidiser is of great importance to the steelmaker, and it is summarised as nearly as possible by the mass action product of the reaction which is known as the "deoxidation constant".

In even the simplest systems, this quantity is not necessarily a true constant for any one temperature. It may change with the residual concentration of the oxidising element due to variation in the activity coefficients of the deoxidiser and the oxygen. (The activity coefficient of oxygen in pure iron is practically constant, but is modified by the presence of a third element.) In practical steelmaking, the "deoxidation constant" is influenced by numerous other factors, among which are the presence of various alloying elements and the assimilation of the primary product of deoxidation in a slag of complex composition. For this reason steelmakers control their deoxidation as well as other refining reactions from largely empirical relationships known to be valid for the production of certain types of steel composition under certain types of slag. Much information is being gained by laboratory-scale investigation and thermodynamic calculation of the simpler equilibria.

A. DEOXIDATION WITH SILICON

There have been several determinations of this equilibrium, most of which have been based on the addition of silicon to oxygen-containing iron melted in a silica crucible. A determination by Gokcen and Chipman [221] involved further measurements, namely those of equilibria between melts of silicon and oxygen in iron and controlled mixtures of hydrogen and water vapour. The mass-action product of the reaction

$$\langle SiO_2 \rangle = [Si]_{Fe} + 2[O]_{Fe} \qquad (\alpha)$$

(i.e. $K_c = c_{Si}\, c_O^2$) was found to have a constant value of $2{\cdot}8 \times 10^{-5}$ at 1600°C, the concentrations being expressed as weight percentages.

The individual Gibbs energy equations from which the mass action constant of eqn. (α) may be approximately derived are the following:

$$\langle SiO_2 \rangle = \{Si\} + (O_2); \qquad \Delta G^\circ = 227{,}700 - 48{\cdot}7T$$

$$(O_2) = 2[O]_{Fe}; \qquad \Delta \bar{G} = -55{,}850 + 12{\cdot}15T + 9{\cdot}15T \log N_O$$

$$\{Si\} = [Si]_{Fe}; \qquad \Delta \bar{G} = -31{,}400 + 3{\cdot}59T + 4{\cdot}575T \log N_{Si}$$

Upon summation, the Gibbs energy of reaction (α) is obtained:

$$\Delta G = 140{,}450 - 32{\cdot}96T \log N_O + 4{\cdot}575T \log N_{Si} \tag{ε}$$

At equilibrium $\Delta G = 0$, and consequently,

$$2 \log N_O + \log N_{Si} = -30{,}700T^{-1} + 7{\cdot}20 \tag{β}$$

At 1600°C

$$2 \log N_O + \log N_{Si} = -9{\cdot}19$$

This result can be transformed into the weight percentage basis c_O and c_{Si} by the use of the general equation for a very dilute solute i, in the solvent j,

$$\log N_i = \log \frac{\text{wt.\%}}{M_i} \frac{M_j}{100}$$

Therefore

$$2 \log c_O + \log c_{Si} = -4{\cdot}58$$

$$c_O^2\, c_{Si} = 2{\cdot}7 \times 10^{-5}$$

This result is approximate because the interaction coefficients of oxygen and silicon $\varepsilon^{Si} = \varepsilon^{o}_{Si}$ in solution in iron have been ignored. However, the concentrations are quite small and during deoxidation the minimum amount of silicon would be added; hence the effect can be ignored for this application.

B. DEOXIDATION WITH ALUMINIUM

The equilibrium constant for the corresponding reaction with aluminium

$$\langle Al_2O_3 \rangle = 2[Al]_{Fe} + 3[O]_{Fe} \tag{γ}$$

can also be calculated from the thermochemical data which are available. The result

$$c_O^3\, c_{Al}^2 = 10^{-13}$$

disagrees, however, with the bulk of the deoxidation constants observed in practice, as may be seen from Table XXIX which is reproduced from a paper by Fitterer [419]. The possible reason for this discrepancy has also been given by Fitterer.

One contributory factor probably is: when aluminium "shot" is added to undeoxidised liquid steel, the reaction in the immediate of the aluminium "particle" proceeds to a very high degree because the reaction with oxygen proceeds at a much higher rate

Table XXIX. Deoxidation constants for aluminium in iron (1600°C)

Investigators	$(\%Al)^2(\%O)^3$
Hessenbruch, 1929	$6{\cdot}2 \times 10^{-8}$
Herty, Fitterer and Byrns (exptl.), 1930	$6{\cdot}3 \times 10^{-9}$
Herty, Fitterer and Byrns (calcd.), 1930	$0{\cdot}19 \times 10^{-9}$
Wentrup and Hieber, 1939	$0{\cdot}11 \times 10^{-9}$
Geller and Dickie, 1943	4×10^{-13}
Hilty and Crafts, 1950	$3{\cdot}0 \times 10^{-9}$
Present calculation	1×10^{-13}

than that at which aluminium dissolves and diffuses through the iron. However, other areas of the steel are essentially unaffected. The second contributory factor appears to be that when a resultant Al_2O_3 particle in a deoxidised area is transmitted to an undeoxidised area by turbulence, etc., oxygen in the undeoxidised area comes out of solution to form FeO which in turn combines with the Al_2O_3 particle to form a spinel, $FeO{\cdot}Al_2O_3$. This would result in a high FeO content as well as a larger deoxidation constant. Both of these factors would tend to explain the discrepancy between the calculated and the experimental "equilibrium" constants. From the thermochemical data, one can estimate that the spinel is, indeed, stable when the oxygen content exceeds approximately 0·01 wt.% at the melting temperature of iron.

However, the discrepancy has been resolved to some extent by Repetylo, Olette and Kozakevitch [419a] who made careful experiments over longer periods under purified argon at 1600°C and demonstrated that the deoxidation constant varies with time approaching a "final" figure of 5×10^{-13}. The phenomenon shown to be responsible for this behaviour is the elimination of the alumina suspension, the coarser particles being eliminated in 7–10 min, whereas a finer suspension ($< 20\ \mu$) remains in the melt even 20 min after the addition of aluminium. Further, Foerster and Richter [419b] have observed the actual concentration of oxygen in killed steel by an e.m.f. method and found that supersaturation can reach a very high value in relatively good agreement with nucleation theories by Volmer.

11. Refining of Iron

High-purity iron is at present required in fairly large quantities for research purposes. The methods available for purification can be classified broadly as chemical, electrolytic and metallurgical. A detailed account of these methods has been given by Hopkins [420]. The metallurgical methods are particularly effective in removing certain non-metallic elements and can, in addition, be expanded to a large laboratory scale fairly easily. It appears timely therefore to discuss quantitatively the thermochemical background of some of the individual steps involved in this purification procedure.

It is, of course, advisable to use fairly pure commercial irons as starting materials, such as the selected Swedish iron quoted by Hopkins [420] which contains, for instance, 0·015% C, 0·016% Si, 0·044% O, 0·01% N, and is reasonably low in sulphur (0·006%) and phosphorus (<0·001%).

The common metallurgical procedure for reducing the carbon content is melting the iron in air so that it becomes saturated with oxygen. Carbon will then be removed as the monoxide according to the equation

$$[C]_{\{Fe\}} + [O]_{\{Fe\}sat} = (CO) \qquad (\alpha)$$

The melt is allowed to solidify and then placed in a vacuum furnace, again heated above the melting point, and the gas pressure reduced to, say, 0·1 mmHg for some time. The Gibbs energy equation for reaction (α) may be obtained from the individual equations, as follows:

$$[C]_{\{Fe\}} = \langle C \rangle; \qquad \Delta\bar{G}_\beta = -6400 - 4{\cdot}574T \log N_C + 4{\cdot}25T$$

$$[O]_{\{Fe\}sat} = \tfrac{1}{2}(O_2); \qquad \Delta\bar{G}_\gamma = 27{,}930 + 3{\cdot}45T$$

$$\langle C \rangle + \tfrac{1}{2}(O_2) = (CO); \qquad \Delta G^\circ_\delta = -26{,}700 - 20{\cdot}95T$$

The Gibbs energies $\Delta\bar{G}_\beta$ and ΔG_γ have been taken from Catterall's tables [421], that of reaction δ from Table E. By addition,

$$\Delta\bar{G}_\alpha = \Delta\bar{G}_\beta + \Delta G_\gamma + \Delta G^\circ_\delta = -5170 - 4{\cdot}574T \log N_C - 13{\cdot}25T$$

By continuous removal of CO from the surface the carbon content could in theory be reduced to virtually nil. In practice, however, 0·001–0·004% C is left in the iron [422]. In a representative run the carbon was 0·0012% after one hour's evacuation at 1550°C. Since the calculation reveals that its equilibrium content should be much lower, it follows that the residual carbon is present for kinetic reasons and an increase in the time of evacuation and/or the surface area should effectively reduce the carbon content further.

During the first stage some of the nitrogen is also removed. Its percentage is further reduced during the next stage when hydrogen of 1 atm pressure is passed over the molten iron. In the earlier work at the National Physical Laboratory [423] the hydrogen was circulated and passed over magnesium turnings at 620°C for intermediate purification. The average figure for nitrogen in the iron product was about

0·0016% N

From the experimental data it was estimated that the hydrogen took up an amount of nitrogen corresponding to 1 mmHg which was to be removed by the magnesium. From Catterall's tables [421] we may take the Gibbs energy of the reaction

$$2[N]_{\{Fe\}} = (N_2) \qquad (\beta)$$

to be

$$\Delta\bar{G}_\varepsilon = -2800 - 9{\cdot}15T \log N_N - 24{\cdot}0T$$

1 mmHg nitrogen corresponds to $\log p_{N_2} = -2{\cdot}88$, and $-4{\cdot}574T \log p_{N_2} = 24{,}000$ (for $T = 1825$°K). Hence

$$9{\cdot}15 \times 1825 \log N_N = -2800 - 24{\cdot}0 \times 1825 - 24{,}000$$

$$\log N_N = -4{\cdot}225$$

The atomic fraction of nitrogen in liquid iron at 1552°C in equilibrium with an N_2 pressure of 1 mmHg is thus $N_N = 0{\cdot}000060$, or, expressed in weight per cent,

0·0015% N

Since this agrees with the average figure observed and given above, it must be concluded that the purification of the hydrogen by magnesium was not effective. Consequently, the hydrogen in the furnace was replaced after some time by fresh hydrogen from the cylinder, and the iron left in this atmosphere for a further period. The nitrogen content of the iron then dropped [424], the average figure for 25 melts being

$$0{\cdot}0008\%\ \text{N}$$

The nitrogen content of the cylinder hydrogen was 0·05% or 0·0005 atm for $p_{H_2} = 1$ atm. If the calculation above is repeated with the new figure, $-4{\cdot}574T \log p_{N_2} = 27{,}600$, the atomic fraction of nitrogen for equilibrium is found to be $N_N = 0{\cdot}000036$ or

$$0{\cdot}0009\%\ \text{N}$$

Probably the cylinder hydrogen contained somewhat less nitrogen than indicated by the suppliers' rough figure.

The oxygen percentage after the hydrogen treatment ranged from 0·0005 to 0·002% [420], a representative figure being 0·0007% O. It is known [422], however, that adsorption on the surface of the machined sample used for oxygen determinations amounts to about 0·0003% O, so that the actual figure for dissolved oxygen was approximately

$$0{\cdot}0004\%\ \text{O}$$

The chemical reaction by which oxygen is removed, is as follows:

$$[\text{O}]_{\{\text{Fe}\}} + (\text{H}_2) = (\text{H}_2\text{O}) \qquad (\gamma)$$

Since p_{H_2} is nearly 1 atm, hydrogen is in its standard state and does not enter into the calculation. The minimum pressure of H_2O depends on the impurities in the hydrogen. Actually, the hydrogen contained <0·05% O and some water vapour. The gas was passed through an excess of silica gel. The oxygen would, of course, have passed through the drying agent, but then catalytically reduced to water at the metal surface. Since the gas was continuously circulated and the silica gel was in the cycle, all the oxygen should have been finally in the form of water vapour of a pressure of about 0·04 mmHg, that is, its pressure in equilibrium with silica gel [425]. In actual fact, however, one would not expect the H_2O pressure to become as low as that because the walls of the furnace and the refractories must have been a constant source of water vapour, since the furnace was, of course, regularly opened after each melt, and gaseous impurities, predominantly water, would have been adsorbed.

In this case therefore, one may use the thermochemical calculation to find the apparent H_2O pressure in the furnace from the figure of 0·0004% for the oxygen content of the iron product.

From pp. 64, 238 and 380 respectively the following two Gibbs energy equations are taken:

$$[\text{O}]_{\{\text{Fe}\}} = \tfrac{1}{2}(\text{O}_2); \qquad \Delta\bar{G}_\eta = 27{,}930 - 4{\cdot}574T \log N_O - 6{\cdot}08T$$

$$(\text{H}_2) + \tfrac{1}{2}(\text{O}_2) = (\text{H}_2\text{O}); \qquad \Delta G^\circ_\theta = -58{,}900 + 13{\cdot}1T$$

By addition,

$$\Delta\bar{G}_\zeta = \Delta\bar{G}_\eta + \Delta G_\theta = -30{,}970 - 4{\cdot}574T \log N_O + 7{\cdot}02T$$

N_O equals 0·000014, and $\Delta\bar{G}_\xi = -4{\cdot}574T \log p_{H_2O}$ is therefore 22,300 for $T = 1825$°K.

This corresponds to a partial pressure of water vapour in the system of 1·6 mmHg. This is rather a high pressure, but possibly some oxygen was passing unchecked through the silica gel in the form of carbon monoxide, the carbon removal not being complete, as was seen above. Some further improvement in this part of the process seems to be possible.

It is seen that a quantitative thermochemical treatment of a refining process may lead to useful conclusions. The example with nitrogen in particular illustrates how a very simple calculation (taking not more than 10 min when the Gibbs energy equations are available) helps considerably in improving experimental techniques. Unfortunately, the thermochemical data for solutions of non-metals in solid or liquid metals other than in iron are, as yet, generally much less accurate.

12. Refining of Lead

In the removal of impurities from blast-furnace-produced lead, there are two steps concerned with the removal of silver and gold which are interesting to analyse thermodynamically. In the first step, zinc is added to the liquid lead, and the temperature is reduced to just above the melting point of lead. A crust separates which contains alloys of zinc and silver or gold. After separation of these metals, the zinc must be removed from the lead, and this is sometimes accomplished by bubbling chlorine through the liquid metal to form, principally, zinc chloride.

To begin the analysis we consider the zinc–silver system. Satisfactory phase diagram and thermodynamic studies have been made of this system so that chemical potentials of silver, as a function of composition across the solid region and in dilute solution in liquid zinc, can be fairly accurately described.

The liquid silver–zinc system can be considered as approximately regular, and the partial heat of solution of silver in dilute solution in zinc is then found to be −3400 cal.

The distribution of silver between liquid lead and zinc has been studied by a radioactive technique by Peterson and Kontrimas [426]. The results for the distribution coefficient K_d show that it is virtually independent of the concentration of silver in either metal from 5×10^{-5} to 5×10^{-2} for the mole fraction of silver. Hence Henry's law may be used for silver in dilute solution in both lead and zinc.

Values of K_d as a function of temperature

T(°K)	K_d	$RT \ln K_d = \Delta\bar{G}^{\text{Ag}}_{\text{Zn}\to\text{Pb}}$
711	55·7 ± 2·1	5680 cal
781	27·5 ± 0·5	5120 cal
821	20·1 ± 1·1	4900 cal

These results were used to calculate the two-term Gibbs energy equation for the transfer, at constant composition of 1 mole of silver, from a dilute solution in liquid zinc to one in liquid lead:

$$\Delta\bar{G}^{\text{Ag}}_{\text{Zn}\to\text{Pb}} = 10{,}900 - 7{\cdot}0T \quad \text{cal/g-atom} \qquad (\alpha)$$

Now as the silver–zinc system can be considered as a regular solution, we may calculate from the results given above that for the dilute solution of silver in zinc, with solid silver as standard state,

$$\Delta \bar{G}_{Zn}^{Ag} = -3400 + RT \ln N_{Ag} \quad \text{cal/g-atom} \qquad (\beta)$$

Hence, combining (α) and (β) we derive a Gibbs energy equation for the dilute solution of silver in liquid lead:

$$\Delta \bar{G}_{Pb}^{Ag} = 7500 - 7{\cdot}0T + RT \ln N_{Ag} \qquad (\gamma)$$

This equation allows the calculation of the chemical potential of silver at various concentrations in liquid lead, and as an example, these have been made for 610°K and for 0·02, 0·2, 2, 20 and 200 oz silver per ton of lead, which represents the range of concentration covered in the desilvering process.

Now, clearly, the solid silver–zinc alloy which is in equilibrium with these solutions at this temperature will have the same silver potential and this is demonstrated in Fig. 114. This shows the integral Gibbs energy of formation of the silver–zinc system at 610°K obtained from the data given above together with tangents which intercept the pure silver ordinate at the potentials for the given silver contents of liquid lead at the same temperature. The points of tangency on the integral silver–zinc curve represents the compositions of the solids in equilibrium with these liquids. Thus, at 200 oz/ton the liquid is in equilibrium with a $\beta + \gamma$ mixture of the silver–zinc phases, whilst at and below 0·2 oz/ton the liquid is in equilibrium with the η solid solution.

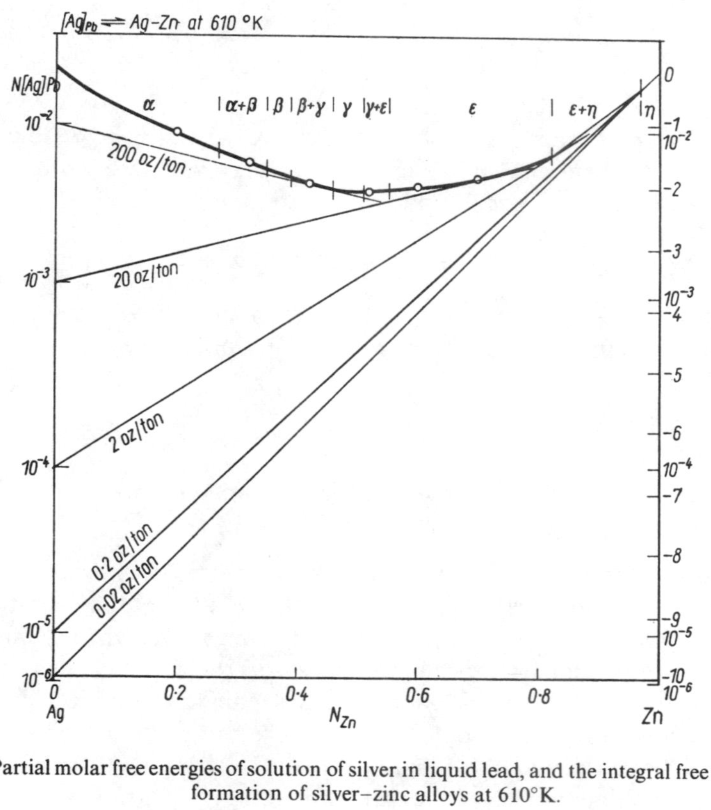

FIG. 114. Partial molar free energies of solution of silver in liquid lead, and the integral free energies of formation of silver–zinc alloys at 610°K.

A. THE REMOVAL OF ZINC FROM LEAD

After the silver and gold have been removed, the lead bullion is left with a zinc content which must be reduced. This can be done by bubbling chlorine through the metal when the zinc content is considerably reduced by preferential chloride formation. The chloride phase which is formed will consist, assuming the process is carried out at or near equilibrium, of a liquid chloride slag containing both zinc and lead, the proportions being determined by the metal activities in the metal phase. The amounts of lead and zinc in the slag can be calculated by considering the equilibrium

$$PbCl_2 + Zn = ZnCl_2 + Pb$$

$$\Delta G^\circ = -15{,}090 \text{ cal} \quad \text{at } 400^\circ\text{C}$$

$$K = \frac{a_{ZnCl_2}\, a_{Pb}}{a_{PbCl_2}\, a_{Zn}} = 7{\cdot}96 \times 10^4 \quad \text{at } 400^\circ\text{C} \tag{δ}$$

Because of the large difference in the standard Gibbs energies of formation of these two halides, the reaction will be displaced to the right and the lead activity can be taken as unity. Lacking quantitative information, we will assume that the halide slag is an ideal liquid mixture of the two component chlorides; then

$$K = \frac{N_{ZnCl_2}}{N_{PbCl_2}} \frac{1}{a[Zn]_{Pb}} \tag{ε}$$

$$\Delta G^\circ = -RT \ln K = RT \ln a[Zn]_{Pb} - RT \ln \frac{N_{ZnCl_2}}{N_{PbCl_2}} \tag{ζ}$$

The dilute solution of zinc in lead shows a positive partial heat of solution of zinc, and an excess partial entropy of +2 cal/deg. We may therefore write for the partial Gibbs energy of solution of zinc in lead, as a function of zinc content,

$$RT \ln a[Zn]_{Pb} = \Delta\bar{G}[Zn]_{Pb} = 5000 - 2T + RT \ln N[Zn]_{Pb} \tag{η}$$

Hence at 400°C, when the slag contains a mole fraction of lead chloride of 0·1, the metal phase has a mole fraction of zinc of 4×10^{-5}. This can be shown by substitution into eqns. (ζ) and (η).

13. The Decarburisation of Iron–Chromium–Carbon Iron–Silicon–Carbon Liquid Alloys

Liquid alloys of iron and chromium are produced by reduction of the oxides in a submerged electric arc furnace with carbon. The temperature of the operation normally reaches about 1700°C, and in order to maintain the integrity of the graphite electrodes an excess of coke is added to the furnace charge. The product always has a high carbon content as a result of this procedure. In refining the alloy it is desirable to reduce the carbon content to about 0·01 wt.%. Typical ferro chromium alloys might contain 10–25 wt.% Cr, and the refining process is required to reduce the carbon content with the minimum loss of chromium from the alloy. Since the oxidation products are substantially carbon monoxide and Cr_2O_3, the equilibrium constant for the reaction

$$\tfrac{2}{3}Cr + (CO) \rightarrow \tfrac{1}{3}Cr_2O_3 + C \qquad K = \frac{a_{Cr_2O_3}^{1/3}\, a_C}{a_{Cr}^{2/3}\, p_{CO}}$$

can be used to calculate the carbon monoxide pressure which would be in equilibrium with a given iron–chromium–carbon alloy and pure solid Cr_2O_3 at the refining temperature, and providing the carbon monoxide pressure can be kept below this level, oxidation of the chromium could not take place unless Cr_2O_3 could be formed in solution in some other oxide ($a_{Cr_2O_3} < 1$).

The equilibrium constant for this reaction varies with temperature according to the equation

$$\log_{10} K = 12{,}580T^{-1} - 9{\cdot}10$$

and has the values $3{\cdot}32 \times 10^{-3}$ at 1900°K and $1{\cdot}55 \times 10^{-3}$ at 2000°K, which straddle the operating temperature range. It follows from the equilibrium constant that when the chromium activity in an alloy is 0·1, the carbon activity is 3×10^{-4} when the system is in equilibrium with pure Cr_2O_3 and 1 atm pressure of carbon monoxide and one-hundredth of this when the carbon monoxide pressure is 10^{-2} atm at 2000°K. The activity of carbon in iron–chromium alloys at these temperatures have been reported by Richardson and Dennis [427]. Chromium forms stable carbides whereas those of iron have very low stability. It might therefore be expected that the chromium–carbon bond strength would be very much stronger than the iron–carbon bond and since the iron–chromium system is almost Raoultian, then chromium should lower the activity of carbon in solution in iron. The interaction coefficient ε_C^{Cr} in iron as solvent has the value $-4{\cdot}3$ and e_C^{Cr}, which is $\partial \log \gamma_C / \partial$ wt.% Cr, is correspondingly $-0{\cdot}020$ at these temperatures. The measurements show that when the atom fraction of chromium has the values 0·1 and 0·23 the activity coefficient of carbon with respect to graphite as standard state is lowered from the value in iron, 0·57, to 0·35 and 0·21 respectively at 1660°C when the carbon content is less than 1 at.%. The presence of chromium in the alloy will therefore approximately double the atom fraction of carbon for a given activity in solution in liquid iron for typical alloy compositions.

If it is possible to reduce the carbon activity only to 3×10^{-4} by oxidation of carbon before Cr_2O_3 is formed, then the best alloy that could be made would contain about 0·04 wt.% C. By lowering the oxygen partial pressure, values much lower than this are theoretically possible. Krivsky [428] has described the use of argon–oxygen mixtures in the removal of carbon in an industrial furnace, and states that by finishing the refining period with practically pure argon it is possible to reach a carbon level of 0·01 wt.% in a typical alloy containing 18·5 wt.% Cr.

The iron–silicon–carbon system has two significant differences from the iron–chromium–carbon system with respect to the oxidation behaviour and the first of these is that silicon has a very strong affinity for iron. Therefore, despite the fact that silicon forms a very stable carbide, the interaction parameter ε_C^{Si} has a value of approximately +10 and silicon *raises* the activity coefficient of carbon in solution in liquid iron.

If we now evaluate the oxidation equilibrium constant for iron–silicon–carbon in the same way as was done for chromium,

$$[\text{Si}]_{\text{Fe}} + 2(\text{CO}) \rightarrow \text{SiO}_2 + 2[\text{C}]_{\text{Fe}}; \quad K = \frac{a_{\text{SiO}_2}\, a_{\text{C}}^2}{a_{\text{Si}}\, p_{\text{CO}}^2}$$

$$\log K = 38{,}100T - 19{\cdot}8$$

The constant has the value of unity at 1924°K which is in the range of steelmaking temperatures, and the activity of carbon is equal to the square root of the silicon activity when p_{CO} is 1 atm and one-tenth of the square root of a_{Si} when p_{CO} is 10^{-1} atm.

The partial Gibbs energy of solution of silicon in liquid iron at a mole fraction of 0·2 is given by the equation

$$\Delta\bar{G}_{\text{Si}} = -26{,}200 + 1{\cdot}54T \text{ (cal)}$$

and hence the activity of silicon is $2{\cdot}1 \times 10^{-3}$ at 1900°K and $3{\cdot}0 \times 10^{-3}$ at 2000°K. The activity of carbon has a value equal to the square roots of these numbers, *ca* 5×10^{-2} when SiO_2, carbon monoxide at 1 atm pressure and an iron–silicon–carbon liquid alloy containing silicon at an atom fraction of 0·2 are in equilibrium in the temperature range 1900–2000°K. This carbon activity corresponds to a carbon content of 1·3 wt.% in the iron–carbon system; when p_{CO} is held at 10^{-2} atm this content falls to about 0·02%. However, this reaction is not the only one to be considered in the oxidation behaviour of iron–silicon–carbon because the formation of gaseous SiO, a product having no parallel in the iron–chromium–carbon system, becomes important at higher temperatures above 1700°C. The reaction would be described by the equilibrium

$$[\text{Si}]_{\text{Fe}} + (\text{CO}) \rightarrow (\text{SiO}) + [\text{C}]_{\text{Fe}}; \qquad K = \frac{a_{\text{C}}\, p_{\text{SiO}}}{a_{\text{Si}}\, p_{\text{CO}}}$$

$$\log K = 1990T - 1{\cdot}94$$

This constant has a value of approximately 0·1 at steelmaking temperatures and hence the two gases will be formed with the same partial pressure when the activity of carbon is one-tenth that of silicon; carbon and silicon will be evolved together at the same rate. No dilution of the oxidising gas with argon would permit further decarburisation without the accompanying silicon loss in this case.

Carbon and silicon have greatly differing activity coefficients in solution in liquid iron, and neglecting the ε_{C}^{Si} and conjugate parameter ε_{Si}^{C} for the present purposes, the atom fraction ratio for a given activity ratio is inversely related to the activity coefficients. The carbon content of an alloy in which CO and SiO are being formed at the same rate will be given approximately by

$$10a_{\text{C}} = a_{\text{Si}}; \qquad 10N_{\text{C}} = \frac{\gamma_{\text{Si}}}{\gamma_{\text{C}}} N_{\text{Si}}$$

Table XXX shows the values at 1873°K for both solutes in iron and in the range of compositions which are of interest in ferro-silicon. The silicon content is usually large compared with the carbon content because of the activity coefficient effect and when N_{Si} is 0·2 (11 wt.%), N_C is only 0·0003 (0·008 wt.%) when carbon and silicon are oxidised at the same rate. As the silicon content increases, the activity coefficient of silicon rises rapidly, and when N_{Si} is 0·5, γ_{Si} is now 0·446 and the activity of carbon must therefore be 0·0223. The effect of silicon on the activity coefficient of carbon is no longer negligible and should be taken into account when attempting to calculate N_C.

Table XXX. Activity coefficients of carbon and silicon dissolved in liquid iron at 1873°K

N_C	γ_C	N_{Si}	γ_{Si}
0·00	0·57	0·00	0·0013
0·05	0·85	0·1	0·0030
0·10	1·37	0·2	0·0090
0·15	2·30	0·3	0·040
0·20	4·12	0·4	0·178
		0·5	0·446

There is reasonable experimental evidence that the interaction coefficient ε_C^{Si} is relatively constant over the range $N_{Si} = 0–0·25$ [429] and therefore the activity coefficient of carbon in the dilute solution range will be given by the equation

$$\begin{aligned}\ln [\gamma_C]_{Fe-Si} &= \ln [\gamma_C]_{Fe} + N_{Si}\varepsilon_C^{Si}\\ &= -0·56 + (0·2 \times 10)\\ &= 1·44\end{aligned}$$

$$[\gamma_C]_{Fe-Si} = 4·21$$

Therefore when N_{Si} is equal to 0·2 the limiting carbon content, when $p_{CO} = p_{SiO}$, is

$$N_C = \frac{0·0223}{4·21} = 0·01 = 0·22 \text{ wt.\% C}$$

Bringing the results for SiO_2 and SiO formation together it can now be concluded that lowering the partial pressure of carbon monoxide in the equilibrium gas by using $Ar–O_2$ mixtures could avoid the formation of SiO_2 as a separate phase, and the carbon level could be reduced below the value predicted for this particular equilibrium. As the carbon content was decreased, however, the partial pressure of SiO would increase until at about 0·2 wt.% C the rate of loss of silicon from the melt would equal that of carbon on an atomic basis.

14. Metal–refractory Interaction

The extent of reaction of metals with oxides is important for the problem of finding suitable refractories for the melting of metals and alloys. The problem is of growing interest as more and more high-melting-point metals are coming to the fore.

As has frequently been mentioned in this monograph, the higher the temperature the less likely are kinetic checks and the more significant are the positions of the thermochemical equilibria. For the high-affinity metals of high melting point, such as titanium, satisfactory refractories in which the molten metal can be kept have not yet been found. But even when the affinity of the metal for non-metals is relatively low, interactions with the refractory are likely to occur and should be carefully assessed.

Let us consider as an example the amount of oxide pure molten iron is likely to dissolve when held in an alumina crucible. The appropriate chemical equation, the

Gibbs energy of which must be found, is as follows:

$$\langle Al_2O_3\rangle = 2[Al]_{Fe} + 3[O]_{Fe} \tag{a}$$

In order to obtain the corresponding Gibbs energy equation, the chemical equation should be broken up as follows:

$$\langle Al_2O_3\rangle = 2\{Al\} + 1\tfrac{1}{2}(O_2); \qquad \Delta G_b \tag{b}$$

$$1\tfrac{1}{2}(O_2) = 3[O]_{Fe}; \qquad 3\Delta\overline{G}_c \tag{c}$$

$$2\{Al\} = 2[Al]_{Fe}; \qquad 2\Delta\overline{G}_d \tag{d}$$

The corresponding Gibbs energy equations are as follows:

$$\Delta G_b^\circ = 400{,}900 - 76{\cdot}2T \tag{α}$$

$$3\Delta G_c^\circ = -83{,}800 + 13{\cdot}7T \log N_O + 18{\cdot}25T \tag{β}$$

$$2\Delta G_d^\circ = -52{,}000 + 9{\cdot}15T \log N_{Al} + 11{\cdot}2T \tag{γ}$$

where N_O and N_{Al} are the atomic fractions of oxygen and aluminium respectively in liquid iron.

When equilibrium is reached in reaction (a), the sum $\Delta G_b^\circ + 3\Delta G_c^\circ + 2\Delta G_d^\circ$ becomes zero, so that

$$0 = 265{,}100 - 46{\cdot}75T + 13{\cdot}7T \log N_O + 9{\cdot}15T \log N_{Al}$$

Since Al_2O_3 dissolves with the atomic proportion $1{\cdot}5N_{Al} = N_O$, the equilibrium concentrations of aluminium and oxygen can be evaluated. For $T = 1873°K$, for instance,

$$\log N_{Al} = -4{\cdot}25 \quad \text{and} \quad \log N_O = -4{\cdot}07$$

corresponding to the weight percentages: $c_{Al} = 0{\cdot}0027\%$ Al and $c_O = 0{\cdot}0024\%$ O.

The concentrations of aluminium and oxygen in pure iron after melting 11 kg ingots in alumina crucibles observed by Hopkins *et al.* [423] at the National Physical Laboratory were about 0·001 to 0·002% each. The difference is probably not due to errors in the calculation but to the fact that equilibrium had not been reached in the large melts. The kinetic delay in the metal-refractory interactions is, indeed, the smelter's best chance for keeping down the impurities, an advantage that is progressively lost with increasing temperature. A particularly smooth surface of the refractory also assists to a considerable extent to cut down the reaction rate at the interface: a fact that is well known to the practical smelter.

Let us consider next the interaction of an alloy containing a baser metal, such as a 10% chromium–nickel alloy, with alumina. Again the alumina will be dissolved to a certain extent in the alloy, according to the equation

$$\langle Al_2O_3\rangle = 2[Al]_{Cr-Ni} + 3[O]_{Cr-Ni} \tag{e}$$

The calculation corresponds to the one above for iron. Since the solubility of oxygen in liquid chromium–nickel alloys is unknown, the Gibbs energies may be taken to be additive, according to

$$\tfrac{1}{2}(O_2) = [O]_{Ni}; \qquad \Delta\overline{G}_f \tag{f}$$

$$\tfrac{1}{2}(O_2) = [O]_{Cr}; \qquad \Delta\overline{G}_g \tag{g}$$

and

$$\tfrac{1}{2}(O_2) = [O]_{Ni-Cr}; \qquad \Delta G_h = 0{\cdot}9\Delta G_f + 0{\cdot}1\Delta G_g \tag{h}$$

for a 10 at.% alloy. This simplification admittedly introduces an additional error, and the total error in the finally calculated concentrations may amount to as much as half a logarithmic unit. In view of the analytical uncertainties at these low concentrations, however, the calculated values still represent useful estimates.

In the present case the apparent concentrations of aluminium and oxygen in the chromium–nickel alloys are obtained as 0·0019% Al and 0·0017% O by weight at 1550°C.

The stability of Cr_2O_3 is much greater than those of Fe_3O_4 and NiO, and although it is still much lower than that of Al_2O_3 it approaches it more closely. This is why the thermochemist must consider another possible equilibrium, namely

$$\langle Al_2O_3\rangle + 2[Cr]_{Cr-Ni} = 2[Al]_{Cr-Ni} + \langle Cr_2O_3\rangle \tag{i}$$

that is, chromic oxide may be precipitated on or even dissolved in the refractory while the equivalent amount of aluminium goes into solution.

Again, eqn. (i) may be broken up into individual equations, the Gibbs energies of which can be assessed.

$$\langle Al_2O_3\rangle = 2\{Al\} + 1\tfrac{1}{2}(O_2); \qquad \Delta G_b = 400{,}900 - 76{\cdot}2T \tag{b}$$

$$2\{Cr\} + 1\tfrac{1}{2}(O_2) = \langle Cr_2O_3\rangle; \qquad \Delta G_j = -276{,}900 + 66{\cdot}4T \tag{j}$$

$$2\{Al\} = 2[Al]_{Ni}; \qquad 2\Delta\bar{G}_k = -76{,}000 + 9{\cdot}15T\log N_{Al} \tag{k}$$

$$2[Cr]_{Cr-Ni} = 2\{Cr\}; \qquad 2\Delta\bar{G}_l = -9{\cdot}15T\log N_{Cr} \tag{l}$$

The Gibbs energy equation for reaction (l) implies that chromium and nickel form ideal solutions. This is only an approximation since recent measurements have shown a deviation from ideality.

When these four chemical equations are added, eqn. (i) results, and

$$\Delta G_b + \Delta G_j + 2\Delta\bar{G}_k + 2\Delta\bar{G}_l = 48{,}000 - 9{\cdot}8T + 9{\cdot}15T\log N_{Al} - 9{\cdot}15T\log N_{Cr}$$

represents the Gibbs energy of eqn. (i). At equilibrium, ΔG_i is zero, and for 1550°C and $N_{Cr} = 0{\cdot}1$, for instance, $\log N_{Al} = -2{\cdot}81$. This corresponds to 0·07% Al by weight in the liquid alloy.

The equilibrium constant of reaction (e) in terms of weight percentages according to the calculation further above is

$$\log K_p = \log[c_{Al}^2 c_O^3] = \log[0{\cdot}0019^2 \times 0{\cdot}0017^3] = -13{\cdot}8$$

In fact, however, another equilibrium, namely reaction (i), must be taken into account. This was shown to produce an aluminium concentration of $c_{Al} = 0{\cdot}07$, and, using the equilibrium constant, we find that

$$c_O = \sqrt[3]{1{\cdot}6 \times 10^{-14}/c_{Al}^2} = 0{\cdot}00015\%$$

so that the oxygen percentage in the chromium–nickel alloy melted in an alumina pot should actually be very low.

The evaluation of reaction (i) shows that the assumption that $1{\cdot}5N_{Al} = N_O$ made to obtain c_{Al} and c_O for reaction (e) was not justified, and this example shows the importance of finding the correct chemical equations representing a particular metal-refractory interaction. In the case of pure iron or nickel as the metal phase the reaction of the metal with the aluminium in the refractory corresponding to eqn. (i) need not be considered because of the lower stabilities of iron and nickel oxides.

The above calculation with chromium–nickel alloys would also apply to more complex chromium alloys provided that the alloying metals have a lower affinity for oxygen than chromium—to Ni–Co–Cr–Mo alloys for instance. In this way, the thermochemist can, in fact, handle fairly complex systems with some confidence provided that an error of, say, half a logarithmic unit in the final concentrations can be tolerated.

If such alloys contain carbon, however, considerable interaction with the refractory oxides must be anticipated, owing to the stability and volatility of the carbon oxides. Carbon in iron would react with an alumina pot in the following manner:

$$\tfrac{1}{3}\langle Al_2O_3\rangle + [C]_{Fe} = (CO) + \tfrac{2}{3}[Al]_{Fe} \qquad \text{(m)}$$

Using the Gibbs energies of formation of $\langle Al_2O_3\rangle$ and (CO) and the Gibbs energies of solution of carbon and aluminium in liquid iron, one obtains at 1600°C for the equilibrium constant

$$p_{CO}N_{Al}^{2/3}/N_C = 0{\cdot}05$$

where p_{CO} is the pressure of carbon monoxide in atmospheres. In a vacuum, and even in an inert gas, carbon monoxide will continuously be removed from the system, and equivalent amounts of aluminium dissolved. Eventually the melt will become virtually carbon-free and contain aluminium instead. Adcock [430] therefore, when making pure carbon-containing iron, used thoria linings or, alternatively, the carbon can be added just before casting. In the second case, use is made of the kinetic delay in reaction (m), in the first of the much lower equilibrium constant for the reaction:

$$\tfrac{1}{2}\langle ThO_2\rangle + [C]_{Fe} = (CO) + \tfrac{1}{2}[Th]_{Fe} \qquad \text{(n)}$$

$$p_{CO}N_{Th}^{1/2}/N_C = 0{\cdot}0002$$

Because of the continuous removal of carbon monoxide in a good vacuum, even here gradual dissolution of thorium is expected, only it takes longer owing to the low equilibrium constant.

Silicon in the iron would act similarly to carbon, but, owing to the low equilibrium constant,

$$p_{SiO}N_{Al}/N_{Si} = 6{\cdot}5 \times 10^{-5}$$

the danger is not serious.

When the oxygen potential is increased over a sample of iron which is held in an alumina crucible, a reaction ensues between metal and container at the point where the spinel phase $FeAl_2O_4$, hercynite, can be formed. This phase has the normal spinel structure in which the ferrous ions occupy tetrahedral and the aluminium ions octahedral holes in a f.c.c. oxide structure. Because of the similarity in structure between the spinel phase and γ-Al_2O_3 which is the same as $FeAl_2O_4$ but with vacancies where some of the Fe^{2+} ions are present in hercynite, there is a measurable range of solid solutions between these phases at temperatures exceeding 1500°C.

The spinel which is formed when iron is oxidised in contact with alumina is therefore that which is saturated with Al_2O_3 and to predict the behaviour of the iron–oxygen–

Al_2O_3 system it is necessary to have thermodynamic information about the Al_2O_3-saturated phase. Results for this system were obtained by Chan *et al.* [431] over the temperature range 700–1600°C using a combination of galvanic cells with solid oxide electrolytes. For the low-temperature region Fe–FeO was used as the reference electrode, and for the high-temperature system Mo–MoO_2 was employed. The electrode containing hercynite was a mixture of Fe, Al_2O_3, and $FeAl_2O_4$. The Gibbs energy equation which was obtained from the study for the reaction

$$2Fe + O_2 + Al_2O_3 \rightarrow 2FeO{\cdot}Al_2O_3 \text{ } (Al_2O_3 \text{ saturated})$$

is of special interest in the present context for liquid iron since a knowledge of this makes it possible to calculate the oxygen content of the liquid iron when the oxidation just begins. For iron in the liquid state ($T \geq 1536$°C) the equation is

$$\Delta G^\circ = -146{,}390 + 36{\cdot}48T$$

Combining this equation with that for oxygen dissolved in liquid iron,

$$(O_2) = 2[O]_{Fe}; \qquad \Delta G^\circ = -55{,}850 + 12{\cdot}15T + 9{\cdot}15T \log N_O$$

we obtain the equation for the Gibbs energy change of the reaction

$$2Fe + 2[O] + Al_2O_3 \rightarrow 2FeO{\cdot}Al_2O_3 \text{ } (Al_2O_3 \text{ saturated})$$

$$\Delta G^\circ = -90{,}540 + 24{\cdot}33T + 9{\cdot}15T \log N_O$$

At equilibrium this is equal to zero since iron, alumina and hercynite are all present at essentially unit activity. At 1873°K the solution of this equation, after conversion from atom fraction to weight percentage, yields an oxygen content close to 0·07 wt.%. The saturation value for iron in contact with pure FeO at this temperature is 0·22 wt.%. The FeO activity in the spinel phase in contact with iron and alumina is therefore 0·07/0·22 or about one-third.

Another popular refractory is magnesia. Here the chemical reactions with molten metals are somewhat different. The solubility of magnesium in iron is virtually nil. On the other hand, the vapour pressure of magnesium is high, so that we have to consider the following chemical reaction:

$$\langle MgO \rangle = (Mg) + [O]_{Fe} \qquad \text{(p)}$$

At 1600°C, the individual Gibbs energies are as follows:

$$\langle MgO \rangle = (Mg) + \tfrac{1}{2}(O_2); \qquad \Delta G_q = 85{,}000 \qquad \text{(q)}$$

$$\tfrac{1}{2}(O_2) = [O]_{Fe}; \qquad \Delta G_c = -16{,}500 + 8560 \log N_O \qquad \text{(c)}$$

Hence

$$-\log p_{Mg} = \log N_O + 8{\cdot}0 \quad \text{at } 1600°\text{C}$$

There is no defined pressure of magnesium. This would evaporate from the reaction zone and condense in the cooler parts of the system. Thus the iron would gradually pick up oxygen up to the saturation point, simultaneously dissolving the magnesia pot. From the above equation one obtains:

at $p_{Mg} =$	10^{-3}	10^{-2}	1	mmHg of magnesium
$c_O =$	0·2	0·02	0·0002	wt.% O

Fischer and Hoffmann [432], who determined the oxygen concentrations in iron after holding the metal for various periods of time in crucibles made of magnesia, found that after 22 hours 0·15% O had accumulated in the iron. For rougher magnesia surfaces the pick-up of oxygen was faster.

Thus although the affinity of magnesium for oxygen is higher than that of aluminium, and the solubility of magnesium in iron less, alumina should be preferred as a container because of the more favourable mechanism of reaction.

Refractories other than oxides can be treated in a similar manner. Roughly, the heat of formation of a refractory compound gives an indication of its suitability, but individual combinations should be considered more fully. Among the sulphides, cerium monosulphide appears to be particularly stable. Among the carbides and silicides, those of zirconium and titanium have highly negative heats of formation. Borides seem to gain some importance as refractory materials, but, unfortunately, these are as yet beyond consideration by the thermochemist because thermochemical data are hardly available.

It is, however, seen that in quite a large number of practical cases the thermochemist may be of assistance in assessing the maximum degree of interaction between a refractory and a molten alloy, although the scarcity of accurate data may affect the accuracy of his computations. Knowing the chemical mechanism of an interaction may often help to reduce the pick-up of impurities; hopeless combinations may be eliminated at an early stage, thus avoiding costly trial runs; and current methods may be checked. If, for instance, a higher pick-up of impurity were found in a particular case than estimated thermochemically one would conclude that side-reactions, with impurities in the refractory for example, take place which could possibly be eliminated.

15. The Oxidation of Alloys of Platinum and of Gold

A thermochemical problem which involves the application of partial molar Gibbs energies arises in the resistance to oxidation of alloys of platinum or gold with copper or nickel. These alloys are used extensively for electrical contacts, on which atmospheric oxidation has a very deleterious effect. Platinum alloys containing 5 and 10% Cu have been found to retain their bright surface at high temperatures, but in Germany during the last war it became necessary to determine whether platinum alloys could be replaced by those of gold with similar resistance to oxidation.

The results of an oxidation test on platinum and gold alloys containing 5 and 10% Cu are shown in Fig. 115. The platinum alloys all oxidise slowly, and those containing 5 and 10% Cu did not oxidise at all above 845° and 950°C respectively. The gold alloys oxidised more rapidly, the oxidation showing hardly any tendency to fall off to their melting points.

The most direct approach to the explanation of this behaviour is to calculate the oxygen pressures in equilibrium with these alloys at various temperatures. It may reasonably be assumed that only the copper in the alloys reacts with atmospheric oxygen. The following reaction mechanism therefore pertains (M = Au, Pt):

$$\begin{array}{rcl} 4\langle Cu\rangle + (O_2) = 2\langle Cu_2O\rangle & & \Delta G_1^\circ \\ 4\langle Cu\rangle = 4[Cu]_{M_xCu_y} & & \Delta \bar{G}_2^\circ \\ \hline 2\langle Cu_2O\rangle = 4[Cu]_{M_xCu_y} + (O_2) & & \Delta \bar{G}_2^\circ - \Delta G_1^\circ \end{array}$$

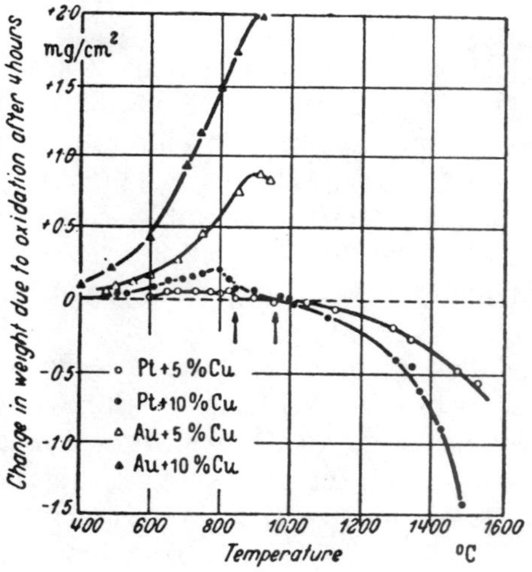

FIG. 115. Rate of oxidation of platinum or gold alloys containing copper.

The dissociation pressure of cuprous oxide in equilibrium with the alloy is obtained from the equation

$$\Delta\overline{G}_2^\circ - \Delta G_1^\circ = -4{\cdot}574T \log p_{O_2}$$

The Gibbs energy of formation of cuprous oxide (ΔG_1°) is given in Table E, the Gibbs energies of solution of copper in its alloys with platinum and gold, respectively, in Table F. The logarithms of the oxygen pressures have been calculated and are plotted against $1/T$ in Fig. 116. The horizontal line in the diagram represents the pressure of oxygen in air. An alloy with a higher oxygen dissociation pressure remains unattacked. Thus the points at which the horizontal line intersects the others in Fig. 116 represent the temperatures above which the alloys will remain bright in air. The temperatures

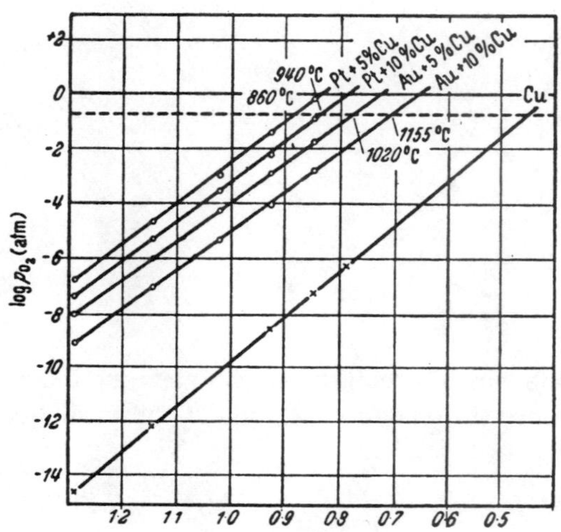

FIG. 116. Oxygen pressures of cuprous oxide in equilibrium with pure copper or with copper alloyed to gold or platinum.

calculated agree closely with those determined in the oxidation test of the platinum–copper alloys; 860°C calculated, 845°C observed, and 940°C calculated, 930–970°C observed, being the respective critical temperatures for the 5% and 10% copper alloys.

The lower dissociation pressures of the gold alloys are in accordance with their higher oxidation rates shown in Fig. 115. There is however, no direct correlation between the two quantities, for thermochemical calculations are concerned with equilibria, not kinetics. All that can be predicted are the temperatures above which the alloy should remain untarnished in air. These temperatures are indicated in Fig. 116.

It must be emphasised that in order to solve the problem completely from a fundamental standpoint at least two further factors must be considered, namely: (1) the ratio of the molecular volume of the oxide layer formed to the molecular (atomic) volume of the parent alloy, and (2) the mechanism of diffusion in the oxide layer. Their introduction would, however, not be appropriate to this monograph. The original paper [433] and a more detailed study of the oxidation of platinum–nickel alloys [434] may be consulted for a fuller discussion which illustrates how metallurgical problems of this general type must be dealt with from the kinetic as well as the thermodynamic standpoint.

16. The Oxidation of Iron–Chromium Alloys

The b.c.c. solid solutions of chromium in iron, the α solid solutions, exist over the whole composition range at temperatures above 1392°C, and below this temperature a closed range of compositions exist which are iron-rich and have the f.c.c. structure (γ solid solutions). The thermodynamic properties of the α solid solutions have been reviewed by Müller and Kubaschewski who obtained the results shown in Table XXXI. These are only quoted for the compositions up to $N_{Cr} = 0{\cdot}25$ because those are all of the results which are needed for the present discussion. More data are available in the original paper [435].

Table XXXI. Partial thermodynamic properties in αFe–Cr alloys at high temperatures

N_{Cr}	$\Delta\bar{H}_{Cr}$	$\Delta\bar{S}_{Cr}$	ΔH_{Fe}	ΔS_{Fe}
0·05	+5415	8·162	15	0·109
0·10	4860	6·584	60	0·233
0·15	4335	5·609	135	0·371
0·20	3840	4·878	240	0·524
0·25	3375	4·274	375	0·700

The other information which is required to complete the thermodynamic properties of this system with respect to oxidation is that for the formation of the oxides. The Gibbs energy equations are as follows:

$$2\langle Fe\rangle + (O_2) \rightarrow 2\langle FeO\rangle \qquad \Delta G^\circ = -124{,}100 + 29{\cdot}90T$$

$$\tfrac{4}{3}\langle Cr\rangle + (O_2) \rightarrow \tfrac{2}{3}\langle Cr_2O_3\rangle \qquad \Delta G^\circ = -178{,}500 + 41{\cdot}40T$$

$$2\langle Fe\rangle + (O_2) + 2\langle Cr_2O_3\rangle \rightarrow 2\langle FeCr_2O_4\rangle \qquad \Delta G^\circ = -151{,}400 + 34{\cdot}7T$$

These data lead to the ready conclusion that the lowest oxygen potential in most of the potential alloy–oxide systems in the ternary Fe–Cr–O system is that of the Cr–Cr_2O_3 system, $FeCr_2O_4$ being the next and the oxides of iron only being formed at much higher oxygen pressures. The results for the oxides are shown in the Ellingham diagram (Fig. 117).

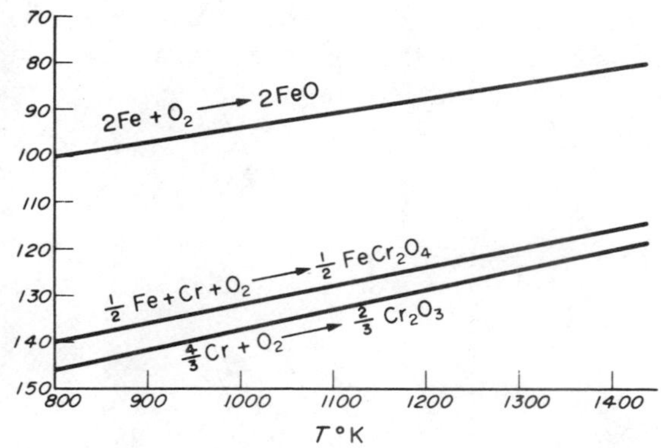

FIG. 117. Ellingham diagram for the oxides FeO, $FeCr_2O_4$ and Cr_2O_3.

The experimental evidence for the oxidation mechanism by Moreau [436] and Rahmel [437] at 1000°C shows that alloys containing greater than 15% Cr are protected from corrosion by formation of $FeCr_2O_4$. Those with less than this chromium content oxidise to form a multilayer oxide comprising a mixture of FeO and $FeCr_2O_4$ in contact with the metal, then Fe_3O_4 containing Cr^{3+} in the octahedral sides and finally Fe_2O_3. There appears to be no role for Cr_2O_3, the most stable oxide, in the reaction mechanism. However, there are a number of points which emerge when a thermodynamic analysis is made and these largely centre around the relative stabilities of Cr_2O_3 and $FeCr_2O_4$. The equilibrium constant for the displacement reaction

$$Fe + \tfrac{4}{3}Cr_2O_3 \rightarrow FeCr_2O_4 + \tfrac{2}{3}Cr$$

provides information concerning the alloy composition ranges in which this reaction goes from left to right or vice versa. When the oxide phases are present at unit activity, the constant simplifies to

$$K = a_{Cr}^{2/3}/a_{Fe}$$

It is soon found by insertion of typical values (the activities of iron and chromium are linked by the fact that N_{Cr} equals $1 - N_{Fe}$) that the reaction only proceeds to the right when chromium is present in the coexisting alloy phase at very high dilution. From the data which have been given for the formation of the oxides, we obtain for the displacement reaction

$$\Delta G^\circ = 13{,}550 - 3{\cdot}35T \qquad K = \frac{a_{FeCr_2O_4}\, a_{Cr}^{2/3}}{a_{Cr_2O_3}^{4/3}\, a_{Fe}}$$

Since iron is present at nearly unit activity when the reaction proceeds from left to right, we may write, as a good approximation for this Gibbs energy,

$$\Delta G^\circ = -\boldsymbol{R}T \ln K = -\tfrac{2}{3}\Delta\mu_{Cr}$$

In order to convert these chemical potentials into the corresponding chromium concentrations in the alloys it is necessary to obtain an equation between the chromium potentials and the chromium atom fraction over a range of temperatures. Using the results quoted above from Müller and Kubaschewski, we calculate the excess partial entropy for chromium at each of the stated compositions. These are then plotted against composition and a reasonable extrapolation produced for the low values of N_{Cr} (Fig. 118). The results fit the equation

$$\Delta\bar{S}^E_{Cr} = 2{\cdot}4 - 3{\cdot}6N_{Cr}$$

but for the region below 1 at.% Cr we shall assume the constant value 2·40. The integral heat of mixing curve shows a strictly regular solution behaviour and can be described by the equation

$$\Delta H^M = 6000N_{Fe}N_{Cr}$$

from the results of Dench [144]. At N_{Cr} equal to 0·01 this yields the partial heat of solution for chromium

$$\Delta\bar{H}_{Cr} = 6000N^2_{Fe} = 5880$$

Hence the equation for the partial Gibbs energy of chromium in solution in iron is

$$\Delta\bar{G}_{Cr} = 5880 - 2{\cdot}4T + 4{\cdot}575T \log N_{Cr} \qquad (N_{Cr} < 0{\cdot}01)$$

From the value for the Gibbs energy change of the displacement reaction, we may now calculate the values of N_{Cr} at equilibrium with iron, Cr_2O_3, and $FeCr_2O_4$ at any temperature. The results are given in Table XXXII, and it must be observed that when the chromium content is higher than these values then chromium displaces

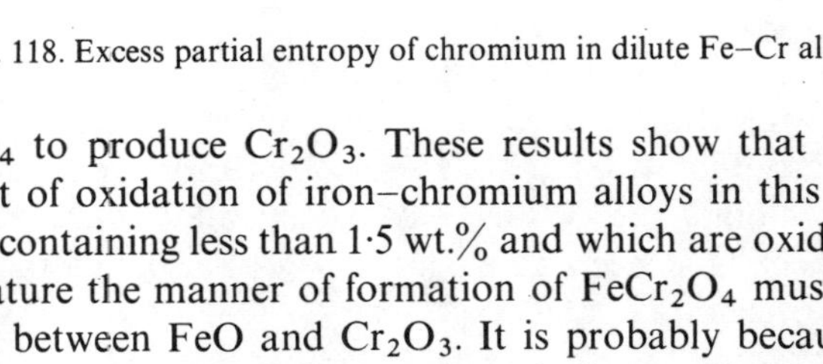

FIG. 118. Excess partial entropy of chromium in dilute Fe–Cr alloys.

iron from $FeCr_2O_4$ to produce Cr_2O_3. These results show that Cr_2O_3 will always be the first product of oxidation of iron–chromium alloys in this temperature range excepting for those containing less than 1·5 wt.% and which are oxidised above 1400°K. Below this temperature the manner of formation of $FeCr_2O_4$ must be the result of a secondary reaction between FeO and Cr_2O_3. It is probably because of this that this compound is formed below the original surface during oxidation at 1000°C. This is demonstrated by the observation that inert markers which were originally placed on the surface of the alloy are found after oxidation between the FeO layer and the FeO–$FeCr_2O_4$ dispersion.

Table XXXII. Chromium contents of Fe–Cr alloys in equilibrium with Cr_2O_3 and iron chromite

T (°K)	ΔG° (displacement)	N_{Cr} (equilibrium)
800	10,870	3×10^{-6}
1000	10,200	$7{\cdot}9 \times 10^{-5}$
1200	9530	$7{\cdot}1 \times 10^{-4}$
1400	8860	$3{\cdot}4 \times 10^{-3}$

CHAPTER V

Thermochemical Data

The Tabulation of the Data

(For references in the text and tables, see pp. 423 to 445)

The tables in this part of the monograph give the heats (enthalpies) of formation, transformation, fusion and evaporation, the standard entropies, the vapour pressures and the heat capacities (molar heats) of the elements, metal halides, chalcogenides, sulphates, nitrides, phosphides, carbides, carbonates, silicides, silicates, and intermetallic compounds. Each value represents the best available up to 1976 in the opinion of the authors.

1. Heats of Formation, Entropies and Heats of Fusion

It has never been the intention of the authors to assess an entirely new set of thermochemical data of substances of actual or potential interest in metallurgy. Firstly, such a task would have exceeded the possible and, secondly, it should be the aim of the assessors of critical tables to agree internationally on the tabulated values as far as possible. Thus, several critical compilations have been perused here.

Values for the standard entropies and heats of formation of key substances (CO_2, H_2O, etc.) have been internationally agreed by CODATA [861], and these have naturally been incorporated in the present tables and denoted "key values".

The National Bureau of Standards provided an excellent compilation in its so-called "Circular 500" [198] as early as 1952. This work has been continuously revised, and the revisions appear as NBS reports, 270 [918], as well as in some special reports [860]. The disadvantage of these tables [918] is that the evaluations cannot be traced in detail to their origin, but in the experience of the authors the NBS is always willing to supply such information for individual substances to the enquirer. A similar compilation is being provided by the Institute for High Temperatures in Moscow [907]. As far as the present authors are aware, the data for the compounds of the elements of the following Groups have so far appeared in print: He, F, O, N, C, B, Zn, Cu, and Fe (Group VIII). The compilation is fully documented.

The American and Russian compilations mentioned present the thermochemical values for compounds and aqueous solutions at room temperature. The latter give in addition heats and entropies of transformation, fusion and evaporation. In contrast, the JANAF Tables [817] provide complete evaluations up to high temperatures for a large number of inorganic compounds. Of particular value are the complete sets for gaseous species, missing information being estimated where necessary. Heat capacities and entropies are often obtained from estimates of the molecular constants.

Critical compilations of individual groups of substances have also been perused, in particular the DCS Reports of the National Physical Laboratory on silicates [902], other double oxides [840], silicides [959], magnesium [963] and alkali metal alloys [971], and III–V compounds [747]. Also from the N.P.L. comes a very good review of the thermochemical properties of metal sulphides, selenides and tellurides by Mills [854]. The Atomic Energy Reviews of the International Atomic Energy Agency similarly provide data for the compounds and alloys of Pu, Nb, Ta, Be, Th and Zr [755], uranium having been dealt with by Rand and Kubaschewski [399]. Earlier reviews by Kelley [121–126] and Brewer [22–27] are still very useful. In all these publications, previous literature is discussed in detail.

The tables compiled by Hultgren and his colleagues [99] are, apart from the corresponding reviews already mentioned, a source for the thermochemical properties of intermetallic compounds.

Where values for the same compound are presented by several compilers, these have been compared. Where they agree, there is no problem. In the case of disagreement, a selection had to be made. The present authors hope that their experience has helped them to make the best choice, with consistency always in their mind.

As may be seen from the list of new references, a large number of original papers have also been consulted. Where the results necessitated a revision of earlier assessments or provided new values, the appropriate changes have been made.

2. Vapour Pressures

Vapour pressure and heat capacities are tabulated separately as formulae which include temperature coefficients.

Kelley has reviewed the vapour pressures of inorganic substances [120] and has obtained from them the heats of evaporation. Brewer [25] has revised critically the vaporisation data of the metal halides and has estimated the boiling points and heats of evaporation not yet determined experimentally. A summary by Stull [239] contains a number of reference and average values of vapour pressures of many inorganic substances. Hultgren and his colleagues have reassessed the vaporisation data of the metals [99].

The present collection is supplemented by many values which they did not consider. It has sometimes been found necessary to adjust some of these earlier formulae to accommodate them to more recently determined information on a neighbouring phase: thus the vapour pressure curve for a liquid may have been modified so as to intersect the curve for the solid phase in a manner in accord with the heat fusion. No note is made of this where the original values are altered only slightly. A number of formulae have been simplified, and the range of validity of others has been enlarged by the introduction of a C_p term, a method which has been adequately described by Kelley in his papers.

Vapour pressures are expressed in millimeters of mercury (Torr). The authors deemed this to be preferable to presentation as Gibbs energies of vaporisation for two reasons. Firstly, for many practical purposes, it is desirable to know the pressure in common, though eventually superseded, units. Secondly, hardly any inorganic substance evaporates as a single species. It may associate or dissociate, or both. Thus, the recorded vapour pressures rarely represent true Gibbs energy values.

Consequently, the recorded normal boiling points mostly refer to the actual boiling points and not to those of the listed species. Many of the vapour pressure equations in Table D have been derived by Second Law evaluations, the ΔC_ps being often estimated. To derive approximate Gibbs energy formulae of evaporation, the figure 2·88 (= log 760) is subtracted from the log p term in column D of Table D, and the formula thus obtained is multiplied by $-4{\cdot}574T$. For a number of gaseous species the authors of the JANAF tables have estimated the standard entropies and heat capacities and then made a Third Law evaluation of the Gibbs energies of evaporation, and obtained the standard heats. Such results may be found in Table A.

3. Heat Capacities

For heat capacity formulae an early critical review, compiled by Kelley [124], is at hand. Some of the formulae have been modified by one of the authors [148]. Experimental results that have been published more recently are also included in the tables, the individual references being given.

Estimated formulae are taken from Kelley's publication or, where no reference is stated, are derived by the authors. They are based on room temperature heat capacities and estimates of the value of the next transition point, as described on pp. 181–3. They should be sufficiently accurate for thermochemical calculations.

4. General

An effort has been made to give the probable limits of accuracy of every value in some tables. It must be emphasised that these limits are not necessarily those quoted by the original authors. It has been the intention to make the limits of accuracy so wide that further determinations will not fall outside them. Some reviewers of the first edition of this monograph have referred to these limits as being very pessimistic, but the authors make no apology for the tendency in this new edition for these limits to be wider rather than narrower. This is believed to be justifiable since quite a number of values re-determined after publication of the first edition now fall outside the limits of error first suggested. It is usual in these days for an observer to estimate and quote faithfully the precision of his particular method of measurement, but frequently no account is taken of such imponderables as the homogeneity of the final product, purity of substances and other factors which are difficult to trace. Further, it is futile to report thermochemical values to four or more figures. A value such as 80,573 cal may be inaccurate even in the hundreds, so the 73 is meaningless. Sometimes the mean is taken of values which differ by more than, say, 200 cal, but the mean quoted is with limits of accuracy less than this amount. The authors therefore request the indulgence of those who see their values quoted with a greater error than originally reported; they believe that their own results are treated no less ruthlessly.

In estimating the accuracy of standard entropies, Kelley [125] has observed these same principles, and the accuracies which he computed have for the most part been accepted.

Values obtained by an approximate method and not checked by other means may sometimes be subject to a very large and quite unknown error. Such values, and those which are doubted by the authors, are enclosed in *parentheses* ( ).

An attempt has also been made to estimate a number of thermochemical values as yet undetermined. This is easier for entropies and heats of fusion and evaporation than for heats of formation, but the latter are, it may be repeated, the most important for free energy calculation. Some of the methods of estimation are described in Part III. The values thus obtained are printed in the following tables in *italics*. They, also, are quoted with the limits of accuracy to be expected, assuming that the particular compound does not behave in an exceptional manner which cannot be predicted in the light of present theoretical knowledge. Occasionally the estimated value is felt to be more reliable than the one determined experimentally. A note is made of these.

It would of course be desirable to discuss each value given in the tables, or at least those that differ from the figures accepted in previous reviews. The authors are also aware of the unfairness of omitting the names of the observers who have done the actual work of measuring the values in favour of those of a few critical reviewers. It was considered, however, that a complete list of references would overload this monograph far too much with material of interest only to a few investigators. The *Selected Values* for instance quote, almost 9000 references, and even the review of a single function, that by Kelley [125] of the heat capacities of inorganic substances, gives 559 references.

Thus earlier literature already assessed in one of the reviews is generally not cited.

The following principles are observed in giving the *reference numbers*. Where values of earlier works differ from those given in the reference, the alteration may be due to a change in the accepted value for one of the components of the system or phase, or one of the values which were necessary to calculate the final figure. This can be confirmed by searching through the references to the various components and comparing their values with those used in the earlier critical evaluation.

References to values which are not considered in the earlier reviews but which either confirm the earlier value or are not considered reliable enough to alter it are placed after the reference to the review. Values which deviate considerably but are regarded as less reliable are referred to in parentheses. Recent values which are an improvement on the earlier ones are cited first. The reference to the critical work is placed afterwards as a source of information on earlier work (in parentheses when it deviates widely from the present value). The order in which references are given is a general indication of the reliance placed on the respective values in evaluation. This also applied to the previous critical evaluations.

In the arrangement of the substances the authors have preferred the use of the alphabetical order of chemical symbols. Any other order, however well founded logically, usually results in considerable waste of time in searching. The compounds of any element are arranged as follows: halides, oxyhalides, oxides, sulphides, sulphates, selenides, tellurides, nitrides, phosphides, arsenides, antimonides, bismuthides, carbides, carbonates, silicides, silicates, intermetallic compounds. The same applies to all tables.

Table A indicates also the lattice structures of the substances considered, and a brief description of some fundamental structures is appended (Table Aa). Emphasis is placed upon isomorphism, as this is considered to be of more immediate importance to the thermochemist than full details of any structure.

It is now well known that many inorganic compounds, particularly those of the transition metals, are stable at compositions other than stoichiometric. Many compounds of metals have considerable ranges of homogeneity. Where information concerning the non-stoichiometric nature of the compounds is available, this is given in the last column of Table A. Earlier information was obtained from a paper by Anderson [8]. If no temperature is assigned to these limits of stability, the limits are still somewhat uncertain.

5. Gibbs Energy Calculations

Change of Gibbs energy can be calculated from heats of formation, entropies and heat capacities by means of eqn. (34a) (p. 19) or for approximate calculations with eqn. (32a), introducing an estimated ΔC_p.

Gibbs energy formulae derived from the application of eqn. (34a) have often been given in the literature [117, 122, 123]. They usually contain five or six terms, as a result of the calculation. As we have already discussed on p. 24, these formulae are unnecessarily cumbersome, and their complexity gives an impression of a degree of accuracy which they do not possess. Those who evaluate such formulae should always endeavour to simplify them, since the small effort required to do this will save an untold amount of arithmetical work for subsequent calculators—to say nothing of giving a more immediate impression of their numerical value.

Equation 36 (p. 22) is quite satisfactory for approximate calculations, while for normal purposes a third term may be introduced which implies a constant value for ΔC_p. If heat capacity measurements are not available, a constant ΔC_p value may be estimated (see pp. 179–85).

Two examples may be given of this contention.

1. M. de Kay-Thompson, in a comprehensive review of Gibbs energies of oxides, gave for the integral free energy of formation of zirconium oxide, for instance,

$$\Delta G^\circ_T = -259{,}000 + 8{\cdot}64T\log T - 5{\cdot}14\times10^{-3}T^2 + 0{\cdot}889\times10^5T^{-1} + 171{\cdot}4T^{1/2} + 20{\cdot}00T \qquad (\alpha)$$

(The original expression has been altered by a slight simplification of the figures and by the introduction of a more recent value of the entropy of ZrO_2.)

Simple evaluation of eqn. (32a), using heat formation and entropies and neglecting heat capacities, gives

$$\Delta G^\circ_T = -258{,}500 + 46{\cdot}5T \qquad (\beta)$$

An estimation of ΔC°_p for oxides formed from the solid metal and gaseous oxygen gives a value 1·5 cal/deg g-atom of oxygen (p. 185). Thus $\Delta C^\circ_p = 3{\cdot}0$, and adding this to eqn. (β) we obtain

$$\Delta G^\circ_T = -259{,}400 - 6{\cdot}9T\log T + 66{\cdot}1T \qquad (\gamma)$$

Lastly, de Kay-Thompson's equation may be simplified, the third, fourth and fifth terms being eliminated by calculation for the arbitrarily chosen temperatures 300°, 700° and 1200°K, to give

$$\Delta G^\circ_T = -260{,}300 - 10{\cdot}15T\log T + 77{\cdot}1T \qquad (\delta)$$

In Table XXXIII a comparison is made of the free energy values obtained from these four formulae for four temperatures. The values given by eqns. (α), (γ) and (δ) do not differ by more than 1000 cal, and are generally much closer. The error of the heat of formation is itself ± 1000 cal, and the errors in the entropies and heat capacities must be added to this. For thermochemical calculations, therefore, eqn. (δ) is no less exact than (α), and even eqn. (γ), which involves an estimated ΔC_p value, is satisfactory. Equation (β) gives values nearly 5000 cal too low, so it is obvious that it can only be used for very rough calculations.

Table XXXIII. Gibbs energies of formation, calculated by different formulae

Formula see text	300°K	500°K	1000°K	1500°K
		(a) $\langle Zr\rangle + (O_2) = \langle ZrO_2\rangle$		
α	−244,700	−235,520	−214,550	−193,660
β	−244,550	−235,250	−212,000	−188,750
γ	−244,680	−235,660	−214,000	−193,100
δ	−244,680	−235,440	−213,650	−192,900
		(b) $\langle Ca\rangle + \frac{1}{2}(O_2) = \langle CaO\rangle$		
ε	−144,230	−139,280	−127,350	
ζ	−144,400	−139,420	−126,940	
η	−144,380	−139,490	−127,680	
θ	−144,220	−139,260	−127,350	
ι	−144,210	−139,490	−127,660	

2. A similar comparison may be made of the Gibbs energy of calcium oxide formation. De Kay-Thompson [117] gives the value

$$\Delta G_T^\circ = -152{,}879 - 0{\cdot}17T\log T - 0{\cdot}66\times10^{-3}T^2 + 0{\cdot}54\times10^5T^{-1} + 85{\cdot}67T^{1/2} + 23{\cdot}85T \qquad (\varepsilon)$$

Simplification of this formula leads to

$$\Delta G_T^\circ = -152{,}220 - 3{\cdot}45T\log T + 35{\cdot}22T \qquad (\theta)$$

The application of eqn. (27a) gives

$$\Delta G_T^\circ = -151{,}900 + 24{\cdot}96T \qquad (\zeta)$$

which together with the estimated heat capacity $\Delta C_p^\circ = 1{\cdot}5$ cal deg (p. 185), gives

$$\Delta G_T^\circ = -152{,}350 - 3{\cdot}45T\log T + 35{\cdot}02T \qquad (\eta)$$

which is almost identical with eqn. (θ).

If ΔC_p° is as small as in this example one can easily express the Gibbs energy by a linear relationship. Instead of using the room temperature values ΔH_{298}° and ΔS_{298}° as in eqn. (ζ) it is better to derive average values ΔH_x and ΔS_x for the temperature range in consideration. This method has been applied to a number of oxides by Richardson and Jeffes [191] who obtained for calcium oxide the value

$$\Delta G_T^\circ = -151{,}320 + 23{\cdot}66T \qquad (\iota)$$

The values obtained from these formulae are compared in Table XXXIII. Again, deviations between eqns. (ε), (η) and (θ) do not exceed the error in the value of the heat of formation (± 300 cal/mole), and even the linear equation (ι) agrees well with the multi-term formulae.

Further examples of the adequacy of two-term formulae are given in Richardson and Jeffes' paper [191].

Table E contains formulae for the free energy of a number of substances which are or may become of metallurgical interest. Further formulae may be calculated in the way described from the data listed in Tables A–D.

Several authors have evaluated free energy equations of inorganic substances. Of these, we have made particular use of the publications by Richardson on metal oxides [191, 193], sulphides [192] and carbides [189], by Pearson and Ende on metal nitrides [184], and by Villa on metal chlorides [255]. Formulae derived by the present authors on the basis of data listed in Table A to D bear no reference number.

Certain binary systems, particularly the oxides, sulphides, nitrides and carbides of transition metals, do not obey Dalton's laws in that stable phases may extend over compositions other than stoichiometric (see p. 27). This means that a compound or the metal may take both or one of the constituent elements into solid or liquid solution. In such cases, it should be clearly stated whether the Gibbs energy equation corresponding to the equation for the chemical reaction refers to stoichiometric compositions or merely to the heterogeneity range between the solubility limits (see example Ti–TiO on p. 42). We have distinguished the two types in the following way. Inverted commas, e.g. $\langle \text{'TiO'} \rangle = \langle \text{'Ti'} \rangle + \frac{1}{2}(O_2)$, indicate that the free energy equation refers to *stoichiometric* compositions, disregarding any mutual solubility. Such values are generally obtained from calorimetric investigations of the pure compounds. On the other hand, bars over the chemical formulae, e.g. $\langle \overline{Cr_2N} \rangle = 2\langle \overline{Cr} \rangle + \frac{1}{2}(N_2)$, refer to reaction between the *saturated solutions*, i.e. Cr_2N deficient in nitrogen and chromium saturated with nitrogen. Such values are generally obtained from equilibrium measurements, but unfortunately the observers have not always determined accurately the compositions of the condensed phases in equilibrium with given gas mixtures, and the chemical reactions are therefore often not well defined. Such Gibbs energy values can be taken only as a rough guide.

6. Data for Metallic Solutions

In Table F, partial and integral thermochemical values of metallic solutions are compiled. The integral values are given for the whole range of composition at the atomic fractions 0·1, 0·3, 0·5, 0·7, 0·9 of the element mentioned second in column 1; and the partial values, in addition, for the atomic fraction 0·0, i.e. infinite dilution.

Table F is subdivided into tables (1, 2, 3,...) arranged in alphabetical order of the solvent metals and, for the sake of clearer presentation, these again are subdivided into one for the partial heats or excess Gibbs energies (α) and another for the excess entropies (β). For a complete description of a system, the values for ΔH, $\Delta \overline{H}_i$ and ΔS^E, $\Delta \overline{S}_i^E$ in Tables α and β must be combined, but where values for excess Gibbs energies (ΔG^E, $\Delta \overline{G}_i^E$) only appear in Table α, the excess entropies are to be taken as zero and are not listed in Table β.

Admittedly, the data presented in Table F suffice for a full thermochemical description only when the system in question is relatively simple. However, a wider tabulation would require too much space in the present monograph, and readers are advised to turn to the source quoted which in most cases are the tables of Hultgren and his colleagues [99]. An earlier source of this nature is a monograph by Kubaschewski and Catterall [307]. Otherwise the original references must be consulted. This is also advisable for those systems where blank spaces indicate incomplete information.

Greek letters after the heat or entropy values in the tables denote the phases to which they refer, but mostly they are omitted when confusion is unlikely to arise.

It should be pointed out that the number of digits in the individual values often stem from a computer evaluation and do *not* reflect their accuracy which varies substantially between, say, 50 and 500 cal/g-atom at $N = 0{\cdot}5$ for the enthalpies and between 0·1 and 0·5 cal/deg g-atom for the entropies. However, the consistency in terms of Gibbs energy for a particular phase is usually much better. In contrast, consistency of the tabulated values with the equilibrium diagrams is not guaranteed in all cases. Work to make all the thermochemical data and the phase boundaries consistent is in progress, system by system, but it is a tedious process and consequently slow.

The temperatures tabulated are those for which the heats and entropies have been calculated. As experimental work continues, it is being increasingly realised that significant deviations from Neumann–Kopp's rule occur. Nevertheless, evaluations accepting this rule, i.e. $\Delta C_p = 0$, lead to appreciable errors only in exceptional cases, so that for most purposes these deviations may continue to be disregarded.

Tables

Further abbreviations used in the tables

A	anion
α	degree of dissociation
b.c.	body centred
b.pt., b.p.	boiling point at 1 atm. pressure
C.No.	co-ordination number
c.p.	close packed
dec.	decomposes, decomposition
def.	deformed, deformation
disprop.	disproportionation
diss.	dissociated, dissociation
dist.	distorted
f.c.	face centered
$\Delta H_{max(min)}$	maximum (*or* minimum) integral heat of formation in a binary system
hetp.	heteropolar
homop.	homopolar
italics	estimated values
K	cation
l.	lattice
$\lambda_{t,f}$	heat of transformation, fusion
r.t.	room temperature
m.pt., m.p.	melting point
n.	net (layer lattice)
sb.pt., s.p.	sublimation point
s.l.	super lattice
sol.	solution
str.	structure
tr.pt.	transformation point
⟨ ⟩ [ ]	The various parentheses and brackets enclosing chemical formulae refer to the state of aggregation, as defined on p. xviii. In the column "m.pt." and "b.pt." angular brackets ⟨ ⟩ around the melting temperature denote melting point under pressure, and around the boiling point denote the theoretical (extrapolated) boiling temperature of the substance, which is not normally attained owing to sublimation. Brackets [ ] around the melting temperature denote peritectic decomposition.

Table A. Heats of formation,

Substance	ΔH°_{298} (kcal/mole)	Ref.	S°_{298} (cal/deg mole)	Ref.
⟨Ag⟩	0		10·20 ± 0·05	125
⟨AgF⟩	−49·2 ± 0·9	907	20·0 ± 3·0	—
⟨AgF_2⟩	−85·9 ± 2·5	907	—	—
⟨AgCl⟩	−30·37 ± 0·02	key value	23·0 ± 0·05	key value
⟨AgBr⟩	−24·06 ± 0·2	907	25·6 ± 0·1	125
⟨$AgBrO_3$⟩	−6·3	907	36·8	907
⟨AgI⟩α	−14·8 ± 0·1	907	27·6 ± 0·4	125
⟨Ag_2O⟩	−7·3 ± 0·1	22	29·1 ± 0·3	125
⟨Ag_2S⟩	−7·6 ± 0·2	307	34·3 ± 0·4	635
⟨Ag_2SO_4⟩	−171·4 ± 0·15	907	47·76 ± 0·1	907
⟨Ag_2Se⟩	−10·4 ± 0·2	854	35·9 ± 0·2	854
⟨Ag_2Te⟩	−8·6 ± 0·1	854	36·7 ± 0·5	854
⟨Ag_3N⟩	+61·0 ± 3·0	81	—	—
⟨$AgNO_3$⟩	−29·76 ± 0·16	907	33·68 ± 0·1	907
⟨AgP_2⟩	−10·7 ± 2·0	307	(21·0)	307
⟨AgP_3⟩	−16·5 ± 2·5	307	(25·2)	307
⟨Ag_2CO_3⟩	−120·8 ± 2·0	138	40·0 ± 0·9	125
⟨Al⟩	0		6·77 ± 0·05	125
(Al)	+78·7 ± 0·5	99	39·3 ± 0·1	99
⟨AlH_3⟩	−2·75 ± 0·15	827	*7·0 ± 1·1*	—
(AlF)	−63·4 ± 0·8	817	51·4 ± 0·1	345
⟨AlF_3⟩	−361·0 ± 0·3	817	15·89 ± 0·1	379
(AlF_3)	−289·04 ± 0·16	817	66·15 ± 0·2	817
(Al_2F_6)	−629·45 ± 4·0	817	92·5 ± 3·0	817
(AlCl)	−12·3 ± 1·0	817	54·45 ± 0·03	817
⟨$AlCl_3$⟩	−168·65 ± 0·2	817	26·12 ± 0·1	817
($AlCl_3$)	−139·7 ± 0·7	817	75·12 ± 1·0	817
(Al_2Cl_6)	−309·64 ± 0·8	817	113·65 ± 1·0	817
(AlBr)	+3·8 ± 3·0	817	57·25	817
⟨$AlBr_3$⟩	−122·15 ± 0·2	797	43·08 ± 0·3	817
($AlBr_3$)	−98·2 ± 0·4	817	83·43 ± 0·3	817
(Al_2Br_6)	−231·3 ± 5·0	(907)	124·4 ± 4·0	(907)
(AlI)	+14·3 ± 2·0	817	*59·24*	817
⟨AlI_3⟩	−73·9 ± 1·5	817	45·3 ± 2·0	817
(AlI_3)	−49·0 ± 1·8	817	86·78	817
(Al_2I_6)	−121·0 ± 3·0	817	*139·6*	817
(AlOF)	−145·0 ± 3·5	903	55·97	817
(AlOCl)	−84·4 ± 2·5	903	59·47 ± 0·7	907
⟨AlOCl⟩	−189·6 ± 0·3	817	*13·0 ± 1·0*	817
⟨$AlCl_3·6H_2O$⟩	−643·7 ± 0·9	(918)	76·03 ± 0·4	905
⟨Al_2O_3⟩	−400·9 ± 1·5	72	12·2 ± 0·1	129, 125
⟨$Al_2O_3·H_2O$⟩	−479·0 ± 2·0	907	16·83 ± 0·07	907
⟨$Al_2O_3·3H_2O$⟩	−618·2 ± 0·8	907	33·5 ± 0·2	907
⟨Al_2S_3⟩	−172·9 ± 4·0	113, 198	*29·5 ± 3·3*	—
⟨Al_2Se_3⟩	−129·0 ± 3·5	213, 113	*37·5*	—
⟨Al_2Te_3⟩	−76·2 ± 0·8	854	*45·0 ± 4·0*	—
⟨$Al_2(SO_4)_3$⟩	−822·6 ± 0·5	907	57·17 ± 0·3	907
⟨AlN⟩	−76·1 ± 0·5	907	4·82 ± 0·03	907
⟨AlP⟩	−39·3 ± 0·7	907, 908	11·3 ± 1·0	907
⟨AlAs⟩	−29·3 ± 1·2	908	*14·4 ± 1·0*	907
⟨AlSb⟩	−12·0 ± 0·1	99	15·53 ± 0·1	99
⟨$AlPO_4$⟩	−414·2 ± 0·4	906	21·7 ± 0·1	54
⟨$Al_6Si_2O_{13}$⟩	+4·1 ± 0·5*	840	65·7 ± 0·1	840
⟨Al_2TiO_5⟩	−2·0 ± 2·0*	840	26·2 ± 0·2	303
⟨Al_4C_3⟩	−51·55 ± 2·0	817	21·2 ± 0·1	817
⟨AlB_2⟩	−16·0 ± 2·5	—	—	—
⟨AlB_{12}⟩	−48·0 ± 3·0	—	—	—

standard entropies and structures

tr.pts. (°C)	m.pt. (°C)	b.pt. (°C)	Structure	Remarks
—	960·8	2200	see Table Aa	
—	435	1150	NaCl (def.)	
—	(690)	—	—	
—	455	1550	NaCl	
—	424	1560	NaCl	
—	—	—	—	
136, 147, 430	554	1500	ZnS	
—	—	—	Cu_2O	
176, 586	830	—	ZnS	
427	660	—	Na_2SO_4	
133	897	—	ZnS	
—	959	—	orthorhomb.	
—	—	—	f.c.c.	
160	210	—	—	
—	—	—	CuP_2	
—	—	—	—	
—	—	—	—	
—	660	2520	Ag	
—	—	—	—	
—	—	—	hexagonal	
—	—	—	—	
455	Sb	1276	rhombohed.	
—	—	—	—	
—	—	—	—	
—	—	—	—	
—	(193)	181	hexagonal	
—	—	—	—	
—	—	—	—	
—	—	—	—	
—	97	255	—	
—	—	—	—	
—	—	—	—	
—	—	—	—	
—	191	385	—	
—	—	—	—	
—	—	—	—	
—	—	—	—	
—	—	—	—	
—	—	—	—	
—	—	—	—	
1000	2050	—	Cr_2O_3	
—	—	—	—	
—	—	—	—	
—	1100	—	hexagonal	
—	950	—	wurtzite	
—	900	—	—	
—	—	—	orthorhomb.	
—	—	—	wurtzite	
—	1740	—	—	
—	1080	—	ZnS	
580, 705	2000	—	hexagonal	
—	1850	—	—	*from oxides, mullite
—	1590	—	Ti_3O_5	*from oxides
—	2200	—	—	
—	[1975]	—	—	
—	[2070]	—	—	

Table A. Heats of formation,

Substance	ΔH°_{298} (kcal/mole)	Ref.	S°_{298} (cal/deg mole)	Ref.
$\langle 9Al_2O_3\cdot 2B_2O_3\rangle$	−24·5 ± 5·3*	840	156·3	840
$\langle Al_4B_2O_9\rangle$	−16·6 ± 1·2*	840	*37·2*	—
$\langle As\rangle$	0	—	8·53 ± 0·1	99
(As_2)	+52·82 ± 1·7	99	57·55	99
(As_4)	+36·64 ± 0·4	99	78·23	99
(AsH_3)	+15·9 ± 4·0	509	53·2 ± 0·05	325
$\{AsF_3\}$	−228·7 ± 0·8	907	43·3 ± 0·1	907
(AsF_3)	−220·2 ± 0·8	907	69·06 ± 0·1	907
(AsF_5)	−295·6 ± 2·0	392	*75·8 ± 2·9*	—
$(AsCl_3)$	−64·8	907	77·97 ± 0·15	907
$(AsBr_3)$	−31·57 ± 1·0	907	86·94 ± 0·1	907
$\langle AsI_3\rangle$	−15·5	907	50·92	907
$\langle As_2O_3\rangle$	−156·15 ± 0·8	380	29·33 ± 0·1	380
$\langle As_2O_5\rangle$	−218·5 ± 1·5	—	25·2 ± 0·4	125
$\langle As_2S_3\rangle$	−40·0 ± 5·0	854	39·1 ± 1·0	854
$\langle AsS\rangle$	−17·0 ± 5·0	854	15·15 ± 0·4	854
$\langle As_2Se_3\rangle$	−24·5 ± 4·0	854	46·5 ± 0·8	854
$\langle As_2Te_3\rangle$	−9·0 ± 1·5	854	54·1 ± 1·5	854
$\langle Au\rangle$	0	—	11·32 ± 0·05	67, 125
(Au)	+88·0 ± 0·8	99	43·115	99
$\langle AuF_3\rangle$	−87·3 ± 8·0	418	27·3 ± 2·0	—
$\langle AuCl\rangle$	−8·3 ± 1·0	224, 198	22·2 ± 1·5	224
$\langle AuCl_3\rangle$	−27·5 ± 1·0	224, 198	35·4 ± 1·5	224
$\langle AuBr\rangle$	−4·4 ± 1·0	198	27·0 ± 1·5	—
$\langle AuI\rangle$	+0·2 ± 1·0	198	28·5 ± 2·0	—
$\langle Au_2O_3\rangle$	(−0·8)	—	(31·15)	—
$\langle AuTe_2\rangle$	−4·45 ± 0·7	753	33·87 ± 0·1	753
$\langle Au_2P_3\rangle$	−23·3 ± 3·5	307	36·0 ± 3·0	307
$\langle AuSb_2\rangle$	−4·65 ± 0·3	307	28·5 ± 0·3	307
$\langle AuSn\rangle$	−7·28 ± 0·3	136, 307	23·45 ± 0·7	307
$\langle AuPb_2\rangle$	−1·5 ± 0·4	136, 307	41·9 ± 0·3	307
$\langle Au_2Al\rangle$	−25·0 ± 1·5	679	—	—
$\langle AuAl\rangle$	−18·5 ± 1·0	679	—	—
$\langle AuAl_2\rangle$	−30·2 ± 1·0	679	—	—
$\langle B\rangle$	0	—	1·4 ± 0·05	105
(B)	136·5 ± 0·8	99	36·65	99
$\{B\}$†	1·0 ± 0·25	907	1·56 ± 0·03	105
(B_2H_6)	7·5 ± 0·3	702	55·3 ± 0·2	702
(BF_3)	−271·4 ± 0·3	692, 704	60·8 ± 0·5	125
(BCl_3)	−96·3 ± 2·0	817	69·3 ± 0·5	125
$\{BBr_3\}$	−57·0 ± 0·3	817	54·7 ± 2·0	817
(BBr_3)	−48·8 ± 0·3	817	77·49	817
$\langle B_2O_3\rangle$	−304·0 ± 0·5	817	12·9 ± 0·1	128
$\{B_2O_3\}$†	−299·5 ± 0·7	817	18·75 ± 0·2	817
$\langle H_3BO_3\rangle$	−261·5 ± 0·4	817, 907	21·21 ± 0·15	106
$\langle B_2S_3\rangle$	−60·3 ± 2·0	854	*22·0 ± 5·0*	854
$\langle BN\rangle$	−60·3 ±.0·3	797	3·54 ± 0·04	907
$\langle BP\rangle$	−27·6 ± 1·4	798	6·4 ± 0·1	907
$\langle B_4C\rangle$‡	−17·1 ± 2·8	—	6·48 ± 0·04	907
$\langle Ba\rangle$	0	—	14·92 ± 0·2	99
(Ba)	+43·5 ± 2·0	99	40·66	99
$\langle BaH_2\rangle$	−42·7 ± 1·0	57	15·4 ± 1·5	702
$\langle BaF_2\rangle$	−288·5	198	23·03 ± 0·1	198
$\langle BaCl_2\rangle$	−205·4 ± 0·6	57	29·55 ± 0·1	684

standard entropies and structures (contd.)

tr.pts. (°C)	m.pt. (°C)	b.pt. (°C)	Structure	Remarks
—	[1950]	—	—	*from oxides
—	[1035]	—	—	*from oxides
—	sb:	603	Table Aa	
—	—	—	—	
—	—	—	—	
−167·6	−116·9	−62·5	—	
—	−6	58	—	
—	—	—	—	
—	−80	−53	—	
—	−16	130	—	
—	31	221	—	
—	141	371	Table Aa	
—	—	—	Table Aa	
—	—	—	—	
267	312	565	—	
—	307	—	—	
—	377	—	—	
—	375	—	—	
—	1063	2860	Ag	
—	—	—	—	
—	—	—	—	
—	288	—	—	
—	288	—	—	
—	—	—	—	
—	—	—	—	
—	—	—	—	
—	464	—	—	
—	—	—	—	
—	460	—	pyrite	
—	418	—	NiAs	
—	[254]	—	$CuAl_2$	
—	624	—	—	
—	[625]	—	—	
—	1060	—	CaF_2	
—	2180	3800	various	
—	—	—	—	
—	—	—	—	†amorph.
—	—	—	—	
−131	−129	−100	—	
—	−107	12·5	—	
—	−47	91	—	
—	—	—	—	
—	450	—	cubic	
—	dec.	—	—	†amorph.
—	563	—	triclinic	
—	2970	—	—	
—	—	—	graphite	
—	2350	—	—	‡$B_{7-4}C$
—	729	(1700)	Na	
—	—	—	—	
—	—	—	—	
—	1290	2215	—	
922	962	—	CaF_2	

Table A. Heats of formation,

Substance	ΔH°_{298} (kcal/mole)	Ref.	S°_{298} (cal/deg mole)	Ref.
⟨$BaBr_2$⟩	−180·5 ± 0·6	57	35·5 ± 5·0	155
⟨BaI_2⟩	−144·5 ± 0·8	57	39·47 ± 0·1	871
⟨BaO⟩†	−132·3 ± 4·5	918	16·83 ± 0·3	125
⟨BaO_2⟩	−151·6 ± 2·8	824	22·25 ± 2·3	824
⟨$Ba(OH)_2$⟩	−225·5 ± 1·8	918	*24·1 ± 2·2*	—
⟨BaS⟩	−106·0 ± 3·0	854	18·7 ± 0·2	854
⟨$BaSO_4$⟩	−352·1 ± 4·0	198	31·6 ± 0·3	125
⟨$BaSe$⟩	−94·0 ± 9·0	854	*21·7 + 2·0*	854
⟨$BaTe$⟩	−64·4 ± 7·0	182	*23·8 ± 3·0*	—
⟨Ba_3N_2⟩	−81·5 ± 7·5	—	*36·4 ± 2·0*	122
⟨$Ba(NO_3)_2$⟩	−237·1	918	51·1	918
⟨Ba_3P_2⟩	−118·0 ± 10·0	—	—	—
⟨Ba_3Sb_2⟩	−175·0 ± 10·0	153	—	—
⟨Ba_3Bi_2⟩	−160·0 ± 10·0	153	—	—
⟨BaC_2⟩	−17·7 ± 3·0	683	21·0 ± 2·0	683
⟨$BaCO_3$⟩	−290·7 ± 5·0	918	26·8 ± 0·5	125
⟨$BaSiO_3$⟩	−38·0 ± 1·0*	342	26·8 ± 1·5	—
⟨Ba_2SiO_4⟩	−64·5 ± 1·0*	342	43·5 ± 2·5	—
⟨$BaTiO_3$⟩	−37·0 ± 2·0*	840	25·8 ± 0·2	840
⟨Ba_2TiO_4⟩	−46·0 ± 3·0*	840	47·0 ± 0·5	840
⟨$BaZrO_3$⟩	−28·6 ± 5·0*	721	26·5 ± 2·0	—
⟨Ba_2Sn⟩	−90·0 ± 8·0	153	30·3 ± 2·5	—
⟨$BaSn_3$⟩	−46·5 ± 4·5	153	*45·0 ± 3·5*	—
⟨Ba_2Pb⟩	−70·0 ± 3·0	153	—	—
⟨$BaPb$⟩	−36·0 ± 2·0	153	—	—
⟨$BaPb_3$⟩	−42·0 ± 2·5	153	—	—
⟨$BaO\cdot Al_2O_3$⟩	−24·0 ± 1·5*	318	35·5 ± 2·2	—
⟨$3BaO\cdot Al_2O_3$⟩	−45·0 ± 2·0*	318	—	—
⟨$BaHfO_3$⟩	−30·7 ± 4·0*	845	*29·2 ± 3·2*	—
⟨$BaMoO_4$⟩	−49·8 ± 2·5*	840	34·5 ± 1·5	840
⟨$BaUO_4$⟩	*−49·0 ± 4·0**	—	*40·3 ± 2·9*	—
⟨Be⟩	0	—	2·27 ± 0·2	755
(Be)	+77·5 ± 1·5	755	32·54	755
⟨BeH_2⟩	−4·55 ± 0·5	755	4·2 ± 1·0	702
(BeF)	−41·8 ± 2·5	755	49·15 ± 0·2	755
⟨BeF_2⟩	−245·4 ± 1·0	755	12·75 ± 0·1	755
($BeCl$)	+2·0 ± 1·0	496	53·0 ± 2·5	496
⟨$BeCl_2$⟩	−118·6 ± 0·8	755	18·12 ± 0·05	755
($BeCl_2$)	−86·1 ± 2·5	755	60·0 ± 1·0	755
(Be_2Cl_4)	(−196·3)	755	92·5 ± 2·0	755
⟨$BeBr_2$⟩	−84·5	755	25·4 ± 2·0	755
⟨BeI_2⟩	−46·0	755	28·8 ± 2·0	755
⟨BeO⟩	−145·4 ± 0·8	755	3·38 ± 0·05	755
⟨$Be(OH)_2$⟩	−216·5 ± 0·5	755	12·6 ± 1·0	—
($BeOH$)	−47·5 ± 7·0	755	*53·28*	817
⟨BeS⟩	−55·8 ± 1·0	755	8·4 ± 1·5	—
⟨$BeSO_4$⟩α	−287·0 ± 0·8	755	18·64 ± 0·2	755
⟨$BeSO_4\cdot 2H_2O$⟩	−435·7	918	39·01	918
⟨Be_3N_2⟩α	−140·9 ± 0·6	755	8·20 ± 0·08	755
⟨Be_2C⟩	−28·0 ± 0·5	755	*3·9 ± 0·8*	755
⟨Be_2SiO_4⟩	−4·7 ± 0·5*	755	15·37 ± 0·2	755
⟨$Be_3B_2O_6$⟩	−13·38 ± 0·5†	755	22·1 ± 3·0	755
⟨$BeAl_2O_4$⟩	−549·4 ± 3·3	755	15·84 ± 0·1	755
⟨$BeAl_6O_{10}$⟩	−1342·6 ± 5·5	755	42·0	755
⟨Bi⟩	0	—	13·55 ± 0·1	471
(Bi)	+50·1 ± 0·5	99	44·67	99
(Bi_2)	+52·6	99	65·4	99

standard entropies and structures (contd.)

tr.pts. (°C)	m.pt. (°C)	b.pt. (°C)	Structure	Remarks
—	854	—	$PbCl_2$	
—	712	—	$BaCl_2$	
—	1920	—	$BaCl_2$	†$BaO_{0\cdot907-1\cdot09}$
—	—	(2750)	$BaCl_2BaCl$	
—	408	—	CaC_2	
—	—	—	—	
—	(2400)	—	NaCl	
1150	1350	—	orthorhomb.	
—	—	—	NaCl	
—	2070	—	NaCl	
—	—	—	pseudo-hexag.	
—	—	—	—	
—	—	—	—	
—	—	—	—	
—	—	—	—	
806, 968	—	—	—	
—	—	—	orthorhomb.	*from oxides
—	—	—	—	*from oxides
5, 120	1760	—	K_2SO_4	*from oxides
120	1615	—	tetragonal	*from oxides
—	1860	—	—	*from oxides
—	—	—	—	
—	—	—	—	
—	700	—	—	
—	928	—	—	
—	—	—	—	
—	617	—	—	*from oxides
330	1830	—	hexagonal	*from oxides
—	—	—	—	*from oxides
—	—	—	—	*from oxides
—	—	—	—	*from oxides
1254	1287	2472	Mg	
—	—	—	—	
—	—	—	—	
227	552	1175	tetragonal	
—	—	—	—	
403	415	532	—	
—	—	—	—	
—	—	—	—	
—	488	521	—	
350	480	482	tetragonal	
—	2580	—	wurtzite	
—	—	—	—	
—	—	—	—	
—	—	—	ZnS	
590, 635	—	—	—	
—	—	—	—	
157	—	—	Sc_2O_3	
—	*2100*	—	CaF_2	
—	—	—	rhombohed.	*from oxides
		—	—	†from ⟨BeO⟩ and {B_2O_3} glass
—	1495	—	—	
—	1870			
—	271	1564	As	
—	—	—	—	
—	—	—	—	

Table A. Heats of formation,

Substance	ΔH°_{298} (kcal/mole)	Ref.	S°_{298} (cal/deg mole)	Ref.
(BiF)	−7·0 ± 5·0	907	58·32 ± 0·5	907
$\langle BiF_3\rangle$	−217·6 ± 4·0	773	29·3 ± 1·4	773
(BiF_3)	−169·5	773	75·92	773
$(BiCl)$	+6·0 ± 2·5	907	60·94 ± 0·08	907
$\langle BiCl_3\rangle$	−90·5 ± 1·2	907	41·0 ± 2·0	907
$\langle BiOCl\rangle$	−88·7 ± 1·0	907	24·5 ± 3·0	907
$\langle BiBr_3\rangle$	−66·0 ± 2·0	772	43·4 ± 1·5	772
$\langle BiI\rangle$	−13·33 ± 0·8	874	29·5 ± 1·0	874
$\langle BiI_3\rangle$	−36·0 ± 1·5	771	53·7 ± 1·5	771
$\langle Bi_2O_3\rangle$	−136·4 ± 0·8	877	36·2 ± 0·7	125
$\langle Bi_2S_3\rangle$	−48·2	854	47·9 ± 2·0	854
$\langle Bi_2(SO_4)_3\rangle$	−608·0 ± 6·0	918	*74·7 ± 5·0*	918
$\langle Bi_2Se_3\rangle$	−33·5 ± 0·8	854	57·3 ± 2·0	854
$\langle Bi_2Te_3\rangle$	−18·7 ± 0·8	523	62·4 ± 2·0	854
$\{Br_2\}$	0	—	36·38 ± 0·01	key value
(Br_2)	+7·39 ± 0·03	key value	58·64 ± 0·05	key value
(Br)	+26·735 ± 0·03	key value	41·80 ± 0·01	key value
Graphite	0	—	1·372 ± 0·005	key value
Diamond	+0·437 ± 0·02	907	0·566 ± 0·01	907
(CH_4)	−17·89 ± 0·1	198	44·5 ± 0·1	125
(C_2H_2)	+54·19	817	48·0	817
(CF_4)	−223·0 ± 0·4	907	62·46 ± 0·03	907
$\{CCl_4\}$	−32·33 ± 0·6	907	51·7 ± 0·2	907
(CCl_4)	−24·6 ± 0·5	907, 918	74·03 ± 0·1	907, 918
(COF_2)	−151·1 ± 1·5	907	61·85 ± 0·05	907
$(COCl_2)$	−53·3 ± 1·5	198	69·1 ± 0·5	125
(CO)	−26·42 ± 0·05	key value	47·22 ± 0·01	key value
(CO_2)	−94·05 ± 0·03	key value	51·07 ± 0·01	key value
(CS)	−57·0 ± 2·5	854	50·3 ± 0·2	125
$\{CS_2\}$	+21·41 ± 0·25	—	36·15 ± 0·12	854
(CS_2)	+28·0 ± 0·2	854	56·83	854
(COS)	−33·08 ± 0·2	817	55·32	817
$\langle Ca\rangle$	0	—	9·95 ± 0·1	125
(Ca)	+42·6 ± 0·4	99	36·99	99
(CaH)	+54·7	918	48·19 ± 0·1	918
$\langle CaH_2\rangle$	−44·5 ± 2·0	57	10·0 ± 1·5	125
$\langle CaF_2\rangle$	−291·5 ± 2·2	918	16·45 ± 0·1	125
$\langle CaCl_2\rangle$	−190·2 ± 0·9	918	25·0	918
$\langle CaOCl_2\rangle$	−178·4	918	*27·0 ± 2·2*	—
$\langle CaBr_2\rangle$	−163·2 ± 3·0	57, 198	32·0 ± 2·5	—
$\langle CaI_2\rangle$	−127·5 ± 2·5	918	(34·0)	918
$\langle CaO\rangle$	−151·6 ± 0·4	198, 529	9·5 ± 0·2	125
$\langle CaO_2\rangle$	−157·5 ± 0·5	138	20·0 ± 4·0	—
$\langle Ca(OH)_2\rangle$	−235·5 ± 2·0	918	19·93 ± 0·1	514
$\langle CaS\rangle$	−113·8 ± 2·4	854	13·5 ± 0·3	125
$\langle CaSO_3\rangle$	*−280·0 ± 9·0*	—	*24·2 ± 1·2*	—
$\langle CaSO_4\rangle$	−342·76 ± 1·0	918	25·5 ± 0·4	125
$\langle CaSO_4\cdot\frac{1}{2}H_2O\rangle$	−376·55 ± 3·5	918	31·2 ± 0·6	918
$\langle CaSO_4\cdot 2H_2O\rangle$	−483·12 ± 3·5	918	46·4 ± 0·3	125
$\langle CaSe\rangle$	−88·0 ± 6·0	854	*16·5 ± 1·5*	854
$\langle CaTe\rangle$	−65·0 ± 8·0	182	19·3 ± 2·0	134
$\langle Ca_3N_2\rangle$	−105·0 ± 3·0	—	*25·8 ± 2·8*	—
$\langle Ca_3P_2\rangle$	−121·0 ± 6·0	918	*29·6 ± 5·0*	—
$\langle Ca_2P_2O_7\rangle$	−797·5 ± 5·0	918	45·25 ± 0·1	55
$\langle Ca_3P_2O_8\rangle$	−984·0 ± 6·0	918	56·4 ± 0·2	918
$\langle CaHPO_4\rangle$	−433·65	918	26·62	918

standard entropies and structures (contd.)

tr.pts. (°C)	m.pt. (°C)	b.pt. (°C)	Structure	Remarks
—	—	—	—	
—	649	—	Table Aa	
—	—	—	—	
—	—	—	—	
—	234	441	—	
—	—	—	—	
—	219	461	—	
291	—	—	—	
—	409	—	AsI_3	
705	824	—	monoclinic	
—	767	—	orthorhomb.	
—	—	—	—	
—	722	—	—	
—	597	—	rhombohed.	
—	−7	61	solid: I_2	
—	—	—	—	
—	—	—	—	
—	—	—	Table Aa	
—	—	—	Table Aa	
—	—	—	—	
—	—	—	—	
−197	−184	−151	—	
−48	−23	77	—	
—	—	—	—	
—	—	—	—	
—	−128	7·5	—	
−212	−205	−192	—	
—	—	−79	—	
—	—	—	—	
—	−112	46	—	
—	—	—	—	
—	−139	−50	—	
448	839	1484	Ag	
—	—	—	—	
—	—	—	—	
780	816	—	$PbCl_2$	
1151	1418	2510	Table A2a	
—	772	(2000)	TiO_2(def.)	
—	—	—	—	
—	742	(1800)	TiO_2(def.)	
—	777	—	CdI_2	
—	2615	(3500)	NaCl	
—	—	—	CaC_2	
—	—	—	CdI_2	
—	—	—	orthorhomb.	
—	—	—	—	
1193	(1400)	—	—	
—	—	—	—	
—	—	—	—	
—	—	—	NaCl	
—	—	—	NaCl	
—	1195	—	pseudo-hexag.	
—	1320	—	—	
1140	1353	—	—	
1100	—	—	—	
—	—	—	—	

Table A. Heats of formation,

Substance	ΔH°_{298} (kcal/mole)	Ref.	S°_{298} (cal/deg mole)	Ref.
$\langle CaV_2O_6\rangle$	−34·3 ± 0·5*	711	42·8 ± 0·5	711
$\langle Ca_2V_2O_7\rangle$	−62·7 ± 0·6*	711	52·7 ± 0·5	711
$\langle Ca_3V_2O_8\rangle$	−77·0 ± 0·8*	711	65·7 ± 0·6	711
$\langle Ca_3Sb_2\rangle$	−174·0 ± 9·0	153	*37·6 ± 4·5*	—
$\langle Ca_3Bi_2\rangle$	−126·0 ± 7·5	153	*42·5*	—
$\langle CaC_2\rangle$	−14·1 ± 2·0	189	16·8 ± 0·5	125
$\langle CaCO_3\rangle$	−288·51 ± 0·7	138	21·2 ± 0·3†	125
	−288·46	—	22·2 ± 0·2‡	125
$\langle CaSi\rangle$	−36·0 ± 2·0	153	—	—
$\langle CaSi_2\rangle$	−36·0 ± 3·0	153	—	—
$\langle Ca_2Si\rangle$	−50·0 ± 3·0	153	—	—
$\langle CaSiO_3\rangle$	−21·3 ± 0·3	193	19·6 ± 0·3	125
$\langle Ca_2SiO_4\rangle$	−32·7 ± 2·0*	840	28·8 ± 0·2	840
$\langle Ca_3SiO_5\rangle$	−27·0 ± 1·5*	131, 193, 352	40·3 ± 0·3	243
$\langle CaMgSiO_4\rangle$	−27·6 ± 1·5*	389	*26·4 ± 1·0*	—
$\langle CaAl_2\rangle$	−51·8 ± 4·0	99	*20·4 ± 2·5*	—
$\langle CaAl_4\rangle$	−52·2 ± 4·0	99	*33·0 ± 4·0*	—
$\langle Ca_3Al_2O_6\rangle$	−2·0 ± 0·8*	840	49·1 ± 0·3	304
$\langle CaAl_2O_4\rangle$	−3·6 ± 0·3*	840	27·25 ± 0·2	840
$\langle CaAl_4O_7\rangle$	−3·0 ± 0·5*	840	42·5 ± 0·3	304
$\langle Ca_2Sn\rangle$	−75·0 ± 4·5	153	*24·0 ± 2·6*	—
$\langle CaSn\rangle$	−38·0 ± 2·5	153	*16·9 ± 1·9*	—
$\langle CaSn_3\rangle$	−43·0 ± 4·0	264	—	—
$\langle Ca_2Pb\rangle$	−51·5 ± 2·5	264	25·2 ± 2·5	147
$\langle CaPb\rangle$	−28·6 ± 1·5	264	*19·3 ± 2·5*	—
$\langle CaO\cdot 2B_2O_3\rangle$	−42·9 ± 1·0*	193	32·2 ± 0·3	125
$\langle CaO\cdot B_2O_3\rangle$	−29·4 ± 1·0*	193	25·1 ± 0·2	125
$\langle 2CaO\cdot B_2O_3\rangle$	−45·8 ± 1·0*	193	34·7 ± 0·2	125
$\langle 3CaO\cdot B_2O_3\rangle$	−60·0 ± 1·0*	193	43·9 ± 0·3	125
$\langle CaTl\rangle$	−39·0 ± 3·0	153, 264	—	—
$\langle CaMg_2\rangle$	−9·4 ± 0·4	840	24·53 ± 0·25	840
$\langle CaOMgO\rangle$	−1·8 ± 0·3*	840	15·85 ± 0·4	840
$\langle CaTiO_3\rangle$	−19·3 ± 0·9*	840	22·4 ± 0·2	125
$\langle Ca_3Ti_2O_7\rangle$	−50·0*	840	56·1 ± 0·5	840
$\langle Ca_4Ti_3O_{10}\rangle$	−71·0*	840	*78·5 ± 2·0*	840
$\langle CaZrO_3\rangle$	−7·3 ± 2·0*	840	*22·4 ± 2·0*	840
$\langle CaHfO_3\rangle$	−7·5*	137	*24·0 ± 2·2*	—
$\langle CaMoO_4\rangle$	−39·8 ± 1·0*	840	29·3 ± 0·2	637
$\langle CaWO_4\rangle$	−34·9 ± 2·5*	840	30·2 ± 0·2	546
$\langle CaUO_4\rangle$	*−32·0 ± 4·0**	—	*34·2 ± 2·7*	—
$\langle Ca_2Fe_2O_5\rangle$	−9·6 ± 1·0*	840	45·1 ± 0·3	378
$\langle CaFe_2O_4\rangle$	−5·0 ± 2·5	840	34·7 ± 0·2	378
$\langle CaZn\rangle$	−17·5 ± 2·0	264	*15·9 ± 1·8*	—
$\langle CaZn_2\rangle$	−22·5 ± 2·5	264	*24·3 ± 2·4*	—
$\langle CaZn_5\rangle$	−33·0 ± 3·0	264	—	—
$\langle CaSnO_3\rangle$	−17·4 ± 0·5*	814	20·0 ± 0·5	814
$\langle CaSnO_4\rangle$	−17·1 ± 0·5*	814	30·65 ± 0·5	814
$\langle Cd\rangle$	0	—	12·38 ± 0·05	99
(Cd)	+26·72 ± 0·15	99	40·065	99
$\langle CdF_2\rangle$	−167·4 ± 0·5	907	20·0 ± 2·0	907
$\langle CdCl_2\rangle$	−93·4 ± 0·3	907	27·55 ± 0·1	907
$\langle CdBr_2\rangle$	−75·35 ± 0·4	907	33·18 ± 0·1	907
$\langle CdI_2\rangle$	−48·8 ± 0·8	907	37·84 ± 0·25	907
$\langle CdO\rangle$	−62·0 ± 0·4	916	13·1 ± 0·2	125, 165
$\langle CdS\rangle$	−35·7 ± 0·5	854	16·5 ± 0·5	165, 125
$\langle CdSO_4\rangle$	−222·0 ± 0·6	860	29·4 ± 0·15	393

standard entropies and structures (contd.)

tr.pts. (°C)	m.pt. (°C)	b.pt. (°C)	Structure	Remarks
—	778	—	—	*from oxides
—	1015	—	—	*from oxides
—	1380	—	—	*from oxides
—	1350	—	—	
—	930	—	—	
447	2300	—	Table Aa	
50	—	—	$BaCO_3$†	†aragonite
—	—	—	$MgCO_3$‡	‡calcite
—	1240	—	orthorhomb.	
—	—	—	rhombohed.	
—	—	—	$PbCl_2$	
1190	1544	—	triclinic	
675, 1420	2130	—	Na_2BeF_4	*from oxides
—	—	—	—	*from oxides
—	—	—	—	*from oxides
—	1080	—	—	
—	777	—	—	
—	(1535)	—	cubic	*from oxides
—	1605	—	—	*from oxides
—	(1535)	—	—	*from oxides
—	1122	—	—	
—	987	—	CaSi	
—	627	—	Ag: sup. str.	
—	1110	—	—	
—	950	—	—	
—	990	—	—	*from oxides
—	1160	—	—	*from oxides
—	1310	—	—	*from oxides
—	1490	—	—	*from oxides
—	—	—	CsCl	
—	714	—	$MgZn_2$	
—	—	—	—	*from oxides
1260	—	—	orthorhomb.	*from oxides
—	(1725)	—	—	*from oxides
—	(1755)	—	—	*from oxides
—	2340	—	—	*from oxides
—	—	—	—	*from oxides
—	—	—	—	*from oxides
546	1580	—	—	*from oxides
752	—	—	—	*from oxides
—	1450	—	—	*from oxides
—	(1216)	—	—	
—	430	—	—	
—	690	—	—	
—	680	—	hexagonal	
—	—	—	—	*from oxides
—	—	—	—	*from oxides
—	321	767	*c.* Mg	
—	—	—	—	
—	1072	1750	CaF_2	
—	569	960	Table Aa	
—	568	863	$CdCl_2$	
322	388	796	Table Aa	
—	Sb	1497	NaCl	
—	1405	—	wurtzite	
—	1000	—	—	

Table A. Heats of formation,

Substance	ΔH°_{298} (kcal/mole)	Ref.	S°_{298} (cal/deg mole)	Ref.
⟨CdSe⟩	−34·6 ± 3·0	854	19·9 ± 0·5	854
⟨$CdSeO_3$⟩	−139·1 ± 1·3	907	28·7 ± 2·0	907
⟨CdTe⟩	−24·33 ± 0·2	99	22·24 ± 0·6	99
⟨Cd_3N_2⟩	38·6 ± 3·0	264	—	—
⟨Cd_3As_2⟩	−9·1 ± 2·0	898	49·5 ± 2·5	898
⟨CdSb⟩†	−3·19 ± 0·3	758	22·85 ± 0·15	758
⟨$CdCO_3$⟩	−179·7 ± 2·0	860	22·1 ± 1·2	860
⟨$CdSiO_3$⟩	−4·8 ± 0·5	840	23·3 ± 0·3	840
⟨$CdTiO_3$⟩	−6·6 ± 0·2*	840	*25·1 ± 1·0*	840
⟨$CdAl_2O_4$⟩	+3·6 ± 0·7*	840	*29·9 ± 4·0*	—
⟨$CdGa_2O_4$⟩	−324·1 ± 1·4	840	*33·4 ± 2·4*	—
⟨$CdWO_4$⟩	−18·9 ± 1·8*	840	*37·0 ± 2·5*	—
⟨Ce⟩	0	—	17·2	918
(Ce)	+101·0 ± 3·0	99	45·81	—
⟨CeH_2⟩	−49·0	918	13·33 ± 0·08	759
⟨CeF_3⟩	*−425·0 ± 15·0*	—	27·55 ± 0·2	—
⟨$CeCl_3$⟩	−252·8 ± 2·0	233, (18)	35·8 ± 1·8	—
⟨CeI_3⟩	−155·3 ± 2·2	918	*51·3 ± 3·5*	—
⟨Ce_2O_3⟩	−435·4 ± 0·9	—	36·0 ± 1·0	—
⟨CeO_2⟩	−260·5 ± 0·4	—	14·9 ± 0·1	—
⟨CeS⟩	−109·1 ± 2·5	854	18·7 ± 0·3	125
⟨Ce_3S_4⟩	−395·0 ± 3·0	854	*61·0 ± 3·0*	854
⟨Ce_2S_3⟩	−284·0 ± 3·0	854	43·1 ± 0·5	854
⟨$Ce_2(SO_4)_3$⟩	−944·5 ± 2·0	767	*68·7 ± 6·0*	—
⟨CeN⟩	−78·0 ± 6·0	264, 198	—	—
⟨Ce_2C_3⟩	−42·2 ± 1·8	756	*41·5 ± 4·0*	—
⟨CeC_2⟩	−23·2 ± 1·7	756	*21·5 ± 2·0*	—
⟨CeB_6⟩	*−84·0 ± 23·0*	—	*17·7 ± 3·5*	—
⟨$CeAlO_3$⟩	−421·4 ± 4·0	843	25·5 ± 3·5	843
⟨$CeAl_4$⟩	−39·0 ± 3·0	264	*38·7*	—
⟨$CeAl_2$⟩	−39·0 ± 4·0	264	*27·1*	—
(Cl_2)	0	—	53·29 ± 0·01	key value
(Cl)	+28·99 ± 0·01	key value	39·45 + 0·01	key value
⟨Co⟩	0	—	7·18 ± 0·1	99
(Co)	+102·4 ± 1·0	99	42·88	99
⟨CoF_2⟩	−165·4 ± 2·4	918	19·59 ± 0·1	918
⟨CoF_3⟩	−193·8 ± 5·0	918	*22·8 ± 2·8*	—
⟨$CoCl_2$⟩	−74·1 ± 0·8	907	26·21 ± 0·05	907
⟨$CoBr_2$⟩	−52·8 ± 2·7	918	*32·0 ± 2·0*	—
⟨CoI_2⟩	−21·2 ± 2·5	918	*36·6 ± 2·5*	—
⟨CoO⟩	−57·1 ± 0·5	349, 22	12·65 ± 0·1	379
⟨Co_3O_4⟩	−216·3 ± 3·0	—	24·5 ± 0·2	379
⟨$Co(OH)_2$⟩	−129·4 ± 1·5	792	22·3 ± 0·4	792
⟨$CoS_{0·89}$⟩	−22·6 ± 1·0	854	12·5 ± 0·8	854
⟨$CoS_{1·33}$⟩	−28·6 ± 2·0	854	*14·7 ± 2·0*	854
⟨CoS_2⟩	−36·6 ± 4·0	854	16·5 ± 1·5	854
⟨$CoSO_4$⟩	−212·0 ± 0·6	860	27·1	860
⟨$CoSeO_3$⟩	−139·1 ± 2·0	907	27·2 ± 2·0	907
⟨Co_3N⟩	*+2·0 ± 4·0*	82	*23·6 ± 1·5*	—
⟨Co_2P⟩	−44·8 ± 3·5	264	*18·5 ± 3·5*	—
⟨CoP⟩	−30·0 ± 5·0	264	*12·0*	—
⟨CoP_3⟩	−48·9 ± 5·0	264	*23·5*	—
⟨$CoAs_2$⟩	−26·6 ± 3·0	139	53·3 ± 3·0	139
⟨Co_2As⟩	−13·5 ± 1·5	139	23·0 ± 2·0	139
⟨Co_3As_3⟩	−27·2 ± 3·0	139	38·3 ± 2·5	139
⟨CoAs⟩	−13·6 ± 1·5	139	15·4 ± 1·0	139
⟨Co_2As_3⟩	−34·4 ± 3·0	139	39·2 ± 3·0	139

standard entropies and structures (contd.)

tr.pts. (°C)	m.pt. (°C)	b.pt. (°C)	Structure	Remarks
—	1264	—	wurtzite	
—	680	—	—	
—	1091	—	ZnS	
—	—	—	SC_2O_3	
578	721	—	tetragonal	
—	456	—	orthorhomb.	†$CdSb_{0\cdot 99-1\cdot 0}$ (250°C)
—	—	—	$MgCO_3$	
—	1785	—	—	
830	—	—	—	*from oxides
—	—	—	—	*from oxides
—	—	—	—	
—	—	—	—	*from oxides
726	798	3424	Mg	
—	—	—	—	
—	—	—	—	
—	1437	—	hexagonal	
—	800	1731	UCl_3	
—	760	—	—	
—	—	—	La_2O_3	
—	—	—	CaF_2	
—	2450	—	NaCl	
—	2050	—	—	
—	1890	—	cubic	
—	—	—	—	
—	2560	—	—	
—	—	—	—	
—	—	—	—	
—	2000	—	—	
—	—	—	—	
1005	1245	—	tetragonal	
—	1460	—	Table Aa	
—	−101	−34	—	
—	—	—	—	
434	1495	2930	Mg	
—	—	—	—	
—	1250	—	rutile	
—	—	—	FeF_3	
—	740	1053	$CdCl_2$	
—	676	—	CdI_2	
—	520	—	CdI_2	
—	1805	—	NaCl	
—	—	—	spinel	
—	—	—	—	
—	[835]	—	cubic	
—	[625]	—	—	
—	—	—	pyrite	
—	—	—	—	
—	659	—	—	
—	—	—	hexag. c.p.†	†metastable
—	—	—	hexag.(def.)	
—	—	—	orthorhomb.	
—	—	—	—	
—	923	—	—	
—	958	—	—	
—	[1014]	—	—	
–	1180	—	CoP	
—	—	—	—	

Table A. Heats of formation,

Substance	ΔH°_{298} (kcal/mole)	Ref.	S°_{298} (cal/deg mole)	Ref.
⟨$CoAs_2$⟩	−22·0 ± 2·5	139	22·2 ± 2·0	139
⟨CoSb⟩‡	−10·0 ± 0·8	264	16·9 ± 1·3	790
⟨Co_2C⟩	4·0 ± 4·0	189	17·8 ± 2·5	189
⟨$CoCO_3$⟩	−170·0 ± 3·5	886	21·2 ± 0·4	833
⟨Co_2Si⟩	−28·0 ± 2·5	840	—	—
⟨CoSi⟩	−22·7 ± 2·0	840	10·2 ± 0·7	840
⟨$CoSi_2$⟩	−23·6 ± 2·5	840	15·3 ± 1·0	840
⟨Co_2B⟩	*−30·0 ± 7·0*	—	*14·3 ± 3·0*	—
⟨CoB⟩	*−22·5 ± 4·0*	—	*7·3 ± 1·6*	—
⟨CoSn⟩	−7·0 ± 1·0	264	*17·1 ± 2·5*	—
⟨CoAl⟩§	−26·4 ± 2·0	264	—	—
⟨Co_2Al_5⟩	−70·0 ± 5·0	264	—	—
⟨$CoAl_4$⟩‖	−38·5 ± 3·0	264	—	—
⟨$CoTiO_3$⟩	−5·9 ± 1·0*	840	23·15	840
⟨$CoWO_4$⟩	−14·8 ± 0·9*	840	*30·2 ± 2·2*	840
⟨$CoAl_2O_4$⟩	−8·0 ± 0·7*	813	23·8 ± 0·7	813
⟨$CoCr_2O_4$⟩	−14·3 ± 0·5*	840, 813	30·3 ± 0·2	840, 813
⟨$CoFe_2O_4$⟩	−6·6 ± 0·6*	840	34·1 ± 2·0	840
⟨Cr⟩	0	—	5·65 ± 0·05	99
(Cr)	+95·0 ± 1·0	99	41·63	99
⟨CrF_2⟩	−186·3 ± 3·0	918	*20·0 ± 3·0*	27
⟨CrF_3⟩	−266·0 ± 5·0	27, 260	22·45 ± 0·1	512
⟨CrF_4⟩	−286·5 ± 6·0	260	—	—
⟨$CrCl_2$⟩	−94·5 ± 3·0	918	27·55 ± 0·1	625
⟨$CrCl_3$⟩	−133·0 ± 4·0	918	29·4 ± 0·1	512
⟨$CrBr_2$⟩	−72·2 ± 5·0	918	*32·2 ± 1·7*	—
⟨CrI_2⟩	−37·8 ± 1·4	295, (198)	—	—
⟨CrI_3⟩	−49·0 ± 1·6	918	*47·7 ± 2·6*	—
(CrO_2Cl_2)	−128·6 ± 3·0	918	78·7	918
⟨Cr_2O_3⟩†	−270·0 ± 2·5	198, 22, 311	19·4 ± 0·2	125
⟨CrO_2⟩	−139·2 ± 2·0	340	*12·2 ± 2·1*	—
⟨CrO_3⟩	−138·5 ± 2·5	198, 22	17·2 ± 2·5	—
⟨CrS⟩	−37·2 ± 2·5	856	*15·3 ± 1·5*	854
⟨Cr_2S_3⟩	(*−87*)	854	(*35·6*)	854
⟨$Cr_2(SO_4)_3$⟩	−695·7 ± 3·5	911	61·85 ± 0·5	912
⟨CrN⟩*	−29·4 ± 0·8	—	78·5 ± 0·8	—
⟨Cr_4C⟩	−23·5 ± 2·0	99	25·3 ± 0·3	125
⟨Cr_7C_3⟩	−54·5 ± 2·5	99	48·0 ± 0·3	125
⟨Cr_3C_2⟩	−26·2 ± 2·0	99	20·4 ± 0·2	125
⟨$Cr(CO)_6$⟩	−222·0	914	*76·5 ± 0·8*	—
⟨Cr_3Si⟩	−22·0 ± 3·8	769	20·6 ± 1·5	769
⟨Cr_5Si_3⟩	−50·5 ± 7·5	769	40·4 ± 3·0	769
⟨CrSi⟩	−12·7 ± 2·0	769	10·5 ± 0·8	769
⟨$CrSi_2$⟩	−19·1 ± 3·0	769	14·0 ± 1·1	769
⟨CrB⟩	*−18·0*	—	*5·75*	—
⟨CrB_2⟩	*−22·5*	—	*6·25*	—
⟨Cr_2N⟩‡ —	−27·3 ± 0·8	—	17·7 ± 1·0	—
⟨Cs⟩	0	—	20·37 ± 0·1	key value
⟨CsF⟩	−132·6 ± 0·5	817	*21·1 ± 2·0*	817
(CsF)	−85·2 ± 0·5	817	58·1 ± 0·01	817
⟨CsCl⟩	−105·8 ± 0·4	817	24·18 ± 0·01	817
(CsCl)	−57·4 ± 1·0	817	61·175	817
⟨CsBr⟩	−94·3 ± 3·0	198	27·1 ± 1·5	177
⟨CsI⟩	−80·5 ± 2·5	198	30·0 ± 2·5	—
⟨Cs_2O⟩	−75·9 ± 3·0	22, 198	*30·5 ± 2·5*	—
⟨Cs_2O_3⟩	−135·0 ± 20·0	22	*56·0 ± 5·0*	—

standard entropies and structures (contd.)

tr.pts. (°C)	m.pt. (°C)	b.pt. (°C)	Structure	Remarks
—	—	—	marcasite	
—	1193	—	NiAs	‡$CoSb_{0\cdot74=0\cdot96}$
—	—	—	orthorhomb.	
—	—	—	$MgCO_3$	
—	1330	—	—	
—	1460	—	—	
—	1327	—	CaF_2	
—	1265	—	—	
—	[1110]	—	—	
525	[936]	—	NiAs(def.)	
—	1628	—	CsCl	§$CoAl_{1\cdot0-1\cdot3}$ (25°C)
—	—	—	—	
—	—	—	—	¶$CoAl_{4\cdot6}$
—	—	—	—	*from oxides
—	—	—	—	*from oxides
—	—	—	—	*from oxides
—	—	—	—	*from oxides
—	—	—	—	*from oxides
—	1857	2672	—	
—	—	—	—	
—	1100	—	—	
—	1100	—	—	
—	—	—	—	
—	815	1300	orthorhomb.	
—	Sb	945	—	
—	842	—	monoclinic	
—	792	—	—	
—	—	—	—	
—	−95	117	—	
—	(2400)	—	Table Aa	†$CrO_{1\cdot5-1\cdot54}$
—	—	—	—	
—	185	—	—	
(200)	1567	—	—	
—	—	—	—	
—	—	—	—	
—	—	—	NaCl	
—	1520	—	cubic	
—	1780	—	hexagonal	
—	1890	—	orthorhombic	
—	151	—	—	
—	1770	—	—	
—	1647	—	—	
—	—	—	—	
—	1457	—	—	
—	—	—	—	
—	—	—	—	
—	—	—	Cr = Mg	‡$CrN_{0\cdot38-0\cdot50}$
—	30	—	Na	
—	703	1270	NaCl	
—	—	—	—	
470	645	1324	Table Aa	
—	—	—	—	
—	639	1300	CsCl	
—	626	1280	CsCl	
—	(490)	—	$CdCl_2$	
—	502	—	$Cs_4(O_2)_3$: cubic	

Table A. Heats of formation,

Substance	ΔH°_{298} (kcal/mole)	Ref.	S°_{298} (cal/deg mole)	Ref.
⟨CsO_2⟩	−76·0 ± 8·0	22	*35·0 ± 3·5*	—
⟨Cs_2S⟩	−81·1 ± 8·0	198	—	—
⟨Cs_2SO_4⟩	−339·4 ± 6·0	198	*49·6 ± 4·1*	—
⟨Cs_2CO_3⟩	−274·2 ± 2·5	919	48·87 ± 0·1	872
⟨Cu⟩	0	—	7·92 ± 0·05	99
(Cu)	+80·5 ± 0·3	99	39·74 ± 0·02	99
⟨CuF_2⟩	−131·2 ± 3·0	812	16·4 ± 1·0	817
⟨CuCl⟩	−32·8 ± 0·7	918	20·6 ± 0·4	918
⟨$CuCl_2$⟩	−52·0 ± 2·0	918	25·83 ± 0·1	625
⟨CuBr⟩	−24·9 ± 0·3	93, 198	23·0 ± 0·2	93, 125
⟨$CuBr_2$⟩	−33·9 ± 2·2	918	*32·0 ± 1·7*	—
⟨CuI⟩†	−16·2 ± 1·0	198	23·1 ± 0·3	125
⟨Cu_2O⟩	−40·0 ± 0·7	22, 198, 165	22·25 ± 0·1	93, 165, 501
⟨CuO⟩	−37·1 ± 0·8	22, 198	10·2 ± 0·05	93, 125
⟨Cu_2S⟩‡	−19·0 ± 0·3	854	28·9 ± 0·5	854
⟨CuS⟩	−12·5 ± 1·0	854	15·9 ± 0·5	125
⟨Cu_2SO_4⟩	−179·2	859	*43·6*	—
⟨$CuSO_4$⟩	−184·25 ± 0·3	907	26·1 ± 0·1	907
⟨Cu_2Se⟩§	−15·6 ± 1·5	854	31·0 ± 1·0	854
⟨CuSe⟩	−10·0 ± 1·0	854	*18·7 ± 1·5*	854
⟨Cu_2Te⟩	−10·0 ± 2·5	854	*32·2 ± 2·0*	854
⟨Cu_3N⟩	−17·8 ± 3·0	198	—	—
⟨Cu_3P⟩	−30·82 ± 4·0	264	*28·5*	—
⟨CuP_2⟩	−21·52 ± 2·5	801	*19·5*	—
⟨Cu_5FeS_4⟩	−90·9	—	*86·6*	—
⟨$CuFeS_2$⟩	−45·5	—	29·87	—
⟨$CuCO_3$⟩	−142·5 ± 2·2	886	21·0	918
⟨$CuAlO_2$⟩*	−2·85 ± 0·5	810	16·0 ± 0·3	810
⟨$CuAl_2O_4$⟩*	+4·4 ± 0·5	810	26·6 ± 0·4	810
⟨$CuGaO_2$⟩*	−1·7 ± 0·5	811	19·9 ± 0·3	811
⟨$CuGa_2O_4$⟩*	+4·1 ± 0·5	811	35·0 ± 0·4	811
⟨$CuFeO_2$⟩	−122·5	—	21·24	—
⟨$CuFe_2O_4$⟩	−231·0	—	35·08	—
⟨Cu_2Sb⟩	−2·74 ± 0·4	99	30·23 ± 0·4	99
⟨$CuCr_2O_4$⟩	+0·75 ± 0·9*	857	31·0 ± 2·2	—
⟨$CuMg_2$⟩	−6·95 ± 0·3	840	22·05 ± 0·3	840
⟨Cu_2Mg⟩	−8·4 ± 0·4	840	23·4 ± 0·4	840
⟨Dy⟩	0	—	17·87 ± 0·1	368
⟨$DyCl_3$⟩	−239·0 ± 2·6	918	*35·1 ± 2·4*	—
⟨Dy_2O_3⟩	−445·3 ± 3·0	808	35·8 ± 0·1	536
⟨Er⟩	0	—	17·5 ± 0·15	412
⟨$ErCl_3$⟩	−229·1 ± 1·0	332	35·1 ± 1·0	332
⟨Er_2O_3⟩	−453·6 ± 1·0	371	36·6 ± 0·1	536
⟨Eu⟩	0		19·3	99
⟨$EuCl_3$⟩	−220·4 ± 1·8	779	*30·0 ± 3·5*	779
⟨Eu_2O_3⟩	−412·4 ± 2·5	—	*35·0*	—
⟨EuS⟩	−106·7 ± 1·8	802	23·6 ± 1·3	802
(F_2)	0	—	48·44 ± 0·1	173, 125
(F)	+18·86 ± 0·6	817	37·92 ± 0·02	817
⟨Fe⟩	0	—	6·52 ± 0·03	99
(Fe)	+99·3 ± 0·3	99	43·11	99
⟨FeF_2⟩	−170·0 ± 2·5	—	20·8 ± 0·1	415
⟨FeF_3⟩	−249·0	817	23·5	817

standard entropies and structures (contd.)

tr.pts. (°C)	m.pt. (°C)	b.pt. (°C)	Structure	Remarks
—	432	—	CaC_2	
—	—	—	—	
—	1019	—	K_2SO_4	
—	—	—	—	
—	1083	2560	Ag	
—	—	—	—	
—	770	—	CaF_2	
408	451	1212	ZnS	
—	—	—	—	
380, 470	489	1355	ZnS	
—	—	—	—	
372, 410	600	1320	ZnS	†$CuI_{1·0-1·0045}$
—	1236	—	Table Aa	
—	—	—	monoclinic	
103, 444	1129	—	ZnS	‡$CuS_{0·5-0·50025}$
—	—	—	hexagonal	
—	—	—	—	
—	—	—	—	
122	1117	—	$\sim CaF_2$	§$CuSe_{0·333-0·355}$
153	[382]	—	CuS	
160, 258, 317, 360, 568	855	—	hexagonal	
—	—	—	ReO_3	
—	1030	—	hexagonal	
—	—	—	monoclinic	
212	—	—	—	
—	557	—	—	
—	—	—	—	
—	—	—	—	*from oxides
—	—	—	—	*from oxides
—	—	—	—	*from oxides
—	—	—	—	*from oxides
818	1197			
202, 522	1085	—	—	
—	[586]	—	—	
—	—	—	tetragonal.	*from oxides
—	566	—	f.c. o'rhomb.	
—	819	—	Table Aa	
1384	1409	2560	Mg	
—	651	—	—	
2200	2380	—	Sc_2O_3	
1470	1522	2860	Mg	
—	774	—	—	
2270	2400	—	Sc_2O_3	
—	—	—	—	
—	817	1600	b.c.c.	
—	623	—	monoclinic	
622, 2240	2320	—	—	
−227·6	−219·6	−188	—	
—	—	—	—	
911, 1392	1536	2860	Na	
—	—	—	—	
—	1100	1837	TiO_2	
—	Sb	930	rhombohed.	

Table A. Heats of formation,

Substance	ΔH°_{298} (kcal/mole)	Ref.	S°_{298} (cal/deg mole)	Ref.
⟨$FeCl_2$⟩	−81·8 ± 0·2	158, 551	28·2 ± 0·5	907
⟨$FeCl_3$⟩	−95·5 ± 0·2	907	34·0 ± 0·4	817
⟨$FeBr_2$⟩	−59·7 ± 0·5	918	33·4 ± 0·7	907
⟨$FeBr_3$⟩	−63·9 ± 2·0	158	*41·5 ± 3·0*	—
⟨FeI_2⟩	−27·8 ± 2·0	907	40·6 ± 0·8	907
⟨FeOCl⟩	−97·65 ± 2·0	406	18·3 ± 1·5	406
⟨$Fe_{0·95}O$⟩*	−63·2 ± 0·3	307	14·05 ± 0·2	307
⟨Fe_3O_4⟩	−266·9 ± 1·0	307	36·2 ± 0·6	307
⟨Fe_2O_3⟩	−196·3 ± 0·8	307	20·9 ± 0·1	503
⟨FeOOH⟩	−133·6	918	14·2 ± 1·2	—
⟨FeS⟩†	−24·0 ± 0·6	907	14·41 ± 0·05	907
⟨FeS_2⟩	−41·0 ± 2·0	854	12·65 ± 0·1	503
⟨$FeSO_4$⟩	−222·0 ± 3·0	817	28·91	817
⟨$Fe_2(SO_4)_3$⟩	(−617·35 ± 2·0)	—	73·5 ± 0·5	796
⟨$FeSe_{0·96}$⟩	−16·0 ± 1·5	854	16·54 ± 1·0	854
⟨$FeSe_{1·14}$⟩	−15·8 ± 0·8	854	20·95 ± 1·0	854
⟨Fe_3Se_4⟩§	−50·7 ± 0·7	854	66·9 ± 3·0	854
⟨$FeTe_{0·9}$⟩	−5·5 ± 0·9	854	19·15 ± 0·1	854
⟨$FeTe_2$⟩	−17·2 ± 1·1	854	23·95 ± 0·15	854
⟨Fe_4N⟩	−2·65 ± 1·0	907	37·2 ± 1·8	—
⟨Fe_3P⟩	−39·2 ± 2·0	307	24·3 ± 3·0	—
⟨Fe_2P⟩	−38·3 ± 2·0	307	17·3 ± 2·0	—
⟨FeP⟩	−33·0 ± 3·0	307	—	—
⟨FeP_2⟩	−52·8 ± 4·0	307	—	—
⟨Fe_3C⟩	+6·0 ± 1·0	99	25·0 ± 1·0	99
⟨$FeCO_3$⟩	−177·0 ± 1·8	918	22·2 ± 0·4	125
($Fe(CO)_5$)	−173·0 ± 2·0	907	106·4 ± 0·8	907
⟨FeSi⟩	−18·38 ± 1·2	769	10·0 ± 0·5	769
⟨Fe_2SiO_4⟩	−8·2 ± 1·5*	840	34·7 ± 0·5	125
⟨Fe_2B⟩	−17·0 ± 4·0	795	*13·54*	—
⟨FeB⟩	−17·0 ± 4·0	795	*6·62*	—
⟨FeTi⟩	−9·7 ± 0·4	308	*12·6 ± 2·1*	—
⟨$FeTiO_3$⟩	−7·1 ± 1·2*	—	25·3 ± 0·3	125
⟨$FeAl_2O_4$⟩	−6·65 ± 0·3*	768	25·5 ± 0·2	768
⟨Fe_2ZnO_4⟩	−1·2 ± 1·0*	840	32·2 ± 0·2	377
⟨$FeMoO_4$⟩	*−13·0 ± 4·5**	840	30·9 ± 0·5	840
⟨$FeWO_4$⟩	*−18·0 ± 3·0**	840	31·5 ± 0·5	840
⟨Fe_2NiO_4⟩	−5·4 ± 2·0*	840	30·1 ± 0·2	377
⟨Fe_2CoO_4⟩	−6·6 ± 0·6*	840	34·1 ± 2·0	840
⟨$FeCo_2O_4$⟩	—	—	30·0 ± 0·2	377
⟨$FeCr_2O_4$⟩	−12·45 ± 0·5*	812	35·1 ± 0·4	812
⟨FeV_2O_4⟩	−5·8 ± 0·5*	812	37·5 ± 0·7	812
⟨Fe_3Mo_2⟩	−1·0 ± 2·0	902	35·0 ± 2·0	902
⟨Ga⟩	0	—	9·8 ± 0·1	1, 125
(Ga)	68·3 ± 2·5	457	40·4 ± 0·2	99
⟨$GaCl_3$⟩	−125·4 ± 1·1	907	*32·3 ± 3·3*	—
⟨$GaBr_3$⟩	−92·4 ± 1·0	27	43·0 ± 2·8	—
⟨GaI_3⟩	−57·2 ± 2·9	907	*48·7 ± 4·0*	—
⟨Ga_2O_3⟩	−258·8 ±.1·2	310, 484, 22	20·23 ± 0·1	1
(Ga_2S)	+5·0 ± 8·0	854	69·3 ± 2·0	854
⟨GaS⟩	−50·0 ± 5·0	854	13·8 ± 1·5	854
⟨Ga_2S_3⟩	−122·8 ± 3·0	—	*33·4 ± 3·5*	—
⟨GaSe⟩	−38·5 ± 2·5	854	16·8 ± 0·5	854
⟨Ga_2Se_3⟩	−97·0 ± 4·0	—	*45·9 ± 3·7*	—
⟨$Ga_2(SeO_4)_3$⟩	−472·8 ± 3·6	906	(72·5)	—
⟨GaTe⟩	−29·5 ± 2·3	854	20·4 ± 0·15	854
⟨Ga_2Te_3⟩	−65·0 ± 2·6	80	53·2 ± 2·5	421

standard entropies and structures (contd.)

tr.pts. (°C)	m.pt. (°C)	b.pt. (°C)	Structure	Remarks
—	677	1012	$CdCl_2$	
—	308	315	AsI_3	
—	688	968	—	
377	691	968	AsI_3	
—	—	—	—	
—	—	—	—	
—	1378	—	NaCl	*$FeO_{1\cdot055-1\cdot19}$ (1400°C)
—	1597	—	spinel	
—	1594	—	—	
—	—	—	—	
138, 325	1195	—	NiAs	†$FeS_{1\cdot00-1\cdot14}$
—	743	—	Table Aa	
—	—	—	—	
—	—	—	—	
457	—	—	tetrag., PbO	
—	—	—	NiAs (hex.)	
—	—	—	monoclinic	§$FeSe_{1\cdot18-1\cdot35}$
—	927	6	—	
—	660	—	—	
—	—	—	Ag¶	¶N interstitial
—	1166	—	tetragonal	
—	1365	—	Table Aa	
—	—	—	CoP	
—	—	—	marcasite	
212	1225	—	orthorhomb.	
—	—	—	$MgCO_3$	
—	—	—	—	
—	1410	—	cubic	
—	1220	—	Mg_2SiO_4	*from oxides
—	1389	—	—	
—	1650	—	—	
—	1317	—	CsCl	
—	1200	—	Cr_2O_3	*from oxides
—	1780	—	spinel	*from oxides
—	1720	—	—	*from oxides
—	—	—	—	*from oxides
—	—	—		*from oxides
—	580	1650	spinel	*from oxides
—	—	—	spinel	*from oxides
—	—	—	spinel	
—	—	—	spinel	*from oxides
—	—	—	—	*from oxides
—	[1370]	—	—	
—	30	2420	Table Aa	
—	—	—	—	
—	78	302	—	
—	122	292	—	
—	213	—	—	
—	1795	—	Cr_2O_3	
—	—	—	—	
—	960	—	ZnS	
—	—	—	ZnS	
—	960	—	ZnS	
—	1020	—	~ZnS	
—	—	—	—	
—	835	—	ZnS	
670	792	—	~ZnS	

Table A. Heats of formation,

Substance	ΔH°_{298} (kcal/mole)	Ref.	S°_{298} (cal/deg mole)	Ref.
⟨GaN⟩	−26·2 ± 2·2	907	*7·1 ± 1·8*	—
⟨GaAs⟩	−19·5 ± 0·6	840	15·35 ± 0·1	907
⟨GaP⟩	−29·2 ± 2·0	908	12·5 ± 0·2	850
⟨GaSb⟩	−10·0 ± 0·3	611, 421	18·48 ± 0·3	591, 421
⟨Gd⟩	0	—	16·24 ± 0·2	99
⟨GdOCl⟩	−235·1 ± 0·9	793	*22·8 ± 2·2*	—
⟨$GdCl_3$⟩	−240·1 ± 1·0	332	34·9 ± 0·6	332
⟨GdI_3⟩	−142·0	318	*51·8*	—
⟨Gd_2O_3⟩	−434·0 ± 4·0	372	36·0 ± 0·1	536
⟨Ge⟩	0	—	7·43 ± 0·05	482, 591
(Ge)	+89·5 ± 0·5	99	40·10	99
(GeH_4)	+21·7 ± 0·5	907	51·9 ± 0·2	907
⟨GeF_2⟩	−157·35 ± 1·0	749	*20·1*	—
(GeF_4)	−284·87 ± 0·3	797	72·35 ± 0·2	907
{$GeCl_4$}	(−130·0)	198	—	—
($GeCl_4$)	−120·6	907	83 ± 0·3	907
($GeBr_4$)	−71·4 ± 2·2	907	94·7 ± 0·4	907
(GeI_4)	−9·0 ± 2·5	907	102·5 ± 0·5	907
(GeO)	−7·33	907	53·5 ± 0·05	907
⟨GeO_2⟩	−138·6 ± 0·3	754	9·5 ± 0·05	754
⟨GeS⟩	−18·2 ± 1·1	854	15·77 ± 0·24	854
(GeS)	+23·2 ± 2·0	854	56·26 ± 0·5	854
⟨GeS_2⟩	−37·5 ± 3·0	854	20·9 ± 0·5	854
⟨GeSe⟩	*−16·5 ± 2·5*	854	18·9 ± 0·3	854
⟨$GeSe_2$⟩	*−27·0 ± 5·0*	854	26·9 ± 1·0	854
⟨GeTe⟩	−7·8 ± 3·0	907	21·3 ± 0·3	854
⟨Ge_3N_4⟩	(−15·6)	264	(37·0)	—
⟨GeP⟩	−6·5 ± 2·5	232	14·6 ± 2·0	232
(H_2)	0	—	31·21 ± 0·02	125
(H)	+52·1 ± 0·5	198	27·4 ± 0·01	125
(HF)	−65·14 ± 0·35	817	41·51	918
(HCl)	−22·063	key value	44·643	key value
(HBr)	−8·695	key value	47·463	key value
(HI)	+6·3 ± 0·19	key value	49·35 ± 0·01	key value
{H_2O}	−68·315	key value	16·718	key value
(H_2O)	−57·795	key value	45·106	key value
{H_2O_2}	−44·87 ± 0·05	907	26·17 ± 0·12	907
(H_2O_2)	−32·59 ± 0·04	907	55·66 ± 0·12	907
(H_2S)	−4·9 ± 0·1	198	49·15 ± 0·1	198
{H_2S_2}	−4·3 ± 0·2	854	48·0 ± 3·0	854
(H_2S_2)	+3·76 ± 0·2	854	63·66	854
(H_2Se)	+7·0 ± 0·3	854	52·3 ± 0·03	854
(H_2Te)	+23·8 ± 0·2	854	54·7 ± 0·5	854
(HCN)	+32·2	907	48·20 ± 0·03	907
⟨Hf⟩	0	—	10·41 ± 0·05	99
(Hf)	+148·0 ± 1·0	99	44·65	99
⟨HfF_4⟩	−461·4 ± 2·5	500	*32·5*	—
⟨$HfCl_4$⟩	−236·7 ± 1·0	505	45·6 ± 0·6	243
⟨HfO_2⟩	−266·0 ± 3·0	97, 22	14·2 ± 0·1	243
⟨HfN⟩	−88·2 ± 0·6	97	12·1 ± 1·5	—
⟨HfC⟩	−52·3 ± 0·7	710	9·45 ± 0·1	99
⟨HfB_2⟩	−80·3 ± 2·0	918	10·2 ± 0·1	918

standard entropies and structures (contd.)

tr.pts. (°C)	m.pt. (°C)	b.pt. (°C)	Structure	Remarks
—	1500	—	wurtzite	
—	1238	—	—	
—	1520	—	—	
—	712	—	ZnS	
1260	1312	3270	Mg	
—	—	—	—	
—	609	—	—	
740	930	—	—	
—	2340	—	Sc_2O_3	
—	937	2834	diamond	
—	—	—	—	
—	—	—	—	
—	—	—	—	
—	—	−36	—	
—	−50	84	—	
—	—	—	—	
—	26	—	—	
—	146	—	—	
—	—	—	—	
1035	1116	—	tetragonal	
—	665	—	orthorhomb.	
—	—	—	—	
—	860	—	—	
—	675	—	—	
—	740	—	—	
—	724	—	—	
—	—	—	Be_2SiO_4	
—	—	—	—	
—	−259	−253	—	
—	—	—	—	
—	−83	19·5	—	
—	−114	−8·5	—	
—	−87	−67	—	
—	−51	−35	—	
—	0	100	—	
—	—	—	—	
—	−0·4	152	—	
—	—	—	—	
—	−86	−60	—	
—	−90	71	—	
—	—	—	—	
—	−66	−41	—	
—	—	—	—	
—	—	—	—	
1740	2230	4600	Mg	
—	—	—	—	
—	—	—	—	
—	—	316	—	
1700	2810	—	monoclinic	
—	3330	—	—	
—	3830	—	—	
—	3370	—	—	

Table A. Heats of formation,

Substance	ΔH°_{298} (kcal/mole)	Ref.	S°_{298} (cal/deg mole)	Ref.
$\{Hg\}$	0	—	18·14 ± 0·03	907
(Hg)	+14·676	99	41·8	99
$\langle HgF\rangle$	−58·7 ± 3·5	441	20·9 ± 2·0	—
$\langle HgF_2\rangle$	−95·0 ± 10·0	22	28·0 ± 5·0	22
$\langle HgCl\rangle$	−31·5 ± 0·3	22, 198	23·0 ± 0·5	125
$\langle HgCl_2\rangle$	−54·55 ± 0·3	907	33·47	907
$\langle HgBr\rangle$	−24·7 ± 0·3	198	26·0 ± 0·5	125
$\langle HgBr_2\rangle$	−40·6 ± 0·3	907	43·0	907
$\langle HgI\rangle$	14·4 ± 0·2	198	28·1	907
$\langle HgI_2\rangle$†	25·2 ± 0·4	198	43·0 ± 1·8	918
$\langle HgO\rangle_{red}$	21·7 ± 0·2	198	16·8 ± 0·1	125
$\langle HgS\rangle_{red}$	12·75 ± 1·0	854	19·7 ± 0·5	854
$\langle Hg_2SO_4\rangle$	−177·6 ± 3·0	918	48·0 ± 0·1	125, 588
$\langle HgSO_4\rangle$	−168·3 ± 3·0	918	*33·5 ± 2·0*	—
$\langle HgSe\rangle$	−10·35 ± 3·3	859	*24·1 ± 1·0*	854
$\langle HgSeO_3\rangle$	−87·3 ± 1·2	907	38·8 ± 8·0	907
$\langle HgTe\rangle$	−7·6 ± 1·0	854	27·0 ± 0·5	854
$\langle Ho\rangle$	0	—	17·93	99
$\langle HoF_3\rangle$	−408·0	918	*26·5 ± 2·8*	—
$\langle HoCl_3\rangle$	−240·3 ± 2·5	918	*35·1 ± 2·3*	—
$\langle Ho_2O_3\rangle$	449·6 ± 2·0	371	37·8 ± 0·1	536
$\langle I_2\rangle$	0	—	27·76 ± 0·02	key value
(I_2)	14·92 ± 0·02	key value	62·277	key value
(I)	25·52 ± 0·02	key value	43·182	key value
$\langle In\rangle$	0	—	13·82 ± 0·2	99
(In)	58·0 ± 0·3	99	41·50	99
$\langle InCl\rangle$	−44·5 ± 2·0	907	22·7 ± 1·5	907
$\langle InCl_2\rangle$	−86·7 ± 4·0	907	29·2 ± 4·0	907
$\langle InCl_3\rangle$	−128·4 ± 2·0	907	*33·7 ± 2·0*	—
$\langle InBr\rangle$	−41·9 ± 2·0	907	27·3 ± 2·2	—
$(InBr)$	−12·3 ± 2·5	(907)	62·03 ± 0·03	907
$\langle InBr_3\rangle$	−98·2 ± 2·0	907	*42·7 ± 3·3*	—
$\langle InI\rangle$	−27·8 ± 2·0	414	31·1 ± 2·0	—
$\langle InI_3\rangle$	−56·1 ± 3·0	907	*48·6 ± 2·5*	—
$\langle In_2O_3\rangle$	−221·3 ± 0·4	907	25·8 ± 0·8	907
$\langle InS\rangle$	−32·0 ± 3·0	854	*16·5 ± 1·5*	854
(InS)	+55·7 ± 3·0	854	60·15	854
$\langle In_5S_6\rangle$	−185·0 ± 15·0	854	*89·5 ± 10·0*	854
$\langle In_2S_3\rangle$	−85·0 ± 5·0	854	39·1 ± 0·6	854
$\langle In_2(SO_4)_3\rangle$	−651·4 ± 0·9	907	72·2 ± 0·5	907
$\langle InSe\rangle$	−28·2 ± 3·0	854	19·5 ± 1·0	854
$\langle In_2Se_3\rangle$	−78·0 ± 4·0	854	*48·1 ± 4·0*	854
$\langle In_2Te\rangle$	−19·0 ± 0·4	725	*37·5 ± 2·5*	—
$\langle InTe\rangle$	−17·2 ± 0·5	725	25·26 ± 0·05	854
$\langle In_2Te_3\rangle$	−45·8 ± 0·7	725	*57·0 ± 2·0*	—
$\langle In_2Te_5\rangle$	−45·8 ± 0·6	725	—	—
$\langle InP\rangle$	−18·0 ±.2·0	908	14·28 ± 0·1	—
$\langle InN\rangle$	−33·0 ± 2·5	915	*10·4 ± 1·5*	—
$\langle InAs\rangle$	−13·8 ± 0·5	840	17·85 ± 0·3	840
$\langle InSb\rangle$	−7·44	—	20·95	—
$\langle Ir\rangle$	0	—	8·49 ± 0·05	99
(Ir)	+160·0 ± 1·5	99	46·25	99
(IrF_6)	−130·0 ± 20·0	907	85·5 ± 0·4	907
$\langle IrCl_3\rangle$	−60·8 ± 4·0	757	27·45 ± 4·0	757

standard entropies and structures (*contd.*)

tr.pts. (°C)	m.pt. (°C)	b.pt. (°C)	Structure	Remarks
—	−38·86	357	Table Aa	
—	—	—	—	
—	570	—	—	
—	645	(650)	—	
—	543	—	tetragonal	
155	280	302	Orthorhom.	
—	407	—	HgCl	
—	238	—	—	
—	—	—	HgCl	
96	256	353	Table Aa	† red
—	—	—	—	
345	825	—	hexagonal	
—	—	—	—	
—	—	—	—	
—	770	—	ZnS	
360	—	—	—	
670	—	—	ZnS	
1428	1470	2700	Mg	
—	1070	1143	—	
—	720	—	—	
2270	2390	—	Sc_2O_3	
—	114	183	Table Aa	
—	—	—	—	
—	—	—	—	
—	157	2073	—	
—	—	—	—	
120	225	591	—	
190	236	(523)	—	
—	583	—	—	
—	285	—	InI	
—	—	—	—	
—	420	—	—	
—	365	(986)	orthorhomb.	
—	207	447	—	
1250	1910	—	Sc_2O_3	
—	[680]	—	—	
—	—	—	—	
—	—	—	—	
387, 827	1090	—	—	
670	—	—	—	
—	660	—	—	
197, 650, 750	890	—	—	
—	—	—	—	
—	695	—	—	
610	670	—	—	
—	—	—	—	
637	1055	—	—	
—	[1200]	—	wurtzite	
—	942	—	—	
—	525	—	ZnS	
—	2443	4430	Ag	
—	—	—	—	
—	—	—	—	
—	—	—	—	

Table A. Heats of formation,

Substance	ΔH°_{298} (kcal/mole)	Ref.	S°_{298} (cal/deg mole)	Ref.
$\langle IrBr_3 \rangle$	(−33·1)	831	*(42·0)*	—
$\langle IrI \rangle$	−11·0 ± 3·0	22	26·0 ± 5·0	22
$\langle IrI_2 \rangle$	−20·0 ± 3·0	22	38·0 ± 5·0	22
$\langle IrO_2 \rangle$	−57·7 ± 3·0	757	13·5 ± 1·5	757
(IrO_3)	+3·2 ± 4·0	—	69·4 ± 2·0	—
$\langle Ir_2S_3 \rangle$	−58·4 ± 1·8	907	23·2	907
$\langle IrS_2 \rangle$	−34·6 ± 1·5	907	14·7	907
$\langle K \rangle$	0	—	15·46 ± 0·05	key value
(K)	+21·3 ± 0·2	817	38·30	key value
(K_2)	+30·37 ± 0·5	817	59·67	817
$\langle KH \rangle$	−13·8 ± 1·2	702	12·3 ± 0·4	702
$\langle KF \rangle$	−135·9 ± 0·1	817	15·9 ± 0·1	269
$\langle KCl \rangle$	−104·37 ± 0·2	817	19·73 ± 0·1	817
$\langle KBr \rangle$	−94·12 ± 0·4	817	22·93 ± 0·1	817
$\langle KI \rangle$	−78·37 ± 0·3	817	25·43 ± 0·1	817
$\langle K_2O \rangle$	−86·8 ± 0·8	817	*22·5 ± 1·5*	—
$\langle KO_2 \rangle$	−68·0 ± 0·8	817	29·3 ± 1·2	817
$\langle KOH \rangle$	−101·5 ± 0·4	817	18·85 ± 0·2	817
$\langle K_2S \rangle$	−102·4 ± 3·5	198	—	—
$\langle K_2SO_3 \rangle$	−272·0 ± 5·0	879	*38·4 ± 4·0*	879
$\langle K_2SO_4 \rangle$	−342·6 ± 1·0	198	42·0 ± 0·5	125
$\langle KAl(SO_4)_2 \rangle$	−589·2 ± 4·0	198	48·9 ± 0·3	125
$\langle K_2Se \rangle$	−89·0 ± 6·0	—	—	—
$\langle K_2HPO_4 \rangle$	−424·0	830	42·0	830
$\langle KH_2PO_4 \rangle$	−373·4	830	32·15	830
$\langle K_3Sb \rangle$	−45·0 ± 8·0	—	—	—
$\langle K_3Bi \rangle$	−54·1 ± 4·0	99	*47·3 ± 4·0*	—
$\langle K_2CO_3 \rangle$	−274·9 ± 1·5	817	37·17	817
$\langle K_2Si_4O_9 \rangle$	−73·9 ± 3·0*	840	*63·5 ± 3·0*	840
$\langle K_2Si_2O_5 \rangle$	−71·5 ± 3·0*	840	*43·5 ± 2·0*	840
$\langle K_2SiO_3 \rangle$	−65·5 ± 3·0*	840	34·9 ± 0·5	840
$\langle K_2CrO_4 \rangle$	−331·2 ± 1·2	803, 896	47·8 ± 0·8	124, 125
$\langle KTl \rangle$	−13·5 ± 2·5	—	—	—
$\langle KHg \rangle$	−13·4 ± 1·0	264	—	—
$\langle KHg_2 \rangle$	−18·5 ± 1·5	264	—	—
$\langle KHg_3 \rangle$	−20·0 ± 2·0	264	—	—
$\langle KHg_4 \rangle$	−21·5 ± 2·0	264	—	—
$\langle KHg_8 \rangle$	−25·0 ± 2·5	264	—	—
$\langle La \rangle$	0	—	13·6 ± 0·5	182, 125
(La)	+103·0	99	43·56	99
$\langle LaH_2 \rangle$	−45·2	918	12·35 ± 0·07	759
$\langle LaF_3 \rangle$	*−426·0*	784	*27·1*	—
$\langle LaCl_3 \rangle$	−255·9 ± 0·8	332	(34·5)	332
$\langle LaBr_3 \rangle$	*−208·0*	784	*43·5*	—
$\langle LaI_3 \rangle$	−157·0 ± 2·5	88	*51·3*	—
$\langle La_2O_3 \rangle$	−428·6 ± 0·5	682	30·6 ± 0·1	494
$\langle LaOCl \rangle$	−243·8 ± 2·0	793	*19·8 ± 2·5*	—
$\langle La_2S_3 \rangle$	−292·0 ± 16·0	854	39·43 ± 0·1	—
$\langle La_2Se_3 \rangle$	−223·0 ± 5·0	854	48·33 ± 0·2	854
$\langle La_2Te_3 \rangle$	−187·5 ± 6·1	854	53·36 ± 0·1	854
$\langle LaN \rangle$	−71·5 ± 4·0	264	*10·6 ± 1·1*	—
$\langle LaAl_2 \rangle$	−36·0 ± 9·0	264	23·60 ± 0·05	777
$\langle LaMg \rangle$	−4·3 ± 1·2	866	22·35 ± 1·5	866
$\langle Li \rangle$	0	—	6·95 ± 0·1	568
(Li)	+38·5 ± 2·0	293	33·15 ± 0·1	293

standard entropies and structures (contd.)

tr.pts. (°C)	m.pt. (°C)	b.pt. (°C)	Structure	Remarks
—	—	—	—	
—	—	—	—	
—	—	—	—	
—	—	—	TiO_2	
—	—	—	—	
—	—	—	—	
—	—	—	—	
—	63·2	770	Na	
—	—	—	—	
—	—	—	—	
—	—	—	—	
—	858	1517	NaCl	
—	772	1437	NaCl	
—	734	1398	NaCl	
—	681	1345	NaCl	
—	—	—	—	
—	509	—	tetragonal	
243	406	1327	—	
146	835	—	CaF_2	
—	*700*	—	—	
586	1069	—	orthorhomb.	
—	—	—	—	
—	—	—	CaF_2	
—	—	—	—	
—	—	—	—	
—	812	—	Na_3Bi	
—	675	—	Na_3Bi	
—	901	—	—	
592	770	—	—	* from oxides
—	1045	—	—	* from oxides
—	976	—	orthorhomb.	* from oxides
666	971	—	—	
—	335	—	—	
—	180	—	triclinic	
—	275	—	orthorhomb.	
—	204	—	—	
—	173	—	—	
—	(70)	—	—	
280, 861	920	3460	Mg	
—	—	—	—	
—	—	—	—	
—	1493	—	—	
—	855	1812	UCl_3	
—	788	—	—	
—	761	1472	$PuBr_3$	
2040, 2210	2320	—	Table Aa	
—	—	—	—	
—	2130	—	Ce_2S_3	
—	1630	—	—	
—	1490	—	—	
—	2450	—	—	
—	1924	—	NaCl	
—	—	—	$MgCu_2$	
—	—	—	Na	
—	181	1324	—	

Table A. Heats of formation,

Substance	ΔH°_{298} (kcal/mole)	Ref.	S°_{298} (cal/deg mole)	Ref.
(Li_2)	+50·5 ± 3·0	293	47·05 ± 0·2	293
$\langle LiH\rangle$	−21·5 ± 0·2	386, 138	4·8 ± 0·1	702
$\langle LiF\rangle$	−147·45 ± 1·5	817	8·5 ± 0·05	36
$\langle LiCl\rangle$	−96·9 ± 2·0	198, 225	14·17 ± 0·1	620
$\langle LiBr\rangle$	−83·4 ± 2·0	198	17·0 ± 2·0	22
$\langle LiI\rangle$	−64·8 ± 2·0	198	*20·5 ± 2·0*	—
$\langle Li_2O\rangle$	−142·6 ± 1·0	198, 104, 552	9·06 ± 0·05	104
$\langle Li_2O_2\rangle$	−151·8 ± 2·0	138	*13·5*	—
$\langle LiOH\rangle$	−116·6 ± 0·3	198, 104	10·2 ± 0·1	11
$\langle Li_2S\rangle$	−106·7 ± 2·0	854	*14·5*	—
$\langle Li_2SO_4\rangle$	−342·8 ± 2·5	138	29·0 ± 6·0	155
$\langle Li_2Se\rangle$	−96·0 ± 9·0	854	*17·0 ± 3·0*	854
$\langle Li_2Te\rangle$	−85·0 ± 5·0	854	*18·5 ± 3·0*	854
$\langle Li_3N\rangle$	−47·0 ± 0·5	264	—	—
$\langle Li_3Sb\rangle$	−43·0 ± 3·0	264	—	—
$\langle Li_3Bi\rangle$	−55·5 ± 4·0	264	—	—
$\langle Li_2CO_3\rangle$	−290·8 ± 2·0	198	21·55	817
$\langle Li_2SiO_3\rangle$	−33·4 ± 2·0*	840	19·2 ± 0·3	840
$\langle Li_2TiO_3\rangle$	−30·9 ± 2·5*	721	21·9 ± 0·1	303
$\langle LiAl\rangle$	*−11·7 ± 2·5*	—	*11·2 ± 1·5*	—
$\langle LiAlO_2\rangle$	−12·9 ± 1·5*	359	12·7 ± 0·1	398
$\langle LiTl\rangle$	−12·8 ± 1·0	264	—	—
$\langle LiFeO_2\rangle$	−4·25 ± 0·5	353	18·0 ± 0·2	398
$\langle Li_2BeF_4\rangle$	−544·25 ± 1·9	755	*29·8 ± 2·0*	817
$\langle LiBeF_3\rangle$	−394·55 ± 2·0	755	*21·3 ± 2·0*	817
$\langle Lu\rangle$	0	—	12·18 ± 0·2	99
$\langle Lu_2O_3\rangle$	−449·8 ± 3·0	371	26·3 ± 0·2	—
$\langle Mg\rangle$	0	—	7·81 ± 0·02	99
(Mg)	+35·0 ± 0·3	99	35·50	99
$\langle MgH_2\rangle$	−18·0 ± 0·7	918	7·43 ± 1·0	918
$\langle MgF_2\rangle$	−268·5 ± 1·5	918	13·68 ± 0·1	918
$\langle MgCl_2\rangle$	−153·3 ± 0·2	918	21·42 ± 0·2	918
$\langle MgBr_2\rangle$	−125·3 ± 1·2	918	28·0	918
$\langle MgI_2\rangle$	−87·0 ± 1·2	918	31·0 ± 2·5	—
$\langle MgO\rangle$	−143·7 ± 0·2	817	6·44 ± 0·1	817
$\langle Mg(OH)_2\rangle$	8·85 ± 0·2*	772	15·1 ± 0·05	828
$\langle MgS\rangle$	−84·0 ± 0·8	854	12·03 ± 0·2	854
$\langle MgSO_4\rangle$	−307·1 ± 3·0	918	21·9 ± 0·2	828
$\langle MgSe\rangle$	−65·2 ± 4·0	—	*15·0 ± 2·0*	—
$\langle MgSeO_3\rangle$	−215·15	918	*22·9 ± 3·2*	—
$\langle MgTe\rangle$	−50·0 ± 5·0	264	*17·8 ± 1·4*	—
$\langle Mg_3N_2\rangle$	−110·3 ± 1·7	892	22·4 ± 1·8	892
$\langle Mg_3P_2\rangle$	*−111·0 ± 20·0*	—	*18·5 ± 6·0*	—
$\langle Mg_3(PO_4)_2\rangle$	−903·6	918	45·22	918
$\langle Mg(NO_3)_2\rangle$	−189·0	918	39·2	918
$\langle Mg_3Sb_2\rangle$	−71·7 ± 4·0	—	32·65 ± 3·5	—
$\langle Mg_3Bi_2\rangle$	−30·3	—	45·85	—
$\langle Mg_2C_3\rangle$	+19·0 ±.8·0	101	*14·0 ± 2·5*	—
$\langle MgC_2\rangle$	+21·0 ± 5·0	794	*14·0 ± 2·5*	—
$\langle MgCO_3\rangle$	−265·7 ± 3·5	860	15·74 ± 0·2	860
$\langle Mg_2Si\rangle$	−18·9 ± 1·2	840	15·25 ± 1·7	840
$\langle Mg_2SiO_4\rangle$	−15·0 ± 0·5*	840	22·75 ± 0·2	828
$\langle MgSiO_3\rangle$	−8·7 ± 0·7*	840	16·2 ± 0·3	828
$\langle Mg_2Ge\rangle$	−27·53 ± 0·4	840	17·43 ± 0·45	840
$\langle Mg_2Sn\rangle$	−19·25 ± 1·3	840	24·25 ± 0·04	840
$\langle Mg_2Pb\rangle$	−11·5 ± 0·7	—	28·5 ± 3·0	—

standard entropies and structures (contd.)

tr.pts. (°C)	m.pt. (°C)	b.pt. (°C)	Structure	Remarks
—	—	—	—	
—	688	—	LiCl	
—	848	1681	NaCl	
—	610	1382	NaCl	
—	550	1310	NaCl	
—	469	1170	NaCl	
—	> 1700	—	CaF_2	
225	dec.	—	—	
—	473	770	SnO	
—	950		CaF_2	
575	859	—	—	
—	1102	—	CaF_2	
—	—	—	—	
—	dec.	—	hexagonal	
—	(960)	—	Na_3Bi	
—	1145	—	BiF_3	
—	726	—	—	
—	1201	—	—	*from oxides
—	1077	—	—	*from oxides
—	720	—	NaTl	
—	1430	—	—	
	508	—	CsCl	
600	—	—	tetragonal	*from oxides
—	472	—	—	
—		—	—	
1600	1663	3395	—	
—	—	—	Sc_2O_3	
—	649	1090	Table Aa	
—	—	—	—	
—	—	—	—	
—	1263	2332	rutile	
—	714	1418	$CdCl_2$	
—	711	—	CdI_2	
—	650	—	CdI_2	
—	2825	3260	NaCl	
—	—	—	CdI_2	*from $\langle MgO \rangle + \{H_2O\}$
—	—	—	NaCl	
1010	1130	—	—	
—	—	—	NaCl	
—	—	—	—	
—	—	—	wurtzite	
550, 788	—	—	Sc_2O_3	
—	—	—	Sc_2O_3	
—	1348	—	—	
—	—	—	—	
925	1245	—	La_2O_3	
702	821	—	La_2O_3	
—	—	—	—	
—	—	—	ThC_2	
—	diss.	—	rhombohed.	
—	1077	—	CaF_2	
—	1890	—	orthorhomb.	*from oxides
—	1580	—	orthorhomb.	*from oxides
—	1115	—	—	
—	775	—	CaF_2	
—	549	—	CaF_2	

Table A. Heats of formation,

Substance	ΔH°_{298} (kcal/mole)	Ref.	S°_{298} (cal/deg mole)	Ref.
$\langle Mg_2TiO_4 \rangle$	—	—	24·8 ± 0·2	243
$\langle MgTiO_3 \rangle$	−5·5 ± 0·5*	—	17·8 ± 0·2	828
$\langle MgTi_2O_5 \rangle$	—	—	30·4 ± 0·2	243
$\langle MgB_2 \rangle$	−22·0	817	8·6 ± 0·05	417
$\langle MgB_4 \rangle$	−25·1	817	12·4 ± 0·1	417
$\langle MgAl_2O_4 \rangle$	−8·5 ± 0·4*	840	19·25 ± 0·1	304
$\langle MgCe \rangle$	−3·85 ± 0·8	866	25·25 ± 0·9	866
$\langle MgZn_2 \rangle$	−7·8 ± 1·0	548	—	—
$\langle MgZn \rangle$	−4·0 ± 0·3	548	—	—
$\langle MgNi_2 \rangle$	−13·5 ± 0·5	548	21·2 ± 0·1	548
$\langle Mg_2Ni \rangle$	−12·4 ± 0·5	548	22·7 ± 1·5	—
$\langle Mg_2Ca \rangle$	−9·7 ± 0·5	548	19·5 ± 2·0	548
$\langle MgFe_2O_4 \rangle$	−3·7 ± 0·4*	840	29·6 ± 0·5	846
$\langle MgMn_2O_4 \rangle$	−2·7 ± 1·0*	840	—	—
$\langle MgCr_2O_4 \rangle$	−10·0 ± 1·0*	840	25·3 ± 0·3	840
$\langle MgMoO_4 \rangle$	−12·9 ± 1·0*	840	28·4 ± 0·2	840
$\langle MgWO_4 \rangle$	−17·5 ± 0·5*	840	24·1 ± 0·2	840
$\langle MgV_2O_7 \rangle$	−19·55 ± 0·6*	711	47·8 ± 0·9	637
$\langle MgV_2O_6 \rangle$	−11·75 ± 0·4*	711	38·4 ± 0·7	637
$\langle Mn \rangle$	0	—	7·65 ± 0·02	99
(Mn)	+67·7 ± 1·0	99	41·49	99
$\langle MnF_2 \rangle$	(−192)	—	22·05 ± 0·3	918
$\langle MnF_3 \rangle$	(−240)	—	*25·0 ± 2·5*	—
$\langle MnCl_2 \rangle$	−115·2 ± 0·5	551	28·25 ± 0·1	453
$\langle MnBr_2 \rangle$	−92·0 ± 2·0	918	*33·0 ± 2·5*	—
$\langle MnI_2 \rangle$	−58·0 ± 3·0	399, 772	36·0 ± 2·5	—
$\langle MnO \rangle$†	−92·0 ± 0·5	876	14·3 ± 0·2	244, 828
$\langle Mn_3O_4 \rangle$‡	−331·4 ± 1·0	876	36·8 ± 0·8	719
$\langle Mn_2O_3 \rangle$§	−228·7 ± 1·2	876	26·4 ± 0·5	303
$\langle MnO_2 \rangle$¶	−124·3 ± 0·5	876	12·7 ± 0·1	828
$\langle Mn_2O_7 \rangle$	−174·1 ± 2·5	756	—	—
$\langle MnS \rangle_{green}$	−51·0 ± 0·5	854	19·2 ± 0·2	854
$\langle MnS_2 \rangle$	−53·5 ± 2·0	854	23·88 ± 0·1	854
$\langle MnSO_4 \rangle$	−254·6 ± 1·0	918	26·8 ± 0·3	125
$\langle MnSe \rangle$	−37·0 ± 2·5	854	21·7 ± 0·5	125
$\langle MnTe \rangle$	−26·6 ± 2·0	714	22·4 ± 0·5	125
$\langle MnTe_2 \rangle$	−30·0 ± 9·0	854	34·66 ± 0·1	854
$\langle Mn_{2\cdot5}N \rangle$†	−24·1 ± 0·8	564	—	—
$\langle Mn_4N \rangle$‡	−30·5 ± 0·8	564, 307	—	—
$\langle MnP \rangle$	−23·0 ± 4·0	327	*12·5 ± 2·2*	—
$\langle MnP_3 \rangle$	−41·6 ± 7·5	—	23·1 ± 4·5	—
$\langle MnAs \rangle$	−13·7 ± 4·0	327	—	—
$\langle Mn_2Sb \rangle$	−7·8 ± 0·8	782	32·7 ± 1·5	782
$\langle MnSb \rangle$	−6·5 ± 0·6	782	22·1 ± 1·0	782
$\langle MnBi \rangle$	−4·7 ± 4·0	327	—	—
$\langle Mn_7C_3 \rangle$	−26·5 ± 2·5	794	57·1 ± 1·0	794
$\langle MnCO_3 \rangle$	−213·7 ± 1·2	772	20·5 ± 0·3	828
$\langle Mn_3Si \rangle$	−19·0 ± 3·0	840	24·8 ± 0·3	707
$\langle Mn_5Si_3 \rangle$	−48·0 ± 9·0	840	57·0 ± 0·8	840
$\langle MnSi \rangle$	−14·5 ± 3·0	840	11·1 ± 0·3	840
$\langle Mn_4Si_7 \rangle$	−73·7 ± 15·0	840	54·5 ± 1·5	840
$\langle Mn_2SiO_4 \rangle$	−11·8 ± 0·7*	301	24·0 ± 3·0	840
$\langle MnSiO_3 \rangle$	−5·9 ± 0·4*	131	24·5 ± 0·5	840
$\langle MnFe_2O_4 \rangle$	−4·8 ± 1·0	681	—	—
$\langle MnSn_2 \rangle$	−6·55 ± 0·7	845	31·28 ± 0·9	845
$\langle MnB \rangle$	−18·0	—	7·75	—
$\langle MnB_2 \rangle$	−22·5	—	8·25	—

standard entropies and structures (contd.)

tr.pts. (°C)	m.pt. (°C)	b.pt. (°C)	Structure	Remarks
—	1830	—	spinel	
—	1840	—	Cr_2O_3	*from oxides
—	—	—	Ti_3O_5	
—	—	—	—	
—	—	—	—	
—	—	—	spinel	*from oxides
—	740	—	CsCl	
—	590	—	Table Aa	
—	—	—	—	
—	1145	—	Table Aa	
—	[760]	—	hexagonal	
—	—	—	—	
—	(1700)	—	—	*from oxides
—	—	—	—	*from oxides
—	2350	—	spinel	*from oxides
—	—	—	—	*from oxides
—	1358	—	—	*from oxides
—	—	—	—	*from oxides
—	710	—	—	*from oxides
710, 1090, 1136	1244	—	Table Aa	
—	—	—	—	
—	856	—	rutile	
—	—	—	monoclinic	
—	650	1231	$CdCl_2$	
—	—	—	CdI_2	
—	638	—	CdI_2	
−56	1875	—	NaCl	†$MnO_{1\cdot0-1\cdot12}$
1172	1565	—	tetragonal	‡$MnO_{1\cdot33-1\cdot41}$
—	—	—	Sc_2O_3	§$MnO_{1\cdot15-1\cdot6}$
250	—	—	orthorhomb.	¶$MnO_{1\cdot96-2\cdot0}$
—	—	—	—	
—	1530	—	various	
—	—	—	pyrites	
—	—	—	—	
−157	—	—	various	
—	1153	—	NiAs	
—	—	—	—	
—	—	—	—	†$MnN_{0\cdot4-0\cdot5}$
—	—	—	—	‡$MnN_{0\cdot2-0\cdot25}$
—	1147	—	CoP	
		—	—	
	935		NiAs	
—	948	—	—	
—	—	—	NiAs	
—	—	—	NiAs	
—	—	—	—	
—	—	—	$MgCO_3$	
—	[1070]	—	Cubic DO_3	
—	1300	—	hexagonal $D8_8$	
—	1275	—	cubic B20	
—	—	—	tetragonal	
—	1345	—	Mg_2SiO_4	*from oxides
—	[1291]	—	triclinic	*from oxides
—	—	—	—	
—	—	—	—	
—	—	—	—	
—	—	—	—	

Table A. Heats of formation,

Substance	ΔH°_{298} (kcal/mole)	Ref.	S°_{298} (cal/deg mole)	Ref.
⟨$MnAl_2O_4$⟩	$-9{\cdot}0 \pm 1{\cdot}0$*	—	$24{\cdot}8 \pm 0{\cdot}5$	—
⟨$MnCr_2O_4$⟩	$-16{\cdot}5 \pm 1{\cdot}0$*	—	$32{\cdot}0 \pm 0{\cdot}5$	—
⟨$MnMoO_4$⟩	$-14{\cdot}6 \pm 2{\cdot}5$*	840	$32{\cdot}5 \pm 2{\cdot}2$	—
⟨Mo⟩	0	—	$6{\cdot}85 \pm 0{\cdot}05$	99
(Mo)	$157{\cdot}3 \pm 1{\cdot}3$	99	43·46	99
(MoF_6)	$-372{\cdot}4 \pm 0{\cdot}3$	817	$83{\cdot}76 \pm 0{\cdot}5$	817
⟨$MoCl_5$⟩	$-126{\cdot}0 \pm 2{\cdot}0$	817	*57·0 ± 3·0*	817
($MoCl_5$)	$-107{\cdot}0 \pm 1{\cdot}3$	817	*95·0 ± 3·0*	817
⟨$MoCl_4$⟩	$-114{\cdot}0 \pm 2{\cdot}0$	817	*53·5 ± 3·0*	817
($MoCl_4$)	$-92{\cdot}0 \pm 3{\cdot}0$	817	*88·9*	817
⟨MoO_2⟩	$-140{\cdot}5 \pm 0{\cdot}4$	817	$11{\cdot}95 \pm 0{\cdot}3$	817
⟨MoO_3⟩	$-178{\cdot}1 \pm 0{\cdot}2$	817	$18{\cdot}59 \pm 0{\cdot}15$	817
(MoO_3)	$-86{\cdot}2 \pm 5{\cdot}0$	817	$66{\cdot}9 \pm 4{\cdot}0$	817
(MoO_2Cl_2)	$-151{\cdot}3 \pm 3{\cdot}5$	817	$80{\cdot}7 \pm 1{\cdot}3$	817
⟨Mo_2S_3⟩	$-98{\cdot}0 \pm 3{\cdot}0$	854	$27{\cdot}0 \pm 3{\cdot}0$	854
⟨MoS_2⟩	$-65{\cdot}8 \pm 1{\cdot}4$	854	$14{\cdot}96 \pm 0{\cdot}25$	854
⟨Mo_2N⟩†	$-16{\cdot}6 \pm 0{\cdot}5$	307	(21·0)	307
⟨Mo_2C⟩‡	$-11{\cdot}0 \pm 0{\cdot}7$	722	$15{\cdot}7 \pm 0{\cdot}2$	722
⟨$Mo(CO)_6$⟩	$-235{\cdot}3 \pm 2{\cdot}5$	358, 416	$78{\cdot}2 \pm 0{\cdot}3$	341
⟨Mo_3Si⟩	$-27{\cdot}8 \pm 2{\cdot}3$	840	$25{\cdot}45 \pm 0{\cdot}2$	—
⟨Mo_5Si_3⟩	$-74{\cdot}1 \pm 5{\cdot}8$	840	$49{\cdot}7 \pm 2{\cdot}2$	840
⟨$MoSi_2$⟩	$-31{\cdot}5 \pm 2{\cdot}3$	840	$15{\cdot}55 \pm 0{\cdot}9$	840
(N_2)	0	—	$45{\cdot}77 \pm 0{\cdot}02$	key value
(N)	112·97	key value	36·613	key value
(NH_3)	$-10{\cdot}98 \pm 0{\cdot}1$	key value	$46{\cdot}05 \pm 0{\cdot}02$	key value
(N_2O)	$19{\cdot}61 \pm 0{\cdot}1$	817	$52{\cdot}55 \pm 0{\cdot}03$	125
(NO)	$21{\cdot}58 \pm 0{\cdot}06$	817	$50{\cdot}35 \pm 0{\cdot}03$	125
(N_2O_3)	$19{\cdot}8 \pm 0{\cdot}2$	817	$73{\cdot}9 \pm 0{\cdot}5$	817
(NO_2)	$7{\cdot}91 \pm 0{\cdot}2$	817	$57{\cdot}34 \pm 0{\cdot}03$	817
(N_2O_4)	$2{\cdot}17 \pm 0{\cdot}4$	817	$72{\cdot}72 \pm 0{\cdot}2$	817
(N_2O_5)	$2{\cdot}7 \pm 0{\cdot}3$	817	$82{\cdot}8 \pm 1{\cdot}0$	817
⟨NH_4F⟩	$-110{\cdot}8 \pm 0{\cdot}6$	907	$17{\cdot}2 \pm 0{\cdot}1$	707
⟨NH_4Cl⟩	$-75{\cdot}1 \pm 0{\cdot}2$	907	$22{\cdot}9 \pm 0{\cdot}5$	907
⟨NH_4Br⟩	$-56{\cdot}56 \pm 0{\cdot}2$	907	$26{\cdot}97 \pm 0{\cdot}1$	—
⟨NH_4I⟩	−48·04	907	—	—
⟨$(NH_4)_2SO_4$⟩	$-282{\cdot}1 \pm 0{\cdot}18$	907	$52{\cdot}6 \pm 0{\cdot}3$	907
⟨Na⟩	0	—	$12{\cdot}25 \pm 0{\cdot}05$	568, 186, 828
(Na)	$26{\cdot}0 \pm 0{\cdot}6$	293	$36{\cdot}7 \pm 0{\cdot}1$	293
(Na_2)	$33{\cdot}7 \pm 1{\cdot}0$	293	$55{\cdot}0 \pm 0{\cdot}2$	293
⟨NaH⟩	$-13{\cdot}5 \pm 0{\cdot}2$	702	$9{\cdot}55 \pm 0{\cdot}1$	702
⟨NaF⟩	$-137{\cdot}52 \pm 0{\cdot}2$	817	$12{\cdot}24 \pm 0{\cdot}02$	817
(NaF)	$-69{\cdot}42 \pm 0{\cdot}5$	817	51·98	817
⟨NaCl⟩	$-98{\cdot}62 \pm 0{\cdot}1$	817	$17{\cdot}24 \pm 0{\cdot}05$	817
(NaCl)	$-43{\cdot}36 \pm 0{\cdot}5$	817	54·9	817
⟨NaBr⟩	$-86{\cdot}45 \pm 0{\cdot}3$	817	$20{\cdot}75 \pm 0{\cdot}06$	817
(NaBr)	$-34{\cdot}4 \pm 0{\cdot}5$	817	57·63	817
⟨NaI⟩	$-68{\cdot}8 \pm 0{\cdot}2$	817	23·5	817
⟨$NaClO_4$⟩	−91·48	817	34·0	817
⟨Na_2O⟩	$-99{\cdot}2 \pm 0{\cdot}7$	817	17·94	817
(NaO)	$+25{\cdot}2 \pm 5{\cdot}0$	867	$54{\cdot}0 \pm 1{\cdot}0$	867
⟨Na_2O_2⟩	$-123{\cdot}0 \pm 1{\cdot}5$	817	$22{\cdot}6 \pm 0{\cdot}3$	243
⟨NaO_2⟩	$-62{\cdot}5 \pm 1{\cdot}0$	—	$27{\cdot}7 \pm 0{\cdot}3$	243
⟨NaOH⟩	$-101{\cdot}8 \pm 0{\cdot}2$	817 (1974)	$15{\cdot}4 \pm 0{\cdot}2$	580
(NaOH)	$-47{\cdot}3 \pm 3{\cdot}0$	817 (1974)	$54{\cdot}57 \pm 0{\cdot}1$	817 (1974)
⟨Na_2S⟩	$-89{\cdot}5 \pm 3{\cdot}0$	854	$19{\cdot}0 \pm 3{\cdot}0$	854

standard entropies and structures (contd.)

tr.pts. (°C)	m.pt. (°C)	b.pt. (°C)	Structure	Remarks
—	1850	—	—	* from oxides (MnO, Al_2O_3)
—	—	—	—	* from oxides
—	—	—	—	* from oxides
—	2620	4600	Na	
—	—	—	—	
—	7·5	34	—	
—	194	264	$NbCl_5$	
—	—	—	—	
—	317	410	—	
—	—	—	—	
—	—	—	monoclinic	
—	800	—	Table Aa	
—	—	—	—	
—	—	—	—	
—	1185	—	hexagonal	
—	—	—	—	
—	—	—	—	† $MoN_{0\cdot49-0\cdot53}$
—	[2430]	—	hexagonal	‡ $MoC_{0\cdot44=0\cdot50}$
—	Sb	102	A_{15}	
—	—	—	A15, Cr_3Si	
—	—	—	D8m, cubic	
—	—	—	C11, tetrag.	
−237·5	−210	−196	—	
—	—	—	—	
—	−78	−33·5	—	
—	—	—	—	
—	—	—	—	
—	—	—	—	
—	—	—	—	
—	—	—	—	
—	—	—	—	
16	—	—	—	
185	520	—	—	
140	542	—	—	
−13	551	—	—	
—	—	—	—	
—	98	882	Table Aa	
—	—	—	—	
—	—	—	—	
—	—	—	NaCl	
—	996	1787	NaCl	
—	—	—	—	
—	801	1465	Table Aa	
—	—	—	—	
—	747	1393	NaCl	
—	—	—	—	
—	660	1304	NaCl	
—	308	—	—	
750, 970	1132	—	CaF_2	
—	—	—	—	
—	675	—	tetragonal	
—	—	—	KO_2	
299	323	(1560)	tetragonal	
—	—	—	—	
—	978	—	CaF_2	

Table A. Heats of formation,

Substance	ΔH°_{298} (kcal/mole)	Ref.	S°_{298} (cal/deg mole)	Ref.
⟨NaS⟩	−48·1 ± 1·7	854	10·7 ± 2·5	854
⟨Na_2S_3⟩	−103·4 ± 5·0	854	24·3 ± 5·0	854
⟨NaS_2⟩	−49·2 ± 2·8	854	20·0 ± 3·3	854
⟨Na_2SO_4⟩	333·5 ± 2·0	—	35·73 ± 0·1	285, 828
⟨Na_2Se⟩	−82·0 ± 3·0	895	—	—
⟨NaSe⟩	−46·4 ± 5·0	854	*15·0 ± 2·5*	854
⟨Na_2Te⟩	−82·0 ± 6·0	917	22·6 ± 4·0	917
⟨NaTe⟩	−41·4 ± 2·5	917	20·0 ± 3·5	917
⟨$NaTe_3$⟩	−50·3 ± 5·0	917	34·7 ± 5·5	917
⟨Na_3As⟩	*−52·0 ± 6·0*	—	—	—
⟨Na_3Sb⟩	−47·2 ± 3·0	264	*42·1 ± 4·0*	—
⟨NaSb⟩	−15·8 ± 1·0	264	*23·83 ± 3·0*	—
⟨Na_3Bi⟩	−45·5 ± 3·0	264	(*38·2*)	—
⟨NaBi⟩	−15·6 ± 1·0	264	(*26·2*)	—
⟨Na_2CO_3⟩	−270·3 ± 0·4	860	33·17 ± 0·2	860
⟨Na_4SiO_4⟩	*−86·0 ± 5·0**	840	46·8 ± 0·6	125
⟨Na_2SiO_3⟩	−55·5 ± 3·5*	921	27·2 ± 0·3	125
⟨$Na_2Si_2O_5$⟩$_\alpha$	−55·1 ± 2·0*	840	39·4 ± 0·6	125
⟨Na_2TiO_3⟩	−51·0 ± 4·0*	—	29·1 ± 0·2	125
⟨$Na_2Ti_2O_5$⟩	−55·5 ± 5·0*	—	41·5 ± 0·3	125
⟨$Na_2Ti_3O_7$⟩	−56·7 ± 5·0*	—	55·9 ± 0·5	125
⟨Na_4Sn⟩	−14·0 ± 1·0	264	—	—
⟨Na_2Sn⟩	−14·4 ± 1·0	264	—	—
⟨NaSn⟩	−12·0 ± 0·8	264	—	—
⟨Na_5Pb_2⟩	−34·3 ± 2·5	307	81·3 ± 9·0	—
⟨NaPb⟩	−11·0 ± 1·0	307	23·6 ± 3·0	—
⟨Na_2Pb_5⟩†	−23·0 ± 2·0	307	98·5 ± 9·0	—
⟨NaTl⟩	−9·0 ± 1·0	264	—	—
⟨$NaCd_2$⟩	−7·0 ± 1·2	840	32·8 ± 2·2	—
⟨Na_3AlF_6⟩	−19·8 ± 0·2*	797	57·0 ± 0·4	379
⟨$NaAlO_2$⟩	−20·9 ± 1·2*	359	16·9 ± 0·2	378
⟨Na_2CrO_4⟩	−317·6 ± 1·9	896	*44·4 ± 1·9*	797
⟨$NaCrO_2$⟩	−24·25 ± 0·5*	840	*19·9 ± 1·6*	—
⟨$NaFeO_2$⟩	*−10·5 ± 2·0*	—	21·1 ± 0·2	—
⟨Na_2MoO_4⟩	*−350·4 ± 9·0*	—	38·1 ± 0·3	840
⟨$NaVO_3$⟩	−38·7 ± 1·5*	711	27·2 ± 0·3	711
⟨$Na_4V_2O_7$⟩	−127·7 ± 4·0*	711	76·1 ± 0·8	711
⟨Na_3VO_4⟩	−85·4 ± 2·5*	711	45·3 ± 0·5	711
⟨Nb⟩	0	—	8·73 ± 0·15	99
(Nb)	+172·4 ± 4·0	99	44·48	99
⟨NbF_5⟩	−433·5 ± 1·0	660, 172	38·3 ± 0·7	445
(NbF_5)	−415·2	755	*78·63*	—
⟨$NbCl_3$⟩	−139·0 ± 3·0	602	35·2 ± 2·5	602
⟨$NbCl_4$⟩	−166·0 ± 2·5	602	44·0 ± 2·5	602
⟨$NbCl_5$⟩	−190·5 ± 1·0	763, 602	54·0 ± 2·5	602
⟨$NbBr_5$⟩	−131·53	752	[illegible]3·0	752
⟨NbI_5⟩	−64·7	752	[illegible]·0	752
⟨$NbOCl_2$⟩	−185·0 ±.2·0	604	—	—
⟨$NbOCl_3$⟩	−210·2 ± 2·5	602	38·0 ± 2·5	602
⟨NbO⟩	−100·3 ± 3·0	817 (1975)	*11·0 ± 2·0*	817
⟨NbO_2⟩	−190·0 ± 2·0	817 (1975)	13·03 ± 0·07	817
⟨Nb_2O_5⟩	−454·0 ± 1·2	817 (1975)	32·82 ± 0·3	303
⟨Nb_2N⟩	−59·4 ± 1·5	755	*16·0 ± 1·2*	755
⟨NbN⟩	−55·9 ± 1·5	755	*9·0 ± 1·0*	755
⟨Nb_2C⟩	−44·5 ± 2·0	755	15·3 ± 0·1	755
⟨NbC⟩	−33·0 ± 1·0	755	8·38 ± 0·1	755

standard entropies and structures (contd.)

tr.pts. (°C)	m.pt. (°C)	b.pt. (°C)	Structure	Remarks
—	480	—	—	
—	353	—	—	
—	185	—	—	
249, 707	884	—	orthorhomb.	
—	900	—	CaF_2	
—	495	—	—	
—	953	—	CaF_2	
—	377	—	—	
—	455	—	—	
—	—	—	Na_3Bi	
—	856	—	Na_3Bi	
—	475	—	monoclinic	
—	848	—	hexagonal	
—	446	—	In, super 1	
450	850	—	—	
—	1120	—	—	* from oxides
—	1088	—	orthorhomb.	* from oxides
678, 707	874	—	orthorhomb.	* from oxides
287	1030	—	—	* from oxides
—	985	—	—	* from oxides
—	1128	—	—	* from oxides
—	405	—	orthorhomb.	
—	477	—	—	
—	576	—	—	
—	400	—	—	
—	368	—	—	
—	320	—	Ag, super 1	† $NaPb_{1\cdot 85=2\cdot 7}$
—	306	—	Table Aa	
—	382	—	—	
568, 880	1012	—	—	* from fluorides
—	1560	—	—	* from oxides
421	797	—	—	
—	—	—	—	* from oxides
—	1072	—	—	* from oxides
485, 593, 642	698	—	—	
—	—	—	—	* from oxides
—	—	—	—	* from oxides
—	—	—	—	* from oxides
—	2468	4750	Na	
—	—	—	—	
—	79	233	monoclinic?	
—	—	—	—	
—	—	—	—	
—	—	455	—	
—	205	247	monoclinic	
—	227	272	orthorhomb.	
—	327	347	—	
—	—	—	—	
—	—	335	—	
—	1935	—	hexagonal	
817, 927	1902	—	*c.* rutile	
—	1510	—	Ta_2O_3	
—	—	—	Co_2N	
1370	2630	—	hexagonal	
—	—	—	Co_2N	
—	—	—	NaCl	

Table A. Heats of formation,

Substance	ΔH°_{298} (kcal/mole)	Ref.	S°_{298} (cal/deg mole)	Ref.
$\langle NbB_2 \rangle$	−60·0 ± 4·0	740	9·0 ± 0·1	740
$\langle NbSi_2 \rangle$	−33·0 ± 11·0	840	*16·7 ± 2·5*	—
$\langle NbFe_2 \rangle$	−14·7 ± 3·0	755	18·0 ± 2·0	755
$\langle NbCr_2 \rangle$	−5·0 ± 0·7	849	20·0 ± 0·05	849
$\langle Nd \rangle$	0	—	17·0 ± 1·0	99
$\langle NdH_2 \rangle$	−46·0	918	14·08 ± 0·08	759
$\langle NdF_3 \rangle$	−399·0 ± 9·0	742	*27·6*	—
$\langle NdCl_3 \rangle$	−245·6 ± 2·0	918	*34·4 ± 2·2*	—
$\langle NdI_3 \rangle$	−150·2 ± 2·5	233, 779	—	—
$\langle NdOCl \rangle$	−241·7 ± 2·0	793	*22·6 ± 2·0*	—
$\langle Nd_2O_3 \rangle$	−432·1 ± 1·2	683	37·9 ± 0·1	846
$\langle Nd_2S_3 \rangle$	−276·0 ± 9·0	854	44·28 ± 0·2	854
$\langle Nd_2(SO_4)_3 \rangle$	(−932·0)	233, 772	68·9	—
$\langle Nd_2Zr_2O_7 \rangle$	−10·54*	832	*62·0 ± 2·7*	—
$\langle Ni \rangle$	0	—	7·14 ± 0·05	125
(Ni)	+102·8 ± 0·5	99	43·52	99
$\langle NiF_2 \rangle$	−157·0 ± 0·9	907	17·6 ± 0·1	356
$(NiCl)$	−55·0 ± 9·0	597	57·5 ± 2·5	597
$\langle NiCl_2 \rangle$	−73·0 ± 0·5	30, 399	23·35 ± 0·2	30, 828
$\langle NiBr_2 \rangle$	−50·7 ± 2·4	918	32·5 ± 2·4	—
$\langle NiI_2 \rangle$	−18·7 ± 2·0	918	36·8 ± 2·4	—
$\langle NiO \rangle$	−57·5 ± 0·5	307	9·1 ± 0·1	379
$\langle Ni_3S_2 \rangle$	−51·6 ± 2·5	854	32·0 ± 0·3	636
$\langle NiS_{0\cdot84} \rangle$	−19·7 ± 1·1	854	11·7 ± 1·0	854
$\langle NiS \rangle$	−22·5 ± 0·9	854	12·65 ± 0·2	636
$\langle NiSO_4 \rangle$	−208·7 ± 0·6	907	24·82 ± 0·25	907
$\langle NiSe_{1\cdot05} \rangle$	−17·9 ± 0·9	854	17·97 ± 1·0	854
$\langle NiSe_{1\cdot143} \rangle$	−19·05 ± 0·6	854	18·44 ± 1·0	854
$\langle NiSeO_3 \rangle$	−135·6 ± 1·6	854	24·7 ± 2·0	854
$\langle NiTe \rangle$	−8·52 ± 0·44	875	19·13 ± 0·35	—
$\langle Ni_3N \rangle$	+0·2 ± 1·0	770	—	—
$\langle Ni_3P \rangle$	−47·83 ± 2·5	—	*25·4 ± 3·2*	—
$\langle Ni_5P_2 \rangle$	−93·5 ± 4·5	—	*44·2 ± 5·1*	—
$\langle Ni_2P \rangle$	−39·3 ± 4·0	—	*18·5 ± 2·8*	—
$\langle Ni_6P_5 \rangle$	−132·9 ± 14·0	—	*66·0 ± 7·5*	—
$\langle NiP_2 \rangle$	−30·9 ± 3·5	—	*17·5 ± 4·0*	—
$\langle NiP_3 \rangle$	−37·7 ± 4·0	—	*23·5 ± 4·0*	—
$\langle NiAs \rangle$	−17·2 ± 0·9	—	*12·4 ± 2·7*	—
$\langle NiSb \rangle$	−20·0 ± 1·2	—	*18·7 ± 3·0*	—
$\langle NiBi \rangle$	−1·85 ± 0·4	—	*21·1 ± 1·0*	—
$\langle NiCO_3 \rangle$	−162·5 ± 3·0	886	20·4 ± 0·4	833
$\langle Ni(CO)_4 \rangle$	−138·3 ± 1·5	907	99·3 ± 0·2	907
$\langle Ni_2Si \rangle$	−34·1 ± 2·6	840	—	—
$\langle Ni_3Si_2 \rangle$	−55·5 ± 4·6	840	—	—
$\langle NiSi \rangle$	−21·4 ± 1·8	840	10·6 ± 0·8	840
$\langle NiSi_2 \rangle$	−22·5 ± 3·0	840	15·6 ± 1·1	840
$\langle Ni_2Ge \rangle$	−26·3 ± 1·4	—	*21·7 ± 2·5*	—
$\langle Ni_3Ti \rangle$	−33·5 ± 1·5	152	25·0 ± 2·7	—
$\langle NiTi \rangle$	−15·9 ± 1·0	152	12·7 ± 1·8	—
$\langle NiTi_2 \rangle$	−20·0 ± 1·5	152	20·0 ± 1·5	—
$\langle Ni_3Sn \rangle$	−24·6 ± 2·6	—	*31·4 ± 4·4*	—
$\langle Ni_3Sn_2 \rangle$	−46·05 ± 1·2	—	*41·5 ± 4·5*	—
$\langle Ni_3Sn_4 \rangle$	−56·3 ± 0·9	—	*61·6 ± 4·5*	—
$\langle Ni_4B_3 \rangle$	−74·5 ± 12·0	—	*27·4 ± 4·5*	—
$\langle NiB \rangle$	−24·0 ± 4·0	—	*7·2 ± 1·5*	—
$\langle Ni_3Al \rangle$	−36·6 ± 2·0	—	27·2 ± 1·0	—

standard entropies and structures (contd.)

tr.pts. (°C)	m.pt. (°C)	b.pt. (°C)	Structure	Remarks
—	—	—	—	
—	—	—	—	
—	1650	—	—	
1600	1740	—	—	
855	1016	3070	Mg	
—	—	—	—	
—	1377	—	—	
—	760	—	UCl_3	
574	785	—	—	
—	—	—	—	
140	2272	—	La_2O_3	
—	2210	—	—	
—	—	—	—	
—	—	—	—	* from oxides
—	1453	2910	Ag	
—	—	—	—	
—	1450	1474	rutile	
—	—	—	—	
—	1009	970	$CdCl_2$	
—	963	919	$CdCl_2$	
750	797	—	$CdCl_2$	
252	1955	—	NaCl	
550	[790]	—	—	
391	[560]	—	—	$NiS_{0\cdot82-0\cdot867}$
389	996	—	NiAs	
—	—	—	—	
320, 660	980	—	—	
(230)	—	—	—	
—	—	—	—	
—	—	—	NiAs	
—	—	—	Mg	
—	[970]	—	cubic b.c.	
1025	1180	—	—	
—	1110	—	Fe_2P	
—	—	—	—	
—	—	—	—	
—	—	—	—	
—	964	—	Table Aa	
—	1160	—	NiAs	
—	[655]	—	—	
—	—	—	—	
—	—	—	—	
1214	1318	—	orthorhomb.	
[845]	—	—	Ag, superstr.	
—	992	—	FeSi	
—	980	—	—	
—	—	—	—	
—	1378	—	hexagonal	
—	1240	—	CsCl	
—	1015	—	f.c.c.	
921	1174	—	$MgCd_3$	
—	1264	—	NiAs	
—	—	—	—	
—	1540	—	—	
—	[1600]	—	—	
—	—	—	$AuCu_3$	

Table A. Heats of formation,

Substance	ΔH°_{298} (kcal/mole)	Ref.	S°_{298} (cal/deg mole)	Ref.
$\langle NiAl\rangle$	−28·3 ± 1·2	—	12·93 ± 0·1	—
$\langle Ni_2Al_3\rangle$	−67·5 ± 4·0	99	*32·6 ± 2·5*	—
$\langle NiAl_3\rangle$	−36·0 ± 2·0	99	26·45 ± 1·7	—
$\langle NiAl_2O_4\rangle$	−1·45 ± 0·3*	840	23·5 ± 1·0	—
$\langle NiTiO_3\rangle$	−4·3 ± 2·1*	840	19·15 ± 0·9	840
$\langle NiWO_4\rangle$	−10·8 ± 1·5*	840	28·2 ± 1·5	840
$\langle NiCr_2O_4\rangle$	−3·25 ± 0·9*	857	29·7 ± 2·0	—
$\langle NiGa_2O_4\rangle$	−1·06 ± 0·3*	—	31·1 ± 0·5	—
$\langle Np\rangle$	0	—	—	—
$\langle NpF_3\rangle$	−360·0 ± 8·0	198	*28·3 ± 2·8*	—
(NpF_6)	*−410·0 ± 25·0.*	—	88·7 ± 0·5	—
$\langle NpCl_3\rangle$	−216·0 ± 5·0	267	38·3 ± 0·5	—
$\langle NpCl_4\rangle$	−238·0 ± 5·0	267	47·7 ± 0·7	—
$\langle NpOCl_2\rangle$	−245·8 ± 1·2	—	*33·5 ± 0·9*	—
$\langle NpO_3\cdot H_2O\rangle$	−332·4	—	*32·8 ± 1·6*	—
$\langle NpO_2\rangle$	−246·0 ± 10·0	22	19·2 ± 0·1	268
(O_2)	0	—	49·0 ± 0·01	125
(O)	59·55 ± 0·2	key value	38·48 ± 0·01	125
$\langle Os\rangle$	0	—	7·8 ± 0·5	125
(Os)	188·4 ± 2·3	99	46·0	99
$\langle OsO_2\rangle$	−70·4 ± 2·3	—	12·4 ± 2·6	—
$\langle OsO_4\rangle$*	−94·1 ± 2·0	907	32·7 ± 2·0	907
(OsO_4)	−80·35 ± 2·1	907	70·18 ± 0·08	907
$\langle OsS_2\rangle$	−24·0 ± 2·5	307	20·1 ± 2·5	307
$\langle OsP_2\rangle$	−31·5 ± 4·5	—	*19·6 ± 3·1*	—
$\langle P\rangle_{white}$	0	—	9·82 ± 0·02	907
$\langle P\rangle_{red}$	−4·15 ± 0·15	907	5·45 ± 0·15	907
(P)	+75·64	907	38·98 ± 0·03	907
(P_2)	+34·38	907	52·1 ± 0·05	907
(P_4)	+14·1 ± 0·2	907	66·9 ± 0·1	907
(PH_3)	+1·3 ± 0·8	907	50·24 ± 0·1	907
(PF_3)	−220·0 ± 8·0	907	65·2 ± 0·5	907
(PF_5)	−381·1 ± 0·4	693, 717	70·0 ± 0·3	907
$\{PCl_3\}$	−76·7 ± 1·0	—	52·22 ± 0·2	—
(PCl_3)	−69·0 ± 1·0	817	74·48 ± 0·1	907
(PCl_5)	−87·7 ± 2·5	907	87·1 ± 0·4	907
(PBr_3)	−31·6 ± 2·0	907	83·2 ± 0·1	907
(PI_3)	−4·3 ± 5·0	—	89·5 ± 0·2	907
$\{POCl_3\}$	−142·8 ± 0·5	907	53·17 ± 0·2	907
$(POCl_3)$	−133·6 ± 0·5	907	77·4 ± 0·2	907
(P_4O_6)	−380·86	907	82·9 ± 0·2	907
$\langle P_2O_5\rangle$	−356·6 ± 2·5	299	27·35 ± 0·1	426
(H_3PO_4)	−305·7 ± 2·5	299	26·4 ± 0·1	772
$\langle P_4S_3\rangle$	−53·5 ± 5·0	854	48·6 ± 0·1	907
(PS)	+53·0 ± 25·0	854	56·26 ± 0·3	854
$\langle P_4S_5\rangle$	−72·8 ± 3·0	854	*60·4 ± 4·0*	854
$\langle P_2S_5\rangle$	*−44·5*	854	45·62 ± 0·1	854
$\langle Pb\rangle$	0	—	15·55 ± 0·1	99
(Pb)	+46·62 ± 0·3	99	41·89	99
$\langle PbF_2\rangle$	−161·8 ± 1·0	817 (1975)	27·0 ± 3·0	817
$\langle PbCl_2\rangle$	−85·9 ± 0·2	817 (1975)	32·5 ± 0·5	817
$(PbCl_2)$	−41·6 ± 1·0	817 (1975)	75·8 ± 0·6	817
$\langle PbBr_2\rangle$	−66·3 ± 0·6	817 (1975)	38·5 ± 0·5	817

standard entropies and structures (contd.)

tr.pts. (°C)	m.pt. (°C)	b.pt. (°C)	Structure	Remarks
—	1639	—	CsCl	
—	1133	—	—	
—	—	—	orthorhomb.	
—	(1830)	—	—	*from oxides ($NiO + Al_2O_3$)
—	—	—	—	*from oxides ($NiO + TiO_2$)
—	—	—	—	*from oxides ($NiO + WO_3$)
—	—	—	—	*from oxides ($NiO + Cr_2O_3$)
—	—	—	—	*from oxides ($NiO + Ga_2O_3$)
280, 577	637	—	—	
—	—	—	—	
—	—	—	—	
—	—	—	UCl_3	
—	—	—	$ThCl_4$	
—	—	—	—	
—	—	—	—	
—	—	—	—	
−250	−219	−183	—	
—	—	—	—	
—	3030	5030	Mg	
—	—	—	—	
—	—	—	—	
—	—	—	monoclinic	*yellow
—	41	131	—	
—	—	—	pyrite	
—	—	—	—	
—	44	—	cubic	
—	593	—	—	
—	—	—	—	
—	—	—	—	
—	—	—	—	
+	−134	−88	—	
+	−151	−101	—	
—	−94	−85	—	
—	−91	75	—	
—	—	—	—	
—	160	159	—	
—	25	174	—	
—	61	—	—	
—	1	107	—	
—	—	—	—	
—	24	175	—	
—	580	605	orthorhomb.	
—	42	—	—	
40	173	—	—	
—	—	—	—	
—	—	—	—	
—	287	—	—	
—	327	1750	Ag	
—	—	—	—	
310	830	1293	CaF_2	
—	501	—	Table Aa	
—	—	—	—	
—	371	912	$PbCl_2$	

Table A. Heats of formation,

Substance	ΔH°_{298} (kcal/mole)	Ref.	S°_{298} (cal/deg mole)	Ref.
$\langle PbI_2 \rangle$	−41·92 ± 0·2	817 (1975)	41·8 ± 0·1	817
$\langle PbO \rangle_{red}$	−52·44 ± 0·2	817 (1974)	15·85 ± 0·2	817
(PbO)	+16·8 ± 1·7	817 (1974)	57·34 ± 0·1	817
$\langle Pb_3O_4 \rangle$	−171·77 ± 1·5	817 (1974)	50·66 ± 1·6	817
$\langle PbO_2 \rangle$	−65·6 ± 0·7	817 (1974)	17·16 ± 0·1	817
$\langle PbS \rangle$	−23·5 ± 0·6	854	21·83 ± 0·4	854
(PbS)	+31·5 ± 1·5	817 (1974)	60·07 ± 0·02	817
$\langle PbSO_4 \rangle$	−219·9 ± 0·4	918	35·5 ± 0·2	485
$\langle PbSe \rangle$	−23·8 ± 0·8	481	24·5 ± 0·5	854
$(PbSe)$	+30·7 ± 2·0	854	63·0 ± 1·0	854
$\langle PbSeO_3 \rangle$	−127·3	907	*30·7*	—
$\langle PbTe \rangle$	−16·5 ± 0·2	782	26·3 ± 0·5	264
$\langle PbCO_3 \rangle$	−167·2 ± 1·3	907	31·3 ± 0·8	125
$\langle PbO \cdot PbCO_3 \rangle$	−195·2	860	48·8	860
$\langle Pb_2SiO_4 \rangle$	−4·0 ± 2·0*	840	44·6 ± 0·5	542
$\langle PbSiO_3 \rangle$	−4·2 ± 0·5*	840	26·2 ± 0·3	542
$\langle PbTiO_3 \rangle$	−286·5 ± 3·3	852	26·75 ± 2·2	852
$\langle Pb_2V_2O_7 \rangle$	−35·1 ± 1·5*	840	—	—
$\langle Pb_3V_2O_8 \rangle$	−40·8 ± 2·0*	840	—	—
$\langle PbMoO_4 \rangle$	−10·0 ± 3·5*	840	39·7 ± 0·5	840
$\langle PbWO_4 \rangle$	−14·0 ± 2·0*	840	40·0 ± 0·5	840
$\langle Pd \rangle$	0	—	9·05 ± 0·05	37, 125
(Pd)	+90·1 ± 0·7	511, 473, 423	39·92	627
$\langle PdF_2 \rangle$	(−112·0)	27	—	—
$\langle PdCl_2 \rangle$	−41·39	435	24·8	435
$(PdCl_2)$	+29·05	435	74·95	435
$\langle PdI_2 \rangle$	−15·2 ± 4·0	918	36·0 ± 2·0	918
$\langle PdO \rangle$	−26·9	—	9·4	—
$\langle Pd_4S \rangle$	−16·5 ± 0·5	854	43·18 ± 0·1	854
$\langle PdS \rangle$	−16·9 ± 1·5	854	*13·5 ± 1·5*	854
$\langle PdS_2 \rangle$	−18·7 ± 3·0	854	21·0 ± 3·0	854
$\langle PdTe \rangle$	−9·0 ± 4·0	854	21·42 ± 0·1	854
$\langle Pr \rangle$	0	—	17·7 ± 0·3	99
(Pr)	85·0 ± 2·0	99	45·34	—
$\langle PrH_2 \rangle$	−47·4	918	13·57 ± 0·1	—
$\langle PrCl_3 \rangle$	−252·6 ± 1·3	918	*(34·5)*	—
$\langle PrI_3 \rangle$	−156·4 ± 4·0	918	*(51·3)*	—
$\langle Pr_2O_3 \rangle$	−436·3 ± 0·6	240	37·9 ± 0·3	—
$\langle PrO_2 \rangle$	−232·9 ± 1·5	—	*19·1 ± 1·1*	—
$\langle PrS \rangle$	−108·0 ± 6·0	854	*18·6 ± 1·2*	854
$\langle Pr_3S_4 \rangle$	−371·5 ± 30·0	854	*61·2 ± 3·0*	854
$\langle PrAl_2 \rangle$	−40·0 ± 9·0	—	27·42 ± 0·05	854
$\langle Pt \rangle$	0	—	9·95 ± 0·05	99
(Pt)	+135·0 ± 0·4	99	45·96	99
$\langle PtCl_2 \rangle$	−26·5 ± 3·0	918	*(31·0)*	—
$\langle PtCl_3 \rangle$	−41·6 ± 3·0	918	*(35·5)*	—
$\langle PtCl_4 \rangle$	−56·6 ± 4·0	918	*49·0 ± 5·0*	—
$\langle PtBr_2 \rangle$	−19·2 ± 3·0	222, 27	*(37·0)*	—
$\langle PtBr_3 \rangle$	−30·5 ± 4·0	222, 27	*(43·0)*	—
$\langle PtBr_4 \rangle$	−33·6 ± 4·0	222, 27	*(60·0)*	—
$\langle PtI_4 \rangle$	−17·4 ± 4·0	918	*67·2 ± 6·0*	—
(PtO_2)	+40·3 ± 2·5	423	61·15 ± 2·5	423
$\langle PtS \rangle$†	−19·86 ± 0·6	907	13·16 ± 0·05	907
$\langle PtS_2 \rangle$	−26·7 ± 0·7	907	17·85 ± 0·1	907
$\langle Pt_5Se_4 \rangle$	−63·4 ± 6·0	881	78·2 ± 5·2	881

standard entropies and structures (contd.)

tr.pts. (°C)	m.pt. (°C)	b.pt. (°C)	Structure	Remarks
—	410	832	CdI_2	
489	886	—	tetragonal	
—	—	—	—	
—	dec.	596	tetragonal	
—	—	—	rutile	
—	1113	—	NaCl	
—	—	—	—	
866	1170	—	$BaSO_4$	
—	1076	—	NaCl	
—	—	—	—	
—	680	—	—	
—	920	—	NaCl	
—	—	—	$BaCO_3$	
—	—	—	—	
—	743	—	—	*from oxides
—	764	—	monoclinic	*from oxides
490	1170	—	$CaTiO_3$, def.	
—	722	—	—	*from oxides
—	960	—	—	*from oxides
—	940	—	—	*from oxides
—	1125	—	—	*from oxides
37, 125	1552	2940	Ag	
627	—	—	—	
—	—	—	—	
—	678	—	orthorhomb.	
—	—	—	—	
—	dec.	—	—	
—	—	—	Table Aa	
—	—	—	—	
—	—	—	—	
—	972	—	—	
—	—	—	—	
795	931	3510	M_2	
—	—	—	—	
—	—	—	—	
—	786	—	UCl_3	
—	738	—	—	
—	—	—	La_2O_3	
—	—	—	CaF_2	
—	2230	—	—	
—	—	—	—	
—	—	—	—	
—	1769	(4100)	Ag	
—	—	—	—	
—	—	—	—	
—	—	—	—	
—	—	—	—	
—	—	—	—	
—	—	—	—	
—	—	—	—	
—	—	—	—	
—	—	—	—	
—	—	—	PdO	† $PtS_{0 \cdot 975-1 \cdot 025}$
—	—	—	CdI_2	
—	—	—	—	

Table A. Heats of formation,

Substance	ΔH°_{298} (kcal/mole)	Ref.	S°_{298} (cal/deg mole)	Ref.
⟨Pu⟩	0	—	13·2 ± 0·5	595
⟨PuH_2⟩	−33·3 ± 1·8	865	*14·3 ± 1·2*	865
⟨PuH_3⟩	−33·0 ± 4·0	865	*15·5 ± 1·2*	865
⟨PuF_3⟩	−371·0 ± 4·0	865	*27·0 ± 3·0*	865
⟨PuF_4⟩	−414·4 ± 5·0	865	38·7 ± 0·5	595
⟨PuF_6⟩	−424·6 ± 7·0	865	53·2 ± 0·7	865
(PuF_6)	−413·0 ± 7·0	865	88·3 ± 0·2	865
⟨$PuCl_3$⟩	−226·8 ± 0·8	595	39·2 ± 3·0	595
⟨PuOCl⟩	−219·7 ± 1·0	595	25·2 ± 3·0	595
⟨$PuBr_3$⟩	−198·8 ± 0·6	865	*45·7 ± 3·0*	595
⟨PuOBr⟩	*−212·4 ± 5·0*	865	*28·5 ± 2·0*	865
⟨PuI_3⟩	(*−143·0*)	865, 595	*53·5 ± 5·0*	865, 595
⟨PuOI⟩	*−197·8 ± 5·0*	865	*30·2 ± 3·0*	865
⟨PuO_2⟩	−252·4 ± 0·4	818	16·35 ± 0·5	818
⟨PuS⟩	*−104·3 ± 5·0*	865	*18·7 ± 1·5*	865
⟨Pu_2S_3⟩	*−235·0 ± 4·5*	865	*46·0 ± 4·0*	865
⟨$Pu(SO_4)_2$⟩	*−526·0 ± 5·0*	865	*39·0 ± 2·5*	865
⟨PuN⟩	−71·5 ± 1·6	—	17·3 ± 0·6	—
⟨$PuC_{0·88}$⟩	−11·1 ± 0·8	766	17·3 ± 0·6	766
⟨$PuC_{1·5}$⟩	−13·2 ± 1·0	766	20·7 ± 1·0	766
⟨PuC_2⟩	−8·25 ± 2·5	—	24·1 ± 1·0	766
⟨Rb⟩	0	—	18·35 ± 0·1	99
⟨RbF⟩	−132·2 ± 1·8	772	*17·6 ± 1·7*	—
⟨RbCl⟩	−102·9 ± 2·0	772	21·9 ± 0·8	906, 155
⟨RbBr⟩	−93·0 ± 2·0	772	26·0 ± 0·2	36
⟨RbI⟩	−78·5 ± 2·0	772	28·2 ± 0·2	36
⟨Rb_2O⟩	−78·9 ± 3·0	772	—	—
⟨Rb_2O_3⟩	−126·0 ± 10·0	22	—	—
⟨RbO_2⟩	−68·0 ± 6·0	22	—	—
⟨RbOH⟩	−98·9 ± 4·0	772	—	—
⟨Rb_2S⟩	−83·2 ± 6·0	772	*31·8 ± 3·0*	854
⟨Rb_2SO_4⟩	−340·5 ± 2·5	772	—	—
⟨Rb_2CO_3⟩	−269·6 ± 4·0	772	*41·8*	861
⟨Re⟩	0	—	8·73 ± 0·1	99
(Re)	+185·2 ± 3·0	99	45·13	627
{ReF_6}	−278·0 ± 12·0	399	—	—
⟨$ReCl_3$⟩	−63·0 ± 2·5	547	29·6 ± 0·1	666
⟨$ReBr_3$⟩	−39·3 ± 2·5	547	35·0 ± 2·5	666
⟨ReO_2⟩	−103·4 ± 1·2	787	*15·0 ± 1·1*	—
⟨ReO_3⟩	−146·0 ± 4·0	284	19·3 ± 2·0	787
⟨Re_2O_7⟩	−298·4 ± 2·5	787	49·55 ± 0·1	284
⟨ReS_2⟩	−42·7 ± 2·5	163	14·5 ± 2·0	854
⟨ReS_3⟩	*−49·8 ± 3·3*	—	*18·0 ± 3·0*	854
⟨Re_2S_7⟩	−107·9 ± 4·0	163	*40·0 ± 5·0*	854
⟨Re_2Te_5⟩	−29·3 ± 6·0	869	60·2 ± 8·0	869
⟨Re_3As_7⟩	−22·8 ± 4·0	—	*80·5 ± 6·0*	—
⟨Re_5Si_3⟩	−37·6 ± 15·0	840	61·15 ± 4·0	840
⟨ReSi⟩	−12·6 ±·5·0	840	13·23 ± 1·0	840
⟨$ReSi_2$⟩	−21·6 ± 7·5	840	17·7 ± 1·6	840
⟨Rh⟩	0	—	7·53 ± 0·05	99
(Rh)	+132·2 ± 1·5	99	44·39	99
⟨$RhCl_3$⟩	−65·8 ± 6·0	435	*30·3 ± 3·5*	435
⟨$RhBr_3$⟩	−50·0 ± 4·0	27	45·0 ± 7·0	27
⟨Rh_2O_3⟩	−91·5 ± 4·0	828	*22·0 ± 3·5*	—
(RhO_2)	−46·8 ± 2·0	831	63·0 ± 2·5	831

standard entropies and structures (contd.)

tr.pts. (°C)	m.pt. (°C)	b.pt. (°C)	Structure	Remarks
122, 205, 318, 452	640	3420	orthorhomb.	
—	—	—	—	
—	—	—	—	
—	1426	—	CeF_3	
—	1037	—	—	
—	52	—	—	
—	—	—	—	
—	767	1790	UCl_3	
—	—	—	orthorhomb.	
—	681	1475	—	
—	—	—	—	
—	780	—	—	
—	—	—	—	
—	2390	—	—	
—	2350	—	—	
—	1730	—	—	
—	—	—	—	
—	2570	—	—	
—	—	—	—	
—	—	—	—	
—	—	—	—	
—	38	688	Na	
—	775	1390	NaCl	
—	715	1381	NaCl	
—	680	1352	NaCl	
—	640	1304	NaCl	
—	>570	—	CaF_2	
—	489	—	cubic†	† Th_3P_4 type
—	412	—	CaC_2	
245	300	—	—	
—	—	—	CaF_2	
649	1050	—	K_2SO_4	
—	840	—	—	
—	3180	5600	Mg	
—	—	—	—	
—	19	34	—	
—	—	—	—	
—	—	—	—	
—	—	—	MoO_2	
—	—	—	cubic†	† approaching c.p.
—	297	361	—	
—	—	—	—	
—	—	—	—	
—	—	—	—	
—	935	—	—	
—	—	—	—	
—	—	—	—	
—	—	—	—	
—	—	—	—	
—	1960	3700	Mg	
—	—	—	—	
—	—	—	—	
—	—	—	—	
—	—	—	Cr_2O_3	
—	—	—	—	

Table A. Heats of formation,

Substance	ΔH°_{298} (kcal/mole)	Ref.	S°_{298} (cal/deg mole)	Ref.
⟨Ru⟩	0	—	6·82 ± 0·1	—
(Ru)	+155·7 ± 2·0	99	44·55	627
⟨RuF_5⟩	−213·4 ± 2·5	723	*38·5 ± 3·5*	—
⟨$RuCl_3$⟩	−60·5 ± 3·0	434	30·5 ± 2·5	434
($RuCl_3$)	+13·4 ± 4·0	434	95·0 ± 5·0	434
($RuCl_4$)	−22·3 ± 4·0	434	89·5 ± 5·0	434
⟨RuO_2⟩	−72·8 ± 0·8	770	14·5 ± 1·1	770
(RuO_3)	−18·7 ± 2·8	907	66·0 ± 3·0	907
(RuO_4)	−44·0 ± 1·8	907	69·47 ± 0·1	907
⟨RuS_2⟩	−49·2 ± 5·0	854	*(13·0)*	854
⟨S⟩†	0	—	7·62 ± 0·05	125
⟨S⟩‡	+0·08 ± 0·01	854	7·88 ± 0·1	854
(S)	+66·2 ± 0·1	key value	40·09	key value
(S_2)	+30·75 ± 0·1	854	54·50	854
(S_4)	45·0 ± 7·0	854	71 ± 1·0	—
(S_6)	24·6 ± 1·5	854	84·58 ± 0·5	854
(S_8)	23·8 ± 3·0	854	101·13	854
(SF_6)	−291·8 ± 0·5	718, 504	69·9 ± 1·0	577
{SCl_4}	−13·6 ± 3·0	198	—	—
{S_2Cl_2}	−13·9 ± 1·0	854	53·0 ± 3·0	854
(S_2Cl_2)	−4·1 ± 2·0	854	77·95 ± 0·8	854
(SCl_2)	−5·4 ± 0·5	854	67·2	854
($SOCl_2$)	−50·75 ± 0·75	907	73·6 ± 0·1	907
(SO_2Cl_2)	−86·2 ± 0·95	907	74·4 ± 0·2	907
(SO)	1·2 ± 0·3	918	53·02 ± 0·1	907
(SO_2)	−70·94 ± 0·05	key value	59·30	key value
{SO_3}	−104·8 ± 0·2	854	29·3 ± 2·0	854
(SO_3)	−94·58 ± 0·2	854	61·35 ± 0·05	854
⟨Sb⟩	0	—	10·88 ± 0·05	99
(Sb)	63·23 ± 0·8	99	43·06	99
(Sb_2)	55·26 ± 0·6	99	60·9	99
(Sb_4)	49·36 ± 0·3	99	83·65	99
(SbH_3)	−34·6 ± 2·5	509	55·65 ± 2·0	509
⟨SbF_3⟩	−218·8 ± 4·0	918	30·4	918
⟨$SbCl_3$⟩	−91·35 ± 0·4	918	43·6 ± 0·9	—
($SbCl_3$)	−74·95 ± 1·0	—	80·6 ± 1·0	870
($SbCl_5$)	−94·25 ± 1·5	918	96·04 ± 0·1	918
⟨SbOCl⟩	−91·0 ± 4·0	907	25·7 ± 2·2	—
⟨$SbBr_3$⟩	−62·0 ± 3·5	907	50·2 ± 1·0	907
⟨SbI_3⟩	−24·0 ± 7·0	907	51·5 ± 0·4	671
⟨Sb_2O_3⟩	−169·35 ± 0·8	907	33·7 ± 1·0	907
⟨SbO_2⟩	−108·45 ± 0·6	907	15·2 ± 1·0	907
⟨Sb_2O_5⟩	−240·8	907	29·9 ± 1·8	907
⟨Sb_2S_3⟩†	−49·0	854	43·5 ± 0·8	854
⟨$Sb_2(SO_4)_3$⟩	−574·2 ± 5·0	918	69·6 ± 5·0	—
⟨Sb_2Se_3⟩	−30·5 ± 1·0	523	50·5 ± 0·5	854
⟨Sb_2Te_3⟩	−13·5 ± 0·4	523	58·8 ± 0·6	854
⟨Sc⟩	0	—	8·28 ± 0·05	99
(Sc)	90·3 ± 1·5	99	41·75	627
⟨ScF_3⟩	−396·0 ± 8·0	784	21·5 ± 2·0	—
⟨$ScCl_3$⟩	−215·0 ± 10·0	—	29·0	—
⟨$ScBr_3$⟩	−170·0 ± 10·0	—	40·0	—
⟨Sc_2O_3⟩	−455·6 ± 1·5	524	18·4 ± 0·1	637
⟨Sc_2S_3⟩	−280·0	854	29·5 ± 4·0	854
⟨ScN⟩	−75·0 ± 4·0	524	7·1 ± 1·0	—

standard entropies and structures (contd.)

tr.pts. (°C)	m.pt. (°C)	b.pt. (°C)	Structure	Remarks
—	2250	4150	Mg	
—	—	—	—	
—	101	272	—	
—	—	—	—	
—	—	—	—	
—	—	—	—	
—	—	—	rutile	
—	—	—	—	
—	—	—	—	
—	—	—	pyrite	
95·5	—	445	Table Aa	† orthorhombic
—	115	445	Table Aa	‡ monoclinic
—	—	—	—	
—	—	—	—	
—	—	—	—	
—	—	—	—	
—	—	—	—	
−179	Sb.	−64	—	
—	−30	—	—	
—	−80	138	—	
—	—	—	—	
—	—	—	—	
—	—	—	—	
—	—	70	—	
—	—	—	—	
—	—	44·5	—	
—	—	—	—	
—		—	—	
—	631	1587	As	
—	—	—	—	
—	—	—	—	
—	—	—	—	
—	—	—	—	
—	291	319	—	
—	73	—	—	
—	—	—	—	
—	2	—	—	
—	dec.	—	—	
—	97	289	—	
—	170	402	AsI_3	
⟨572⟩	656	1456	orthorhomb.	
—	—	—	cubic	
—	—	—	cubic	
—	550	—	Bi_2S_3	† black
—	—	—	—	
—	615	—	—	
—	619	—	—	
1335	1539	2930	Mg	
—	—	—	—	
—	—	—	—	
—	966	967	AsI_3	
—	Sb	929	—	
—	<2400	—	Table Aa	
—	—	—	—	
—	(2650)	—	NaCl	

Table A. Heats of formation,

Substance	ΔH°_{298} (kcal/mole)	Ref.	S°_{298} (cal/deg mole)	Ref.
$\langle Se \rangle$	0	—	10·1 ± 0·1	854
(Se)	56·25	854	24·21 ± 0·05	854
(Se_2)	33·3 ± 0·5	854	58·2 ± 0·9	854
$\{SeF_4\}$	−204·8	854	44·3	854
(SeF_4)	−194·0 ± 9·0	854	70·85	854
(SeF_6)	−266·95 ± 0·2	854	74·92	854
$\{SeCl_2\}$	−20·0 ± 1·0	854	59·0 ± 5·0	854
(Se_2Cl_2)	−5·2 ± 2·0	854	84·56	854
$(SeCl_2)$	−8·0 ± 3·0	854	70·65	854
$\langle SeCl_4 \rangle$	−45·1 ± 1·5	854	46·5 ± 4·0	854
(SeO)	14·9 ± 5·0	854	55·9	854
$\langle SeO_2 \rangle$	−53·8 ± 0·5	854	15·94 ± 0·4	854
(SeO_2)	−25·8 ± 4·0	854	63·31	854
$\langle Si \rangle$	0	—	4·5 ± 0·05	125
(Si)	108·9 ± 1·2	99	40·12	99
(Si_2)	142·0 ± 1·8	99	54·92	99
(Si_3)	147·0 ± 2·5	99	60·63	99
(SiH_4)	−7·3 ± 0·8	578	48·7 ± 0·1	125
(Si_2H_6)	19·1 ± 0·4	907	65·55 ± 0·4	907
$(SiCl_2)$	−40·0 ± 1·0	817	67·36 ± 1·0	817
$(SiCl_4)$	−158·4 ± 0·4	817	79·07 ± 0·1	817
$(SiBr_2)$	−10·2 ± 2·0	907	72·5 ± 0·5	907
$\{SiBr_4\}$	−110·3 ± 1·0	907	—	—
$(SiBr_4)$	−100·3 ± 1·2	907	90·2 ± 0·3	907
(SiI_2)	19·6 ± 1·5	—	75·8 ± 0·6	907
$\langle SiI_4 \rangle$	−47·6 ± 1·5	890	63·44 ± 1·2	—
(SiO)	−23·5 ± 2·5	215	50·55 ± 0·1	125
$\langle SiO_2 \rangle$	−217·6 ± 0·8	817	9·91 ± 0·05	817
(SiS)	+25·3 ± 3·0	817	53·46 ± 0·5	817
$\langle SiS_2 \rangle$	−51·0 ± 5·0	817	19·2 ± 1·0	—
$(SiSe)$	+48·5 ± 5·0	854	56·2 ± 0·3	854
$\langle SiSe_2 \rangle$	−35·0 ± 9·0	854	22·5 ± 3·0	854
$\langle Si_3N_4 \rangle$	−178·0 ± 3·0	817	27·0 ± 2·0	817
$\langle SiP \rangle$	−14·8 ± 3·5	—	7·8 ± 2·3	—
$\langle SiC \rangle$	−16·0 ± 1·5	840	3·95 ± 0·05	125
$\langle Sm \rangle$	0	—	16·62 ± 0·1	534
(Sm)	+49·6 ± 1·5	405	43·72	99
$\langle SmCl_2 \rangle$	−195·6 ± 1·1	383	30·5 ± 1·5	—
$\langle SmCl_3 \rangle$	−245·2	918	34·8	—
$\langle SmOF \rangle$	−274·6 ± 2·5	920	22·5 ± 4·0	920
$\langle Sm_2O_3 \rangle$	−437·9 ± 2·0	—	36·1 ± 0·1	536
$\langle SmC_2 \rangle$	−23·4 ± 2·0	—	18·6 ± 2·0	—
$\langle Sn \rangle_{white}$	0	—	12·24 ± 0·1	99
$\langle Sn \rangle_{grey}$	+0·47 ± 0·03	99	10·55 ± 0·1	99
(Sn)	+72·0 ± 0·5	99	40·24	99
(SnH_4)	+38·9 ± 0·8	907	54·65 ± 0·1	907
$\langle SnCl_2 \rangle$	−79·1	907	31·5 ± 1·5	—
$\{SnCl_4\}$	−126·4 ± 3·5	907	61·9 ± 1·0	907
$(SnCl_4)$	−116·9	907	87·2 ± 0·3	907
$\langle SnBr_2 \rangle$	−62·15 ± 1·6	907	35·8 ± 2·8	—
$(SnBr_4)$	−83·2	907	98·6 ± 0·5	907
$\langle SnI_2 \rangle$	−34·7 ± 4·5	907	(40·1)	—
(SnI_4)	—	—	106·7 ± 0·4	907
$\langle SnO \rangle$	68·4 ± 0·3	98	13·5 ± 0·5	125
(SnO)	+5·0 ± 2·3	907	55·45 ± 0·05	907

standard entropies and structures (contd.)

tr.pts. (°C)	m.pt. (°C)	b.pt. (°C)	Structure	Remarks
(106)	220	685	Table Aa	
—	—	—	—	
—	—	—	—	
—	10	108	—	
—	—	—	—	
—	−35	Sb	—	
—	−85	—	—	
—	—	—	—	
—	—	—	—	
—	305	—	—	
—	—	—	—	
—	340	330	—	
—	—	—	—	
—	1412	3280	diamond	
—	—	—	—	
—	—	—	—	
—	—	—	—	
−210	−185	−112	—	
—	−131	−14·5	—	
—	—	—	—	
—	−70	58	—	
—	—	—	—	
—	5	153	—	
—	—	—	—	
—	—	—	—	
—	123	301	SnI_4	
—	(1610)	—	—	
574, 806	1722	—	—	
—	—	—	—	
—	1090	—	orthorhomb.	
—	—	—	—	
—	—	—	—	
—	—	—	—	
—	—	—	—	
—	—	—	—	
918	1072	1800	—	
—	—	—	—	
—	740	—	$PbCl_2$	
—	678	—	—	
524, 922	—	—	—	
922	2350	—	monoclinic	
—	—	—	—	
—	232	2605	Table Aa	
—	—	—	diamond	
—	—	—	—	
—	—	—	—	
—	247	670	orthorhomb.	
—	−33	112	—	
—	—	—	—	
—	232	641	—	
—	30	208	—	
—	320	718	—	
—	145	349	Table Aa	
—	—	—	Table Aa	
—	—	—	—	

Table A. Heats of formation,

Substance	ΔH°_{298} (kcal/mole)	Ref.	S°_{298} (cal/deg mole)	Ref.
$\langle SnO_2 \rangle$	−138·8 ± 0·5	907	12·5 ± 0·3	907
$\langle SnS \rangle$	−25·9 ± 0·8	724, 321	18·4 ± 0·2	133
$\langle Sn_3S_4 \rangle$	−88·5 ± 4·0	854	58·2 ± 4·0	854
$\langle Sn_2S_3 \rangle$	−63·0 ± 5·0	854	39·3 ± 3·6	854
$\langle SnS_2 \rangle$	−36·7 ± 5·0	854	20·9 ± 0·2	133
$\langle SnSO_4 \rangle$	−242·5 ± 7·5	—	33·12 ± 0·2	905
$\langle Sn(SO_4)_2 \rangle$	−389·4 ± 7·5	918	35·8	—
$\langle SnSe \rangle$	−21·2 ± 1·5	854	20·6 ± 1·5	—
$\langle SnSe_2 \rangle$	−29·8 ± 1·0	—	28·2 ± 2·0	—
$\langle SnTe \rangle$	−14·5 ± 0·35	762	23·6 ± 0·6	—
$\langle Sr \rangle$	0	—	12·5 ± 0·6	125
(Sr)	+39·2 ± 0·4	817	39·23	817
$\langle SrH_2 \rangle$	−43·0 ± 1·4	57	11·9 ± 1·6	702
$\langle SrF_2 \rangle$	−290·9 ± 0·7	817	19·63 ± 0·1	817
$\langle SrCl_2 \rangle$	−198·1 ± 0·6	817	27·45 ± 0·1	817
$\langle SrBr_2 \rangle$	−171·6 ± 0·6	817	34·3 ± 1·0	817
$\langle SrI_2 \rangle$	−134·2 ± 0·5	817	38·03 ± 0·2	817
$\langle SrO \rangle$	−141·5 ± 0·9	817	13·27 ± 0·1	817
$\langle SrO_2 \rangle$	−151·4 ± 3·6	918	14·1 ± 2·6	—
$\langle SrS \rangle$	−108·1 ± 8·0	198	16·5 ± 1·0	—
$\langle SrSO_4 \rangle$	−345·3 ± 5·0	198	28·2 ± 2·0	125
$\langle SrSe \rangle$	−95·0 ± 9·0	854	19·3 ± 1·5	854
$\langle SrTe \rangle$	−62·0 ± 8·0	—	—	—
$\langle Sr_3N_2 \rangle$	−93·4 ± 5·0	198	29·5 ± 2·5	—
$\langle SrC_2 \rangle$	−20·2 ± 4·0	683	17·0 ± 2·0	683
$\langle SrCO_3 \rangle$	−291·55 ± 2·0	918	23·2 ± 0·4	125
$\langle SrSiO_3 \rangle$	−31·2 ± 0·5*	342	23·0 ± 1·8	840
$\langle Sr_2SiO_4 \rangle$	−50·0 ± 0·8*	342	35·8 ± 2·0	840
$\langle SrTiO_3 \rangle$	−32·0 ± 1·7*	840	25·9 ± 0·3	840
$\langle Sr_2TiO_4 \rangle$	−38·0 ± 2·0*	840	38·0 ± 0·3	247
$\langle SrHfO_3 \rangle$	−18·7*	844	27·0 ± 2·7	—
$\langle SrAl_2O_4 \rangle$	−559·0 ± 4·0*	—	26·0 ± 6·0	—
$\langle SrMoO_4 \rangle$	−50·7 ± 2·1*	840	30·8 ± 1·2	840
$\langle SrWO_4 \rangle$	−44·5 ± 4·5*	840	32·0 ± 1·2	840
$\langle Ta \rangle$	0	—	9·92 ± 0·04	99
(Ta)	+186·8 ± 0·8	99	44·24	99
$\langle TaF_5 \rangle$	−454·9 ± 1·8	755	41·0 ± 3·2	—
(TaF_5)	−435·5 ± 4·0	755	84·35 ± 4·0	755
$\langle TaCl_3 \rangle$	−132·2 ± 1·8	603	37·0 ± 2·0	601
$\langle TaCl_4 \rangle$	−169·5 ± 1·9	601	46·0 ± 2·0	601
$\langle TaCl_5 \rangle$	−205·3 ± 0·5	755	56·0 ± 2·0	601
$(TaCl_4)$	−136·7 ± 2·0	755	91·5 ± 2·5	755
$(TaCl_5)$	−182·4 ± 2·0	755	102·6 ± 2·5	755
$(TaOCl_3)$	−187·5 ± 3·3	606	87·0 ± 3·0	606
$\langle TaBr_5 \rangle$	−143·0 ± 1·5	79	—	—
$\langle TaI_5 \rangle$	—	—	—	—
$\langle Ta_2O_5 \rangle$	−489·0 ±.1·0	817	34·2 ± 0·3	125
$\langle Ta_2N \rangle$	−65·2 ± 1·8	755	17·8 ± 2·2	—
$\langle TaN \rangle$	−60·3 ± 3·0	755	10·2 ± 0·4	755
$\langle Ta_2C \rangle$	−48·5 ± 3·4	755	20·0 ± 1·2	—
$\langle TaC \rangle$	−34·2 ± 0·8	755	10·13 ± 0·2	125
$\langle Ta_2Si \rangle$	−30·0 ± 6·0	840	25·2 ± 2·8	840
$\langle Ta_5Si_3 \rangle$	−80·0 ± 8·0	840	67·1 ± 6·5	840
$\langle TaSi_2 \rangle$	−28·5 ± 3·0	840	18·0 ± 1·6	840
$\langle TaCr_2 \rangle$	−6·45 ± 0·6	849	21·05 ± 0·05	849

standard entropies and structures (contd.)

tr.pts. (°C)	m.pt. (°C)	b.pt. (°C)	Structure	Remarks
425	(2000)	(2500)	rutile	
602	880	1276	NaCl, def.	
—	710	—	—	
—	—	—	—	
—	765	—	CdI_2	
—	—	—	—	
—	—	—	—	
—	880	—	—	
—	675	—	—	
—	806	—	NaCl	
555	768	1381	Ag	
—	—	—	—	
—	—	—	—	
(1148, 1211)	1477	2486	CaF_2	
729	875	2058	CaF_2	
645	657	2146	—	
—	538	1908	—	
—	2460	—	NaCl	
—	—	—	CaC_2	
—	—	—	NaCl	
—	2000	—	$BaSO_4$	
—	—	—	NaCl	
—	—	—	NaCl	
—	1030	—	Ba_3N_2	
—	—	—	—	
924	—	—	—	
—	1580	—	$BaCO_3$	*from oxides
—	—	—	—	*from oxides
—	1910	—	—	*from oxides
—	[1855]	—	—	*from oxides
—	—	—	—	*from oxides
659	—	—	—	*from oxides
—	—	—	—	*from oxides
—	1535	—	—	*from oxides
—	3015	5450	Na	
—	—	—	—	
—	97	—	—	
—	—	—	—	
—	—	—	—	
—	—	—	—	
—	217	234	—	
—	—	—	—	
—	—	—	—	
—	—	—	—	
—	263	347	—	
—	496	545	—	
—	1900	—	orthorhomb.	
—	(2700)	—	Co_2N	
—	2950	—	hexagonal	
—	—	—	—	
—	4000	—	NaCl	
—	—	—	—	
—	—	—	Ta_5Si_3	
—	—	—	hexagonal	
1530	2020	—	—	

Table A. Heats of formation,

Substance	ΔH°_{298} (kcal/mole)	Ref.	S°_{298} (cal/deg mole)	Ref.
⟨$TaFe_2$⟩	$-13{\cdot}8 \pm 2{\cdot}0$	755	$25{\cdot}5 \pm 1{\cdot}5$	755
⟨TaB_2⟩	$-50{\cdot}0$	892	10·6	892
⟨Tb⟩	0	—	$17{\cdot}52 \pm 0{\cdot}2$	—
⟨$TbCl_3$⟩	$-238{\cdot}3 \pm 2{\cdot}3$	918	$35{\cdot}3 \pm 1{\cdot}8$	—
⟨Tb_2O_3⟩	$-436{\cdot}8 \pm 2{\cdot}0$	240	—	—
⟨Tc⟩	0	—	*7·5 ± 0·5*	—
⟨TcO_2⟩	$-103{\cdot}5 \pm 3{\cdot}0$	284	*14·0 ± 1·2*	—
⟨TcO_3⟩	$-129{\cdot}0 \pm 6{\cdot}0$	22	*17·0 ± 1·5*	—
⟨Tc_2O_7⟩	$-266{\cdot}0 \pm 4{\cdot}0$	284	*44·0 ± 2·5*	—
⟨Te⟩	0	—	$11{\cdot}83 \pm 0{\cdot}05$	99
(Te)	$+50{\cdot}6 \pm 0{\cdot}6$	854	43·64	854
(Te_2)	$+38{\cdot}33 \pm 0{\cdot}5$	854	61·87	854
⟨TeF_6⟩	$-327{\cdot}2 \pm 0{\cdot}8$	718	$80{\cdot}25 \pm 0{\cdot}2$	718
⟨$TeCl_4$⟩	$-77{\cdot}4 \pm 2{\cdot}0$	854	$48{\cdot}0 \pm 5{\cdot}0$	854
⟨$TeBr_4$⟩	$-46{\cdot}6 \pm 2{\cdot}2$	907	$58{\cdot}2 \pm 5{\cdot}0$	—
⟨TeO_2⟩	$-77{\cdot}3 \pm 1{\cdot}0$	854	$17{\cdot}7 \pm 1{\cdot}5$	854
(TeO_2)	$-14{\cdot}2 \pm 2{\cdot}0$	854	$65{\cdot}7 \pm 2{\cdot}0$	854
⟨Th⟩	0	—	$12{\cdot}76 \pm 0{\cdot}05$	75
(Th)	$+142{\cdot}7 \pm 2{\cdot}5$	755	45·42	755
⟨ThH_2⟩	$-34{\cdot}3 \pm 1{\cdot}0$	755	$14{\cdot}2 \pm 1{\cdot}2$	755
⟨ThF_4⟩	$-504{\cdot}6 \pm 2{\cdot}0$	755	$33{\cdot}95 \pm 0{\cdot}1$	309
(ThF_4)	$-424{\cdot}3 \pm 3{\cdot}0$	755	$81{\cdot}7 \pm 3{\cdot}0$	755
⟨$ThCl_4$⟩	$-283{\cdot}6 \pm 0{\cdot}4$	755	$45{\cdot}5 \pm 1{\cdot}0$	755
($ThCl_4$)	$-233{\cdot}0 \pm 3{\cdot}0$	755	$93{\cdot}4 \pm 2{\cdot}0$	755
⟨$ThBr_4$⟩	$-230{\cdot}8 \pm 1{\cdot}2$	755	$54{\cdot}5 \pm 1{\cdot}5$	755
⟨ThI_4⟩	(−159)	755	(63·5)	755
⟨$ThOF_2$⟩	$-401{\cdot}0 \pm 4{\cdot}0$	755	$24{\cdot}2 \pm 2{\cdot}5$	755
⟨$ThOCl_2$⟩	$-294{\cdot}8 \pm 1{\cdot}2$	755	$28{\cdot}0 \pm 2{\cdot}5$	755
⟨$ThOBr_2$⟩	$-270{\cdot}0 \pm 4{\cdot}0$	755	$31{\cdot}3 \pm 2{\cdot}5$	755
⟨$ThOI_2$⟩	$-237{\cdot}2 \pm 1{\cdot}0$	755	$34{\cdot}6 \pm 3{\cdot}0$	755
⟨ThO_2⟩	$-293{\cdot}2 \pm 1{\cdot}0$	22	$15{\cdot}6 \pm 0{\cdot}03$	317
(ThO)	$-6{\cdot}8 \pm 0{\cdot}5$	755	$57{\cdot}37 \pm 0{\cdot}1$	755
(ThO_2)	$-120{\cdot}5 \pm 5{\cdot}0$	755	$61{\cdot}6 \pm 3{\cdot}0$	755
⟨$Th(SO_4)_2$⟩	$-607{\cdot}4 \pm 4{\cdot}0$	569	$39{\cdot}5 \pm 5{\cdot}0$	755
⟨ThS⟩	$-95{\cdot}5 \pm 1{\cdot}5$	755	$16{\cdot}68 \pm 0{\cdot}05$	755
⟨Th_2S_3⟩	$-258{\cdot}9 \pm 3{\cdot}0$	755	$42{\cdot}0 \pm 3{\cdot}0$	755
⟨ThS_2⟩	$-150{\cdot}0 \pm 9{\cdot}0$	755	$23{\cdot}0 \pm 0{\cdot}5$	755
⟨ThN⟩	$-90{\cdot}5 \pm 2{\cdot}5$	755	$13{\cdot}7 \pm 0{\cdot}2$	755
⟨Th_3N_4⟩	$-312{\cdot}0 \pm 6{\cdot}0$	755	$43{\cdot}7 \pm 0{\cdot}5$	755
⟨Th_2N_2O⟩	$-308{\cdot}0 \pm 5{\cdot}0$	755	$29{\cdot}7 \pm 1{\cdot}5$	755
⟨ThP⟩	$-86{\cdot}3 \pm 9{\cdot}0$	820	$16{\cdot}8 \pm 1{\cdot}5$	755
⟨Th_3P_4⟩	$-285{\cdot}5 \pm 26{\cdot}0$	820	$59{\cdot}0 \pm 7{\cdot}0$	755
⟨$ThC_{0{\cdot}97}$⟩	$-30{\cdot}0 \pm 2{\cdot}0$	755	$14{\cdot}1 \pm 0{\cdot}2$	755
⟨$ThC_{1{\cdot}94}$⟩	$-28{\cdot}0 \pm 2{\cdot}5$	755	$16{\cdot}8 \pm 0{\cdot}1$	755
⟨Th_3Si_2⟩	$-66{\cdot}8 \pm 12{\cdot}0$	751	$39{\cdot}75 \pm 7{\cdot}0$	751
⟨ThSi⟩	$-30{\cdot}1 \pm 4{\cdot}0$	751	$15{\cdot}0 \pm 2{\cdot}7$	751
⟨Th_3Si_5⟩	$-114{\cdot}1 \pm 20{\cdot}0$	751	$55{\cdot}4 \pm 5{\cdot}0$	751
⟨$ThSi_2$⟩	$-40{\cdot}8 \pm 7{\cdot}0$	751	$21{\cdot}3 \pm 2{\cdot}6$	751
⟨$ThAl_3$⟩	$-26{\cdot}7 \pm 4{\cdot}5$	755	$23{\cdot}2 \pm 3{\cdot}0$	—
⟨$ThRe_2$⟩	$-41{\cdot}6 \pm 1{\cdot}2$	884	$29{\cdot}55 \pm 0{\cdot}8$	884
⟨$ThMg_2$⟩	$-7{\cdot}5 \pm 1{\cdot}2$	864	$22{\cdot}1 \pm 2{\cdot}2$	864
⟨Ti⟩	0	—	$7{\cdot}32 \pm 0{\cdot}05$	855, 828
(Ti)	$+112{\cdot}3 \pm 1{\cdot}5$	52, 450, 15	43·07	—

standard entropies and structures (contd.)

tr.pts. (°C)	m.pt. (°C)	b.pt. (°C)	Structure	Remarks
—	—	—	—	
—	3100	—	—	
1290	1360	3220	—	
510	582	—	—	
—	—	—	Sc_2O_3	
—	2200	(4250)	—	
—	—	—	MoO_2	
—	disp.	—	—	
—	119	312	—	
—	450	988	Se	
—	—	—	—	
—	—	—	—	
−74	−39	—	—	
—	224	416	—	
—	380	421	—	
—	733	—	rutile	
—	—	—	—	
1360	1750	4975	Ag	
—	—	—	—	
—	—	—	—	
—	1110	1782	monoclinic	
—	—	—	—	
407	770	942	tetrag. b.c.	
—	—	—	—	
420	679	905	$ThCl_4$	
—	566	853	—	
—	—	—	—	
—	—	—	—	
—	—	—	—	
—	—	—	—	
—	3370	—	CaF_2	
—	—	—	—	
—	—	—	—	
—	—	—	—	
—	—	—	NaCl	
—	2000	—	hexagonal	
—	1917	—	$PbCl_2$	
—	2820	—	—	
—	dec.	—	NaCl	
—	—	—	—	
—	2950	—	—	
—	—	—	—	
—	(2500)	—	—	
1430, 1490	2610	—	—	
—	1900	—	—	
—	(1780)	—	—	
—	1745	—	—	
—	(1850)	—	—	
—	1120	—	—	
—	—	—	—	
—	—	—	—	
882	1670	3290	Mg	
—	—	—	—	

Table A. Heats of formation,

Substance	ΔH°_{298} (kcal/mole)	Ref.	S°_{298} (cal/deg mole)	Ref.
$\langle TiF_3\rangle$	−343·1 ± 9·0	817	21·0 ± 3·0	817
$\langle TiF_4\rangle$	−394·2 ± 0·9	817	32·0 ± 0·1	477
(TiF_4)	−370·8 ± 1·1	817	75·24 ± 0·5	817
$\langle TiCl_2\rangle$	−123·2 ± 4·0	817	20·88 ± 1·0	817
$\langle TiCl_3\rangle$	−172·5 ± 1·0	817	33·4 ± 0·3	817
$(TiCl_3)$	−128·9 ± 1·5	817	75·7 ± 1·0	817
$\{TiCl_4\}$	−192·2 ± 0·9	817	60·33	817
$(TiCl_4)$	−182·4 ± 0·9	817	84·8 ± 0·7	817
$\langle TiBr_3\rangle$	−131·4 ± 1·5	856	41·0 ± 2·5	856
$\langle TiBr_4\rangle$	−148·0 ± 0·5	762, 407, 197	57·2 ± 0·5	856
$\langle TiI_2\rangle$	−57·6 ± 6·0	84	33·0 ± 4·0	—
(TiI_2)	−2·2 ± 8·0	84	77·3 ± 4·0	—
$\langle TiI_4\rangle$	−89·8 ± 2·0	817	58·8 ± 1·6	817
(TiI_4)	−66·27 ± 2·0	817	103·48 ± 2·0	817
$\langle TiOCl\rangle$	−184·2 ± 3·8	608	17·5 ± 2·0	608
$\langle TiO\rangle$	(−129·7)	817 (1975)	8·3 ± 0·5	828
$\langle Ti_2O_3\rangle$	−363·5 ± 2·5	817 (1975)	18·46 ± 0·1	817 (1975)
$\langle Ti_3O_5\rangle$	−587·8 ± 1·8	817 (1975)	30·93 ± 0·3	817 (1975)
Rutile	−225·8 ± 0·4	817 (1975)	12·03 ± 0·05	817 (1975)
$\langle TiS\rangle$	−65·0 ± 7·0	854	13·5 ± 2·0	854
(TiS)	+79·0 ± 8·0	854	58·9 ± 0·5	854
$\langle Ti_5Si_3\rangle$	−138·5 ± 15·0	—	52·1 ± 3·4	—
$\langle TiSi\rangle$	−31·0 ± 3·5	99	11·7 ± 1·8	—
$\langle TiSi_2\rangle$	−32·1 ± 5·0	99	14·6 ± 2·2	—
$\langle TiN\rangle$†	−80·4 ± 0·8	788, 264	7·24 ± 0·1	828
$\langle TiC\rangle$‡	−43·9 ± 1·5	788, 264	5·8 ± 0·2	828
$\langle TiB_2\rangle$	−77·4 ± 0·9	700	6·81 ± 0·11	817
$\langle TiAl\rangle$§	−17·4 ± 0·6	839	12·5 ± 0·7	888
$\langle TiAl_3\rangle$	−34·0 ± 1·0	839	22·6 ± 1·5	—
$\langle Tl\rangle$	0	—	15·34 ± 0·05	99
(Tl)	+43·25 ± 0·3	99	43·23	99
$\langle TlF\rangle$	−77·8 ± 1·1	846	22·87 ± 0·05	846
$\langle TlF_3\rangle$	−136·9 ± 2·5	418	—	—
$\langle TlCl\rangle$	−48·8 ± 0·4	918	26·6 ± 0·3	918
$\langle TlCl_3\rangle$	−75·3 ± 2·5	918	36·4 ± 1·8	—
$\langle TlBr\rangle$	−41·27 ± 0·2	907	29·3 ± 0·1	907
$\langle TlI\rangle$	−29·56 ± 0·2	907	30·52 ± 0·05	907
$\langle Tl_2O\rangle$	−40·0 ± 1·8	907	32·1 ± 2·7	—
$\langle Tl_2O_3\rangle$	−93·3 ± 1·4	907	32·8 ± 3·3	—
$\langle Tl_2S\rangle$	−22·7 ± 1·0	854	38·0 ± 3·0	854
$\langle Tl_2SO_4\rangle$	−223·15 ± 0·2	907	48·0 ± 3·3	—
$\langle Tl_2Se\rangle$†	−22·5 ± 0·5	854	41·5 ± 1·5	854
$\langle TlSe\rangle$	−14·65 ± 0·3	854	24·5 ± 0·6	854
$\langle Tl_2Te\rangle$‡	−19·2 ± 2·8	854	41·6 ± 2·6	854
$\langle Tm\rangle$	0	—	(17·7)	99
$\langle TmCl_3\rangle$	−235·8 ± 2·5	918	35·1	—
$\langle U\rangle$	0	—	12·02 ± 0·1	99
(U)	125·0 ± 2·0	99	47·72	99
$\langle UH_3\rangle_\beta$	−30·4 ± 0·2	399	15·2 ± 0·1	399
$\langle UF_3\rangle$	−357·1 ± 1·2	797	28·0 ± 2·0	—
$\langle UF_4\rangle$	−453·7 ± 1·2	797	36·3 ± 0·1	399
$\langle UF_5\rangle$	−491·5 ± 0·6	399	45·0 ± 3·0	399
$\langle UF_6\rangle$	−522·4 ± 0·4	—	54·4 ± 0·5	399
(UF_6)	−510·7 ± 0·5	399	90·4 ± 0·5	399
$\langle UCl_3\rangle$	−213·5 ± 2·0	399	38·0 ± 0·2	399

standard entropies and structures (contd.)

tr.pts. (°C)	m.pt. (°C)	b.pt. (°C)	Structure	Remarks
—	Sb	1037	—	
—	Sb	285	—	
—	—	—	—	
—	Sb	1308	CdI_2	
−53	Sb	831	—	
—	—	—	—	
—	−24	136	—	
—	—	—	—	
—	—	—	—	
—	38	232	SnI_4	
—	—	—	—	
—	—	—	SnI_4	
—	—	—	—	
106	155	—	SnI_4	
—	920	—	—	
991	2020	—	NaCl	
200	1842	—	Cr_2O_3	
180	1780	—	monoclinic	
—	1860	—	Table Aa	
—	1930	—	—	
—	—	—	—	
—	—	—	—	
—	—	—	—	
—	—	—	—	
—	2950	—	NaCl	† $TiN_{0\cdot42-1\cdot0}$
—	3150	—	NaCl	‡ $TiC_{0\cdot6-11\cdot0}$
—	3225	—	—	
—	(1460)	—	AuCu	§ $TiAl_{0\cdot38-1\cdot4}$
—	(1340)	—	tetragonal	
234	304	1473	Mg	
—	—	—	—	
83	322	—	—	
—	—	—	—	
—	431	820	CsCl	
—	—	—	—	
—	460	825	CsCl	
178	442	833	CsCl	
—	579	—	Cu_2O	
—	834	—	Sc_2O_3	
—	457	—	CdI_2, def.	
430	632	—	K_2SO_4	
—	390	—	—	† $Tl_{1\cdot7-2\cdot0}Se$
—	330	—	—	
—	453	—	—	‡ $Tl_{1\cdot9-2\cdot0}Te$
1510	1545	1950	—	
—	845	—	—	
668, 775	1132	4135	orthorhomb.	
—	—	—	—	
—	—	—	cubic	
—	—	—	CeF_4	
—	1036	—	ThF_4	
—	—	—	tetragonal	
—	Sb	57	—	
—	—	—	—	
—	836	(1660)	hexagonal	

Table A. Heats of formation,

Substance	ΔH°_{298} (kcal/mole)	Ref.	S°_{298} (cal/deg mole)	Ref.
$\langle UCl_4 \rangle$	−251·3 ± 1·0	399	47·4 ± 0·3	399
$\langle UCl_5 \rangle$	−261·5 ± 2·0	399	58·0 ± 1·5	399
$\langle UCl_6 \rangle$	−270·7 ± 3·0	399	68·3 ± 0·4	399
$\langle UBr_3 \rangle$	−272·3 ± 2·0	399	45·0 ± 2·0	399
$\langle UBr_4 \rangle$	−197·5 ± 1·0	399	56·0 ± 2·0	399
$\langle UI_3 \rangle$	−90·0 ± 9·0	910	53·5 ± 5·0	—
$\langle UI_4 \rangle$	−102·0 ± 9·0	910	63·5 ± 5·0	—
$\langle UO_2F_2 \rangle$	−399·0 ± 5·0	399	32·4 ± 0·3	125
$\langle UOCl_2 \rangle$	−260·0 ± 3·0	399	33·1 ± 0·2	367
$\langle UOCl_3 \rangle$	−284·2 ± 3·0	399	42·0 ± 3·0	399
$\langle UO_2Cl_2 \rangle$	−297·0 ± 1·1	399	36·0 ± 0·1	399
$\langle UOBr_2 \rangle$	−240·0 ± 3·0	399	37·7 ± 0·2	367
$\langle UOBr_3 \rangle$	−235·9 ± 2·5	399	49·0 ± 3·0	399
$\langle UO_2Br_2 \rangle$	−276·6 ± 3·0	399	40·5 ± 2·5	399
$\langle UO_2 \rangle$†	−259·0 ± 0·6	399	18·6 ± 0·1	109
$\langle U_4O_9 \rangle$	−107·8 ± 4·0	399	80·3 ± 0·1	399
$\langle U_3O_8 \rangle$‡	−854·1 ± 2·0	399	67·5 ± 0·1	399
$\langle UO_3 \rangle$	−294·0	399	23·6 ± 0·1	109
$\langle UO_3{\cdot}H_2O \rangle$	−367·5 ± 6·0	399	*32·1 ± 1·2*	—
$\langle US \rangle$	−76·5 ± 3·5	847	18·64 ± 0·1	854
$\langle U_2S_3 \rangle$	−196·0 ± 12·0	—	45·0 ± 2·5	—
$\langle US_2 \rangle$	−110·0 ± 9·0	—	26·4 ± 0·1	854
$\langle U(SO_4)_2 \rangle$	−542·0 ± 17·0	399	38·5 ± 3·5	—
$\langle USe \rangle$	−65·9 ± 5·0	854	23·07 ± 0·1	854
$\langle UN \rangle$	−70·4 ± 1·0	399	14·97 ± 0·1	670
$\langle U_2N_3 \rangle$	−169·4 ± 2·5	399	30·8 ± 0·3	670
$\langle UBi \rangle$	−17·6 ± 4·0	99	23·2 ± 4·0	—
$\langle U_3Bi_4 \rangle$	−63·7 ± 9·0	99	81·9 ± 12·0	—
$\langle UBi_2 \rangle$	−24·5 ± 6·0	99	36·3 ± 6·0	—
$\langle UC \rangle$	−21·7 ± 1·0	—	14·08	426
$\langle U_2C_3 \rangle$	−49·0 ± 3·0	399	33·07	426
$\langle UC_{1\cdot9} \rangle$	−23·0 ± 2·0	—	16·3 ± 0·1	426
$\langle U_3Si \rangle$	−22·0 ± 3·0	—	40·0 ± 4·0	—
$\langle U_3Si_2 \rangle$	−40·8 ± 2·5	399	47·2 ± 2·0	399
$\langle USi \rangle$	−20·2 ± 1·0	399	15·9 ± 1·0	399
$\langle U_3Si_5 \rangle$	−84·7 ± 4·0	399	55·3 ± 3·2	399
$\langle USi_2 \rangle$	−31·0 ± 0·6	399	19·6 ± 1·2	399
$\langle USi_3 \rangle$	−31·2 ± 0·7	399	25·4 ± 1·5	399
$\langle U_5Ge_3 \rangle$	−56·2 ± 4·0	399	89·6 ± 3·3	399
$\langle UGe \rangle$	−14·7 ± 1·0	399	21·6 ± 0·8	399
$\langle U_3Ge_5 \rangle$	−57·3 ± 4·0	399	84·1 ± 3·2	399
$\langle UGe_2 \rangle$	−20·9 ± 1·5	399	31·2 ± 1·2	399
$\langle UGe_3 \rangle$	−25·5 ± 2·0	399	40·8 ± 2·0	399
$\langle UB_2 \rangle$	−39·3 ± 4·0	786	13·17 ± 0·05	786
$\langle UB_4 \rangle$	58·7	399	17·0 ± 2·0	399
$\langle UB_{12} \rangle$	103·5	399	33·4	399
$\langle UAl_2 \rangle$	−23·6 ± 1·3	775	25·5 ± 3·0	399
$\langle UAl_3 \rangle$	−27·3 ± 5·0	775	32·5 ± 3·0	399
$\langle UAl_4 \rangle$	−31·0 ± 5·0	775	39·0 ± 4·0	399
$\langle UCd_{11} \rangle$	−10·9 ± 2·2	819	139·4 ± 3·3	819
$\langle UFe_2 \rangle$	−7·7 ± 0·5	750	25·0 ± 4·0	399
$\langle URh_3 \rangle$	−62·0 ± 3·5	805	35·4 ± 3·3	805
$\langle URu_3 \rangle$	−52·0 ± 4·0	805	25·9 ± 4·0	805
$\langle V \rangle$	0	—	6·92 ± 0·1	817 (1975)
(V)	+123·2 ± 1·9	817 (1975)	43·54 ± 0·2	817 (1975)
$\langle VF_3 \rangle$	−275·0 ± 11·0	—	23·18 ± 0·1	918

standard entropies and structures (contd.)

tr.pts. (°C)	m.pt. (°C)	b.pt. (°C)	Structure	Remarks
—	590	789	ThCl	
—	—	—	—	
—	—	—	hexagonal	
—	727	—	UCl_3	
—	519	777	—	
—	—	—	$PuBr_3$	
—	506	—	—	
—	—	—	—	
—	—	—	—	
—	—	—	—	
—	—	—	—	
—	—	—	—	
—	—	—	—	
—	—	—	—	
—	2830	—	CaF_2	† $UO_{2\cdot0-2\cdot24}$ (1100°C)
—	—	—	cubic	
—	—	—	orthorhomb.	‡ $UO_{2\cdot61-2\cdot67}$
—	—	—	hexagonal	
—	—	—	—	
—	2460	—	—	
—	2030	—	—	
1350	1850	—	—	
—	—	—	—	
—	—	—	—	
—	2850	—	NaCl	
—	—	—	Sc_2O_3	
—	[1450]	—	NaCl	
—	[1150]	—	Th_3P_4	
—	[1010]	—	Cu_2Sb	
—	2100	—	—	
—	—	—	—	
1825	2180	—	CaC_2	
—	—	—	—	
—	1665	—	tetragonal	
—	—	—	orthorhomb.	
—	—	—	—	
—	1700	—	TiB_2	
—	1510	—	$AuCu_3$	
—	—	—	—	
—	—	—	—	
—	—	—	—	
—	—	—	—	
—	—	—	$AuCu_3$	
—	—	—	—	
—	2260	—	—	
—	—	—	—	
—	—	—	$C_{Te}Al_2$	
—	—	—	$AuCu_3$	
—	—	—	orthorhomb.	
—	—	—	—	
—	1200	—	—	
—	—	—	—	
—	[1850]	—	—	
—	1902	3350	Na	
—	—	—	—	
—	—	—	—	

Table A. Heats of formation,

Substance	ΔH°_{298} (kcal/mole)	Ref.	S°_{298} (cal/deg mole)	Ref.
⟨VF_4⟩	-335.4 ± 5.0	918	29.0 ± 2.7	—
(VF_5)	-342.7 ± 5.0	918	76.67 ± 0.1	918
⟨VCl_3⟩	-134.0 ± 3.5	505	31.3 ± 0.3	828
{VCl_4}	-136.2 ± 0.8	505	61.53 ± 0.5	—
(VCl_4)	-126.8 ± 1.0	—	84.0 ± 2.5	—
⟨VBr_2⟩	-38.0 ± 6.0	617	30.0 ± 3.0	617
⟨VBr_3⟩	-107.0 ± 6.0	617	34.0 ± 3.0	617
(VBr_4)	-94.0 ± 5.0	617	80.0 ± 3.0	617
⟨VI_2⟩	-63.0 ± 5.0	631	35.0 ± 3.0	631
⟨VI_3⟩	-67.0 ± 4.0	631	48.5 ± 3.0	631
⟨VO⟩	-103.2 ± 1.5	817 (1975)	9.32 ± 0.2	244
⟨V_2O_3⟩	-291.3 ± 1.5	817 (1975)	23.44 ± 0.3	817 (1975)
⟨VO_2⟩	-170.5 ± 0.9	817 (1975)	12.37 ± 0.2	817 (1975)
⟨V_2O_5⟩	-370.6 ± 1.5	817 (1975)	31.2 ± 0.4	817 (1975)
⟨$VN_{0.465}$⟩	-31.6 ± 1.5	817 (1975)	6.38 ± 0.04	817 (1975)
⟨VN⟩	-51.9 ± 2.5	310	8.91 ± 0.1	125
⟨$VC_{0.88}$⟩	-24.35 ± 0.5	Mah	6.61 ± 0.08	Mah
⟨VC⟩†	-24.1 ± 0.8	—	6.6 ± 0.12	—
⟨V_3Si⟩	-36.0 ± 7.5	840	24.25 ± 2.7	—
⟨V_5Si_3⟩	-110.5 ± 18.0	840	49.9 ± 3.0	—
⟨VSi_2⟩	-30.0 ± 9.0	840	14.0 ± 1.2	840
⟨W⟩	0	—	7.8 ± 0.1	99
(W)	$+203.0 \pm 3.0$	99	41.55	99
(WF_6)	-411.5 ± 0.6	817	81.5	817
⟨WCl_5⟩	-122.6 ± 9.0	817	52.0	817
(WCl_5)	-98.6 ± 8.0	817	69.9	817
(WCl_6)	-117.6 ± 6.0	817	100.2	817
⟨WOF_4⟩	-333.3 ± 15.0	817 (1975)	42.0 ± 2.0	817 (1975)
(WOF_4)	-319.5 ± 15.0	817 (1975)	80.0 ± 2.0	817 (1975)
(WO_2I_2)	-102.8 ± 2.5	799	90.1 ± 2.0	799
⟨WO_2⟩	-140.95 ± 1.5	546, 22	12.08 ± 0.4	817
⟨WO_3⟩	-201.45 ± 0.7	564, 22	18.14 ± 0.3	817
⟨WS_2⟩	-48.44 ± 4.0	307	20.0 ± 3.0	307
⟨W_5Si_3⟩	-32.3 ± 5.0	840	54.9 ± 3.3	840
⟨WSi_2⟩	-22.2 ± 2.2	840	15.3 ± 1.2	840
⟨W_2C⟩	-6.3 ± 0.6	—	—	—
⟨WC⟩†	-9.1 ± 2.5	772	10.0 ± 1.0	655
⟨$W(CO)_6$⟩	-226.3 ± 2.0	358, 416	—	—
⟨Y⟩	0	—	10.63 ± 0.1	703
(Y)	101.5 ± 2.0	339	42.87	790
⟨YF_3⟩	-410.7 ± 3.5	202	26.2 ± 1.5	—
⟨YCl_3⟩	-232.7 ± 1.0	332	(32.7)	332
⟨YI_3⟩	-147.7 ± 7.5	918	(49.5)	—
⟨Y_2O_3⟩	-455.4 ± 1.5	371	23.7 ± 0.1	—
⟨YN⟩	-71.5 ± 5.0	176	9.0 ± 1.2	—
⟨YRe_2⟩	-32.45 ± 1.2	884	26.3 ± 1.0	884
⟨Yb⟩	0	—	14.3 ± 0.06	99
⟨$YbCl_2$⟩	-191.1 ± 2.1	NBS 270	31.2 ± 1.5	—
⟨$YbCl_3$⟩	-229.4 ± 1.8	734	35.3 ± 2.8	—
⟨Yb_2O_3⟩	-433.7 ± 1.5	371	31.8 ± 0.1	—
⟨Zn⟩	0	—	9.95 ± 0.05	125
(Zn)	$+31.17$	176	38.45	991
⟨ZnF_2⟩	-182.7 ± 0.5	402	17.6 ± 0.1	415
⟨$ZnCl_2$⟩	-99.2 ± 0.25	907	26.64 ± 0.1	907

standard entropies and structures (contd.)

tr.pts. (°C)	m.pt. (°C)	b.pt. (°C)	Structure	Remarks
—	—	—	—	
—	—	—	—	
—	—	—	AsI_3	
—	−26	160	—	
—	—	—	—	
—	—	—	—	
—	—	—	—	
—	—	—	—	
—	—	—	—	
—	—	—	—	
—	1790	—	f.c.c.	
—	2070	—	Cr_2O_3	
67	1545	—	monoclinic	
—	670	[1690]	orthorhomb.	
—	[2600]	—	—	
—	2350	—	NaCl	
—	[2345]	—	NaCl	
—	—	—	—	† hypothetical
—	1935	—	—	
—	2010	—	—	
—	1680	—	—	
—	3410	5555	Na	
—	—	—	—	
—	3	17	—	
—	253	288	—	
—	—	—	—	
230	282	—	rhombohed.	
—	106	—	—	
—	—	—	—	
—	—	—	—	
—	[1724]	—	monoclinic	
777	1472	1837	monoclinic	
—	—	—	$MoSi_2$	
—	—	—	—	
—	2150	—	$MoSi_2$	
—	—	—	—	
—	—	—	hexagonal	† $WC_{0.98-1.01}$ (2500°C)
—	—	—	—	
1480	1526	3340	Mg	
—	—	—	—	
1077	1155	—	—	
—	721	(1510)	—	
—	—	—	—	
1057	2704	(4300)	Sc_2O_3	
—	>2670	—	NaCl	
—	—	—	—	
760	824	1194	—	
—	—	—	—	
—	854	—	—	
—	2439	—	Sc_2O_3	
—	420	907	Mg	
—	—	—		
—	950	1505	rutile	
—	317	732	$CdCl_2$	

Table A. Heats of formation,

Substance	ΔH°_{298} (kcal/mole)	Ref.	S°_{298} (cal/deg mole)	Ref.
⟨$ZnBr_2$⟩	−78·8 ± 0·6	907	32·5 ± 1·0	907
⟨ZnI_2⟩	−49·75 ± 0·2	907	38·6 ± 1·0	907
⟨ZnO⟩	−83·8 ± 0·1	907	10·43 ± 0·1	907
⟨ZnS⟩	−49·05 ± 0·7	907	13·78 ± 0·1	907
(ZnS)	+48·67 ± 3·2	907	57·0 ± 0·4	907
⟨$ZnSO_4$⟩	−234·55 ± 0·4	907	26·42 ± 0·1	907
⟨$ZnSe$⟩	−38·0 ± 2·0	854	16·8 ± 1·0	854
⟨$ZnSeO_3$⟩	−155·9 ± 0·75	907	23·5 ± 2·0	907
⟨$ZnTe$⟩	−28·5 ± 0·2	99	18·68 ± 0·4	99
⟨Zn_3N_2⟩	−5·3 ± 2·0	264	33·5 ± 2·8	—
⟨Zn_3As_2⟩	−32·2 ± 2·5	898	39·4 ± 2·9	898
⟨Zn_3P_2⟩	−38·0 ± 3·5	821	36·0 ± 2·5	821
⟨ZnP_2⟩	−24·3 ± 3·0	821	14·4 ± 4·5	821
⟨$ZnSb$⟩	−4·55 ± 0·3	99	19·75 ± 0·4	99
⟨$ZnCO_3$⟩	−195·5 ± 0·3	907	19·7 ± 0·3	125
⟨Zn_2SiO_4⟩	−7·8 ± 0·5*	840	31·4 ± 0·3	243
⟨Zn_2TiO_4⟩	−0·8 ± 0·5*	816	35·4 ± 0·4	816
⟨$ZnTiO_3$⟩	−1·6 ± 0·4*	816	22·7 ± 0·4	816
⟨$ZnFeO_4$⟩	−1·2 ± 1·0*	840	36·64 ± 0·1	907
⟨$ZnWO_4$⟩	−10·0 ± 0·5*	840	34·5 ± 2·5	840
⟨$ZnAl_2O_4$⟩	−10·75 ± 0·4*	815	20·8 ± 0·5	815
⟨$ZnCr_2O_4$⟩	−15·0 ± 0·3*	815	27·8 ± 0·3	815
⟨Zr⟩	0	—	9·32 ± 0·05	—
(Zr)	143·67 ± 2·0	—	43·32 ± 1·0	—
(ZrF_2)	−140·0 ± 10·0	—	—	—
(ZrF_3)	−265·0 ± 10·0	—	—	—
⟨ZrF_4⟩	−456·8 ± 0·5	—	25·0 ± 0·1	—
(ZrO)	20·0 ± 10·0	—	—	—
⟨$ZrCl_4$⟩	−234·4 ± 0·3	—	43·5 ± 0·6	—
($ZrCl_4$)	−207·9 ± 1·0	—	88·0 ± 1·0	—
⟨$ZrBr_4$⟩	−181·6 ± 1·0	—	—	—
($ZrBr_4$)	−155·0 ± 2·0	—	—	—
⟨ZrI_4⟩	−115·9 ± 1·0	—	—	—
⟨ZrO_2⟩	−263·1 ± 0·5	—	12·12(±0·4)	—
(ZrF_4)	−400·1 ± 1·0	—	76·0 ± 1·0	—
⟨ZrN⟩	−88·0 ± 0·6	—	9·29 ± 0·3	—
⟨$ZrC_{0·96}$⟩	−48·27 ± 0·6	—	7·93 ± 0·5	—
⟨Zr_2Si⟩	−81·0 ± 3·0	—	—	—
⟨Zr_5Si_3⟩	−146·7	—	—	—
⟨Zr_6Si_5⟩	−203·3	—	—	—
⟨$ZrSi$⟩	−35·3	—	—	—
⟨$ZrSi_{1·62}$⟩	−35·8	—	—	—
⟨$ZrSiO_4$⟩	−486·4 ± 2·0	—	20·2	—
⟨ZrB_2⟩	−77·4 ± 1·5	—	8·59	—
⟨ZrH_2⟩	−40·5(±0·5)	—	8·40	—
⟨ZrD_2⟩	−41·82 ± 0·5	—	9·18	—

standard entropies and structures (contd.)

tr.pts. (°C)	m.pt. (°C)	b.pt. (°C)	Structure	Remarks
—	402	670	orthorhomb.	
—	442	—	—	
—	1975	—	wurtzite	
1175	1775	—	Table Aa	
—	—	—	—	
+	—	—	—	
—	1525	—	ZnS	
—	621	—	—	
—	1290	—	ZnS	
—	—	—	Sc_2O_3	
190, 646, 692	945	—	—	
—	—	—	—	
—	980	—	—	
—	548	—	CdSb	
—	—	—	M_2CO_3	
—	1512	—	$BeSiO_4$	*from oxides
—	1549	—	spinel	*from oxides TiO_2(Rut). ZnO
—	decomp.	—	—	*from oxides
—	1590	—	—	*from oxides
—	—	—	—	*from oxides
—	—	—	—	*from oxides ZnO, $Al_2O_3(\alpha)$
—	—	—	—	*from oxides
—	1857	4305	Mg	
— —	—	—		
—	—	—	—	
—	—	—	—	
—	932	sub 904	ThF_4	
—	—	—	—	
—	437	sub 335	SnI_4	
—	—	—	—	
—	450	sub 360	—	
—	—	—	—	
—	499	sub 428	—	
1175, 2295	2680	—	monoclinic	
—	—	—	—	
—	2950	—	NaCl	
—	3420	—	$CuAl_2$	
—	2110	—	$CuAl_2$	
—	2150	—	Ti_5Si_3	
—	—	—	—	
—	2090	—	hexagonal	
—	2430	—	—	
—	—	—	tetragonal	
—	3200	—	—	
—	sub	908	—	
—	—	—	—	

Table Aa. Structures

Substance	Description	Atoms or molecules per cell	Coordination	Remarks
Elements				
Ag	cubic face centred	4A	12	metallic latice
As	rhombohedral	2A	3	homop. layers
diamond	cubic	8A	4	homop. lattice
graphite	Widely separated layers 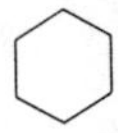	4A	3	homop. layers
Ga	orthorhombic	8A	1 + 2 + 2 + 2	homop.—metallic
Hg	rhombohedral, dist. face centred cubic	1A	6 + 6	metallic (+homop.)
I	orthorhombic, similar to Ga	8A	1	I_2 molecules
In	tetragonal, dist. face centred cubic	2A	4 + 8	metallic (+homop.)
Mg	hexagonal close packed	2A	12	metallic
α—Mn	cubic, dist. body centred	58A	12 − 16	
Na	cubic body centred	2A	8	metallic
S	rhombic	128A	2	S_8 molecules
S	monoclinic	48A	2	homop. chains
Se	hexagonal	3A	2	homop. chains
Sn, white	tetragonal, dist. diamond str.	4A	6	homop.—metallic
U	orthorhombic, dist. Mg str.	4A	12	metallic
AB-types				
CsCl	cubic body centred	1M	K8l, A8l	hetp. (+metallic)
NaCl	simple cubic	4M	K6l, A6l	hetp. to homop.
NaTl	superlattice of CsCl, Na and Tl form separate diamond str.	8M	K4l, A4l	hetp., homop. and metallic, mixed
NiAs	hexagonal	2M	K6l, A6l	Ni atoms form chains C.N.2
PdO	tetragonal, similar to SnO	2M	K4l, A4l	(homop.)
SnO	tetragonal	2M	K4l, A4l	(homop.)
ZnS blende*	cubic face centred	4M	K4l, A4l	ionic
ZnS wurtzite	hexagonal	2M	K4l, A4l	ionic
AB_2-types				
CaC_2	tetragonal	2M	Ca(4−2)l, C(4−2)l	
CaF_2	cubic	4M	K8l, A4l	ionic
$CdCl_2$	rhombohedral	1M	K6n, A3n	layer lattice
CdI_2	hexagonal	1M	K6n, A3n	layer lattice
Co_2P	Co hexagonal cl. p., P interstitial			
Cu_2O	cubic	2M	K2l, A4l	
Fe_2P	hexagonal	3M	complicated	
FeS_2 pyrites	cubic	4M	K6l, A6l	
marcasite	orthorhombic	2M	Fe6l, S6l	Fe atoms forming chains, S atoms forming S_2 islands
HgI_2 red	tetragonal	2M	Hg4l, I2l	C.N. of I atoms app. 12, ionic, layer lattice
$HgCl_2$	orthorhombic	4M	Hg2l, Cl,l	molecular layer str.
$MgCu_2(CeAl_2)$	cubic	8M	Mg12l, Cu6l	metallic
$MgNi_2$	hexagonal	8M	Mg12l, Ni6	metallic
$MgZn_2$	hexagonal cl. p.	4M	Mg12l, Zn6l	metallic

*quoted as ZnS in Table A.

Table Aa. Structures (contd.)

Substance	Description	Atoms or molecules per cell	Coordination	Remarks
MoS_2	hexagonal	2M	Mo6l, S3l	layer lattice of Mo
$PbCl_2$	orthorhombic	4M	{Pb(3 + 1)l, Cl(2 or 1)	
α, β-quartz	hexagonal	3M	Si4l, O2l	
β-cristobalite	cubic	8M	Si4l, O2l	
β-tridymite	hexagonal, similar to quartz	4M	Si4l, O2l	
TiO_2 rutile	tetragonal	2M	Ti6l, O3l	Ti chains O:C.N.1l
A_2B_3-types				
As_2O_3	cubic	16M	As3, O2	As_4O_6 molecules
Cr_2O_3	rhombohedral	2M	Cr6l, O4l	ionic, O: cl.p.
La_2O_3	hexagonal	1M	{La(4 + 3)l, O(2 + 2)l	ionic (+metallic)
Sc_2O_3	cubic, dist. CaF_2	16M	Sc6l, O4l	ionic, Sc: C.N.12
AB_3-types				
AlF_3	rhombohedral } similar	2M	K6l, A2l	ionic, C.N:F(8 + 2)
AsI_3	rhombohedral } similar	2M	K6l, A2l	ionic, C.N.F(8 + 2)
BiF_3	cubic	4M	K8l, A4l	ionic, Bi:C.N 12 F:C.N. 4 or 8
Co_3N	Co hexagonal cl.p., N interstitial			
MoO_3	orthorhombic	4M	Mo6l, O1, 2 or 3	O:C.N.12
ReO_3	cubic	1M	Re6l, O2l	ionic
SnI_4	cubic	8M	Sn4l, Il	Sn_4 molecules

Table B. Heats and temperatures of transformation

Substance	θ_t (°C)	λ_t (kcal/mole)	Ref.	θ_f (°C)	λ_f (kcal/mole)	Ref.
Ag	—	—	—	960·8	2·65 ± 0·1	147
AgCl	—	—	—	455	3·1 ± 0·1	124
AgBr	—	—	—	(430)	2·2 ± 0·4	124
AgI	147	(1·47)	124	558	2·25 ± 0·5	124
Ag_2S	176; 586	1·4 ± 0·4; —	192	830	2·7 ± 0·5	307
Ag_2SO_4	430	3·75	364	660	(4·3)	364
Ag_2Se	133	2·19 ± 0·1	—	—	—	907
$AgNO_3$	160	0·6	124	210	2·75 ± 0·2	124
AgCd	211; 443	—; 0·46 ± 0·2	264	—	—	—
$AgCd_2$	—	—	—	[578]	6·05 ± 0·3	264
Al	—	—	—	659	2·5 ± 0·03	147
AlF_3	454	0·15	314	—	—	—
Al_2Cl_6	—	—	—	⟨193⟩	17·0 ± 0·6	124
Al_2Br_6	—	—	—	97	5·4 ± 0·3	124
Al_2I_6	—	—	—	191	8·0 ± 0·6	124
Al_2O_3	—	—	—	2051	25·7	938
AlSb	—	—	—	1060	19·6 ± 0·5	907
AsF_3	—	—	—	−6	2·49 ± 0·03	125
AsF_5	—	—	—	−80	2·7 ± 0·4	198
$AsCl_3$	—	—	—	−16	2·4 ± 0·4	121
$AsBr_3$	—	—	—	31	2·8 ± 0·3	198
AsI_3	—	—	—	142	2·2 ± 0·6	25
As_4O_6	−33	1·35 ± 0·2	380	309	8·8 ± 1·0	198
As_2S_3	177	0·07	—	312	6·85	854
As_2Se_3	—	—	—	377	9·75 ± 1·0	854
As_2Te_3	—	—	—	375	11·2 ± 1·0	854
Au	—	—	—	1063	3·05 ± 0·1	147
AuSn	—	—	—	418	6·12 ± 0·2	264
$AuPb_2$	—	—	—	[254]	5·7 ± 0·4	124
AuCd	267 (o—d)	0·24	264	627	4·3 ± 0·2	264
B	—	—	—	2180	5·4	817
BF_3	−131	0·018	125	−129	1·014	125
B_2O_3	—	—	—	450	5·3 ± 0·3	198
B_2S_3	—	—	—	563	11·5	854
Ba	370	0·15 ± 0·08	147	729	1·83 ± 0·1	147
BaF_2	—	—	—	1290	6·8 ± 1·0	394
$BaCl_2$	922	4·1	474	962	4·0 ± 0·1	531
$BaBr_2$	—	—	—	854	7·5 ± 0·2	474
BaI_2	—	—	—	712	6·35 ± 0·1	474
BaO	—	—	—	1925	*1·38 ± 2·0*	—
$Ba(OH)_2$	—	—	—	408	3·4 ± 0·2	198
$BaSO_4$	1150	—	—	1350	9·7 ± 0·8	121
$BaCO_3$	806; 968	4·5; 0·7	154	dec.	—	—
$BaTiO_3$	5; 120	0·016; 0·047	14	1760	—	—
Be	1254	0·61	—	1287	2·92	755
BeF_2	227	0·053	—	552	1·135	755
$BeCl_2$	403	1·63 ± 0·15	—	415	*2·07 ± 0·06*	755
$BeBr_2$	—	—	—	488	(4·5)	25
BeI_2	—	—	—	480	*5·0*	755
BeO	—	—	—	2580	19·3 ± 1·0	755
$BeSO_4$	590, 635	0·266; 4·67	—	—	—	755

Table B. Heats and temperatures of transformation (contd.)

Substance	θ_t (°C)	λ_t (kcal/mole)	Ref.	θ_f (°C)	λ_f (kcal/mole)	Ref.
Bi	—	—	—	271·3	2·6 ± 0·05	147
$BiCl_3$	—	—	—	233	5·7 ± 0·2	—
$BiBr_3$	—	—	—	218	5·2 ± 0·2	—
BiI_3	—	—	—	409	9·35	924
Bi_2O_3	705	13·6 ± 3·0	—	824	*14·3*	933
Bi_2S_3	—	—	—	767	*18·7*	—
Bi_2Te_3	—	—	—	587	28·8 ± 0·6	443, 523
Bi_3Tl_2	—	—	—	213	8·7 ± 0·5	264
Br_2	—	—	—	−7·3	2·52 ± 0·1	23
C	—	—	—	3800	25 ± 2*	447
CCl_4	−48	1·1 ± 0·05	125	−23	0·60 ± 0·02	125
CBr_4	47	1·6 ± 0·05	387	90	0·95 ± 0·05	25
$COCl_2$	—	—	—	−128	1·37 ± 0·05	125
CS_2	—	—	—	−112	1·05 ± 0·1	125
COS	—	—	—	−139	1·13 ± 0·1	125
Ca	464	0·06	—	843	2·0 ± 0·1	147
CaF_2	1151	1·14 ± 0·2	125	1418	7·1 ± 0·1	125
$CaCl_2$	—	—	—	772	6·8 ± 0·1	125
$CaBr_2$	—	—	—	742	6·9 ± 0·1	531
CaO	—	—	—	2615	*(19·0)*	—
$CaSO_4$	1193	—	—	—	(6·7)	121
$Ca_2P_2O_7$	1140	1·6 ± 0·2	55	1353	24·1 ± 0·5	55
CaC_2	447	1·33 ± 0·2	124	2300	—	—
$CaCO_3$	50*	0·045 ± 0·02	126	dec.	—	—
$CaTiSiO_5$	—	—	—	1400	29·6 ± 1·0	305
$CaSiO_3$	1190	(1·3)	—	1540	(13·4)§	121
Ca_2SiO_4	675; 1420	1·06; 0·78	262	2130	—	—
$CaTiO_3$	1260	(0·55)	198	—	—	—
$CaO{\cdot}2B_2O_3$	—	—	—	990	27·1 ± 0·4	124
$CaO{\cdot}B_2O_3$	—	—	—	1160	17·7 ± 0·4	124
$2CaO{\cdot}B_2O_3$	531	1·1 ± 0·2	124	1310	24·1 ± 0·4	124
$3CaO{\cdot}B_2O_3$	—	—	—	1490	35·5 ± 0·6	124
$Ca_2Fe_2O_5$	—	—	—	[1480]	36·1 ± 1·2	348
$CaFe_2O_4$	—	—	—	[1240]	25·9 ± 1·0	348
$Ca_3P_2O_8$	1100	3·7 ± 0·3	124	—	—	—
$CaMgSi_2O_6$	—	—	—	1392	30·7	902
$CaAl_2Si_2O_8$	—	—	—	1553	39·9	902
$CaUO_4$	752	0·22	931	—	—	—
Cd	—	—	—	321	1·53 ± 0·04	147
$CdCl_2$	—	—	—	568	7·2 ± 0·2	632
$CdBr_2$	—	—	—	567	8·0 ± 0·2	632
CdI_2	—	—	—	390	4·95 ± 0·3	632
CdSb	—	—	—	456	7·7 ± 0·3	264
$CdSeO_3$	—	—	—	680	14·0 ± 2·0	907
Ce	725	0·7	99	795	1·25 ± 0·3	99
CeF_3	—	—	—	1437	13·5 ± 2·5	932
$CeCl_3$	—	—	—	817	12·8 ± 0·4	927
$CeBr_3$	—	—	—	732	12·4 ± 0·4	927
CeI_3	—	—	—	760	12·2	927
Cl_2	—	—	—	−101	1·53 ± 0·08	125

*at 48 Kbar

Table B. Heats and temperatures of transformation (contd.)

Substance	θ_t (°C)	λ_t (kcal/mole)	Ref.	θ_f (°C)	λ_f (kcal/mole)	Ref.
Co	430	0·11	—	1495	3·70 ± 0·1	99
$CoCl_2$	—	—	—	740	14·1 ± 3·0	208
Co_2Si	—	—	769	1330	*16·5 ± 1·5*	264
CoSi	—	—	769	1460	16·54	—
$CoSi_2$	—	—	—	1327	23·9	—
$CoWO_4$	713	0·445	—	940	—	—
Cr	—	—	—	1857	*5·0 ± 0·6*	—
$CrCl_2$	—	—	—	815	7·7 ± 1·5	198
CrO_3	—	—	—	198	3·4 ± 0·3	935
Cr_5Si_3	—	—	—	1647	64·25	769
$CrSi_2$	—	—	—	1457	30·55	769
Cs	—	—	—	29·8	0·50 ± 0·01	147
CsF	—	—	—	703	5·2 ± 0·2	—
CsCl	470	0·9	—	645	3·8	817
CsBr	—	—	—	635	5·65 ± 0·2	474
CsI	—	—	—	621	5·7 ± 0·2	474
Cu	—	—	—	1083	3·1 ± 0·1	147
CuBr	380; 465	1·4; 0·7	91	488	*(2·3)*	25
CuCl	407	1·45 ± 0·07	—	423	1·75 ± 0·1	939
CuI	—	—	—	588	*(2·6)*	25
CuF_2	—	—	—	770	9·4	817
Cu_2O	—	—	—	1236	13·58	817
Cu_2S	103; 350	0·92; 0·20	124	1130	2·6 ± 0·5	190
Cu_2Se	122	1·63 ± 0·08	—	—	—	930
Cu_2Te	160, 258, 317, 360, 568	0·053; 0·455; 0·23; 0·585; 0·477	—	—	—	930
$CuFeO_2$	818	0·09	—	1197	15·38	—
$CuFe_2O_4$	402; 522	0·18	—	1085	—	—
$DyCl_3$	—	—	—	651	6·1	927
$ErCl_3$	—	—	—	776	7·8	927
Fe	(760); 914; 1391	1·22; 0·16; 0·2	—	1536	3·29 ± 0·1	925
$FeCl_2$	—	—	—	677	10·3 ± 0·1	124
$FeCl_3$	—	—	—	304	10·3	817
$FeBr_3$	377	0·1	—	691	*12·0*	817
FeI_2	375	0·2	—	590	10·7	817
$Fe_{0·95}O$	—	—	—	1378	7·4 ± 0·2	42, 45
Fe_3O_4	—	—	—	1597	33·0 ± 2·0	198
Fe_2O_3	(680); (780)	0·16; —	42	1594	—	—
FeS	138; 325	0·57; 0·12	41	1195	7·73 ± 0·2	41
Fe_3C	190	0·18	124	1227	12·3 ± 1·0	124
$Fe(CO)_5$	—	—	—	−20	3·15 ± 0·1	559
Fe_2SiO_4	—	—	—	1220	22·0 ± 1·0	181
$FeTiO_3$	—	—	—	1470	21·7 ± 0·5	124
Ga	—	—	—	29·8	1·336 ± 0·01	147
$GaBr_3$	—	—	—	122	*(2·8)*	—
$GaCl_3$	—	—	—	78	2·75	907
GaI_3	—	—	—	212	*(3·9)*	—
$GdCl_3$	—	—	—	602	9·7	927
GdI_3	740	0·14	—	930	12·9	927

Table B. Heats and temperatures of transformation (*contd.*)

Substance	θ_t (°C)	λ_t (kcal/mole)	Ref.	θ_f (°C)	λ_f (kcal/mole)	Ref.
Ge	—	—	—	940	8·8 ± 0·3	614
$GeCl_4$	—	—	—	−49	—	—
$GeBr_4$	—	—	—	26	*2·9 ± 0·5*	—
GeO_2	1033	—	—	1116	*10·5 ± 2·0*	—
GeS	—	—	—	615	5·1 ± 1·5	622
HF	—	—	—	−83	0·94 ± 0·01	—
H_2O	—	—	—	0	1·436 ± 0·003	125
H_2O_2	—	—	—	−0·5	2·92 ± 0·1	64
H_2S	−170; −147	0·37; 0·11	125	−85·5	0·586 ± 0·01	125
H_2S_2	—	—	—	−90	1·8 ± 0·3	198
H_2Se	−191; −101	0·31; 0·27	198	−66	0·60	198
H_2Te	—	—	—	−49	(1·6)	—
Hf	1750	*(1·65)*	—	2230	*(5·75)*	—
HfO_2	1700	2·5	—	2810	25·0	—
Hg	—	—	—	−39	0·55 ± 0·005	30, 147
$HgCl_2$	—	—	—	278	4·65 ± 0·1	632
$HgBr_2$	—	—	—	238	4·3 ± 0·1	713
HgF_2	—	—	—	645	5·5	817
HgI_2	130	0·65 ± 0·1	124	250	4·5 ± 0·5	25
HgS	345	1·0	198	825	—	—
HoF_3	1070	—	—	1143	13·46 ± 0·15	937
$HoCl_3$	—	—	—	720	7·3	927
I_2	—	—	—	114	3·77 ± 0·08	124
In	—	—	—	157	0·78 ± 0·02	147
InCl	120	1·65	—	225	2·2	907
InBr	—	—	—	275	5·8	907
InI	—	—	—	365	5·35 ± 1·1	907
In_2S_3	390; 830	0·26; 0·96	—	—	—	854
InTe	—	—	—	695	8·6 ± 0·2	907
In_2Te_3	—	—	—	670	19·5 ± 0·3	907
InP	637	0·09	—	1055	15·0 ± 2·5	907
InAs	—	—	—	942	18·4 ± 2·0	748
InSb	—	—	—	525	11·8	748
Ir	—	—	—	2454	*(6·3)*	—
IrF_6	0·4	—	1·7	44	(1·2)	25
K	—	—	—	63·5	0·571 ± 0·005	147
KF	—	—	—	857	6·75 ± 0·3	124
KCl	—	—	—	772	6·35 ± 0·1	474
KBr	—	—	—	740	6·1 ± 0·1	474
KI	—	—	—	685	5·75 ± 0·1	474
KOH	249	1·51	—	400	2·24 ± 0·3	817
K_2CO_3	—	—	—	901	6·6	817
K_2SO_4	583	2·14 + 0·2	124	1069	8·8 ± 0·2	364
La	868	*0·68*	—	920	*2·03 ± 0·3*	—
$LaCl_3$	—	—	—	855	13·0 ± 0·2	474
$LaBr_3$	—	—	—	783	13·0 ± 1·0	474
LaF_3	—	—	—	1493	12·0 ± 0·2	937
LaI_3	—	—	—	778	13·4	927
Li	—	—	—	180·5	0·70 ± 0·03	147, 290

Table B. Heats and temperatures of transformation (contd.)

Substance	θ_t (°C)	λ_t (kcal/mole)	Ref.	θ_f (°C)	λ_f (kcal/mole)	Ref.
LiF	—	—	—	848	6·4 ± 0·15	288
LiCl	—	—	—	610	4·75 ± 0·1	474
LiBr	—	—	—	550	4·22 ± 0·2	474
LiI	—	—	—	469	3·5 ± 0·1	474
LiOH	—	—	—	473	5·0 ± 0·1	326
Li_2SO_4	575	6·8	364	859	(3·05)	364
LiH	—	—	—	689	5·4	817
Li_2O	—	—	—	1570	14·0	902
Li_2CO_3	350; 410	0·134; 0·535	—	720	10·7	817
$Li_2Si_2O_5$	936	0·225	—	1034	12·86	817
$LiBO_2$	—	—	—	844	8·09	817
Li_3AlF_6	475; 575; 705	0·5; 0·3; 0·1	—	785	20·6	817
Li_2BeF_4	—	—	—	472	10·6	755
Mg	—	—	—	650	2·1 ± 0·1	147
MgF_2	—	—	—	1263	13·9 ± 0·2	124
$MgCl_2$	—	—	—	714	10·3 ± 0·3	124
$MgBr_2$	—	—	—	710	(8·3)	121
MgO	—	—	—	2825	18·5	817
$MgSO_4$	—	—	—	1130	(3·5)	121
Mg_3N_2	550; 788	0·11; 0·22	124	dec.	—	—
$Mg_3(PO_4)_2$	—	—	—	1348	29·0	817
Mg_3Sb_2	930	—	—	1230	*36·5 ± 4·0*	—
Mg_3Bi_2	700	—	—	823	*25·5 ± 2·5*	—
Mg_2Si	—	—	—	1100	*20·5 ± 2·5*	—
$MgSiO_3$	630; 985	0·16; 0·39	—	1577	*18·0 ± 5·0*	817
Mg_2SiO_4	—	—	—	1890	*17·0 ± 5·0*	817
Mg_2Sn	—	—	—	778	*11·5 ± 1·0*	145
Mg_2Pb	—	—	—	550	9·4 ± 0·5	264
$MgTiO_3$	—	—	—	1840	*21·6*	817
$MgZn_2$	—	—	—	590	10·2 ± 1·0	264
Mn	720; 1090; 1136	0·53; 0·53; 0·43	—	1244	3·5	99
MnF_2	—	—	—	856	5·5	—
$MnCl_2$	—	—	—	650	9·0 ± 0·2	124
MnO	—	—	—	1875	13·0 ± 1·5	—
Mn_3O_4	1172	5·0 ± 0·3	124	1560	—	—
MnS	—	—	—	1530	6·25 ± 1·0	854
Mn_3C	1040	3·6 ± 0·3	124	—	—	—
Mn_5Si_3	—	—	—	1300	41·4	769
MnSi	—	—	—	1275	14·2	769
Mn_2SiO_4	—	—	—	1345	21·42	902
Mo	—	—	—	2620	**9·35 ± 0·3**	—
MoF_6	−9	1·96	—	17·5	1·0 ± 0·1	449, 445
$MoOF_4$	—	—	—	97	1·0 ± 0·1	449
$MoCl_5$	—	—	—	194	*(8·0)*	25
MoO_3	—	—	—	795	11·57	—
N_2	−237·5	0·0547	125	−210	0·172	125
NH_3	—	—	—	−78	1·352 ± 0·01	125
Na	—	—	—	97·8	0·63 ± 0·01	147
NaF	—	—	—	992	8·0 ± 0·1	314
NaCl	—	—	—	801	6·7 ± 0·1	474
NaBr	—	—	—	750	6·25 ± 0·1	474
NaI	—	—	—	660	5·65 ± 0·1	474

Table B. Heats and temperatures of transformation (contd.)

Substance	θ_t (°C)	λ_t (kcal/mole)	Ref.	θ_f (°C)	λ_f (kcal/mole)	Ref.
NaOH	293	1·52 ± 0·08	289	320	1·52 ± 0·08	289
Na_2O	750; 970	0·42; 2·85	—	1132	11·4	817
Na_2SO_4	249, 707	2·58; 0·08	—	884	5·5	817
Na_2CO_3	450	0·165	—	850	7·09	817
$Na_2Si_2O_5$	678; 707	0·1; 0·15	817	874	8·5 ± 0·4	124
Na_2SiO_3	—	—	—	1088	12·5 ± 0·3	124
Na_2TiO_3	287	0·4 ± 0·1	124	1030	16·8 ± 0·4	124
$Na_2Ti_2O_5$	—	—	—	985	26·2 ± 0·5	124
$Na_2Ti_3O_7$	—	—	—	1128	37·1 ± 0·5	124
Na_5Pb_2	—	—	—	400	12·0 ± 0·6	264
NaPb	—	—	—	368	3·9 ± 0·3	264
$NaHg_2$	—	—	—	350	6·3 ± 0·4	264
Na_2WO_4	588	(8·4)	198	695	(5·7)	121
Na_4SiO_4	—	—	—	1120	13·8	902
$NaBO_2$	—	—	—	966	8·66	817
Na_3AlF_6	565; 880	1·97; 0·09	—	1012	25·64	817
Na_2CrO_4	421	2·29 ± 0·15	—	797	5·79 ± 0·2	926
Na_2MoO_4	445; 592; 642	5·2; 0·5; 1·98	—	689	5·12	926
Nb	—	—	—	2468	7·0 ± 1·0	—
NbF_5	—	—	—	78	2·9 + 0·1	445
$NbCl_5$	—	—	—	205	6·9 ± 1·0	—
NbO_2	820; 930	0·72	—	1902	15·0	—
Nb_2O_5	—	—	—	1512	24·6	—
Nd	862	0·713	—	1024	1·705	99
NdF_3	—	—	—	1377	13·08 ± 0·2	937
$NdCl_3$	—	—	—	759	11·6	474
NdI_3	574	3·3	—	787	9·9	927
$NdBr_3$	—	—	—	684	10·8 ± 0·2	474
Ni	*c.* 358	0·14 ± 0·02	147	1455	4·1 ± 0·08	147
$NiCl_2$	—	—	—	⟨1030⟩	18·5 ± 0·4	41
$NiBr_2$	—	—	—	963	*13·0 ± 2·5*	—
Ni_3S_2	550	—	—	·790	(5·8)	121
$Ni(CO)_4$	—	—	—	−25	3·3 ± 0·01	—
Ni_2Si	—	—	—	1318	12·0 ± 0·6	264
NiSi	—	—	—	992	10·6 ± 0·6	264
Ni_3Sn_2	—	—	—	1264	27·6 ± 0·8	264
$NiO{\cdot}Fe_2O_3$	*c.* 582	0·85 ± 0·1	17	—	—	—
NpF_6	—	—	—	54·4	4·2 ± 0·2	567
O_2	−250; −229	0·0224; 0·178	125	−219	0·106	125
OsF_6	−0·4	1·97	449	33·4	1·76	449
OsO_4	white	—	—	42	(2·3)	120
	yellow	—	—	31	3·41	907
P_4, white	—	—	—	44	0·63	907
PCl_3	—	—	—	−91	1·08 ± 0·2	—
$POCl_3$	—	—	—	2	3·2 ± 0·6	120
P_4O_6	—	—	—	24	(3·4)	121
P_4O_{10}	o'rhomb.	—	—	570	(11·5)	198
	—	—	—	⟨420⟩	—	—
H_3PO_4	—	—	—	43	3·1	817
P_2S_5	—	—	—	287	4·91	854

Table B. Heats and temperatures of transformation (contd.)

Substance	θ_t (°C)	λ_t (kcal/mole)	Ref.	θ_f (°C)	λ_f (kcal—mole)	Ref.
Pb	—	—	—	327	1·15 ± 0·03	288, 147
PbF_2	340; 452	0·61; 0·65	—	818	4·16	817
$PbCl_2$	—	—	—	499	5·5	939
$PbBr_2$	—	—	—	370	4·96	817
PbI_2	—	—	—	410	3·87	817
PbO	489	0·18	—	886	6·57 ± 0·3	817
PbS	—	—	—	1119	8·7	922
$PbSO_4$	866	(4·1)	198	1090	(9.6)	121
PbSe	—	—	—	1076	11·8	922
PbTe	—	—	—	920	13·7	922
Pb_4SiO_6	—	—	—	[725]	25·9 ± 3·0	902
Pb_2SiO_4	—	—	—	743	12·8 ± 3·0	902
$PbSiO_3$	—	—	—	764	8·25 ± 2·0	902
$PbTiO_3$	*c.* 490	1·15 ± 0·1	226	—	—	—
Pd	—	—	—	1552	*4·0 ± 0·1*	—
$PdCl_2$	—	—	—	679	4·4 + 2·0	—
Pr	789	—	—	935	2·7 ± 0·5	121
$PrCl_3$	—	—	—	786	12·1 ± 0·2	474
$PrBr_3$	—	—	—	693	11·3 ± 0·2	474
PrI_3	—	—	—	738	12·5 ± 0·4	474
Pt	—	—	—	1769	*5·3 ± 0·3*	—
Pu	122; 205; 318; 451; 476	0·80; 0·15; 0·13; 0·02; 0·44	595	640	0·67 ± 0·03	595
PuF_3	—	—	—	1426	7·9 ± 1·0	185
PuF_6	—	—	—	52	4·45	755
$PuCl_3$	—	—	—	760	15·2 ± 1·5	185
$PuBr_3$	—	—	—	681	14·0 ± 1·5	185
PuI_3	—	—	—	777	*12·0*	755
Rb	—	—	—	39	0·525 ± 0·01	147
RbF	—	—	—	775	5·5 ± 0·4	929
RbCl	—	—	—	715	(4·4)	25
RbBr	—	—	—	680	(3·7)	25
RbI	—	—	—	640	(3·0)	25
Re	—	—	—	3180	*8·0 ± 1·0*	410
ReF_6	−3	2·1	—	19	1·0 ± 0·1	—
ReF_7	—	—	—	48	1·8 ± 0·2	566
$ReOF_4$	—	—	—	108	3·25	449
$ReOF_5$	30	1·34	—	41	1·2	449
Re_2O_7	—	—	—	296	(15·3)	120
S	95·5	0·096	124	119	0·40 ± 0·03	198
SO_2	—	—	—	−75·5	1·77 ± 0·03	125
SO_3	α	—	—	17	0·5	—
	β	—	—	32·5	2·5	—
	γ	—	—	⟨62⟩	6·1	—
Sb_2	—	—	—	630·5	9·5 ± 0·2	147
Sb_4	—	—	—	630·5	19·0 ± 0·4	147
$SbCl_3$	—	—	—	73	3·1 ± 0·3	121
$SbCl_5$	—	—	—	3	2·4 ± 0·3	25
$SbBr_3$	—	—	—	97	3·5 ± 0·3	25
SbI_3	—	—	—	170	*4·2 ± 0·6*	—

Table B. Heats and temperatures of transformation (*contd.*)

Substance	θ_t (°C)	λ_t (kcal/mole)	Ref.	θ_f (°C)	λ_f (kcal—mole)	Ref.
SbF_3	—	—	—	291	*5·1 ± 0·6*	—
Sb_2O_3	—	—	—	656	13·1	907
Sb_2S_3	26	0·013	—	550	11·45	854
Sb_2Se_3	—	—	—	615	12·85 ± 0·5	854
Sb_2Te_3	—	—	—	619	23·65 ± 0·4	854
Sc	1334	*0·5*	—	1538	*(4·0)*	—
$ScCl_3$	—	—	—	967	16·1	927
Se	106	0·0	—	220	1·4 ± 0·2	854
SeF_6	—	—	—	−35	2·0 ± 0·3	25
Si	—	—	—	1410	12·1 ± 0·4	391
SiH_4	−210	0·147	125	−185	0·16 ± 0·01	125
$SiCl_4$	—	—	—	−70	1·85 ± 0·1	125
SiI_4	—	—	—	122	*3·5 ± 0·7*	—
SiO_2	574; 806	0·175; 0·6	—	1722	2·6 ± 0·8	902
Sm	917	0·74	624	1072	2·13 ± 0·1	624
Sm_2O_3	922	0·25	587	—	—	—
Sn	13	0·50 ± 0·03	147	232	1·69 ± 0·03	147
$SnCl_4$	—	—	—	−34	2·2 ± 0·3	125
$SnBr_2$	—	—	—	232	(1·72)	121
$SnBr_4$	—	—	—	30	2·85 ± 0·5	218
SnI_4	—	—	—	145	4·6 ± 0·2	25
$SnCl_2$	—	—	—	247	(3·47)	936
SnI_2	—	—	—	320	4·5	936
SnS	602	0·16	—	881	7·55 ± 0·2	934
SnTe	—	—	—	806	10·8	922
SnO_2	410, 540	(0·45), (0·30)	198	—	—	—
Sr	589	0·20	—	770	2·4	99
SrF_2	—	—	—	1477	7·09 ± 0·15	817
$SrCl_2$	729	0·65	—	875	3·65 ± 0·2	531
$SrBr_2$	645	(2·9)	—	657	2·5 ± 0·3	474
SrI_2	—	—	—	538	4·7 ± 0·2	474
SrO	—	—	—	2460	*16·7 ± 2·0*	—
$SrCO_3$	924	(4·7)	154	—	—	—
$SrAl_2O_4$	659	0·46	—	—	—	923
Ta	—	—	—	3015	*5·9 ± 0·7*	—
TaF_5	—	—	—	97	4·5	755
$TaCl_5$	—	—	—	216	7·4 ± 0·8	755
$TaBr_5$	—	—	—	269	10·9 ± 0·5	—
TaI_5	—	—	—	496	(1·1)	2
Ta_2O_5	—	—	—	1900	36·12	755
$TbCl_3$	510	3·4	—	782	4·65	927
Tc_2O_7	—	—	—	119	11·5 ± 1·7	—
Te_2	—	—	—	450	8·36 ± 0·3	147
TeF_4	—	—	—	130	6·35 ± 1·0	110
TeF_6	−74	0·5	198	−38	2·1 ± 0·4	198
$TeCl_4$	—	—	—	224	4·5 ± 0·1	124
TeO_2	—	—	—	733	6·95 ± 0·2	—

Table B. Heats and temperatures of transformation (contd.)

Substance	θ_t (°C)	λ_t (kcal/mole)	Ref.	θ_f (°C)	λ_f (kcal—mole)	Ref.
Th	1360	0·86	—	1750	3·33	755
ThF_4	—	—	—	1110	*10·5*	755
$ThCl_4$	—	—	—	770	10·5	755
$ThBr_4$	420	1·0	—	679	*13·0*	755
ThI_4	—	—	—	566	11·5	755
$ThC_{1·94}$	1430; 1490	*0·5; 2·5*	—	—	—	755
Ti	882	0·80 ± 0·07	408	1667	*3·5 ± 0·5*	—
$TiCl_4$	—	—	—	−25	2·24 ± 0·1	125
$TiBr_4$	—	—	—	38	3·1 ± 0·2	197
TiO	991	0·82 ± 0·2	124	2020	*14·0 ± 2·0*	—
TiI_4	106	2·37	—	155	4·74	817
Ti_2O_3	200	0·215	—	1839	26·4	817
TiO_2	—	—	—	1870	16·0	902
Tl	234	0·09 ± 0·01	147	304	1·03 ± 0·03	147
TlCl	—	—	—	429	3·8 ± 0·3	463
TlBr	—	—	—	460	3·9 ± 0·2	463
TlF	83	0·08	—	322	3·32	463
TlI	178	0·22	—	442	3·52 ± 0·1	907
Tl_2O	—	—	—	579	7·24 ± 0·2	907
Tl_2SO_4	430	~0	—	632	5·5 ± 0·5	364
U	668; 775	0·70; 1·15	124	1130	*3·0 ± 0·7*	—
UF_4	—	—	—	1036	10·2	399
UF_6	—	—	—	⟨64⟩	4·59 ± 0·05	160, 179
UCl_4	—	—	—	590	10·7	399
UBr_4	—	—	—	519	13·2	399
UI_4	—	—	—	506	16·9 ± 1·8	399
UFe_2	—	—	—	1200	16·2 ± 0·9	—
V	—	—	—	1902	4·0 ± 0·6	—
VO_2	72	1·03 ± 0·2	124	[1360]	13·6 ± 0·3	124
V_2O_5	—	—	—	670	15·6 ± 0·4	124
W	—	—	—	3410	*(8·4)*	23
WF_6	−8·2	1·4	198	1	0·42 ± 0·1	449
WCl_5	—	—	—	240	(4·2)	—
WCl_4	230	1·73	—	281	1·54	236
WCl_6	177; 230	1·0; 3·77	—	282	1·6	817
WO_3	780	—	—	1472	17·55	817
WOF_4	—	—	—	105	2·26	449
Y	1485	1·2	437	1530	2·75 ± 0·1	437
Y_2O_3	1057	0·31	587	2439	—	—
YF_3	1077	7·75 ± 0·2	—	1155	6·7 ± 0·2	937
YCl_3	—	—	—	721	7·5	927
Zn	—	—	—	419·5	1·74 ± 0·03	147
ZnF_2	—	—	—	875	10·0 + 3·0	907
$ZnCl_2$	—	—	—	318	2·45 ± 0·2	463
$ZnBr_2$	—	—	—	402	3·74 ± 0·2	463
ZnS	1020	(3·2)	—	—	—	—
Zn_3Sb_2	405; 455	—	—	565	(22·0)	264
ZnSb	—	—	—	549	7·35	99
Zr	852	0·9 ± 0·1	132	1857	*4·6 ± 0·7*	—
ZrO_2	1175, 2295	1·42	—	2680	*20·8*	902
$ZrCl_4$	—	—	—	⟨437⟩	9·0 ± 2·5	584

Table C1. *Heat capacities of the elements*
$C_p = a + bT + cT^{-2} + dT^2$

Element	C_p (cal/deg mole)				Remarks	Range (°K)	Ref.
	a	b × 10^3	c × 10^{-5}	d × 10^6			
⟨Ag⟩	5·09	2·04	0·36	—	—	298–m.p.	124
{Ag}	7·30	—	—	—	—	m.p.–1600	124
⟨Al⟩	4·94	2·96	—	—	—	298–m.p.	124
{Al}	7·6	—	—	—	—	m.p.–2400	99
⟨As⟩	5·54	1·32	—	—	—	298–sb.pt.	124
⟨Au⟩	5·66	1·24	—	—	—	298–m.p.	124
{Au}	7·00	—	—	—	—	m.p.–1600	124
⟨B⟩	4·735	1·38	−2·20	—	—	298–1700	944
{B}	7·50	—	—	—	liquid	m.p.–3500	99
{B}	3·835	2·39	−1·50	—	amorphous	298–1240	644
⟨Ba⟩α	—	—	—	—	—	—	—
⟨Ba⟩β	−1·36	19·2	—	—	—	673–m.p.	148
{Ba}	11·5	—	—	—	—	m.p.–1125	148
⟨Be⟩α	5·07	1·36	−1·405	+0·23	—	298–1527	755
⟨Be⟩β	7·70	—	—	—	—	1527–m.p.	755
{Be}	7 04	—	—	—	—	m.p.–2800	755
(Be)	4·97	—	—	—	—	298–1900	755
	4·53	0·14	+6·51	—	—	1900–3000	755
⟨Bi⟩	4·49	5·40	—	—	—	298–m.p.	124
{Bi}	4·78	1·47	5·05	—	—	m.p.–820	99
(Bi)	4·97	—	—	—	—	298-1500	—
(Bi_2)	8·97	0·18	—	—	—	298-2000	—
(Br)	4·75	0·356	+0·10	—	—	298–2000	817
(Br_2)	8·93	0·11	−0·31	—	—	298–2000	817
⟨C⟩	0·026	9·307	−0·354	−4·155	graphite	298–1100	817
	5·841	0·104	−7·559	—	—	1100–4000	817
⟨C⟩	2·18	3·16	−1·48	—	diamond	298–1200	124
⟨Ca⟩α	6·064	−1·736	—	+5·67	—	298–720	99
⟨Ca⟩β	−0·086	9·86	—	—	—	720–m.p.	99
{Ca}	7·00	—	—	—	—	m.p.–1757	99
⟨Cd⟩	5·31	2·94	—	—	—	298–m.p.	124
{Cd}	7·10	—	—	—	—	m.p.–1100	124
⟨Ce⟩α	5·613	2·485	—	+0·97	—	298–999	99
⟨Ce⟩β	8·99	—	—	—	—	999–m.p.	99
{Ce}	9·01	—	—	—	—	m.p.–1500	99
(Cl_2)	8·82	0·06	−0·68	—	—	298–3000	124
⟨Co⟩α	5·11	3·42	−0·21	—	—	440–650	124
⟨Co⟩β	3·30	5·86	—	—	—	718–1400	124
{Co}	9·65	—	—	—	—	m.p.–1900	258
⟨Cr⟩	5·84	2·36	−0·88	—	—	298–m.p.	124
{Cr}	9·40	—	—	—	—	m.p.–	124

Table C1. Heat capacities of the elements (contd.)

$C_p = a + bT + cT^{-2} + dT^2$

Element	C_p(cal/deg mole)				Remarks	Range (°K)	Ref.
	a	b × 10^3	c × 10^{-5}	d × 10^6			
{Cs}	7·62	—	—	—	—	m.p.–330	360
⟨Cu⟩	5·41	1·50	—	—	—	298–m.p.	124
{Cu}	7·50	—	—	—	—	m.p.–1600	124
(F_2)	8·29	0·44	−0·80	—	—	298–2000	124
⟨Fe⟩αδ	8·873	1·474	—	—	$-56{\cdot}92T^{-1/2}$	298–1809	925
⟨Fe⟩γ	5·85	2·02	—	—	—	1187–1674	925
{Fe}	10·0	—	—	—	—	m.p.–1873	124
⟨Ga⟩	6·236	—	—	—	—	300	749
{Ga}	6·30	—	+0·30	—	—	m.p.–1200	99
⟨Ge⟩	5·16	1·40	—	—	—	298–m.p.	99
{Ge}	6·60	—	—	—	—	m.p.–1600	99
(H_2)	6·52	0·78	0·12	—	—	298–3000	124
⟨Hf⟩α	5·607	1·82	—	—	—	298–1000	515
	6·315	1·696	—	—	—	1000–2023	99
{Hg}	7·26	−2·74	—	+2·427	—	298–630	99
(Hg)	6·61	—	—	—	—	298–b.p.	124
(I)	4·874	0·096	+0·067	—	—	298–2000	817
(I_2)	8·94	0·136	−0·15	—	—	298–2000	817
⟨In⟩	5·81	2·50	—	—	—	298–m.p.	124
{In}	7·24	−0·33	—	—	—	m.p.–800	99
⟨Ir⟩	5·56	1·42	—	—	—	298–1800	124
⟨K⟩	6·04	3·12	—	—	—	298–m.p.	124
{K}	8·886	−4·57	—	+2·94	—	m.p.–1037	817
⟨La⟩	6·17	1·60	—	—	—	298–800	124
⟨Li⟩	3·33	8·21	—	—	—	298–m.p.	99
{Li}	5·85	1·31	2·07	−0·467	—	m.p.–1200	99
⟨Mg⟩	5·33	2·45	−0·103	—	—	293–m.p.	148
{Mg}	7·80	—	—	—	—	m.p.–1150	99
⟨Mn⟩α	5·70	3·38	−0·375	—	—	298–990	944
⟨Mn⟩β	8·33	0·66	—	—	—	990–1360	944
⟨Mn⟩γ	6·03	3·56	−0·443	—	—	298–1410	944
⟨Mn⟩δ	11·10	—	—	—	—	1410–1517	944
{Mn}	11·0	—	—	—	—	m.p.–b.pt.	944
⟨Mo⟩	5·961	0·626	-0·43	—	$1{\cdot}913{\times}10^{-10}T^3$	298-2893	—
{Mo}	9·64	—	—	—	—	2893-3383	—
(N_2)	6·66	1·02	—	—	—	298–2500	124
⟨Na⟩	19·71	−88·27	—	150·0	—	298–m.p.	99
{Na}	8·965	−4·594	—	2·542	—	m.p.–1200	99

Table C1. Heat capacities of the elements (contd.)

$C_p = a + bT + cT^{-2} + dT^2$

Element	C_p(cal/deg mole)				Remarks	Range (°K)	Ref.
	a	b × 10^3	c × 10^{-5}	d × 10^6			
⟨Nb⟩α	6·12	-0·027	—	0·496	$0.317\times10^5T^{-2}$	298-1400	
⟨Nb⟩β	9·40	-3·014	—	1·288	$-13{\cdot}123\times10^5T^{-2}$	1400-2741	—
{Nb}	9·98	—	—	—	—	2741-3159	—
⟨Nd⟩α	3·503	6·434	+1·07	—	—	298–1128	944
⟨Nd⟩β	10·65	—	—	—	—	1128–m.p.	944
{Nd}	11·66	—	—	—	—	m.p.–1400	944
⟨Ni⟩	4·06	7·04	—	—	—	298–633	124
	6·00	1·80	—	—	—	633–m.p.	124
{Ni}	9·20	—	—	—	—	m.p.–2000	124
(O_2)	7·16	1·0	−0·40	—	—	298–3000	124
⟨Os⟩	5·63	0·92	—	—	—	298–2000	944
⟨P⟩wh	4·57	3·78	—	—	white	298–317	817
⟨P⟩re	4·05	3·56	—	—	red	298–870	817
{P}	6·29	—	—	—	—	317–870	817
(P_2)	8·675	+0·19	−0·995	—	—	298–2000	817
(P_4)	19·56	+0·16	−3·21	—	—	298–2000	817
⟨Pb⟩	5·63	2·33	—	—	—	298–m.p.	288
{Pb}	7·75	−0·74	—	—	—	m.p.–1200	288
⟨Pr⟩α	6·21	0·3	—	+3·11	—	298–1068	99
⟨Pr⟩β	9·19	—	—	—	—	1068–m.p.	99
{Pr}	10·27	—	—	—	—	m.p.–1500	99
⟨Pt⟩	5·81	1·27	0·02	-0·01	—	298-2000	—
⟨Pu⟩α	5·91	5·8	—	—	—	298–395	99
⟨Pu⟩β	5·21	7·05	—	—	—	395–480	99
⟨Pu⟩γ	2·98	11·1	—	—	—	480–588	99
⟨Pu⟩δ	9·00	—	—	—	—	588–753	99
⟨Pu⟩ε	8·40	—	—	—	—	753–913	99
<Rb>	7·27	—	—	—	—	298–m.p.	124
⟨Re⟩	5·66	1·3	—	—	—	298–2000	99
<Rh>	5·49	2·06	—	—	—	298–1900	124
⟨Ru⟩	5·28	1·1	—	+0·3	—	298–2000	99
{S}	107·5	−229·4	−49·92	145·1	—	388–718	944
(S)	5·40	5·50	—	—	—	m.p.–b.p.	124
(S_2)	8·72	0·16	−0·90	—	—	298–3000	854
(S_6)	31·49	0·16	−4·11	—	—	298–2000	854
⟨Sb⟩	5·51	1·74	—	—	—	298–m.p.	124
{Sb}	7·50	—	—	—	—	m.p.–1300	124
(Sb_2)	8·935	—	−0·22	—	—	298–2000	99
(Sb_4)	19·85	—	−0·44	—	—	298–2000	99

Table C1. Heat capacities of the elements (contd.)

$$C_p = a + bT + cT^{-2} + dT^2$$

Element	C_p(cal/deg mole) a	b × 10^3	c × 10^{-5}	d × 10^6	Remarks	Range (°K)	Ref.
⟨Sc⟩α	6·02	0·16	—	+1·28	—	298–1608	99
⟨Sc⟩β	10·57	—	—	—	—	1608–1812	99
⟨Se⟩	4·275	6·0	—	—	hexag.	298–m.p.	854
{Se}	8·4	—	—	—	—	m.p.–800	854
(Se)	5·13	0·36	−0·22	—	—	298–2000	854
(Se_2)	10·66	−0·635	−0·598	—	—	298–2000	854
⟨Si⟩	5·72	0·59	−0·99	—	—	298–m.p.	124
{Si}	6·12	—	—	—	—	m.p.–1873	391
(Sm)	12·57	—	—	—	—	m.p.–1398	629
⟨Sn⟩wh	5·16	4·34	—	—	white	298–m.p.	99
{Sn}	8·29	−2·2	—	—	—	510–810	99
⟨Ta⟩	6·65	−0·52	−0·45	+0·47	—	298–3250	755
{Ta}	10·0	—	—	—	—	3250–4000	755
⟨Te⟩	4·58	5·25	—	—	—	273–m.p.	148
{Te}	9·0	—	—	—	—	m.p.–873	148
⟨Th⟩α	6·09	1·87	−0·122	—	—	298–1633	755
⟨Th⟩β	3·75	2·86	—	—	—	1633–m.p.	755
{Th}	11·0	—	—	—	—	m.p.–3000	755
⟨Ti⟩α	5·28	2·4	—	—	—	298–1155	944
⟨Ti⟩β	4·74	1·90	—	—	—	1155–m.p.	944
⟨Tl⟩α	3·74	6·04	+0·67	—	—	298–507	944
⟨Tl⟩β	5·00	5·00	—	—	—	507–m.p.	944
{Tl}	7·2	—	—	—	—	m.p.–1760	944
⟨U⟩α	2·61	8·95	+1·17	—	—	298–941	399
⟨U⟩β	10·0	—	—	—	—	941–1048	399
⟨U⟩γ	9·1	—	—	—	—	1048–1405	399
⟨V⟩	4·90	2·58	+0·2	—	—	298–m.p.	99
{V}	11·35	—	—	—	—	m.p.–2600	941
⟨W⟩	5·475	1·12	—	—	—	298-2500	—
⟨Y⟩α	5·72	1·805	+0·08	—	—	298–1752	99
⟨Y⟩β	8·37	—	—	—	—	1752–m.p.	99
{Y}	9·51	—	—	—	—	m.p.–2360	904
⟨Zn⟩	5·35	2·40	—	—	—	298–m.p.	124
{Zn}	7·50	—	—	—	—	m.p.–1200	124
⟨Zr⟩α	5·25	2·78	—	—	—	298–1135	817
⟨Zr⟩β	5·55	1·11	—	—	—	1135–m.p.	817

Table C2. Heat capacities of the compounds
$C_p = a + bT + cT^{-2}$ (cal/deg mole)

Substance	C_p in cal/deg mole			Remarks	Temp. range (°K)	Ref.
	a	b × 10^3	c × 10^{-5}			
$\langle AgCl\rangle$	14·88	1·0	−2·70	—	298–m.p.	124
$\{AgCl\}$	16·0	—	—	—	m.p.–900	124
$\langle AgBr\rangle$	7·93	15·40	—	—	298–m.p.	124
$\{AgBr\}$	14·9	—	—	—	m.p.–900	124
$\langle AgI\rangle\alpha$	5·82	24·10	—	—	298–423	124
$\langle AgI\rangle\beta$	13·5	—	—	—	423–600	124
$\langle Ag_2O\rangle$	14·18	9·75	−1·0	—	298–500	128
$\langle Ag_2S\rangle\alpha$	10·13	26·40	—	—	298–452	124
$\langle Ag_2S\rangle\beta$	21·64	—	—	—	452–850	124
$\langle Ag_2SO_4\rangle$	*23·1*	*27·9*	—	—	298–m.p.	124
$\langle Ag_2Se\rangle\alpha$	23·29	0·39	−4·00	—	298–406	854
$\langle Ag_2Se\rangle\beta$	20·4	—	—	—	406–698	124
$\langle Ag_3Sb\rangle$	19·53	16·0	—	—	298–700	124
$\langle AgNO_3\rangle\alpha$	8·76	45·2	—	—	298–433	124
$\langle AgNO_3\rangle\beta$	25·5	—	—	—	433–483	124
$\{AgNO_3\}$	30·6	—	—	—	483–600	124
$\langle Ag_2CO_3\rangle$	18·97	25·85	—	—	298–450	138
$\langle Ag_2CrO_4\rangle$	*31·6*	*16·0*	*−2·13*	—	298–500	817
(AlF)	8·9	—	−1·45	—	298–2000	77
$\langle AlF_3\rangle\alpha$	17·27	10·96	−2·30	—	298–727	314
$\langle AlF_3\rangle\beta$	20·93	3·0	—	—	727–1400	314
$(AlCl)$	9·0	—	−0·68	—	298–2000	77
$\langle AlCl_3\rangle$	13·25	28·00	—	—	273–m.p.	124
$\{AlCl_3\}$	31·2	—	—	—	m.p.–500	124
$(AlCl_3)$	*19·8*	—	*−2·64*	—	298–1800	77, 255
$(AlBr)$	8·905	0·14	−0·39	—	298–3000	817
$\langle AlBr_3\rangle$	29·5	—	—	—	m.p.–500	124
$\{AlBr_3\}$	29·87	—	—	—	370–550	817
(Al_2Br_6)	43·5	0·1	−3·12	—	298–2000	817
$\langle AlB_{12}\rangle$	*50·5*	*27·5*	*−20·4*	—	298–m.p.	—
$\langle AlI_3\rangle$	16·88	22·66	—	—	298–m.p.	124
$\{AlI_3\}$	29·0	—	—	—	m.p.–500	124
(Al_2O)	13·18	0·373	−2·21	—	298–2000	944
(AlO)	8·45	0·334	−1·10	—	298–2000	944
$\langle Al_2O_3\rangle$	25·48	4·25	−6·82	—	298–1800	480, 129
$\langle Al_2O_3\cdot 3H_2O\rangle$	17·30	91·2	—	—	298–425	124
$(AlOF)$	14·26	0·30	−3·27	—	298–2500	817
$(AlOCl)$	14·04	0·83	−2·08	$-0{\cdot}2\times10^{-6}\,T^2$	298–2500	817
$\langle Al_2S_3\rangle$	*24·42*	*8·62*	—	—	298–1370	—
$\{Al_2S_3\}$	*37·5*	—	—	—	1370–1800	—
$\langle Al_2(SO_4)_3\rangle$	87·55	14·96	−26·68	—	298–1100	124
$\langle AlN\rangle$	*8·22*	*4·05*	*−2·00*	—	298–1500	—
$\langle Al_4C_3\rangle$	36·97	6·866	−10·02	—	298–1800	942
$\langle Al_2SiO_5\rangle$	41·50	6·24	−13·85	$-0{\cdot}15\times10^{-6}\,T^2$	*298–1600	902
$\langle Al_2SiO_5\rangle$	42·54	5·71	−13·85	kyanite	298–1500	902
$\langle Al_2SiO_5\rangle$	40·57	6·87	−12·15	sillimanite	298–1500	902
$\langle Al_6Si_2O_{13}\rangle$	55·83	151·5	−13·35	$-92{\cdot}25\times10^{-6}\,T^2$	†298–600	902
	120·38	8·39	−55·0	$-0{\cdot}6\times10^{-6}\,T^2$	†600–2123	902
$\langle Al_2TiO_5\rangle$	43·63	5·30	−11·21	—	298–1800	280
$\{AsF_3\}$	30·3		—	—	298–b.p.	125
(AsF_3)	18·18		−2·66	—	298–1000	124
(AsF_5)	31·3		−7·1	—	298–2500	124

* Andalusite. † Mullite.

Table C2. Heat capacities of the compounds (contd.)
$C_p = a + bT + cT^{-2}$ (cal/deg mole)

Substance	C_p in cal/deg mole			Remarks	Temp. range (°K)	Ref.
	a	$b \times 10^3$	$c \times 10^{-5}$			
$\{AsCl_3\}$	31·9	—	—	—	298–371	124
$(AsCl_3)$	19·62	0·24	−1·42	—	298–1000	124
$\langle As_2O_3\rangle$	14·3	42·0	—	claudetite	298–m.p.	380
$\langle As_2O_5\rangle$	10·15	59·0	—	—	298–550	—
$\langle As_2S_3\rangle$	25·25	8·71	—	—	298–m.p.	—
$\langle As_2Te_3\rangle$	23·0	25·0	—	—	298–m.p.	763
$\langle AuSb_2\rangle$	17·12	4·64	—	—	298–628	124
$\langle AuSn\rangle$	11·13	3·8	—	—	298–m.p.	99
$\{AuSn\}$	14·5	—	—	—	m.p.–910	146
$\langle AuZn\rangle$	11·51	3·56	—	—	298–m.p.	124
$\{AuZn\}$	13·6	—	—	—	m.p.–1200	124
(BF_3)	12·44	6·70	−2·12	—	298–1000	124
(BCl_3)	16·86	2·86	−2·44	—	298–1000	124
(BBr_3)	17·83	2·04	−1·95	—	298–1000	124
$\langle B_2O_3\rangle$	13·63	17·45	−3·36	—	298–m.p.	128
$\{B_2O_3\}$	30·50	—	—	—	m.p.–1800	124
$\langle BN\rangle$	8·10	3·52	−5·51	cubic	298–1200	572
$\langle BP\rangle$	*5·25*	*6·7*	—	—	298–1100	—
$\langle B_4C\rangle$	22·99	5·40	−10·72	—	298–1373	124
$\langle H_3BO_3\rangle$	19·44	—	—	—	298	106
$\langle BaF_2\rangle$	*20·75*	*0·9*	*−3·6*	—	298–m.p.	—
$\{BaF_2\}$	*22·5*	—	—	—	m.p.–1900	—
$\langle BaCl_2\rangle\alpha$	22·20	0·76	−4·0	—	298–1198	531, 198
$\langle BaCl_2\rangle\beta$	26·61	—	—	—	1198–m.p.	531
$\{BaCl_2\}$	24·96	—	—	—	m.p.–1339	531
$\langle BaBr_2\rangle$	15·96	6·22	—	—	487–1126	531
$\{BaBr_2\}$	31·00	—	—	—	1126–1237	531
$\langle BaO\rangle$	12·74	1·04	−1·984	—	298–1270	154
$\langle Ba(OH)_2\rangle$	*17·5*	*13·5*	—	—	298–750	—
$\langle BaSO_4\rangle$	33·80	—	−8·43	—	298–1300	124
$\langle Ba(NO_3)_2\rangle$	30·05	35·7	−4·01	—	298–850	124
$\langle BaC_2\rangle$	*17·6*	*0·9*	*−2·31*	—	298–1500	—
$\langle BaCO_3\rangle\alpha$	20·77	11·70	−2·86	—	298–1040	154
$\langle BaCO_3\rangle\beta$	37·0	—	—	—	1079–1241	154
$\langle BaCO_3\rangle\gamma$	39·0	—	—	—	1241–	154
$\langle BaSiO_3\rangle$	*29·16*	*1·70*	*−7·46*	—	298–1300	902
$\langle BaAl_2O_4\rangle\alpha$	34·25	17·66	−10·97	—	298–600	923
$\langle BaAl_2O_4\rangle\beta$	33·03	7·56	—	—	600–1630	923
$\langle BaMoO_4\rangle$	32·33	6·80	−6·00	—	298–1700	—
$\langle BaUO_4\rangle$	29·04	9·62	—	—	298–1084	842
$\langle BaTiO_3\rangle$	29·03	2·04	−4·58	—	298–1800	43
$\langle Ba_2TiO_4\rangle$	43·00	1·60	−6·96	—	298–1800	43
(BeF_2)	12·23	2·02	−1·59	—	298–1000	755
	14·78	0·04	−7·83	—	1000–3000	755
$\langle BeF_2\rangle\alpha$	15·56	5·15	−4·20	—	298–500	755
$\langle BeF_2\rangle\beta$	1·24	18·59	+13·14	—	500–823	755
$\{BeF_2\}$	13·81	7·105	−4·51	—	823–1500	755
$(BeCl_2)$	13·89	0·83	−1·67	$-0{\cdot}18 \times 10^{-6}\,T^2$	298–3000	755
$\langle BeCl_2\rangle\alpha$	18·31	2·95	−3·29	—	298–688	755
$\langle BeCl_2\rangle\beta$	15·69	4·99	−2·01	—	298–676	755
$\{BeCl_2\}$	29·02	—	—	—	688–805	755
$\langle BeO\rangle$	9·94	2·44	−4·15	$-0{\cdot}32 \times 10^{-6}\,T^2$	298–2835	755

Table C2. Heat capacities of the compounds (*contd.*)

$C_p = a + bT + cT^{-2}$ (cal/deg mole)

Substance	C_p in cal/deg mole			Remarks	Temp. range (°K)	Ref.
	a	b × 10^3	c × 10^{-5}			
$\langle Be(OH)_2\rangle\beta$	*4·56*	*36·75*	*−0·1*	—	298–450	817
$\langle BeS\rangle$	*9·97*	*1·94*	*−2·30*	—	298–1800	755
$\langle BeSO_4\rangle\alpha$	26·96	−2·21	−6·62	$+18{\cdot}17 \times 10^{-6}\,T^2$	298–863	755
$\langle BeSO_4\rangle\beta$	38·73	—	—	—	863–908	755
$\langle BeSO_4\rangle\gamma$	69·32	−12·56	−171·51	$+2{\cdot}03 \times 10^{-6}\,T^2$	908–1500	755
$\langle Be_3N_2\rangle$	12·89	24·75	−4·22	—	298–430	755
$\langle Be_2C\rangle$	27·37	3·60	−14·23	—	430–1200	755
$\langle Be_2SiO_4\rangle$	29·65	18·77	−10·84	$-8{\cdot}0 \times 10^{-6}\,T^2$	298–1833	755
$\langle BeAl_2O_4\rangle$	29·26	22·62	−8·93	$-9{\cdot}36 \times 10^{-6}\,T^2$	298–1200	755
	37·58	4·10	—	—	1200–2143	755
$\langle BeAl_6O_{10}\rangle$	88·45	18·57	−27·75	$-2{\cdot}77 \times 10^{-6}\,T^2$	298–2150	755
	9·17	10·76	−2·01	$-2{\cdot}29 \times 10^{-6}\,T^2$	298–2400	817
$\langle BiI_3\rangle$	9·55	26·3	0·71	—	298–682	924
$\{BiI_3\}$	37·7	—	—	—	682–819	924
$\langle Bi_2O_3\rangle\alpha$	24·74	8·00	—	—	298–800	124
$\langle Bi_2O_3\rangle\beta$	*35·0*	—	—	—	978–1097	—
$\langle Bi_2S_3\rangle$	*26·25*	*9·8*	—	—	298–1050	—
$\langle Bi_2Te_3\rangle$	36·0	13·5	−3·12	—	373–m.p.	443
(CH_4)	5·65	11·44	−0·46	—	298–1500	124
(C_2H_2)	10·43	7·565	−1·794	$-1{\cdot}51 \times 10^{-6}\,T^2$	298–2000	944
(CF_4)	25·18	0·31	−23·04	$+4{\cdot}04 \times 10^{-6}\,T^2$	298–2000	944
(CCl_4)	24·90	0·48	−4·74	—	298–2000	817
$\{CCl_4\}$	32·0	—	—	—	298–b.p.	120
$\langle CBr_4\rangle\alpha$	31·7	—	—	—	295–320	387
$\langle CBr_4\rangle\beta$	33·0	—	—	—	320–m.p.	387
$\{CBr_4\}$	36·7	—	—	—	m.p.–450	124
(CBr_4)	25·03	0·60	−3·03	—	298–1000	124
(CO)	6·79	0·98	−0·11	—	298–2500	124
(CO_2)	10·55	2·16	−2·04	—	298–2500	124
(COS)	11·33	2·18	−1·83	—	298–1800	124
$\{CS_2\}$	18·4	—	—	—	298–b.p.	124
(CS_2)	12·45	1·60	−1·80	—	298–1800	124
$\langle CaH_2\rangle$	*7·1*	*5·85*	—	—	298–1053	—
(CaH)	7·68	0·74	−0·70	—	298–2000	124
$\langle CaF_2\rangle\alpha$	14·30	7·28	0·47	—	298–1424	124
$\langle CaF_2\rangle\beta$	25·81	2·50	—	—	1424–m.p.	124
$\{CaF_2\}$	23·90	—	—	—	m.p.–1800	124
$\langle CaCl_2\rangle$	17·18	3·04	−0·60	—	600–m.p.	124
$\{CaCl_2\}$	24·70	—	—	—	m.p.–1700	124
$\langle CaBr_2\rangle$	13·96	7·86	—	—	434–m.p.	531
$\{CaBr_2\}$	27·38	—	—	—	m.p.–1123	531
$\langle CaO\rangle$	11·86	1·08	−1·66	—	298–1177	154
$\langle Ca(OH)_2\rangle$	25·17	2·855	−4·533	—	298–1000	944
$\langle CaS\rangle$	*10·8*	*1·85*	—	—	298–2000	—
$\langle CaSO_4\rangle$	16·78	23·60	—	—	298–1400	124
$\langle CaSO_4{\cdot}\frac{1}{2}H_2O\rangle$	16·95	39·0	—	—	298–450	—
$\langle Ca_3N_2\rangle$	*33·17*	*3·70*	*−6·29*	—	298–1468	—
$\langle Ca_3P_2O_8\rangle\alpha$	48·24	39·08	−5·00	—	298–1373	124
$\langle Ca_3P_2O_8)\beta$	79·00	—	—	—	1373–2000	124
$\langle Ca_2P_2O_7)\alpha$	53·03	14·76	−11·16	—	298–1413	124
$\langle Ca_2P_2O_7\rangle\beta$	76·11	—	—	—	1413–1626	55
$\{Ca_2P_2O_7\}$	96·85	—	—	—	1626–1700	55
$\langle CaHPO_4\rangle$	*33·08*	*13·17*	*−9·65*	—	298–1000	—
$\langle CaC_2\rangle\alpha$	16·40	2·84	−2·07	—	298–720	124

Table C2. Heat capacities of the compounds (contd.)

$C_p = a + bT + cT^{-2}$ (cal/deg mole)

Substance	C_p in cal/deg mole			Remarks	Temp. range (°K)	Ref.
	a	b × 10^3	c × 10^{-5}			
$\langle CaC_2\rangle\beta$	15·40	2·00	—	—	720–1275	124
$\langle CaCO_3\rangle\alpha, \beta$	24·98	5·24	−6·20	calcite	298–1200	124
$\langle CaO{\cdot}MgO{\cdot}2SiO_2\rangle$	52·87	7·84	−15·74	diopside	298–1600	124
$\langle CaSiO_3\rangle$	25·85	3·94	−5·65	pseudowollastonite	298–1700	902
$\langle CaSiO_3\rangle$	26·64	3·60	−6·52	wollastonite	298–1450	124
$\beta\langle Ca_2SiO_4\rangle$	36·25	8·83	−7·24	—	298–1200	(243)
$\langle CaTiSiO_5\rangle$	42·39	5·54	−9·63	—	298–1670	305
$\{CaTiSiO_5\}$	66·8	—	—	—	1670–1811	305
$\langle CaTiO_3\rangle\alpha$	30·47	1·36	−6·69	—	298–1530	124
$\langle CaTiO_3\rangle\beta$	32·03	—	—	—	1530–1800	124
$\langle CaO{\cdot}2B_2O_3\rangle$	51·34	19·16	−17·16	—	298–m.p.	124
$\{CaO{\cdot}2B_2O_3\}$	106·30	—	—	—	m.p.–1800	124
$\langle CaO{\cdot}B_2O_3\rangle$	31·02	9·76	−8·07	—	298–m.p.	124
$\{CaO{\cdot}B_2O_3\}$	61·70	—	—	—	m.p.–1700	124
$\langle 2CaO{\cdot}B_2O_3\rangle\alpha$	43·75	11·50	−10·69	—	298–804	124
$\langle 2CaO{\cdot}B_2O_3\rangle\beta$	52·29	2·40	—	—	804–m.p.	124
$\{2CaO{\cdot}B_2O_3\}$	68·20	—	—	—	m.p.–1900	124
$\langle 3CaO{\cdot}B_2O_3\rangle$	56·44	10·42	−13·02	—	298–m.p.	124
$\{3CaO{\cdot}B_2O_3\}$	94·00	—	—	—	m.p.–1900	124
$\langle Ca_3Al_2O_6\rangle$	62·28	4·58	−12·01	—	298–1800	279
$\langle CaAl_2O_4\rangle$	36·0	5·96	−7·96	—	298–1800	279
$\langle CaAl_4O_7\rangle$	66·09	5·48	−17·8	—	298–1800	279
$\langle CaFe_2O_4\rangle$	39·42	4·76	−3·66	—	298–1510	348
$\{CaFe_2O_4\}$	54·90	—	—	—	1510–1800	348
$\langle Ca_2Fe_2O_5\rangle$	59·42	—	-11·68	—	298–1750	348
$\{Ca_2Fe_2O_5\}$	74·20	—	—	—	1750–1850	348
$\langle CaWO_4\rangle$	26·48	10·94	—	—	298–1073	940
$\langle CaUO_4\rangle\alpha$	27·63	11·9	—	—	293–1025	842
$\langle CaUO_4\rangle\beta$	27·01	12·58	—	—	1025–1134	842
$\langle CdF_2\rangle$	*14·35*	5·5	—	—	298–m.p.	—
$\langle CdCl_2\rangle$	11·3	21·9	—	—	298–842	—
$\{CdCl_2\}$	26·3	—	—	—	m.p.–1250	632
$\langle CdBr_2\rangle$	19·11	5·04	−2·03	—	298–m.p.	632
$\{CdBr_2\}$	24·3	—	—	—	m.p.–1150	632
$\langle CdI_2\rangle\alpha$	16·15	9·0	—	—	298–595	632
$\langle CdI_2\rangle\beta$	21·5	—	—	—	595–661	632
$\{CdI_2\}$	24·4	—	—	—	m.p.–1100	632
$\langle CdO\rangle$	11·53	1·525	−1·17	—	298–1500	856
$\langle CdS\rangle$	10·65	3·30	—	—	298–1200	854
$\langle CdSO_4\rangle$	*18·48*	*18·5*	—	—	298–m.p.	122
$\langle CdTe\rangle$	12·55	4·54	−1·76	—	298–m.p.	99
$\langle CdCO_3\rangle$	*10·3*	*31·5*	—	—	298–700	—
$\langle CdTiO_3\rangle$	*27·75*	*2·3*	*4·35*	—	298–1600	—
$\langle CeH_2\rangle$	*8·40*	*4·6*	—	—	298–1200	—
$\langle CeF_3\rangle$	*19·2*	*4·0*	—	—	298–m.p.	—
$\{CeF_3\}$	32·0	—	—	—	m.p.–2420	932
$\langle CeBr_3\rangle$	22·64	5·86	−0·07	—	298–m.p.	927
$\{CeBr_3\}$	36·5	—	—	—	m.p.–1150	927
$\langle Ce_2O_3\rangle$	25·77	9·9	−2·2	—	298–1200	884
$\langle CeO_2\rangle$	15·49	4·23	−1·815	—	298–1250	841
$\langle CeN\rangle$	*11·1*	*1·65*	*−1·73*	—	298–2000	—
$\langle Cl_2C_3\rangle$	*29·25*	*2·85*	*−4·8*	—	293–1500	—
$\langle ClC_2\rangle$	*16·4*	*2·2*	*−2·2*	—	298–1500	—

Table C2. Heat capacities of the compounds (*contd.*)

$C_p = a + bT + cT^{-2}$ (cal/deg mole)

Substance	C_p in cal/deg mole			Remarks	Temp. range (°K)	Ref.
	a	b × 10^3	c × 10^{-5}			
⟨CoF_2⟩	*15·3*	*3·74*	—	—	298–m.p.	—
⟨CoF_3⟩	*20·55*	*5·83*	−0·61	—	298–1200	—
⟨$CoCl_2$⟩	14·41	14·60	—	—	298–1000	124
⟨CoO⟩	11·54	2·04	0·4	—	298–1800	544
⟨Co_3O_4⟩	30·84	17·08	−5·72	—	298–1000	544
⟨CoS⟩	*10·6*	*2·51*	—	—	273–1373	124, 122
⟨$CoSO_4$⟩	30·09	9·91	—	—	298–1000	944
⟨$CoSi$⟩	11·75	2·89	−1·80	—	298–1733	769
{$CoSi$}	20·88	—	—	—	m.p.–1900	769
⟨$CoSi_2$⟩	16·935	4·46	−2·37	—	298–m.p.	769
{$CoSi_2$}	27·75	—	—	—	m.p.–1900	769
⟨$CoWO_4$⟩α	27·6	11·59	—	—	298–986	940
⟨$CoWO_4$⟩β	29·25	10·02	—	—	986–1073	940
⟨$CrCl_2$⟩	*15·25*	*5·96*	—	—	298–m.p.	—
{$CrCl_2$}	*24·0*	—	—	—	m.p.–	49
⟨$CrCl_3$⟩	19·44	7·03	—	—	298–s.p.	42
⟨Cr_2O_3⟩	28·53	2·20	−3·74	—	350–1800	124
⟨CrO_3⟩	*19·73*	*5·18*	*−4·18*	—	298–m.p.	—
{CrO_3}	*30·0*	—	—	—	m.p.–551	935
⟨$Cr_2(SO_4)_3$⟩	85·58	19·0	−21·45	—	298–1200	—
⟨Cr_2N⟩	15·24	6·8	—	—	273–800	204
⟨CrN⟩	9·84	3·9	—	—	273–800	204
⟨Cr_3C_2⟩	30·03	5·58	−7·46	—	298–1500	124
⟨Cr_7C_3⟩	56·96	14·54	−10·12	—	298–1700	124
⟨Cr_4C⟩	29·35	7·40	−5·02	—	298–1700	124
⟨$Cr(CO)_6$⟩	48·7	18·0	—	—	293–329	219
⟨Cr_5Si_3⟩	47·46	11·78	−6·12	—	298–1300	769
⟨$CrSi$⟩	12·43	2·09	−2·01	—	298–1700	769
⟨$CrSi_2$⟩	15·68	5·38	−1·855	—	298–1730	769
{$CrSi_2$}	21·5	—	—	—	1730–1900	769
⟨CrB⟩	10·12	3·83	−2·40	—	298–1200	572
⟨CrB_2⟩	9·63	10·7	—	—	298–1200	572
⟨CsF⟩	*11·3*	*2·71*	—	—	298–m.p.	124
⟨$CsCl$⟩α	12·78	1·23	−0·46	—	293–743	540
⟨$CsCl$⟩β	0·805	17·61	−0·89	—	743–m.p.	540
{$CsCl$}	13·86	4·28	—	—	m.p.–1170	540
⟨CsI⟩	7·06	10·36	+1·93	—	298–m.p.	540
{CsI}	−4·29	20·5	—	—	m.p.–1170	540
⟨Cs_2CO_3⟩	*27·59*	*16·57*	−*2·62*	—	298–m.p.	—
⟨$CuCl$⟩α	8·63	10·0	—	—	298–680	939
⟨$CuCl$⟩β	15·46	—	—	—	680–696	939
{$CuCl$}	13·90	—	—	—	696–1000	939
⟨$CuCl_2$⟩	15·42	12·00	—	—	298–m.p.	124
⟨$CuBr$⟩	13·1	—	—	—	300	93
⟨CuI⟩	*12·1*	*2·86*	—	—	298–m.p.	124
⟨CuI_2⟩	*20·1*	—	—	—	298–m.p.	124
⟨Cu_2O⟩	14·90	5·70	—	—	298–1200	124
⟨CuO⟩	9·27	4·80	—	—	298–1250	124
⟨Cu_2S⟩α	19·50	—	—	—	298–376	124
⟨Cu_2S⟩β	23·25	—	—	—	376–623	124
⟨Cu_2S⟩γ	20·30	—	—	—	623–1000	124
⟨CuS⟩	*10·6*	*2·64*	—	—	273–1273	124
⟨$CuSO_4$⟩	18·77	17·20	—	—	298–900	124

Table C2. Heat capacities of the compounds (contd.)
$C_p = a + bT + cT^{-2}$ (cal/deg mole)

Substance	C_p in cal/deg mole			Remarks	Temp. range (°K)	Ref.
	a	b × 10^3	c × 10^{-5}			
⟨$CuO\cdot CuSO_4$⟩	38·37	11·56	−7·42	—	298–1200	124
⟨Cu_2Se⟩α	14·0	18·5	—	—	298–395	930
⟨Cu_2Se⟩β	20·1	—	—	—	395–800	930
⟨Cu_2Te⟩α	14·3	12·8	—	—	298–433	930
⟨Cu_2Te⟩β	14·45	12·8	—	—	433–531	930
⟨Cu_2Te⟩γ	27·0	—	—	—	531–590	930
⟨Cu_2Te⟩δ	32·0	—	—	—	590–633	930
⟨Cu_2Te⟩ε	26·12	−4·1	—	—	633–841	930
⟨Cu_2Te⟩ζ	20·9	—	—	—	841–950	930
⟨Cu_3N⟩	21·7	—	—	—	273–373	124
⟨Cu_2Cd_3⟩	18·84	30·40	—	—	298–m.p.	124
{Cu_2Cd_3}	41·50	—	—	—	m.p.–1000	124
⟨Cu_5Cd_8⟩	58·78	54·40	—	—	298–m.p.	124
{Cu_5Cd_8}	106·00	—	—	—	m.p.–1000	124
⟨Cu_2Mg⟩	14·96	7·76	—	—	273–873	124
⟨$CuFeO_2$⟩α	23·42	1·8	−4·3	—	298–1091	826
⟨$CuFeO_2$⟩β	21·87	3·6	—	—	1091–m.p.	826
{$CuFeO_2$}	30·28	—	—	—	m.p.–1600	826
⟨$CuFe_2O_4$⟩	33·37	28·15	−5·6	—	298–675	826
	54·3	—	—	—	675–795	826
	39·68	9·8	—	—	795–1358	825
{$CuFe_2O_4$}	54·0	—	—	—	1358–1500	825
⟨$DyCl_3$⟩	22·60	4·30	−0·34	—	298–m.p.	927
{$DyCl_3$}	34·60	—	—	—	m.p.–1000	—
⟨$ErCl_3$⟩	22·84	4·2	−0·25	—	298–1049	780
⟨Eu_2O_3⟩α	29·60	6·48	−2·08	monoclinic	298–895	587
⟨Eu_2O_3⟩β	31·06	4·16	—	monoclinic	895–1802	587
⟨Eu_2O_3⟩	31·90	4·38	−3·04	cubic	298–1350	587
⟨$FeCl_2$⟩	18·94	2·08	−1·17	—	298–950	124
{$FeCl_2$}	24·40	—	—	—	950–1110	124
⟨$FeCl_3$⟩	29·56	—	−6·11	—	298–m.p.	245
⟨$Fe_{0\cdot947}O$⟩	11·66	2·00	−0·67	—	298–m.p.	42
{$Fe_{0\cdot947}O$}	16·30	—	—	—	m.p.–1800	42
⟨FeO⟩	12·38	1·62	−0·38	—	298–1200	124
⟨Fe_3O_4⟩α	21·88	48·2	—	—	298–900	42
⟨Fe_3O_4⟩β	48·00	—	—	—	900–1800	42
⟨Fe_2O_3⟩α	23·49	18·6	−3·55	—	298–950	42
⟨Fe_2O_3⟩β	36·0	—	—	—	950–1050	42
⟨Fe_2O_3⟩γ	31·70	1·76	—	—	1050–1750	42
⟨FeS⟩α	5·19	26·40	—	—	298–411	41
⟨FeS⟩β	17·40	—	—	—	411–598	41
⟨FeS⟩γ	12·20	2·38	—	—	598–m.p.	41
{FeS}	17·00	—	—	—	m.p.–1500	41
⟨FeS_2⟩	17·88	1·32	−3·05	—	298–1000	41
⟨Fe_4N⟩	*26·84*	*8·16*	—	—	273–1000	124
⟨Fe_2N⟩	*14·91*	*6·09*	—	—	273–1000	124
⟨Fe_3C⟩α	19·64	20·00	—	—	273–463	124
⟨Fe_3C⟩β	25·62	3·00	—	—	463–1500	124
⟨$FeSi$⟩	10·72	4·30	—	—	298–900	124
{$Fe(CO)_5$}	57·5	—	—	—	298–382	907
($Fe(CO)_5$)	47·26	9·4	−8·9	—	298–1000	907
⟨Fe_2SiO_4⟩	36·51	9·36	−6·70	—	298–1490	181

Table C2. Heat capacities of the compounds (contd.)

$C_p = a + bT + cT^{-2}$ (cal/deg mole)

Substance	C_p in cal/deg mole			Remarks	Temp. range (°K)	Ref.
	a	b × 10³	c × 10⁻⁵			
$\{Fe_2SiO_4\}$	57·5	—	—	—	1490–1724	181
$\langle FeTiO_3\rangle$	27·87	4·36	−4·79	—	298–1640	124
$\{FeTiO_3\}$	47·60	—	—	—	1640–1800	124
$\langle Fe_2TiO_4\rangle$	33·34	15·08	−3·40	—	298–1600	280
$\langle Fe_2TiO_5\rangle$	46·03	5·26	−7·41	—	298–1700	280
$\langle FeW_2\rangle$	28·29	9·90	−1·60	—	298–1250	99
$\langle FeCr_2O_4\rangle$	38·96	5·34	−7·62	—	298–1800	124
$\langle Ga_2O_3\rangle$	26·98	3·69	−5·02	—	298–1800	856
$\langle GaN\rangle$	*9·1*	*2·15*	—	—	298–1773	—
$\langle GaP\rangle$	*10·0*	*1·63*	—	—	298–1790	—
$\langle GaAs\rangle$	10·80	1·45	—	—	298–1238	747
$\langle GaSb\rangle$	10·90	3·0	—	—	298–985	747
$\langle GdCl_3\rangle$	20·67	8·20	+0·34	—	298–m.p.	927
$\{GdCl_3\}$	33·345	—	—	—	m.p.–1000	927
$\langle GdI_3\rangle\alpha$	24·29	1·80	−1·00	—	298–1013	927
$\langle GdI_3\rangle\beta$	30·64	—	—	—	1013–m.p.	927
$\{GdI_3\}$	37·25	—	—	—	m.p.–1300	927
$\langle Gd_2O_3\rangle$	27·28	3·54	−2·54	monoclinic	298–1852	587
$\langle Gd_2O_3\rangle$	28·72	2·84	−3·88	cubic	298–1550	587
(GeF_4)	23·33	2·0	−3·91	—	298–1000	256
$(GeBr_4)$	25·66	0·16	−1·22	—	298–1500	124
$\langle GeO_2\rangle$	15·92	2·77	−4·24	tetragonal	298–1500	754
$\langle GeO_2\rangle$	*16·47*	*2·35*	*−4·23*	hexagonal	298–1500	—
$\langle GeTe\rangle$	*11·65*	*3·06*	—	—	298–997	—
(HF)	6·43	0·82	0·26	—	298–2000	124
(HCl)	6·34	1·10	0·26	—	298–2000	124
(HBr)	6·25	1·40	0·26	—	298–1600	124
(HI)	6·29	1·42	0·22	—	298–2000	124
$\{H_2O\}$	18·03	—	—	—	273–373	124
(H_2O)	7·17	2·56	0·08	—	298–2500	124
(H_2S)	7·81	2·96	−0·46	—	298–2000	854
$\{H_2S_2\}$	22·0	—	—	—	273–b.p.	31
$\langle HfF_4\rangle$	31·9	0·75	−9·0	—	298–1100	540
$\langle HfCl_4\rangle$	31·47	—	−2·38	—	298–485	181
$\langle HfO_2\rangle$	17·39	2·08	−3·48	—	298–1800	181
$\langle HfC\rangle$	12·0	1·18	−2·4	—	298–2500	—
$\langle HfB_2\rangle$	15·57	4·40	−4·50	—	298–1600	572
$\langle HgCl\rangle$	*11·05*	*3·70*	—	—	273–m.p.	124
$\langle HgCl_2\rangle$	*15·28*	*10·4*	—	—	273–553	124
$\langle HgI\rangle$	*11·4*	*4·61*	—	—	273–563	124
$\langle HgI_2\rangle\alpha$	18·50	—	—	—	298–403	124
$\langle HgI_2\rangle\beta$	20·20	—	—	—	403–m.p.	124
$\{HgI_2\}$	25·00	—	—	—	m.p.–600	124
$\langle HgO\rangle$	*9·00*	*6·0*	—	—	—	117
$\langle HgS\rangle$	*10·46*	*3·72*	—	red	298–618	854
$\langle HgS\rangle$	*10·52*	*3·63*	—	black	298–1000	854
$\langle HoF_3\rangle$	25·54	0·115	−5·04	—	298–1343	901
$\langle HoF_3\rangle$	39·04	−3·93	−57·14	—	1343–1416	901
$\{HoF_3\}$	22·85	0·04	−0·85	—	m.p.–1843	901

Table C2. Heat capacities of the compounds (*contd.*)
$C_p = a + bT + cT^{-2}$ (cal/deg mole)

Substance	C_p in cal/deg mole			Remarks	Temp. range (°K)	Ref.
	a	b × 10^3	c × 10^{-5}			
⟨$HoCl_3$⟩	22·84	3·10	−0·23	—	298–993	927
{$HoCl_3$}	35·53	—	—	—	993–1100	927
⟨Ho_2O_3⟩	30·08	1·66	−2·75	—	298–1621	909
⟨In_2O_3⟩	*29·6*	*1·9*	*−5·51*	—	298–1600	854
⟨$IrCl_3$⟩	*20·3*	*4·5*	*−1·0*	—	298–1080	—
⟨IrO_2⟩	*17·33*	*1·93*	*−3·68*	—	298–1400	—
⟨Ir_2S_3⟩	*26·0*	*9·68*	*−2·52*	—	298–1200	—
⟨KF⟩	11·02	3·12	—	—	298–m.p.	124
⟨KCl⟩	9·89	5·20	0·77	—	298–m.p.	124
{KCl}	16·00	—	—	—	m.p.–1200	124
(KCl)	8·94	0·18	−0·24	—	298–2500	817
⟨KBr⟩	12·84	2·50	−2·84	—	600–1000	39
⟨KI⟩	21·10	−8·38	−10·38	—	600–1000	39
⟨K_2O⟩	22·86	−1·18	−2·64	$+5·66 \times 10^{-6}\ T^2$	298–1100	902
⟨K_2SO_4⟩α	28·77	23·80	−4·26	—	298–856	124
⟨K_2SO_4⟩β	33·60	13·40	—	—	856–m.p.	124
{K_2SO_4}	47·80	—	—	—	m.p.–1700	124
⟨K_2CO_3⟩	19·19	26·06	—	—	630–m.p.	533
{K_2CO_3}	36·95	10·64	—	—	m.p.–1250	533
⟨LaF_3⟩	24·32	0·873	0·708	—	298–950	901
	−12·37	27·46	+101·9	—	950–m.p.	901
⟨LaI_3⟩	23·22	4·735	−0·653	—	298–m.p.	927
{LaI_3}	36·27	—	—	—	m.p.–1200	927
⟨La_2O_3⟩	28·86	3·08	−3·28	—	298–1171	16
⟨LaN⟩	*11·1*	*1·65*	*−1·73*	—	298–1800	—
⟨LiF⟩	9·14	5·19	—	—	298–m.p.	288
{LiF}	15·50	—	—	—	m.p.–1170	288
⟨LiCl⟩	*11·0*	*3·40*	—	—	298–m.p.	124
⟨Li_2O⟩	14·94	6·08	−3·38	—	298–1045	325
⟨LiOH⟩	11·99	8·24	−2·27	—	298–m.p.	325
{LiOH}	20·74	—	—	—	m.p.–880	325
⟨Li_3N⟩	11·73	23·00	—	—	298–880	124
⟨Li_2CO_3⟩	3·82	42·72	—	—	560–m.p.	533
{Li_2CO_3}	30·85	13·55	—	—	m.p.–1150	533
⟨Li_2SiO_3⟩	28·39	9·86	−6·49	$-1·46 \times 10^{-6}\ T^2$	298–m.p.	902
($LiAlF_4$)	27·50	5·03	−6·37	$-1·54 \times 10^{-6}\ T^2$	298–2200	817
⟨Li_3AlF_6⟩α	49·22	26·25	−7·72	—	298–748	817
⟨Li_3AlF_6⟩β	68·0	—	—	—	748–848	817
⟨Li_3AlF_6⟩γ	70·5	—	—	—	848–978	817
⟨Li_3AlF_6⟩δ	73·0	—	—	—	978–1058	817
{Li_3AlF_6}	86·0	—	—	—	1058–2000	817
⟨Li_3BeF_4⟩	21·99	35·36	−0·24	—	298–m.p.	902
{Li_2BeF_4}	55·5	—	—	—	m.p.–2000	902
⟨$LiBeF_3$⟩	*18·43*	*20·29*	*−2·09*	—	298–633	817
⟨MgH_2⟩	6·50	11·8	−1·4	—	298–620	817
⟨MgF_2⟩	16·93	2·52	−2·20	—	298–m.p.	124
{MgF_2}	22·60	—	—	—	m.p.–1800	124
⟨$MgCl_2$⟩	18·90	1·42	−2·06	—	298–m.p.	124
{$MgCl_2$}	22·10	—	—	—	m.p.–1500	124
⟨MgO⟩	11·71	0·75	−2·80	—	298–3098	902

Table C2. *Heat capacities of the compounds* (*contd.*)

$C_p = a + bT + cT^{-2}$ (cal/deg mole)

Substance	C_p in cal/deg mole a	b × 10^3	c × 10^{-5}	Remarks	Temp. range (°K)	Ref.
$\langle MgS \rangle$	*10·30*	*1·97*	—	—	298–2000	854
$\langle Mg_3N_2 \rangle \alpha$	22·81	7·30	—	—	298–823	944
$\langle Mg_3N_2 \rangle \beta$	29·60	—	—	—	823–1061	944
$\langle Mg_3N_2 \rangle \gamma$	29·54	—	—	—	1061–1300	944
$\langle Mg_3(PO_4)_2 \rangle$	29·025	80·255	+0·26	$-26{\cdot}0 \times 10^{-6}\, T^2$	298–1621	817
$\{Mg_3(PO_4)_2\}$	113·445	—	—	—	1621–200	817
$\langle MgCO_3 \rangle$	18·62	13·80	−4·16	—	298–750	124
$\langle Mg_2Si \rangle$	17·52	3·58	−2·11	—	298–873	241
$\langle MgSiO_3 \rangle$	22·0	7·9	−4·25	clinoenstatite	298–903	902
$\langle MgSiO_3 \rangle$	28·76	—	—	rhomb. enstatite	903–1258	902
$\langle MgSiO_3 \rangle$	29·26	—	—	proto-enstatite	1258–1850	902
$\{MgSiO_3\}$	35·0	—	—	—	m.p.–2300	902
$\langle Mg_2SiO_4 \rangle$	35·81	6·54	−8·52	—	298–1808	181
$\{Mg_2SiO_4\}$	49·0	—	—	—	m.p.–2500	902
$\langle Mg_2V_2O_7 \rangle$	*39·95*	*28·9*	—	—	298–1200	977
$\langle MgV_2O_6 \rangle$	*30·2*	*31·0*	—	—	298–1100	977
$\langle MgTiO_3 \rangle$	28·29	3·28	−6·53	—	298–1800	124
$\langle Mg_2TiO_4 \rangle$	35·96	8·54	−6·89	—	298–1800	316
$\langle MgTi_2O_5 \rangle$	40·68	9·20	−7·35	—	298–1800	316
$\langle MgAl_2O_4 \rangle$	36·80	6·40	−9·78	—	298–1800	279
$\langle Mg_2Ge \rangle$	17·55	4·3	−0·76	—	298–m.p.	963
$\langle MgZn_2 \rangle$	15·55	7·20	—	—	298–800	124
$\langle MgCu_2 \rangle$		additive		—	273–650	241
$\langle MgAg \rangle$		additive		—	298–773	241
$\langle MgCr_2O_4 \rangle$	40·02	3·56	−9·58	—	298–1800	124
$\langle MgNi_2 \rangle$	15·67	7·30	—	—	298–900	124
$\langle MnF_2 \rangle$	*14·79*	*5·70*	*−0·47*	—	298–m.p.	—
$\langle MnF_3 \rangle$	*19·6*	*7·4*	—	—	298–1000	—
$\langle MnCl_2 \rangle$	18·04	3·16	−1·37	—	298–m.p.	124
$\{MnCl_2\}$	22·60	—	—	—	m.p.–1200	124
$\langle MnO \rangle$	11·11	1·94	−0·88	—	298–1800	124
$\langle Mn_3O_4 \rangle \alpha$	34·64	10·82	−2·20	—	298–1445	124
$\langle Mn_3O_4 \rangle \beta$	50·20	—	—	—	1445–1800	11
$\langle Mn_2O_3 \rangle$	24·73	8·38	−3·23	—	298–1350	315
$\langle MnO_2 \rangle$	16·60	2·44	−3·88	—	298–780	124
$\langle MnS \rangle$	11·40	1·80	—	—	298–1803	41
$\{MnS\}$	16·00	—	—	—	1803–2000	124
$\langle MnS_2 \rangle$	16·66	4·22	−1·04	—	250–350	854
$\langle MnSO_4 \rangle$	29·26	8·92	−7·04	—	298–1100	124
$\langle Mn_4N \rangle$	22·16	30·5	—	—	298–800	124
$\langle Mn_5N_2 \rangle$	30·55	38·40	—	—	298–800	124
$\langle Mn_3C \rangle$	25·26	5·60	−4·07	—	298–1310	124
$\langle Mn_7C_3 \rangle$	*58·95*	*13·10*	*−9·5*	—	298–1500	—
$\langle Mn_3Si \rangle$	24·11	12·45	−3·52	—	298–950	769
$\langle Mn_5Si_3 \rangle$	48·13	12·94	−4·68	—	298–m.p.	769
$\langle MnSi \rangle$	11·79	3·05	−1·53	—	298–m.p.	769
$\langle Mn_{0\cdot37}Si_{0\cdot63} \rangle$	6·37	4·085	−1·16	—	298–1425	769
$\langle MnCo_3 \rangle$	21·99	9·30	−4·69	—	298–700	124
$\langle MnSiO_3 \rangle$	26·42	3·88	−6·16	—	298–1500	124
$\langle Mn_2SiO_4 \rangle$	38·02	4·66	−7·44	—	298–m.p.	902
$\{Mn_2SiO_4\}$	58·1	—	—	—	m.p.–1800	902
$\langle MnWO_4 \rangle$	26·0	12·26	—	—	298–1073	940
$\langle MoO_3 \rangle$	20·07	5·90	−3·68	—	298–1808	181
$\langle Mo_2N \rangle$	11·19	13·80	—	—	298–800	124

Table C2. Heat capacities of the compounds (contd.)
$C_p = a + bT + cT^{-2}$ (cal/deg mole)

Substance	C_p in cal/deg mole			Remarks	Temp. range (°K)	Ref.
	a	b × 10^3	c × 10^{-5}			
⟨Mo_2C⟩	16·50	2·30	−2·74	—	298–2200	722
⟨$Mo(CO)_6$⟩	49·05	37·0	—	—	293–351	219
⟨Mo_3Si⟩	20·52	5·42	+0·076	—	298–2200	769
⟨Mo_5Si_3⟩	43·82	8·37	−2·87	—	298–2200	769
⟨$MoSi_2$⟩	16·22	2·86	−1·57	—	298–2200	769
⟨Mo_2B⟩	18·42	1·3	—	—	298–800	572
⟨MoB⟩	9·77	3·07	−1·13	—	298–1200	572
⟨MoB_2⟩	7·92	13·0	—	—	600–1200	572
(NH_3)	8·92	4·46	−1·55	—	298–2400	—
(N_2O)	10·92	2·06	−2·04	—	298–200	124
(NO)	7·03	0·92	−0·14	—	298–2500	124
⟨NaF⟩	10·40	3·88	−0·33	—	298–m.p.	314
⟨$NaCl$⟩	10·98	3·90	—	—	298–m.p.	124
{$NaCl$}	16·0	—	—	—	m.p.–1300	124
⟨$NaBr$⟩	11·87	2·10	—	—	298–550	124
⟨Na_2O⟩α	13·26	16·78	−0·99	$-7·3 \times 10^{-6}\ T^2$	298–1023	902
⟨Na_2O⟩β	19·67	3·05	—	—	1023–1243	902
⟨Na_2O⟩γ	20·28	2·56	—	—	1243–m.p.	902
{Na_2O}	25·0	—	—	—	m.p.–2200	902
⟨$NaOH$⟩	17·15	−26·5	—	$+56·35 \times 10^{-6}\ T^2$	298–568	289
⟨$NaOH$⟩β	20·55	—	—	—	568–m.p.	289
{$NaOH$}	21·38	−1·40	—	—	m.p.–980	289
⟨Na_2S⟩	18·10	3·0	—	—	298–m.p.	81
{Na_2S}	21·52	0·22	—	—	m.p.–2000	817
⟨Na_2SO_4⟩α	23·50	31·74	—	—	298–450	285
⟨Na_2SO_4⟩γ	29·06	19·34	—	—	514–1157	285
{Na_2SO_4}	47·18	—	—	—	1157–1850	285
⟨Na_2CO_3⟩	13·98	54·4	−3·13	—	298–500	138
{Na_2CO_3}	33·99	10·7	—	—	m.p.–1210	533
⟨Na_4SiO_4⟩	38·86	17·74	—	—	298–m.p.	902
{Na_4SiO_4}	62·0	—	—	—	m.p.–1800	902
⟨Na_2SiO_3⟩	31·14	9·60	−6·47	—	298–m.p.	124
{Na_2SiO_3}	42·80	—	—	—	m.p.–1800	124
⟨$Na_2Si_2O_5$⟩	44·38	16·86	−10·67	—	298–m.p.	124
{$Na_2Si_2O_5$}	62·35	—	—	—	m.p.–1800	124
⟨$NaBO_2$⟩	12·10	12·83	—	—	298–m.p.	817
{$NaBO_2$}	35·0	—	—	—	m.p.–2000	817
⟨Na_3AlF_6⟩α	45·95	29·46	−2·78	—	298–845	314
⟨Na_3AlF_6⟩β	52·15	15·86	—	—	845–1300	314
{Na_3AlF_6}	93·4	—	—	—	1300–1400	314
⟨Na_2TiO_3⟩α	25·18	20·72	—	—	298–560	124
⟨Na_2TiO_3⟩β	25·95	17·00	—	—	560–1303	124
{Na_2TiO_3}	46·90	—	—	—	1303–1600	124
⟨$Na_2Ti_2O_5$⟩	49·32	7·06	−4·60	—	298–1258	124
{$Na_2Ti_2O_5$}	68·50	—	—	—	1258–1600	774
⟨$Na_2Ti_3O_7$⟩	63·46	10·64	−5·64	—	298–1401	124
{$Na_2Ti_3O_7$}	94·15	—	—	—	1401–1700	124
⟨Na_2CrO_4⟩α	24·15	33·46	—	—	298–694	774
⟨Na_2CrO_4⟩β	35·84	12·33	—	—	694–1070	774
{Na_2CrO_4}	48·9	—	—	—	1070–1550	774
⟨$NbCl_4$⟩	*31·9*	—	*−2·9*	—	298–600	755
⟨$NbCl_5$⟩	*38·0*	—	*−3·0*	—	298–477	755
⟨NbO⟩	10·04	2·35	−0·78	—	298–1700	488

Table C2. Heat capacities of the compounds (*contd.*)

$C_p = a + bT + cT^{-2}$ (cal/deg mole)

Substance	C_p in cal/deg mole			Remarks	Temp. range (°K)	Ref.
	a	b × 10³	c × 10⁻⁵			
$\langle NbO_2\rangle\alpha$	14·68	6·16	−2·42	—	298–1640	755
$\langle NbO_2\rangle\beta$	21·28	—	—	—	1040–1700	755
$\langle Nb_2O_5\rangle$	38·76	3·54	−7·32	—	298–1700	488
$\{Nb_2O_5\}$	57·90	—	—	—	m.p.–1810	755
$\langle NbN\rangle$	8·69	5·40	—	—	298–600	124
$\langle Nb_2C\rangle$	15·88	3·0	−2·05	—	298–1703	488
$\langle NbC_{0\cdot75}\rangle$	8·95	2·25	−1·25	—	298–1763	488
$\langle NbC\rangle$	10·79	1·73	−2·15	—	298–1790	488
$\langle Nb_5Si_3\rangle$	45·21	7·36	−3·60	—	298–2000	769
$\langle NbSi_2\rangle$	15·10	3·67	−0·67	—	298–2000	769
$\langle NbB_2\rangle$	11·23	9·21	−2·25	—	298–1400	572
$\langle NbCr_2\rangle$	17·75	5·68	−1·76	—	298–1500	849
$\langle NdF_3\rangle$	17·92	8·75	+4·85	—	298–1650	901
$\{NdF_3\}$	44·09	−1·06	−28·5	—	1650–1873	901
$\langle NdCl_3\rangle$	18·65	14·6	—	—	298–m.p.	780
$\{NdCl_3\}$	35·0	—	—	—	m.p.–1250	780
$\langle NdI_3\rangle\alpha$	21·72	8·56	−0·15	—	298–847	927
$\langle NdI_3\rangle\beta$	28·06	—	—	—	847–1060	927
$\{NdI_3\}$	37·22	—	—	—	1060–1200	927
$\langle Nd_2O_3\rangle\alpha$	27·67	7·12	−2·84	—	298–1395	587
$\langle Nd_2O_3\rangle\beta$	37·20	—	—	—	1395–1795	587
$\langle NiF_2\rangle$	*15·0*	*4·3*	—	—	298–1723	—
$\langle NiCl_2\rangle$	17·50	3·16	−1·19	—	298–m.p.	41
$\{NiCl_2\}$	24·00	—	—	—	m.p.–1336	41
$\langle NiO\rangle\alpha$	−4·99	37·58	+3·89	—	298–525	544
$\langle NiO\rangle\beta$	13·88	—	—	—	525–565	544
$\langle NiO\rangle\gamma$	11·18	2·02	—	—	565–1800	544
$\langle NiS\rangle$	*9·3*	*6·4*	—	—	298–670	—
$\langle NiSO_4\rangle$	*30·1*	9·92	—	—	298, 200	124
$\langle NiTe\rangle$	11·57	3·30	—	—	298–700	124
$\langle Ni_2Si\rangle$	15·8	3·29	—	—	298–1582	124
$\langle NiSi\rangle$	11·65	1·47	−1·56	—	298–1265	769
$\langle Ni_{0\cdot35}Si_{0\cdot65}\rangle$	5·98	0·88	−0·86	—	298–1200	769
$\langle NiB\rangle$	*10·26*	*3·5*	*−2·69*	—	298–1300	—
$\langle Ni_3B_4\rangle$	*37·28*	*11·74*	*−9·03*	—	298–1300	—
$\langle NiAl_2O_4\rangle$	*38·05*	*5·58*	*−7·35*	—	295–1900	—
$\langle NiWO_4\rangle$	24·38	15·7	—	—	298–1120	978
$\langle NiTiO_3\rangle$	*27·51*	*3·82*	*−4·38*	—	298–1750	—
$\langle Ni_3Sn\rangle$	20·78	10·20	—	—	298–900	124
$\langle OsO_2\rangle$	16·72	2·48	−3·39	—	298–1200	787
(OsO_4)	20·55	4·88	−3·82	—	298–1000	124
(PH_3)	9·11	2·86	−1·71	—	298–2000	124
(PF_3)	17·18	1·92	−3·88	—	298–2000	124
(PCl_3)	19·15	0·74	−1·91	—	298–1000	124
(PCl_5)	31·42	0·20	−4·27	—	298–2000	817
$(POCl_3)$	24·95	0·62	−4·13	—	298–2000	817
(PBr_3)	19·81	—	−1·43	—	298–1000	124
$\langle P_2O_5\rangle$	17·90	38·8	−3·73	—	298–700	817
(PS)	8·85	0·12	−0·43	—	298–2000	854
$\langle PbF_2\rangle$	16·5	4·10	—	—	293–1097	124
$\langle PbCl_2\rangle$	16·1	4·0	—	—	298–m.p.	939

Table C2. Heat capacities of the compounds (contd.)
$C_p = a + bT + cT^{-2}$ (cal/deg mole)

Substance	C_p in cal/deg mole a	$b \times 10^3$	$c \times 10^{-5}$	Remarks	Temp. range (°K)	Ref.
$\{PbCl_2\}$	28·2	—	—	—	m.p.–950	939
$(PbCl_2)$	13·53	0·23	—	—	298–2000	944
$\langle PbBr_2 \rangle$	18·59	2·20	—	—	298–m.p.	124
$\{PbBr_2\}$	27·60	—	—	—	m.p.–900	124
$\langle PbI_2 \rangle$	18·00	4·70	—	—	298–m.p.	124
$\{PbI_2\}$	32·40	—	—	—	m.p.–800	124
$\langle PbO \rangle_{red}$	10·95	3·75	−1·0	—	298–1000	755
$\langle PbO \rangle_{yellow}$	9·05	6·40	—	—	298–1000	124
$\langle PbO \rangle$	11·05	2·74	−0·81	—	1000–1159	755
$\{PbO\}$	15·54	—	—	—	1159–1700	755
$\langle Pb_3O_4 \rangle$	32·80	7·80	—	—	298–1000	817
$\langle PbO_2 \rangle$	*13·26*	*7·44*	—	—	298–1000	—
$\langle PbS \rangle$	10·66	3·92	—	—	298–900	124
(PbS)	8·92	—	−0·49	—	298–2000	124
$\langle PbSO_4 \rangle$	17·67	24·2	—	—	298–1139	—
$\langle PbCO_3 \rangle$	*12·39*	*28·6*	—	—	298–800	124
$\langle PbSiO_3 \rangle$	17·84	26·47	−2·74	$-12·94 \times 10^{-6}\,T^2$	298–m.p.	902
$\{PbSiO_3\}$	31·1	—	—	—	m.p.–1800	902
$\langle Pb_2SiO_4 \rangle$	30·52	19·73	−3·27	—	298–600	902
	47·71	−0·37	−21·77	—	600–m.p.	902
$\{Pb_2SiO_4\}$	45·19	—	—	—	m.p.–1800	902
$\langle PbTiO_3 \rangle$	18·7	23·3	—	—	300–850	226
$\langle Pb_{0·37}Tl_{0·63} \rangle$	4·93	4·5	—	—	298–m.p.	148
$\langle PdO \rangle$	10·83	1·68	−0·30	$+0·09 \times 10^{-6}\,T^2$	298–1200	—
$\langle PrF_3 \rangle$	21·81	4·88	1·99	—	298–m.p.	901
$\{PrF_3\}$	28·65	1·00	25·91	—	m.p.–1863	901
$\langle PrCl_3 \rangle$	20·6	11·4	—	—	298–m.p.	780
$\{PrCl_3\}$	32·0	—	—	—	m.p.–1250	780
$\langle PrI_3 \rangle$	21·3	9·7	—	—	298–m.p.	780
$\{PrI_3\}$	34·2	—	—	—	m.p.–1250	780
$\langle Pr_2O_3 \rangle$	23·76	12·5	—	—	298–1070	841
$\langle PrO_{1·83} \rangle$	15·88	4·36	−1·55	—	298–1172	124
(PtO_2)	*13·25*	*0·5*	*−2·75*	—	298–2000	—
$\langle PtS \rangle$	*11·95*	*2·6*	*−2·12*	—	298–1100	—
$\langle PtS_2 \rangle$	*14·07*	*7·08*	*−0·4*	—	298–1000	—
$\langle PtSb_2 \rangle$	15·27	5·06	—	—	298–900	124
(PuF_6)	37·18	0·29	−5·65	—	298–2200	865
$\langle PuO_2 \rangle$	22·18	0·21	−4·94	—	298–m.p.	836
$\langle PuC_{0·88} \rangle$	13·08	1·14	−3·23	—	298–1400	595
$\langle PuC_{1·5} \rangle$	8·99	9·23	−0·01	$-1·59 \times 10^{-6}\,T^2$	298–2200	865
$\langle PuCl_2 \rangle$	14·38	6·46	−2·75	$-1·34 \times 10^{-6}\,T^2$	298–2525	865
$\langle RbF \rangle$	7·97	9·2	+1·21	—	298–m.p.	540
$\{RbF\}$	−11·30	+0·83	+350·7	—	m.p.–1200	540
$\langle RbCl \rangle$	*11·5*	*2·49*	—	—	283–990	124
$\{RbCl\}$	*15·3*	—	—	—	990–	—
$\langle RbBr \rangle$	*11·6*	*2·55*	—	—	283–945	124
$\langle RbI \rangle$	11·6	2·63	—	—	283–911	124
(ReF_6)	*37·1*	*0·31*	*−7·65*	—	298–2000	—
$\langle ReCl_3 \rangle$	*25·21*	*6·6*	*−4·56*	—	298–750	—
$\langle Re_2O_7 \rangle$	*29·15*	*44·0*	*−2·25*	—	298–570	—
$\langle Re_5Si_3 \rangle$	*45·60*	*10·8*	*−3·36*	—	298–1500	769

Table C2. Heat capacities of the compounds (*contd.*)

$C_p = a + bT + cT^{-2}$ (cal/deg mole)

Substance	C_p in cal/deg mole			Remarks	Temp. range (°K)	Ref.
	a	b × 10^3	c × 10^{-5}			
$\langle Rh_2O_3\rangle$	20·74	13·8	—	—	291–973	274
(SO)	7·70	0·84	−0·65	—	298–2000	124
(SO_2)	10·38	2·54	−1·42	—	298–1800	124
(SO_3)	13·70	6·42	−3·12	—	298–1200	124
$(SOCl_2)$	17·75	1·72	−2·09	—	298–1100	—
(SO_2Cl_2)	12·84	19·0	—	—	298–500	124
$\langle SbCl_3\rangle$	*10·3*	*57·1*	—	—	273–346	124
$\langle SbBr_3\rangle$	*17·2*	*29·3*	—	—	273–370	124
$\langle SbI_3\rangle$	17·0	21·2	—	—	298–m.p.	924
$\{SbI_3\}$	34·3	—	—	—	m.p.–b.p.	924
$\langle Sb_2O_3\rangle$	*19·10*	*17·1*	—	—	273–930	124
$\langle SbO_2\rangle$	*11·3*	*8·1*	—	—	298–1198	124
$\langle Sb_2O_5\rangle$	*10·95*	*57·57*	—	—	298–500	944
$\langle Sb_2S_3\rangle$	*24·34*	*14·47*	—	—	298–823	854
$\langle Sb_2Te_3\rangle$	27·5	10·0	—	—	298–893	979
$\{Sb_2Te_3\}$	49·5	—	—	—	892–950	979
$\langle ScF_3\rangle$	*23·75*	*1·77*	*−2·25*	—	298–1250	—
$\langle ScCl_3\rangle$	22·86	3·68	−1·74	—	298–1240	927
$\{ScCl_3\}$	34·28	—	—	—	1240–1350	927
$\langle Sc_2O_3\rangle$	*23·17*	*5·64*	—	—	298–2500	—
$\langle ScN\rangle$	*10·95*	*1·3*	*−2·2*	—	298–2000	—
(SeF_4)	24·15	0·88	−6·58	—	298–2000	854
(SeO)	8·2	0·5	−0·8	—	298–2000	124
$\langle SeO_2\rangle$	16·63	0·93	−2·64	—	298–613	854
(SeO_2)	12·63	0·74	−2·37	—	298–2000	854
(SiF_4)	21·86	3·17	−4·7	—	298–1000	257
$(SiCl_2)$	13·76	0·09	−1·35	—	298–2200	817
$\{SiCl_4\}$	35·0	—	—	—	298–350	817
$(SiCl_4)$	24·25	1·64	−2·75	—	298–1000	124
$\langle SiI_4\rangle$	19·59	20·9	—	—	298–m.p.	706
$\{SiI_4\}$	35·25	9·87	—	—	m.p.–573	706
(SiI_4)	25·57	0·24	−1·44	—	298–1000	124
$\langle SiO_2\rangle$	10·49	0·24	−1·44	α-quartz	298–847	817
$\langle SiO_2\rangle$	14·08	2·4	—	β-quartz	847–1079	817
$\langle SiO_2\rangle$	4·28	21·06	—	α-cristobalite	298–523	124
$\langle SiO_2\rangle$	17·39	0·31	−9·90	β-cristobalite	523–1995	817
$\langle SiO_2\rangle$	3·27	24·80	—	α-tridymite	298–390	124
$\langle SiO_2\rangle$	13·64	2·64	—	β-tridymite	390–2000	124
$\{SiO_2\}$	13·38	3·68	−3·45	glass	298–2000	124
$\{SiO_2\}$	20·6	—	—	liquid	—	817
$\langle Si_3N_4\rangle$	16·86	23·6	—	—	298–900	124
$\langle SiC\rangle$	12·14	0·47	−11·76	$+1{\cdot}96 \times 10^8\ T^{-3}$	298–3260	944
$\langle SmCl_3\rangle$	*19·66*	*11·4*	*+0·18*	—	298–m.p.	—
$\langle Sm_2O_3\rangle\alpha$	30·75	4·64	−4·30	monoclinic	298–1195	587
$\langle Sm_2O_3\rangle\beta$	36·90	—	—	monoclinic	1195–1798	587
$\langle Sm_2O_3\rangle$	30·64	5·08	−3·96	cubic	298–1150	587
(SnH_4)	*12·38*	*9·0*	*−2·7*	—	298–800	—
$\langle SnCl_2\rangle$	*16·2*	*9·26*	—	—	298–520	124
$\{SnCl_4\}$	39·5	—	—	—	298–400	907

Table C2. Heat capacities of the compounds (*contd.*)
$C_p = a + bT + cT^{-2}$ (cal/deg mole)

Substance	C_p in cal/deg mole			Remarks	Temp. range (°K)	Ref.
	a	b × 10³	c × 10⁻⁵			
$(SnCl_4)$	25·57	0·20	−1·87	—	298–1000	124
$\langle SnI_4\rangle$	19·4	36·0	—	—	298–m.p.	124
$\{SnI_4\}$	40·1	—	—	—	m.p.–450	124
$\langle SnO\rangle$	9·55	*3·50*	—	—	298–1273	124
(SnO)	8·42	0·32	−0·84	—	298–2000	124
$\langle SnO_2\rangle$	17·66	2·40	−5·16	—	298–1500	124
$\langle SnS\rangle\alpha$	8·53	7·48	0·9	—	298–875	315
$\langle SnS\rangle\beta$	9·78	3·74	—	—	875–m.p.	315
$\langle SnS_2\rangle$	15·51	4·20	—	—	298–1000	315
(SnS)	8·83	0·08	−0·55	—	298–2000	124
$\langle SrF_2\rangle$	*15·5*	*4·3*	—	—	298–m.p.	—
$\langle SrCl_2\rangle\alpha$	15·32	9·4	—	—	298–1002	531
$\langle SrCl_2\rangle\beta$	32·35	—	—	—	1002–m.p.	531
$\{SrCl_2\}$	26·66	—	—	—	m.p.–1350	531
$\langle SrBr_2\rangle\alpha$	17·95	3·22	—	—	298–918	780
$\langle SrBr_2\rangle\beta$	27·5	—	—	—	918–930	780
$\{SrBr_2\}$	27·8	—	—	—	930–1200	780
$\langle SrI_2\rangle$	*18·6*	*3·05*	—	—	298–m.p.	124
$\langle SrO\rangle$	12·34	1·12	−1·81	—	298–1270	154
$\langle SrCO_3\rangle\alpha$	21·42	8·56	−3·40	—	298–1197	154
$\langle SrCO_3\rangle\beta$	34·0	—	—	—	1197–1325	154
$\langle SrAl_2O_4\rangle\alpha$	42·35	1·18	−12·67	—	298–932	764
$\langle SrAl_2O_4\rangle\beta$	34·92	7·0	—	—	932–1600	764
$\langle SrWO_4\rangle$	28·84	8·63	—	—	298–1100	883
$\langle SrTiO_3\rangle$	28·23	2·04	−4·58	—	298–1800	43
$\langle Sr_2TiO_4\rangle$	38·45	3·84	−4·67	—	298–1800	43
$\langle TaCl_4\rangle$	*31·9*	—	*−2·0*	—	298–700	755
$(TaCl_4)$	25·8	—	−2·0	—	298–700	755
$\langle TaCl_5\rangle$	*38·0*	—	*−3·0*	—	298–489	755
$(TaCl_5)$	*31·6*	—	*−3·7*	—	298–600	755
Ta_2O_5	37·0	6·56	−5·92	—	298–1800	755
$\langle Ta_2N\rangle$	16·85	4·22	−1·69	—	298–3000	755
$\langle TaN\rangle$	*13·21*	*0·65*	*−3·02*	—	298–3360	—
$\langle Ta_2C\rangle$	15·88	3·33	−2·05	—	298–3775	755
$\langle TaC\rangle$	10·35	2·71	−2·10	—	298–4270	755
$\langle Ta_5Si_3\rangle$	*42·95*	*9·35*	*−2·13*	—	298–2000	—
$\langle TaSi_2\rangle$	17·51	1·84	−2·17	—	298–2200	769
$\langle TaB\rangle$	7·82	5·8	—	—	500–1200	572
$\langle TaB_2\rangle$	14·21	4·49	−3·60	—	298–3370	755
$\langle TaFe_2\rangle$	16·0	6·3	—	—	298–1300	755
$(TaCr_2\rangle$	17·65	5·45	−1·72	—	298–m.p.	849
$\langle TeCl_4\rangle$	33·2	—	—	—	298–m.p.	124
$\{TeCl_4\}$	53·2	—	—	—	m.p.–550	124
$\langle TeO_2\rangle$	15·58	3·48	−1·20	—	298–m.p.	571
$\{TeO_2\}$	26·92	0·52	—	—	m.p.–1146	571
(TeO_2)	13·09	0·58	−2·83	—	298–2000	854
$\langle ThF_4\rangle$	*26·75*	*5·85*	*−1·81*	—	298–1383	755
$\langle ThCl_4\rangle$	30·35	3·24	−2·18	—	298–m.p.	980
$\langle ThI_4\rangle$	*31·0*	*3·1*	*−1·47*	—	298–m.p.	755
$\langle ThO_2\rangle$	16·65	2·13	−2·24	—	298–2500	317, 522
$\langle Th(SO_4)_2\rangle$	25·0	55·2	—	—	623–897	569
$\langle ThN\rangle$	*11·34*	*2·28*	*−1·14*	—	298–2000	755

Table C2. Heat capacities of the compounds (contd.)
$C_p = a + bT + cT^{-2}$ (cal/deg mole)

Substance	C_p in cal/deg mole			Remarks	Temp. range (°K)	Ref.
	a	b × 10^3	c × 10^{-5}			
$\langle Th_3N_4 \rangle$	*39·33*	*6·24*	*−5·33*	—	298–2000	755
$\langle ThC_{1\cdot94} \rangle$	*15·17*	*2·89*	*−2·21*	—	298–1700	755
$\{TiCl_4\}$	35·7	—	—	—	285–b.p.	—
$(TiCl_4)$	25·45	0·24	−2·36	—	298–2000	124
$\{TiBr_4\}$	32·75	15·66	—	—	m.p.–423	197
$(TiBr_4)$	25·77	—	−1·58	—	>298	197
$\langle TiI_4 \rangle \alpha$	17·7	41·34	—	—	298–379	124
$\langle TiI_4 \rangle \beta$	35·4	—	—	—	379–428	124
$\{TiI_4\}$	37·4	—	—	—	428–650	124
(TiI_4)	25·82	—	−0·80	—	298–2500	817
$\langle TiO \rangle \alpha$	10·57	3·60	−1·86	—	298–1264	124
$\langle TiO \rangle \beta$	11·85	3·00	—	—	1264–1800	124
$\langle Ti_2O_3 \rangle \alpha$	7·31	53·52	—	—	298–473	124
$\langle Ti_2O_3 \rangle \beta$	34·68	1·30	−10·20	—	473–1800	124
$\langle Ti_3O_5 \rangle \alpha$	35·47	29·5	—	—	298–450	124
$\langle Ti_3O_5 \rangle \beta$	41·60	8·0	—	—	450–1400	124
$\langle TiO_2 \rangle$	17·97	0·28	−4·35	rutile	298–1800	124
$\langle TiO_2 \rangle$	17·83	0·50	−4·23	anatase	298–1300	124
$\langle TiS_2 \rangle \alpha$	8·08	27·34	—	—	298–420	245
$\langle TiS_2 \rangle \beta$	14·99	5·14	—	—	420–1010	245
$\langle TiN \rangle$	11·91	0·94	−2·96	—	298–1800	124
$\langle TiC \rangle$	11·83	0·80	−3·58	—	298–1800	124
$\langle TiB_2 \rangle$	13·48	6·18	−4·17	$-0{\cdot}8 \times 10^{-6}\,T^2$	298–3190	817
$\langle TlF \rangle \alpha$	12·31	2·62	—	—	298–355	463
$\langle TlF \rangle \beta$	5·29	17·78	—	—	355–595	463
$\{TlF\}$	16·08	—	—	—	m.p.–850	463
$\langle TlCl \rangle$	12·00	2·00	—	—	298–700	463
$\{TlCl\}$	14·2	—	—	—	700–850	463
$\langle TlBr \rangle$	11·07	4·95	—	—	298–m.p.	463
$\{TlBr\}$	25·25	−9·04	—	—	m.p.–950	463
$\langle TlI \rangle \alpha$	11·56	3·32	—	—	298–451	463
$\langle TlI \rangle \beta$	7·72	11·26	—	—	451–m.p.	463
$\{TlI\}$	17·2	—	—	—	m.p.–950	463
(TlI)	9·23	—	−0·3	—	298–1600	907
$\langle Tl_2O_3 \rangle$	31·52	0·85	−5·32	—	298–m.p.	856
$\langle UF_4 \rangle$	25·7	7·00	−0·06	—	298–1309	399
$\langle UF_6 \rangle$	12·6	92·0	—	—	273–337	399
(UF_6)	22·3	28·5	—	—	273–400	399
$\langle UCl_3 \rangle$	20·8	7·75	1·05	—	298–900	399
$\langle UCl_4 \rangle$	27·2	8·57	−0·79	—	298–800	399
$\{UCl_4\}$	25·8	14·4	—	—	890–920	399
$\langle UBr_4 \rangle$	31·4	4·92	−3·15	—	350–750	399
$\langle UI_4 \rangle$	34·8	2·38	−4·72	—	380–720	399
$\{UI_4\}$	39·6	—	—	—	820–870	399
$\langle UO_2 \rangle$	19·20	1·62	−3·96	—	298–1500	124
$\langle U_3O_8 \rangle$	67·5	8·83	−11·94	—	298–900	399
$\langle UO_3 \rangle$	22·09	2·54	−2·97	—	298–900	124, 244
$\langle UO_2F_2 \rangle$	*25·34*	*5·23*	*−1·02*	—	298–1200	—
$\langle US \rangle$	12·63	1·56	−0·905	—	298–2000	847
$\langle U_2S_3 \rangle$	*29·8*	*3·85*	*−0·9*	—	298–2000	—
$\langle US_2 \rangle$	17·16	2·30	—	—	298–1833	854
$\langle UC \rangle$	13·40	1·02	−1·46	—	298–2073	—
$\langle U_2C_3 \rangle$	29·9	3·06	−3·71	—	298–2050	—

Table C2. Heat capacities of the compounds (contd.)

$C_p = a + bT + cT^{-2}$ (cal/deg mole)

Substance	C_p in cal/deg mole			Remarks	Temp. range (°K)	Ref.
	a	b × 10^3	c × 10^{-5}			
$\langle UC_2 \rangle$	16·5	2·04	−2·25	—	298–2050	—
$\langle UB_2 \rangle$	25·58	−9·62	−8·48	$+5·86 \times 10^{-6}\ T^2$	298–2300	789
$\langle VCl_2 \rangle$	17·25	2·72	−0·71	—	298–1300	124
$\langle VCl_3 \rangle$	22·99	3·92	−1·68	—	298–900	124
$\langle VO \rangle$	11·32	3·22	−1·26	—	298–1700	315
$\langle VO_2 \rangle \alpha$	14·96	—	—	—	298–345	124
$\langle VO_2 \rangle \beta$	17·85	1·70	−3·95	—	345–m.p.	124
$\{VO_2\}$	25·5	—	—	—	m.p.–1900	124
$\langle V_2O_3 \rangle$	29·35	4·76	−5·42	—	298–1800	124
$\langle V_2O_5 \rangle$	46·54	−3·90	−13·22	—	298–m.p.	124
$\{V_2O_5\}$	45·60	—	—	—	m.p.–1500	124
$\langle VN \rangle$	10·94	2·10	−2·21	—	298–1600	124
$\langle VC_{0·88} \rangle$	8·69	3·18	−1·7	—	298–2000	848
$\langle V_3Si \rangle$	22·41	4·37	−1·66	—	298–1400	769
$\langle VSi_2 \rangle$	17·08	2·78	−2·25	—	298–1950	769
$\langle WO_3 \rangle$	17·48	6·79	—	—	298–1550	124
$\langle WC \rangle$	10·37	2·06	−2·23	$-0·24 \times 10^{-6}\ T^2$	298–2500	99
$\langle W(CO)_6 \rangle$	39·30	62·6	—	—	293–372	219
$\langle WS_2 \rangle$	*16·55*	*3·4*	*−1·0*	—	298–1300	—
$\langle W_5Si_3 \rangle$	42·94	9·36	−2·13	—	298–2200	769
$\langle WSi_2 \rangle$	16·21	2·64	−1·46	—	298–2200	769
$\langle W_2B \rangle$	18·46	1·44	−3·14	—	298–1200	572
$\langle WB \rangle$	13·89	−0·78	−5·02	—	298–1200	572
$\langle W_2B_3 \rangle$	39·27	2·08	−16·72	—	298–1200	572
$\langle YF_3 \rangle$	23·76	1·78	−1·36	orthorhombic	298–1350	901
$\langle YF_3 \rangle$	−76·35	50·86	+673·5	—	1350–1428	901
$\{YF_3\}$	31·96	−0·006	−0·115	—	1428–1873	901
$\langle YCl_3 \rangle$	25·03	0·77	−2·90	—	298–m.p.	927
$\{YCl_3\}$	32·44	—	—	—	m.p.–1100	927
$\langle YN \rangle$	*10·9*	*1·55*	*−1·75*	—	298–2200	—
$\langle YbCl_3 \rangle$	*22·63*	*2·23*	*−0·45*	—	298–m.p.	—
$\langle Yb_2O_3 \rangle$	*30·75*	*4·65*	*−4·1*	—	298–1800	—
$\langle ZnCl_2 \rangle$	14·5	5·5	—	—	298–m.p.	463
$\{ZnCl_2\}$	24·1	—	—	—	m.p.–1000	463
$\langle ZnBr_2 \rangle$	12·6	10·4	—	—	298–m.p.	463
$\{ZnBr_2\}$	27·2	—	—	—	m.p.–1000	463
$\langle ZnO \rangle$	11·71	1·22	−2·18	—	298–1600	124
$\langle ZnS \rangle$	12·16	1·24	−1·36	—	298–1200	124
(ZnS)	9·08	—	−0·47	—	298–1500	907
$\langle ZnSO_4 \rangle$	*21·9*	*18·2*	—	—	298–m.p.	—
$\langle ZnTe \rangle$	*11·11*	*2·61*	—	—	298–m.p.	—
$\langle Zn_3N_2 \rangle$	19·00	22·5	—	—	273–700	204
$\langle ZnCO_3 \rangle$	*9·3*	*33·0*	—	—	298–	126
$\langle Zn_2SiO_4 \rangle$	*34·63*	*8·83*	*−7·24*	—	298–1785	902
$\langle Zn_2TiO_4 \rangle$	39·82	5·54	−7·69	—	298–1800	280
$\langle ZnFe_2O_4 \rangle$	37·15	10·65	−3·82	—	298–1000	17
$\langle ZnWO_4 \rangle$	27·08	9·78	—	—	298–1110	978
$\langle ZrCl_4 \rangle$	31·92	—	−2·91	—	298–s.p.	132
$\langle ZrBr_4 \rangle$	25·5	15·1	—	—	298–s.p.	120
$\langle ZrO_2 \rangle \alpha$	16·64	1·80	−3·36	—	298–1478	132

Table C2. Heat capacities of the compounds (contd.)

$C_p = a + bT + cT^{-2}$ (cal/deg mole)

Substance	C_p in cal/deg mole			Remarks	Temp. range (°K)	Ref.
	a	b × 10^3	c × 10^{-5}			
⟨ZrO_2⟩β	17·80	—	—	—	1478–1850	132
⟨ZrN⟩	11·10	1·68	−1·72	—	298–1700	132
⟨$ZrSiO_4$⟩	31·48	3·92	−8·08	—	298–1800	132
⟨ZrB_2⟩	15·79	4·22	−3·51	—	298–1200	254

Table D. Vapour pressures and
$\log p = A \cdot T^{-1} + B \log T$

Substance	log *p* (mmHg)				Remarks	Temp. range (°K)	Ref.
	A	B	C × 10^3	D			
⟨Ag⟩	−14,900	−0·85	—	12·20	—	298–m.p.	99
{Ag}	−14,400	−0·85	—	11·70	—	m.p.–b.p.	99
⟨AgCl⟩	−11,830	−0·30	−1·02	12·39	—	298–m.p.	120
{AgCl}	−11,320	−2·55	—	17·34	—	m.p.–b.p.	120
{AgBr}	−12,400	−2·97	—	19·33	—	m.p.–b.p.	20
{AgI}	−10,250	−3·52	—	20·09	—	m.p.–b.p.	120
{Al}	−16,380	−1·0	—	12·32	—	m.p.–b.p.	446
⟨AlF_3⟩	−16,700	−3·02	—	23·27	—	298–s.p.	646
⟨Al_2Cl_6⟩	−6360	3·77	−6·12	9·66	—	298–s.p.	120
⟨Al_2Br_6⟩	−5280	−1·75	−4·08	20·81	—	298–m.p.	120
{Al_2Br_6}	−5280	−12·59	—	46·70	—	m.p.–b.p.	120
⟨Al_2I_6⟩	−7150	+0·12	−4·96	17·76	—	298–m.p.	120
{Al_2I_6}	−6760	−11·89	—	46·67	—	m.p.–b.p.	120
{Am}	−13,700	−1·0	—	13·97	—	1103–1453	355
⟨AmF_3⟩	−24,600	−7·05	—	36·87	—	1100–1300	355
⟨As_4⟩	−6160	—	—	9·82	—	600–900	58, 120
{AsF_3}	−4150	−18·26	—	61·38	—	265–292	25
{AsF_5}	−1088	—	—	7·72	—	m.p.–b.p.	120
{$AsCl_3$}	−2660	−5·83	—	24·76	—	m.p.–b.p.	120
⟨As_4O_6⟩	−5282	—	—	10·91	claudetite	373–573	539
⟨As_4O_6⟩	−5452	—	—	11·468	arsenolite	488–573	539
⟨As_2S_3⟩$_x$	−5100	—	—	4·67	—	450–600	92
⟨Au⟩	−19,820	−0·306	−0·16	10·81	—	298–m.p.	120, 58
{Au}	−19,280	−1·01	—	12·38	—	m.p.–b.p.	120
⟨B⟩	−29,900	−1·0	—	13·88	—	1000–m.p.	589, 409
{BCl_3}	−2115	−7·04	—	27·56	—	m.p.–b.p.	120
{BBr_3}	−2710	−7·04	—	28·36	—	m.p.–b.p.	120
{BI_3}	−3342	−5·4	—	24·31	—	m.p.–b.p.	6
⟨Ba⟩	−9730	—	—	7·83	—	750–m.p.	159
{Ba}	−9340	—	—	7·42	—	m.p.–	159
{BaF_2}	−20,330	−5·03	—	28·04	—	m.p.–b.p.	25
⟨BaO⟩	−21,900	—	—	9·99	—	1200–1700	120
⟨Be⟩	−16,730	—	0·145	9·065	—	1000–1557	89
{Be}	−17,000	−0·775	—	11·90	—	1557–2670	(89), (58)
{BeF_2}	−13,000	−3·98	—	24·56	—	m.p.–b.p.	217, 498
⟨$BeCl_2$⟩	−7870	−5·03	—	27·15	—	298–m.p.	25
{$BeCl_2$}	−7220	−5·03	—	26·28	—	m.p.–b.p.	25
⟨Be_2Cl_4⟩	−8970	−7·65	—	37·0	—	298–m.p.	25
⟨$BeBr_2$⟩	−7650	−5·03	—	27·15	—	298–m.p.	25
{$BeBr_2$}	−6570	−5·03	—	25·63	—	m.p.–b.p.	25
⟨Be_2Br_4⟩	−8320	−7·65	—	35·9	—	298–s.p.	25
⟨BeI_2⟩	−7000	−5·03	—	26·5	—	298–m.p.	25
{BeI_2}	−5800	−5·03	—	24·96	—	m.p.–b.p.	25
⟨Be_2I_4⟩	−8520	−7·65	—	35·9	—	298–m.p.	25

evaporation constants
+ CT + D (mmHg)

Substance	b.pt. (°C)	L_e at b.pt. (kcal/mole)	sb.pt. (°C)	L_s at m.pt. or sb.pt.	L_e at 25°C (kcal/mole)	L_s at 25°C (kcal/mole)	Ref. and remarks
Ag	2147	61·6 ± 1·5	—	66·3 ± 1·0	65·2 ± 1·5	67·9 ± 1·0	99
AgCl	1564	42·5 ± 4·0	—	51·2 ± 3·0	50·3 ± 4·0	53·6 ± 4·0	120
AgBr	1560	*45·9 ± 4·0*	—	—	55·0 ± 5·0	—	20
AgI	1505	34·5 ± 5·0	—	—	44·8 ± 5·0	—	120
Al	2450	69·5 ± 2·0	—	75·6 ± 2·0	—	76·9 ± 3·0	446
AlF_3	—	—	1280	67·0 ± 3·0	—	74·5 ± 4·0	637
Al_2Cl_6	⟨160⟩	9·7 ± 0·3	180	26·7 ± 0·8	—	28·8 ± 0·8	120
Al_2Br_6	255	10·9 ± 0·5	—	20·3 ± 0·6	16·7 ± 0·7	21·5 ± 0·8	120
Al_2I_6	385	15·4 ± 0·5	—	27·9 ± 1·5	23·9 ± 1·5	30·7 ± 1·5	120
Am	2600	57·0 ± 4·0	—	—	62·2 ± 3·5	—	355
As_4	—	—	622	27·3 ± 2·5	—	29·0 ± 2·5	23
AsF_3	58	7·1 ± 0·5	—	—	7·3 ± 0·5	—	25
AsF_5	−53	5·0 ± 0·2	—	—	—	—	25
$AsCl_3$	130	7·5 ± 0·4	—	—	8·7 ± 0·5	—	198
$AsBr_3$	221	10·3 ± 1·8	—	14·6 ± 1·7	11·8 ± 1·5	14·6 ± 1·7	25
AsI_3	424	14·2 ± 2·0	—	—	—	—	25
As_4O_6	459	14·2 ± 1·0	—	22·3 ± 1·0*	—	30·5 ± 1·5†	120 *monocl. † ortho-rhomb.
Au	2950	82·0 ± 2·5	—	88·7 ± 1·5	87·7 ± 2·0	90·5 ± 2·0	120, 58
B	—	—	—	134·0 ± 3·0	—	136 ± 5·0	99
BF_3	−101	4·6 ± 0·1	—	—	—	—	25
BCl_3	13	5·7 ± 0·2	—	—	—	—	120
BBr_3	91	7·3 ± 0·5	—	—	8·2 ± 0·5	—	120
BI_3	210	10·1 ± 0·6	—	—	12·1 ± 0·7	—	6
Ba	(1700)	—	—	(44·3)	(44·4)	(46·0)	159, 120
BaF_2	2382	64·5	—	80·0	83·3	90·1	—
$BaCl_2$	(1830)	*(50·0)*	—	—	—	—	25
BaO	—	—	—	(90·0)	—	—	120
Be	2472	69·9	—	—	—	77·5 ± 2·0	755
BeF_2	1175	47·65	—	—	—	55·15	755
$BeCl_2$	532	25·0	—	—	—	32·5 ± 2·0	755
$BeBr_2$	(511)	23·9	—	—	—	30·3 ± 2·5	755
BeI_2	(482)	(23·0)	—	—	—	30·3 ± 2·5	755

Table D. Vapour pressures and
$\log p = \mathrm{A} \cdot T^{-1} + \mathrm{B} \log T$

Substance	log *p* (mmHg)				Remarks	Temp. range (°K)	Ref.
	A	B	C × 10³	D			
$\{\mathrm{Bi}\}$	−10,400	−1·26	—	12·35	—	m.p.–b.p.	350
$\{\mathrm{Bi_2}\}$	−10,730	−3·02	—	18·1	—	m.p.–b.p.	350
$\langle \mathrm{BiCl_3} \rangle$	−6200	—	—	12·83	—	298–m.p.	467
$\{\mathrm{BiCl_3}\}$	−5980	−7·04	—	31·38	—	m.p.–b.p.	120
$\{\mathrm{BiBr_3}\}$	−6190	−7·04	—	31·40	—	m.p.–b.p.	120
$\{\mathrm{Br_2}\}$	−2200	−4·15	—	19·96	—	298–390	362
$\{\mathrm{CCl_4}\}$	−2400	−5·30	—	23·60	—	m.p.–b.p.	120
$\langle \mathrm{CBr_4} \rangle$	−2841	—	—	9·386	monoclin.	295–319	444
$\langle \mathrm{CBr_4} \rangle$	−2579	—	—	8·567	cubic	321–329	444
$\{\mathrm{CS_2}\}$	−1865	−3·13	—	16·55	—	m.p.–b.p.	120
$\langle \mathrm{Ca} \rangle$	−9350	−1·39	—	12·82	—	298–m.p.	99
$\{\mathrm{Ca}\}$	−8920	−1.39	—	12·45	—	m.p.–b.p.	99
$\langle \mathrm{CaF_2} \rangle \alpha$	−23,600	−4·525	—	27·41	—	298–1424	—
$\langle \mathrm{CaF_2} \rangle \beta$	−23,350	−4·525	—	27·23	—	1424–m.p.	612, 442
$\{\mathrm{CaF_2}\}$	−21,800	−4·525	—	26·31	—	m.p.–b.p.	25
$\{\mathrm{CaCl_2}\}$	−13,570	—	—	9·22	—	1110–1281	520
$\langle \mathrm{Cd} \rangle$	−5908	−0·232	−0·284	9·717	—	298–594	58, 120
$\{\mathrm{Cd}\}$	−5819	−1·257	—	12·287	—	594–1050	58, 120
$\{\mathrm{CdF_2}\}$	−16,170	−5·03	—	27·50	—	m.p.–b.p.	25
$\langle \mathrm{CdCl_2} \rangle$	−9270	−2·11	—	17·46	—	298–m.p.	177
$\{\mathrm{CdCl_2}\}$	−9183	−5·04	—	25·907	—	m.p.–b.p.	418
$\langle \mathrm{CdBr_2} \rangle$	−8250	−2·5	—	18·15	—	298–m.p.	25
$\{\mathrm{CdBr_2}\}$	−7150	−2·5	—	16·85	—	m.p.–b.p.	25
$\langle \mathrm{CdI_2} \rangle$	−7530	−2·5	—	18·01	—	298–m.p.	25
$\{\mathrm{CdI_2}\}$	−6720	−2·5	—	16·79	—	m.p.–b.p.	25
$\langle \mathrm{CdS} \rangle$	−11,460	−2·5	—	16·06	—	298–1200	92
$\{\mathrm{Ce}\}$	−20,304	—	—	8·207	—	1611–2038	464
$\langle \mathrm{CeF_3} \rangle$	−20,460	—	—	12·09	—	1374–1637	984
$\langle \mathrm{CeCl_3} \rangle$	−18,750	−7·05	—	36·38	—	298–m.p.	618
$\langle \mathrm{CeBr_3} \rangle$	−18,000	−7·05	—	36·49	—	298–m.p.	618
$\langle \mathrm{Co} \rangle$	−22,209	—	−0·223	10·817	—	298–m.p.	52
$\langle \mathrm{CoCl_2} \rangle$	−14,150	−5·03	—	30·10	—	298–m.p.	208
$\{\mathrm{CoCl_2}\}$	−11,050	−5·03	—	27·06	—	m.p.–b.p.	208
$\langle \mathrm{Cr} \rangle$	−20,680	−1·31	—	14·56	—	298–m.p.	235, 556, 570
$\langle \mathrm{CrCl_2} \rangle$	−14,000	−0·62	−0·58	15·14	—	298–m.p.	49
$\{\mathrm{CrCl_2}\}$	−13,800	−5·03	—	27·70	—	m.p.–b.p.	49
$\langle \mathrm{CrCl_3} \rangle$	−13,950	−0·73	−0·77	17·49	—	298–s.p.	49
$\langle \mathrm{CrI_2} \rangle$	−16,080	−3·53	—	25·92	—	298–m.p.	3
$\{\mathrm{CrO_2Cl_2}\}$	−3340	−9·08	—	34·94	—	m.p.–b.p.	120
$\langle \mathrm{CrO_3} \rangle$	−10,300	—	—	20·14	—	448–468	492
$\langle \mathrm{Cr(CO)_6} \rangle$	−3738	—	—	11·75	—	274–311	982

evaporation constants (contd.)
+ *C**T* + D (mmHg)

Substance	b.pt. (°C)	L_e at b.pt. (kcal/mole)	sb.pt. (°C)	L_s at m.pt. or sb.pt.	L_e at 25°C (kcal/mole)	L_s at 25°C (kcal/mole)	Ref. and remarks
Bi	1680	42·8 ± 2·0	—	48·9 ± 1·5	46·9 ± 1·5	49·5 ± 1·5	350
Bi_2	1790	36·7 ± 3·0	—	51·0 ± 2·5	47·3 ± 2·5	52·5 ± 2·5	350
$BiCl_3$	441	17·3 ± 1·5	—	—	23·2 ± 1·5.	29·7 ± 2·5	120, 467
$BiBr_3$	461	18·0 ± 1·5	—	—	24·1 ± 1·5	—	120
Bi_2O_3	(1890)	—	—	—	—	—	—
Br_2	61	7·3 ± 0·1	—	—	7·6 ± 0·1	—	362, 120
C	—	—	(4350)	—	—	170·5 ± 3	23, 298, 283
C_2	—	—	—	—	—	195 ± 10	283, 281
CCl_4	77	7·28 ± 0·1	—	—	7·88 ± 0·1	13·2 ± 0·5	25
CBr_4	190	10·6 ± 0·5	—	11·7 ± 0·5	12·2 ± 0·5	13·0 ± 0·3	218, 198
$COCl_2$	7·5	5·83 ± 0·05	—	—	—	—	125
CS_2	46	6·5 ± 0·1	—	—	6·65 ± 0·1	—	120
COS	−50	4·43 ± 0·02	—	—	—	—	125
Ca	1483	36·0 ± 1·0	—	39·7 ± 0·5	40·1 ± 1·0	42·1 ± 0·5	50, 330, 398
CaF_2	2510	74·6 ± 4·0	—	91·6 ± 3·0	97·0 ± 4·0	105·2 ± 3·0	25, 612, 442
$CaCl_2$	(2000)	—	—	70·0 ± 2·0	—	73·3 ± 2·5	520
$CaBr_2$	*(1800)*	—	—	—	—	—	—
CaO	—	—	—	—	—	149·5 ± 8·0	22
Cd	765	23·9 ± 0·3	—	26·3 ± 0·4	25·5 ± 0·5	26·75 ± 0·5	120
CdF_2	1750	52·0 ± 4·0	—	63·5 ± 6·0	69·0 ± 6·0	—	25
$CdCl_2$	961	29·6 ± 0·8	—	38·9 ± 1·0	39·0 ± 2·0	41·2 ± 2·0	347, 120
$CdBr_2$	863	27·0 ± 2·0	—	33·5 ± 2·5	31·2 ± 4·0	36·2 ± 3·0	25
CdI_2	796	25·4 ± 2·0	—	31·1 ± 3·0	29·3 ± 3·0	32·9 ± 3·0	25
Ce	(3470)	—	—	—	(92·9)	—	—
$CeCl_3$	1731	—	—	70·7 ± 1·5	—	81·6 ± 2·0	618
$CeBr_3$	1562	—	—	68·3 ± 1·5	—	78·2 ± 2·0	618
Cl_2	−34·1	4·88 ± 0·05	—	—	—	—	125
Co	(2900)	—	—	98·4 ± 3·0	—	101·5 ± 4·0	52
$CoCl_2$	1025	37·6 ± 2·5	—	54·6 ± 2·5	47·6 ± 4·0	61·7 ± 4·0	208, (120)
Cr	2690	81·7 ± 1·5	—	88·8 ± 0·8	—	94·8 ± 0·7	235, 570, 150
$CrCl_2$	1300	47·0 ± 2·0	—	59·0 ± 1·5	59·9 ± 2·5	63·0 ± 2·5	25
$CrCl_3$	—	—	945	56·8 ± 2·5	—	63·0 ± 2·5	25
CrI_2	—	—	—	65·6 ± 1·5	—	71·4 ± 2·5	3
CrO_2Cl_2	117	8·3 ± 0·8	—	—	9·9 ± 0·7	—	120
$Cr(CO)_6$	—	—	151	17·2 ± 2·5	—	—	198

Table D. Vapour pressures and
$\log p = \mathrm{A} \cdot T^{-1} + \mathrm{B} \log T$

Substance	log p (mmHg)				Remarks	Temp. range (°K)	Ref.
	A	B	C × 10³	D			
{Cs}	−4075	−1·45	—	11·38	—	280–1000	58
⟨CsF⟩	−10,930	−2·12	—	17·51	—	298–m.p.	594
{CsF}	−9950	−2·84	—	18·62	—	m.p.–b.p.	—
⟨CsCl⟩	−10,800	−3·02	—	19·99	—	700–m.p.	38, 177, 25
{CsCl}	−9815	−3·52	—	20·38	—	m.p.–b.p.	25, 38
⟨CsBr⟩	−10,950	−3·02	—	20·02	—	700–m.p.	38, 25
{CsBr}	−10,080	−3·52	—	20·56	—	m.p.–b.p.	25, 38
⟨CsI⟩	−10,420	−3·02	—	19·70	—	600–m.p.	38, 25
{CsI}	−9678	−3·52	—	20·35	—	m.p.–b.p.	25, 38
⟨Cu⟩	−17,770	−0·86	—	12·29	—	298–m.p.	85
{Cu}	−17,520	−1·21	—	13·21	—	m.p.–b.p.	85
{CuCl}	−10,170	—	—	8·04	—	1000–1900	25
⟨Cu_3Cl_3⟩	−8156	—	—	11·235	—	548–658	619
⟨Cu_3Br_3⟩α	−7674	—	—	9·693	—	578–658	619
⟨Cu_3Br_3⟩β	−6756	—	—	8·30	—	670–704	619
⟨Cu_3Br_3⟩γ	−6309	—	—	7·787	—	—	619
⟨Cu_3I_3⟩α	−9463	—	—	11·14	—	629	619
⟨Cu_3I_3⟩β	−8351	—	—	9·41	—	643–670	619
⟨Cu_3I_3⟩γ	−7853	—	—	8·678	—	684–770	619
⟨Fe⟩	−21,080	−2·14	—	16·89	—	298–m.p.	52
{Fe}	−19,710	−1·27	—	13·27	—	m.p.–b.p.	52
{$FeCl_2$}	−9475	−5·23	—	26·53	—	m.p.–b.p.	322, 120
⟨Fe_2Cl_6⟩	−10,754	−12·64	—	55·898	$+1{\cdot}335 \times 10^5 T^{-2}$	298–m.p.	210
⟨FeI_2⟩	−12,180	−5·03	—	29·59	—	298–m.p.	621, 421
{FeI_2}	−8750	−5·536	—	27·185	—	m.p.–b.p.	421
{$Fe(CO)_5$}	−2075	—	—	8·42	—	298–b.p.	559, 120
{Ga}	−14,330	−0·844	—	11·42	—	298–m.p.	234
{$GaCl_3$}	−4886	−6·44	—	29·14	—	m.p.–b.p.	25
{$GaBr_3$}	−4700	−6·44	—	28·69	—	m.p.–b.p.	25
⟨Ge⟩	−20,150	−0·91	—	13·28	—	298–m.p.	215
{Ge}	−18,700	−1·16	—	12·87	—	m.p.–b.p.	215
{$GeCl_4$}	−2940	−9·08	—	34·27	—	m.p.–b.p.	120
{$GeBr_4$}	−3690	−9·05	—	35·00	—	m.p.–b.p.	120
⟨GeI_4⟩	−4920	−4·02	—	22·73	—	298–m.p.	108
⟨GeS⟩	−8350	—	—	10·78	—	500–m.p.	469
{H_2O}	−2900	−4·65	—	22·613	—	m.p.–b.p.	—
{H_2O_2}	−3560	−7·04	—	29·68	—	m.p.–	120
{H_2S_2}	−1850	—	—	8·26	—	273–b.p.	31

evaporation constants (contd.)
$+ CT + D$ (mmHg)

Substance	b.pt. (°C)	L_e at b.pt. (kcal/mole)	sb.pt. (°C)	L_s at m.pt. or sb.pt.	L_e at 25°C (kcal/mole)	L_s at 25°C (kcal/mole)	Ref. and remarks
Cs	700	15·9 ± 0·5	—	—	18·3 ± 0·5	18·8 ± 0·5	58, 120
CsF	1210	37·2 ± 1·5	—	45·9 ± 1·0	43·0 ± 1·5	48·7 ± 1·0	594
CsCl	1300	38·2 ± 2·0	—	48·6 ± 2·0	47·1 ± 2·5	52·3 ± 2·5	177, 120
CsBr	1300	36·0 ± 2·5	—	46·2 ± 2·0	44·9 ± 2·0	—	120
CsI	1280	35·9 ± 2·0	—	46·0 ± 2·0	44·7 ± 2·0	—	120
Cu	2570	73·3 ± 1·5	—	79·1 ± 1·0	79·4 ± 1·5	80·7 ± 1·5	85, 120
CuCl	1690	39·6	—	—	—	—	25
Cu_3Cl_3	1595	8·0	—	—	—	—	25
CuBr	1318	(33·4)	—	—	—	—	25
"Cu_3Br_3"	1730	(8·6)	—	—	—	—	25
CuI	1207	(31·1)	—	—	—	—	25
Fe	3070	81·3 ± 3·0	—	88·7 ± 2·5	89·4 ± 4·0	95·2 ± 4·0	52, 120
$FeCl_2$	1012	30·0 ± 2·0	—	44·0 ± 2·0	40·2 ± 2·5	47·5 ± 3·0	120, 322
Fe_2Cl_6	315	14·5 ± 2·0	—	32·7 ± 3·0	21·7 ± 3·0	—	459
Fe_2Br_6	—	—	—	—	—	34·5 ± 3·0	296
FeI_2	935	26·7 ± 6·0	—	47·0 ± 6·0	36·7 ± 6·0	52·7 ± 6·0	421, 621
$Fe(CO)_5$	109	7·8 ± 0·5	—	—	9·5 ± 0·3	—	559, 120
Ga	2420	64·6 ± 3·0	—	68·3 ± 2·5	67·0 ± 3·0	68·4 ± 3·0	457
$GaCl_3$	302	15·0 ± 2·0	—	—	—	—	25
Ga_2Cl_6	201	10·5 ± 1·0	—	—	—	—	25
$GaBr_3$	314	14·0 ± 2·0	—	—	—	—	25
Ga_2Br_6	292	12·0 ± 1·5	—	—	—	—	25
GaI_3	349	16·2 ± 1·5	—	—	—	—	25
Ga_2I_6	462	19·7 ± 2·5	—	—	—	—	25
Ge	2870	78·3 ± 3·0	—	90·0 ± 2·5	84·8 ± 3·0	91·7 ± 3·5	215, 300
$GeCl_4$	84	7·1 ± 0·7	—	—	8·1 ± 0·7	—	120
$GeBr_4$	189	8·6 ± 0·7	—	14·4 ± 1·2	11·5 ± 0·8	14·4 ± 1·2	120
GeI_4	dec.	—	—	—	—	20·1 ± 2·0	108
GeO_2	—	—	—	—	—	—	—
GeS	760	34·7 ± 3·0	—	40·5 ± 2·5	38·4 ± 4·0	43·5 ± 3·0	411, 622
HF	19·5	*	—	—	—	—	*89·5 cal/g
H_2O	100	9·82 ± 0·03	—	—	10·51 ± 0·01	—	125
H_2O_2	158	10·3 ± 0·8	—	—	12·2 ± 0·8	—	120
H_2S	−60·4	4·46 ± 0·05	—	—	—	—	125, 198
H_2S_2	71	—	—	—	8·54 ± 0·8	—	31

Table D. Vapour pressures and
$\log p = A \cdot T^{-1} + B \log T$

Substance	log p (mmHg) A	B	C × 10³	D	Remarks	Temp. range (°K)	Ref.
$\langle Hf\rangle\alpha$	−32,000	−0·5	—	11·81	—	298–2023	585
$\langle Hf\rangle\beta$	−31,630	−0·5	—	11·63	—	2023–m.p.	585
$\{Hf\}$	−29,830	—	—	9·20	—	m.p.–b.p.	—
$\langle HfCl_4\rangle$	−5197	—	—	11·71	—	476–681	584
$\langle HfI_4\rangle\alpha$	−10,700	—	—	19·56	—	575–597	236
$\langle HfI_4\rangle\beta$	−7360	—	—	13·97	—	598–645	236
$\{Hg\}$	−3305	−0·795	—	10·355	—	298–b.p.	30, 46
$\langle HgCl_2\rangle$	−4580	−2·0	—	16·39	—	298–m.p.	203, 177
$\langle HgBr_2\rangle$	−4500	0·05	−1·51	11·47	—	298–m.p.	120, 177
$\{HgBr_2\}$	−4370	−5·03	—	24·18	—	m.p.–b.p.	120
$\langle HgI_2\rangle$	−5690	−6·47	—	30·27	—	298–m.p.	120
$\{HgI_2\}$	−4620	−5·53	—	25·72	—	m.p.–b.p.	120
$\langle I_2\rangle$	−3578	−2·51	—	17·715	—	298–m.p.	23
$\{I_2\}$	−3205	−5·18	—	23·65	—	m.p.–b.p.	23, 120
$\{In\}$	−12,580	−0·45	—	9·79	—	m.p.–b.p.	7
$\{InCl\}$	−4640	—	—	8·03	—	m.p.–b.p.	194
$\langle InCl_3\rangle$	−8270	—	—	13·62	—	500–s.p.	194
$\langle InBr\rangle$	−6470	−2·01	—	16·31	—	298–m.p.	346
$\{InBr_2\}$	−4480	—	—	7·84	—	m.p.–b.p.	194
$\langle InBr_3\rangle$	−5670	—	—	11·67	—	500–s.p.	194
$\langle InI\rangle$	−6730	−1·97	—	15·74	—	298–m.p.	346
$\langle Ir\rangle$	−35,070	−0·7	—	13·18	—	298–m.p.	585, 511
$\{IrF_6\}$	−1657	—	—	7·952	—	m.p.–b.p.	449
$\{K\}$	−4770	−1·37	—	11·58	—	350–1050	58
$\langle KF\rangle$	−12,930	−2·06	—	17·30	—	298–m.p.	594, 177
$\{KF\}$	−11,570	−2·32	—	16·90	—	m.p.–b.p.	—
$\langle KCl\rangle$	−12,230	−3·0	—	20·34	—	298–m.p.	168, 120, 19
$\{KCl\}$	−10,710	−3·0	—	18·91	—	m.p.–b.p.	168, 120
$\langle KBr\rangle$	−11,110	−2·0	—	16·60	—	298–m.p.	276
$\{KBr\}$	−10,180	−3·0	—	18·67	—	m.p.–b.p.	120, 276
$\langle KI\rangle$	−11,000	−2·0	—	16·99	—	298–m.p.	276
$\{KI\}$	−10,050	−3·52	—	20·41	—	m.p.–b.p.	120
$\{KOH\}$	−10,230	−5·03	—	25·42	—	m.p.–b.p.	120
$\langle La\rangle$	−22,120	−0·33	—	10·39	—	298–m.p.	—
$\{La\}$	−21,530	−0·33	—	9·89	—	m.p.–b.p.	339, 464, 454
$\langle LaCl_3\rangle$	−19,040	−7·05	—	36·20	—	298–m.p.	618
$\langle LaBr_3\rangle$	−18,780	−7·05	—	36·83	—	298–m.p.	618
$\langle LaI_3\rangle$	−18,390	−7·05	—	37·00	—	298–m.p.	618
$\{Li\}$	−8415	−1·0	—	11·34	—	m.p.–b.p.	290, 167
$\{LiF\}$	−14,560	−4·02	—	23·56	—	m.p.–b.p.	120
$\{LiCl\}$	−10,760	−4·02	—	22·30	—	m.p.–b.p.	120

evaporation constants (contd.)
+ C*T* + D (mmHg)

Substance	b.pt. (°C)	L_e at b.pt. (kcal/mole)	sb.pt. (°C)	L_s at m.pt. or sb.pt.	L_e at 25°C (kcal/mole)	L_s at 25°C (kcal/mole)	Ref. and remarks
H_2Se	−41	4·6 ± 0·3	—	—	—	—	198
H_2Te	−2	5·65 ± 0·4	—	—	—	—	120
Hf	4450	136·4 ± 6·0	—	142·2 ± 3·0	—	146·0 ± 4·0	585
$HfCl_4$	—	—	316	23·8 ± 1·0	—	27·0 ± 1·5	584
$HfBr_4$	—	—	322	(24·0)	—	(26·0)	25
Hg	357	14·13 ± 0·1	—	—	14·65 ± 0·1	—	30, 46
$HgCl_2$	304	14·1 ± 0·4	—	18·6 ± 0·5	16·8 ± 0·5	19·9 ± 0·5	120, 177
$HgBr_2$	319	14·1 ± 0·5	—	19·8 ± 0·2	17·0 ± 0·5	20·5 ± 0·5	120, 177
HgI_2	354	14·3 ± 0·5	—	19·8 ± 0·4	17·9 ± 0·7	21·8(α)	120
I_2	183	9·96 ± 0·2	—	14·45 ± 0·1	11·57 ± 0·3	14·9 ± 0·1	23, 24, 124
In	2062	55·4 ± 2·0	—	57·9 ± 2·0	—	58·1 ± 2·0	7
InCl	608	20·4 ± 1·5	—	—	23·4 ± 1·5	—	194
$InCl_2$	(485)	(42·1)*	—	—	—	—	194, *disprop.?
$InCl_3$	—	—	498	37·8 ± 2·0	—	—	194
InBr	660	22·7 ± 1·5	—	27·4 ± 1·0	25·3 ± 1·5	28·4 ± 1·5	346
$InBr_2$	631	19·7 ± 1·5	—	—	24·0 ± 2·0	—	194
$InBr_3$	—	—	371	25·9 ± 2·0	—	—	194
InI	770	23·1 ± 2·0	—	28·3 ± 1·5	26·1 ± 2·0	29·6 ± 1·5	346
Ir	4390	146·3 ± 3·0	—	156·4 ± 2·0	—	160·0 ± 2·0	511, 585, 589
IrF_6	53	7·4 ± 1·0	—	—	—	9·0 ± 1·0	449
K	779	18·9 ± 0·5	—	—	21·0 ± 0·7	21·5 ± 0·7	120
KF	1510	44·6 ± 2·0	—	54·4 ± 1·5	51·4 ± 1·5	57·8 ± 1·5	594, 177
KCl	1407	39·0 ± 0·5	—	49·7 ± 0·5	47·2 ± 0·8	54·1 ± 0·8	276, 177, 120
KBr	1383	37·2 ± 0·7	—	48·7 ± 0·6	46·6 ± 1·0	51·2 ± 0·8	276, 120
KI	1330	34·7 ± 1·5	—	46·5 ± 1·5	43·5 ± 2·0	49·2 ± 2·0	276, 120
KOH	1330	30·9 ± 1·5	—	—	43·9 ± 1·5	—	120
La	3420	96·1 ± 2·0	—	99·7 ± 1·5	—	101·0 ± 1·5	339, 464, 454
$LaCl_3$	1812	—	—	71·3 ± 1·5	—	82·9 ± 2·0	618
$LaBr_3$	1586	—	—	71·1 ± 1·5	—	81·8 ± 2·0	618
LaI_3	1472	—	—	69·7 ± 1·5	—	80·0 ± 2·0	618
Li	1329	35·3 ± 2·0	—	38·3 ± 2·0	37·9 ± 2·5	38·6 ± 2·5	290, 167
LiF	1681	51·0 ± 2·0	—	63·5 ± 3·0	64·2 ± 3·0	66·5 ± 2·5	120
LiCl	1382	36·0 ± 1·5	—	47·0 ± 2·0	46·8 ± 1·8	51·0 ± 2·5	120, 177

Table D. Vapour pressures and
$\log p = A \cdot T^{-1} + B \log T$

Substance	log p (mmHg) A	B	C × 10³	D	Remarks	Temp. range (°K)	Ref.
{LiBr}	−10,170	−3·52	—	20·55	–	m.p.–b.p.	120
{LiI}	−11,110	−3·52	—	21·70	–	m.p.–b.p.	120
⟨Mg⟩	−7780	−0·855	—	11·41	—	298–m.p.	58, 253
{Mg}	−7550	−1·41	—	12·79	—	m.p.–b.p.	58, 253
⟨MgF_2⟩	−20,600	−2·11	—	19·06	—	298–1536	—
{MgF_2}	−18,150	−3·9	—	23·17	—	1536–2605	—
{$MgCl_2$}	−10,840	−5·03	—	25·53	—	m.p.–b.p.	120
{$MgBr_2$}	*−10,930*	*−5·03*	—	*26·07*	—	m.p.–b.p.	25
{MgI_2}	*−8090*	*−5·03*	—	*25·18*	—	m.p.–b.p.	25
⟨Mn⟩	−14,920	−1·96	—	16·19	α, β, γ, δ	298–m.p.	99
{Mn}	−14,520	−3·02	—	19·24	—	m.p.–b.p.	99
⟨MnF_2⟩	−17,400	−3·02	—	22·06	—	298–m.p.	432
{$MnCl_2$}	−10,606	−4·33	—	23·68	—	m.p.–b.p.	322, (120)
⟨Mo⟩	−34,700	−0·236	−0·145	11·66	—	298–m.p.	99
{MoF_5}	−2772	—	—	8·58	—	m.p.–b.p.	449
{MoF_6}	−1500	—	—	7·76	—	m.p.–b.p.	449, 120
⟨$MoOF_4$⟩	−2854	—	—	9·21	—	313–m.p.	449
{$MoOF_4$}	−2671	—	—	8·716	—	m.p.–b.p.	449
⟨$MoCl_5$⟩	−5210	—	—	13·1	—	298–m.p.	33
⟨MoO_3⟩	−15,230	−4·02	—	27·16	—	298–m.p.	340, 250
{MoO_3}	−12,480	−4·02	—	24·60	—	m.p.–b.p.	419, 239
{Na}	−5780	−1·18	—	11·50	—	298–b.p.	312, 58
⟨NaF⟩	−14,960	−2·01	—	17·53	—	298–m.p.	594, 177
{NaF}	−13,500	−2·52	—	17·93	—	m.p.–b.p.	177, 120
⟨NaCl⟩	−12,440	−0·90	−0·46	14·31	—	298–m.p.	120, 276, 168
{NaCl}	−11,530	−3·48	—	20·77	—	m.p.–b.p.	120, 168
⟨NaBr⟩	−12,100	−3·0	—	20·39	—	298–m.p.	168
{NaBr}	−10,500	−3·0	—	18·81	—	m.p.–b.p.	120
{NaBr}	−9687	—	—	8·957	—	1143–1403	20
{NaI}	−10,740	−3·52	—	20·96	—	m.p.–b.p.	120
{NaOH}	−7520	—	—	7·43	—	1280–1700	120
⟨Nb⟩	−37,650	+0·715	−0·166	8·94	—	298–m.p.	99
⟨NbF_5⟩	−4900	—	—	14·397	—	298–m.p.	111
{NbF_5}	−2780	—	—	8·37	—	m.p.–b.p.	111
⟨$NbCl_4$⟩	−6870	—	—	12·30	—	577–651	322a
⟨$NbCl_5$⟩	−4370	—	—	11·51	—	403–m.p.	206, 2
{$NbCl_5$}	−2870	—	—	8·37	—	m.p.–b.p.	206, 2
⟨$NbOCl_3$⟩	−5330	—	—	11·67	—	298–s.p.	602
{$NbBr_5$}	−4085	—	—	9·33	—	m.p.–b.p.	2
⟨$NdCl_3$⟩	−18,220	−7·05	—	36·27	—	298–m.p.	618
⟨$NdBr_3$⟩	−17,650	−7·05	—	36·51	—	298–m.p.	618
⟨NdI_3⟩	−17,490	−7·05	—	36·61	—	298–m.p.	618
⟨Ni⟩	−22,500	−0·96	—	13·60	—	298–m.p.	107
{Ni}	−22,400	−2·01	—	16·95	—	m.p.–b.p.	—
⟨NiF_2⟩	−14,650	−3·02	—	20·28	—	298–m.p.	478

evaporation constants (*contd.*)
$+ CT + D$ (mmHg)

Substance	b.pt. (°C)	L_e at b.pt. (kcal/mole)	sb.pt. (°C)	L_s at m.pt. or sb.pt.	L_e at 25°C (kcal/mole)	L_s at 25°C (kcal/mole)	Ref. and remarks
LiBr	1310	35·4 ± 1·5	—	45·6 ± 2·0	44·4 ± 1·8	—	120
LiI	1170	40·8 ± 3·0	—	50·3 ± 3·5	48·8 ± 3·0	—	120
Mg	1105	30·5 ± 1·5	—	34·0 ± 1·5	33·6 ± 2·0	35·0 ± 2·0	58, 120, 253
MgF_2	2332	65·3 ± 1·5	—	87·5 ± 1·5	83·8 ± 1·5	93·0 ± 1·5	497
$MgCl_2$	1418	32·7 ± 2·0	—	50·0 ± 2·0	46·6 ± 2·5	—	120
$MgBr_2$	*1230*	(*35·0*)	—	—	—	—	25
Mn	2060	52·7 ± 2·0	—	62·3 ± 1·0	—	67·1 ± 1·5	99
$MnCl_2$	1231	35·6 ± 2·0	—	—	46·0 ± 3·5	—	322, (120)
Mo	4650	141·0 ± 5·0	—	151·8 ± 3·0	—	158·7 ± 3·0	99
MoF_5	214	12·4 ± 0·5	—	—	—	—	449
MoF_6	34	6·65 ± 0·3	—	—	6·85 ± 0·2	7·8 ± 0·2	449, 432
$MoOF_4$	186	12·1 ± 0·5	—	—	—	—	449
$MoCl_5$	268	*15·0* ± 3·5	—	23·5 ± 3·0	—	—	33
MoO_3	1100	46·0 ± 2·5	—	61·0 ± 2·5	54·6 ± 2·5	67·2 ± 2·5	419, 340
N_2	−196	1·333 ± 0·01	—	—	—	—	125
NH_3	−33·5	5·58 ± 0·02	—	—	—	—	125
Na	882	23·7 ± 0·4	—	26·2 ± 0·5	25·7 ± 0·6	26·35 ± 0·6	312, 58
NaF	1710	51·8 ± 2·0	—	63·3 ± 2·0	60·2 ± 2·5	67·2 ± 2·0	177, 594
NaCl	1465	40·7 ± 0·2	—	52·5 ± 0·5	50·5 ± 0·7	56·1 ± 0·8	276, 177, 120
NaBr	1393	38·0 ± 1·0	—	49·2 ± 1·5	45·5 ± 2·0	53·5 ± 1·5	168, 177, 120
NaI	1304	38·1 ± 1·0	—	48·6 ± 1·5	47·1 ± 1·5	51·6 ± 1·5	120
NaOH	1390	(34·5)	—	(36·5)	—	—	120
NbF_5	233	12·5 ± 3·0	—	22·4 ± 2·5	14·3 ± 3·0	23·0 ± 3·0	111, (120)
$NbCl_5$	250	13·1 ± 0·8	—	20·0 ± 1·0	—	—	206, 2
$NbOCl_3$	—	—	335	24·4 ± 1·0	—	26·1 ± 1·5	602
$NbBr_5$	361	18·7 ± 1·0	—	—	—	—	2
$NdCl_3$	1674	—	—	68·9 ± 1·5	—	79·2 ± 2·0	618
$NdBr_3$	1495	—	—	67·4 ± 1·5	—	76·6 ± 2·0	618
NdI_3	1437	—	—	65·4 ± 1·5	—	75·95 ± 2·0	618
Ni	2920	89·6 ± 4·0	—	99·7 ± 3·0	—	102·6 ± 3·0	107, 120
NiF_2	(1740)	— —	—	56·7 ± 2·0	—	65·3 ± 2·0	478

Table D. Vapour pressures and
$\log p = A \cdot T^{-1} + B \log T$

Substance	log p (mmHg) A	B	C × 10^3	D	Remarks	Temp. range (°K)	Ref.
⟨$NiCl_2$⟩	−13,300	−2·68	—	21·88	—	298–s.p.	322, 120
⟨$NiBr_2$⟩	−13,110	−1·71	−0·35	16·68	—	298–s.p.	207
{$Ni(CO)_4$}	−1530	—	—	7·73	—	298–b.p.	120
⟨NpF_6⟩	−2892	−2·7	—	18·48	—	273–327	567
{NpF_6}	−1913	−2·347	—	14·613	—	328–350	567
⟨Os⟩	−39,880	—	—	10·365	—	2157–2592	451
{OsF_5}	−3429	—	—	9·75	—	m.p.–b.p.	449
⟨OsF_6⟩	−1858	—	—	8·726	—	273–m.p.	449
{OsF_6}	−1473	—	—	7·47	—	m.p.–b.p.	449
⟨OsO_4⟩	−2955	—	—	9·64	yellow	273–329	120
⟨OsO_4⟩	−2580	—	—	10·70	white	273–315	120
{OsO_4}	−2065	—	—	8·01	—	m.p.–b.p.	120
⟨P_4⟩	−3530	−3·5	—	19·09	white	298–m.p.	44, 120
⟨P_4⟩	−7240	—	—	12·85	black	298–m.p.	120
⟨P_4⟩	−5600	—	—	11·00	violet	298–m.p.	120
{P_4}	−2740	—	—	7·84	—	m.p.–b.p.	120
{PCl_3}	−2370	−5·14	—	22·74	—	273–b.p.	120
⟨PCl_5⟩	−3520	—	—	11·035	—	373–432	25
⟨P_4O_{10}⟩	−4940	—	—	10·70	hexagonal	298–631	120
⟨P_4O_{10}⟩α	−8250	—	—	12·55	o'rhombic	298–842	120
{$POCl_3$}	−1830	—	—	7·72	—	273–b.p.	120
{Pb}	−10,130	−0·985	—	11·16	—	m.p.–b.p.	77, 58, 120
{PbF_2}	−11,800	−5·03	—	26·48	—	m.p.–b.p.	120
⟨$PbCl_2$⟩	−9890	−0·95	−0·91	15·36	—	298–m.p.	120
{$PbCl_2$}	−10,000	−6·65	—	31·60	—	m.p.–b.p.	120, 177
⟨$PbBr_2$⟩	−9320	−2·08	−0·34	18·44	—	298–m.p.	100, 120
{$PbBr_2$}	−9540	−6·76	—	31·67	—	m.p.–b.p.	120
⟨PbI_2⟩	−9340	−2·35	−0·32	19·68	—	298–m.p.	120, (177)
{PbI_2}	−10,000	−9·21	—	39·80	—	m.p.–b.p.	120
⟨PbO⟩	−13,480	−0·92	−0·35	14·36	—	298–m.p.	120
⟨PbS⟩	−13,300	−0·81	−0·43	14·85	—	298–1200	120, 252, 92
⟨Pd⟩	−19,800	−0·755	—	11·82	—	298–m.p.	423, 511, 473
{Pd}	−17,500	+1·0	—	4·81	—	m.p.–b.p.	—
{Po}	−5810	−1·0	—	10·70	—	m.p.–b.p.	351
{Pr}	−17,190	—	—	8·10	—	1425–1692	99
⟨$PrCl_3$⟩	−18,490	−7·05	—	36·31	—	298–m.p.	618
⟨$PrBr_3$⟩	−17,800	−7·05	—	36·53	—	298–m.p.	618
⟨PrI_3⟩	−17,470	−7·05	—	36·66	—	298–m.p.	618
⟨Pt⟩	−28,460	−1·27	—	14·33	—	298–m.p.	58
{Pt}	−27,890	−1·77	—	15·71	—	m.p.–b.p.	58

evaporation constants (*contd.*)
+ CT + D (mmHg)

Substance	b.pt. (°C)	L_e at b.pt. (kcal/mole)	sb.pt. (°C)	L_s at m.pt. or sb.pt.	L_e at 25°C (kcal/mole)	L_s at 25°C (kcal/mole)	Ref. and remarks
$NiCl_2$	—	—	970	53·8 ± 2·0	—	59·1 ± 2·0	322, 120
$NiBr_2$	—	—	919	53·7 ± 1·5	—	58·9 ± 1·5	207
Ni_3S_2	—	—	—	—	—	—	—
$Ni(CO)_4$	42	7·0 ± 0·2	—	—	7·0 ± 0·3	—	120
NpF_6	55	7·2 ± 0·2	—	11·45 ± 0·2	7·35 ± 0·2	11·6 ± 0·2	567
O_2	−183	1·63 ± 0·01	—	—	—	—	125
Os	5030	—	—	—	—	189·0 ± 3·0	451
OsF_5	226	15·7 ± 3·0	—	—	—	—	449
OsF_6	47·5	6·72	—	—	—	—	449
OsO_4	130	9·45 ± 0·7	—	11·8 ± 0·7	—	—	120
P_4	280	12·4 ± 0·7	—	13·9 ± 0·5	13·4 ± 0·5	14·05 ± 0·5	44, 120
	—	—	453	33·1 ± 3·0	—	—	120
	—	—	413	25·6 ± 3·0	—	—	120
PCl_3	75	7·3 ± 0·3	—	—	7·7 ± 0·3	—	120, 25
PCl_5	166*	—	—	14·0 ± 2·5	—	—	25, *dec.
PBr_3	174	9·5 ± 0·4	—	—	—	—	25
PI_3	(230)	10·5 ± 0·7	—	—	—	—	25
P_4O_6	175	10·4 ± 2·0	—	—	—	—	120
P_4O_{10}	600	(26·0)	—	(37·8)	—	—	120
	—	—	360	(17·6)	—	—	198
$POCl_3$	105	8·4 ± 0·5	—	11·6 ± 0·8	8·4 ± 0·5	—	120
Pb	1740	42·5 ± 0·5	—	46·4 ± 0·6	45·8 ± 0·7	46·9 ± 0·7	77, 58, 120
PbF_2	1293	38·3 ± 3·0	—	—	51·0 ± 5·0	—	120
$PbCl_2$	952	30·3 ± 1·2	—	42·0 ± 1·2	42·6 ± 1·2	45·0 ± 1·5	347, 120
$PbBr_2$	914	27·7 ± 2·0	—	39·3 ± 2·0	39·6 ± 2·0	41·3 ± 2·0	120
PbI_2	872	24·6 ± 1·0	—	38·6 ± 1·0	40·3 ± 1·5	41·2 ± 1·5	120
PbO	—	— —	—	—	—	—	—
Pd	2940	86·4 ± 2·5	—	87·8 ± 0·7	—	90·1 ± 0·7	473, 511, 423
$PdCl_2$	(*1200*)	—	—	34·1 ± 2·0	—	—	582
Po	965	24·1 ± 2·0	—	—	26·0 ± 2·0	—	351
Pr	(3020)	—	—	—	—	—	—
$PrCl_3$	1709	—	—	69·75 ± 1·5	—	80·4 ± 2·0	618
$PrBr_3$	1522	—	—	67·9 ± 1·5	—	77·25 ± 2·0	618
PrI_3	1422	—	—	65·8 ± 1·5	—	75·7 ± 2·0	618
Pt	4170	112·1 ± 6·0	—	125·8 ± 4·0	126·6 ± 5·0	130·2 ± 5·0	58, 473, 511

Table D. Vapour pressures and
$\log p = \mathrm{A} \cdot T^{-1} + \mathrm{B} \log T$

Substance	log *p* (mmHg) A	B	C × 10^3	D	Remarks	Temp. range (°K)	Ref.
{Pu}	– 17,590	—	—	7·90	—	1392–1793	99
⟨PuF_3⟩	– 24,950	– 7·05	—	36·91	—	298–m.p.	355, 185
{PuF_3}	– 23,500	– 6·45	—	34·47	—	m.p.–200	185
⟨$PuCl_3$⟩	– 18,270	– 5·34	—	32·60	—	298–m.p.	185
{$PuCl_3$}	– 15,490	– 6·45	—	31·76	—	m.p.–b.p.	185
⟨$PuBr_3$⟩	– 17,460	– 5·34	—	31·32	—	298–m.p.	185
{$PuBr_3$}	– 15,030	– 6·45	—	32·34	—	m.p.–b.p.	185
{Rb}	– 4688	– 1·76	—	13·07	—	813–1258	658
⟨RbF⟩	– 12,600	– 2·66	—	19·59	—	298–m.p.	594
{RbF}	– 11,230	– 2·66	—	18·26	—	m.p.–b.p.	—
⟨RbCl⟩	– 11,670	– 3·0	—	20·157	—	298–m.p.	168
{RbCl}	– 10,300	– 3·0	—	18·77	—	m.p.–b.p.	168, 120
⟨RbBr⟩	– 11,510	– 3·0	—	20·155	—	298–m.p.	168
{RbBr}	– 10,220	– 3·0	—	18·805	—	m.p.–b.p.	168, 120
{RbI}	– 10,280	– 3·52	—	20·64	—	m.p.–b.p.	120
⟨Re⟩	– 40,800	– 1·16	—	14·20	—	298–3000	410
{ReF_5}	– 3037	—	—	9·025	—	m.p.–b.p.	449
{ReF_6}	– 1489	—	—	7·73	—	m.p.–b.p.	449
⟨ReF_7⟩	– 2206	– 1·47	—	13·045	—	259–m.p.	566
{ReF_7}	– 244·3	+ 9·91	—	– 21·585	—	m.p.–b.p.	566
⟨$ReOF_4$⟩	– 3888	—	—	11·88	—	323–m.p.	449
{$ReOF_4$}	– 3206	—	—	10·09	—	m.p.–b.p.	449
⟨$ReOF_5$⟩β	– 1959	—	—	8·62	—	303–m.p.	449
{$ReOF_5$}	– 1679	—	—	7·73	—	m.p.–b.p.	449
{ReO_2F_3}	– 3437	—	—	10·36	—	m.p.–b.p.	449
⟨Re_2O_7⟩	– 6775	—	—	13·67	—	413–478	492
⟨Rh⟩	– 29,360	– 0·88	—	13·50	—	298–m.p.	473, 511, 585
⟨Ru⟩	– 32,770	—	—	10·38	—	1940–2377	451
⟨$RuCl_3$⟩	– 16,750	– 4·63	—	30·53	—	298–1000	434
⟨S_2⟩	– 6750	—	—	11·32	—	298–m.p.	21, 120
{S_2}	– 6975	– 1·53	– 1·0	16·22	—	m.p.–b.p.	21, 122
{S_x}	– 4830	– 5·0	—	23·88	total *p*	m.p.–b.p.	21
⟨SO_3⟩α	– 2680	—	—	11·44	—	273–m.p.	120
⟨SO_3⟩β	– 2860	—	—	11·97	—	273–m.p.	120
⟨SO_3⟩γ	– 3610	—	—	14·00	—	273–m.p.	120
{SO_3}	– 2230	—	—	9·90	—	m.p.–b.p.	120
{SO_2Cl_2}	– 1660	—	—	7·65	—	m.p.–b.p.	120
⟨Sb_x⟩	– 10,320	—	—	10·59	total *p*	500–m.p.	99
{Sb_x}	– 6500	—	—	6·37	total *p*	m.p.–b.p.	99
⟨Sb_4⟩	– 11,560	– 3·52	—	22·40	—	298–m.p.	Richards
⟨$SbCl_3$⟩	– 3460	3·88	– 5·6	2·81	—	298–m.p.	120
{$SbCl_3$}	– 3770	– 7·04	—	29·48	—	m.p.–b.p.	120
{$SbCl_5$}	– 2530	—	—	8·56	—	m.p.–350	120
{$SbBr_3$}	– 4760	– 4·54	—	23·68	—	m.p.–b.p.	983

evaporation constants (contd.)
+ CT + D (mmHg)

Substance	b.pt. (°C)	L_e at b.pt. (kcal/mole)	sb.pt. (°C)	L_s at m.pt. or sb.pt.	L_e at 25°C (kcal/mole)	L_s at 25°C (kcal/mole)	Ref. and remarks
Pu	(3230)	82·1 ± 8·0	—	81·0 ± 7·0	—	84·1 ± 8·0	99
PuF_3	(2120)	(76·6)	—	96·6 ± 2·5	103 ± 4·0	108·5 ± 4·0	185
PuF_6	62	7·2 ± 0·8	—	—	—	—	595
$PuCl_3$	1790	44·4 ± 2·5	—	72·8 ± 2·5	67·0 ± 3·0	80·4 ± 3·5	185
$PuBr_3$	1475	46·2 ± 3·0	—	69·9 ± 2·5	64·9 ± 3·0	76·7 ± 3·5	185
Rb	673	18·1 ± 1·0	—	20·9 ± 1·0	20·4 ± 1·0	20·9 ± 1·0	658
RbF	1390	42·5 ± 1·5	—	52·1 ± 1·5	49·5 ± 1·5	55·7 ± 1·5	594
RbCl	1381	39·6 ± 2·0	—	(48·6)	49·1 ± 2·0	—	177, 120, 168
RbBr	1352	37·0 ± 1·0	—	(44·7)	45·0 ± 2·0	—	168, 120
RbI	1304	36·0 ± 1·0	—	(43·0)	44·9 ± 2·0	—	120
Re	(5650)	—	—	178·0 ± 2·5	—	186·1 ± 2·0	410, 99
ReF_6	33·7	6·8 ± 0·2	—	—	6·9 ± 0·3	7·85 ± 0·2	566, 449
ReF_7	74	8·55 ± 0·2	—	9·15 ± 0·2	7·0 ± 0·2	9·2 ± 0·2	566
$ReOF_4$	172	14·6 ± 0·5	—	17·8 ± 0·5	—	—	449
$ReOF_5$	73	7·7 ± 0·4	—	8·95 ± 0·5	—	—	449
ReO_2F_3	185	15·7 ± 0·5	—	—	—	—	449
Re_2O_7	362	18·0 ± 3·5	—	33·4 ± 3·5	—	—	120
Rh	3760	—	—	129·4 ± 2·0	—	132·8 ± 2·5	511, 473
S	444·6*	—	—	—	—	—	* S_x
S_2	625	25·4 ± 1·0	—	30·7 ± 1·0	30·5 ± 1·0	31·05 ± 1·0	21, 122
S_4	625	22·9 ± 3·0	—	31·5 ± 3·0	30·7 ± 3·0	31·75 ± 3·0	21
S_6	527	15·8 ± 2·5	—	26·9 ± 2·5	25·95 ± 2·5	27·45 ± 2·5	21
S_8	490	15·1 ± 2·5	—	29·1 ± 2·5	27·95 ± 2·5	30·0 ± 2·5	21
S_2F_{10}	29	7·0 ± 0·2	—	—	7·0 ± 0·2	—	25
S_2Cl_2	138	8·7 ± 0·1	—	—	—	—	25
SO_2	−10	5·96 ± 0·05	—	—	—	—	125
SO_3	43	—	—	11·76	—	—	198
SO_2Cl_2	69	7·5 ± 0·2	—	—	7·8 ± 0·2	—	120
$SOCl_2$	76	7·4 ± 0·3	—	—	7·7 ± 0·3	—	120
Sb_2	1675	39·4 ± 0·8	—	56·0 ± 0·8	49·3 ± 1·0	—	321, 23
Sb_4	—	—	—	46·6 ± 1·0	33·6 ± 1·0	50·8 ± 2·0	321, 23
$SbCl_3$	220	10·4 ± 0·3	—	15·5 ± 0·3	13·1 ± 0·4	15·8 ± 0·4	120
$SbCl_5$	dec.	—	—	—	11·7 ± 2·5	—	120
$SbBr_3$	280	(14·1)	—	—	15·1 ± 2·5	—	239

Table D. Vapour pressures and $\log p = A \cdot T^{-1} + B \log T$

Substance	log p (mmHg) A	B	C × 10³	D	Remarks	Temp. range (°K)	Ref.
$\{SbI_3\}$	−3600	—	—	8·25	—	474–674	239
$\langle Sb_4O_6\rangle$	−10,360	—	—	12·195	cubic	742–839	120
$\langle Sb_4O_6\rangle$	−9625	—	—	11·312	orthorhombic	742–914	120
$\{Sb_4O_6\}$	−3900	—	—	5·137	—	929–1073	120
$\langle Sc\rangle\beta$	−19,700	−1·0	—	13·07	—	1607–m.p.	554, 623, 339
$\langle ScCl_3\rangle$	−14,200	—	—	14·37	—	1065–1233	25
$\langle ScBr_3\rangle$	−13,780	—	—	14·35	—	1042–1200	25
$\langle ScI_3\rangle$	−13,340	—	—	14·17	—	1010–1180	25
$\{Se_x\}$	−4990	—	—	8·09	mainly (Se_6)	m.p.–b.p.	28
$\{SeF_4\}$	−2457	—	—	9·44	—	m.p.–b.p.	183
$\langle SeO_2\rangle_x$	−4995	—	—	10·98	—	420–492	492
$\{Si\}$	−20,900	−0·565	—	10·78	—	m.p.–b.p.	502, 468
$\langle Si\rangle$	−23,550	−0·565	—	12·35	—	298–m.p.	—
$\{SiCl_4\}$	−1572	—	—	7·64	—	273–333	119
$\{SiI_4\}$	−3863	−5·0	—	23·38	—	m.p.–b.p.	6
$\langle Sm\rangle$	−11,170	−1·56	—	13·76	—	298–m.p.	405, 517
$\{Sn\}$	−15,500	—	—	8·23	—	505–b.p.	99
$\{SnCl_4\}$	−1925	—	—	7·865	—	298–b.p.	112, 86
$\{SnBr_4\}$	−3510	−6·5	—	27·63	—	303–b.p.	112, 218
$\langle SnI_4\rangle$	−3990	—	—	10·08	—	298–m.p.	218
$\{SnI_4\}$	−2975	—	—	7·666	—	m.p.–b.p.	112, 218, 25
$\langle SnS\rangle$	−10,460	—	—	9·97	—	890–m.p.	321
$\langle Sr\rangle$	−9450	−1·31	—	13·08	—	813–m.p.	398
$\{Sr\}$	−9000	−1·31	—	12·63	—	m.p.–b.p.	398, 120
$\{SrF_2\}$	−21,660	−5·03	—	28·04	—	m.p.–b.p.	25
$\langle SrO\rangle$	−27,500	−2·0	—	17·74	dec?	298–1600	169
$\langle Ta\rangle$	−40,800	—	—	10·29	—	298–m.p.	99
$\langle TaCl_4\rangle$	−6600	—	—	11·71	—	—	615
$\langle TaCl_5\rangle$	−4654	—	—	12·197	—	—	601
$\{TaCl_5\}$	−2865	—	—	8·54	—	m.p.–b.p.	601
$\langle TaBr_5\rangle$	−7320	−7·04	—	34·85	—	298–m.p.	273, 2
$\{TaBr_5\}$	−3260	—	—	8·14	—	m.p.–b.p.	273, 2
$\langle TaI_5\rangle$	−6660	−7·04	—	31·61	—	298–m.p.	2
$\{TaI_5\}$	−3955	—	—	7·72	—	m.p.–b.p.	2
$\langle Tc_2O_7\rangle$	−7205	—	—	18·28	—	298–m.p.	284
$\{Tc_2O_7\}$	−3570	—	—	9·00	—	m.p.–b.p.	284

evaporation constants (contd.)
$+ CT + D$ (mmHg)

Substance	b.pt. (°C)	L_e at b.pt. (kcal/mole)	sb.pt. (°C)	L_s at m.pt. or sb.pt.	L_e at 25°C (kcal/mole)	L_s at 25°C (kcal/mole)	Ref. and remarks
SbI_3	400	16·4 ± 2·0	—	20·6 ± 3·0	—	—	239
Sb_4O_6	1425	—	—	(44·0)	—	—	120
Sb_4S_6	—	—	—	(51·2)	—	—	252, (92) *298°K
Sc	(2870)	—	—	86·3 ± 3·0	—	89·9 ± 4·5	554, 623, 339
$ScCl_3$	967	—	—	65·0 ± 3·0	—	—	25
$ScBr_3$	—	—	929	63·0 ± 3·0	—	—	25
ScI_3	—	—	909	61·0 ± 3·0	—	—	25
Se_6	695	(21·5)	—	—	—	—	28, 205
SeF_4	102	—	—	—	11·3 ± 1·0	—	183
SeF_6	—	—	−46	6·4 ± 0·5	—	—	25
Si	3280	91·6 ± 2·5	—	105·8 ± 2·0	95·3 ± 2·5	107·8 ± 2·5	502, 468
SiH_4	−112	2·96 ± 0·05	—	—	—	—	125
SiF_4	−97	4·5 ± 0·3	−95	6·15 ± 0·3	—	—	25
$SiCl_4$	58	6·8 ± 0·4	—	—	7·2 ± 0·4	—	119, 25
$SiBr_4$	153	9·05 ± 0·2	—	—	—	—	25
SiI_4	301	12·0 ± 0·6	—	—	14·75 ± 0·7	—	6
Sm	1800	39·4 ± 4·0	—	45·9 ± 1·5	—	49·6 ± 1·5	405, 517
Sm_2O_3	—	—	—	—	—	—	—
Sn	2623	70·8 ± 2·0	—	71·9 ± 1·5	—	72·2 ± 1·5	215, 282
$SnCl_2$	652	19·5 ± 2·0	—	—	25·8 ± 3·0	—	25
$SnCl_4$	115	8·1 ± 0·4	—	—	9·3 ± 0·3	—	112, 86
$SnBr_2$	639	23·3 ± 2·0	—	—	30·0 ± 3·0	—	25
$SnBr_4$	207	9·8 ± 0·8	—	—	12·2 ± 1·2	—	112, 218
SnI_2	715	23·8 ± 2·0	—	—	34·0 ± 3·0	—	25
SnO_2	—	—	—	—	—	—	—
Sr	1350	36·9 ± 3·0	—	40·6 ± 3·0	40·4 ± 3·0	—	398, 120
SrF_2	2480	71·5 ± 5·0	—	86·5 ± 5·0	96·0 ± 6·0	(100·0)	25
SrO	—	—	—	—	—	124·8 ± 4·0	169
Ta	—	—	—	187·8 ± 2·5	—	186·9 ± 1·5	99
$TaCl_4$	—	—	—	—	—	32·5 ± 2·0	601
$TaCl_5$	234	12·0 ± 0·8	—	20·9 ± 1·0	—	22·3 ± 1·5	601
$TaBr_5$	347	14·9 ± 0·7	—	25·8 ± 1·0	—	29·25 ± 1·0	2, 273
TaI_5	545	17·9 ± 1·5	—	19·5 ± 1·5	—	25·2 ± 2·0	2
Tc_2O_7	312	14·0 ± 1·5	—	30·2 ± 1·5	21·2 ± 1·0	31·7 ± 2·0	—

Table D. Vapour pressures and
$\log p = A \cdot T^{-1} + B \log T$

Substance	log p (mmHg) A	B	C × 10^3	D	Remarks	Temp. range (°K)	Ref.
⟨Te_2⟩	−9175	−2·71	—	19·68	—	298–m.p.	—
{Te_2}	−7830	−4·27	—	22·29	—	m.p.–b.p.	563, 28, 294
⟨TeF_4⟩	−3174	—	—	9·093	—	298–m.p.	110
{TeF_4}	−1787	—	—	5·640	—	m.p.–467	110
⟨TeF_6⟩	−1460	—	—	9·13	—	194–241	120
{$TeCl_2$}	−3350	—	—	8·51	—	m.p.–b.p.	25
⟨TeO_2⟩	−13,940	−3·52	—	23·51	—	298–m.p.	250
⟨Th⟩	−30,200	−1·0	—	12·95	—	298–m.p.	466
⟨$ThCl_4$⟩	−12,900	—	—	14·30	—	974–1043	25
{$ThCl_4$}	−7980	—	—	9·57	—	1043–1186	25
⟨$ThBr_4$⟩	−9630	—	—	11·73	—	903–951	25
{$ThBr_4$}	−7550	—	—	9·56	—	955–1126	25
{ThI_4}	−6890	—	—	9·09	—	856–1107	25
⟨ThO_2⟩	−31,600	—	—	10·1	—	2000–2800	339
⟨Ti⟩β	−24,400	−0·91	—	13·18	—	1155–m.p.	450, 52, 15
{Ti}	−23,200	−0·66	—	11·74	—	m.p.–b.p.	—
⟨TiF_4⟩	−5332	−2·57	—	19·51	—	298–s.p.	369
⟨$TiCl_2$⟩	−15,230	−2·51	—	19·36	—	298–m.p.	145
{$TiCl_2$}	−13,110	−2·51	—	17·93	—	m.p.–b.p.	145
{$TiCl_4$}	−2919	−5·788	—	25·129	—	298–b.p.	211
{$TiBr_4$}	−3621	—	—	11·26	—	275–311	982
⟨TiI_2⟩	−12,500	−1·51	—	16·90	—	298–1000	84
{TiI_4}	−3054	—	—	7·576	—	430–643	15
⟨Tm⟩	−12,550	—	—	9·18	—	807–1219	99
{Tl}	−9300	−0·892	—	11·10	—	700–1800	58
⟨TlF⟩	−7710	−2·18	—	17·66	—	298–m.p.	344
⟨TlCl⟩	−7370	−2·11	—	16·49	—	298–m.p.	344
{TlCl}	−6650	−2·62	—	16·92	—	m.p.–b.p.	344, 120
⟨TlBr⟩	−7420	−2·0	—	16·18	—	298–m.p.	344
{TlBr}	−6840	−3·02	—	18·26	—	m.p.–b.p.	344, 120
⟨TlI⟩	−7270	−2·01	—	15·85	—	298–m.p.	344
{TlI}	−6890	−3·02	—	18·20	—	m.p.–b.p.	344, 120
⟨U⟩	−25,580	−2·62	—	18·58	—	298–m.p.	399
{U}	−24,090	−1·26	—	13·20	—	m.p.–b.p.	399
⟨UF_4⟩	−16,400	−3·02	—	22·60	—	298–m.p.	399
{UF_4}	−15,300	−5·03	—	28·05	—	m.p.–b.p.	399
⟨UF_6⟩	−3312	−5·53	—	26·843	—	273–m.p.	399
{UF_6}	−1502	—	—	7·50	—	m.p.–b.p.	399
⟨UCl_4⟩	−11,350	−3·02	—	23·21	—	298–m.p.	399
{UCl_4}	−9950	−5·53	—	28·96	—	m.p.–b.p.	399
⟨UCl_6⟩	−4000	—	—	10·20	—	298–450	399
⟨UBr_3⟩	−16,420	−3·02	—	22·95	—	298–m.p.	399
{UBr_3}	−15,000	−5·03	—	27·54	—	m.p.–b.p.	399
⟨UBr_4⟩	−10,800	−3·02	—	23·15	—	298–m.p.	399
{UBr_4}	−8770	−5·53	—	27·93	—	m.p.–b.p.	399

evaporation constants (*contd.*)
$+ CT + D$ (mmHg)

Substance	b.pt. (°C)	L_e at b.pt. (kcal/mole)	sb.pt. (°C)	L_s at m.pt. or sb.pt.	L_e at 25°C (kcal/mole)	L_s at 25°C (kcal/mole)	Ref. and remarks
Te_2	998	25·0 ± 2·0	—	38·0 ± 2·0	33·3 ± 2·0	40·3 ± 2·5	28, 563, 294
Te_2F_{10}	54	—	—	—	9·45 ± 1·0	—	56
TeF_4	—	—	—	14·5 ± 1·0	—	—	110
TeF_6	—	—	−39	6·5 ± 0·4	—	—	198
$TeCl_2$	322	15·3 ± 1·5	—	—	—	—	25
$TeCl_4$	392	16·8 ± 1·5	—	24·3 ± 1·5	23·4 ± 1·5	27·5 ± 1·6	120
TeO_2	—	—	—	56·8 ± 3·0	—	61·7 ± 3·5	250
Th	(4850)	—	—	133·4 ± 2·5	—	137·6 ± 2·0	466
$ThCl_4$	922	36·5 ± 4·0	—	—	—	—	25
$ThBr_4$	857	34·5 ± 4·0	—	44·5 ± 4·0	—	—	25
ThI_4	837	31·5 ± 3·0	—	—	—	—	25
Ti	3285	101·7 ± 2·5	—	108·2 ± 1·0	—	112·1 ± 1·0	15, 52, 450
TiF_4	—	—	—	21·5 ± 0·5	—	22·9 ± 0·5	369
$TiCl_3$	(750)	(33·0)	283	(38·0)	—	(41·6)	—
$TiCl_4$	137	8·65 ± 0·5	—	—	9·92 ± 0·6	—	211
$TiBr_4$	233	10·5 ± 0·4	—	16·1 ± 0·6	13·1 ± 0·6	16·2 ± 0·6	369, 218, 197
TiI_2	—	—	1170	53·2 ± 6·0	—	56·3 ± 6·0	84
TiI_4	377	13·4 ± 1·0	—	20·5 ± 1·5	17·6 ± 1·5	21·6 ± 1·5	15
Tl	1460	39·7 ± 0·8	—	42·4 ± 0·6	41·9 ± 0·8	43·2 ± 0·8	120, 58
TlF	(700)	(27·7)	—	32·7 ± 1·0	—	34·0 ± 1·5	344
TlCl	816	24·75 ± 1·0	—	30·7 ± 1·0	28·9 ± 1·2	32·4 ± 1·0	344, 120
TlBr	825	24·7 ± 1·2	—	31·0 ± 1·0	29·5 ± 1·2	32·7 ± 1·2	344, 120
TlI	845	24·8 ± 1·5	—	30·3 ± 1·2	29·7 ± 1·5	—	344, 120
Tl_2SO_4	—	—	—	—	—	—	—
U	3930	99·7 ± 4·0	—	—	109·4 ± 3·0	115·2 ± 3·0	99
UF_4	1457	53·0 ± 4·0	—	—	—	—	399
UF_6	⟨51⟩	—	57	11·5 ± 0·1	7·25 ± 0·2	11·85 ± 0·1	160, 166, 179
UCl_4	789	33·8 ± 3·0	—	—	—	—	399
UBr_4	777	28·5 ± 3·0	—	—	—	—	399

Table D. Vapour pressures and
$\log p = A \cdot T^{-1} + B \log T$

Substance	log p (mmHg) A	B	C × 10³	D	Remarks	Temp. range (°K)	Ref.
⟨UI₄⟩	−12,330	−3·52	—	26·62	—	298–m.p.	399
{UI₄}	−9310	−5·53	—	28·57	—	m.p.–b.p.	399
⟨V⟩	−26,900	+0·33	−0·265	10·12	—	298–m.p.	99
{VF₅}	−2423	—	—	10·43	—	m.p.–b.p.	35
⟨VCl₂⟩	−9720	—	—	8·61	—	1183–1373	583
{VCl₄}	−2875	−6·07	—	25·56	—	298–m.p.	229
{VOCl₃}	−1921	—	—	7·70	—	298–b.p.	142, 63
{V₂O₅}	−7100	—	—	5·05	*	m.p.–1500	187
⟨W⟩	−44,000	+0·50	—	8·76	—	298–m.p.	99
{WF₆}	−1380	—	—	7·635	—	m.p.–b.p.	449, 120
⟨WCl₄⟩α	−3996	—	—	9·615	—	458–503	236
⟨WCl₄⟩β	−3588	—	—	8·795	—	503–554	236
{WCl₄}	−3253	—	—	8·195	—	555–600	236
⟨WCl₅⟩	−3670	—	—	9·50	—	413–m.p.	221
{WCl₅}	−2760	—	—	7·72	—	m.p.–b.p.	221
⟨WCl₆⟩α	−4580	—	—	10·73	—	425–t.p.	130
⟨WCl₆⟩β	−4080	—	—	9·73	—	t.p.–m.p.	130
{WCl₆}	−3050	—	—	7·87	—	m.p.–b.p.	221, 130
{WOF₄}	−3125	—	—	9·69	—	m.p.–b.p.	449
⟨WOF₄⟩	−3605	—	—	10·96	—	298–m.p.	449
⟨WO₃⟩	−24,600	—	—	15·63	—	1000–m.p.	250
⟨Y⟩	−22,230	−0·66	—	11·835	—	298–m.p.	—
{Y}	−22,280	−1·97	—	16·13	—	m.p.–b.p.	339
⟨Zn⟩	−6850	−0·755	—	11·24	—	298–m.p.	251, 120
{Zn}	−6620	−1·255	—	12·34	—	m.p.–b.p.	58
{ZnF₂}	−13,650	−5·03	—	26·90	—	m.p.–b.p.	25
{ZnCl₂}	−8415	−5·035	—	26·42	—	(693)–883	541
{Zn₂Cl₄}	−4700	—	—	5·89	—	(693)–883	541
{ZnBr₂}	−7785	−5·03	—	26·11	—	(671)–835	541
{Zn₂Br₄}	−5613	—	—	7·915	—	(671)–835	541
⟨ZnI₂⟩	−6450	−1·76	—	14·70	—	298–m.p.	177
⟨ZnS⟩	−13,980	—	—	8·98	—	970–1280	92
⟨Zr⟩β	−31,820	−0.50	—	11·78	—	1125–m.p.	230
{Zr}	−30,300	—	—	9·38	—	m.p.–b.p.	—
⟨ZrF₄⟩	−14,700	−5·03	—	30·80	—	298–s.p.	324, 62
⟨ZrCl₄⟩	−5400	—	—	11·765	—	480–689	584
⟨ZrBr₄⟩	−6780	−1·76	−1·65	19·60	—	298–s.p.	120
⟨ZrI₄⟩	−7680	−2·164	−1·344	20·87	—	298–s.p.	120

* Apparent vapour pressures. V_2O_5 loses oxygen with increasing temperature.

evaporation constants (*contd.*)
+ CT + D (mmHg)

Substance	b.pt. (°C)	L_e at b.pt. (kcal/mole)	sb.pt. (°C)	L_s at m.pt. or sb.pt.	L_e at 25°C (kcal/mole)	L_s at 25°C (kcal/mole)	Ref. and remarks
V	(3350)	—	—	119·9 ± 6·0	—	121·9 ± 7·0	51
VF_5	48	11·1 ± 1·0	—	—	11·1 ± 1·0	—	35
VCl_4	160	7·9 ± 0·4	—	—	9·5 ± 0·3	—	229
$VOCl_3$	127	8·0 ± 0·6	—	—	9·2 ± 0·6	—	63, 142
V_2O_5	dec.	—	—	(76·4)	(70·8)	—	187
W	(5500)	197·0 ± 5·0	—	205·0 ± 3·0	—	203·0 ± 3·0	99
WF_6	17	6·35 ± 0·5	—	—	6·15 ± 0·4	—	120, 449
WCl_4	—	—	—	—	—	—	—
WCl_5	298	12·6 ± 3·0	—	16·8 ± 2·5	—	—	221
WCl_6	338	13·9 ± 2·0	—	18·7 ± 3·0	—	—	221, 130
WOF_4	186	14·25	—	16·5	—	—	449
$WOCl_4$	220	11·5 ± 2·0	—	—	—	—	141, 221
WO_3	—	—	—	(109·0)	—	(122·0)	250
Y	3300	87·8 ± 3·0	—	97·4 ± 2·0	—	101·5 ± 2·0	339
Y_2O_3	—	—	—	—	—	—	—
Zn	907	27·3 ± 0·4	—	30·25 ± 0·4	29·5 ± 0·7	30·9 ± 0·7	251, 120
ZnF_2	1500	44·0 ± 3·0	—	—	59·0 ± 5·0	—	25
$ZnCl_2$	732	28·5 ± 3·0	—	37·1 ± 1·5	35·6 ± 2·0	38·0 ± 1·5	177, 120
$ZnBr_2$	650	(23·5)	—	29·8 ± 1·5	—	31·3 ± 1·5	177
ZnI_2	(730)	—	—	27·3 ± 1·5	—	28·6 ± 2·0	177
Zr	4400	—	—	143·4 ± 2·0	—	146·2 ± 2·5	230
ZrF_4	—	—	908	55·5 ± 3·5	—	64·3 ± 4·0	324, 62
$ZrCl_4$	[279]	—	334	24·7 ± 1·0	—	27·9 ± 1·5	584
$ZrBr_4$	—	—	357	25·8 ± 2·5	—	29·3 ± 2·5	120
ZrI_4	—	—	431	29·0 ± 2·5	—	33·3 ± 2·5	120
ZrO_2	—	—	—	—	—	—	—

Table E. Standard Gibbs energies of reaction
$\Delta G^\circ_T = A + BT \log T + CT$

Reaction	ΔG° in cal A	B	C	± kcal	Temp. range (°K)	Ref.
$2\langle Ag_2S\rangle_\alpha = 4\langle Ag\rangle + (S_2)$	44,800	—	−22·1	1	298–452	192
$2\langle Ag_2S\rangle_\beta = 4\langle Ag\rangle + (S_2)$	41,980	—	−16·52	1	452–1115	192
$2\{Ag_2S\} = 4\{Ag\} + (S_2)$	51,800	—	−24·33	1·5	1234–1500	192
$2\langle Al\rangle + (Cl_2) = 2(AlCl)$	−21,360	11·75	−76·5	3	298–933	255
$2\{Al\} + (Cl_2) = 2(AlCl)$	−25,860	11·8	−71·9	3	933–2000	255
$\langle Al\rangle + 1\frac{1}{2}(Cl_2) = (AlCl_3)$	−138,160	2·48	4·75	5	453–933	255
$\{Al\} + 1\frac{1}{2}(Cl_2) = (AlCl_3)$	−140,400	2·50	7·05	8	>933	255
$2\langle Al\rangle + 3(Cl_2) = (Al_2Cl_6)$	−304,440	4·95	4·23	9	453–933	255
$\langle Al_2O_3\rangle = 2\langle Al\rangle + 1\frac{1}{2}(O_2)$	400,810	3·98	−87·64	3	298–923	—
$\langle Al_2O_3\rangle = 2\{Al\} + 1\frac{1}{2}(O_2)$	405,760	3·75	−92·22	4	923–1800	—
$2\langle AlN\rangle = 2\langle Al\rangle + (N_2)$	154,000	—	−44·5	16	298–923	—
$\langle Al_4C_3\rangle = 4\langle Al\rangle + 3\langle C\rangle$	51,600	—	−10·0	2	298–932	655
$\langle Al_4C_3\rangle = 4\{Al\} + 3\langle C\rangle$	63,700	—	−23·0	2	932–2000	655
$\{Al_2Se_3\} + 4\{Al\} = 3(Al_2Se)$	168,300	—	−97·1	5	1300–1500	294
$2\langle BaO\rangle = 2\langle Ba\rangle + (O_2)$	271,600	—	−46·4	8	298–983	—
$2\langle BaO\rangle = 2\{Ba\} + (O_2)$	278,000	—	−53·2	10	983–1600	—
$\langle BaO\rangle + \langle SiO_2\rangle = \langle BaSiO_3\rangle$	−26,800	—	+0·1	3	298–1600	193
$\langle Ba_3N_2\rangle = 3\langle Ba\rangle + (N_2)$	87,000	—	−57·4	9	298–1000	—
$\langle BeF_2\rangle = \langle Be\rangle + (F_2)$	243,200	15·4	−81·9	5	298–818	—
$\{BeF_2\} = \langle Be\rangle + (F_2)$	237,280	19·63	−87·0	5	818–1455	—
$(BeF_2) = \{Be\} + (F_2)$	178,000	5·65	−2·0	5	>1557	—
$(BeF_2) + \langle Be\rangle = 2(BeF)$	91,500	—	−44·3	—	1425–1675	498
$(BeCl_2) + \{Be\} = 2(BeCl)$	89,100	—	−39·4	5	1573–1723	496
$2\langle BeO\rangle = 2\langle Be\rangle + (O_2)$	286,900	3·32	−56·1	10	298–1557	—
$2\langle BeO\rangle = 2\{Be\} + (O_2)$	291,900	3·32	−59·32	13	1557–2000	—
$\langle Be_3N_2\rangle = 3\langle Be\rangle + (N_2)$	134,700	—	−40·6	12	298–1000	123
$\frac{2}{3}\langle BiCl_3\rangle = \frac{2}{3}\langle Bi\rangle + (Cl_2)$	62,400	12·05	−66·6	3	298–503	255
$\frac{2}{3}\{BiCl_3\} = \frac{2}{3}\{Bi\} + (Cl_2)$	63,570	17·5	−83·4	4	544–714	255
$\frac{2}{3}(BiCl_3) = \frac{2}{3}\{Bi\} + (Cl_2)$	45,330	−3·96	+3·42	5	714–1500	255
$(BiCl_3) + 2\{Bi\} = 3(BiCl)$	72,000	—	−54·0	5	873–973	462
$(BiBr_3) + 2\{Bi\} = 3(BiBr)$	66,700	—	−54·0	5	870–980	462
$\langle Bi_2S_3\rangle + 3(H_2) = 2\langle Bi\rangle + 3(H_2S)$	24,500	27·73	−114·7	3	298–544	212
$\langle Bi_2O_3\rangle = 2\{Bi\} + 1\frac{1}{2}O_2$	137,890	—	−63·43	1	773–1093	947
$\langle C\rangle + 2(H_2) = (CH_4)$	−16,520	+12·25	−15·62	1	298–1200	189
$\langle C\rangle + 2(F_2) = (CF_4)$	−220,000	—	+36·0	5	298–1500	—
$\langle C\rangle + 2(Cl_2) = (CCl_4)$	−26,250	−5·16	49·32	1	350–800	255
$\langle C\rangle + \frac{1}{2}(O_2) = (CO)$	−26,700	—	−20·95	1	298–2500	191
$\langle C\rangle + (O_2) = (CO_2)$	−94,200	—	−0·2	1	298–2000	191
$\langle C\rangle + \frac{1}{2}(S_2) = (CS)$	59,000	—	−22·75	7	1873–2073	361
$\langle C\rangle + (S_2) = (CS_2)$	−3,100	—	−1·73	1	298–1600	192
$2(CO) + (S_2) = 2(COS)$	−45,720	—	+37·4	3	298–1500	192
graphite = diamond	+310	—	1·13	0·2	298–1500	198
$\langle CaF_2\rangle\alpha = \langle Ca\rangle + (F_2)$	293,300	7·7	−64·4	5	298–1123	—
$\langle CaF_2\rangle\alpha = \{Ca\} + (F_2)$	297,600	16·9	−98·35	6	1123–1424	—
$\langle CaF_2\rangle\beta = \{Ca\} + (F_2)$	285,000	—	−35·0	7	1424–1756	—
$\langle CaCl_2\rangle = \langle Ca\rangle + (Cl_2)$	189,900	—	−34·0	3	298–1055	255
$\{CaCl_2\} = \{Ca\} + (Cl_2)$	195,400	24·44	−112·6	5	1120–1900	255
$2\langle CaO\rangle = 2\langle Ca\rangle + (O_2)$	302,650	—	−47·32	3	298–1124	191
$2\langle CaO\rangle = 2\{Ca\} + (O_2)$	307,100	—	−51·28	3	1124–1760	191
$2\langle CaO\rangle = 2(Ca) + (O_2)$	380,200	—	−93·24	5	1760–2500	191

Table E. Standard Gibbs energies of reaction (contd.)

$$\Delta G^\circ_T = A + BT \log T + CT$$

Reaction	ΔG° in cal			± kcal	Temp. range (°K)	Ref.
	A	B	C			
$2\langle CaS\rangle = 2\langle Ca\rangle_\alpha + (S_2)$	258,870	—	−45·62	2	298–673	192
$2\langle CaS\rangle = 2\langle Ca\rangle_\beta + (S_2)$	259,100	—	−45·92	2	673–1124	192
$2\langle CaS\rangle = 2\{Ca\} + (S_2)$	263,560	—	−49·88	3	1124–1760	192
$2\langle CaS\rangle = 2(Ca) + (S_2)$	336,710	—	−91·44	5	1760–2000	192
$\langle Ca_3N_2\rangle = 3\langle Ca\rangle + (N_2)$	105,000	—	−50·0	11	298–1100	—
$\langle Ca\rangle_\alpha + 2\langle C\rangle = \langle CaC_2\rangle$	−13,600	—	−5·9	3	298–720	189
$\langle Ca\rangle_\beta + 2\langle C\rangle = \langle CaC_2\rangle$	−11,620	—	−8·64	3	720–1123	189
$\{Ca\} + 2\langle C\rangle = \langle CaC_2\rangle$	−13,700	—	−6·80	3	1123–1963	189
$(Ca) + 2\langle C\rangle = \langle CaC_2\rangle$	−51,210	—	+12·3	5	1963–2200	—
$\langle CaCO_3\rangle = \langle CaO\rangle + (CO_2)$	40,250	—	−34·4	1	449–1150	370
$\langle Ca\rangle + \langle Si\rangle = \langle CaSi\rangle$	−36,000	—	−0·5	4	298–1123	—
$\{Ca\} + \langle Si\rangle = \{CaSi\}$	−25,750	—	−6·58	5	1520–1717	—
$2\{Ca\} + \langle Si\rangle = \{Ca_2Si\}$	−42,600	—	−4·65	5	1200–1717	—
$\langle CaO\rangle + \langle SiO_2\rangle = \langle CaSiO_3\rangle_\alpha$	−21,300	—	+0·12	1	298–1483	193
$\langle CaO\rangle + \langle SiO_2\rangle = \langle CaSiO_3\rangle_\beta$	−19,900	—	−0·82	2	1483–1813	193
$2\langle CaO\rangle + \langle SiO_2\rangle = \langle Ca_2SiO_4\rangle$	−30,200	—	−1·2	2·5	298–1700	—
$2\langle CdO\rangle = 2(Cd) + (O_2)$	174,500	—	−99·9	—	1242–1379	491
$\langle CeCl_3\rangle = \langle Ce\rangle + 1\frac{1}{2}(Cl_2)$	255,200	18·4	−114·9	4	298–900	476
$\langle CeO_2\rangle = \langle Ce\rangle + (O_2)$	260,900	5·0	−65·25	3	298–1000	—
$2\langle CeS\rangle = 2\langle Ce\rangle + (S_2)$	267,000	—	−40·0	10	298–2200	192
$2\langle CeN\rangle = 2\langle Ce\rangle + (N_2)$	156,000	—	−57·5	17	298–1000	123
$\langle CeC\rangle = \langle Ce\rangle + \langle C\rangle$	27,900	—	−8·2		815–940	465
$2(CoCl_3) = 2(CoCl_2) + (Cl_2)$	33,800	—	−36·3	2	900–1100	208
$\langle CoCl_2\rangle = \langle Co\rangle + (Cl_2)$	78,700	6·9	−54·12	3	298–1000	
$\{CoCl_2\} = \langle Co\rangle + (Cl_2)$	72,200	9·2	−54·52	4	1000–1323	
$(CoCl_2) = \langle Co\rangle + (Cl_2)$	31,700	−13·8	+47·8	4	1323–1786	—
$\langle CoF_2\rangle = \langle Co\rangle + (F_2)$	158,000	—	−33·7	6	298–1450	—
$2\langle CoO\rangle = 2\langle Co\rangle + (O_2)$	111,800	—	−33·8	2	298–1400	—
$\langle Co_3O_4\rangle = 3\langle CoO\rangle + \frac{1}{2}(O_2)$	43,800	—	−35·4	3	298–1300	—
$\frac{1}{4}\langle Co_9S_8\rangle = \frac{9}{4}\langle Co\rangle + (S_2)$	79,240	—	−39·81	2	298–1048	192
$\frac{1}{2}\langle Co_9S_8\rangle + (S_2) = \frac{3}{2}\langle Co_3S_4\rangle$	−51,640	—	+34·03	3	600–750	196
$3\langle CoS_2\rangle = \langle Co_3S_4\rangle + (S_2)$	50,160	—	−50·41	3	600–900	196
$\langle Co_2C\rangle = 2\langle Co\rangle + \langle C\rangle$	−3950	—	+2·08	5	298–1200	189
$\langle CrCl_2\rangle = \langle Cr\rangle + (Cl_2)$	93,900	8·7	−55·0	3	298–1088	255
$\{CrCl_2\} = \langle Cr\rangle + (Cl_2)$	86,900	8·36	−47·5	3	1088–1577	255
$(CrCl_2) = \langle Cr\rangle + (Cl_2)$	27,700	−6·26	+37·0	5	1577–2100	255
$\frac{2}{3}\langle CrCl_3\rangle = \frac{2}{3}(Cr\rangle + (Cl_2)$	87,400	7·42	−58·4	5	298–1200	255
$\langle Cr_2O_3\rangle = 2\langle Cr\rangle + 1\frac{1}{2}(O_2)$	267,750	—	−62·1	2	298–2100	191
$2\langle\overline{Cr_2N}\rangle = 4\langle\overline{Cr}\rangle + (N_2)$	51,900	11·5	−66·0	5	298–1400	307
$4\langle CrN\rangle = 2\langle\overline{Cr_2N}\rangle + (N_2)$	64,000	11·5	−83·2	5	298–1400	307
$\frac{1}{6}\langle Cr_{23}C_6\rangle = \frac{23}{6}\langle Cr\rangle + \langle C\rangle$	+16,380	—	+1·54	3	298–1673	189
$\frac{23}{27}\langle Cr_7C_3\rangle = \frac{7}{27}\langle Cr_{23}C_6\rangle + \langle C\rangle$	10,050	—	2·85	3	298–1673	189
$\frac{7}{5}\langle Cr_3C_2\rangle = \frac{3}{5}\langle Cr_7C_3\rangle + \langle C\rangle$	3200	—	0·20	3	298–1673	189
$(CuCl) + \frac{1}{2}(H_2) = \langle Cu\rangle + (HCl)$	−43,660	−4·17	+31·81	2	298–1300	25
$(Cu_3Cl_3) + \frac{3}{2}(H_2) = 3\langle Cu\rangle + 3(HCl)$	−4500	—	−9·4	2	1300	25
$\langle Cu_2O\rangle = 2\langle Cu\rangle + \frac{1}{2}(O_2)$	40,500	3·92	−29·5	1	298–1356	—
or	39,800	—	−16·88		—	—
$\langle Cu_2O\rangle = 2\{\text{'Cu'}\} + \frac{1}{2}(O_2)$	46,700	3·92	−34·1	1·5	1356–1503	—
$2\langle CuO\rangle = \langle Cu_2O\rangle + \frac{1}{2}(O_2)$	34,950	6·1	−44·3	0·5	298–1300	—
$\langle Cu_2S\rangle_\alpha = 2\langle Cu\rangle + \frac{1}{2}(S_2)$	35,655	4·34	−27·31	1	298–376	307
$\langle Cu_2S\rangle_\alpha = \langle Cu_2S\rangle_\beta$	−490	−8·63	+23·53	1	—	307
$\langle Cu_2S\rangle_\gamma = 2\langle Cu\rangle + \frac{1}{2}(S_2)$	34,150	6·22	−28·74	1	623–1360	307

Table E. Standard Gibbs energies of reaction (contd.)

$\Delta G^\circ_T = A + BT \log T + CT$

Reaction	ΔG° in cal A	B	C	± kcal	Temp. range (°K)	Ref.
$\langle FeCl_2\rangle = \langle Fe\rangle_\alpha + (Cl_2)$	82,770	6·98	−50·84	1	298–950	255
$\{FeCl_2\} = \langle Fe\rangle + (Cl_2)$	68,450	—	−15·22	1	950–1300	255
$(FeCl_2) = \langle Fe\rangle_\gamma + (Cl_2)$	25,250	−23·0	+89·65	4	1300–1812	255
$2\langle FeCl_3\rangle = 2\langle FeCl_2\rangle + (Cl_2)$	25,900	—	−38·7	1	400–500	210
$2(FeCl_3) = (Fe_2Cl_6)$	−32,550	—	+31·65	1	778–978	210
$\langle FeF_2\rangle = \langle Fe\rangle + (F_2)$	167,000	—	−31·8	6	298–1350	—
$2(FeBr_3) = 2(FeBr_2) + (Br_2)$	44,600	—	−33·7	2	750	421
$\langle FeI_2\rangle = \langle Fe\rangle + (I_2)$	45,770	6·9	−51·2	5	298–867	—
$\{FeI_2\} = \langle Fe\rangle + (I_2)$	40,600	11·5	−58·6	6	867–1208	—
$2(FeI_3) = 2(FeI_2) + (I_2)$	32,000	—	−32·0	2	700–1000	421
$\langle \overline{FeO}\rangle = \langle Fe\rangle + \frac{1}{2}(O_2)$	63,310	—	−15·62	3	298–1642	191
$\{\overline{FeO}\} = \{Fe\} + \frac{1}{2}(O_2)$	55,620	—	−10·83	3	1808–2000	191
$\langle Fe_3O_4\rangle = 3\langle \overline{FeO}\rangle + \frac{1}{2}(O_2)$	74,620	—	−29·9	3	298–1642	191
$3\langle Fe_2O_3\rangle = 2\langle Fe_3O_4\rangle + \frac{1}{2}(O_2)$	59,620	—	−33·62	8	298–1460	191
$\langle Fe\rangle + 2(H_2O) = (Fe(OH)_2) + (H_2)$	42,700	—	−7·31	2	1573–1733	436
$2\langle FeS\rangle_\alpha = 2\langle Fe\rangle_\alpha + (S_2)$	74,320	—	−31·18	1	298–412	192
$2\langle FeS\rangle_\beta = 2\langle Fe\rangle_\alpha + (S_2)$	71,820	—	−25·12	1	412–1179	192
$2\langle FeS\rangle_\beta = 2\langle Fe\rangle_\gamma + (S_2)$	72,140	—	−25·48	2	1179–1261	192
$2\langle \overline{FeS_2}\rangle = 2\langle \overline{FeS}\rangle + (S_2)$	86,700	—	−90·0	3	600–1100	192
$\langle Fe_4N\rangle = 4\langle Fe\rangle + \frac{1}{2}(N_2)$	200	−11·62	24·85		298–950	123
$\langle Fe_3P\rangle = 3\langle Fe\rangle + \frac{1}{2}(P_2)$	51,000	—	−11·3	8	298–1439	191
$3\langle FeCO_3\rangle = \langle Fe_3O_4\rangle + 2(CO_2) + (CO)$	49,206	—	-117·94	14	298–700	191
$2\langle Fe\rangle + (O_2) + \langle SiO_2\rangle = \langle Fe_2SiO_4\rangle$	−135,300	—	+34·4	3	298–1478	193
$2\langle Fe\rangle + (O_2) + 2\langle TiO_2\rangle = 2\langle FeTiO_3\rangle$	−129,220	—	+25·86	2	1173–1400	193
$2\langle Fe\rangle + (O_2) + 2\langle Cr_2O_3\rangle = 2\langle FeCr_2O_4\rangle$	−131,600	—	+24·2	3	1173–1700	193
$2\langle Fe\rangle + (O_2) + 2\langle Al_2O_3\rangle = 2\langle FeAl_2O_4\rangle$	−150,000	—	+41·8	8	1100–1400	193
$\langle Ga_2O_3\rangle = 2\{Ga\} + 1\frac{1}{2}O_2$	+259,340	—	-78·11	1	820–1073	945
$2\langle GeI_2\rangle = \langle Ge\rangle + (GeI_4)$	32,500	18·4	−94·6	3	298–643	108
$(GeO) = \frac{1}{2}\langle Ge\rangle + \frac{1}{2}\langle GeO_2\rangle$	−54,600	−6·9	+62·0	4	298–860	108
$\langle GeO_2\rangle = \langle Ge\rangle + O_2$	+137,330	—	-44·86	1	600–1173	—
$2(HF) = (H_2) + (F_2)$	130,200	—	+1·9		298–2500	—
$\frac{1}{2}(H_2) + \frac{1}{2}(Cl_2) = (HCl)$	−21,770	0·99	−5·22	0·6	298–2100	255
$(H_2) + \frac{1}{2}(O_2) = (H_2O)$	−57,250	4·48	−2·21	0·3	298–2500	—
or	−58,900	—	13·1	1	298–2500	191
$(H_2) + \langle S\rangle = (H_2S)$	−3320	11·72	−44·55	0·6	298–392	122
$2(H_2) + (S_2) = 2(H_2S)$	−40,210	7·25	−1·21	0·7	298–1750	122
or	−43,160	—	23·61	1	298–1800	192
$\frac{1}{2}(N_2) + 1\frac{1}{2}(H_2) = (NH_3)$	−10,400	7·1	3·79	0·5	298–1000	123
or	−12,050	—	+26·7	2	298–1000	184
$\langle In_2O_3\rangle + 3(H_2) = 2\{In\} + 3(H_2O)$	44,960	3·35	−49·63	3	430–1123	238
$\langle In_2O_3\rangle = 2\{In\} + 1\frac{1}{2}O_2$	+215,550	—	-72·63	1	800–1123	946
$\langle InS\rangle + (H_2) = \{In\} + (H_2S)$	41,900	2·75	−41·4	3	298–1000	335
$\langle IrO_2\rangle = \langle Ir\rangle + (O_2)$	52,200	12·65	−77·25	4	298–1366	274
$\langle Ir\rangle + 1\frac{1}{2}(O_2) = (IrO_3)$	4150	—	+10·8		1473–1673	423
$\frac{2}{3}\langle \overline{Ir_2S_3}\rangle = \frac{4}{3}\langle Ir\rangle + (S_2)$	64,500	—	−44·66	5	1000–1600	192
$\langle KCl\rangle = \{K\} + \frac{1}{2}(Cl_2)$	104,900	—	−24·0	1	298–1043	255
$\{KCl\} = (K) + \frac{1}{2}(Cl_2)$	116,600	—	−35·2	1·5	1043–1680	255
$(KCl) = (K) + \frac{1}{2}(Cl_2)$	66,000	−16·1	46·84	2	1680–2300	255
$\langle K_2O\rangle = 2\langle K\rangle + \frac{1}{2}(O_2)$	86,600	4·6	−44·6	3	298–	—

Table E. Standard Gibbs energies of reaction (contd.)

$\Delta G^\circ_T = A + BT \log T + CT$

Reaction	ΔG° in cal			±	Temp. range	Ref.
	A	B	C	kcal	(°K)	
$\langle La_2O_3\rangle = 2\langle La\rangle + 1\frac{1}{2}(O_2)$	447,200	9·2	−96·9	4	298–1000	—
$\langle LaN\rangle = \langle La\rangle + \frac{1}{2}(N_2)$	72,100	—	−25·0	9	298–1000	123
$2\langle LiOH\rangle = \langle Li_2O\rangle + (H_2O)$	34,940	—	−25·98	3	648–795	48
$\{Li_2CO_3\} = \{Li_2O\} + (CO_2)$	77,950	—	−69·0		m.p.–1125	532
$\langle MgF_2\rangle = \langle Mg\rangle + (F_2)$	267,200	—	−40·2	3	298–923	—
$\langle MgF_2\rangle = \{Mg\} + (F_2)$	269,300	—	−42·5	3	923–1276	—
$\langle MgCl_2\rangle = \langle Mg\rangle + (Cl_2)$	154,440	9·88	−68·18	1	298–923	255
$\{MgCl_2\} = \{Mg\} + (Cl_2)$	147,850	13·58	−72·77	1·5	987–1376	255
$\{MgCl_2\} = (Mg) + (Cl_2)$	184,100	20·71	−121·52	1·5	1380–1691	255
$(MgCl_2) = (Mg) + (Cl_2)$	134,500	−2·29	−17·94	5	1691–2000	255
$\langle MgO\rangle = \langle Mg\rangle + \frac{1}{2}(O_2)$	144,350	2·95	−33·95	1·5	298–923	191a
$\langle MgO\rangle = \{Mg\} + \frac{1}{2}(O_2)$	145,350	0·24	−26·95	1·5	923–1380	191a
$\langle MgO\rangle = (Mg) + \frac{1}{2}(O_2)$	181,600	7·37	−75·7	3	1380–2500	191a
$\langle Mg(OH)_2\rangle = \langle MgO\rangle + (H_2O)$	11,000	—	−24·0		653–923	420
$\langle MgS\rangle = \langle Mg\rangle + \frac{1}{2}(S_2)$	99,650	—	−22·8	5	298–923	192
$\langle MgS\rangle = \{Mg\} + \frac{1}{2}(S_2)$	101,800	—	−25·65	5	923–1380	192
$\langle MgS\rangle = (Mg) + \frac{1}{2}(S_2)$	134,350	—	−48·75	5	1380–2000	192
$\langle Mg_3N_2\rangle_\alpha = 3\langle Mg\rangle + (N_2)$	109,600	—	−47·41	3	298–823	184
$\langle Mg_3N_2\rangle_\beta = 3\{Mg\} + (N_2)$	115,970	—	−54·30	3	923–1061	184
$\langle Mg_3N_2\rangle_\gamma = 3\{Mg\} + (N_2)$	115,750	—	−54·09	3	1061–1300	184
$\langle MgCO_3\rangle = \langle MgO\rangle + (CO_2)$	28,100	—	−40·6	4	298–1000	191
$\langle MgO\rangle + \langle SiO_2\rangle = \langle MgSiO_3\rangle$	−8900	—	+1·1	1	298–1600	193
$2\langle MgO\rangle + \langle SiO_2\rangle = \langle Mg_2SiO_4\rangle$	−15,120	—	0·0	2	298–1700	193
$\langle MnCl_2\rangle = \langle Mn\rangle_\alpha + (Cl_2)$	116,300	8·37	−56·0	1	298–923	—
$\{MnCl_2\} = \langle Mn\rangle_\beta + (Cl_2)$	107,970	9·11	−49·15	1	993–1373	—
$\{MnCl_2\} = \langle Mn\rangle_\delta + (Cl_2)$	105,930	4·07	−31·84	1·5	1400–1504	—
$(MnCl_2) = \{Mn\} + (Cl_2)$	53,450	−16·4	+68·2	2	1517–	—
$\langle MnO\rangle = \langle Mn\rangle + \frac{1}{2}(O_2)$	91,950	—	−17·4	3	298–1500	191
$\langle MnO\rangle = \{Mn\} + \frac{1}{2}(O_2)$	95,400	—	−19·7	3	1500–2050	191
$\langle MnS\rangle = \langle Mn\rangle_\alpha + \frac{1}{2}(S_2)$	64,000	—	−15·32	1·5	298–1000	192
$\langle MnS\rangle = \langle Mn\rangle_\beta + \frac{1}{2}(S_2)$	64,540	—	−15·86	1·5	1000–1374	192
$\langle MnS\rangle = \langle Mn\rangle_\delta + \frac{1}{2}(S_2)$	65,510	—	−16·56	1·5	1410–1517	192
$\langle MnS\rangle = \{Mn\} + \frac{1}{2}(S_2)$	69,010	—	−18·86	2	1517–1803	192
$\{MnS\} = \{Mn\} + \frac{1}{2}(S_2)$	62,770	—	−15·40	2	1803–2000	192
$\langle\text{'}Mn_3C\text{'}\rangle = 3\langle Mn\rangle + \langle C\rangle$	3330	—	+0·26	3	298–1010	189
$\langle MnSiO_3\rangle = \langle MnO\rangle + \langle SiO_2\rangle$	5920	—	−3·0	4	298–1600	193
$(MoI_2) = \langle Mo\rangle + 2(I)$	25,000	—	9·5	2	1050–1500	278
$\langle MoO_2\rangle = \langle Mo\rangle + (O_2)$	140,500	4·6	−56·8	6	298–1300	307
$\langle MoO_3\rangle = \langle MoO_2\rangle + \frac{1}{2}(O_2)$	38,700	—	−19·5	3	298–1300	307
$2\langle\overline{Mo_2N}\rangle = 4\langle\overline{Mo}\rangle + (N_2)$	34,400	9·2	−57·9	8	298–1300	307
$\frac{2}{3}\langle Mo_2S_3\rangle = \frac{4}{3}\langle Mo\rangle + (S_2)$	85,700	—	−36·41	2	1123–1473	626
$\langle NaF\rangle = \{Na\} + \frac{1}{2}(F_2)$	137,000	—	−24·9	3	271–1156	—
$\langle NaCl\rangle = \{Na\} + \frac{1}{2}(Cl_2)$	99,000	—	−23·6	1·5	298–1073	255
$\{NaCl\} = (Na) + \frac{1}{2}(Cl_2)$	114,300	—	−35·8	2	1183–1738	—
$\langle Na_2S\rangle = 2\{Na\} + \frac{1}{2}(S_2)$	105,250	—	−31·45	5	371–1187	192
$\langle Na_2SiO_3\rangle = \langle Na_2O\rangle + \langle SiO_2\rangle$	55,550	—	+1·40	8	298–1361	193
	43,080	—	10·55	10	1361–1600	193
$2\langle NbCl_4\rangle = \langle NbCl_3\rangle + (NbCl_5)$	28,300	—	−47·0	1	475–590	206
$2\langle NbO\rangle = 2\langle\text{'}Nb\text{'}\rangle + (O_2)$	198,700	11·5	−77·0	4	298–	—
$2\langle NbO_2\rangle = 2\langle NbO\rangle + (O_2)$	185,100	11·5	−79·5	4	298–	—
$2\langle\overline{Nb_2O_5}\rangle = 4\langle NbO_2\rangle + (O_2)$	149,700	11·5	−68·9	5	298–	—
$\langle\overline{NbC}\rangle = \langle Nb\rangle + \langle C\rangle$	31,100	—	−0·4		1180–1370	656

Table E. Standard Gibbs energies of reaction (contd.)

$\Delta G_T^\circ = A + BT \log T + CT$

Reaction	ΔG° in cal A	B	C	± kcal	Temp. range (°K)	Ref.
$\langle NdCl_3\rangle + (H_2O) = \langle NdOCl\rangle + 2(HCl)$	20,910	—	−30·31	3	785	137
$\langle NiCl_2\rangle + (H_2) = \langle Ni\rangle + 2(HCl)$	29,075	6·15	−58·53	0·8	298–1260	30
$\langle NiO\rangle = \langle Ni + \frac{1}{2}O_2\rangle$	56,010	—	−20·37	2	298–1725	191
$\langle NiO\rangle = \{Ni\} + \frac{1}{2}(O_2)$	62,650	—	−25·98	3	1725–2200	191
$\langle Ni_3C\rangle = 3\langle Ni\rangle + \langle C\rangle$	−8110	—	+1·70	3	298–1000	189
$\langle Ni_3S_2\rangle = 3\langle Ni\rangle + (S_2)$	79,240	—	−39·01	2	650–800	196
$2\langle \overline{NiS}\rangle = 2\langle Ni\rangle + (S_2)$	69,960	—	−34·41	3	670–850	196
$(P_4) = 2(P_2)$	53,860	4·35	−50·05	3	298–1200	237
$(PCl_5) = (PCl_3) + (Cl_2)$	22,850	4·37	−56·22	4	298–1000	237
$\langle PbCl_2\rangle = Pb + (Cl_2)$	85,000	—	−34·3	2	298–771	255
$\{PbCl_2\} = \{Pb\} + (Cl_2)$	83,600	14·12	−73·3	3	771–1224	255
$(PbCl_2) = \{Pb\} + (Cl_2)$	37,800	−16·3	+57·92	3	1224–1500	255
$2\langle PbO\rangle = 2\langle Pb\rangle + (O_2)$	105,700	6·9	−68·86	4	298–600	—
$2\langle PbO\rangle = 2\{Pb\} + (O_2)$	107,500	—	−52·6	4	600–760	—
$2\langle PbO\rangle = 2\{Pb\} + (O_2)$	106,600	—	−51·4	4	760–1150	—
$2\langle PbS\rangle = 2\langle Pb\rangle + (S_2)$	72,710	—	−34·17	3	298–600	192
$2\langle PbS\rangle = 2\{Pb\} + (S_2)$	75,160	—	−38·25	3	600–1380	192
$\langle PbSe\rangle = \langle Pb\rangle + \{Se\}$	25,340	—	−3·8	2	473–573	481
$\langle PdCl_2\rangle = \langle Pd\rangle + (Cl_2)$						
$\{PdCl_2\} = \langle Pd\rangle + (Cl_2)$						
$\langle PdO\rangle = \langle Pd\rangle + \frac{1}{2}(O_2)$	22,400	5·75	−36·9	2	298–1133	274
$\langle PrCl_3\rangle + (H_2O) = \langle PrOCl\rangle + 2(HCl)$	21,530	—	−30·07	3	785	137
$(PtO_2) = \langle Pt\rangle + (O_2)$	−39,270	—	+0·93	2	1373–1823	423
$2\langle \overline{PtS}\rangle = 2\langle \overline{Pt}\rangle + (S_2)$	66,100	—	−43·6	5	1000–1700	192
$2\langle PtS_2\rangle = 2\langle \overline{PtS}\rangle + (S_2)$	43,750	—	−43·8	5	700–1100	192
$\langle ReS_2\rangle = \langle Re\rangle + (S_2)$	64,200	9·2	−68·6	5	298–1500	—
$(RhO_2) = \langle Rh\rangle + (O_2)$	−45,140	—	+4·94	1	1473–1773	423
$(RuCl_4) = \langle Ru\rangle + 2(Cl_2)$	21,650	−4·83	−8·4	2	298–1400	434
$\langle RuCl_3\rangle = \langle Ru\rangle + 1\frac{1}{2}(Cl_2)$	61,800	10·35	−86·5			434
$\langle RuO_2\rangle = \langle Ru\rangle + (O_2)$	73,600	6·9	−62·4		298–1850	605
$(S_2) = \frac{1}{2}(S_4)$	−14,200	—	+15·94	2·5	298–1300	21
$(S_2) = \frac{1}{3}(S_6)$	−21,240	—	+23·1	1	298–1300	21
$(S_2) = \frac{1}{4}(S_8)$	−23,040	—	+25·3	2	298–1300	21
$(S_2) = 2(S)$	+77,250	—	−29·7	10	298–2000	221
$2(SO) = (S_2) + (O_2)$	30,800	—	+2·8	7	298–2000	—
$(S_2) + 2(O_2) = 2(SO_2)$	−173,240	—	34·62	1	298–2000	192
$(S_2) + 3(O_2) = 2(SO_3)$	−218,440	—	77·34	3	318–1800	192
$\langle Sb_2O_3\rangle = 2\langle Sb\rangle + 1\frac{1}{2}O_2$	+164,340	—	−57·62	1	773–900	947
$(SiCl_4) = \langle Si\rangle + 2(Cl_2)$	155,600	3·64	−43·9	3	298–1000	
$\langle Si\rangle + (SiCl_4) = 2(SiCl_2)$	83,000	12·0	−89·4	3	298–1800	209
$\langle SiO_2\rangle + \langle Si\rangle = 2(SiO)$	169,600	12·9	−124·3	3	298–1700	p. 221
$\langle SiO_2\rangle = \langle Si\rangle + (O_2)$	215,600	—	−41·5	3	700–1700	—
$\langle SiO_2\rangle = \{Si\} + (O_2)$	227,700	—	−48·7	3	1700–2000	—
$\langle Si_3N_4\rangle = 3\langle Si\rangle + 2(N_2)$	177,000	5·76	−96·3	6	298–1686	590

Table E. Standard Gibbs energies of reaction (contd.)

$\Delta G^\circ_T = A + BT \log T + CT$

Reaction	ΔG° in cal			$\pm$ kcal	Temp. range (°K)	Ref.
	A	B	C			
$\langle Si_3N_4\rangle = 3\{Si\} + 2(N_2)$	209,000	—	−96·8	1	1686–1973	590
$\langle SiC\rangle = \langle Si\rangle + \langle C\rangle$	14,000	1·3	−5·68	2·5	298–1686	—
$\langle SiC\rangle = \{Si\} + \langle C\rangle$	27,100	2·73	−18·1	2	1686–2000	502, 468, 189
$(SiS) = \langle Si\rangle + \frac{1}{2}S_2$	12,740	—	−19·18	2	863–1680	948
$\{SnCl_2\} = \{Sn\} + (Cl_2)$	79,600	—	−28·3	4	520–925	—
$(SnCl_4) = \{Sn\} + 2(Cl_2)$	122,500	—	−36·0	2·5	500–1200	—
$\langle SnO_2\rangle = \{Sn\} + (O_2)$	140,180	—	−51·52	1	770–980	433
$(TaCl_5) = (TaCl_4) + \frac{1}{2}(Cl_2)$	33,600	—	−31·3	—	—	615
$2\langle TaCl_4\rangle = \langle TaCl_3\rangle + (TaCl_5)$	27,000	—	−48·3	—	—	615
$3\langle TaCl_3\rangle = 2(TaCl_4) + (TaCl_5)$	33,400	—	−32·3	—	—	615
$\langle Ta_2O_5\rangle + 3(TaCl_5) = 5(TaOCl_3)$	99,000	—	−95·0		854–1026	606
$\langle Ta_2O_5\rangle = 2\langle Ta\rangle + 2\frac{1}{2}(O_2)$	491,000	20·7	−168·4	5	298–2000	307
$2\langle TaN\rangle = 2\langle 'Ta'\rangle + (N_2)$	117,800	13·8	−79·7	7	298–2240	307
$\langle \overline{TaC}\rangle = \langle Ta\rangle + \langle C\rangle$	34,900	—	−0·5	—	1250–1400	656
$\langle TeO_2\rangle = \{Te\} + O_2$	+75,120	—	−41·18	1	773–973	947
$\langle ThO_2\rangle = \langle Th\rangle + (O_2)$	293,400	1·61	−47·7	5	298–1800	—
$\langle Th(SO_4)_2\rangle = \langle Th\rangle + (S_2) + 4(O_2)$	645,150	33·2	−331·05	—	—	569
$\langle Th_3N_4\rangle = 3\langle Th\rangle + 2(N_2)$	310,400	—	−89·7	20	298–2000	123
$\langle \overline{ThC_2}\rangle = \langle Th\rangle + 2\langle C\rangle$	45,000	—	−2·6	10	298–2300	—
$2\langle TiCl_3\rangle = \langle TiCl_2\rangle + (TiCl_4)$	40,500	18·4	−96·4	4	298–1000	—
$(TiCl_4) = \langle Ti\rangle + 2(Cl_2)$	180,700	1·8	−34·65	3	298–1700	—
$2\langle TiI_2\rangle = \langle Ti\rangle + (TiI_4)$	50,600	13·8	−76·5	8	298–1000	84
$\langle 'TiO'\rangle = \langle 'Ti'\rangle + \frac{1}{2}(O_2)$	122,300	—	−21·3	4	600–2000	149
$\langle Ti_2O_3\rangle = 2\langle 'TiO'\rangle + \frac{1}{2}(O_2)$	114,150	—	−19·05	4	298–2000	149
$2\langle Ti_3O_5\rangle = 3\langle Ti_2O_3\rangle + \frac{1}{2}(O_2)$	88,500	—	−19·7	3	700–2000	149
$3\langle 'TiO_2'\rangle = \langle Ti_3O_5\rangle + \frac{1}{2}(O_2)$	73,000	—	−23·0	3	298–2123	149
$2\langle 'TiN'\rangle = 2\langle Ti\rangle_\alpha + (N_2)$	160,500	—	−44·4	4	298–1155	184
$2\langle 'TiN'\rangle = 2\langle Ti\rangle_\beta + (N_2)$	161,700	—	−45·54	4	1155–1500	184
$\langle 'TiC'\rangle = \langle 'Ti'\rangle + \langle C\rangle$	43,750	—	−2·41	3	298–1155	189
$\langle 'TiC'\rangle = \langle 'Ti'\rangle + \langle C\rangle$	44,600	—	−3·16	3	1155–2000	189
$\langle UF_4\rangle = \langle U\rangle + 2(F_2)$	448,500	—	−67·4	—	298–1309	399
$\{UF_4\} = \langle U\rangle_\gamma + 2(F_2)$	433,100	—	−55·6	—	1309–1405	399
$\{UF_4\} = \{U\} + 2(F_2)$					1405–1730	399
$\langle UF_6\rangle = \langle UF_4\rangle + (F_2)$	73,000	—	−30·4	—	298–337	399
$(UF_6) = \langle UF_4\rangle + (F_2)$	62,300	—	+2·8	—	298–1309	399
$(UF_6) = \{UF_4\} + (F_2)$	74,900	—	−7·1	—	1309–1730	399
$\langle UCl_4\rangle = \langle U\rangle + 2(Cl_2)$	253,100	14·3	−112·8	—	298–861	399
$\{UCl_4\} = \langle U\rangle + 2(Cl_2)$	236,700	—	−52·0	—	861–1060	399
$(UCl_6) = \langle UCl_4\rangle + (Cl_2)$	−2600	—	+1·5	—	450–900	399
$\langle UBr_4\rangle = \langle U\rangle + 2(Br_2)$	211,000	—	−70·9	—	298–792	399
$\{UBr_4\} = \langle U\rangle + 2(Br_2)$	194,000	—	−49·2	—	792–1050	399
$\langle UO_2\rangle = \langle U\rangle + (O_2)$	258,000	—	−40·0	—	298–1405	399
$\langle UO_2\rangle = \{U\} + (O_2)$	269,700	15·4	−97·0	—	1405–2000	399
$2\langle U_4O_9\rangle = 8\langle UO_2\rangle + (O_2)$	86,700	23·0	−105·3	—	298–1395	399
$\langle U_3O_8\rangle = 3\langle UO_2\rangle + (O_2)$	75,800	—	−34·0	—	298–1395	399
$\langle UN\rangle = \langle U\rangle + \frac{1}{2}(N_2)$	70,000	—	−20·5	—	298–1405	399
$\langle UN\rangle = \{U\} + \frac{1}{2}(N_2)$	71,200	—	−21·5	—	1405–2000	399
$\langle UC\rangle = \langle U\rangle + \langle C\rangle$	21,600	—	+1·5	1·5	298–m.p.	—

Table E. Standard Gibbs energies of reaction (contd.)

$\Delta G_T^\circ = A + BT \log T + CT$

Reaction	ΔG° in cal A	B	C	± kcal	Temp. range (°K)	Ref.
$\langle UC\rangle = \{U\} + \langle C\rangle$	24,600	—	−1·2	1·5	m.p.–2500	—
$\langle UC_2\rangle = \{U\} + 2\langle C\rangle$	27,700	—	−2·6	2	m.p.–2050	—
$2\langle VCl_3\rangle = \langle VCl_2\rangle + (VCl_4)$	38,000	—	−39·6	—	698–928	581
$2\langle VBr_3\rangle + (Br_2) = 2(VBr_4)$	19,200	—	−32·9	—	523–873	617
$2\langle \text{'}VO\text{'}\rangle = 2\langle V\rangle + (O_2)$	205,900	—	−35·9	6	900–1800	4
$2\langle V_2O_3\rangle = 4\langle \text{'}VO\text{'}\rangle + (O_2)$	175,300	—	−40·5	5	823–1385	4
$4\langle \text{'}VO_2\text{'}\rangle = 2\langle V_2O_3\rangle + (O_2)$	102,800	—	−33·5	4	1020–1180	4
$2\langle \text{'}V_2O_5\text{'}\rangle = 4\langle VO_2\rangle + (O_2)$	64,480	—	−45·6	3	298–943	4
$2\{\text{'}V_2O_5\text{'}\} = 4\langle VO_2\rangle + (O_2)$	84,300	—	−40·3	4	943–1800	4
$\langle \overline{VC}\rangle = \langle V\rangle + \langle C\rangle$	24,100	—	−1·5	—	1180–1370	656
$\langle WO_2\rangle = \langle W\rangle + (O_2)$	138,500	—	−36·6	5	298–1500	—
$\langle \text{'}WO_3\text{'}\rangle = \langle W\rangle + 1\frac{1}{2}(O_2)$	201,500	10·2	−91·7	5	298–1400	191a
$\langle WS_2\rangle = \langle W\rangle + (S_2)$	62,360	—	−23·0	10	298–1400	192
$\langle WC\rangle = \langle W\rangle + \langle C\rangle$	9000	—	−0·4	3	298–2000	—
$\langle ZnCl_2\rangle = \langle Zn\rangle + (Cl_2)$	101,385	12·85	−75·45	3	298–586	255
$\{ZnCl_2\} = Zn + (Cl_2)$	93,950	—	−27·35	5	586–1005	255
$(ZnCl_2) = (Zn) + (Cl_2)$	93,800	9·44	−38·6	5	1180–1800	—
$\langle ZnO\rangle = \langle Zn\rangle + \frac{1}{2}(O_2)$	84,100	6·9	−44·14	1·5	298–693	—
$\langle ZnO\rangle = (Zn) + \frac{1}{2}(O_2)$	115,420	10·35	−82·38	2	1170–2000	—
$\langle ZnS\rangle = \langle Zn\rangle + \frac{1}{2}(S_2)$	63,650	4·6	−36·68	4	298–693	—
$\langle ZnS\rangle = (Zn) + \frac{1}{2}(S_2)$	94,970	8·05	−74·92	5	1120–2000	—
$2\langle ZnO\rangle + \langle SiO_2\rangle = \langle Zn_2SiO_4\rangle$	−7130	—	+0·23	2	300–1300	135
$\langle ZrCl_4\rangle = \langle Zr\rangle + 2Cl_2$	+233,250	—	−69·46	1	298–700	—
$\langle \text{'}ZrO_2\text{'}\rangle_\alpha = \langle \text{'}Zr\text{'}\rangle_\alpha + (O_2)$	259,940	4·33	−59·12	4	298–1143	307
$\langle \text{'}ZrO_2\text{'}\rangle_\beta = \langle \text{'}Zr\text{'}\rangle_\beta + (O_2)$	260,200	6·44	−65·99	4	1478–2138	307
$2\langle \text{'}ZrN\text{'}\rangle = 2\langle Zr\rangle_\alpha + (N_2)$	174,000	—	−44·61	3	298–1135	184
$2\langle \text{'}ZrN\text{'}\rangle = 2\langle Zr\rangle_\beta + (N_2)$	175,850	—	−46·22	3	1135–1500	184
$\langle \text{'}ZrC\text{'}\rangle = \langle Zr\rangle + \langle C\rangle$	44,100	—	−2·2	3	298–2200	—

Table F. Thermochemical data of
(N_2 is the atomic fraction of the second
F1α. **Ag.** Silver alloys: partial and integral

System	Components	Thermo-chemical function	Temp. (°K)	$N_2 =$ 0·0	0·1
⟨Ag–Al⟩	⟨Ag⟩⟨Al⟩	$\Delta\bar{G}^E_{Al}$	722	−5600	−5620
{Ag–Al}	{Ag} {Al}	ΔH	1273	0	−750
	{Al}	$\Delta\bar{H}_{Al}$	1273	−8500	−6553
⟨Ag–Au⟩	⟨Ag⟩⟨Au⟩	ΔH	800	0	−430
	⟨Au⟩	$\Delta\bar{H}_{Au}$	800	−4850	−3796
{Ag–Au}	{Ag} {Au}	ΔH	1350	0	−443
	{Au}	$\Delta\bar{H}_{Au}$	1350	−4920	−3985
{Ag–Bi}	{Ag} {Bi}	ΔH	1000	0	—
	{Bi}	$\Delta\bar{H}_{Bi}$	1000	—	—
⟨Ag–Cd⟩	⟨Ag⟩⟨Cd⟩	ΔH	673	0	−603
	⟨Cd⟩	$\Delta\bar{H}_{Cd}$	673	−6400	−5597
{Ag–Cd}	{Ag} {Cd}	ΔG^E	1223	0	−521
	{Cd}	$\Delta\bar{G}^E_{Cd}$	1223	—	−4536
{Ag–Cu}	{Ag} {Cu}	ΔH	1423	0	+455
	{Cu}	$\Delta\bar{H}_{Cu}$	1423	+5500	+3750
{Ag–Ga}	{Ag} {Ga}	ΔH	1000	0	—
	{Ga}	$\Delta\bar{H}_{Ga}$	1000	—	—
{Ag–Ge}	{Ag} {Ge}	ΔH	1250	0	−163
	{Ge}	$\Delta\bar{H}_{Ge}$	1250	−3000	−510
⟨Ag–Hg⟩	⟨Hg⟩	$\Delta\bar{G}^E_{Hg}$	500	−2930	−2320
{Ag–In}	{Ag} {In}	ΔH	1100	0	—
	{In}	$\Delta\bar{H}_{In}$	—	−4500	—
⟨Ag–Mg⟩	⟨Ag⟩⟨Mg⟩	ΔH	773	0	−1110
	⟨Mg⟩	$\Delta\bar{H}_{Mg}$	773	−11200	−9185
⟨Ag–Mn⟩	⟨Mn⟩β	$\Delta\bar{G}^E_{Mn}$	1150	+200	+818
{Ag–Pb}	{Ag} {Pb}	ΔH	1273	0	+252
	{Pb}	$\Delta\bar{H}_{Pb}$	1273	+2500	+2500
⟨Ag–Pd⟩	⟨Ag⟩⟨Pd⟩	ΔH	1200	0	−664
	⟨Pd⟩	$\Delta\bar{G}^E_{Pd}$	1200	−5860	−4080
{Ag–S}	$\frac{1}{2}(S_2)$	$\Delta\bar{H}_S$	1400	−17370	−20530
⟨Ag–Sb⟩	⟨Sb⟩	$\Delta\bar{G}^E_{Sb}$	600	−814	−1336
{Ag–Sb}	{Ag} {Sb}	ΔH	1250	0	−504
	{Sb}	$\Delta\bar{H}_{Sb}$	1250	−6000	−4022
{Ag–Si}	{Ag} {Si}	ΔH	1473	0	−62
	{Si}	$\Delta\bar{H}_{Si}$	1473	−1193	+24
{Ag–Sn}	{Ag} {Sn}	ΔH	1250	0	−536
	{Sn}	$\Delta\bar{H}_{Sn}$	900	—	—
	{Sn}	$\Delta\bar{G}^E_{Sn}$	1250	+990	−334
{Ag–Tl}	{Ag} {Tl}	ΔH	—	—	+173
	{Tl}	$\Delta\bar{H}_{Tl}$	975	—	—
⟨Ag–Zn⟩	⟨Ag⟩⟨Zn⟩	ΔH	873	0	−400
	⟨Zn⟩	$\Delta\bar{H}_{Zn}$	873	−2826	−3397
{Ag–Zn}	{Ag} {Zn}	ΔG^E	1023	—	—
	{Zn}	$\Delta\bar{G}^E_{Zn}$	1023	—	—

F1β. **Ag.** Silver alloys: partial and integral

System	Components	Thermo-chemical function	Temp. (°K)	$N_2 =$ 0·0	0·1
{Ag–Al}	{Ag} {Al}	ΔS^E	1273	0 0	−0·013
	{Al}	$\Delta\bar{S}^E_{Al}$	1273	−0·314	0·00
⟨Ag–Au⟩	⟨Ag⟩⟨Au⟩	ΔS^E	800	0	−0·122
	⟨Au⟩	$\Delta\bar{S}^E_{Au}$	800	−1·356	−1·103
{Ag–Au}	{Ag} {Au}	ΔS^E	1350	0	−0·148
	{Au}	$\Delta\bar{S}^E_{Au}$	1350	−1·584	−1·376

binary metallic systems
element named in the first column)
excess Gibbs energies of solution in cal/g-atom

0·3	0·5	0·7	0·9	Ref.
−3130	—	—	—	99
−1530	−942	−270	+67	99
−2146	+1320	+540	+97	99
−967	−1111	−900	−370	99
−2137	−1009	−334	−34	99
−1033	−1230	−1033	−443	99
−2410	−1230	−443	−49	99
+167	+292	+409	+250	99
+558	+674	+438	+65	99
−1458	—	−1540ε	—	99
−3684	—	—	—	99
−1092	−1134	−810	−286	99
−2070	−710	−132	0	99
+921	+1014	+829	+352	99
+1788	+900	+343	+38	99
−868	−638	−204	+55	99
−863	+350	+411	+77	99
+94	+534	+642	+329	99
+1750	+1310	+492	+61	99
−294	—	—	—	99
−1350	−988	−472	−122	99
−1973	+426	+157	+22	99
—	−4400	—	—	99
—	—	—	—	99
+1710	—	—	—	99
+696	+885	+711	+274	99
+1913	+880	+220	+11	99
−1317	−1200	−780	−250	99
−1140	+230	+210	+21	99
−29,450	—	—	—	307
—	—	—	—	99
−630	−15	+240	+157	99
+1531	+1060	+352	+40	99
+187	+628	+856	+509	967
+1672	+1565	+855	+167	967
−640	−227	+59	+83	99
—	+600	+190	+42	99
−1233	−1025	−552	−77	99
+450	+594	+547	+280	99
+1220	+735	+328	+64	99
−2025	−952β	−930ε	—	99
−2248	−1885β	−50ε	—	99
−1020	−1104	−860	−320	99
−2600	−1005	−227	−8	99

excess entropies of solution in cal/deg g-atom

0·3	0·5	0·7	0·9	Ref.
+0·018	+0·372	+0·479	+0·274	99
+0·312	+1·298	+0·416	+0·069	99
−0·287	−0·343	−0·289	−0·125	99
−0·672	−0·345	−0·125	−0·014	99
−0·367	−0·468	−0·418	−0·190	99
−0·947	−0·541	−0·216	−0·026	99

Table F. Thermochemical data of

(N_2 is the atomic fraction of the second

F1β. **Ag** Silver alloys: partial and integral

System	Components	Thermo-chemical function	Temp. (°K)	N_2 = 0·0	0·1
{Ag–Bi}	{Ag} {Bi}	ΔS^E	1000	0	—
	{Bi}	$\Delta \bar{S}^E_{Bi}$	1000	—	—
⟨Ag–Cd⟩	⟨Ag⟩⟨Cd⟩	ΔS^E	673	0	−0·085
	⟨Cd⟩	$\Delta \bar{S}^E_{Cd}$	673	−1·486	−0·514
{Ag–Cu}	{Ag} {Cu}	ΔS^E	1423	0	+0·104
	{Cu}	$\Delta \bar{S}^E_{Cu}$	1423	+1·430	+0·737
{Ag–Ga}	{Ag} {Ga}	ΔS^E	1000	0	—
	{Ga}	$\Delta \bar{S}^E_{Ga}$	1000	—	—
{Ag–Ge}	{Ag} {Ge}	ΔS^E	1250	0	−0·006
	{Ge}	$\Delta \bar{S}^E_{Ge}$	1250	−0·640	+0·424
{Ag–In}	{Ag} {In}	ΔS^E	1100	0	—
	{In}	$\Delta \bar{S}^E_{In}$	1100	—	—
⟨Ag–Mg⟩	⟨Ag⟩⟨Mg⟩	ΔS^E	773	0	+0·133
	⟨Mg⟩	$\Delta \bar{S}^E_{Mg}$	773	+1·682	+3·186
{Ag–Pb}	{Ag} {Pb}	ΔS^E	1273	0	+0·161
	{Pb}	$\Delta \bar{S}^E_{Pb}$	1273	+2·12	+1·334
⟨Ag–Pd⟩	⟨Ag⟩⟨Pd⟩	ΔS^E	1200	0	−0·203
{Ag–S}	$\frac{1}{2}(S_2)$	$\Delta \bar{S}^E_S$	1400	−11·2	−12·33
{Ag–Sb}	{Ag} {Sb}	ΔS^E	1250	0	+0·068
	{Sb}	$\Delta \bar{S}^E_{Sb}$	1250	+1·012	+0·536
{Ag–Si}	Si	$\Delta \bar{S}^E_{Si}$	1473	−1·29	−0·715
{Ag–Sn}	{Ag} {Sn}	ΔS^E	1250	0	−0·16
	{Sn}	$\Delta \bar{S}^E_{Sn}$	900	—	—
{Ag–Tl}	{Ag} {Tl}	ΔS^E	975	0	—
	{Tl}	$\Delta \bar{S}^E_{Tl}$	975	—	—
⟨Ag–Zn⟩	⟨Ag⟩⟨Zn⟩	ΔS^E	873	0	−0·096
	⟨Zn⟩	$\Delta \bar{S}^E_{Zn}$	873	−0·03	+0·006

F2α. **Al.** Aluminium alloys: partial and integral heats or

System	Components	Thermo-chemical function	Temp. (°K)	N_2 = 0·0	0·1
⟨Al–Ag⟩	⟨Ag⟩	$\Delta \bar{G}^E_{Ag}$	722	+706	—
{Al–Ag}	{Al} {Ag}	ΔH	1273	0	+67
	{Ag}	$\Delta \bar{H}_{Ag}$	1273	+1750	−203
{Al–Au}	{Al} {Au}	ΔH	1338	0	−1845
	{Au}	$\Delta \bar{H}_{Au}$	1338	−16,665	−19,624
{Al–Be}	{Be}	$\Delta \bar{H}_{Be}$	1600	—	—
{Al–Bi}	{Al} {Bi}	ΔH	—	0	+532
	{Bi}	$\Delta \bar{H}_{Bi}$	1173	+5310	—
{Al–Cu}	{Al} {Cu}	ΔH	1373	0	−459
	{Cu}	$\Delta \bar{H}_{Cu}$	1373	−4225	−4850
{Al–Fe}	{Fe}	$\Delta \bar{H}_{Fe}$	971	−30,700	—
{Al–Ga}	{Al} {Ga}	ΔH	1023	0	+71
	{Ga}	$\Delta \bar{H}_{Ga}$	1023	+832	+600
{Al–Ge}	{Al} {Ge}	ΔH	1200	0	−272
	{Ge}	$\Delta \bar{H}_{Ge}$	1200	−2760	−2658
{Al–In}	{Al} {In}	ΔH	1173	0	+584
	{In}	$\Delta \bar{H}_{In}$	1173	+6800	+5006
{Al–Mg}	{Al} {Mg}	ΔH	1073	0	−312
	{Mg}	$\Delta \bar{H}_{Mg}$	1073	−3480	−2833
{Al–Pb}	{Al} {Pb}	ΔH	>1650	0	+924
	{Pb}	$\Delta \bar{H}_{Pb}$	1200	+9970	—
{Al–Si}	{Al} {Si}	ΔH	—	0	−244
	{Si}	$\Delta \bar{H}_{Si}$	—	−2500	−2175

binary metallic systems (contd.)
element named in the first column)
excess entropies of solution in cal/deg g-atom (contd.)

0·3	0·5	0·7	0·9	Ref.
+0·172	+0·097	+0·132	+0·072	99
−0·107	+0·066	+0·168	+0·001	99
−0·201	—	−0·348ε	—	99
−1·117	—	—	—	99
+0·154	+0·121	+0·081	+0·030	99
+0·109	+0·024	+0·020	+0·001	99
+0·361	+0·481	+0·622	+0·404	99
+0·470	+0·921	+0·685	+0·120	99
+0·227	+0·466	+0·467	+0·208	99
+1·264	+0·802	+0·276	+0·016	99
−0·871	−0·607	−0·302	−0·104	99
−1·346	+0·337	+0·022	−0·001	99
—	−0·090	—	—	99
—	—	—	—	99
+0·375	+0·438	+0·305	+0·093	99
+0·895	+0·345	+0·006	−0·01	99
−0·38	−0·438	−0·405	−0·167	99
−17·28	—	—	—	307
+0·31	+0·606	+0·544	+0·23	99
+2·004	+0·820	+0·232	+0·027	99
−0·025	+0·24	+0·215	+0·07	967
−0·043	+0·189	+0·206	+0·046	99
—	+0·334	+0·034	+0·021	99
+0·14	+0·133	+0·089	+0·053	99
+0·21	+0·06	+0·028	+0·025	99
+0·109	+0·376β	+0·255ε	—	99
+0·419	−0·370β	+0·712ε	—	99

excess Gibbs energies of solution in cal/g-atom

0·3	0·5	0·7	0·9	Ref.
—	—	−527ζ	−14	99
−270	−942	−1531	−749	99
−2158	−3205	−1267	−104	99
−5795	−8070	−7147	−2916	99
−17,924	−10,092	−3194	−205	99
—	+5112	+2450	+380	99
+1512	+1708	+1167	+440	99
—	—	+154	+31	99
−1445	−2163	−1980	−800	99
−4675	−3270	−680	−65	99
—	—	—	—	99
+148	+157	+120	+49	99
+290	+113	+40	+3	99
−729	−928	−773	−300	99
−1960	−997	−274	−9	99
+1241	+1365	+1087	+455	99
+2506	+1137	+401	+48	99
−719	−806	−614	−230	99
−1580	−641	−157	−8	99
+2130	+2236	+1540	+584	99
—	—	—	—	99
−585	−742	−662	−300	—
−1500	−860	−336	−42	—

Table F. Thermochemical data of

(N_2 is the atomic fraction of the second

F2α. **Al.** Aluminium alloys: partial and integral heats or

System	Components	Thermo-chemical function	Temp. (°K)	$N_2 =$ 0·0	0·1
{Al–Sn}	{Al} {Sn}	ΔH	973	0	+464
	{Sn}	$\Delta\bar{H}_{Sn}$	973	+5845	+3680
⟨Al–Zn⟩	⟨Al⟩⟨Zn⟩	ΔH	653	0	+326
	⟨Zn⟩	$\Delta\bar{H}_{Zn}$	653	+3742	+2850
{Al–Zn}	{Al–Zn}	ΔH	1000	0	+225
	{Zn}	$\Delta\bar{H}_{Zn}$	1000	+2538	+1995

F2β. **Al.** Aluminium alloys: partial and integral

System	Components	Thermo-chemical function	Temp. (°K)	$N_2 =$ 0·0	0·1
{Al–Ag}	{Al} {Ag}	ΔS^E	1273	0	+0·274
	{Ag}	$\Delta\bar{S}^E_{Ag}$	1273	+3·515	+2·122
{Al–Au}	{Al} {Au}	ΔS^E	1338	0	+0·326
	{Au}	$\Delta\bar{S}^E_{Au}$	1338	+5·635	+1·428
{Al–Be}	{Be}	$\Delta\bar{S}^E_{Be}$	1600	—	—
{Al–Bi}	{Bi}	$\Delta\bar{S}^E_{Bi}$	1173	−1·524	—
{Al–Cu}	{Al} {Cu}	ΔS^E	1373	0	+0·275
	{Cu}	$\Delta\bar{S}^E_{Cu}$	1373	+3·238	+2·344
{Al–Ga}	{Al} {Ga}	ΔS^E	1023	0	+0·043
	{Ga}	$\Delta\bar{S}^E_{Ga}$	1023	+0·52	+0·35
{Al–Ge}	{Al} {Ge}	ΔS^E	1200	0	+0·087
	{Ge}	$\Delta\bar{S}^E_{Ge}$	1200	+1·176	+0·622
{Al–In}	{Al} {In}	ΔS^E	1173	0	+0·095
	{In}	$\Delta\bar{S}^E_{In}$	1173	+1·20	+0·74
{Al–Mg}	{Al} {Mg}	ΔS^E	1073	0	+0·001
	{Mg}	$\Delta\bar{S}^E_{Mg}$	1073	+0·30	−0·255
{Al–Si}	{Al} {Si}	ΔS^E	—	0	+0·054
	{Si}	$\Delta\bar{S}^E_{Si}$	—	+0·61	+0·455
{Al–Sn}	{Al} {Sn}	ΔS^E	973	0	+0·160
	{Sn}	$\Delta\bar{S}^E_{Sn}$	973	+2·245	+1·107
⟨Al–Zn⟩	⟨Al⟩⟨Zn⟩	ΔS^E	653	0	+0·115
	⟨Zn⟩	$\Delta\bar{S}^E_{Zn}$	635	+1·426	+1·009
{Al–Zn}	{Al} {Zn}	ΔS^E	1000	0	+0·081
	{Zn}	$\Delta\bar{S}^E_{Zn}$	1000	+1·007	+0·65

F3α. **Au.** Gold alloys: partial and integral heats or

System	Components	Thermo-chemical function	Temp. (°K)	$N_2 =$ 0·0	0·1
⟨Au–Ag⟩	⟨Ag⟩⟨Au⟩	ΔH	800	0	−371
	⟨Ag⟩	$\Delta\bar{H}_{Ag}$	800	−4040	−3404
{Au–Ag}	{Au} {Ag}	ΔH	1350	0	−443
	{Ag}	$\Delta\bar{H}_{Ag}$	1350	−4920	−3985
{Au–Al}	{Au} {Al}	ΔH	1338	0	−2916
	{Al}	$\Delta\bar{H}_{Al}$	1338	−30,724	−27,304
{Au–Bi}	{Au} {Bi}	ΔH	973	0	—
	{Bi}	$\Delta\bar{H}_{Bi}$	973	—	—
⟨Au–Cd⟩	⟨Au⟩⟨Cd⟩	ΔH	700	0	−1431α
	⟨Cd⟩	$\Delta\bar{H}_{Cd}$	700	−15,600	−13,203
{Au–Cd}	{Au} {Cd}	ΔG^E	1000	0	—
	{Cd}	$\Delta\bar{G}^E_{Cd}$	1000	—	—
⟨Au–Co⟩	⟨Co⟩β	$\Delta\bar{H}_{Co}$	1150	+9590	+8877
⟨Au–Cu⟩	⟨Au⟩⟨Cu⟩	ΔH	800	0	−290
	⟨Cu⟩	$\Delta\bar{H}_{Cu}$	800	−2780	−2956

binary metallic systems (*contd.*)
element named in the first column)
excess Gibbs energies of solution in cal/g-atom (contd.)

0·3	0·5	0·7	0·9	Ref.
+900	+966	+762	+316	99
+1690	+766	+273	+33	99
+715	+825	—	—	99
+1504	+905	—	—	99
+516	+614	+526	+234	99
+1178	+631	+145	+36	99

excess entropies of solution in cal/deg g-atom

0·3	0·5	0·7	0·9	Ref.
+0·48	+0·372	+0·018	−0·013	99
+0·627	−0·553	−0·109	−0·015	99
+0·124	−0·028	+0·044	+0·127	99
−0·991	−0·058	+0·23	+0·056	99
—	+2·502	+1·151	+0·178	99
—	—	−0·074	+0·019	99
0·615	+0·812	+0·918	+0·491	99
+1·505	+1·141	+0·871	+0·092	99
+0·083	+0·080	+0·056	+0·021	99
+0·14	+0·037	+0·013	+0·001	99
+0·131	+0·104	+0·07	+0·026	99
+0·098	+0·023	+0·011	+0·001	99
+0·164	+0·142	+0·088	+0·032	99
+0·198	+0·031	0	+0·007	99
−0·106	−0·198	−0·195	−0·086	99
−0·532	−0·322	−0·109	−0·008	99
+0·112	+0·060	−0·013	−0·053	—
+0·10	−0·13	−0·14	−0·03	—
+0·27	+0·29	+0·25	+0·12	99
+0·468	+0·265	+0·128	+0·03	99
+0·254	+0·314	—	—	99
+0·51	+0·524	—	—	99
+0·171	+0·226	+0·212	+0·08	99
+0·405	+0·311	+0·106	−0·005	99

excess Gibbs energies of solution in cal/g-atom

0·3	0·5	0·7	0·9	Ref.
−899	−1111	−967	−429	99
−2220	−1214	−466	−55	99
−1033	−1230	−1033	−443	99
−2411	−1230	−443	−49	99
−7147	−8070	−5795	−1845	99
−16,370	−6047	−597	+130	99
+100	+150	+120	+45	99
+423	+167	+29	+2	99
−3393α_2	−4440β	−3660ε	—	99
−9390α_2	−4855β	−4200ε	—	99
—	−3153	−2574	−1036	99
—	−3173	−936	−77	99
—	—	—	—	99
−854	−1221	−1189	−440	99
−2610	−1765	−632	+29	99

Table F. Thermochemical data of
(N_2 is the atomic fraction of the second

F3α. **Au.** Gold alloys: partial and integral heats or

System	Components	Thermo-chemical function	Temp. (°K)	N_2 = 0·0	0·1
{Au–Cu}	{Au}{Cu}	ΔH	1550	0	−343
	{Cu}	$\Delta\bar{H}_{Cu}$	1550	−3725	−3163
⟨Au–Fe⟩	⟨Au⟩⟨Fe⟩α	ΔH	1123	0	+528
	⟨Fe⟩α	$\Delta\bar{H}_{Fe}$	1123	+6090	+4415
{Au–Fe}	{Au}{Fe}	ΔH	1473	0	+1400
	{Fe}	$\Delta\bar{H}_{Fe}$	1473	15,500	12,795
{Au–Ga}	{Au}{Ga}	ΔH	1030	0	−1885*
	{Ga}	$\Delta\bar{H}_{Ga}$	1030	−22,400*	−15,600*
{Au–Ge}	{Au}{Ge}	ΔH	1423	0	−510
⟨Au–Hg⟩	⟨Hg⟩	$\Delta\bar{H}_{Hg}$	500	−1270	−530
{Au–Mn}	{Au}{Mn}	ΔG^E	1535	0	−837
	{Mn}	$\Delta\bar{G}^E_{Mn}$	1535	−8500	−8667
⟨Au–Ni⟩	⟨Au⟩⟨Ni⟩	ΔH	1150	0	+500
	⟨Ni⟩	$\Delta\bar{H}_{Ni}$	1150	+5140	+4844
{Au–Pb}	{Au}{Pb}	ΔH	1200	0	−104
	{Pb}	$\Delta\bar{H}_{Pb}$	1200	—	−728
⟨Au–Pd⟩	⟨Au⟩⟨Pd⟩	ΔH	—	0	−975
	⟨Pd⟩	$\Delta\bar{H}_{Pd}$	—	−11,140	−7710
⟨Au–Pt⟩	⟨Au⟩⟨Pt⟩	ΔH	—	0	0
	⟨Pt⟩	$\Delta\bar{H}_{Pt}$	—	−635	+510
{Au–Sn}	{Au}{Sn}	ΔH	823	0	—
	{Sn}	$\Delta\bar{H}_{Sn}$	823	—	—
{Au–Te}	⟨Au⟩{Te}	ΔH	737	0	—
{Au–Tl}	{Tl}	$\Delta\bar{H}_{Tl}$	973	—	—
{Au–Zn}	{Au}{Zn}	ΔH	1080	—	—
	{Zn}	$\Delta\bar{H}_{Zn}$	1080	—	—

* Supercooled.

F3β. **Au.** Gold alloys: partial and integral

System	Components	Thermo-chemical function	Temp. (°K)	N_2 = 0·0	0·1
⟨Au–Ag⟩	⟨Au⟩⟨Ag⟩	ΔS^E	800	0	−0·122
	⟨Ag⟩	$\Delta\bar{S}^E_{Ag}$	800	−1·388	−1·120
{Au–Ag}	{Au}{Ag}	ΔS^E	1350	0	−0·190
	{Ag}	$\Delta\bar{S}^E_{Ag}$	1350	−2·163	−1·658
{Au–Al}	{Au}{Al}	ΔS^E	1338	0	+0·127
	{Al}	$\Delta\bar{S}^E_{Al}$	1338	+2·00	+0·77
{Au–Bi}	{Au}{Bi}	ΔS^E	973	0	—
	{Bi}	$\Delta\bar{S}^E_{Bi}$	973	—	—
⟨Au–Cd⟩	⟨Au⟩⟨Cd⟩	ΔS^E	700	0	−0·687α
		$\Delta\bar{S}^E_{Cd}$	700	−8·357	−5·716
⟨Au–Co⟩	⟨Co⟩β	$\Delta\bar{S}^E_{Co}$	1150	+2·00	+3·524
⟨Au–Cu⟩	⟨Au⟩⟨Cu⟩	ΔS^E	800	0	+0·047
	⟨Cu⟩	$\Delta\bar{S}^E_{Cu}$	800	+0·612	+0·360
{Au–Cu}	{Au}{Cu}	ΔS^E	1550	0	+0·113
	{Cu}	$\Delta\bar{S}^E_{Cu}$	1550	+1·306	+0·964
⟨Au–Fe⟩	⟨Au⟩⟨Fe⟩α	ΔS^E	1123	0	+0·41
	⟨Fe⟩α	$\Delta\bar{S}^E_{Fe}$	1123	+5·20	+3·03
{Au–Fe}	{Au}{Fe}	ΔS^E	1473	0	+1·09
	{Fe}	$\Delta\bar{S}^E_{Fe}$	1473	+12·763	+9·35
{Au–Ga}	{Au}{Ga}	ΔS^E	1400	0	0·0
	{Ga}	$\Delta\bar{S}^E_{Ga}$	1400	−1·06	0·0
{Au–Ge}	{Au}{Ge}	ΔS^E	1423	0	−0·09
⟨Au–Hg⟩	⟨Hg⟩	$\Delta\bar{S}^E_{Hg}$	500	−2·00	−1·14

binary metallic systems (*contd.*)
element named in the first column)
excess Gibbs energies of solution in cal/g-atom (contd.)

0·3	0·5	0·7	0·9	Ref.
−839	−1044	−915	−408	99
−2090	−1156	−449	−53	99
+1167	+1566	—	—	99
+2870	+2310	—	—	99
+3360	+4100	—	—	99
+8153	+4380	—	—	99
−3750	−3820	−2720	−968	976
−6730	−2290	−495	−29	976
−1080	−1120	−830	−300	957
—	—	—	—	99
−2706	−4205	−3407	−1120	99
−9628	−5670	−437	+52	99
+1347	+1807	+1608	+643	99
+3768	+2235	+715	+28	99
−189	−167	−73	+62	99
−272	−22	+131	+82	99
−1830	−1780	−1132	−344	958
−3085	−765	+14	+34	958
+265	+635	+795	+455	958
+1550	+1500	+740	+107	958
−2467	−2766	−2090	−760	99
−5422	−2223	−444	−22	99
—	+1010	+659	+253	965
—	+49	−3	−2	99
−4230	−5436	−4487	−1747	99
−12,000	−6020	−1440	−94	99

excess entropies of solution in cal/deg g-atom

0·3	0·5	0·7	0·9	Ref.
−0·287	−0·343	−0·289	−0·125	99
−0·672	−0·341	−0·122	−0·013	99
−0·418	−0·468	−0·367	−0·148	99
−0·890	−0·396	−0·121	−0·011	99
+0·044	−0·028	+0·124	+0·326	99
−0·392	+0·002	+0·603	+0·203	99
+0·245	+0·298	+0·244	+0·106	99
+0·61	+0·295	+0·09	+0·015	99
−1·586α_2	−1·576β	−1·51ε	—	99
−3·81α_2	−1·884β	−3·69ε	—	99
—	—	—	—	99
+0·081	+0·070	+0·031	+0·06	99
+0·094	−0·034	+0·023	+0·053	99
+0·238	+0·254	+0·189	+0·071	99
+0·469	+0·181	+0·044	+0·003	99
+0·71	+0·76	—	—	99
+1·107	+0·773	—	—	99
+2·305	+2·463	—	—	99
+4·553	+1·76	—	—	99
+0·06	+0·27	+0·30	+0·06	976
+0·78	+0·625	+0·245	+0·03	976
−0·13	+0·03	+0·13	+0·16	957
—	—	—	—	99

Table F. Thermochemical data of
(N_2 is the atomic fraction of the second
F3β. **Au.** Gold alloys: partial and integral

System	Components	Thermo-chemical function	Temp. (°K)	$N_2 =$ 0·0	0·1
⟨Au–Ni⟩	⟨Au⟩⟨Ni⟩	ΔS^E	1150	0	+0·139
	⟨Ni⟩	$\Delta \bar{S}^E_{Ni}$	1150	+1·213	+1·516
{Au–Pb}	{Au} {Pb}	ΔS^E	1200	0	+0·153
	{Pb}	$\Delta \bar{S}^E_{Pb}$	1200	—	+1·377
⟨Au–Pd⟩	⟨Au⟩⟨Pd⟩	ΔS^E	—	0	−0·27
	⟨Pd⟩	$\Delta \bar{S}^E_{Pd}$	—	−3·0	−2·43
⟨Au–Pt⟩	⟨Au⟩⟨Pt⟩	ΔS^E	—	0	−0·29
	⟨Pt⟩	$\Delta \bar{S}^E_{Pt}$	—	−3·50	−2·41
{Au–Sn}	{Au} {Sn}	ΔS^E	823	0	—
	{Sn}	$\Delta \bar{S}^E_{Sn}$	823	—	—
{Au–Tl}	{Tl}	$\Delta \bar{S}^E_{Tl}$	973	—	—
{Au–Zn}	{Au} {Zn}	ΔS^E	1080	0	—
	{Zn}	$\Delta \bar{S}^E_{Zn}$	1080	—	—

F4α. **Co.** Cobalt alloys: partial and integral heats or

System	Components	Thermo-chemical function	Temp. (°K)	0·0	0·1
{Co–Al}	{Al}	ΔG^E_{Al}	1873	−19,754	−16,395
⟨Co–Cr⟩	⟨Co⟩fcc, ⟨Cr⟩bcc	ΔH	1473	0	+75
	⟨Cr⟩	$\Delta \bar{H}_{Cr}$	1473	+670	+965
⟨Co–Cu⟩	⟨Cu⟩	ΔG^E_{Cu}	1300	(9100)	—
⟨Co–Fe⟩	⟨Co⟩γ⟨Fe⟩γ	ΔH	1473	0	−105
(f.c.c.)	⟨Fe⟩γ	$\Delta \bar{H}_{Fe}$	1473	−1085	−996
{Co–Fe}	{Co} {Fe}	ΔH	1873	0	−219
	{Fe}	$\Delta \bar{H}_{Fe}$	1873	—	−1987
{Co–Ge}	⟨Co⟩ {Ge}	ΔH	1287	0	—
⟨Co–Mn⟩	⟨Co⟩⟨Mn⟩β	ΔG^E	1023	0	−511
	⟨Mn⟩β	$\Delta \bar{G}^E_{Mn}$	1023	−5875	−4455
⟨Co–Mo⟩	⟨Mo⟩	$\Delta \bar{G}^E_{Mo}$	1350	+4000	+3420
Co–Ni				this	system is nearly ideal in
⟨Co–Pt⟩	⟨Co⟩⟨Pt⟩	ΔG^E	1273	0	−1067
{Co–Si}	{Co} {Si}	ΔH	1873	0	−3470
	{Si}	$\Delta \bar{H}_{Si}$	1873	−35,000	−34,100
{Co–Sn}	{Co} {Sn}	ΔH	1773	0	−380

F4β. **Co.** Cobalt alloys: partial and integral

System	Components	Thermo-chemical function	Temp. (°K)	0·0	0·1
⟨Co–Cr⟩	⟨Co⟩⟨Cr⟩	ΔS^E	1473	0	+0·064
	⟨Cr⟩	$\Delta \bar{S}^E_{Cr}$	1473	+0·42	+0·705
⟨Co–Fe⟩	—	—	—	—	—
{Co–Fe}	—	—	—	—	—
{Co–Si}	{Co} {Si}	ΔS^E	1873	0	−0·61
	{Si}	$\Delta \bar{S}^E_{Si}$	1873	−6·06	−6·29

F5α. **Cr.** Chromium alloys: partial and integral heats or

System	Components	Thermo-chemical function	Temp. (°K)	0·0	0·1
⟨Cr–Co⟩	⟨Co⟩fcc, ⟨Cr⟩bcc	ΔH	1473	0	+715
	⟨Co⟩fcc	$\Delta \bar{H}_{Co}$	1473	(8100)	+6080

binary metallic systems (*contd.*)
element named in the first column)
excess entropies of solution in cal/deg g-atom (contd.)

0·3	0·5	0·7	0·9	Ref.
+0·458	+0·668	+0·561	+0·161	99
+1·513	+0·853	+0·127	−0·039	99
+0·357	+0·425	+0·396	+0·261	99
+0·833	+0·425	+0·324	+0·093	99
−0·63	−0·75	−0·63	−0·27	958
−1·47	−0·75	−0·27	−0·03	958
−0·58	−0·55	−0·35	−0·105	958
−0·95	−0·225	—	—	958
+0·498	+0·313	+0·183	+0·093	99
−0·027	−0·152	+0·092	+0·012	99
—	+0·638	+0·209	+0·02	99
−0·143	−0·493	−0·507	−0·177	99
−1·57	−1·13	−0·17	+0·01	99

excess Gibbs energies of solution in cal/g-atom

0·3	0·5	0·7	0·9	Ref.
−10,207	—	—	—	99
+510	—	+1700	+715	953
+3140	—	—	+120	953
—	—	—	—	99
−270	−314	−233	−85	974
−677	−260	−53	−4	974
−508	−580	−456	−182	974
−1140	−500	−146	−14	974
—	−596	−1060	−395	966
−1127	−1340	−2485	−1190	99
−2482	−1490	−693	−36	99
—	—	—	—	99
the solid and liquid states				960
−2491	−2965	−2491	−1067	99
−9310	−11,150	−8170	−2890	954
−24,400	−8800	−1400	−30	954
−550	−250	+130	+200	99

excess entropies of solution in cal/deg g-atom

0·3	0·5	0·7	0·9	Ref.
+0·426	—	+1·479	+0·905	953
+2·635	—	—	+0·545	953
—	—	—	—	—
—	—	—	—	—
−1·73	−2·20	−1·72	−0·64	954
−4·78	−2·15	−0·44	−0·02	954

excess Gibbs energies of solution in cal/g-atom

0·3	0·5	0·7	0·9	Ref.
+1700	—	+510	+75	953
—	—	−618	−24	953

Table F. Thermochemical data of
(N_2 is the atomic fraction of the second

F5α. **Cr.** Chromium alloys: partial and integral heats or

System	Components	Thermo-chemical function	Temp. (°K)	$N_2 =$ 0·0	0·1
⟨Cr–Fe⟩α	⟨Cr⟩⟨Fe⟩	ΔH	1550	0	+540
	⟨Fe⟩	$\Delta\bar{H}_{Fe}$	1550	+6000	+4860
{Cr–Fe}	{Cr}{Fe}	ΔH	1750–2150	0	+455
	{Fe}	$\Delta\bar{H}_{Fe}$	1750–2150	+5018	+4100
⟨Cr–Mo⟩	⟨Cr⟩⟨Mo⟩	ΔH	1400	0	+745
⟨Cr–Ni⟩	⟨Cr⟩⟨Ni⟩	ΔH	—	0	+1163
	⟨Ni⟩	$\Delta\bar{H}_{Ni}$	—	+13,872	+9614
{Cr–Ni}	{Cr}{Ni}	ΔG^E	—	0	−337
	{Ni}	$\Delta\bar{G}^E_{Ni}$		—	−625
⟨Cr–V⟩	⟨Cr⟩⟨V⟩	ΔH	1550	0	−437
	⟨V⟩	$\Delta\bar{H}_V$	1550	−5210	−3594

F5β. **Cr.** Chromium alloys: partial and integral

System	Components	Thermo-chemical function	Temp. (°K)	$N_2 =$ 0·0	0·1
⟨Cr–Co⟩	⟨Cr⟩bcc ⟨Co⟩fcc	ΔS^E	1473	0	+0·905
	⟨Co⟩	$\Delta\bar{S}^E_{Co}$	1473	(9·1)	+4·14
⟨Cr–Fe⟩α	⟨Cr⟩⟨Fe⟩	ΔS^E	1550	0	+0·266
	⟨Fe⟩	$\Delta\bar{S}^E_{Fe}$	1550	—	+2·115
{Cr–Fe}	{Cr}{Fe}	ΔS^E	1750–2150	0	+0·266
	{Fe}	$\Delta\bar{S}^E_{Fe}$	1750–2150	—	+2·115
⟨Cr–Mo⟩	⟨Cr⟩⟨Mo⟩	ΔS^E	1200	0	+0·205
⟨Cr–Ni⟩	⟨Cr⟩⟨Ni⟩	ΔS^E	—	0	+0·662
	⟨Ni⟩	$\Delta\bar{S}^E_{Ni}$	—	—	+5·211
⟨Cr–V⟩	⟨Cr⟩⟨V⟩	ΔS^E	1550	0	−0·027
	⟨V⟩	$\Delta\bar{S}^E_V$	1550	−0·429	−0·123

F6α. **Cu.** Copper alloys: partial and integral heats or

System	Components	Thermo-chemical function	Temp. (°K)	$N_2 =$ 0·0	0·1
{Cu–Ag}	{Cu}{Ag}	ΔH	1423	0	+352
	{Ag}	$\Delta\bar{H}_{Ag}$	1423	+3900	+2762
{Cu–Al}	{Cu}{Al}	ΔH	1373	0	−800
	{Al}	$\Delta\bar{H}_{Al}$	1373	−8625	−7415
⟨Cu–Au⟩	⟨Cu⟩⟨Au⟩	ΔH	800	0	−440
	⟨Au⟩	$\Delta\bar{H}_{Au}$	800	−3920	−4660
{Cu–Au}	{Cu}{Au}	ΔH	1550	0	−408
	{Au}	$\Delta\bar{H}_{Au}$	1550	−4625	−3600
{Cu–Bi}	{Cu}{Bi}	ΔH	1200	0	—
	{Bi}	$\Delta\bar{H}_{Bi}$	1200	—	—
⟨Cu–Fe⟩	⟨Fe⟩⟨Cu⟩	ΔH	1323	0	+934
{Cu–Fe}	{Cu}{Fe}	ΔH	1823	0	+804
	{Fe}	$\Delta\bar{H}_{Fe}$	1823	+9300	+6978
{Cu–Ge}	{Cu}{Ge}	ΔH	1423	0	−510
{Cu–In}	{Cu}{In}	ΔH	1073	0	—
	{In}	$\Delta\bar{H}_{In}$	1073	—	(N_2 = 0·2) −1427

binary metallic systems (*contd.*)
element named in the first column)
excess Gibbs energies of solution in cal/g-atom (contd.)

0·3	0·5	0·7	0·9	Ref.
+1260	+1500	+1260	+540	951
+2940	+1500	+540	+60	951
+1050	+1245	+1045	+460	951
+2430	+1240	+460	+60	951
+1555	+1725	+1345	+495	952
+2148	+1520	+336	−210	968
+2461	−1929	−1123	−210	968
−410	−653	−742	−335	968
−853	−1225	−503	−35	968
−779	−455	−235	−203	99
−718	+928	−790	−27	99

excess entropies of solution in cal/deg g-atom

0·3	0·5	0·7	0·9	Ref.
+1·479	—	+0·426	+0·064	953
—	—	+0·524	+0·065	953
+0·595	+0·688	+0·554	+0·222	951
+1·375	+0·607	+0·207	+0·024	951
+0·595	+0·688	+0·554	+0·222	951
+1·375	+0·607	+0·207	+0·024	951
+0·42	+0·46	+0·33	+0·087	952
+1·163	+0·967	+0·429	+0·087	968
+1·529	−0·504	−0·291	+0·014	968
+0·030	+0·242	+0·213	+0·004	99
+0·526	+0·864	−0·460	−0·010	99

excess Gibbs energies of solution in cal/g-atom

0·3	0·5	0·7	0·9	Ref.
+829	+1014	+921	+455	99
+1965	+1126	+549	+88	99
−1980	−2163	−1445	−459	99
−5012	−1055	−60	+29	99
−1189	−1221	−854	−290	99
−2487	−677	−101	+7	99
−915	−1044	−839	−343	99
−2002	−931	−303	−30	99
+1224	+1280	+1052	+468	99
+2128	+1055	+451	+77	99
+2300	+2880	+2537	+1138	950
+1790	+2132	+1917	+950	99
+4028	+2282	+1152	+178	99
−1070	−1120	−830	−300	961
−749	−240	+23	+66	99
+465	+829	+233	+40	99

Table F. Thermochemical data of

(N_2 is the atomic fraction of the second

F6α. **Cu.** Copper alloys: partial and integral heats or

System	Components	Thermo-chemical function	Temp. (°K)	$N_2 =$ 0·0	0·1
{Cu–Mg}	{Cu} {Mg}	ΔH	1100	0	—
	{Mg}	$\Delta \bar{H}_{Mg}$	1100	—	($N_2 = 0{\cdot}2$) −6419
⟨Cu–Mn⟩γ	⟨Cu⟩⟨Mn⟩γ	ΔH	1100	0	+240
	⟨Mn⟩γ	$\Delta \bar{G}^E_{Mn}$	1100	−880	−81
{Cu–Mn}	{Cu} {Mn}	ΔG^E	1500	0	−113
	{Mn}	$\Delta \bar{G}^E_{Mn}$	1500	−2000	−356
⟨Cu–Ni⟩	⟨Cu⟩⟨Ni⟩	ΔH	773	0	+130
	⟨Ni⟩	$\Delta \bar{H}_{Ni}$	773	+1480	+1285
{Cu–Ni}	{Cu} {Ni}	ΔG^E	—	0	+270
	{Ni}	$\Delta \bar{G}^E_{Ni}$	—	+2970	+2450
{Cu–Pb}	{Cu} {Pb}	ΔH	1473	0	+707
	{Pb}	$\Delta \bar{H}_{Pb}$	1473	+8620	+5790
⟨Cu–Pd⟩	⟨Cu⟩⟨Pd⟩	ΔH	1350	0	−1005
	⟨Pd⟩	$\Delta \bar{H}_{Pd}$	1350	−10,900	−9274
⟨Cu–Pt⟩	⟨Cu⟩⟨Pt⟩	ΔH	1350	0	−1228
	⟨Pt⟩	$\Delta \bar{H}_{Pt}$	1350	−14,760	−10,178
{Cu–Sb}	{Cu} {Sb}	ΔH	1190	—	—
	{Sb}	$\Delta \bar{H}_{Sb}$	1190	—	(−7000)
{Cu–Si}	{Cu} {Si}	ΔG^E	1760	0	−1380
	{Si}	$\Delta \bar{G}^E_{Si}$	1760	−14,470	−12,260
{Cu–Sn}	{Cu} {Sn}	ΔH	1400	0	−666
	{Sn}	$\Delta \bar{H}_{Sn}$	1400	−8000	−5233
{Cu–Tl}	{Cu} {Tl}	ΔH	1573	0	+647
	{Tl}	$\Delta \bar{H}_{Tl}$	1573	+6730	+6164
⟨Cu–Zn⟩	⟨Cu⟩⟨Zn⟩	ΔH	773	0	−627
	⟨Zn⟩	$\Delta \bar{H}_{Zn}$	773	−5500	−6853
{Cu–Zn}	{Zn}	$\Delta \bar{G}^E_{Zn}$	—	—	—
{Cu–O}	$\frac{1}{2}(O_2)$.	$\Delta \bar{H}_O$	1673	−20,700	−22,800

F6β. **Cu.** Copper alloys: partial and integral

System	Components	Thermo-chemical function	Temp. (°K)	$N_2 =$ 0·0	0·1
{Cu–Ag}	{Cu} {Ag}	ΔS^E	1423	0	+0·030
	{Ag}	$\Delta \bar{S}^E_{Ag}$	1423	+0·323	+0·293
{Cu–Al}	{Cu} {Al}	ΔS^E	1373	0	+0·491
	{Al}	$\Delta \bar{S}^E_{Al}$	1373	+5·838	+4·086
⟨Cu–Au⟩	⟨Cu⟩⟨Au⟩	ΔS^E	800	0	+0·059
	⟨Au⟩	$\Delta \bar{S}^E_{Au}$	800	+1·27	+0·114
{Cu–Au}	{Cu} {Au}	ΔS^E	1550	0	+0·071
	{Au}	$\Delta \bar{S}^E_{Au}$	1550	+0·726	+0·682
{Cu–Bi}	{Cu} {Bi}	ΔS^E	1200	0	—
	{Bi}	$\Delta \bar{S}^E_{Bi}$	1200	—	—
⟨Cu–Fe⟩	⟨Fe⟩⟨Cu⟩	ΔS^E	1323	0	+0·27
{Cu–Fe}	{Cu} {Fe}	ΔS^E	1823	0	+0·038
	{Fe}	$\Delta \bar{S}^E_{Fe}$	1823	+0·411	+0·359
{Cu–Ge}	{Cu} {Ge}	ΔS^E	1423	0	−0·12
{Cu–In}	{Cu} {In}	ΔS^E	1073	0	—
	{In}	$\Delta \bar{S}^E_{In}$	1073	—	($N_2 = 0{\cdot}2$) −0·095
{Cu–Mg}	{Cu} {Mg}	ΔS^E	1100	0	—
	{Mg}	$\Delta \bar{S}^E_{Mg}$	1100	—	($N_2 = 0{\cdot}2$) −1·792

binary metallic systems (contd.)
element named in the first column)
excess Gibbs energies of solution in cal/g-atom (contd.)

0·3	0·5	0·7	0·9	Ref.
−2285	−2465	−1933	−908	99
−4520	−1873	−772	−208	99
+608	+764	+1009	+720	99
+730	+872	+621	—	99
+97	+476	+722	+513	99
+1416	+1335	+872	+205	99
+340	+445	+465	+285	949
+880	+600	+405	+90	949
+680	+835	+735	+315	949
+1570	+835	+315	+35	949
+1422	+1607	+1338	+584	99
+2817	+1545	+573	+73	99
−2440	−2557	−2108	−904	99
−5385	−2057	−891	−100	99
−2482	−2650	−1979	−689	99
−4682	−2169	−237	−5	99
−1344	−697	−151	+82	99
−66	+1093	+465	+81	99
−2820	−2810	−1880	−640	954
−4880	−1420	−120	−10	954
−934	−475	−97	+52	99
+252	+681	+311	+49	99
+1661	+2050	+1582	+543	99
+4287	+2031	+259	−11	99
−1754	—	—	—	99
−4150	—	—	—	99
—	−868	−151	+1	99
−35,100	—	—	—	955

excess entropies of solution in cal/deg g-atom

0·3	0·5	0·7	0·9	Ref.
+0·081	+0·121	+0·154	+0·105	99
+0·224	+0·219	+0·174	+0·035	99
+0·918	+0·812	+0·615	+0·275	99
+1·028	+0·484	+0·234	+0·044	99
+0·031	+0·070	+0·081	+0·047	99
+0·049	+0·175	+0·076	+0·012	99
+0·189	+0·254	+0·238	+0·113	99
+0·526	+0·327	+0·138	+0·018	99
+0·448	+0·435	+0·370	+0·179	99
+0·624	+0·335	+0·188	+0·043	99
+0·63	+0·75	+0·63	+0·27	950
+0·110	+0·162	+0·185	+0·131	99
+0·355	+0·241	+0·207	+0·043	99
−0·12	+0·02	+0·13	+0·13	961
−0·057	+0·160	+0·132	+0·085	99
+0·550	+0·401	−0·006	+0·003	99
−0·687	−0·772	−0·698	−0·449	99
−1·372	−0·725	−0·522	−0·185	99

Table F. Thermochemical data of
(N_2 is the atomic fraction of the second
F6β. **Cu.** Copper alloys: partial and integral

System	Components	Thermo-chemical function	Temp. (°K)	$N_2 =$ 0·0	0·1
⟨Cu–Mn⟩γ	⟨Cu⟩⟨Mn⟩γ	ΔS^E	1100	0	+0·259
⟨Cu–Ni⟩	⟨Cu⟩⟨Ni⟩	ΔS^E	—	0	−0·085
	⟨Ni⟩	$\Delta\bar{S}^E_{Ni}$	—	−0·77	−0·70
{Cu–Pb}	{Cu}{Pb}	ΔS^E	1473	0	+0·173
	{Pb}	$\Delta\bar{S}^E_{Pb}$	1473	+2·55	+1·096
⟨Cu–Pd⟩	⟨Cu⟩⟨Pd⟩	ΔS^E	1350	0	−0·150
	⟨Pd⟩	$\Delta\bar{S}^E_{Pd}$	1350	−1·023	−1·887
⟨Cu–Pt⟩	⟨Cu⟩⟨Pt⟩	ΔS^E	1350	0	−0·096
	⟨Pt⟩	$\Delta\bar{S}^E_{Pt}$	1350	−1·254	−0·718
{Cu–Sb}	{Cu}{Sb}	ΔS^E	1190	—	—
	{Sb}	$\Delta\bar{S}^E_{Sb}$	1190	—	(0·35)
{Cu–Sn}	{Cu}{Sn}	ΔS^E	1400	0	+0·251
	{Sn}	$\Delta\bar{S}^E_{Sn}$	1400	+4·006	+1·477
{Cu–Tl}	{Cu}{Tl}	ΔS^E	1573	0	+0·062
	{Tl}	$\Delta\bar{S}^E_{Tl}$	1573	+0·305	+0·850
⟨Cu–Zn⟩	⟨Cu⟩⟨Zn⟩	ΔS^E	773	0	−0·020
	⟨Zn⟩	$\Delta\bar{S}^E_{Zn}$	773	+1·30	−1·415

F7α. **Fe.** Iron alloys: partial and integral heats or

System	Components	Thermo-chemical function	Temp. (°K)	0·0	0·1
{Fe–Al}	{Fe}{Al}	ΔH	1873	0	−1460
	{Al}	$\Delta\bar{H}_{Al}$	1873	−15,350	−13,870
⟨Fe–Au⟩	⟨Fe⟩α⟨Au⟩	ΔH	1123	0	—
	⟨Au⟩	$\Delta\bar{H}_{Au}$	1123	—	—
{Fe–Au}	{Fe}{Au}	ΔH	1383	0	—
⟨Fe–Co⟩	⟨Fe⟩γ⟨Co⟩γ	ΔH	1473	0	−85
(fcc)	⟨Co⟩γ	$\Delta\bar{H}_{Co}$	1473	−923	−845
{Fe–Co}	{Fe}{Co}	ΔH	1873	0	−182
	{Co}	$\Delta\bar{H}_{Co}$	1873	—	−1698
⟨Fe–Cr⟩α	⟨Fe⟩⟨Cr⟩	ΔH	1550	0	+540
	⟨Cr⟩	$\Delta\bar{H}_{Cr}$	1550	+6000	+4860
{Fe–Cr}	{Fe}{Cr}	ΔH	—	0	+460
	{Cr}	$\Delta\bar{H}_{Cr}$	—	+5165	+4060
⟨Fe–Cu⟩	⟨Fe⟩γ⟨Cu⟩	ΔH	1323	0	+1138
{Fe–Cu}	{Fe}{Cu}	ΔH	1823	0	+950
	{Cu}	$\Delta\bar{H}_{Cu}$	1823	11,370	+7897
{Fe–Ge}	⟨Fe⟩{Ge}	ΔH	1287	0	—
⟨Fe–Ir⟩γ	⟨Fe⟩γ⟨Ir⟩	ΔG^E	1473	0	−1266
	⟨Ir⟩	$\Delta\bar{G}^E_{Ir}$	1473	−13,460	−11,760
⟨Fe–Mn⟩γ	⟨Fe⟩γ⟨Mn⟩γ	ΔH	1450	0	−314
	⟨Mn⟩γ	$\Delta\bar{H}_{Mn}$	1450	−3346	−2952
{Fe–Mn}	{Fe}{Mn}	ΔG^E	1863	0	+95
	{Mn}	$\Delta\bar{G}^E_{Mn}$	1863	+1050	+851
⟨Fe–Ni⟩	⟨Fe⟩γ⟨Ni⟩	ΔH	1200	0	−230
(fcc)	⟨Ni⟩	$\Delta\bar{H}_{Ni}$	1200	—	−1030
{Fe–Ni}	{Fe}{Ni}	ΔH	1873	0	−245
	{Ni}	$\Delta\bar{H}_{Ni}$	1873	−2400	−2333
⟨Fe–Pd⟩	⟨Fe⟩γ⟨Pd⟩	ΔH	1273	0	+407
	⟨Pd⟩	$\Delta\bar{H}_{Pd}$	1273	+8630	+217
{Fe–Pd}	{Fe}{Pd}	ΔH	1873	0	+246
	{Pd}	$\Delta\bar{H}_{Pd}$	1873	+5250	+1590

binary metallic systems (*contd.*)
element named in the first column)
excess entropies of solution in cal/g-atom (contd.)

0·3	0·5	0·7	0·9	Ref.
+0·493	+0·432	+0·485	—	99
−0·190	−0·231	−0·153	−0·010	949
−0·47	−0·185	+0·04	+0·043	949
+0·234	+0·252	+0·225	+0·110	99
+0·277	+0·282	+0·124	+0·021	99
−0·639	−0·846	−0·987	−0·533	99
−2·141	−1·419	−0·830	−0·111	99
−0·179	−0·209	−0·125	+0·039	99
−0·350	−0·328	+0·265	+0·048	99
+0·278	+0·579	+0·570	+0·282	99
+1·587	+0·975	+0·348	+0·052	99
+0·400	+0·545	+0·491	+0·245	99
+1·018	+0·622	+0·277	+0·052	99
+0·288	+0·463	+0·374	+0·109	99
+1·121	+0·657	+0·010	−0·016	99
−0·353	—	—	—	99
−0·944	—	—	—	99

excess Gibbs energies of solution in cal/g-atom

0·3	0·5	0·7	0·9	Ref.
−4529	—	—	—	99
−10,545	—	—	—	99
—	+1450	+1150	+530	981
—	+822	+437	+96	981
—	+1030	+860	+390	956
−233	−314	−270	−105	974
−667	−370	−90	−6	974
−456	−580	−508	−219	974
−1180	−661	−237	−23	974
+1260	+1500	+1260	+540	951
+2940	+1500	+540	+60	951
+1045	+1245	+1050	+455	951
+2410	+1250	+460	+50	951
+2537	+2880	+2300	+934	950
+1917	+2132	+1789	+804	99
+3702	+1982	+830	+118	99
—	+2140	+1160	+440	966
−3032	−3347	−2565	−1010	99
−6723	−2577	−756	−71	99
−1482	−1653	−1456	−796	99
−2291	−1578	−769	−112	99
+221	+263	+221	+95	99
+515	+263	+95	+11	99
−465	−885	−1000	−496	972
−2110	−1430	−940	−80	972
−700	−1100	−1160	−580	972
−2324	−1724	−916	−49	972
−678	−2257	−2873	−1510	99
−6233	−5640	−2418	−284	99
+464	+530	+478	+219	99
+932	+550	+274	+28	99

Table F. Thermochemical data of

(N_2 is the atomic fraction of the second

F7α. **Fe.** Iron alloys: partial and integral heats or

System	Components	Thermo-chemical function	Temp. (°K)	$N_2 =$ 0·0	0·1
⟨Fe–Pt⟩	⟨Fe⟩γ⟨Pt⟩	ΔG^E	1123	0	−1368γ
	⟨Pt⟩	$\Delta \bar{G}^E_{Pt}$	1123	−18,120	−11,744
{Fe–Pt}	{Fe} {Pt}	ΔG^E	1880	0	−1780
	{Pt}	$\Delta \bar{G}^E_{Pt}$	1880	−16,000	−18,643
{Fe–S}	(S_2)	$\Delta \bar{H}_{S_2}$	—	−60,500	−60,100
{Fe–Si}	{Fe} {Si}	ΔH	1873	0	−3050
	{Si}	$\Delta \bar{H}_{Si}$	1873	−31,400	−29,900
{Fe–Sn}	{Fe} {Sn}	ΔG^E	1820	0	+367
	{Sn}	$\Delta \bar{G}^E_{Sn}$	1820	+3724	+3606
⟨Fe–V⟩	⟨Fe⟩α⟨V⟩	ΔH	1600	0	−311
	⟨V⟩	$\Delta \bar{H}_V$	1600	−3090	−2990
{Fe–V}	{Fe} {V}	ΔG^E	2193	0	−225
	{V}	$\Delta \bar{G}^E_V$	2193	−2550	−2395

F7β. **Fe.** Iron alloys: partial and integral

System	Components	Thermo-chemical function	Temp. (°K)	$N_2 =$ 0·0	0·1
{Fe–Al}	{Fe} {Al}	ΔS^E	1873	0	−0·27
	{Al}	$\Delta \bar{S}^E_{Al}$	1873	−2·523	−2·81
{Fe–Au}	{Fe} {Au}	ΔS^E	1383	—	—
⟨Fe–Co⟩	—	—	—	—	—
{Fe–Co}	—	—	—	—	—
⟨Fe–Cr⟩α	⟨Fe⟩⟨Cr⟩	ΔS^E	1550	0	+0·222
	⟨Cr⟩	$\Delta \bar{S}^E_{Cr}$	1550	+2·45	+2·01
{Fe–Cr}	{Fe} {Cr}	ΔS^E	1750–2150	0	+0·222
	{Cr}	$\Delta \bar{S}^E_{Cr}$	1750–2150	+2·45	+2·01
⟨Fe–Cu⟩	⟨Fe⟩γ⟨Cu⟩	ΔS^E	1323	0	+0·27
	⟨Cu⟩	$\Delta \bar{S}^E_{Cu}$	1323	+3·0	+2·43
{Fe–Cu}	{Fe} {Cu}	ΔS^E	1823	0	+0·038
	{Cu}	$\Delta \bar{S}^E_{Cu}$	1823	+1·76	+0·917
⟨Fe–Mn⟩γ	⟨Fe⟩γ⟨Mn⟩γ	ΔS^E	1450	0	−0·296
	⟨Mn⟩γ	$\Delta \bar{S}^E_{Mn}$	1450	−3·17	−2·76
⟨Fe–Ni⟩γ	⟨Fe⟩γ⟨Ni⟩	ΔS^E	1200	0	−0·059
	⟨Ni⟩	$\Delta \bar{S}^E_{Ni}$	1200	—	+0·292
{Fe–Ni}	{Fe} {Ni}	ΔS^E	1873	0	−0·051
	{Ni}	$\Delta \bar{S}^E_{Ni}$	1873	−0·509	−0·509
⟨Fe–Pd⟩γ	⟨Fe⟩γ⟨Pd⟩	ΔS^E	1273	0	+0·613
	⟨Pd⟩	$\Delta \bar{S}^E_{Pd}$	1273	+8·553	+4·045
{Fe–Pd}	{Fe} {Pd}	ΔS^E	1873	0	−0·029
	{Pd}	$\Delta \bar{S}^E_{Pd}$	1873	+0·753	−0·513
{Fe–S}	(S_2)	$\Delta \bar{S}^E_{S_2}$	—	−12·75	−11·70
{Fe–Si}	{Fe} {Si}	ΔS^E	1873	0	−0·50
	{Si}	$\Delta \bar{S}^E_{Si}$	1873	−4·186	−5·67
⟨Fe–V⟩	⟨Fe⟩α⟨V⟩	ΔS^E	1600	0	−0·127
	⟨V⟩	$\Delta \bar{S}^E_V$	1600	+3·4	+2·065

F8α. **Mg.** Magnesium alloys: partial and integral heats or

System	Components	Thermo-chemical function	Temp. (°K)	$N_2 =$ 0·0	0·1
{Mg–Al}	{Ag} {Al}	ΔH	1073	0	−230
	{Al}	$\Delta \bar{H}_{Al}$	1073	−2358	−2227
{Mg–Bi}	{Mg} {Bi}	ΔH	1100	0	−1850
	{Bi}	$\Delta \bar{H}_{Bi}$	1100	—	−16,700

binary metallic systems (*contd.*)
element named in the first column)
excess Gibbs energies of solution in cal/g-atom (contd.)

0·3	0·5	0·7	0·9	Ref.
-3342γ	$-4344\gamma_2$	$-3276\gamma_3$	-1591γ	99
−8970	$-4520\gamma_2$	$-855\gamma_3$	0	99
−5140	−6972	—	—	99
−14,943	−8283	—	—	99
−59,200	−57,400	—	—	307
−7980	−9050	−6700	−2450	959
−19,400	−6650	−1340	−90	959
+1013	—	—	—	99
+2965	—	—	—	99
−858	−1253	−1323	−529	973
−2532	−2040	−526	−66	973
−660	−1071	−1240	−654	972
−2070	−1850	−1080	−120	972

excess entrophies of solution in cal/deg g-atom

0·3	0·5	0·7	0·9	Ref.
−0·84	—	—	—	99
−2·706	—	—	—	99
—	—	+0·60	+0·38	956
—	—	—	—	—
—	—	—	—	—
+0·554	+0·688	+0·595	+0·266	951
+1·36	+0·77	+0·26	+0·05	951
+0·554	+0·688	+0·595	+0·266	951
+1·36	+0·77	+0·26	+0·05	951
+0·63	+0·75	+0·63	+0·27	950
+1·47	+0·75	+0·27	+0·03	950
+0·110	+0·162	+0·185	+0·131	99
+0·133	+0·084	+0·005	+0·002	99
−0·757	−1·017	−0·983	−0·491	99
−2·053	−1·35	−0·64	−0·093	99
−0·069	−0·194	−0·248	−0·134	972
−0·427	−0·269	−0·313	−0·061	972
−0·154	−0·264	−0·296	−0·178	972
−0·561	−0·420	−0·277	−0·042	972
+0·817	+0·689	+0·461	+0·164	99
+0·667	+0·21	+0·07	+0·001	99
−0·112	−0·135	−0·098	−0·034	99
−0·304	−0·109	−0·013	−0·001	99
−10·29	−16·78	—	—	307
−1·65	−2·06	−1·71	−0·666	959
−4·71	−2·145	−0·61	−0·04	959
−0·331	−0·386	−0·263	−0·032	972
+0·553	−0·342	−0·125	−0·042	972

excess Gibbs energies of solution in cal/g-atom

0·3	0·5	0·7	0·9	Ref.
−614	−806	−720	−312	99
−1681	−972	−350	−32	99
−4500	−4530	−3000	−750	962
−8370	−2610	+690	+100	962

Table F. Thermochemical data of

(N_2 is the atomic fraction of the second

F8α. **Mg.** Magnesium alloys: partial and integral heats or

System	Components	Thermo-chemical function	Temp. (°K)	$N_2 =$ 0·0	0·1
{Mg–Ca}	{Mg} {Ca}	ΔH	1110	0	(−1180)
	{Ca}	$\Delta \bar{H}_{Ca}$	1110	—	—
⟨Mg–Cd⟩	⟨Mg⟩ ⟨Cd⟩	ΔH	543	0	−397
	⟨Cd⟩	$\Delta \bar{H}_{Cd}$	543	−4030	−3870
{Mg–Cd}	{Mg} {Cd}	ΔH	923	0	−484
	{Cd}	$\Delta \bar{H}_{Cd}$	923	−5572	−4227
{Mg–Cu}	{Mg} {Cu}	ΔH	1100	0	−908
		$\Delta \bar{H}_{Cu}$	1100	−11,760	−7206
{Mg–In}	{Mg} {In}	ΔH	923	0	−671
	{In}	$\Delta \bar{H}_{In}$	923	−7090	−6306
{Mg–Li}	{Mg} {Li}	ΔH	1000	0	−363
	{Li}	$\Delta \bar{H}_{Li}$	1000	−3020	−3990
{Mg–Ni}	{Ni}	$\Delta \bar{G}^E_{Ni}$	1000	−10,023	−5058
{Mg–Pb}	{Mg} {Pb}	ΔH	973	0	−945
	{Pb}	$\Delta \bar{H}_{Pb}$	973	−9500	−9133
{Mg–Sb}	{Mg} {Sb}	ΔH	1100	0	−5250
		$\Delta \bar{H}_{Sb}$	1100	—	−43,900
{Mg–Sn}	{Mg} {Sn}	ΔG^E	1073	0	−1607
	{Sn}	$\Delta \bar{G}^E_{Sn}$	1073	−18,955	−13,568
{Mg–Tl}	{Mg} {Tl}	ΔH	923	0	−662
	{Tl}	$\Delta \bar{H}_{Tl}$	923	−6350	−6350
{Mg–Zn}	{Mg} {Zn}	ΔH	923	0	−500
	{Zn}	$\Delta \bar{H}_{Zn}$	923	−5200	−4400

F8β. **Mg.** Magnesium alloys: partial and integral

System	Components	Thermo-chemical function	Temp. (°K)	$N_2 =$ 0·0	0·1
{Mg–Al}	{Mg} {Al}	ΔS^E	1073	0	−0·086
	{Al}	$\Delta \bar{S}^E_{Al}$	1073	−0·923	−0·786
{Mg–Bi}	{Mg} {Bi}	ΔS^E	1100	0	−1·005
	{Bi}	$\Delta \bar{S}^E_{Bi}$	1100	—	−9·525
{Mg–Ca}	{Mg} {Ca}	ΔS^E	1110	0	(−0·31)
	{Ca}	$\Delta \bar{S}^E_{Ca}$	1110	—	—
⟨Mg–Cd⟩	⟨Mg⟩ ⟨Cd⟩	ΔS^E	543	0	−0·08
	⟨Cd⟩	$\Delta \bar{S}^E_{Cd}$	543	−0·758	−0·834
{Mg–Cd}	{Mg} {Cd}	ΔS^E	923	0	−0·137
	{Cd}	$\Delta \bar{S}^E_{Cd}$	923	−1·925	−0·956
{Mg–Cu}	{Mg} {Cu}	ΔS^E	1100	0	−0·45
	{Cu}	$\Delta \bar{S}^E_{Cu}$	1100	−6·904	−2·829
{Mg–In}	{Mg} {In}	ΔS^E	923	0	+0·053
	{In}	$\Delta \bar{S}^E_{In}$	923	+1·643	−0·315
{Mg–Li}	{Mg} {Li}	ΔS^E	1000	0	−0·12
	{Li}	$\Delta \bar{S}^E_{Li}$	1000	−0·827	−1·388
{Mg–Pb}	{Mg} {Pb}	ΔS^E	973	0	+0·136
	{Pb}	$\Delta \bar{S}^E_{Pb}$	973	+3·446	−0·150
{Mg–Sb}	{Mg} {Sb}	ΔS^E	1100	0	−2·61
	{Sb}	$\Delta \bar{S}^E_{Sb}$	1100	—	−17·40
{Mg–Tl}	{Mg} {Tl}	ΔS^E	923	0	−0·029
	{Tl}	$\Delta \bar{S}^E_{Tl}$	923	+0·433	−0·45
{Mg–Zn}	{Mg} {Zn}	ΔS^E	923	0	−0·252
	{Zn}	ΔS^E_{Zn}	923	−2·04	−1·85

binary metallic systems (*contd.*)
element named in the first column)
excess Gibbs energies of solution in cal/g-atom (contd.)

0·3	0·5	0·7	0·9	Ref.
−1900	−1710	−1120	−400	963
−2360	−610	−100	−15	963
−1066	−1322	−1008	−340	99
−2875	−1230	−182	+17	99
−1102	−1341	−1123	−460	99
−2605	−1401	−441	−36	99
−1933	−3131	−2865	—	99
−4642	−3056	−1327	—	99
−1620	−1642	−1132	−432	99
−3435	−755	−220	−28	99
−1102	−1283	−857	−261	99
−3002	−786	−11	+26	99
—	—	—	—	99
−2290	−2244	−1470	−495	99
−4555	−931	−69	0	99
−12,600	−11,000	−5100	−1650	962
−25,700	+12,200	+600	−20	962
−3290	−3385	−2410	−852	99
−6095	−2070	−393	−7	99
−2660	−3007	−2412	−1080	99
−3092	−700	0	+28	99
−1160	−1525	−1400	−690	963
−2820	−1920	−810	−155	963

excess entropies of solution in cal/deg g-atom

0·3	0·5	0·7	0·9	Ref.
−0·195	−0·198	−0·106	+0·001	99
−0·396	−0·073	+0·076	+0·029	99
−2·265	−2·025	−1·185	−0·455	962
−2·74	−2·075	+0·23	−0·115	962
−0·64	−0·575	−0·375	−0·13	963
−0·88	−0·18	−0·03	0	963
−0·254	−0·31	−0·145	+0·022	99
−0·796	−0·142	+0·183	+0·051	99
−0·225	−0·250	−0·210	−0·090	99
−0·361	−0·250	−0·090	−0·010	99
−0·698	−0·772	−0·687	—	99
−1·109	−0·819	−0·394	—	99
−0·169	−0·098	+0·084	+0·097	99
−0·737	+0·447	+0·226	+0·065	99
−0·338	−0·183	−0·198	−0·238	99
−0·599	+0·740	+0·621	+0·104	99
−0·149	−0·138	−0·055	−0·047	99
−0·876	+0·065	−0·015	−0·028	99
−5·36	−2·94	−0·40	+0·05	962
−0·70	+6·19	+0·97	+1·01	962
−0·052	+0·174	+0·282	+0·183	99
+0·53	+0·63	+0·325	+0·052	99
−0·515	−0·665	—	−0·285	963
−1·29	−0·86	−0·32	−0·05	963

Table F. Thermochemical data of

(N_2 is the atomic fraction of the second

F9α. **Na.** Sodium alloys: partial and integral heats or

System	Components	Thermo-chemical function	Temp. (°K)	N_2 = 0·0	0·1
{Na–Bi}	{Na} {Bi}	ΔH	1173	0	−3980
	{Bi}	$\Delta\bar{H}_{Bi}$	1173	—	−43,600
{Na–Cd}	{Na} {Cd}	ΔH	673	0	−34
	{Cd}	$\Delta\bar{H}_{Cd}$	673	−55	−644
{Na–Cs}	{Na} {Cs}	ΔH	384	0	+145
	{Cs}	$\Delta\bar{H}_{Cs}$	384	+1170	+855
{Na–Ga}	{Na} {Ga}	ΔG^E	823	0	+262
	{Ga}	$\Delta\bar{G}^E_{Ga}$	823	—	+1150
{Na–Hg}	{Na} {Hg}	ΔH	648	0	−1250
	{Hg}	$\Delta\bar{H}_{Hg}$	648	−11,800	−12,880
{Na–In}	{Na} {In}	ΔH	713	0	−448
	{In}	$\Delta\bar{H}_{In}$	713	−4200	−4665
{Na–K}	{Na} {K}	ΔH	384	0	+76
	{K}	$\Delta\bar{H}_K$	384	+870	+650
{Na–Pb}	{Na} {Pb}	ΔH	700	0	−1614
	{Pb}	$\Delta\bar{H}_{Pb}$	700	−15,500	−16,500
{Na–Rb}	{Na} {Rb}	ΔH	384	0	+143
	{Rb}	$\Delta\bar{H}_{Rb}$	384	+1560	+1115
{Na–Sn}	{Na} {Sn}	ΔH	773	0	−1600
{Na–Tl}	{Na} {Tl}	ΔH	673	0	−783
	{Tl}	ΔH_{Tl}	673	−8000	−7606

F9β. **Na.** Sodium alloys: partial and integral

System	Components	Thermo-chemical function	Temp. (°K)	N_2 = 0·0	0·1
{Na–Bi}	{Na} {Bi}	ΔS^E	1173	0	−1·165
	{Bi}	$\Delta\bar{S}^E_{Bi}$	1173	0	−15·70
{Na–Cd}	{Na} {Cd}	ΔS^E	673	0	−0·16
	{Cd}	$\Delta\bar{S}^E_{Cd}$	673	−1·73	−1·594
{Na–Cs}	{Na} {Cg}	ΔS^E, $\Delta\bar{S}^E$	384	0·0	0·0
{Na–Hg}	{Na} {Hg}	ΔS^E	648	0	−0·66
	{Hg}	$\Delta\bar{S}^E_{Hg}$	648	−5·22	−7·64
{Na–In}	{Na} {In}	ΔS^E	713	0	−0·50
	{In}	$\Delta\bar{S}^E_{In}$	713	−5·60	−4·492
{Na–K}	{Na} {K}	ΔS^E	384	0	0·00
	{K}	$\Delta\bar{S}^E_K$	384	0·0	0·00
{Na–Pb}	{Na} {Pb}	ΔS^E	700	0	−0·703
	{Pb}	$\Delta\bar{S}^E_{Pb}$	700	−5·814	−8·14
{Na–Rb}	{Na} {Rb}	ΔS^E, $\Delta\bar{S}^E$	384	0·0	0·0
{Na–Sn}	{Na} {Sn}	ΔS^E	773	0	−0·785
{Na–Tl}	{Na} {Tl}	ΔS^E	673	0	−0·388
	{Tl}	$\Delta\bar{S}^E_{Tl}$	673	−4·306	−3·492

F10α. **Ni.** Nickel alloys: partial and integral heats or

System	Components	Thermo-chemical function	Temp. (°K)	N_2 = 0·0	0·1
⟨Ni–Al⟩	⟨Ni⟩ ⟨Al⟩	ΔH	298	0	−3650
	⟨Ni⟩ ⟨Al⟩	ΔG^E	1273	0	−3025
	⟨Al⟩	$\Delta\bar{G}^E_{Al}$	1273	−34,640	−26,230
⟨Ni–Au⟩	⟨Ni⟩ ⟨Au⟩	ΔH	1150	0	+643
	⟨Au⟩	$\Delta\bar{H}_{Au}$	1150	+6620	+6174
{Ni–Au}	{Ni} {Au}	ΔH	1369	0	—
Ni–Co				this system is nearly ideal in	

binary metallic systems (*contd.*)
element named in the first column)
excess Gibbs energies of solution in cal/g-atom

0·3	0·5	0·7	0·9	Ref.
−9880	−7260	−4360	−1075	971
−2030	−1910	+750	+250	971
−291	−750	−1101	−739	99
−1597	−1890	−1296	−160	99
+238	+219	+148	+52	971
+378	+88	+16	+3	971
+300	+72	−210	−208	971
−227	−698	−480	−82	971
−3870	−6080	−6070	−2010	99
−12,900	−10,130	−1900	0	99
−1373	−2025	−1766	−672	99
−4350	−3000	−461	−32	99
+162	+177	+134	+51	971
+330	+135	+36	+2	971
−3916	−3883	−2688	−931	99
−6636	−2065	−325	0	99
+284	+294	+217	+80	971
+503	+198	+49	+3	971
−4180	−4630	−2890	−845	971
−2106	−2735	−2188	−804	99
−5826	−2847	−598	−8	99

excess entropies of solution in cal/deg g-atom

0·3	0·5	0·7	0·9	Ref.
−3·195	−1·445	−0·315	+0·435	971
+4·77	+0·80	+1·65	+0·37	971
−0·50	−0·902	−1·108	−0·692	99
−1·808	−1·724	−1·09	−0·12	99
0·0	0·0	0·0	0·0	971
−2·51	−4·46	−4·74	−1·01	99
−9·59	−8·89	−0·29	+0·27	99
−1·19	−1·56	−1·272	−0·435	99
−2·973	−2·203	−0·079	0·00	99
0·00	−0·05	−0·11	−0·11	971
−0·13	−0·19	−0·16	−0·095	971
−1·774	−1·409	−0·753	−0·199	99
−1·667	+0·10	+0·231	+0·027	99
0·0	0·0	0·0	0·0	971
−2·055	−2·22	−1·225	−0·505	971
−0·917	−1·064	−0·642	−0·12	99
−2·206	−0·571	+0·205	+0·07	99

excess Gibbs energies of solution in cal/g-atom

0·3	0·5	0·7	0·9	Ref.
—	−14,050	—	—	99
—	−11,205	—	—	99
—	−9815	—	—	99
+1608	+1807	+1347	+500	99
+3692	+1378	+309	+18	99
—	+514	+432	+185	99
the solid and liquid states				960

MT-N

Table F. Thermochemical data of

(N_2 is the atomic fraction of the second

F10α. **Ni.** Nickel alloys: partial and integral heats or

System	Components	Thermo-chemical function	Temp. (°K)	N_2 = 0·0	0·1
⟨Ni–Cr⟩	⟨Ni⟩⟨Cr⟩	ΔH	—	0	−210
	⟨Cr⟩	$\Delta \bar{H}_{Cr}$	—	−4200	−210
{Ni–Cr}	{Ni}{Cr}	ΔG^E	—	0	−335
	{Cr}	$\Delta \bar{G}^E_{Cr}$	—	−3675	−3037
⟨Ni–Cu⟩	⟨Ni⟩⟨Cu⟩	ΔH	—	0	+285
	⟨Cu⟩	$\Delta \bar{H}_{Cu}$	—	+3940	+2070
{Ni–Cu}	{Ni}{Cu}	ΔG^E	—	0	+315
	{Cu}	$\Delta \bar{G}^E_{Cu}$	—	+3500	+2830
⟨Ni–Fe⟩γ	⟨Ni⟩⟨Fe⟩γ	ΔH	1200	0	−496
	⟨Fe⟩γ	$\Delta \bar{H}_{Fe}$	1200	−5800	−4200
{Ni–Fe}	{Ni}{Fe}	ΔH	1873	0	−580
	{Fe}	$\Delta \bar{H}_{Fe}$	—	−7700	−5413
{Ni–Ge}	⟨Ni⟩{Ge}	ΔH	1287	0	—
⟨Ni–Mn⟩γ	⟨Ni⟩⟨Mn⟩β	ΔH	1050	0	−1295
	⟨Mn⟩β	$\Delta \bar{H}_{Mn}$	1050	−14,375	−11,685
⟨Ni–Pd⟩	⟨Ni⟩⟨Pd⟩	ΔH	1273	0	+131
	⟨Pd⟩	$\Delta \bar{H}_{Pd}$	1273	+2024	+730
{Ni–Pd}	{Ni}{Pd}	ΔH	1873	0	+163
	{Pd}	$\Delta \bar{H}_{Pd}$	1873	+2790	+994
⟨Ni–Pt⟩	⟨Ni⟩⟨Pt⟩	ΔH	298	0	−651
	⟨Ni⟩⟨Pt⟩	ΔG^E	1625	0	−852
	⟨Pt⟩	$\Delta \bar{G}^E_{Pt}$	1625	−9780	−7423
{Ni–Si}	{Ni}{Si}	ΔH	1873	0	−4490
	{Si}	$\Delta \bar{H}_{Si}$	1873	−45,000	−44,700
{Ni–Sn}	{Ni}{Sn}	ΔH	1773	0	−2500
⟨Ni–Zn⟩	⟨Ni⟩⟨Zn⟩	ΔG^E	1100	0	−873
	⟨Zn⟩	$\Delta \bar{G}^E_{Zn}$	1100	−9314	−8167

F10β. **Ni.** Nickel alloys: partial and integral

System	Components	Thermo-chemical function	Temp. (°K)	N_2 = 0·0	0·1
⟨Ni–Au⟩	⟨Au⟩⟨Ni⟩	ΔS^E	1150	0	+0·161
	⟨Au⟩	$\Delta \bar{S}^E_{Au}$	1150	+1·14	+1·957
⟨Ni–Cr⟩	⟨Ni⟩⟨Cr⟩	ΔS^E	—	0	+0·087
	⟨Cr⟩	$\Delta \bar{S}^E_{Cr}$	—	−0·675	+0·740
⟨Ni–Cu⟩	⟨Ni⟩⟨Cu⟩	ΔS^E	—	0	−0·01
	⟨Cu⟩	$\Delta \bar{S}^E_{Cu}$	—	+1·123	−0·485
⟨Ni–Fe⟩	⟨Ni⟩⟨Fe⟩γ	ΔS^E	1200	0	−0·134
	⟨Fe⟩γ	$\Delta \bar{S}^E_{Fe}$	1200	−1·2	−0·788
{Ni–Fe}	{Ni}{Fe}	ΔS^E	1873	0	−0·178
	{Fe}	$\Delta \bar{S}^E_{Fe}$	1873	−2·2	−1·40
⟨Ni–Mn⟩γ	⟨Ni⟩⟨Mn⟩β	ΔS^E	1050	0	−0·166
	⟨Mn⟩β	$\Delta \bar{S}^E_{Mn}$	1050	−1·700	−1·652
⟨Ni–Pd⟩	⟨Ni⟩⟨Pd⟩	ΔS^E	1273	0	+0·166
	⟨Pd⟩	$\Delta \bar{S}^E_{Pd}$	1273	+1·875	+1·475
{Ni–Pd}	{Ni}{Pd}	ΔS^E	1873	0	−0·003
	{Pd}	$\Delta \bar{S}^E_{Pd}$	1873	+0·455	−0·259
⟨Ni–Pt⟩	⟨Ni⟩⟨Pt⟩	ΔS^E	1625	0	+0·124
{Ni–Si}	{Ni}{Si}	ΔS^E	1873	0	−0·65
	{Si}	$\Delta \bar{S}^E_{Si}$	1873	−5·50	−7·48

* Assuming ΔC_p (298–1625°K) = 0.

binary metallic systems (*contd.*)
element named in the first column)
excess Gibbs energies of solution in cal/g-atom (contd.)

0·3	0·5	0·7	0·9	Ref.
+336	+1520	+2148	+1163	968
+3740	+4970	+2014	+224	968
−742	−653	−410	−337	968
−1300	−81	−220	−305	968
+465	+445	+340	+130	949
+605	+290	+110	+3	949
+735	+835	+680	+270	949
+1715	+835	+300	+30	949
−1000	−885	−465	−230	972
−1140	−340	+240	−140	972
−1160	−1100	−700	−245	972
−1732	−475	−4	−13	972
—	−13,090	−8650	−2950	966
−3100	−3397	−1632	—	99
−7057	−1830	+1120	—	99
+84	−128	−282	−202	99
−526	−664	−384	−67	99
+255	+286	+236	+101	99
+463	+268	+98	+12	99
−1867	−2214	−1730	−762	99
−1860	−2060	−1601	−631	99
−3887	−1675	−492	−42	99
−12,640	−13,230	−8780	−3050	954
−33,500	−5600	−820	−40	954
−4400	−4100	−2500	−1100	99
−2156	−2996β	—	—	—
−5468	−3485β	—	—	—

excess entropies of solution in cal/deg g-atom

0·3	0·5	0·7	0·9	Ref.
+0·561	+0·668	+0·458	+0·139	99
+1·575	+0·483	+0·006	−0·014	99
+0·429	+0·967	+1·163	+0·662	968
+2·108	+2·438	+1·006	+0·326	968
−0·153	−0·231	−0·19	−0·085	949
−0·61	−0·235	−0·07	−0·016	949
−0·248	−0·194	0·069	−0·059	972
−0·21	−0·12	+0·085	−0·099	972
−0·296	−0·264	−0·154	−0·051	972
−0·34	−0·10	+0·02	0·0	972
−0·55	−0·494	+0·291	—	99
−1·526	+0·80	+1·50	—	99
+0·376	+0·427	+0·33	+0·118	99
+0·827	+0·367	+0·079	−0·002	99
−0·067	−0·085	−0·067	−0·026	99
−0·214	−0·079	−0·02	−0·001	99
−0·004	−0·095	−0·079	−0·081	99*
−2·45	−2·77	−1·84	−0·55	954
−8·00	−1·54	−0·04	+0·08	954

Table F. Thermochemical data of
(N_2 is the atomic fraction of the second

F11α. **Pb.** Lead alloys: partial and integral heats or

System	Components	Thermo-chemical function	Temp. (°K)	N_2 = 0·0	0·1
{Pb–Ag}	{Pb} {Ag}	ΔH	1273	0	+274
	{Ag}	$\Delta \bar{H}_{Ag}$	1273	+2800	+2640
{Pb–Al}	{Pb} {Al}	ΔH	—	0	+584
{Pb–Au}	{Pb} {Au}	ΔH	1200	0	+62
	{Au}	$\Delta \bar{H}_{Au}$	1200	+53	−114
{Pb–Bi}	{Pb} {Bi}	ΔH	700	0	−80
	{Bi}	$\Delta \bar{H}_{Bi}$	700	−840	−780
{Pb–Cd}	{Pb} {Cd}	ΔH	773	0	+205
	{Cd}	$\Delta \bar{H}_{Cd}$	773	+2230	+1886
{Pb–Cu}	{Pb} {Cu}	ΔH	1473	0	+584
	{Cu}	$\Delta \bar{H}_{Cu}$	1473	+6600	+5184
{Pb–Ga}	{Pb} {Ga}	ΔH	923	0	+391
		$\Delta \bar{H}_{Ga}$	923	+4830	+3180
{Pb–Hg}	{Pb} {Hg}	ΔH	600	0	−34
	{Hg}	$\Delta \bar{H}_{Hg}$	600	−555	−173
⟨Pb–In⟩	⟨In⟩⟨Pb⟩	ΔH	315	0	+100
{Pb–In}	{Pb} {In}	ΔH	673	0	+78
	{In}	$\Delta \bar{H}_{In}$	673	+956	+711
{Pb–K}	{Pb} {K}	ΔH	848	0	−1232
	{K}	$\Delta \bar{H}_{K}$	848	−12,316	−12,168
{Pb–Mg}	{Pb} {Mg}	ΔH	973	0	−495
	{Mg}	$\Delta \bar{H}_{Mg}$	973	−5000	−4950
{Pb–Na}	{Pb} {Na}	ΔH	700	0	−931
	{Na}	$\Delta \bar{H}_{Na}$	700	−9200	−9330
{Pb–Pt}	{Pb} {Pt}	ΔG^E	1273	0	−725
	{Pt}	$\Delta \bar{G}^E_{Pt}$	1273	−7958	−6610
{Pb–Sb}	{Pb} {Sb}	ΔH	905	0	0
	{Sb}	$\Delta \bar{H}_{Sb}$	905	+32	−18
{Pb–Sn}	{Pb} {Sn}	ΔH	1050	0	+130
	{Sn}	$\Delta \bar{H}_{Sn}$	1050	+1500	+1118
⟨Pb–Tl⟩	⟨Pb⟩⟨Tl⟩β	ΔH	523	0	−112
	⟨Tl⟩β	$\Delta \bar{H}_{Tl}$	523	−1170	−1075
{Pb–Tl}	{Pb} {Tl}	ΔH	773	0	−78
	{Tl}	$\Delta \bar{H}_{Tl}$	773	−820	−744
{Pb–Zn}	{Pb} {Zn}	ΔH	—	0	+535
	{Zn}	$\Delta \bar{H}_{Zn}$	—	—	+5010

F11β. **Pb.** Lead alloys: partial and integral

System	Components	Thermo-chemical function	Temp. (°K)	0·0	0·1
{Pb–Ag}	{Pb} {Ag}	ΔS^E	1273	0	+0·093
	{Ag}	$\Delta \bar{S}^E_{Ag}$	1273	+0·792	+1·025
{Pb–Au}	{Pb} {Au}	ΔS^E	1200	0	+0·261
	{Au}	$\Delta \bar{S}^E_{Au}$	1200	+2·345	+1·769
{Pb–Bi}	{Pb} {Bi}	ΔS^E	700	0	+0·02
	{Bi}	$\Delta \bar{S}^E_{Bi}$	700	+0·26	+0·20
{Pb–Cd}	{Pb} {Cd}	ΔS^E	773	0	+0·046
	{Cd}	$\Delta \bar{S}^E_{Cd}$	773	+0·466	+0·449
{Pb–Cu}	{Pb} {Cu}	ΔS^E	1473	0	+0·110
	{Cu}	$\Delta \bar{S}^E_{Cu}$	1473	+1·334	+0·918
{Pb–Hg}	{Pb} {Hg}	ΔS^E	600	0	−0·10
	{Hg}	$\Delta \bar{S}^E_{Hg}$	600	−1·18	−0·875
{Pb–In}	{Pb} {In}	ΔS^E	673	0	+0·059
	{In}	$\Delta \bar{S}^E_{In}$	673	+0·826	+0·49

binary metallic systems (*contd.*)
element named in the first column)
excess Gibbs energies of solution in cal/g-atom

0·3	0·5	0·7	0·9	Ref.
+711	+885	+696	+252	99
+1855	+891	+174	+1	99
+1593	+2236	+2129	+924	99
−73	−167	−189	−104	99
−550	−311	−153	−35	99
−210	−265	−215	−80	99
−550	−265	−70	−5	99
+502	+635	+585	+297	99
+1262	+751	+353	+69	99
+1338	+1607	+1422	+707	99
+3121	−1670	+824	+143	99
+788	+916	+857	+450	970
+1595	+1018	+576	+103	970
−18	+47	+111	+92	99
+156	+230	+174	+37	99
+250	+310	+270β	0	99
+186	+227	+195	+85	99
+449	+239	+90	+10	99
−3432	−4722	−4229	−1487	99
−9859	−7000	−1252	+62	99
−1470	−2244	−2289	−945	99
−4738	−3556	−1318	−35	99
−2688	−3883	−3916	−1614	99
−8200	−5700	−2750	+40	99
−1745	−2141	—	—	99
−4292	−2205	—	—	99
−8	−16	−12	−2	99
−45	−22	+3	+2	99
+285	+327	+274	+120	99
+615	+312	+122	+15	99
−313	−463	−494	—	99
−950	−700	−414	—	99
−200	−252	−211	−88	99
−520	−268	−86	−8	99
+1360	+1800	+1745	+930	161
+3605	+2340	+1140	+180	161

excess entropies of solution in cal/deg g-atom

0·3	0·5	0·7	0·9	Ref.
+0·305	+0·438	+0·375	+0·161	99
+1·003	+0·531	+0·152	+0·031	99
+0·396	+0·425	+0·357	+0·153	99
+0·565	+0·425	+0·153	+0·017	99
+0·05	+0·05	+0·04	+0·03	99
+0·10	+0·04	+0·03	+0·01	99
+0·124	+0·170	+0·174	+0·111	99
+0·345	+0·235	+0·143	+0·042	99
+0·225	+0·252	+0·234	+0·173	99
+0·461	+0·223	+0·216	+0·071	99
−0·23	−0·278	−0·234	−0·11	99
−0·543	−0·285	−0·098	−0·025	99
+0·119	+0·123	+0·084	+0·031	99
+0·221	+0·076	+0·009	0·0	99

Table F. Thermochemical data of
(N_2 is the atomic fraction of the second

F11β. **Pb.** Lead alloys: partial and integral

System	Components	Thermo-chemical function	Temp. (°K)	N_2 = 0·0	0·1
{Pb–K}	{Pb} {K}	ΔS^E	848	0	−0·447
	{K}	$\Delta \bar{S}^E_K$	848	−4·50	−4·415
{Pb–Mg}	{Pb} {Mg}	ΔS^E	973	0	−0·047
	{Mg}	ΔS^E_{Mg}	973	−0·925	−0·216
{Pb–Na}	{Pb} {Na}	ΔS^E	700	0	−0·199
	{Na}	$\Delta \bar{S}^E_{Na}$	700	−1·714	−2·231
{Pb–Sb}	{Pb} {Sb}	ΔS^E	905	0	+0·045
	{Sb}	$\Delta \bar{S}^E_{Sb}$	905	+0·533	+0·383
{Pb–Sn}	{Pb} {Sn}	ΔS^E	1050	0	−0·184
	{Sn}	$\Delta \bar{S}^E_{Sn}$	1050	−2·385	−1·401
⟨Pb–Tl⟩	⟨Pb⟩⟨Tl⟩β	ΔS^E	523	0	−0·138
	⟨Tl⟩β	$\Delta \bar{S}^E_{Tl}$	523	−1·663	−1·157
{Pb–Tl}	{Pb} {Tl}	ΔS^E	773	0	−0·055
	{Tl}	$\Delta \bar{S}^E_{Tl}$	773	−0·608	−0·49
{Pb–Zn}	{Pb} {Zn}	ΔS^E	—	0	+0·185
	{Zn}	$\Delta \bar{S}^E_{Zn}$	—	+2·2	+1·745

F12α. **Pt.** Platinum alloys: partial and integral heats or

System	Components	Thermo-chemical function	Temp. (°K)	0·0	0·1
⟨Pt–Au⟩	⟨Pt⟩⟨Au⟩	ΔH	—	0	+455
	⟨Au⟩	$\Delta \bar{H}_{Au.}$	—	+5690	+3585
⟨Pt–Co⟩	⟨Pt⟩⟨Co⟩	ΔG^E	1273	0	−1067
	⟨Co⟩	$\Delta \bar{G}^E_{Co}$	1273	−11,860	−9607
⟨Pt–Cr⟩ *	⟨Pt⟩⟨Cr⟩	ΔG^E	1773	0	−3140
	⟨Cr⟩	ΔG^E_{Cr}	1773	−35,100	−28,100
⟨Pt–Cu⟩	⟨Pt⟩⟨Cu⟩	ΔH	1350	0	−690
	⟨Cu⟩	$\Delta \bar{H}_{Cu}$	1350	−6870	−6844
⟨Pt–Fe⟩	⟨Pt⟩⟨Fe⟩γ	ΔG^E	1123	0	−1590
	⟨Fe⟩γ	$\Delta \bar{G}^E_{Fe}$	1123	−15,914	−15,914
⟨Pt–Ni⟩	⟨Pt⟩⟨Ni⟩	ΔG^E	1625	0	−631
	⟨Pt⟩⟨Ni⟩	ΔH	298	0	−762
	⟨Ni⟩	ΔG^E_{Ni}	1625	−6700	−5926
⟨Pt–Pd⟩	⟨Pt⟩⟨Pd⟩	ΔH	—	0	−292
	⟨Pd⟩	$\Delta \bar{H}_{Pd}$	—	−3020	−2805

F12β. **Pt.** Platinum alloys: partial and integral

System	Components	Thermo-chemical function	Temp. (°K)	0·0	0·1
⟨Pt–Au⟩	⟨Pt⟩⟨Au⟩	ΔS^E	—	0	−0·105
	⟨Au⟩	$\Delta \bar{S}^E_{Au}$	—	−0·90	−1·15
⟨Pt–Cu⟩	⟨Pt⟩⟨Cu⟩	ΔS^E	1350	0	+0·039
	⟨Cu⟩	$\Delta \bar{S}^E_{Cu}$	1350	+0·926	−0·042
⟨Pt–Ni⟩	⟨Pt⟩⟨Ni⟩	ΔS^E	1625	0	−0·081
⟨Pt–Pd⟩	⟨Pt⟩⟨Pd⟩	ΔS^E	—	0	−0·164
		$\Delta \bar{S}^E_{Pd}$	—	−1·70	−1·57

* ΔH measured at 298°K; $\Delta Cp = 0$, assumed.

binary metallic systems (*contd.*)
element named in the first column)
excess entropies of solution in cal/deg g-atom (contd.)

0·3	0·5	0·7	0·9	Ref.
−1·311	−2·123	−2·339	−0·814	99
−4·141	−5·00	−0·914	+0·067	99
−0·055	−0·138	−0·149	+0·136	99
−0·149	−0·340	+0·162	+0·168	99
−0·753	−1·409	−1·774	−0·703	99
−3·049	−2·917	−1·82	+0·123	99
+0·095	+0·106	+0·091	+0·043	99
+0·194	+0·099	+0·048	+0·007	99
−0·304	−0·244	−0·135	−0·035	99
−0·312	+0·01	+0·035	+0·007	99
−0·298	−0·358	−0·329	—	99
−0·654	−0·340	−0·226	—	99
−0·119	−0·130	−0·10	−0·062	99
−0·236	−0·098	−0·039	−0·035	99
+0·475	+0·66	+0·685	+0·405	161
+1·32	+0·94	+0·49	+0·07	161

excess Gibbs energies of solution in cal/g-atom

+795	+632	+265	0	958
+930	−160	−285	−60	958
−2490	−2965	−2490	−1067	99
−5811	−2965	−1067	−119	99
−7260	—	—	—	964
−16,650	—	—	—	964
−1980	−2650	−2482	−1228	99
−6044	−3130	−1540	−234	99
$-3276\gamma_3$	$-4344\gamma_2$	−3342	−1368	99
$-8924\gamma_3$	$-4166\gamma_2$	−929	−450	99
−1600	−2060	−1860	−852	99
−1730	−2214	−1867	−651	99
−4190	−2445	−990	−122	99
−773	−1032	−960	−451	958
−2130	−1310	−550	−70	958

excess entropies of solution in cal/deg g-atom

−0·35	−0·55	−0·58	−0·29	958
−1·205	−0·875	−0·41	−0·057	958
−0·125	−0·209	−0·179	−0·096	99
−1·034	−0·089	−0·106	−0·027	99
−0·079	−0·095	−0·004	+0·124	99*
−0·432	−0·595	−0·533	−0·250	958
−1·19	−0·725	−0·305	−0·04	958

References to Chapters I, II, III and IV

1. C. B. Griffith and M. W. Mallett, *J. Amer. Chem. Soc.* 1953 **75** 1823.
2. C. B. Alcock, *Nature* 1966 **209** 351.
3. J. H. McAteer and H. Seltz, *J. Amer. Chem. Soc.* 1936 **58** 2081.
4. O. Kubaschewski and O. Huchler, *Z. Elektrochem.* 1948 **52** 170.
5. E. A. Guggenheim, *Trans. Faraday Soc.* 1937 **33** 151, *Proc. Roy. Soc.* (A) 1935 **148** 304.
6. C. W. Bale and A. D. Pelton, *Metall. Trans.* 1974 **5** 2323.
7. W. Biltz and F. Meyer, *Z. anorg. allg. Chem.* 1928 **176** 23.
8. J. Lumsden, *Thermodynamics of Alloys*, Inst. Metals, 1952.
9. M. N. Dastur and J. Chipman, *J. Metals* 1949 **1** 441.
10. L. S. Darken and R. W. Gurry, *J. Amer. Chem. Soc.* 1845 **67** 1398; 1946 **68** 798.
11. K. K. Kelley, *US Bur. Mines, Bull.* No. 393, 1936.
12. O. Kubaschewski, A. Cibula and D. C. Moore, *Gases and Metals*, Metals and Metallurgy Trust, Iliffe, 1970, p. 17.
13. *Disc. Faraday Soc.* No. 4, 1948.
14. M. Rey, *Compt. rend.* 1949 **228** 473, 545.
15. E. Scheil, *Z. Elektrochem.* 1943 **49** 242.
16. M. Rey, see ref. 13, p. 257.
17. W. Slough, *Metallurgical Chemistry*, Proc. Symp. Brunel Univ. and NPL, 1971, HMSO, 1972, p. 311.
18. O. J. Kleppa, *J. Amer. Chem. Soc.* 1949 **71** 3275; 1950 **72** 3346; 1951 **73** 385; 1952 **74** 6037, 6052.
19. G. Scatchard, *Trans. Faraday Soc.* 1937 **33** 160.
20. O. Kubaschewski, *Trans. Faraday Soc.* 1949 **45** 931.
21. O. Kubaschewski and R. Hörnle, *Z. Metalk.* 1951 **42** 129.
22. O. Kubaschewski, *Metal Chemistry*, NPL Sympos. No. 9, 1958, HMSO, 1959, paper 3C.
23. C. Wagner, *Thermodynamics of Alloys*, Addison-Wesley, Cambridge, Mass., 1952.
24. L. S. Darken, *J. Amer. Chem. Soc.* 1950 **72** 2909.
25. See ref. 23.
26. R. Schuhmann, *Acta met.* 1950 **3** 219.
27. F. D. Richardson, *Physical Chemistry of Melts in Metallurgy*, Academic Press, London, 1974.
28. G. W. Toop, *Trans. Met. Soc. AIME* 1965 **233** 850.
29. F. Kohler, *Monatsh. Chemie* 1960 **91** 738.
30. E. Bonnier and R. Caboz, *Compt. rend.* 1960 **250** 527.
31. I. Ansara, *Metallurgical Chemistry*, Proc. Symp. Brunel Univ. and NPL 1971, HMSO, 1972, p. 403.
32. P. J. Spencer, F. H. Hayes and O. Kubaschewski, *Rev. chim. minér.* 1972 **9** 13.
33. K. Wohl, *Trans. AICE* 1946 **42** 215.
34. P. J. Spencer, F. H. Hayes and L. Elford, *Proc. Sympos. Chem. Metall. Iron and Steel*, Iron and Steel Inst., 1973, p. 322.
35. O. Kubaschewski and J. F. Counsell, *Monatsh. Chemie* 1971 **102** 1924.
36. C. B. Alcock and F. D. Richardson, *Acta met.* 1958 **6** 385; 1960 **8** 882.
37. K. T. Jacob and C. B. Alcock, *Acta met.* 1973 **21** 1011.
38. K. T. Jacob and C. B. Alcock, *Acta met.* 1972 **20** 221.
39. J. F. Martin, F. Müller and O. Kubaschewski, *Trans. Faraday Soc.* 1970 **66** 1065.
40. R. Hultgren, R. L. Orr, P. D. Anderson and K. K. Kelley, *Selected Values of the Thermodynamic Properties of Metals and Alloys*, Wiley, New York, 1973.
41. H. A. Skinner (ed.) *Experimental Thermochemistry*, vol. 2, Interscience, New York–London, 1962.
42. F. E. Wittig, *Metal Chemistry*, NPL Sympos. No. 9, 1958, HMSO, 1959, paper 1A.

43. W. C. HAGEL, G. M. POUND and R. F. MEHL, *Acta met.* 1965 **4** 37.
44. J. A. HALL, (a) *Fundamentals of Thermometry*; (b) *Practical Thermometry*, Institute of Physics, London, 1953.
45. British Standards Institution, *Bomb Calorimeter Thermometers*, BSS No. 791, London, 1938.
46. C. R. BARBER, *J. Sci. Instr.* 1950 **27** 47.
47. C. G. MAIER, *J. Phys. Chem.* 1930 **43** 2866.
48. O. KUBASCHEWSKI and A. WALTER, *Z. Elektrochem.* 1939 **45** 630, 732.
49. F. NEUMANN, *Z. anorg. Chem.* 1925 **145** 193.
50. E. LANGE and J. MONHEIM, *Z. phys. Chem.* 1930 **149** 51.
51. A. FERRIER and M. OLETTE, *Compt. rend.* 1962 **254** 4293; G. T. FURUKAWA, T. B. DOUGLAS, R. E. MCCOSKEY and D. C. GINNINGS, *J. Res. Nat. Bur. Stand.* 1956 **57** 67.
52. W. A. ROTH, *Thermochemie*, Berlin: Göschen, 1947.
53. E. R. VAN ARTSDALEN and K. P. ANDERSON, *J. Amer. Chem. Soc.* 1951 **73** 579.
54. O. KUBASCHEWSKI and W. WEBER, *Z. Elektrochem.* 1950 **54** 200.
55. G. W. MURPHY, *Rev. Sci. Instr.* 1949 **20** 372.
56. F. WEIBKE, *Z. anorg. Chem.* 1937 **232** 289.
57. R. S. JESSUP, *J. Res. Nat. Bur. Standards* 1938 **21** 475.
58. W. G. MARSHALL and J. E. RAYNER, *Fuel* 1949 **28** 243.
59. W. A. ROTH, *Z. Elektrochem.* 1935 **41** 112; 1939 **45** 325.
G. BECKER and W. A. ROTH, *Z. phys. Chem.* (A) 1932 **159** 1.
60. W. A. ROTH, U. WOLFF and O. FRITZ, *Z. Elektrochem.* 1940 **46** 42.
61. O. J. KLEPPA and T. YOKOKAWA, *J. Amer. Chem. Soc.* 1964 **86** 2749.
62. W. A. ROTH and O. DOEPKE, *Z. angew. Chem.* 1929 **42** 982.
63. L. D. MCGRAW, H. SELTZ and P. SNYDER, *J. Amer. Chem. Soc.* 1947 **69** 329.
64. G. HUFF, E. SQUITIERI and P. SNYDER, *J. Amer. Chem. Soc.* 1948 **70** 3380.
65. W. P. WHITE, *Amer. J. Sci.* 1919 **47** 2.
66. H. SIEMONSEN, *Z. Elektrochem.* 1939 **45** 637.
67. B. NEUMANN, C. KRÖGER and H. KUNZ *Z. anorg. Chem.* 1934 **218** 379.
68. W. KLEMM and I. SCHNICK, *Z. anorg. Chem.* 1936 **226** 353.
69. H. VON WARTENBERG, *Z. anorg. Chem.* 1942 **249** 100.
70. G. H. BRODIE, H. W. JENNINGS and A. HAYES, *Trans. Amer. Soc. Steel Treat.* 1926 **10** 615.
71. H. SIEMONSEN, *Z. Elektrochem.* 1951 **55** 327.
72. H. FEDER, *et al.*, Proc. IAEA Sympos., *Thermodyn. Nucl. Mat.*, Vienna, 1962, 155.
73. B. NEUMANN, C. KRÖGER and H. HAEBLER, *Z. anorg. Chem.* 1931 **196** 65.
74. R. JUZA and H. HAHN, *Z. anorg. Chem.* 1939 **241** 175.
75. J. THOMSEN, *Thermochemistry*, London, 1908.
76. C. H. SHOMATE and E. H. HUFFMANN, *J. Amer. Chem. Soc.* 1943 **65** 1625.
77. E. HOHMANN and H. BOMMER, *Z. anorg. Chem.* 1941 **248** 383.
78. T. J. BAKER, *Proc. Roy. Soc.* 1901 **68** 9.
79. M. HERSCHKOWITSCH, *Z. phys. Chem.* 1898 **27** 123.
80. W. BILTZ, W. WAGNER, H. PIEPER and W. HOLVERSCHEIDT, *Z. anorg. Chem.* 1924 **134** 25.
81. W. A. ROTH and P. CHALL, *Z. Elektrochem.* 1928 **34** 185.
82. B. NEUMANN, C. KRÖGER and H. KUNZ *Z. anorg. Chem.* 1932 **207** 133.
83. H. VON WARTENBERG, *Z. anorg. Chem.* 1943 **252** 141.
84. W. BILTZ and C. HAASE, *Z. anorg. Chem.* 1923 **129** 141.
85. W. A. ROTH, H. UMBACH and P. CHALL, *Arch. Eisenhüttenwes.* 1930 **4** 87.
86. W. A. ROTH, H. BERENDT and G. WIRTHS, *Z. Elektrochem.* 1941 **47** 185.
87. W. A. ROTH and H. TROITZSCH, *Arch. Eisenhüttenwes.* 1932/33 **6** 79.
88. W. BILTZ, G. ROHLFFS and H. U. VON VOGEL, *Z. anorg. Chem.* 1934 **220** 113.
89. N. PARRAVANO and P. DE CESARIS, *Gazz. chim. Ital.* 1917 **47** 144.
G. NAESER, *Mitt. KWI Eisenforsch. Düsseldorf* 1934 **16** 1.
90. R. FRICKE and E. DÖNGES, *Z. anorg. Chem.* 1942 **250** 202.
91. F. E. YOUNG, *J. Amer. Chem. Soc.* 1945 **67** 257.
92. C. H. SHOMATE, *J. Amer. Chem. Soc.* 1943 **65** 785.
93. R. FRICKE and W. ZERRWECK, *Z. Elektrochem.* 1937 **43** 52.
R. FRICKE and F. BLASCHKE, *Z. Elektrochem.* 1940 **46** 46.
94. W. F. GIAUQUE, *J. Amer. Chem. Soc.* 1949 **71** 3192.
95. M. A. FINEMAN and W. E. WALLACE, *J. Amer. Chem. Soc.* 1948 **70** 4165.
T. M. BUCK, W. E. WALLACE and R. M. RULON, *J. Amer. Chem. Soc.* 1952 **74** 136.
96. W. BILTZ and G. HOHORST, *Z. anorg. Chem.* 1922 **121** 1.
97. C. A. KRAUS and J. A. RIDDERHOF, *J. Amer. Chem. Soc.* 1834 **56** 79.
98. J. B. TAYLER, *Phil. Mag.* (5) 1900 **50** 37.

99. R. Hultgren et al., *Selected Values of the Thermodynamic Properties of Binary Alloys*, Amer. Soc. Met., Metals Park, Ohio (1973).
100. O. J. Kleppa, *Acta met.* 1960 **8** 435.
101. T. W. Richards and G. S. Forbes, *Z. phys. Chem.* 1907 **58** 683.
T. W. Richards and F. Daniels, *J. Amer. Chem. Soc.* 1919 **41** 1732.
102. J. C. Southard, *Ind. Eng. Chem.* 1940 **32** 442.
103. D. R. Torgeson and T. G. Sahama, *J. Amer. Chem. Soc.* 1948 **70** 2156.
104. O. J. Kleppa, *Thermochimie*, Coll. Int. CNRS, No. 201, 1972, p. 119.
105. J. L. Holm and O. J. Kleppa, *Amer. Mineral.* 1966 **51** 1608.
106. A. Navrotsky and O. J. Kleppa, *J. Inorg. Nucl. Chem.* 1968 **30** 479.
107. A. Navrotsky and O. J. Kleppa, *Inorg. Chem.* 1969 **8** 756.
108. F. Müller and O. J. Kleppa, *J. Inorg. Nucl. Chem.* 1973 **35** 2673.
109. J. H. Awberry and E. Griffiths, *Proc. Phys. Soc.* 1926 **38** 378.
110. F. M. Jaeger and E. Rosenbohm, *Recl. Trav. chim. Pays–Bas* 1928 **47** 513; 1932 **51** 1.
111. F. M. Jaeger, E. Rosenbohm and J. A. Bottema, *Kon. Akad. Wet. Amst. Proc.* 1932 **35** 347; *Recl. Trav. chim. Pays-Bas* 1933 **52** 61.
112. F. M. Jaeger and E. Rosenbohm, *Kon. Akad. Wetensch. Amst. Proc.* 1927 **30** 905.
113. J. C. Southard, *J. Amer. Chem. Soc.* 1941 **63** 3142.
114. M. Olette and A. Ferrier, *Metal Chemistry*, NPL Sympos. No. 9, 1958, HMSO, 1959, paper 4H.
115. V. Ya. Chekhovskoi, A. E. Sheindlin and B. Ya. Berezin, *High Temp.–High Press.* 1970 **2** 301.
116. B. Ya. Berezin, V. Ya. Chekhovskoi and A. E. Sheindlin, *High Temp.–High Press.* 1971 **3** 287.
117. H. Zeumer and W. A. Roth, *Z. physik. Chem.* (A) 1935 **173** 365.
118. O. Kubaschewski, *High Temp.–High Press.* 1972 **4**, Appendix, p. 12.
119. O. Kubaschewski and W. A. Dench, *Acta met.* 1955 **3** 339.
120. O. Kubaschewski, *Trans. Faraday Soc.* 1958 **54** 814.
121. F. Körber, W. Oelsen, W. Middel and H. Lichtenberg, *Stahl Eisen* 1936 **56** 1401.
122. W. Seith and O. Kubaschewski, *Z. Elektrochem.* 1937 **43** 743.
123. R. L. McKisson and L. A. Bromley, *J. Amer. Chem. Soc.* 1951 **73** 314.
124. H. von Wartenberg and R. Schütte, *Z. anorg. allg. Chem.* 1933 **211** 222.
125. W. Mielenz and H. von Wartenberg, *Z. anorg. allg. Chem.* 1921 **116** 267.
126. A. Bertram and W. A. Roth, *Z. physik. Chem.* (A) 1937 **178** 227.
127. O. Kubaschewski and O. von Goldbeck, NPL Rep., June 1950.
128. P. Gross, C. Hayman and D. L. Levi, *Trans. Faraday Soc.* 1955 **51** 626; 1954 **50** 477.
129. M. Kawakami, *Sci. Rep. Tôhoku Univ.* 1927 **16** 915 and *Z. anorg. Chem.* 1927 **167** 345.
130. F. E. Wittig and F. Huber, *Z. Elektrochem.* 1956 **60** 1181.
131. O. J. Kleppa, *J. Phys. Chem.* 1955 **59** 175.
132. L. B. Ticknor and M. B. Bever, *J. Metals* 1952 **4** 941.
133. R. L. McKisson and L. A. Bromley, *J. Metals* 1952 **4** 33.
134. W. Fischer and W. Biltz, *Z. anorg. Chem.* 1928 **176** 81.
135. J. Precht, *Ann. Phys.* (4) 1906 **21** 596.
136. H. Schröter, *Z. Metallk.* 1940 **32** 425.
137. D. C. Ginnings and R. J. Corruccini, *J. Res. Nat. Bur. Stand.* 1947 **38** 583.
138. T. Holmberg, *Soc. Sci. Fennica Comm. Phys.-Math.* 1938 **9** No. 17.
139. R. Hultgren, P. Newcombe, R. L. Orr and L. Warner, *Met. Chem.* NPL: Sympos. No. 9. 1958, HMSO, 1959, paper 1H.
140. G. T. Furukawa, D. C. Ginnings, R. E. McCoskey and R. A. Nelson, *J. Res. Nat. Bur. Stand.* 1951 **46** 195.
141. H. Moser, *Phys. Z.* 1936 **37** 737.
142. C. Sykes and F. W. Jones, *J. Inst. Met.* 1936 **59** 257.
N. Swindells and C. Sykes, *Proc. Roy. Soc.* (A) 1938 **168** 237.
143. I. Backhurst, *J. Iron Steel Inst.* 1958 **189** 124.
144. W. A. Dench, *Trans. Faraday Soc.* 1963 **59** 1279.
145. W. A. Dench and O. Kubaschewski, *J. Iron Steel Inst.* 1963 **201** 140.
146. T. W. Richards and H. S. Davis, *J. Amer. Chem. Soc.* 1917 **39** 341.
147. D. A. MacInnes and J. M. Braham, *J. Amer. Chem. Soc.* 1917 **39** 2110.
148. O. Kubaschewski and H. Villa, *Z. Elektrochem.* 1949 **53** 32.
149. A. Cezairliyan, *High Temp.–High Press.* 1969 **1** 517.
150. A. Cezairliyan, M. S. Morse, H. A. Berman and C. W. Beckett, *J. Res., Nat. Bur. Stand.* 1970 **74A** 65.
151. R. Parker, *Trans. Met. Soc. AIME* 1965 **233** 1545.
152. T. G. Kollie, *Rev. Sci. Instr.* 1967 **38** 1452.
153. A. Knappwost, *Z. Elektrochem.* 1943 **49** 1.
154. W. Oelsen, K. H. Rieskamp and O. Oelsen, *Arch. Eisenhüttenw.* 1955 **26** 253.

155. E. CALVET and H. PRAT, *Traité de Microcalorimetrie*, Masson, Paris, 1956.
156. M. LAFFITTE, *Metallurgical Chemistry*, Proc. Symp. Brunel Univ. and NPL, 1971, HMSO, 1972, p. 3.
157. P. GERDANIAN, *Thermochimie*, Proc. Int. Symp., Marseille, 1971, CNRS, 1972, p. 259.
158. G. BOUREAU, Thèse présentée a L'Université de Paris-Sud (1974).
159. O. KUBASCHEWSKI and R. HULTGREN, *Experimental Thermochemistry* (ed. H. A. Skinner), Interscience: New York, 1962 **2** 343.
160. H. KLINKHARDT, *Ann. Physik.* 1927 **84** 167.
161. F. E. WITTIG, *Z. Elektrochem.* 1850 **54** 288.
162. G. TAMMANN, *Z. anorg. Chem.* 1905 **43** 215, **47** 289.
G. D. ROOS, *Z. anorg. Chem.* 1916 **94** 329.
163. E. JENCKEL and H. WILSING, *Z. Elektrochem.* 1949 **53** 4.
W. PLATO, *Z. phys. Chem.* 1906 **55** 721.
164. P. W. BRIDGMAN, *Proc. Amer. Acad.* 1911 **47** 348; *Phys. Rev.* 1935 **48** 893.
165. J. NYSTRÖM, *Arkiv Fysik* 1949 **1** 359; 1950 **2** 151.
G. BORELIUS, *Bull. Soc. Chim.* 1949, p. 160.
166. R. FRICKE, *Z. Elektrochem.* 1948 **52** 72.
167. J. L. MARGRAVE (ed.), *The Characterization of High Temperature Vapors*, Wiley: New York, 1967.
168. R. A. RAPP (ed), *Physicochemical Measurements in Metals Research*, Interscience, 1970.
169. C. H. M. JENKINS, *Proc. Roy. Soc.* (A) 1926 **110** 227.
170. W. FISCHER, R. GEWEHR and H. WINGCHEN, *Z. anorg. allg. Chem.* 1939 **242** 161.
171. L. S. BROOKS, *J. Amer. Chem. Soc.* 1952 **74** 227.
172. R. E. MACHOL and E. F. WESTRUM, *J. Amer. Chem. Soc.* 1958 **80** 2950.
173. L. L. HIRST and A. R. OLSON, *J. Amer. Chem. Soc.* 1929 **51** 2398.
174. F. E. POINDEXTER, *Phys. Rev.* 1926 **28** 208.
175. A. W. HERBENAR, C. A. SIEBERT and O. S. DUFFENDACK, *J. Metals* 1950 **2** 323.
176. P. A. RICE and D. V. RAGONE, *J. Chem. Phys.* 1965 **42** 701; 1966 **45** 4141.
177. R. F. BREBRICK and A. J. STRAUSS, *J. Chem. Phys.* 1964 **40** 3230; **41** 197.
178. A. S. PASHINKIN and A. V. NOVOSELOVA, *Russ. J. Inorg. Chem.* 1959 **4** 1229.
179. Y. M. NESTEROVA, A. S. PASHINKIN and A. V. NOVOSELOVA, *Russ. J. Inorg. Chem.* 1961 **6** 1031.
180. R. HARGREAVES, *J. Inst. Metals* 1939 **64** 115.
181. P. HERASYMENKO, *Acta met.* 1956 **4** 1.
182. K. L. KOMAREK and M. SILVER, *Thermodynamics of Nuclear Materials*, Proc. IAEA Sympos., Vienna, 1962, p. 749.
183. F. D. RICHARDSON and L. E. WEBB, *Trans. IMM* 1955 **64** 529.
184. W. BILTZ and R. JUZA, *Z. anorg. Chem.* 1930 **190** 161.
185. M. BODENSTEIN, *Z. Elektrochem.* 1908 **14** 544.
F. M. G. JOHNSON, *Z. phys. Chem.* 1908 **61** 457.
186. C. ROBERT, *Helv. Phys. Acta* 1936 **9** 405.
187. H. HARALDSEN, *Skr. norske Vidensk Akad. I. Math.-naturw. Kl.* 1932 No. **9**.
188. W. FRANKE, R. JUZA, K. MEISEL and W. BILTZ, *Z. anorg. Chem.* 1934 **218** 346.
189. W. BILTZ and M. HEIMBRECHT, *Z. anorg. Chem.* 1939 **241** 349.
190. W. FISCHER and W. BILTZ, *Z. anorg. Chem.* 1927 **166** 290; *ibid.* 1928 **176** 81.
191. R. BRUNNER, *Z. Elektrochem.* 1932 **38** 55.
192. A. SIEVERTS, *Z. phys. Chem.* 1907 **60** 129.
193. M. WEINSTEIN and J. F. ELLIOTT, *Trans. Met. Soc. AIME* 1963 **227** 382.
194. J. FISCHER, *Z. anorg. allg. Chem.* 1934 **219** 1.
195. O. RUFF and B. BERGDAHL, *Z. anorg. allg. Chem.* 1919 **106** 76.
196. W. FISCHER and O. RAHLFS, *Z. anorg. allg. Chem.* 1932 **205** 1.
197. W. LEITGEBEL, *Z. anorg. allg. Chem.* 1931 **202** 305.
198. G. GATTOW and A. SCHNEIDER, *Angew. Chem.* 1959 **71** 189.
199. E. BAUER and R. BRUNNER, *Helv. chim. acta* 1934 **17** 958.
200. J. L. BARTON and H. BLOOM, *J. Phys. Chem.* 1956 **60** 1413; 1959 **63** 1785.
201. C. B. ALCOCK and G. W. HOOPER, *Proc. Roy. Soc.* (A) 1960 **254** 551.
202. U. MERTENS, *J. Phys. Chem.* 1959 **63** 443.
203. H. von WARTENBERG, *Z. Elektrochem.* 1913 **19** 482.
204. K. JELLINEK and G. A. ROSNER, *Z. physik. Chem.* (A) 1929 **143** 51; 1931 **152** 67.
205. A. SCHNEIDER and E. K. STOLL, *Z. Elektrochem.* 1941 **47** 519.
206. W. FISCHER and R. GEWEHR, *Z. anorg. allg. Chem.* 1932 **209** 17.
207. J. P. MORRIS and G. R. ZELLARS, *J. Metals* 1956 **8** 1086.
208. J. B. BOOKEY, *J. Iron Steel Inst.* 1952 **172** 61.
209. J. A. KITCHENER and S. IGNATOWICZ, *Trans. Faraday Soc.* 1951 **47** 1278.
210. H. SCHÄFER and A. TEBBEN, *Z. anorg. allg. Chem.* 1960 **304** 317.

211. A. SCHNEIDER and U. ESCH, *Z. Elektrochem.* 1943 **49** 55.
212. R. SCHENCK, *Z. anorg. Chem.* 1927 **164** 145.
R. SCHENCK, F. KURZEN and H. WESSELKOCK, *Z. anorg. Chem.* 1931 **203** 159.
213. R. SCHENCK, K. MEYER and K. MAYER, *Z. anorg. Chem.* 1939 **243** 17, 259.
214. P. H. EMMETT, S. B. HENDRICKS and S. BRUNAUER, *J. Amer. Chem. Soc.* 1930 **52** 1456.
215. E. D. EASTMAN and R. M. EVANS, *J. Amer. Chem. Soc.* 1924 **46** 888.
216. P. H. EMMETT and J. F. SHULTZ, *J. Amer. Chem. Soc.* 1929 **51** 3249.
217. Z. SHIBATA and H. KITAGAWA, *J. Fac. Sci. Hokk. Univ.* Ser. III 1938 **2** 223.
218. E. WHALLEY, *J. Chem. Phys.* 1951 **19** 509.
219. O. KUBASCHEWSKI and O. VON GOLDBECK, *Trans. Faraday Soc.* 1949 **45** 948.
220. R. N. PEASE and R. S. COOK, *J. Amer. Chem. Soc.* 1926 **48** 1199.
221. N. A. GOKCEN and J. CHIPMAN, *J. Metals* 1952 **4** 171.
222. M. N. DASTUR and N. A. GOKCEN, *J. Metals* 1949 **1** 665.
223. A. MCLEAN, *High Temp.–High Press.* 1974 **6** 21.
224. H. VON WARTENBERG and S. AOYAMA, *Z. Elektrochem.* 1927 **33** 144.
225. G. GRUBE and M. FLAD, *Z. Elektrochem.* 1939 **45** 835.
G. GRUBE and K. RATSCH, *Z. Elektrochem.* 1939 **45** 838.
226. G. GRUBE and M. FLAD, *Z. Elektrochem.* 1942 **48** 377.
227. P. CHICHE, *Ann. Chimie* 1952 **7** 361.
228. H. PÉLABON, *Ann. Chim. Phys.* (7) 1902 **25** 365.
229. F. G. KEYES and W. A. FELSING, *J. Amer. Chem. Soc.* 1920 **42** 246.
230. M. WATANABE, *Sci. Rep. Tôhoku Univ.* 1933 **22** 407, 902.
231. R. SCHENCK and P. VON DER FORST, *Z. anorg. allg. Chem.* 1939 **241** 145.
232. C. W. SHERMAN, H. I. ELVANDER and J. CHIPMAN, *J. Metals* 1950 **2** 334; 1952 **4** 597.
233. T. ROSENQVIST, *J. Metals* 1949 **1** 451.
234. A. H. JOUNIAUX, *Compt. rend.* 1899 **129** 283; 1901 **132** 1270; 1903 **136** 1006.
235. F. ISHIKAWA and M. WATANABE, *Sci. Rep. Tôhoku Univ.* 1933 **22** 393.
236. K. JELLINEK and K. UHLOHT, *Z. anorg. allg. Chem.* 1926 **151** 157.
237. K. JELLINEK and K. UHLOHT, *Z. physik. Chem.* 1926 **119** 161.
238. K. JELLINEK and A. RUDAT, *Z. anorg. allg. Chem.* 1928 **175** 281.
239. K. SCHWERDTFEGER and R. T. TURKDOGAN, in *Physicochemical Measurements in Metals Research* (A. Rapp, ed.), Interscience, 1970 **I** 321.
240. E. W. DEWING and F. D. RICHARDSON, *Trans. Faraday Soc.* 1959 **55** 611.
241. H. H. KELLOGG, *Trans. Met. Soc. AIME* 1964 **230** 1622.
242. T. R. INGRAHAM and H. H. KELLOGG, *Trans. Met. Soc. AIME* 1963 **227** 1419.
243. M. KNUDSEN, *Ann. Physik.* 1909 **28** 75; 1909 **29** 179.
244. I. LANGMUIR, *Phys. Rev.* (2) 1913 **2** 329.
245. A. W. SEARCY and R. D. FREEMAN, *J. Chem. Phys.* 1954 **22** 762.
246. E. A. GULBRANSEN and K. F. ANDREW, *Trans. Electrochem. Soc.* 1942 **81** 327.
247. O. KUBASCHEWSKI and T. G. CHART, *J. Chem. Thermodyn.* 1974 **6** 467.
248. H. M. SCHADEL and C. E. BIRCHENALL, *J. Metals* 1950 **2** 1134.
H. M. SCHADEL, G. DERGE and C. E. BIRCHENALL, *J. Metals* 1950 **2** 1282.
249. O. KUBASCHEWSKI and G. HEYMER, *Acta met.* 1960 **8** 416.
O. KUBASCHEWSKI, G. HEYMER and W. A. DENCH, *Z. Elektrochem.* 1960 **64** 801.
250. P. FESCHOTTE and O. KUBASCHEWSKI, *Trans. Faraday Soc.* 1964 **60** 1941.
251. J. W. WARD, R. N. R. MULFORD and M. KAHN, *J. Chem. Phys.* 1967 **47** 1710.
252. J. W. WARD, R. N. R. MULFORD and R. L. BIVINS, *J. Chem. Phys.* 1967 **47** 1718.
253. E. K. STORMS, *High Temp. Sci.* 1969 **1** 456.
254. J. BERKOWITZ and W. A. CHUPKA, *Trans. NY Acad. Sci.* 1960 **79** 1073.
255. A. KANT, *J. chem. Phys.* 1966 **44** 2450.
256. G. R. BELTON and R. J. FRUEHAN, *J. Phys. Chem.* 1967 **71** 1403.
257. A. NECKEL and S. WAGNER, *Ber. Bunsenges. Phys. Chem.* 1969 **73** 210.
258. G. SODECK, P. ETNER and A. NECKEL, *High Temp. Sci.* 1970 **2** 311.
259. C. B. ALCOCK, R. SRIDHAR and R. C. SVEDBERG, *J. Chem. Thermodyn.* 1970 **2** 255.
M. CARBONEL, C. BERGMAN and M. LAFFITTE, *Thermochimie, Colloques Internationaux du CNR,* 1972, No. 201, 311.
259a. M. CARBONEL, C. BERGMAN and M. LAFFITTE, *Thermochimie, Colloques Internationaux du CNR,* 1972, No. 201, 311.
260. J. P. HAGER, S. M. HOWARD and J. H. JONES, *Met. Trans.* 1973 **4** 2383.
261. F. E. STAFFORD, *High Temp.-High Press.* 1971 **3** 213.
262. A. L. MARSHALL, R. W. DORNTE and F. J. NORTON, *J. Amer. Chem. Soc.* 1937 **59** 1161.
263. E. A. GULBRANSEN and K. F. ANDREW, *J. Electrochem. Soc.* 1950 **97** 383.

264. L. H. Dreger and J. L. Margrave, *J. Phys. Chem.* 1960 **64** 1323.
265. R. P. Burns, *J. Chem. Phys.* 1966 **44** 3307.
266. E. K. Rideal and R. Littlewood, *Trans. Faraday Soc.* 1956 **52** 1598.
267. E. G. Wolff and C. B. Alcock, *Trans. Brit. Ceram. Soc.* 1962 **61** 667.
268. P. Gross, C. S. Campbell, P. J. C. Kent and D. Levi, *Disc. Faraday Soc.* No. **4** 1948 206.
269. C. B. Alcock and P. Grieveson, *J. Inst. Metals* 1962 **90** 304.
270. W. H. Rodebush and A. O. Dixon, *Phys. Rev.* 1925 **26** 85.
271. W. Leitgebel, *Z. anorg. allg. Chemie* 1922 **202** 305.
272. K. C. Mills and K. Kinoshita, *J. Chem. Thermodynamics* 1973 **5** 129.
273. K. C. Mills, K. Kinoshita and P. Grieveson, *J. Chem. Thermodynamics* 1972 **4** 581.
274. G. R. Zellars, S. L. Payne, J. P. Morris and R. L. Kipp, *Trans. AIME* 1959 **215** 181.
275. R. Speiser, A. J. Jacobs and J. W. Spretnak, *Trans. AIME* 1959 **215** 185.
276. M. Volmer, *Z. physik. Chem., Bodenstein Festb.* 1931 863.
277. J. N. Pratt and A. T. Aldred, *J. Sci. Instrum.* 1959 **36** 465.
278. D. L. Hildenbrand and W. F. Hall, *J. Phys. Chem.* 1962 **66** 754.
279. A. W. Searcy and R. D. Freeman, *J. Amer. Chem. Soc.* 1955 **76** 5229.
280. M. Peleg and C. B. Alcock, *J. Sci. Instrum.* 1966 **43** 558.
281. K. Kiukkola and C. Wagner, *J. Electrochem. Soc.* 1957 **104** 379.
282. H. Schmalzried, *Z. Elektrochem.* 1962 **66** 572.
283. B. C. H. Steele and C. B. Alcock, *Trans. AIME* 1965 **233** 1359.
284. R. A. Rapp and D. A. Shores, *Physicochemical Measurements in Metals Research* (R. A. Rapp, ed.), Interscience, 1970 **2** 123.
285. K. T. Jacob and C. B. Alcock, *Met. Trans. B* 1975 **6** 215.
286. G. G. Charette and S. N. Flengas, *J. Electrochem. Soc.* 1968 **115** 796.
287. R. A. Rapp, *Trans. AIME* 1963 **227** 371.
288. W. A. Dench and O. Kubaschewski, *High Temp.-High Press.* 1969 **1** 357.
289. F. E. Rizzo, R. E. Bidwell and D. F. Frank, *Trans. AIME* 1967 **239** 593.
290. W. A. Fischer, *Z. Naturforsch.* 1967 **22A** 1575.
291. K. Goto, T. Ito and M. Someno, *Trans. AIME* 1969 **245** 1662.
292. R. J. Ruka, J. E. Bauerle and L. Dykstra, *J. Electrochem. Soc.* 1968 **115**, 497.
293. N. Cusack and P. Kendall, *Proc. Phys. Soc.* (*London*) 1958 **72** 898.
294. R. A. Rapp and F. Maak, *Acta met.* 1962 **10** 63.
295. T. N. Rezukhina and Z. V. Proshina, *Russ. J. Phys. Chem.* 1962 **36** 333.
296. A. Kubik and C. B. Alcock, *Metal Sci. J.* 1967 **1** 19.
297. Y. D. Tretyakov and R. A. Rapp. *Trans. AIME* 1969 **242** 1235.
298. S. Zador and C. B. Alcock, *J. Chem. Thermodyn.* 1970 **2** 9.
299. F. Haber and A. Moser, *Z. Elektrochem.* 1905 **11** 593.
300. C. B. Alcock and J. C. Chan, *Can. Met. Quart.* 1972 **11** 559.
301. A. W. Smith, F. W. Meszaros and C. D. Amata, *J. Amer. Ceram. Soc.* 1965 **49** 240.
302. H. R. Larson and J. F. Elliott, *Trans. AIME* 1967 **239** 1713.
303. C. B. Alcock and T. N. Belford, *Trans. Faraday Soc.* 1964 **60** 822.
304. C. Gatellier and M. Olette, *C. r. Acad. Sci.* 1968 **266** 1133.
305. R. J. Fruehan, L. J. Martonik and E. T. Turkdogan, *Trans. AIME* 1967 **239** 1276.
306. G. G. Charette and S. N. Flengas, *Can. Met. Quart.* 1968 **7** 191.
307. M. L. Kapoor and M. G. Frohberg, *Arch. Eisenhüttenw.* 1971 **42** 5.
308. Z. Kozuka and C. S. Samis, *Met. Trans.* 1970 **1** 871.
309. Y. Matsushita and K. Goto, *Tetsu-to-Hagane* (overseas) 1964 **4** 128.
310. R. W. Ure, *J. Chem. Phys.* 1957 **26** 1363.
311. T. N. Rezukhina, T. F. Sislova, L. I. Holokhonova and E. G. Ippolitov, *J. Chem. Thermodyn.* 1974 **6** 883.
312. H. Kleykamp, *Ber. Bunsenges. Phys. Chem.* 1969 **73** 354.
313. M. Kanno, *J. Nucl. Mat.* 1974 **51** 24.
314. J. J. Egan, *J. Phys. Chem.* 1964 **68** 978.
315. S. Aronson, *Compounds of Interest in Nuclear Reactor Technology*, 1964, AIME Special Report No. 13 247.
316. S. Aronson and A. Auskern, *Thermodynamics,* Vol. I, 1966, IAEA, Vienna, p. 165.
317. T. N. Rezukhina and B. S. Pokarev, *J. Chem. Thermodyn.* 1971 **3** 369.
318. M. S. Whittingham and R. A. Huggins, *J. Chem. Phys.* 1971 **54** 414.
319. T. L. Francis, F. E. Phelps and G. Maczura, *Bull. Amer. Ceram. Soc.* 1971 **54** 414.
320. B. V. Joglekar, P. S. Nicholson and W. W. Smeltzer, *Can. Met. Quart.* 1973 **12** 155.
321. I. Wynne-Jones and L. J. Miles, *Proc. Brit. Ceram. Soc.* 1971 **19** 161.
322. D. J. Fray and B. Savory, *J. Chem. Thermodyn.* 1975 **7** 485.

323. K. HAUFFE and A. L. VIERK, *Z. Elektrochem.* 1949 **53** 151.
324. T. W. DAKIN and D. T. EWING, *J. Amer. Chem. Soc.* 1940 **62** 2280.
325. R. G. BATES, *J. Amer. Chem. Soc.* 1939 **61** 1040.
326. T. W. RICHARDS and G. S. FORBES, Publ. Carnegie Inst. No. 56 and *Z. phys. Chem.* 1907 **58** 683.
327. N. ELLIOT and D. M. YOST, *J. Amer. Chem. Soc.* 1934 **56** 1057 2797.
328. R. MÜLLER, H. KUMPFMILLER, E. PINTER and B. VON SEEBACH, *Monatsh. Chem.* 1933 **63** 317.
329. T. W. RICHARDS and C. P. SMYTH, *J. Amer. Chem. Soc.* 1922 **44** 524.
C. WAGNER and G. ENGELHARDT, *Z. phys. Chem.* (A) 1932 **159** 241.
330. H. SELTZ, *Trans. Electrochem. Soc.* 1940 **77** 233.
331. R. W. MILLAR, *J. Amer. Chem. Soc.* 1927 **49** 3003.
331a. J. F. ELLIOTT and J. CHIPMAN, *Trans Faraday Soc.* 1951 **47** 139.
332. G. N. LEWIS and F. G. KEYES, *J. Amer. Chem. Soc.* 1912 **34** 119.
G. N. LEWIS and C. A. KRAUS, *J. Amer. Chem. Soc.* 1910 **32** 1459.
333. K. HAUFFE, *Z. Elektrochem.* 1940 **46** 348.
334. E. WARBURG, *Wied. Ann. Phys.* 1884 **21** 622.
335. A. L. VIERK, *Z. Elektrochem.* 1950 **54** 436.
336. T. W. RICHARDS and J. B. CONANT, *J. Amer. Chem. Soc.* 1922 **44** 601.
337. A. ÖLANDER, *Z. Metallk.* 1937 **29** 361.
338. F. WEIBKE and U. VON QUADT, *Z. Elektrochem.* 1939 **45** 715.
339. F. WEIBKE and H. MATTHES, *Z. Elektrochem.* 1941 **47** 421.
340. A. WACHTER, *J. Amer. Chem. Soc.* 1932 **54** 4609.
341. O. KUBASCHEWSKI, *Z. Elektrochem.* 1936 **42** 5.
342. R. LORENZ and H. VELDE, *Z. anorg. Chem.* 1929 **183** 81.
343. G. GRUBE and E. A. RAU, *Z. Elektrochem.* 1934 **40** 352.
344. F. OPPENHEIMER, *Z. anorg. Chem.* 1930 **189** 297.
345. B. A. ROSE, G. J. DAVIS and H. J. T. ELLINGHAM, *Disc. Faraday Soc.* 1948 No. **4** 154.
346. F. WEIBKE and I. EFINGER, *Z. Elektrochem.* 1940 **46** 61.
347. C. WAGNER and A. WERNER, *J. Electrochem. Soc.* 1963 **110** 326.
348. *JANAF Thermochemical Tables*, US Dept. of Commerce, National Bureau of Standards, 1965–8.
349. H. H. KELLOGG, *Proc. 1st Conf. on Thermodynamic Properties of Materials* (1967) (G. R. Fitterer, ed.) University of Pittsburgh, p. 357.
350. O. KUBASCHEWSKI and H. ÜNAL, *High Temp.–High Press.* 1977 **9** 361.
351. K. K. KELLEY, US Bureau of Mines, Bull. No. 371, 1934.
352. P. SCHÜBEL, *Z. anorg. Chem.* 1914 **87** 81.
353. R. MEZAKI, E. W. TILLEUX, D. W. BARNES and J. L. MARGRAVE, *Thermodynamics of Nuclear Materials*, 1962, Proc. IAEA Symposium Vienna, p. 775.
354. F. FÖRSTER and G. TSCHENDTKE, *Z. Metallkunde* 1940 **32** 191.
355. G. TAMMANN, *Z. physik. Chem.* 1913 **85** 273.
356. O. KUBASCHEWSKI, *Z. Elektrochem.* 1941 **47** 475.
O. KUBASCHEWSKI, *Z. physik. Chem.* 1943 **192** 292.
357. O. KUBASCHEWSKI and F. WEIBKE, *Z. Metallkunde* 1938 **30** 325.
358. I. JOHNSON, *Cpds. of Interest Nucl. Reactor Tech.*, Met. Soc. AIME 1964, p. 171.
359. W. M. LATIMER, *J. Amer. Chem. Soc.* 1951 **73** 1480.
360. K. K. KELLEY, US Bureau of Mines, Bull No. 350, 1932.
361. O. KUBASCHEWSKI, *High Temp.–High Press.* 1972 **4** 1.
362. F. D. RICHARDSON, *J. Iron Steel Inst.* 1949 **163** 382.
363. W. L. WORRELL, *J. Phys. Chem.* 1960 **68** 954.
364. F. GRONVOLD and E. F. WESTRUM, *Inorg. Chem.* 1962 **1** 36.
365. C. KRÖGER, *Z. anorg. Chem.* 1934 **218** 396.
366. P. SÜE, *J. Chim. Phys.* 1945 **42** 45.
367. F. TROMBE, *Comptes Rend.* 1944 **218** 457.
368. A. F. KAPUSTINSKI, *Izv. est. Akad. Nauk SSSR Otdel Khim Nauk* 1948 pp. 568, 589.
369. W. GORDON and W. J. O. THOMAS, *J. Chem. Phys.* 1956 **24** 439.
370. L. BREWER, L. A. BROMLEY, P. W. GILLES and N. L. LOFGREN, *Nat. Nucl. Sci. Ser. Div. IV* 1949 **19B** 76.
371. O. KUBASCHEWSKI, *Z. Elektrochem.* 1941 **47** 623.
372. T. W. RICHARDS, *Z. physik. Chem.* 1902 **40** 597; 1904 **49** 15.
373. O. KUBASCHEWSKI, *Z. Elektrochem.* 1941 **47** 623.
374. O. KUBASCHEWSKI, *Metal Chemistry*, NPL Symposium No. 9, 1958, HMSO, 1959, paper 3C.
375. O. KUBASCHEWSKI and G. HEYMER, *Trans. Faraday Soc.* 1960 **56** 473.
376. A. SCHNEIDER and G. HEYMER, *Metal Chemistry*, NPL. Symposium No. 9, 1958, HMSO, 1959, paper 4A.
377. C. B. ALCOCK, J. B. CORNISH and P. GRIEVESON, *Thermodynamics,* Proc. IAEA Symposium, Vienna, 1966, **I** 211.

378. R. WALTER, *Z. Metallkund* 1921 **13** 225.
379. G. MASING, *Z. Anorg. Chem.* 1909 **62** 265.
380. O. KUBASCHEWSKI, unpublished work.
381. G. TAMMANN and H. O. VON SAMSON-HIMMELSTJERNA, *Z. anorg. Chem.* 1934 **216** 288.
382. T. G. CHART and O. KUBASCHEWSKI, *Metallurgical Chemistry*, Proc. Symp. Brunel Univ. and NPL Symposium, 1971, HMSO, 1972, p. 567.
383. M. H. RAND and O. KUBASCHEWSKI, *The Thermochemical Properties of Uranium Compounds*, Oliver & Boyd, Edinburgh, 1963.
384. J. HARPER and A. E. WILLIAMS, *Extraction and Refining of the Rarer Metals*, Inst. Min. Metall. 1957, p. 143.
385. H. H. KELLOGG, *Trans. AIME* 1950 **188** 862.
386. C. F. GOODEVE, *Disc. Faraday Soc.* 1948 **4** 9.
387. A. SCHNEIDER and E. HESSE, *Z. Elektrochem.* 1940 **46** 279.
387a. A. SCHNEIDER, *Met. u. Erz* 1942 **39** 272, 292.
387b. F. RITTER and G. JAECKEL, *Chem. Technologie* (Ed. K. Winnacker and L. Küchler), C. Hansen: München, 1961 **5** 106.
388. A. SCHNEIDER and H. EBERT, *Z. Metallk.* 1950 **41** 208.
389. L. M. PIDGEON and J. A. KING, *Disc. Faraday Soc.* 1948 **4** 197.
390. J. CHIPMAN, *J. Amer. Chem. Soc.* 1961 **83** 1762.
391. C. B. ALCOCK and J. H. E. JEFFES, *Trans. Inst. Min. Metall.* 1967 **76** C246.
392. J. H. E. JEFFES and C. B. ALCOCK, *J. Materials Science* 1968 **3** 635.
393. N. C. TOMBS and A. J. E. WELCH, *J. Iron Steel Inst.* 1952 **172** 69.
394. G. GRUBE and H. SPEIDEL, *Z. Elektrochem.* 1949 **53** 339, 541.
395. H. F. RAMSTAD and F. D. RICHARDSON, *Trans. Met. Soc. AIME* 1961, **221** 1021.
396. H. SCHÄFER and R. HÖRNLE, *Z. anorg. allg. Chem.* 1950 **263** 261.
397. O. KUBASCHEWSKI and T. G. CHART, *J. Chem. Thermodyn.* 1974 **6** 467.
398. P. V. GEL'D and M. I. KOCHNEV, *Zh. Priklad. Khim.* 1948 **21** 1249.
399. H. VON WARTENBERG, *Z. Elektrochem.* 1949 **53** 343.
400. E. ZINTL, *Z. anorg. Allg. Chem.* 1940 **245** 1.
401. K. HACK, Diplomarbeit, 1976. Lehrstuhl für theoretische Hüttenkunde, Tech. Hochschule, Aachen.
402. O. KUBASCHEWSKI and L. E. H. STUART, *J. Chem. Eng. Data* 1967 **12** 418.
403. M. B. PANISH, *et al*, *J. Phys. Chem.* 1958 **62** 980.
404. G. R. FITTERER and L. A. PUGLIESE, *Met. Trans.* 1970 **1** 1997.
405. F. N. MAZANDARANY and R. D. PEHLKE, *Met. Trans.* 1973 **4** 2067.
406. E. L. FLORES-MAGON and M. VIKRAM RAO, *Met. Trans.* 1971 **2** 2471.
407. G. GRUBE and M. FLAD, *Z. Elektrochem.* 1942 **48** 377.
408. R. J. FRUEHAN, *Trans. Met. Soc. AIME* 1968 **242** 2007.
409. S. W. GILBY and G. R. ST. PIERRE, *Trans. Met. Soc. AIME* 1969 **245** 1749.
410. O. KUBASCHEWSKI and T. G. CHART, *J. Inst. Metals* 1965 **93** 329.
411. R. O. WILLIAMS, *Trans. Met. Soc. AIME* 1958 **212** 497.
412. A. F. COOK and F. W. JONES, *J. Iron Steel Inst.* 1943 **148** 217.
413. L. KAUFMAN and H. BERNSTEIN, *Computer Calculation of Phase Diagrams*, Academic Press, N. Y. 1970.
414. I. ANSARA, *Metallurgical Chemistry*, Proc. Sympos. Brunel Univ. and NPL, 1971, HMSO, 1972, p. 403.
415. O. KUBASCHEWSKI and J. F. COUNSELL, *Monatsh. Chemie* 1971 **102** 1724.
416. E. RAUB and G. WÖRWAG, *Z. Metallkunde* 1955 **46** 513.
417. J. F. COUNSELL, E. B. LEES and P. J. SPENCER, *Metallurgical Chemistry*, Proc. Sympos. Brunel Univ. and NPL, 1971, HMSO, 1972, p. 451.
418. J. F. COUNSELL, E. B. LEES and P. J. SPENCER, *Metal. Sci. J.* 1971 **5** 210.
419. G. R. FITTERER, *Proc. Open Hearth Conf. Pittsburgh*, 1957 **40** 281.
419a. O. REPETYLO, M. OLETTE and P. KOZAKEVITCH, *J. Metals* 1967 (May), p. 45.
419b. F. FOERSTER and H. RICHTER, *Radex Rundschau* 1969 (2), p. 518.
420. B. E. HOPKINS, *Metall. Rev.* 1956 **1** 117.
421. O. KUBASCHEWSKI and J. A. CATTERALL, *Thermochemical Data of Alloys*, Pergamon Press, 1956.
422. H. A. SLOMAN and T. E. ROONEY, *J. Iron Steel Inst.* 1943 **148** 327P.
423. B. E. HOPKINS, G. C. H. JENKINS and H. E. N. STONE, *J. Iron Steel Inst.* 1951 **168** 377.
424. W. P. REES, *J. Iron Steel Inst.* 1958 **188** 351.
425. E. W. WASHBURN and E. N. BUNTING, *Bur. Stand. J. Res.* 1934 **12** 239.
426. D. T. PETERSON and R. KONTRIMAS, *J. Phys. Chem.* 1960 **64** 362.
427. F. D. RICHARDSON and W. E. DENNIS, *J. Iron Steel Inst.* 1953 **175** 257.
428. W. A. KRIVSKY, *Met. Trans.* 1973 **4** 1439.
429. O. NYQUIST, K. W. K. LANGE and J. CHIPMAN, *Trans. Met. Soc. AIME* 1962 **224** 714.

430. F. Adcock, *J. Iron Steel Inst.* 1937, *No.* 1, 281P.
431. J. C. Chan, C. B. Alcock and K. T. Jacob, *Can. Met. Quart.* 1973, **12** 439.
432. W. A. Fischer and A. Hoffmann, *Arch. Eisenhüttenw.* 1958 **29** 339.
433. O. Kubaschewski, *Z. Elektrochem.* 1943 **49** 446.
434. O. Kubaschewski and O. von Goldbeck, *J. Inst. Metals* 1949 **76** 255.
435. F. Müller and O. Kubaschewski, *High Temp.–High Press.* 1969, **1** 543.
436. J. Moreau, *Compt. Rend.* 1953 **236** 85.
437. A. Rahmel, *Z. Elektrochem.* 1962 **66** 363.

References to Chapter V

1. G. B. Adams, H. L. Johnston and E. C. Kerr, C_p: Ga, Ga_2O_3 (15–300°K), *J. Amer. Chem. Soc.* 1952 **74** 4784, 4788.
2. K. M. Alexander and F. Fairbrother, Vapour pressures of $TaCl_5$, $TaBr_5$, TaI_5, $NbCl_5$, $NbBr_5$. *J. Chem. Soc.* 1949 2472 S223.
3. T. L. Allen, Vapour pressure of CrI_2 (943–1054°K), *J. Amer. Chem. Soc.* 1956 **78** 5476.
4. N. P. Allen, O. Kubaschewski and O. von Goldbeck, Thermodynamics of the V–O system, *J. Elektrochem. Soc.* 1951 **98** 417.
5. R. L. Altman, Free energy of the reactions: $\langle MgO\rangle = (MgO)$; $\langle MgO\rangle = (Mg) + \frac{1}{2}(O_2)$; $4\langle MgO\rangle + 2\{Al\} = \langle MgAl_2O_4\rangle + (Mg)$, *J. Phys. Chem.* 1963 **67** 366.
6. H. C. Andersen and L. H. Belz, Vapour pressures of BI_3 and SiI_4, *J. Amer. Chem. Soc.* 1953 **75** 4828.
7. J. S. Anderson, Vapour pressure of In, *J. Chem. Soc.* 1943, 141.
8. J. S. Anderson, Non-stoichiometric compounds, *Ann. Rep. Chem. Soc.* 1946 **43** 104.
9. L. D. Armstrong and H. Grayson-Smith, C_p of Mn, Co and Cr, *Can. J. Res.* 1950 **28** 51.
10. R. F. Barrow and A. D. Caunt, Ultra-violet absorption spectra of gaseous alkali fluorides, *Proc. Roy. Soc.* (A) 1953 **219** 120.
11. T. W. Bauer, H. L. Johnston and E. C. Kerr, C_p (16–335°K) of LiOH and $LiOH{\cdot}H_2O$, *J. Amer. Chem. Soc.* 1950 **72** 4740.
12. A. Baur and A. Lecocq, ΔG (dissociation): $Be(OH)_2$, *Compt. rend.* 1963 **257** 1445.
13. C. E. Birchenall and C. H. Cheng, Vapour pressures of Zn and Cd over Zn–Ag and Cd–Ag alloys, *J. Metals* 1949 **1** 428.
14. H. Blattner, W. Känzig and W. Merz, Properties of $BaTiO_3$, *Helv. Phys. Acta* 1949 **22** 35.
15. J. M. Blocher and I. E. Campbell, Vapour pressures of TiI_4 and Ti, *J. Amer. Chem. Soc.* 1947 **69** 2100; 1949 **71** 4040.
16. J. O. Blomeke and W. T. Ziegler, Heat contents (30–900°C) of La_2O_3, Nd_2O_3 and Pr_6O_{11}, *J. Amer. Chem. Soc.* 1951 **73** 5099.
17. L. Bochirol, C_p: $NiO{\cdot}Fe_2O_3$, $CoO{\cdot}Fe_2O_3$ and $ZnO{\cdot}Fe_2O_3$, *Compt. rend.* 1951 **232** 1473.
18. H. Bommer and E. Hohmann, ΔH (solution) of the chlorides of the rare earth metals, *Z. anorg. allg. Chem.* 1941 **248** 371.
19. R. S. Bradley and P. Volans, Vapour pressure of KCl (438–597°C), *Proc. Roy. Soc.* 1953 **217** 508.
20. H. Bloom, J. O'M. Bockris, N. E. Richards and R. G. Taylor, Vapour pressure measurements: AgCl, AgBr, $CdBr_2$, CdI_2, NaBr, $PbBr_2$, $ZnCl_2$, $ZnBr_2$, *J. Amer. Chem. Soc.* 1958 **80** 2044.
21. H. Braune, S. Peter and V. Neveling, Dissociation of sulphur vapour, *Z. Naturforsch.* 1951 **6a** 32.
22. L. Brewer, Critical review of the thermochemical properties of the metal oxides, *Chem. Rev.* 1953 **52** 1.
23. L. Brewer, Review of thermochemical properties of the elements, *Natl. Nuclear Sci. Ser.*, Div. IV, **19B**, p. 13.
24. L. Brewer, Critical review of the thermochemical properties of the common gases, *ibid.* p. 60.
25. L. Brewer, Critical review of the fusion and evaporation data of metal halides, *ibid.* p. 193.
26. L. Brewer, L. A. Bromley, P. W. Gilles and N. L. Lofgren, Critical review of thermochemical properties of nitrides, carbides and sulphides, *ibid.* p. 40.
27. L. Brewer, L. A. Bromley, P. W. Gilles and N. L. Lofgren, Critical review of the thermochemical properties of metal halides, *ibid.* p. 76.
28. L. S. Brooks, Vapour pressures of Se and Te. Equilibrium: $2Cu + (H_2S) = Cu_2S + (H_2)$, *J. Amer. Chem. Soc.* 1952 **74** 227; 1953 **75** 2471.
29. T. M. Buck, W. E. Wallace and R. M. Rulon, ΔH (solution in HCl. aq) of Mg–Cd alloys, *J. Amer. Chem. Soc.* 1952 **74** 136.
30. R. H. Busey and W. F. Giauque, C_p (15–330°K) of $NiCl_2$ and Hg; equilibrium: $NiCl_2 + (H_2) = Ni + 2(HCl)$, *J. Amer. Chem. Soc.*, 1952 **744** 443; 1953 **75** 806, 1791.
31. K. H. Butler and O. Maass, Properties of H_2S_2, *J. Amer. Chem. Soc.* 1930 **52** 2184.
32. A. D. Caunt and R. F. Barrow, ΔH (solution) of TIF, *Trans. Faraday Soc.* 1950 **46** 154.

33. W. J. CHILDS, J. E. CLINE, W. M. KISNER and J. WULFF, Vapour pressure of $MoCl_5$, *Trans. Amer. Soc. Metals* 1951 **43** 105.
34. J. CHIPMAN, Activities in liquid metallic solutions, *Disc. Faraday Soc.* 1948 No. **4**, 23.
35. H. C. CLARK and H. J. EMELÉUS, Vapour pressure of VF_5, *J. Chem. Soc.* 1957, 2119.
36. K. CLUSIUS, J. GOLDMANN and A. PERLICK, C_p (10–270°K): LiF, NaCl, KCl, KBr, KI, RbBr, RbI, *Z. Naturforsch.* 1949 **4a** 424.
37. K. CLUSIUS and L. SCHACHINGER, C_p of Pd, In and Co at low temperatures, *Z. Naturforsch.* 1947 **2a** 90.
38. G. E. COGIN and G. E. KIMBALL, Vapour pressures of solid NaBr, NaI, KI, CsCl, CsBr and CsI, *J. Chem. Phys.* 1948 **16** 1035.
39. C. B. COOPER, Heat capacities of KBr and KI (300–700°K), *J. Chem. Phys.* 1953 **21** 777.
40. L. A. COSGROVE and P. E. SNYDER, Heat contents (25–1100°C) of MoO_3; ΔH (combustion) of Be, *J. Amer Chem. Soc.* 1953 **75** 1227, 3102.
41. J. P. COUGHLIN, Heat contents: MnS (25–1530°C), FeS (25–1230°C), FeS_2 (25–730°C); $NiCl_2$ (25–1260°C), *J. Amer. Chem. Soc.* 1950 **72** 5445; 1951 **73** 5314.
42. J. P. COUGHLIN, E. G. KING and K. R. BONNICKSON, Heat contents (25–1530°C): Fe oxides, *J. Amer. Chem. Soc.* 1951 **73** 3891.
43. J. P. COUGHLIN and R. L. ORR, Heat contents (25–1530°C): $BaTiO_3$, $SrTio_3$, Ba_2TiO_4, *J. Amer. Chem. Soc.* 1953 **75** 530.
44. F. S. DAINTON and H. M. KIMBERLEY, Vapour pressure of white phosphorus, *Trans. Faraday Soc.* 1950 **46** 912.
45. L. S. DARKEN and R. W. GURRY, Thermodynamics of the system Fe–O, *J. Amer. Chem. Soc.* 1945 **67** 1398; 1946 **68** 798.
46. T. M. DAUPHINÉE, Vapour pressure of Hg, *J. Chem. Phys.* 1951 **19** 389.
47. W. DE SORBO, C_p (15–300°K) of Se, cryst, *J. Chem. Phys.* 1953 **21** 1144.
48. W. E. DITMARS and H. L. JOHNSTON, ΔG (formation) of LiOH, *J. Amer. Chem. Soc.* 1953 **75** 1791.
49. H. A. DOERNER, Equilibria involving Cr chlorides, US Bur Mines, Techn. Paper No. 577, 1937.
50. P. E. DOUGLAS, Vapour pressure of Ca, *Proc. Phys. Soc.* 1954 **67** 783.
51. J. W. EDWARDS, H. L. JOHNSTON and P. E. BLACKBURN, Vapour pressures of Ta, V and Mo, *J. Amer. Chem. Soc.* 1951 **73** 4727; 1952 **74** 1539.
52. J. W. EDWARDS, H. L. JOHNSTON and W. E. DITMARS, Vapour pressures: Fe (1356–1519°K), Co (1363–1522°K); Ti (1587–1764°K), *J. Amer. Chem. Soc.* 1951 **73** 4729; 1953 **75** 2467.
53. W. EGGERSGLUESS, A. G. MONROE and W. G. PARKER, ΔH (combustion) of boron, *Trans. Faraday Soc.* 1949 **45** 661.
54. E. P. EGAN and Z. T. WAKEFIELD, C_p (3–300°K): $AlPO_4$, *J. Phys. Chem.*, 1960 **64** 1953.
55. E. P. EGAN and Z. T. WAKEFIELD, C_p (10–1700°K): $Ca_2P_2O_7$, *J. Amer. Chem. Soc.* 1957 **79** 558.
56. W. D. ENGLISH and J. N. DALE, Vapour pressure of Te_3F_{10}, *J. Chem. Soc.* 1953, 2498.
57. P. EHRLICH, K. PEIK and E. KOCH, ΔH (solution in aqueous acids); hydrides, halides and hydride halides of alkaline earth metals, *Z. anorg. allg. Chem.* 1963 **324** 113.
58. A. EUCKEN, Vapour pressures of metals—assessment, *Metallwirtsch.* 1936 **15** 63.
59. D. F. EVANS and R. E. RICHARDS, ΔH (solution in NaOH·aq) of $GeBr_4$, GeI_4 and GeO_2, *J. Chem. Soc.* 1952, 1292.
60. M. W. EVANS, ΔH (formation) of Ce sulphides, Natl. Nuclear Energy Ser., Div. IV, **19B**, 1950, p. 312.
61. L. EYRING and E. F. WESTRUM, ΔH (solution in 6N HCl) of Th, $ThCl_4$ and Th_2S_3, *J. Amer. Chem. Soc.* 1950 **72** 5555; 1953 **75** 4802.
62. W. FISCHER in FIAT Reports of German Science, 1939–1946, Inorganic Chemistry IV, p. 195.
63. H. FLOOD, J. F. GORRISON and R. VEIMO, Vapour pressure of $VOCl_3$, *J. Amer. Chem. Soc.* 1937 **59** 2494.
64. W. T. FOLEY and P. A. GIGUÈRE, L_f of H_2O_2, *Can. J. Chem.* 1951 **29** 123.
65. H. GABRIEL and C. ALVAREZ-TOSTADO, ΔH (hydrolysis) of SiS_2 and $SiSe_2$, *J. Amer. Chem. Soc.* 1952 **74** 262.
66. G. GATTOW and A. SCHNEIDER, ΔH (formation, direct): Cu–Se, *Z. anorg. allg. Chem.* 1956 **286** 296.
67. T. H. GEBALLE and W. F. GIAUQUE, C_p (15–300°K) of Au, *J. Amer. Chem. Soc.* 1952 **74** 2368.
68. B. C. GERSTEIN, M. GRIFFEL, *et al.*, S_{298}: Ho, *J. Chem. Phys.* 1957 **27** 394.
69. YA. I. GERASSIMOV, E. V. KRUGLOVA and N. N. ROZENBLYUM, Dissociation pressures of the higher Sn sulphides, *J. Gen. Chem. USSR* 1937 **7** 1520.
70. P. A. GIGUÈRE, Evaluation of spectroscopic data of H_2O_2, *Can. J. Res.* 1950 **28** 485.
71. P. W. GILLES and J. L. MARGRAVE, Dissociation of F_2, *J. Chem. Phys.* 1953 **21** 381.
72. O. GLEMSER and H. SCHRÖDER, ΔH: $Mn_2O_7 = \alpha Mn_2O_3 + 2(O_2)$, *Z. anorg. allg. Chem.* 1953 **271** 293.
73. N. A. GOKCEN, Equilibrium: $MoO_2 + 2(H_2) = Mo + 2(H_2O)$, *J. Metals* 1953 **5** 1019.
74. E. S. GREINER, L_f of Ge, *J. Metals* 1952 **4** 1044.
75. M. GRIFFEL and R. E. SKOCHDOPOLE, C_p (18–300°K): Th, *J. Amer. Chem. Soc.* 1953 **75** 5250.
76. P. GROSS, C. HAYMAN and D. L. LEVI, ΔH (reaction) $\frac{3}{2}PbF_2 + Al = AlF_3 + \frac{3}{2}Pb$ and $PbF_3 + Mg = MgF_2 + Pb$, *Trans. Faraday Soc.* 1954 **50** 477.

77. P. Gross, C. S. Campbell, P. J. C. Kent and D. L. Levi, Equilibria involving Al monohalides, *Disc. Faraday Soc.* 1948, No. **4**, 206.
78. P. Gross, C. Hayman and D. L. Levi, ΔH (formation): $TiBr_4$, $TiCl_4$, $ZrCl_4$, *Trans. Faraday Soc.* 1957 **53** 1285, 1601.
79. P. Gross, C. Hayman, D. L. Levi and G. L. Wilson, ΔH_f (directly): $NbCl_5$, $TaCl_5$, $NbBr_5$, $TaBr_5$, *Trans. Faraday Soc.* 1960 **56** 318; 1962 **58** 890.
80. H. Hahn and F. Burow, ΔH (combustion): In–S, In–Se, In–Te, Ga–S, Ga–Se, Ga–Te, *Angew. Chem.* 1956 **68** 382.
81. H. Hahn and E. Gilbert, Properties of Ag_3N, *Z. anorg. allg. Chem.* 1949 **258** 77.
82. H. Hahn and A. Konrad, ΔH (decomposition) of Ni_3N, *Z. anorg. allg. Chem.* 1951 **264** 181.
83. P. Herasymenko, Dissociation pressures: Ag–Cd, *Acta Metallurg.* 1956 **4** 1.
84. H. Herczog and L. M. Pidgeon, Equilibria involving Ti iodides, *Can. J. Chem.* 1956 **34** 1687.
85. H. N. Hersh, Vapour pressure of Cu, *J. Amer. Chem. Soc.* 1953 **75** 1529.
86. W. Hieber and E. Reindl, Vapour pressure and ΔH (formation) of $SnCl_4$, *Z. Elektrochem* 1940 **46** 559.
87. W. Hieber and A. Woerner, Thermochemistry of the ferrous halides and their CO complexes, *Z. Elektrochem.* 1934 **40** 287.
88. E. Hohmann and H. Bommer, Heats of solution of the rare earth metals and their iodides, *Z. anorg. allg. Chem.* 1941 **248** 383.
89. R. B. Holden, R. Speiser and H. L. Johnston, Vapour pressure of Be (900–1280°C), *J. Amer. Chem. Soc.* 1948 **70** 3897.
90. C. E. Holley, E. J. Huber and E. H. Meierkord, ΔH (combustion): In, *J. Amer. Chem. Soc.* 1952 **74** 1084.
91. S. Hoshimo, Transformations of CuBr, *J. Phys. Soc. Japan* 1952 **7** 560.
92. C. M. Hsiao and A. W. Schlechten, Apparent vapour pressures of several metal sulphides, *J. Metals* 1952 **4** 65.
93. J. H. Hu and H. L. Johnston, C_p (15–300°K): Cu_2O, CuBr, CuO, *J. Amer. Chem. Soc.* 1951 **73** 4550; 1952 **74** 4771; 1953 **75** 2471.
94. J. H. Hu, D. White and H. L. Johnston, S_{298} of HF, *J. Amer. Chem. Soc.* 1953 **75** 1232.
95. E. J. Huber and C. E. Holley, ΔH (combustion): La, Ce, Nd, *J. Amer. Chem. Soc.* 1952 **74** 5530; 1953 **75** 3594, 5645.
96. E. J. Huber, C. E. Holley and E. H. Meierkord, ΔH (combustion): U, Th, *J. Amer. Chem. Soc.* 1952 **74** 3406.
97. G. L. Humphrey, ΔH (combustion): Ti oxides, TiN, TiC; Hf and HfN, *J. Amer. Chem. Soc.* 1951 **73** 1587, 2261; 1953 **75** 2806.
98. G. L. Humphrey and C. J. O'Brien, ΔH (combustion) of Sn and SnO, *J. Amer. Chem. Soc.* 1953 **75** 2805.
99. R. Hultgren, R. L. Orr, P. D. Anderson and K. K. Kelley, *Selected Values of Thermodynamic Properties of Metals and Alloys*. J. Wiley: New York (1963).
100. J. L. Hyde, Concerning $PbBr_2$, *J. Amer. Chem. Soc.* 1951 **73** 1860.
101. F. Irmann, Properties of Mg carbides, *Helv. Chim. Acta* 1948 **31** 1584, 2263.
102. E. S. Itskevich and G. P. Strelkov, C_p (1·5–300°K): $CdBr_2$, *Zh. fiz. Khim.* 1960 **34** 1312.
103. H. H. G. Jellinek, Slag equilibria involving BiF_3, *Trans. Faraday Soc.* 1944 **40** 1.
104. H. L. Johnston and T. W. Bauer, Thermochemical data for Li_2O and LiOH, *J. Amer. Chem. Soc.* 1951 **73** 1119.
105. H. L. Johnston, H. N. Hersh and E. C. Kerr, S_{298} of B and B_2O_3, *J. Amer. Chem. Soc.* 1951 **73** 1112.
106. H. L. Johnston and E. C. Kerr, C_p (17–300°K): H_3BO_3, *J. Amer. Chem. Soc.* 1950 **72** 4733.
107. H. L. Johnston and A. L. Marshall, Vapour pressures of Ni and NiO, *J. Amer. Chem. Soc.* 1940 **62** 1382.
108. W. L. Jolly and W. M. Latimer, Vapour pressure of GeI_4. Equilibrium: $Ge + GeO_2 = 2(GeO)$, *J. Amer. Chem. Soc.* 1952 **74** 5754, 5757.
109. W. M. Jones, J. Gordon and E. A. Long, C_p (15–300°K): U, UO_2, UO_3, *J. Chem. Phys.* 1952 **20** 695.
110. J. H. Junkins, H. A. Bernhardt and E. J. Barber, Properties of TeF_4, *J. Amer. Chem. Soc.* 1952 **74** 5749.
111. J. H. Junkins, R. L. Farrer, E. J. Barber and H. A. Bernhardt, Properties of NbF_5, *J. Amer. Chem. Soc.* 1952 **74** 3464.
112. A. Kabesh and R. S. Nyholm, Vapour pressures of $SnCl_4$, $SnBr_4$ and SnI_4, *J. Chem. Soc.* 1951, 3245.
113. A. F. Kapustinski and Yu. M. Golutwin, ΔH (formation): Al_2S_3, Al_2Se_3, Al_2Te_3, *Izvest. Akad. Nauk SSSR, Otdel. Khim. Nauk* 1951, 192.
114. A. F. Kapustinski, I. A. Makolkin and L. I. Krishtalik, Equilibria involving H_2Se, *J. Phys. Chem. USSR* 1947 **21** 125.
115. A. F. Kapustinski and O. Ya. Samoilov, ΔH (hydrolysis): BCl_3, *Izvest. Akad. Nauk SSSR, Otdel. Khim. Nauk* 1952, 218.
116. M. Kawakami, Heat capacities of Fe–Ni alloys, *Sci. Rep. Tôhoku Univ.* 1926 **15** 251.

117. M. DE KAY-THOMPSON, Total and Free Energies of Formation of the Oxides of 32 Metals, *Electrochem. Soc.*, New York; 1942.
118. L. I. KAZARNOVSKAYA and I. A. KAZARNOVSKI, ΔH (solution) of KO_2, *Zh. fiz. Khim.* 1951 **25** 293.
119. K. KEARBY, Vapour pressure of $SiCl_4$, *J. Amer. Chem. Soc.* 1936 **58** 374.
120. K. K. KELLEY, Critical evaluations of vapour pressures and heats of evaporation of inorganic substances, US Bur. Mines, Bull. No. 383, 1935.
121. K. K. KELLEY, Evaluation of the heats of fusion of metals and metal compounds data on freezing point lowering, US Bur. Mines, Bull. No. 393, 1936.
122. K. K. KELLEY, Evaluation of free energy data of sulphur, sulphides and sulphates, US Bur. Mines, Bull. No. 406, 1937.
123. K. K. KELLEY, Evaluation of free energy data of metal carbides and nitrides, US Bur. Mines, Bull. No. 407, 1937.
124. K. K. KELLEY, Critical evaluation of high-temperature heat capacities of inorganic compounds, US Bur. Mines, Bull. No. 476, 1949.
125. K. K. KELLEY, The entropies of inorganic substances: revision of data, US Bur. Mines, Bull No. 477, 1950.
126. K. K. KELLEY and C. T. ANDERSON, Correlation of thermochemical data of metal carbonates, US Bur. Mines, Bull. No. 384, 1935.
127. J. C. R. KELLY and P. E. SNYDER, C_p (60–300°K): NaOH, *J. Amer. Chem. Soc.* 1951 **73** 4114.
128. E. C. KERR, H. N. HERSH and H. L. JOHNSTON, C_p (18–300°K) of B_2O_3, cryst, *J. Amer. Chem. Soc.* 1950 **72** 4738.
129. E. C. KERR, H. L. JOHNSTON and N. C. HALLETT, C_p (20–295°K) of Al_2O_3, *J. Amer. Chem. Soc.* 1950 **72** 4740.
130. J. A. A. KETELAAR, G. W. VAN OOSTERHOUT and P. B. BRAUN, Vapour pressure of WCl_6, *Rec. Trav. Chim. Pays-Bas* 1943 **62** 597.
131. E. G. KING, ΔH (solution in HF·aq) of Zn_2SiO_4, βCa_2SiO_4, Ca_3SiO_5; $MnSiO_3$, Fe_2SiO_4, quartz, and MnO, *J. Amer. Chem. Soc.* 1951 **73** 656; 1952 **74** 4446.
132. E. G. KING and J. P. COUGHLIN, High-temperature heat contents of Zr and Zr compounds, *J. Amer. Chem. Soc.* 1950 **72** 2262.
133. E. G. KING and S. S. TODD, C_p (52–296°K): SnS, SnS_2, *J. Amer. Chem. Soc.* 1953 **75** 3023.
134. V. A. KIREEV, Calculations and estimations of S_{298}: monatomic gases, some metasilicates, and sulphides, *Amer. Chem. Abstr.* 1946 **40** 5611; 1947 **41** 3685.
135. J. A. KITCHENER and S. IGNATOWICZ, Equilibria involving Zn_2SiO_4, *Trans. Farad Soc.* 1951 **47** 1278.
136. O. J. KLEPPA, ΔH (formation): Ag and Au alloys, *J. Phys. Chem.* 1956 **60** 846, 852, 858.
137. O. KNACKE and K. E. PRESCHER, Thermochem. properties of PbO, *Z. Erzbergbau Metallhüttenw.* 1964 **17** 28.
138. K. KOBAYASHI, $\bar{C}_p$ of AgCl (30–500°C), $Ca(OH)_2$ (30–400°C), Ag_2O (44–230°C), Ag_2CO_3 (30–160°C) and Na_2CO_3, *Sci. Rep. Tôhoku Univ.* 1950 **34** 112, 153; 1951 **35** 173; 1952 **35** 283, **36** 21.
139. M. I. KOCHNEV, Dissociation pressures of Co–As alloys, *Doklady Akad. Nauk SSSR* 1950 **70** 433.
140. F. KÖRBER, W. OELSEN and H. LICHTENBERG, Heat contents and heats of formation: Fe–Ni–Al, Fe–Co–Al, Cu–Ni–Al, Fe–Al–Si, Cu–Mn–Al, *Mitt. KWI Eisenforsch.* 1937 **19** 131.
141. A. V. KOMANDIN and D. N. TARASENKOV, Equilibria involving WO_2Cl_3 and $WOCl_4$, *Zh. obchei Khim.* 1940 **14** 1333.
142. A. V. KOMANDIN and M. L. VLODAVETS, Vapour pressure of $VOCl_3$, *Zh. fiz Khim.* 1952 **26** 1291.
143. T. KOOTZ, Solubility of N_2 in Fe, *Arch. Eisenhüttenwesen* 1941 **15** 77.
144. C. W. KOTHEN and H. L. JOHNSTON, C_p (15–305°K) of Ti, *J. Amer. Chem. Soc.* 1953 **75** 3101.
145. O. KUBASCHEWSKI, Physicochem. properties of Ti compounds, *Angew. Chem.* 1960 **72** 255.
146. O. KUBASCHEWSKI, $\bar{C}_p$ and L_f of Au and Ag alloys, *Z. phys. Chem.* (A) 1943 **192** 292.
147. O. KUBASCHEWSKI, L_f and L_t of metals, *Z. Elektrochem.* 1950 **54** 275.
148. O. KUBASCHEWSKI, C_p's of some metals, *Z. Metallkunde* 1950 **41** 445.
149. O. KUBASCHEWSKI and W. A. DENCH, Free energies of dissociation in the Ti–O_2 system, *J. Inst. Metals* 1953 **82** 87.
150. O. KUBASCHEWSKI, W. A. DENCH and G. HEYMER, Diss. pressure of Cr–Ni alloys, *Z. Elektrochem.* 1960 **64** 601.
151. O. KUBASCHEWSKI and O. HUCHLER, E.m.f.'s: Ag/Ag–Au, *Z. Elektrochem.* 1948 **52** 170.
152. O. KUBASCHEWSKI, H. VILLA and W. A. DENCH, Equilibria involving Ti chlorides, *Trans. Faraday Soc.* 1956 **52** 214.
153. O. KUBASCHEWSKI and H. VILLA, ΔH (formation) of intermetallic compounds of Ba and Ca, *Z. Elektrochem.* 1949 **53** 32.
154. J. J. LANDER, Heat contents: SrO (405–1266°K), BaO (420–1300°K), $SrCO_3$ (455–1324°K), $BaCO_3$ (431–1311°K) and CaO (563–1177°K), *J. Amer. Chem. Soc.* 1951 **73** 5794.
155. W. M. LATIMER, Estimation of S_{298} of solid compounds of metals, *J. Amer. Chem. Soc.* 1951 **73** 1480.
156. F. LAVES, *Taschenbuch f. Physiker u. Chemiker* (ed. by J. d'Ans and E. Lax), Springer, Berlin, 1943;

Crystal structures of inorganic substances, *Metals Reference Book* (ed. by C. J. Smithells), Butterworth, London, 1949.
157. M. LEVIN and H. SCHOTTKY, Heat contents of Bi–Cd and Fe–C alloys, *Ferrum* 1912 **10** 193.
158. J. C. M. LI and N. W. GREGORY, ΔH (solution): $FeCl_2$, $FeCl_3$, $FeBr_2$, $FeBr_3$, *J. Amer. Chem. Soc.* 1952 **74** 4670.
159. J. A. M. VAN LIEMPT, Vapour pressure of Ba, *Rec. Trav. Chim. Pays-Bas* 1936 **55** 468.
160. M. W. MALLETT and B. G. KOEHL, Diss. pressures in the tantalum-hydrogen system, *J. Electrochem. Soc.* 1962 **109** 611.
161. J. LUMSDEN, *Thermodynamics of Alloys*, Inst. Metals, London, 1952.
162. C. L. MCCABE and R. G. HUDSON, Vapour pressure of Mn, and dissociation pressure of Mn_7C_3, *J. Metals* 1957 **9** 17.
163. J. E. MCDONALD and J. W. COBBLE, ΔH (combustion): ReS_2, Re_2S_7, J. Phys. Chem. 1962 **66** 791.
164. A. D. MCQUILLAN, Solubility of H_2 in Ti, *Proc. Roy. Soc.* 1950 **204** 309.
165. I. A. MAKOLKIN, *Acta Phys. Chim. USSR* 1940 **13** 347; E.m.f. measurements on CdS, ZnS, MoS_2; HgS, Ag_2S, Cu_2O, CdO, Ag_2O, *J. Phys. Chem. USSR* 1942 **16** 13.
166. A. D. MAH, ΔH (combustion): Al, Mo, MoO_2, *J. Phys. Chem.* 1958 **62** 124.
167. M. MAUCHERAT, Vapour pressure of Li, *J. Phys. Radium* 1939 **10** 441.
168. J. E. MAYER and I. H. WINTNER, Measurements of low vapour pressures of alkali halides, *J. Chem. Phys.* 1938 **6** 301.
169. G. E. MOORE, H. W. ALLISON and J. D. STRUTHERS, Vapour pressure of SrO, *J. Chem. Phys.* 1950 **18** 1572.
170. M. A. MOSEMANN and K. S. PITZER, Heat contents of the various modifications of SiO_2, *J. Amer. Chem. Soc.* 1941 **63** 2348.
171. H. D. MULDER and F. C. SCHMIDT, Reactions of Na_2Se and Na_2Se_2 in liquid ammonia, *J. Amer. Chem. Soc.* 1951 **73** 5575.
172. O. E. MYERS and A. P. BRADY, ΔH (solution in NaOH·aq, etc.): MoF_6, NbF_5, *J. Phys. Chem.* 1960 **64** 591.
173. G. M. MURPHY and E. RUBIN, S_{298} of F_2, *J. Chem. Phys.* 1952 **20** 1179.
174. B. F. NAYLOR, Heat contents of Cu–Mn alloys, US Bur. Mines, Rep. Inv. No. 3835, 1946.
175. J. J. NEELY, C. E. TEETER and J. B. TRICE, C_p: Be_2C, *J. Amer. Ceram. Soc.* 1950 **33** 363.
176. B. NEUMANN, C. KRÖGER and H. KUNZ, ΔH (combustion) of metal nitrides, *Z. anorg. allg. Chem.* 1934 **218** 379.
177. K. NIWA, Vapour pressures of the iodides, bromides and chlorides of alkali metals and Tl, Pb, Zn, Cd, Te and Hg, *J. Fac. Sci. Hokkaido Univ.* 1938 **2** 201; 1940 **3** 17, 75.
178. H. NOWOTNY, FIAT Reports of German Science 1939–1946, Crystal structures of intermetallic phases, *General Metallurgy*, p. 3.
179. F. J. OETTING and N. W. GREGORY, Heat content (60–500°C): $FeCl_2$, FeI_2, *J. Phys. Chem.* 1961 **65** 138, 173.
180. M. OLETTE and A. FERRIER, Heat contents of solid and liquid iron, *Compt. rend.* 1962 **254** 2322.
181. R. L. ORR, Heat contents: Mg_2SiO_4, Fe_2SiO_4 (298–1800°K); HfO_2 (298–1804°K), $HfCl_4$ (298–486°K); Nb_2O_5 and Ta_2O_5 (298–1800°K), *J. Amer. Chem. Soc.* 1953 **75** 528, 1229, 2808.
182. D. H. PARKINSON, F. E. SIMON and F. H. SPEDDING, C_p (2–130°K): Ce, Pr, Nd, La, *Proc. Roy. Soc.* 1951 **207** 137.
183. R. D. PEACOCK, Properties of SeF_4, *J. Chem. Soc.* 1953, 3617.
184. J. PEARSON and U. J. C. ENDE, ΔG (formation) of metal nitrides, *J. Iron Steel Inst.* 1953 **175** 52.
185. T. E. PHIPPS, G. W. SEARS, R. L. SEIFERT and O. C. SIMPSON, Vapour pressures of Pu halides, *J. Chem. Phys.* 1950 **18** 713.
186. G. L. PICKARD and F. E. SIMON, C_p of Pd, Na (2–25°K), and Hg (4–90°K), *Proc. Phys. Soc.* 1948 **61** 1.
187. A. YU. POLYAKOV, Vapour pressure of V_2O_5, *Zh. fiz. Khim.* 1946 **20** 1021.
188. B. D. POLLOCK, Diss. pressure (1430–1670°K): Be_2C, *J. Phys. Chem.* 1959 **63** 587.
189. F. D. RICHARDSON, ΔG (formation) of metal carbides: critical review, *J. Iron Steel Inst.* 1953 **175** 53.
190. F. D. RICHARDSON and J. E. ANTILL, Thermodynamic properties of Cu_2S and its mixtures with Na_2S, *Trans. Faraday Soc.* 1955 **51** 22.
191. F. D. RICHARDSON and J. H. E. JEFFES, ΔG (formation) of metal oxides: assessment, *J. Iron Steel Inst.* 1948 **160** 261; 1949 **163** 397.
191a. O. KUBASCHEWSKI, discussion of this paper: 1949 **163** 150.
192. F. D. RICHARDSON and J. H. E. JEFFES, ΔG (formation) of metal sulphides: assessment, *J. Iron Steel Inst.* 1952 **171** 165.
193. F. D. RICHARDSON, J. H. E. JEFFES and G. WITHERS, Thermodynamics of double oxides, *J. Iron Steel Inst.* 1950 **166** 213.
194. C. ROBERT, Vapour pressures of In halides, *Helv. Phys. Acta* 1936 **9** 405.
195. P. ROCQUET and M. F. ANCEY-MORET, ΔH (solution in HF·aq): SiS_2, *Bull. Soc. Chim. France* 1954, 1038.
196. T. ROSENQVIST, A thermodynamic study of the Fe, Co and Ni sulphides, *J. Iron Steel Inst.* 1954 **176** 36.

197. F. D. ROSSINI, P. A. COWIE, F. O. ELLISON and C. C. BROWNE, *Properties of Ti compounds*, Washington, 1956.
198. F. D. ROSSINI, D. D. WAGMAN, W. H. EVANS, S. LEVINE and I. JAFFEE, Selected values of chemical thermodynamic properties, Nat. Bur. Standards, 1952.
199. W. A. ROTH, FIAT Reports of German Science 1939–1946, ΔH (formation) of inorganic substances, *Inorganic Chemistry* IV, 1948, p. 214.
200. W. A. ROTH, Re-evaluation of ΔH_{298} of components of boron, *Z. Naturforsch.* 1946 **1** 575.
201. W. A. ROTH and H. TROITZSCH, ΔH (formation) of Si compounds, *Z. anorg. allg. Chem.* 1949 **260** 337.
202. E. RUDZITIS, H. M. FEDER and W. N. HUBBARD, ΔH (combustion in F_2): Y., *J. Phys. Chem.* 1965 **69** 2305.
203. R. RUF and W. D. TREADWELL, Vapour pressure of $HgCl_2$ (10–60°C), *Helv. chim. Acta* 1954 **37** 1941.
204. S. SATOH, Heat contents of Cr nitrides, Th_3N_4, Li_3N and Zn_3N_2, *Sci. Pap. Inst. Phys. Chem. Res.* 1938 **34** 1001; 1939 **35** 182.
205. T. SATO and H. KANEKÔ, Vapour pressure of Se, *Technol. Rep. Tôhoku Univ.* 1950 **14** 45.
206. H. SCHÄFER, L. BAYER and H. LEHMANN, The $NbCl_3$–$NbCl_5$ system, *Z. anorg. allg. Chem.* 1952 **269** 76.
207. H. SCHÄFER and H. JACOB, Vapour pressure of $NiBr_2$, *Z. anorg. allg. Chem.* 1956 **286** 56.
208. H. SCHÄFER and K. KREHL, Equilibria involving Co chlorides, *Z. anorg. allg. Chem.* 1952 **268** 25.
209. H. SCHÄFER and J. NICKL, Equilibrium: $(SiCl_4) + Si = 2(SiCl_2)$, *Z. anorg. allg. Chem.* 1953 **274** 250.
210. H. SCHÄFER and A. OEHLER, Equilibrium: $2FeCl_3 = 2FeCl_2 + (Cl_2)$, *Z. anorg. allg. Chem.* 1953 **271** 206.
211. H. SCHÄFER and F. ZEPPERNICK, Vapour pressure of $TiCl_4$, *Z. anorg. allg. Chem.* 1953 **272** 274.
212. R. SCHENCK and P. VON DER FORST, Equilibria with sulphides. Revision of earlier work, *Z. anorg. allg. Chem.* 1939 **241** 148.
213. A. SCHNEIDER and G. GATTOW, ΔH (combustion) of Se, Al and Al_2Se_3. Cu_2Se: Priv. Comm., *Z. anorg. allg. Chem.* 1954 **277** 37.
214. A. SCHNEIDER and G. ZINTL, ΔH (combustion) Te, *Z. anorg. allg. Chem.* 1961 **308** 290.
215. A. W. SEARCY and R. D. FREEMAN, Vapour pressure of Sn and Ge, *J. Amer. Chem. Soc.* 1954 **76** 5229; *J. Chem. Phys.* 1955 **23** 88.
216. A. W. SEARCY and A. G. THARP, Diss. pressures of molybdenum silicides, *J. Phys. Chem.* 1960 **64** 1538.
217. K. A. SENSE and R. W. STONE, Vapour pressures in the system BeF_2–NaF, *J. Phys. Chem.* 1958 **62** 453.
218. S. SEKI, Vapour pressures of $TiBr_4$, CBr_4, $SnBr_4$, SnI_4, *J. Chem. Soc. Japan* 1941 **62** 789.
219. K. A. SHAFIROV and S. M. SKUTAROV, $\bar{C}_p$ of $Cr(CO)_6$, $Mo(CO)_6$ and $W(CO)_6$, *Amer. Chem. Abstr.* 1955 **49** 5099.
220. L. SHARTSIS and E. S. NEWMAN, ΔH (formation): PbO–B_2O_3 and PbO–SiO_2 glasses, *J. Res. Nat. Bur. Stand.* 1948 **40** 471.
221. S. A. SHCHUKAREV and G. I. NOVIKOV, Vapour pressures: WCl_6, WCl_5, $WOCl_4$, *Zh. neorg. Khim.* 1956 **1** 357.
222. S. A. SHCHUKAREV, M. A. ORANSKAYA, T. A. TOLMACHEVA, L. V. KOMANDROVSKAYA and T. S. SHEMYAKINA, *Zh. neorg. Khim.* 1956 **1** 8, 17.
223. S. A. SHCHUKAREV, M. A. ORANSKAYA, T. A. TOLMACHEVA and A. K. YAKHKIND, *p*: VCl_4, *Zh. neorg. Khim.* 1956 **1** 30.
224. S. A. SHCHUKAREV, M. A. ORANSKAYA and V. M. TSINTSIUS, Dissociation: Au–Cl, *Zh. neorg. Khim.* 1956 **1** 881.
225. Z. SHIBATA and I. MORI, Equilibrium: $CoO + (H_2) = Co + (H_2O)$, *Z. anorg. allg. Chem.* 1933 **212** 305.
226. G. SHIRANE and E. SAWAGUCHI, C_p of $PbTiO_3$, *Phys. Rev.* 1951 **81** 440.
227. H. SIEMONSEN, Calorimetric bomb: ΔH_{298} of $BeCl_2$, $CaCl_2$, $SrCl_2$, $BaCl_2$ and $AlCl_3$, *Z. Elektrochem.* 1951 **55** 327.
228. H. SIEMONSEN and U. SIEMONSEN, ΔH (formation) of LiCl and $ZrCl_4$, *Z. Elektrochem.* 1952 **56** 643.
229. J. H. SIMONS and M. G. POWELL, Thermochemical data of VCl_4, *J. Amer. Chem. Soc.* 1945 **67** 76.
230. G. B. SKINNER, J. W. EDWARDS and H. L. JOHNSTON, Vapour pressure of Zr, *J. Amer. Chem. Soc.* 1951 **73** 174.
231. H. A. SKINNER and N. B. SMITH, ΔH (hydrolysis) of BCl_3, liq., *Trans. Faraday Soc.* 1953 **49** 601.
232. C. J. SMITHELLS, Solubility of gases in metals (p. 379). Dissociation pressures of sulphides and phosphides (p. 451), *Metals Reference Book*, Butterworth, London, 1949.
233. F. H. SPEDDING and C. F. MILLER, ΔH (solution) in HCl·aq: Ce, Nd, $CeCl_3$ $NdCl_3$, *J. Amer. Chem. Soc.* 1952 **74** 4195.
234. R. SPEISER and H. L. JOHNSTON, Vapour pressure of Ga, *J. Amer. Chem. Soc.* 1953 **75** 1469.
235. W. STEINER and O. KRISEMENT, ΔH (solution in liquid tin): Fe–Co alloys, *Arch. Eisenhüttenw.* 1962 **33** 877.
236. F. D. STEVENSON, C. E. WICKS and F. E. BLOCK, Vapour pressures of WCl_4 and $HfCl_4$, US Bureau of Mines, Rep. Inv. No. 6367, 1964.
237. D. P. STEVENSON and D. M. YOST, Thermodynamics of phosphorus and its compounds, *J. Chem. Phys.* 1941 **9** 403.
238. M. F. STUBBS, J. A. SCHUFLE and A. J. THOMPSON, Equilibrium: $In_2O_3 + 3(H_2) = 2In + 3(H_2O)$ (823–1123°K), *J. Amer. Chem. Soc.* 1952 **74** 6201.
239. D. R. STULL, Vapour pressures of inorganic substances: review, *Ind. Eng. Chem.* 1947 **39** 540.

240. C. T. STUBBLEFIELD, H. EICK and L. EYRING, ΔH (solution in $HNO_3 \cdot aq$): Tb oxides, *J. Amer. Chem. Soc.* 1956 **78** 3877.
241. G. TAMMANN, Heat capacities of intermetallic compounds, *Lehrbuch der Metallographie*, Voss, Leipzig, 1921, p. 283.
242. D. N. TARASENKOV and A. V. KOMANDIN, Vapour pressures of $NbCl_5$ and $TaCl_5$, *Zh. obshchei Khim.* 1940 **14** 1319.
243. S. S. TODD, C_p (50–300°K): Ca_2SiO_4, Ca_3SiO_5, Zn_2SiO_4; Mg_2TiO_4, $Mg_2Ti_2O_5$; NaO_2, KO_2, Na_2O_2; HfO_2, $HfCl_4$, *J. Amer. Chem. Soc.* 1951 **73** 3277; 1952 **74** 4669; 1953 **75** 1227, 3035.
244. S. S. TODD and K. R. BONNICKSON, C_p (54–300°K): FeO, MnO, VO, *J. Amer. Chem. Soc.* 1951 **73** 3894.
245. S. S. TODD and J. P. COUGHLIN, C_p (50–300°K) and heat contents at higher temperatures: $FeCl_3$ and TiS_2, *J. Amer. Chem. Soc.* 1951 **73** 4184; 1952 **74** 525.
246. S. S. TODD and E. G. KING, C_p (50–300°K): Fe_2TiO_4, Fe_2TiO_5, *J. Amer. Chem. Soc.* 1953 **75** 4547.
247. S. S. TODD and R. E. LORENSEN, C_p (50–300°K): Ba_2TiO_4, Sr_2TiO_4, *J. Amer. Chem. Soc.* 1952 **74** 3764.
248. W. TRZBIATOWSKI and J. TERPILOWSKI, E.M.F., Ag–Mg, *Bull. Acad. Polon. Sci.* 1955 **3** 391.
249. E. E. UNDERWOOD and B. L. AVERBACH, Dissociation pressure of βAg–Zn, *J. Metals* 1951 **3** 1198.
250. K. UYENO, Vapour pressures of WO_3, MoO_3, CdO and TeO_2, *J. Chem. Soc. Japan* 1941 **62** 990.
251. J. E. VANCE and C. I. WHITMAN, Vapour pressure of Zn, *J. Chem. Phys.* 1951 **19** 744.
252. B. K. VESELOVSKI, Vapour pressures of PbS, CdS, ZnS and Sb_2S_3, *J. Appl. Chem. USSR* 1942 **15** 422.
253. A. A. VECHER and Y. I. GERASSIMOV, E.m.f. of cell: $\langle Cu\rangle/[Cu^+]/\langle Cu\text{–}Sb\rangle$, *Zh. fiz. Khim.* 1958 **32** 2835.
254. R. H. VALENTINE, T. F. JAMBOIS and J. L. MARGRAVE, Heat content (400–1125°K): ZrB_3, *J. Chem. Eng. Data* 1964 **9** 182.
255. H. VILLA, ΔG (formation) of metal chlorides: assessment, *J. Soc. Chem. Ind.* 1950, p. 59.
256. F. L. VOELZ, S_{298} and C_p (298–1000°K) from spectroscopic data of GeF_4, gas, *J. Chem. Phys.* 192 **20** 1662.
257. F. L. VOELZ, A. G. MEISTER and F. F. CLEVELAND, C_p of SiF_4 (100–1000°K), *J. Chem. Phys.* 1951 **19** 1084.
258. O. VOLLMER, R. KOHLHAAS and M. BRAUN, C_p (near m.pt.): Fe, Co, Ni, *Z. Naturforsch.* 1966 **21a** 181.
259. B. E. WALKER, J. A. GRAND and R. R. MILLER, Heat contents: Al_2O_3 (65–650°C), $MoSi_2$ (110–373°C), *J. Phys. Chem.* 1956 **60** 231.
260. H. VON WARTENBERG, Calorimetric determination of ΔH (formation) of PbF_4, $CrCl_3$, CrF_4, CrF_3, MgF_2, BeS and CaS, *Z. anorg. allg. Chem.* 1940 **244** 337; 1942 **249** 100; 1943 **252** 136.
261. H. VON WARTENBERG, ΔH (formation) of CF_4, COF_2 and COF_3, *Z. anorg. allg. Chem.* 1949 **258** 356.
262. H. VON WARTENBERG, ΔH (solution in $HCl \cdot aq$): La, La_2O_3, $La(OH)_3$, *Z. anorg. allg. Chem.* 1959 **299** 227.
263. M. WATANABE, Equilibrium: $2Co_3O_4 = 6CoO + (O_2)$, *Bull. Inst. Phys. Chem. Res. Tokyo* 1933 **12** 255.
264. F. WEIBKE and O. KUBASCHEWSKI, *Thermochemie der Legierungen*, Springer, Berlin, 1943.
265. E. F. WESTRUM, ΔH (solution in 6M $HCl \cdot aq$) of $PuBr_3$, Nat. Nucl. Energy Ser., Div. IV, **14B**, Pt. II (1949), p. 926.
265a. E. F. WESTRUM, C_p (5–307°K): ZrF_4, *J. Chem. Eng. Data* 1965 **10** 140.
266. E. F. WESTRUM and L. EYRING, ΔH (solution in $HCl \cdot aq$) of Pu; ΔH (precipitation) of PuF_3, Nat. Nucl. Energy Ser., p. 908.
267. E. F. WESTRUM and L. EYRING, ΔH (solution in $HCl \cdot aq$) of Np, *J. Amer. Chem. Soc.* 1952 **74** 2045.
268. E. F. WESTRUM, J. B. HATCHER and D. W. OSBORNE, C_p (10–315°K): NpO_2, *J. Chem. Phys.* 1953 **21** 419.
269. E. F. WESTRUM and K. S. PITZER, Thermodynamics of the system KHF_2–KF–HF, *J. Amer. Chem. Soc.* 1949 **71** 1940.
270. E. F. WESTRUM and H. P. ROBINSON, ΔH (solution in $HCl \cdot aq$) of Pu, $PuCl_3$ and PuOCl, Nat. Nucl. Energy Ser., Div. IV, **14B**, Pt. II (1949), pp. 914, 930.
271. E. WICKE and H. FRIZ, ΔH_0 of F_2, *Naturwiss.* 1952 **39** 522.
272. H. WISE, Dissociation of F_2, *J. Chem. Phys.* 1952 **20** 927.
273. E. L. WISEMAN and N. W. GREGORY, Vapour pressure of $TaBr_5$, *J. Amer. Chem. Soc.* 1949 **71** 2344.
274. L. WÖHLER and N. JOCHUM, $\bar{C}_p$'s and equilibria with H_2 of the oxides of Rh, Pd, Cu, *Z. Phys. Chem.* (A) 1933 **167** 176.
275. R. W. G. WYCKOFF, *Crystal Structures*, Interscience, London, 1948.
276. B. H. ZIMM and J. E. MAYER, Vapour pressures of some alkali halides, *J. Chem. Phys.* 1944 **12** 362.
277. C. B. ALCOCK and T. N. BELFORD, EMF measurements involving Pb oxides, *Trans. Faraday Soc.* 1964 **60** 822.
278. T. L. ALLEN and D. M. YOST, Dissociation equilibria of (MoI_3) and (ThI_4), *J. Chem. Phys.* 1954 **22** 855.
279. K. R. BONNICKSON, Heat contents (25–1500°C): Ca aluminates, *J. Phys. Chem.* 1955 **59** 220.
280. K. R. BONNICKSON, Heat contents (25–1530°C): Al_2TiO_5, Fe_2TiO_5, Fe_2TiO_4, Zn_2TiO_4, *J. Am. Chem. Soc.* 1955 **77** 2152.
281. L. BREWER, P. W. GILLES and F. A. JENKINS, Vapour pressure of graphite, *J. Phys. Chem.* 1948 **16** 797.
282. L. BREWER and L. F. PORTER, Vap. press. of Sn and MgO, *J. Chem. Phys.* 1953 **21** 2012; 1954 **22** 1867.
283. W. A. CHUPKA and M. G. INGHRAM, Vapour pressure of graphite and spectroscopic data, *J. Phys. Chem.* 1955 **59** 97.
284. J. W. COBBLE, W. T. SMITH and G. E. BOYD, ΔH (formation) of Re- and Tc-oxides. Vapour pressure of Tc oxides, *J. Amer. Chem. Soc.* 1953 **75** 5773, 5777, 5783.

285. J. P. COUGHLIN, Heat contents (25°–1570°C): Na_2SO_4, *J. Amer. Chem. Soc.* 1955 **77** 868.
286. D. J. A. DEAR and D. D. ELEY, ΔH (solution): AlI_3, *J. Chem. Soc.* 1954, 4684.
287. W. DE SORBO, C_p (13–70°K): Sb, *Acta metall.* 1953 **1** 503.
288. T. B. DOUGLAS and J. L. DEVER, C_p (0–895°C): Pb and LiF, *J. Amer. Chem. Soc.* 1954 **76** 4824, 4826.
289. T. B. DOUGLAS and J. L. DEVER, C_p (0–900°C): NaOH and $MoSi_2$, *J. Res. Nat. Bur. Stand.* 1954 **53** 81, 91.
290. T. B. DOUGLAS, J. F. EPSTEIN, J. L. DEVER and W. H. HOWLAND, Properties of Li, *J. Amer. Chem. Soc.* 1955 **77** 2144.
291. A. S. DWORKIN, D. J. SASMOR and E. R. VAN ARTSDALEN, BN: ΔH (combustion) and C_p (20–300°K), *J. Chem. Phys.* 1954 **22** 837.
292. A. S. DWORKIN, D. J. SASMOR and E. R. VAN ARTSDALEN, C_p (15–300°K): CdI_2, *J. Amer. Chem. Soc.* 1955 **77** 1304.
293. W. H. EVANS, R. JACOBSEN, T. R. MUNSON and D. D. WAGMAN, Thermodynamic properties of the alkali metals, *J. Res. Nat. Bur. Stand.* 1955 **55** 83.
294. G. GATTOW and A. SCHNEIDER, ΔH (combustion): Bi_2Se_3; equil. $Al_2Se_3 + 4Al = 3(Al_2Se)$. Vap. pressures: Zn (550–650°C), Se (400–500°C), Te (600–750°C), *Angew. Chem.* 1955 **67** 306; 1956 **68** 520; 1959 **71** 189.
295. N. W. GREGORY and T. R. BURTON, ΔH (solutions in KCl·aq) of Cr iodides, *J. Amer. Chem. Soc.* 1953 **75** 6054.
296. N. W. GREGORY and R. O. MACLAREN, Equilibrium measurements on Fe bromides, *J. Phys. Chem.* 1955 **59** 110.
297. P. GROSS and D. L. LEVI, Equilibria involving Ti halides, *Int. Congr. Pure Appl. Chem., Paris,* 1957.
298. M. HOCH, P. E. BLACKBURN, D. P. DINGLEDY and H. L. JOHNSTON, Vapour pressure of graphite (1900–2500°C), *J. Phys. Chem.* 1955 **59** 97.
299. W. S. HOLMES, ΔH (combustion): phosphorus, *Trans. Faraday Soc.* 1962 **58** 1916.
300. R. E. HONIG, ΔH (evaporation): Ge_{1-4}, *J. Chem. Phys.* 1954 **22** 1610.
301. J. H. E. JEFFES, F. D. RICHARDSON and J. PEARSON, ΔH (solution in HF·aq) of Mn_2SiO_4, MnO, SiO_2, MnS, *Trans. Faraday Soc.* 1954 **50** 364.
302. R. S. JESSUP, R. E. MCCOSKEY and R. H. NELSON, ΔH (formation) of CF_4, *J. Amer. Chem. Soc.* 1955 **77** 244.
303. E. G. KING, C_p(51–298°K): Mn_2O_3, Nb_2O_5; Al_2TiO_5, $Ca_3Ti_2O_7$, Li_2TiO_3 and Zn_2TiO_4, *J. Amer. Chem. Soc.* 1954 **76** 3289; 1955 **77** 2150.
304. E. G. KING, C_p (51–298°K) of Ca aluminates, *J. Phys. Chem.* 1955 **59** 218.
305. E. G. KING, R. L. ORR and K. R. BONNICKSON, C_p (50–1670°K): $CaTiSiO_5$, *J. Amer. Chem. Soc.* 1954 **76** 4320.
306. P. KOSFTAD, W. E. WALLACE and L. J. HYVÖNEN, Diss. pressures in the systems tantalum-hydrogen and vanadium-hydrogen, *J. Amer. Chem. Soc.* 1959 **81** 5015, 5019.
307. O. KUBASCHEWSKI and J. A. CATTERALL, *Thermodynamic Data of Alloys*, Pergamon Press, 1956.
308. M. F. LANTRATOV and E. V. TSARENKO, E.m.f. measurements on liquid K-Tl and Na-Ga alloys, *Zh. fiz. Khim.* 1959 **33** 1792; *Zh. priklad Khim.* 1961 **34** 2435.
309. H. R. LOHR, D. W. OSBORNE and E. F. WESTRUM, C_p(5–300°K): ThF_4, *J. Amer. Chem. Soc.* 1954 **76** 3837.
310. A. D. MAH, ΔH (combustion): Ba, Sr; Al_4C_3, US Bur. Mines, Rep. Inv. No. 6171 (1963); No. 6415 (1964).
311. A. D. MAH, ΔH (combustion): Cd, Cr, *J. Amer. Chem. Soc.* 1954 **76** 3363.
312. M. M. MAKANSI, C. H. MUENDEL and W. A. SELKE, Vapour pressure of Na (620–1130°C), *J. Phys. Chem.* 1955 **59** 40.
313. E. NEALE and L. T. D. WILLIAMS, ΔH (formation): PCl_3, $SOCl_2$, SO_2Cl_2, *J. Chem. Soc.* 1954, 2156.
314. C. J. O'BRIEN and K. K. KELLEY, Heat contents: NaF, AlF_3, Na_3AlF_6, *J. Amer. Chem. Soc.* 1957 **79** 5616.
315. R. L. ORR, Heat contents: Mn_2O_3, VO; SnS, SnS_2, *J. Amer. Soc.* 1954 **76** 857; *J. Phys. Chem.* 1958 **62** 124.
316. R. L. ORR and J. P. COUGHLIN, Heat contents (25–1530°C): Mg_2TiO_4, $MgTi_2O_5$, *J. Amer. Chem. Soc.* 1952 **74** 3186.
317. D. W. OSBORNE and E. F. WESTRUM, C_p (13–300°K): ThO_2, *J. Chem. Phys.* 1953 **21** 1884.
318. R. B. PEPPLER and E. S. NEWMAN, ΔH (solution in HCl·aq): Ba aluminates and their hydrates, *J. Res. Bur. Stand. Nat.* 1951 **47** 439.
319. G. R. St. PIERRE, W. T. EBIHARA, M. J. POOL and R. SPEISER, Equilibria involving tungsten oxides, *Trans. Met. Soc. AIME* 1962 **224** 259.
320. T. N. REZUKHINA, V. A. LEVITSKI and P. OSHEGOV, ΔG (formation) of $FeAl_2O_4$ (e.m.f.), *Zh. fiz. Khim.* 1963 **37** 687.
321. A. W. RICHARDS, Equilibria involving SnS, *Trans. Faraday Soc.* 1955 **51** 1193.
322. H. SCHÄFER, Vapour pressure: $NiCl_2$ (700–800°C), $FeCl_2$ (710–840°C), $MnCl_2$ (725–945°C), $CoCl_2$ (650–800°C), *Z. anorg. allg. Chem.* 1955 **278** 300.
323. H. SCHÄFER and L. BAYER, Properties of Nb chlorides, *Z. anorg. allg. Chem.* 1954 **277** 140.
324. K. A. SENSE, M. J. SNYDER and R. B. FILBERT, Vapour pressure of ZrF_4 (620–890°C), *J. Phys. Chem.* 1954 **58** 995.

325. R. H. SHERMAN and W. F. GIAUQUE, Thermodynamic properties of AsH_3, *J. Amer. Chem. Soc.* 1955 **77** 2154.
326. C. H. SHOMATE and A. J. COHEN, Heat contents: Li_2O (25–780°C), LiOH (25–630°C), *J. Amer. Chem. Soc.* 1955 **77** 285.
327. S. A. SHCHUKAREV, M. P. MOROZOVA and T. A. STOLYAROVA, ΔH (combustion): MnP, MnAs, MnSb, MnBi, *Zh. obshchei Khim.* 1961 **31** 1773.
328. H. A. SKINNER and N. B. SMITH, ΔH (hydrolysis) of BBr_3, *Trans. Faraday Soc.* 1955 **51** 19.
329. D. SMITH, A. S. DWORKIN and E. R. VAN ARTSDALEN, ΔH (combustion): B_4C, *J. Amer. Chem. Soc.* 1955 **77** 2654.
330. W. T. SMITH, G. D. OLIVER and J. W. COBBLE, C_p (20–300°K) of Re, *J. Amer. Chem. Soc.* 1953 **75** 5785.
331. F. H. SPEDDING and A. H. DAANE, Melting points of rare earth metals, *J. Metals* 1954 **6** 505.
332. F. H. SPEDDING and J. P. FLYNN, ΔH (solution in HCl·aq) of rare earth metals and their chlorides, *J. Amer. Chem. Soc.* 1954 **76** 1474.
333. J. E. SPICE, L. A. K. STAVELEY and G. A. HARROW, Thermodynamic properties of $Ni(CO)_4$, *J. Chem. Soc.* 1955, 100.
334. C. G. STEVENS and E. T. TURKDOGAN, ΔH (solution in HF·aq) of Na_2O, SiO_2 and $Na_2Si_2O_5$, *Trans. Faraday Soc.* 1955 **51** 356.
335. A. J. THOMPSON, M. F. STUBBS and J. A. SCHUFLE, Reduction equilibria of In sulphides with H_2, *J. Amer. Chem. Soc.* 1954 **76** 341.
336. D. H. TOMLIN, Vapour pressure of Ca, *Proc. Phys. Soc.* 1954 **67** 787.
337. A. C. VICTOR and T. B. DOUGLAS, Heat contents of MgO and BeO (ice calorimeter), *J. Res. Nat. Bur. Stand.* (A) 1963 **67** 325.
338. R. J. ACKERMANN, P. W. GILLES and R. J. THORN, Vapour pressure of UO_2, *J. Chem. Phys.* 1956 **25** 1089.
339. R. J. ACKERMANN and E. G. RAUH, Effusion measurements: Sc, Y, La, *J. Chem. Phys.* 1962 **36** 448.
340. S. M. ARIYA, S. A. SHCHUKAREV and V. B. GLUSHKOVA, Properties of CrO_2; vapour pressures of MoO_3 and CrO_3, *Zh. obshchei Khim.* 1953 **23** 1241, 2063.
341. D. N. ASTROV, E. S. ITSKEVICH and K. A. SHAFIROV, C_p(10–302°K): $Mo(CO)_6$, *Zh. fiz. Khim.* 1955 **29** 424.
342. R. BARANY, E. G. KING and S. S. TODD, ΔH (solution in HF·aq): Sr and Ba silicates, *J. Amer. Chem. Soc.* 1957 **79** 3639.
343. E. F. BARROW, P. G. DODSWORTH, G. DRUMMOND and E. A. N. S. JEFFRIES, Vapour pressure of GeS (305–380°K), *Trans. Faraday Soc.* 1955 **51** 1480.
344. R. F. BARROW, E. A. N. S. JEFFRIES and J. M. SWINSTEAD, Vapour pressures: TIF, TlCl, TlBr, TlI, *Trans. Faraday Soc.* 1955 **51** 1650.
345. R. F. BARROW, J. W. C. JOHNS and F. J. SMITH, Spectroscopic properties of AlF, *Trans. Faraday Soc.* 1956 **52** 913.
346. R. F. BARROW, A. C. P. PUGH and F. J. SMITH, Vapour pressures: InBr, InI, *Trans Faraday Soc.* 1955 **51** 1657.
347. J. L. BARTON AND H. BLOOM, Vapour pressures: NaCl, KCl, $CdCl_2$, $PbCl_2$, *J. Phys. Chem.* 1956 **60** 1413.
348. K. R. BONNICKSON, Heat contents (298–1800°K): $CaFe_2O_4$, $Ca_2Fe_2O_5$, $MgFe_2O_4$, *J. Amer. Chem. Soc.* 1954 **76** 1480.
349. B. J. BOYLE, E. G. KING and K. C. CONWAY, ΔH (combustion) of Co, *J. Amer. Chem. Soc.* 1954 **76** 3835.
350. E. BRACKETT and L. BREWER, Thermodynamics of Bi, Priv. comm. 1957.
351. L. S. BROOKS, Vapour pressure of Po (438–745°C), *J. Amer. Chem. Soc.* 1955 **77** 3211.
352. S. BRUNAUER, D. L. KANTRO and C. H. WEISE, ΔH (solution in HF·HNO_3·aq): Ca_3SiO_5, Ca_2SiO_4, CaO, *J. Phys. Chem.* 1956 **60** 771.
353. B. R. BURGAN, R. C. HALL and R. F. HEHEMANN, Dissociation pressure: Hg–Pb, *J. Inst. Metals* 1951 **80** 413.
354. R. H. BUSEY, C_p (15–300°K): Re_2O_7, *J. Amer. Chem. Soc.* 1956 **78** 3262.
355. S. C. CARNIGLIA and B. B. CUNNINGHAM, Vapour pressures of Am, AmF_3 and PuF_3, *J. Amer. Chem. Soc.* 1955 **77** 1451, 1502.
356. E. CATALANO and J. W. STOUT, C_p (12–300°K): NiF_2, *J. Chem. Phys.* 1955 **23** 1284.
357. K. CLUSIUS and C. C. LOSA, C_p (10–273°K): Rh, Ir, Ta, *Z. Naturforsch.* 1955 **10a** 545, 939.
358. F. A. COTTON, A. K. FISCHER and G. WILKINSON, ΔH (combustion): $Cr(CO)_6$, $Mo(CO)_6$, $W(CO)_6$, *J. Amer. Chem. Soc.* 1956 **78** 5168.
359. J. P. COUGHLIN, ΔH (solution in HCl·aq): CaO–Al_2O_3; $LiAlO_2$, $NaAlO_2$ etc., *J. Amer. Chem. Soc.* 1956 **78** 5479; **79** 2397.
360. T. M. DAUPHINEE, D. L. MARTIN and H. PRESTON-THOMAS, C_p (20–330°K): K, Rb, Cs, *Proc. Roy. Soc.* 1955 **233** 214.
361. C. J. B. FINCHAM and R. A. BERGMAN, Thermodynamic properties of CS, *J. Metals* 1957 **9** 690.
362. J. FISCHER and J. BINGLE, Vapour pressure (24–116°C): Br_2, *J. Amer. Chem. Soc.* 1955 **77** 6511.
363. A. K. FISCHER, F. A. COTTON and G. WILKINSON, ΔH (combustion): $Ni(CO)_4$, *J. Amer. Chem. Soc.* 1957 **79** 2044.

364. H. F. FISCHMEISTER, L_t and L_f of sulphates, reviewed, *Z. phys. Chem.* (NF) 1956 **7** 91.
365. P. W. GILLES and J. L. MARGRAVE, ΔH (solution in aqua): Na_2O_2, NaO_2, KO_2, *J. Phys. Chem.* 1956 **60** 1331.
366. D. C. GINNINGS, T. B. DOUGLAS and A. F. BALL, Heat contents of Be (0–900°C), *J. Amer. Chem. Soc.* 1951 **73** 1236.
367. E. GREENBERG and E. F. WESTRUM, C_p (10–340°K): $UOCl_2$, $UOBr_2$, *J. Amer. Chem. Soc.* 1956 **78** 5144.
368. M. GRIFFEL, R. E. SKOCHDOPOLE and F. H. SPEDDING, C_p (15–330°K): Gd; Dy, *Phys. Rev.* 1954 **93** 657; *J. Chem. Phys.* 1956 **25** 75.
369. E. H. HALL, J. M. BLOCHER and I. E. CAMPBELL, Vapour pressures of TiF_4 and $TiBr_4$, Priv. comm. 1957.
370. K. J. HILL and E. R. S. WINTER, Equilibrium: $CaCO_3 = CaO + CO_2$, *J. Phys. Chem.* 1956 **60** 1361.
371. E. J. HUBER, E. L. HEAD and C. E. HOLLEY, ΔH (combustion): Er, Ho, Y, Lu, Dy, Tm, Yb, *J. Phys. Chem.* 1956 **60** 1582; 1957 **61** 497, 1021; 1960 **64** 379, 1763.
372. E. J. HUBER and C. E. HOLLEY, ΔH (combustion): Gd, *J. Amer. Chem. Soc.* 1955 **77** 1444.
373. E. J. HUBER, C. O. MATTHEWS and C. E. HOLLEY, ΔH (combustion): Sm, *J. Amer. Chem. Soc.* 1955 **77** 6493.
374. G. L. HUMPHREY, ΔH (combustion): TaC, TiC, *J. Amer. Chem. Soc.* 1954 **76** 978.
375. R. JUZA and W. UPHOFF, ΔH (neutralisation): $LiOH{\cdot}aq + H_2S{\cdot}aq$, *Z. anorg. allg. Chem.* 1956 **287** 113.
376. S. KACHI, E.M.F.: Ag–Mg, *Sci. Rep. Res. Inst. Tôhoku Univ.* 1955 **7** 351.
377. E. G. KING, C_p (53–295°K): spinels, *J. Phys. Chem.* 1956 **60** 410.
378. E. G. KING, C_p (51–298°K): $CaFe_2O_4$, $Ca_2Fe_2O_5$, $MgFe_2O_4$; $LiAlO_2$, $LiFeO_2$, $NaFeO_2$, $NaAlO_2$; GeO_2, *J. Amer. Chem. Soc.* 1954 **76** 5849; 1955 **77** 3189; 1965 **80** 1799.
379. E. G. KING, C_p (54–296°K): Na_3AlF_6, NaF, AlF_3; NiO, CoO, Co_3O_4, *J. Amer. Chem. Soc.* 1957 **79** 2056, 2399.
380. H. J. KIRSCHNING, K. PLIETH and I. N. STRANSKI, E.M.F.: $As_2O_3 + 3H_2 = 2As + 3H_2O$; C_p (20–400°K): As_2O_3. (As_2O_3 = claudetite, arsenolite), *Z. Kristallogr.* 1954 **106** 172.
381. F. KRAUSS and H. WARNCKE, C_p (180–1160°C): Ni, *Z. Metallkunde* 1955 **46** 61.
382. C. KRÖGER and W. JANETZKO, Calorimetry of slow processes involving silicates, *Z. anorg. allg. Chem.* 1956 **284** 83.
383. G. R. MACHLAN, C. T. STUBBLEFIELD and L. EYRING, ΔH (solution in HCl·aq): $SmCl_2$, $YbCl_2$, *J. Amer. Chem. Soc.* 1955 **77** 2975.
384. A. D. MAH and B. J. BOYLE, ΔH (combustion): NbC, ZrC, *J. Amer. Chem. Soc.* 1955 **77** 6512.
385. A. D. MAH and N. L. GELLERT, ΔH (combustion): NbN, TaN, ZrN, *J. Amer. Chem. Soc.* 1956 **78** 3261.
386. C. E. MESSER, L. G. FASOLINO and C. E. THALMEYER, ΔH (solution in aqua:) Li, Na, K, LiH, NaH, KH, *J. Amer. Chem. Soc.* 1955 **77** 4524.
387. J. G. MARSHALL, L. A. STAVELEY and K. R. HART, Specific heat of NH_4Cl and CBr_4, *Trans. Faraday Soc.* 1956 **52** 19.
388. C. A. NEUGEBAUER and J. L. MARGRAVE, ΔH (formation): AlN, *Z. anorg. allg. Chem.* 1957 **290** 82.
389. K. J. NEUVONEN, ΔH (solution in HF·aq): $Ca_3Mg(SiO_4)_2$, $CaMgSiO_4$, *Amer. J. Sci.*, Browen Vol. II, 1952, 373.
390. E. S. NEWMAN and R. H. OFFMAN, ΔH (solution): $6CaO{\cdot}2Al_2O_3{\cdot}Fe_2O_3$, $2CaO{\cdot}Fe_2O_3$, *J. Res. Nat. Bur. Stand.* 1956 **56** 313.
391. M. OLETTE, Heat contents (1100–1600°C): Si, *Compt. rend.* 1957 **244** 1033.
392. P. A. G. O'HARE and W. N. HUBBARD, ΔH (combustion in F_2): As, *J. Phys. Chem.* 1965 **69** 4358.
393. M. N. PAPADOPOULOS and W. F. GIAUQUE, C_p (15–300°K): $CdSO_4$, $CdSO_4{\cdot}H_2O$, *J. Amer. Chem. Soc.* 1955 **77** 2740.
394. G. PETIT and A. CREMIEU, L_f (cryoscopic) of fluorides, *Compt, rend.* 1956 **243** 360.
395. T. E. PHIPPS, G. W. SEARS, R. L. SEIFERT and O. C. SIMPSON, Vapour pressure of Pu, *Amer. Chem. Abstr.* 1956 **50** 14295.
396. J. C. PLATTEEUW and G. MEYER, Equilibrium: SnO_2–Sn–CO_2–CO, *Trans. Faraday Soc.* 1956 **52** 1066.
397. M. M. POPOV and D. M. GINZBURG, Heat contents: Na_2CO_3, Na_2SO_4 NaOH, *J. Gen. Chem. USSR* 1956 **26** 971.
398. YU. A. PRIZELKOV and A. N. NESMEYANOV, Vapour pressure: Ca-45 (475–670°C), Sr-89 (400–600°C), *Dokl. Akad. Nauk SSSR* 1954 **95** 1207.
399. M. H. RAND and O. KUBASCHEWSKI, *The Thermochemical Properties of Uranium Compounds*, Oliver & Boyd, Edinburgh, 1963.
400. F. D. RICHARDSON and L. E. WEBB, Thermodynamics of PbO–SiO_2, *Trans. Inst. Min. Met.* 1955 **64** 529.
401. D. A. ROBINS and I. JENKINS, ΔH (formation) of silicides of transition metals, *Acta Metallurg*, 1955 **3** 598.
402. E. RUDZITIS, R. TERRY, H. M. FEDER and W. N. HUBBARD, ΔH (fluorination): Zn, *J. Phys. Chem.* 1964 **68** 617.
403. K. SANO, K. OKAJIMA and S. TATSUO, E.M.F. Sn–Zn/Zn, *Mem. Fac. Eng. Nagoya Univ.* 1953 **5** 299.
404. T. SATOW, S. KACHI and K. IWASE, Equilibria involving Co–Fe, *Sci. Rep. Res. Inst. Tôhoku Univ.* 1956 **8** 503.
405. W. R. SAVAGE, D. E. HUDSON and F. H. SPEDDING, Vapour pressure (789–833°K): Sm, *J. Chem. Phys.* 1959 **30** 221.

406. H. SCHÄFER, F. E. WITTIG and M. JORI, Thermodynamic properties of FeOCl, *Z. anorg. allg. Chem.* 1956 **287** 61.
407. H. L. SCHLÄFER and H. H. SCHMIDTKE, ΔH (formation): $TiBr_4$, *Z. phys. Chem.* (NF) 1957 **11** 296.
408. T. H. SCHOFIELD, L_t: α–β Ti, *J. Inst. Metals* 1956 **85** 68.
409. A. W. SEARCY and C. E. MYERS, Vapour pressures: B (2115–2413°K), B_2O_3 (1500–1566°K), *J. Phys. Chem.* 1957 **61** 957.
410. E. M. SHERWOOD, D. M. ROSENBAUM, J. M. BLOCHER and I. E. CAMPBELL, Vapour pressure of Re (2220–2725°C), *J. Electrochem. Soc.* 1955 **102** 650.
411. E. SHEMAZAKI and T. WADA, Vapour pressure of GeS (338–399°C), *Bull. Chem. Soc. Japan* 1956 **29** 294.
412. R. E. SKOCHDOPOLE, M. GRIFFEL and F. H. SPEDDING, C_p (15–320°K) of Er, *J. Chem. Phys.* 1955 **23** 2258.
413. W. I. SMIRNOVA and B. F. ORMONT, *Dokl. Akad. Nauk SSSR* 1955 **100** 127.
414. F. J. SMITH and R. F. BARROW, ΔH (solution in $KBr \cdot Br_2 \cdot aq$): In, InBr, InI, I, *Trans. Faraday Soc.* 1955 **51** 1478.
415. J. W. STOUT and E. CATALANO, C_p (11–300°K): ZnF_2, FeF_2, CoF_2, *J. Chem. Phys.* 1955 **23** 2013.
416. K. A. STAFAROV and T. N. REZUKHINA, ΔH (combustion): $Cr(CO)_6$, $Mo(CO)_6$ $W(CO)_6$, *Amer. Chem. Abstr.* 1955 **49** 2173.
417. R. M. SWIFT and D. WHITE, C_p (20–300°K): MgB_2, MgB_4, *J. Amer. Chem. Soc.* 1957 **79** 3641.
418. A. A. WOOLF, ΔH (hydrolysis): AuF_3, ThF_2, *J. Chem. Soc.* 1954, 4694.
419. A. N. ZELIKMAN, N. N. GOROVITS and T. E. PROSENKOVA, Vapour pressure of MoO_3 (820–1000°C), *Zh. neorg. Khim.* 1956 **1** 632.
420. G. M. ZHABROVA and B. M. KADENATZI, Dissociation: $Mg(OH)_2 = MgO + (H_2O)$, *Zh. obshchei Khim.* 1954 **24** 1135.
421. A. S. ABASSOW, A. V. NIKOLSKAYA, Y. I. GERASSIMOV and V. P. VASIL'EV, E.m.f. measurements: systems Ga–Sb, Ga–Te, In–As, *Dokl. Akad. Nauk SSSR* 1964 **156** 118, 1140, 1399.
422. K. P. ABRAHAM, M. W. DAVIES, J. L. BARTON and F. D. RICHARDSON, Equilibrium of Pt–Mn alloys with MnO_{1+x}, *Acta met.* 1960 **8** 888.
423. C. B. ALCOCK and G. W. HOOPER, Volatility in oxygen: Pt, Rh, Ru, Ir, Pd, *Proc. Roy. Soc.* (A) 1960 **254** 551.
424. A. T. ALDRED and J. N. PRATT, Dissociation pressures (900–1080°C): Ag–Pb system, *Trans. Faraday Soc.* 1961 **57** 611.
425. A. T. ALDRED and K. M. MYLES, Diss. pressures (1450–1650°K): V–Cr system, *Trans. Met. Soc. AIME* 1964 **230** 736.
426. R. J. L. ANDON, J. F. COUNSELL, J. F. MARTIN, and H. J. HEDGER, C_p (11–330°K): UC, U_2C_3, $UC_{1\cdot 91}$, *Trans. Faraday Soc.* 1964 **60** 1030.
427. R. J. L. ANDON, J. F. COUNSELL, H. MCKERREL and J. F. MARTIN, C_p (12–324°K): P_2O_5, *Trans. Faraday Soc.* 1963 **59** 2702.
428. I. BACKHURST, Adiabatic calorimeter. C_p: Ti, CrFe, steels, *J. Iron Steel Inst.* 1958 **189** 124.
429. J. D. BAIRD and J. TAYLOR, Equilibrium (1400–1580°C): $SiO_2 + 3C = SiC + 2CO$, *Trans. Faraday Soc.* 1958 **54** 526.
430. R. BARANY, ΔH (solution): $CaWO_4$, $CaMoO_4$, $MgMoO_4$, US Bur. Mines, Rep. Inv. No. 6143, 1962.
431. I. R. BARTKY and W. F. GIAUQUE, C_p (14–300°K): TlCl, *J. Amer. Chem. Soc.* 1959 **81** 4169.
432. R. G. BAUTISTA and J. L. MARGRAVE, Vapour pressure (887–983°K): MnF_2, *J. Phys. Chem.* 1963 **67** 1564.
433. T. N. BELFORD and C. B. ALCOCK, E.m.f. measurements on Sn–SnO_2 system, *Trans. Faraday Soc.* 1965 **61** 443.
434. W. E. BELL, M. C. GARRISON and U. MERTEN, Diss. pressure: $RuCl_3$; equilibria involving Ru chlorides (650–1500°C), *J. Phys. Chem.* 1960 **64** 145; 1961 **65** 510.
435. W. E. BELL, M. TAGAMI and U. MERTEN, Thermodyn. properties of $PdCl_2$ (transpiration technique), *J. Phys. Chem.* 1961 **65** 517.
436. G. R. BELTON and F. D. RICHARDSON, Fuming of iron, *Trans. Faraday Soc.* 1962 **58** 1562.
437. J. R. BERG, F. H. SPEDDING and A. H. DAANE, quoted by 99. Heat contents (273–1948°K): Y.
438. L. R. BIDWELL and R. SPEISER, E.m.f. measurements (700–1200°C): Ni–Pd system, *Acta met.* 1965 **13** 61.
439. M. BIGORGNE, Spectroscopic study of $Ni(CO)_4$, *Compt. rend.* 1958 **246** 1685.
440. J. L. BILLS and F. A. COTTON, ΔH (solution in H_2O_2–HF): GeO_2, Ge, *J. Phys. Chem.* 1964 **68** 802.
441. J. N. BLOCHER and E. H. HALL, Reaction pressures: $\frac{1}{2}\langle Hg_2F_2\rangle + \langle TiF_3\rangle = (TiF_4) + \{Hg\}$, *J. Phys. Chem.* 1959 **63** 127.
442. G. D. BLUE, J. W. GREEN, R. G. BAUTISTA and J. L. MARGRAVE, By vacuum microbalance and mass spectrometer: $\langle CaF_2\rangle = (CaF_2)$; (Ca) + (AlF) = (Al) + (CaF), *J. Phys. Chem.* 1963 **67** 877.
443. G. F. BOLLING, Heat contents: Bi_2Te_3, *J. Chem. Phys.* 1960 **33** 305.
444. R. S. BRADLEY and T. DRURY, Vapour pressure (22–56°C): CBr_4, *Trans. Faraday Soc.* 1959 **55** 1844.
445. A. P. BRADY, O. E. MYERS and J. K. CLAUSS, C_p (50–300°K): MoF_6, NbF_5, *J. Phys. Chem.* 1960 **64** 588.
446. L. BREWER and A. W. SEARCY, Vapour pressure (1383–1468°K): Al, *J. Amer. Chem. Soc.* 1951 **73** 5308.
447. F. P. BUNDY, Graphite: p (9–125 kbar), *J. Chem. Phys.* 1963 **38** 618.
448. D. L. BURK, I. ESTERMANN and S. A. FRIEDBERG, C_p (10–200°K): Hf, *Z. physik. Chem.* (NF) 1958 **16** 183.

449. G. H. CADY and G. B. HARGREAVES, Vapour pressures of some transition metal fluorides, *J. Chem. Soc.* 1961, 1563, 1568.
450. L. G. CARPENTER and W. N. MAIR, Evaporation of Ti (1650–1810°K), *Proc. Phys. Soc.* (B) 1951 **64** 57.
451. N. J. CARRERA, R. F. WALKER and E. R. PLANTE, Vapour pressures of Ru and Os (Langmuir method), *J. Res. Nat. Bur. Stand.* 1964 **68A** 325.
452. R. G. CAVELL and H. C. CLARK, Heats of hydrolysis: VF_4, VF_5, *Trans. Faraday Soc.* 1963 **59** 2706.
453. R. C. CHISHOLM and J. W. STOUT, C_p (11–320°K): $MnCl_2$, $CoCl_2$, *J. Chem. Phys.* 1962 **36** 972.
454. W. A. CHUPKA, M. G. INGHRAM and R. F. PORTER, Vapour pressure (1723–1873°K): La, *J. Chem. Phys.* 1956 **24** 792.
455. K. CLUSIUS and U. PIESBERGEN, C_p (11–272°K): Ru, *Z. Naturforsch.* 1959 **14a** 23.
456. C. N. COCHRAN and L. M. FOSTER, Heats of reaction involving silica, *J. Phys. Chem.* 1962 **66** 360.
457. C. N. COCHRAN and L. M. FOSTER, Vapour pressure (1180–1383°K): Ga, *J. Electrochem. Soc.* 1962 **109** 144.
458. R. COLIN and J. DROWART, Thermodyn. study of SnSe and SnTe using a mass-spectrometer, *Trans. Faraday Soc.* 1964 **60** 673.
459. C. M. COOK, Fe_2Cl_6: vapour pressure and heat content, *J. Phys. Chem.* 1962 **66** 219.
460. J. P. COUGHLIN, ΔH (solution): $AlCl_3 \cdot 6H_2O$, *J. Phys. Chem.* 1958 **62** 419.
461. J. P. COUGHLIN, ΔH (solution): NaF, Na_3AlF_6, *J. Amer. Chem. Soc.* 1958 **80** 1802.
462. D. CUBICCIOTTI, Equilibria involving Bi halides, *J. Phys. Chem.* 1960 **64** 791, 1506.
463. D. CUBICCIOTTI and H. EDING, Heat contents (300–1000°K): $ZnCl_2$, $ZnBr_2$; Tl monohalides, *J. Chem. Phys.* 1964 **40** 978; *J. Chem. Eng. Data* 1965 **10** 343.
464. A. H. DAANE and F. H. SPEDDING, Vapour pressures (1611–2038°K): La, Ce, 134th Meeting, *Amer. Chem. Soc.*, 1958.
465. E. A. DANCY, J. H. EVERETT and C. L. MCCABE, Equilibration of CeC and PrC with CH_4/H_2 mixtures, *Trans. Met. Soc. AIME* 1962 **224** 1095.
466. A. J. DARNELL, W. A. MCCOLLUM and T. A. MILNE, Vapour pressure (1757–1956°K): Th, *J. Phys. Chem.* 1960 **64** 341.
467. A. J. DARNELL and S. J. YOSIM, Equilibria involving Bi chlorides, *J. Phys. Chem.* 1959 **63** 1813.
468. S. G. DAVIS, D. F. ANTHROP and A. W. SEARCY, Vapour pressure of Si and dissociation pressure of SiC, *J. Chem. Phys.* 1961 **34** 659.
469. V. I. DAVYDOV and N. P. DIEV, Vapour pressure of GeS, *J. Inorg. Chem.* (*Russ.*) 1957 **2** 2003.
470. G. DE MARIA, J. DROWART and M. G. INGHRAM, Mass-spectrometric study of In_2, Sb_2 and InSb, *J. Chem. Phys.* 1959 **31** 1076.
471. W. DESORBO, C_p (15–90°K): Bi, W, *J. Phys. Chem.* 1958 **62** 965.
472. E. W. DEWING and F. D. RICHARDSON, Free energies of the reactions: $S_2 + 2SO_2 = 4SO$ and $2SO_2 = SO + SO_3$, *Trans. Faraday Soc.* 1958 **54** 679.
473. L. H. DREGER and J. L. MARGRAVE, Vapour pressures (Langmuir method): Pt (1571–1786°K), Pd (1220–1640°K), Rh (1803–2068°K), *J. Phys. Chem.* 1960 **64** 1323; 1961 **65** 2106.
474. A. S. DWORKIN and M. A. BREDIG, Heat contents of halides of alkali, alkaline earth and rare earth metals, *J. Phys. Chem.* 1960 **64** 269; 1963 **67** 697, 2499.
475. B. H. ECKSTEIN and E. R. VAN ARTSDALEN, ΔH (combustion): B, *J. Amer. Chem. Soc.* 1958 **80** 1352.
476. J. J. EGAN, W. MCCOY and J. BRACKER, E.m.f. measurements pertaining to the formation of $MgCl_2$, $CeCl_3$, $ThCl_4$, etc., *IAEA Conf., Vienna*, 1962, p. 163.
477. R. D. EULER and E. F. WESTRUM, C_p (6–300°K): TiF_4, *J. Chem. Phys.* 1961 **65** 132.
478. M. FARBER, R. T. MEYER and J. L. MARGRAVE, Vapour pressure (1026–1350°K): NiF_2, *J. Phys. Chem.* 1958 **62** 883.
479. H. M. FEDER, W. N. HUBBARD, S. S. WISE and J. L. MARGRAVE, Combination of data involving HF, *J. Phys. Chem.* 1963 **67** 1148.
480. A. FERRIER and M. OLETTE, Heat contents (1123–2103°K): Al_2O_3, *Compt. rend.* 1962 **254** 4293.
481. C. B. FINCH and J. B. WAGNER, E.m.f., $\langle Pb \rangle / [Pb^{++}] / \{PbSe, Se\}$, *J. Electrochem. Soc.* 1960 **107** 932.
482. P. FLUBACHER, A. J. LEADBETTER and J. A. MORRISON, C_p (3–300°K): Si, Ge, *Phil. Mag.* 1959 **4** 273.
483. D. R. FREDERICKSON, R. L. NUTTAL, H. E. FLOTOW and W. N. HUBBARD, ΔH (combustion): ZrH_2, ZrD_2, *J. Phys. Chem.* 1963 **67** 1506.
484. S. N. GADZHILEV and K. A. SHAFIROV, Special calorimetric determination: TeO_2, *Izv. Akad. Nauk. Azerb. SSR*, Ser. Fiz. Mat. i Tekh. Nauk, 1962 [1] 47.
485. K. GALLAGHER, G. E. BRODALE and T. E. HOPKINS, C_p(15–300°K) = $PbSO_4$, *J. Phys. Chem.* 1960 **64** 687.
486. G. GATTOW, ΔH (combustion): various modifications of Se, *Z. anorg. allg. Chem.* 1962 **317** 245.
487. G. GATTOW and H. SCHRÖDER, L_t: α–δ, Bi_2O_3, *Z. anorg. allg. Chem.* 1962 **318** 176.
488. P. V. GEL'D and F. G. KUSENKO, Thermodyn. properties of Nb compounds, *Isvest. Akad. Nauk SSSR* 1960 (No. 2) 79.
489. H. GEOFFRAY, A. FERRIER and M. OLETTE, Heat contents (25–1920°C): Ni, *Compt. rend.* 1963 **256** 139.
490. Y. I. GERASSIMOV and A. V. NIKOLSKAYA, E.m.f. measurements on Bi_2Te_3 and Sb_2Te_3, *Proc. 4th Conf. Semicond. Materials*, 1961, p. 30.

491. O. GLEMSER and U. STÖCKER, Equilibrium: $\langle CdO \rangle = (Cd) + \frac{1}{2}(O_2)$, *Ber. Bunsenges. Phys. Chem.* 1963 **67** 505.
492. O. GLEMSER *et al.*, Vapour pressures: CrO_3, Re_2O_7, SeO_2, *Z. anorg. allg. Chem.* 1964 **333** 25.
493. P. GOLDFINGER and M. JEUNEHOMME, *Trans. Faraday Soc.* 1963 **59** 2851.
494. H. W. GOLDSTEIN, E. F. NELISON, P. N. WALSH and D. WHITE, C_p (16–300°K): Y_2O_3, La_2O_3, Nd_2O_3, *J. Phys. Chem.* 1959 **63** 1445.
495. W. D. GOOD, ΔH (formation): aqueous fluorosilic acid, *J. Phys. Chem.* 1962 **66** 380.
496. M. A. GREENBAUM, M. L. ARIN, M. WONG and M. FARBER, Equilibrium: $(BeCl_2) + \{Be\} = 2(BeCl)$, *J. Phys. Chem.* 1964 **68** 791.
497. M. A. GREENBAUM, H. C. KO, M. WONG and M. FARBER, Vapour pressure (1273–1513°K): MgF_2, *J. Phys. Chem.* 1964 **68** 965.
498. M. A. GREENBAUM *et al.*, Vapour pressures and equilibria involving BeF_2, *J. Phys. Chem.* 1963 **67** 36, 703, 1191.
499. E. GREENBERG, J. L. SETTLE, H. M. FEDER and W. N. HUBBARD, ΔH (fluorination): Zr, *J. Phys. Chem.* 1961 **65** 1186.
500. E. GREENBERG, J. L. SETTLE and W. N. HUBBARD, ΔH (fluorination): Ti, Hf, *J. Phys. Chem.* 1962 **66** 1345.
501. L. V. GREGOR, C_p (3–21°K): Cu_2O, *J. Phys. Chem.* 1962 **66** 1645.
502. P. GRIEVESON and C. B. ALCOCK, Private communication.
503. F. GRÖNVOLD and E. F. WESTRUM, $C°$ (5–350°K): Fe_2O_3; FeS_2, $FeSe_2$, $NiSe_2$, *J. Amer. Chem. Soc.* 1959 **81** 1780; *Inorg. Chem.* 1962 **1** 36.
504. P. GROSS, Fulmer Research Inst., Private communication.
505. P. GROSS and C. HAYMAN, ΔH (formation, directly): $HfCl_4$, VCl_4, *Trans Faraday Soc.* 1964 **60** 45.
506. P. GROSS, C. HAYMAN and D. L. LEVI, ΔH (reaction): $\frac{3}{2}PbF_2 + Al + 3NaF = \frac{3}{2}Pb + Na_3AlF_6$, *Symp. Phys. Chem. Process Metallurgy*, AIME, Pittsburgh, 1959.
507. S. R. GUNN, ΔH (reaction): H_2Se–H_2Te, *J. Phys. Chem.* 1964 **68** 949.
508. S. R. GUNN and L. R. GREEN, ΔH (hydrolysis): B_2H_6, BCl_3, *J. Phys. Chem.* 1960 **64** 61.
509. S. R. GUNN, W. L. JOLLY and L. G. GREEN, ΔH (decomposition): AsH_3, SbH_3, *J. Phys. Chem.* 1960 **64** 1334.
510. E. H. HALL, J. M. BLOCHER and I. E. CAMPBELL, Vapour pressure of TiF_4, *J. Electrochem. Soc.* 1958 **105** 275.
511. R. F. HAMPSON and R. F. WALKER, Vapour pressures: Pt, Ir, Rh; Pd (Langmuir method), *J. Res. Nat. Bur. Stand.* 1961 **65** 289; 1962 **66** 177.
512. W. N. HANSEN and M. GRIFFEL, C_p (15–300°K): CrF_3, $CrCl_3$, *J. Chem. Phys.* 1958 **28** 902.
513. S. B. HARTLEY and J. C. MCCOUBREY, ΔH (combustion): P_4O_6, *Nature* 1963 **198** 476.
514. W. E. HATTON, D. L. HILDENBRAND, G. C. SINKE and D. R. STULL, C_p (19–330°K): $Ca(OH)_2$, *J. Amer. Chem. Soc.* 1959 **81** 5028.
515. D. T. HAWKINS, M. ONILLON and R. L. ORR, Heat content (298–1346°K): Hf, *J. Chem. Eng. Data* 1963 **8** 628.
516. D. D. HAWKINS and R. L. ORR, Heat contents (298–1615°K): Nb, *J. Chem. Eng. Data* 1964 **9** 505.
517. C. C. HERRICK, Vapour pressure (928–1160°K): Sm, *J. Less Comm. Met.* 1964 **7** 330.
518. C. C. HERRICK, Vapour pressure of In, *Trans. Met. Soc. AIME* 1964 **230** 1439.
519. D. L. HILDENBRAND, W. R. KRAMER, R. A. MCDONALD and D. R. STULL, C_p (20–300°K): Br_2, *J. Amer. Chem. Soc.* 1958 **80** 4129.
520. D. L. HILDENBRAND and N. D. POTTER, Vapour pressure (1110–1281°K): $CaCl_2$, *J. Phys. Chem.* 1963 **67** 2231.
521. C. HIRAYAMA, Thermochemical assessment: chalcogenides of Ge, Sn and Pb, *J. Chem. Eng. Data* 1964 **9** 65.
522. M. HOCH and H. L. JOHNSTON, Heat contents (1100–2660°C): Ta, W, *J. Phys. Chem.* 1961 **65** 855.
523. B. W. HOWLETT, S. MISRA and M. B. BEVER, ΔH (solution in liquid Bi): Sb_2S_3, Bi_2Se_3, Sb_2Te_3, Bi_2Te_3, *Trans. Met. Soc. AIME* 1964 **230** 1367.
524. E. J. HUBER, G. C. FITZGIBBON, E. L. HEAD and C. E. HOLLEY, ΔH (solution, HCl·aq): Sc, Sc_2O_3; ΔH (combustion): Sc, *J. Phys. Chem.* 1963 **67** 1731.
525. E. J. HUBER, G. C. FITZGIBBON and C. E. HOLLEY, ΔH (combustion): Eu, *J. Phys. Chem.* 1964 **68** 2720.
526. E. J. HUBER, E. L. HEAD, C. E. HOLLEY, E. K. STORMS and N. H. KRIKORIAN, ΔH (combustion): Nb_2N, Nb carbides, *J. Phys. Chem.* 1961 **65** 1846.
527. E. J. HUBER, E. L. HEAD, C. E. HOLLEY and E. L. BOWMAN, ΔH (combustion): Ta carbides, *J. Phys. Chem.* 1963 **67** 793.
528. E. J. HUBER and C. E. HOLLEY, ΔH (combustion): Th, U, Pu carbides, IAEA Conf. *Thermodyn. Nucl. Materials*, Vienna, 1962, 581.
529. E. J. HUBER and C. E. HOLLEY, *J. Phys. Chem.* 1956 **60** 498.
530. G. J. JANZ and J. GOODKIN, L_f: $HgBr_2$, *J. Phys. Chem.* 1959 **63** 1975.
531. G. J. JANZ, F. J. KELLY and J. L. PERANO, Heat contents: $CaBr_2$, $SrCl_2$, $BaCl_2$, $BaBr_2$, *Trans. Faraday Soc.* 1963 **59** 2718.
532. G. J. JANZ and M. R. LORENZ, Dissociation pressures: Li_2CO_3, Na_2CO_3, *J. Chem. Eng. Data* 1964 **9** 94.

533. G. J. Janz, E. Neuenschwander and F. J. Kelly, Heat contents (25–975°C): Li_2CO_3, K_2CO_3, Na_2CO_3, *Trans. Faraday Soc.* 1963 **59** 841.
534. L. D. Jennings, E. D. Hill and F. H. Spedding, C_p (13–350°K): Sm, *J. Chem. Phys.* 1959 **31** 1240.
535. W. H. Johnson and A. A. Gilliland, ΔH (formation, directly): $BeCl_2$, *J. Res Nat. Bur. Stand.* (A) 1961 **65** 59.
536. B. H. Justice and E. F. Westrum, C_p (10–350°K): Gd_2O_3, Sm_2O_3, Yb_2O_3, La_2O_3, Nd_2O_3, Dy_2O_3, Ho_2O_3, Er_2O_3. *J. Phys. Chem.* 1963 **67** 339, 345, 659.
537. R. Juza, G. Wainoff and W. Uphoff, ΔH (solution): $AlPO_4$, *Z. anorg. allg. Chem.* 1958 **296** 157.
538. P. B. Kantor, R. M. Krasovitskaya and A. N. Kisel, Heat contents (600–2200°K): Be, *Fiz. Metallov i Metalloved.* 1960 **10** 835.
539. I. Karutz and I. N. Stranski, Vapour pressure of As_2O_3, *Z. anorg. allg. Chem.* 1957 **292** 330.
540. C. E. Kaylor, G. E. Walden and D. F. Smith, Heat contents: CsCl, CsI; RbF, HfF_4, *J. Phys. Chem.* 1960 **64** 276; *J. Amer. Chem. Soc.* 1959 **81** 4172.
541. F. J. Keneshea and D. Cubicciotti, Vapour pressures of $ZnCl_2$ and $ZnBr_2$, *J. Chem. Phys.* 1964 **40** 191.
542. E. G. King, C_p (51–298°K): Ca double silicates; Ga_2O_3, GeO_2, MoO_2, NbO_2; Pb oxides; $PbSiO_3$, Pb_2SiO_4, $CdSiO_3$, *J. Amer. Soc.* 1957 **79** 5437; 1958 **80** 1799, 2400; 1959 **81** 799.
543. E. G. King and A. U. Christensen, C_p (54–1450°K): Mo_3Si, *J. Phys. Chem.* 1958 **62** 499.
544. E. G. King and A. U. Christensen, Heat contents: CoO, Co_3O_4, NiO, *J. Amer. Chem. Soc.* 1958 **80** 1800.
545. E. G. King and A. U. Christensen, Heat contents: NbO_2, US Bur. Mines, Rep. Inv. No. 5789, 1961.
546. E. G. King and W. W. Weller, C_p (51–298°K): $MeWO_4$, US Bur. Mines, Rep. Inv. No. 5791, 1961.
547. J. P. King and J. W. Cobble, ΔH (oxidation by basic hypochlorite): $ReCl_3$, $ReBr_3$, *J. Amer. Chem. Soc.* 1960 **82** 2111.
548. R. C. King and O. J. Kleppa, ΔH (solution in liquid tin): some Laves phases, *Acta met.* 1964 **12** 87.
549. W. Kischio, ΔH (solution in 20% HCl): AlAs, *Z. anorg. allg. Chem.* 1964 **328** 187.
550. O. J. Kleppa, ΔH (mixing) of binary alloys of mercury, *Acta met.* 1960 **8** 435.
551. M. F. Koehler and J. P. Coughlin, ΔH (solution): $FeCl_2$, $FeCl_3$, $MnCl_2$, *J. Phys. Chem.* 1959 **63** 605. **63** 605.
552. V. P. Kolesov, S. M. Skutarov and I. D. Zaikin, ΔH (solution in H_2O): Li_2O; ΔH (solution in HF·aq): BeO, *Zh. neorg. Khim.* 1959 **4** 1233, 1237.
553. A. N. Kornilov, V. Y. Leonidov and S. M. Skutarov, ΔH (combustion): Nb, Ta, *Dokl. Akad. Nauk SSSR*, 1962 **144** 355.
554. O. H. Krikorian, Vapour pressure (1478–1797°K): Sc, *J. Phys. Chem.* 1963 **67** 1586.
555. O. Kubaschewski, Thermochemical data of alloys, *Landolt-Börnstein*, 6th edition, vol. II, 4 (1961), p. 807.
556. O. Kubaschewski and G. Heymer, Dissociation pressures in the system Cr–Fe, *Acta met.* 1960 **8** 416.
557. F. G. Kusenko and P. V. Geld, ΔH (combustion): Nb, NbO, NbO_2, NbC, *Izvest. Sibir. Otdel. Akad. Nauk SSSR* 1960 (2) 46.
558. F. Z. Kuznetsov, T. N. Rezukhina and A. N. Gulbenko, *Zh. fis. Khim.* (Engl. trsl.) 1960 **34** 1010.
559. A. J. Leadbetter and J. E. Spice, C_p (20–290°K) and p (20–33°C): $Fe(CO)_5$, *Can. J. Chem.* 1959 **37** 1923.
560. L. S. Levinson, Heat contents: NbC (1289–2778°K), TaC (1296–2843°K), *J. Chem. Phys.* 1963 **39** 1550.
561. C. E. Lowell and W. S. Williams, High temp. calorimetry: TiC, TiB_2, *Rev. Sci. Instr.* 1961 **32** 1120.
562. C. E. Lundin, A. S. Yamamoto and J. F. Nachman, Dissociation pressures in the Pr–Nd system, *Acta met.* 1965 **13** 149.
563. R. E. Machol and E. F. Westrum, Vapour pressure (m.p. – 855°C): Te_x, *J. Amer. Chem. Soc.* 1958 **80** 2950.
564. A. D. Mah, ΔH (combustion): Mn_5N_2, Mn_4N; NbO_2, Nb_2N, Ta_2N; W, WO_2, *J. Amer. Chem. Soc.* 1958 **80** 2954, 3872; 1959 **81** 1582.
565. A. D. Mah, ΔH (combustion): Bi, Ga, Sb, VN, VC, US Bur. Mines, Rep. Inv. No. 5676, 1961; 5965, 5972, 1962; 6177, 1963.
566. J. G. Malm and H. Selig, Vapour pressures of ReF_6 and ReF_7, *J. Inorg. Nucl. Chem.* 1961 **20** 189.
567. J. G. Malm, B. Weinstock and F. E. Weaver, Vapour pressure (0–77°C): NbF_6, *J. Phys. Chem.* 1958 **62** 1506.
568. D. L. Martin, C_p (20–320°K): Li, Na, *Proc. Roy. Soc.* 1960 **254** 433, 444.
569. S. W. Mayer, B. B. Owens, T. H. Rutherford and R. B. Serrins, Equilibria involving $Th(SO_4)_2$, *J. Phys. Chem.* 1960 **64** 911.
570. C. L. McCabe, R. G. Hudson and H. W. Paxton, Diss. pressure in the Cr–Fe system, *Trans. Met. Soc. AIME* 1958 **212** 102.
571. R. Mezaki and J. L. Margrave, Heat contents: TeO_2 (298–1146°K), Na_2TeO_4 (298–804°K), *J. Phys. Chem.* 1962 **66** 1713.
572. R. Mezaki, E. W. Tilleux, D. W. Barnes and J. L. Margrave, Heat contents (300–1200°K): numerous borides of transition metals, and BN, IAEA Conf., *Thermodynamics Nucl. Mat*, Vienna, 1962, p. 775.
573. R. L. Montgomery, ΔH (solution in HCl·aq): Ce, $CeCl_3$, US Bur. Mines, Rep. Inv. No. 6146.
574. A. G. Morachevski, A. F. Alabyshev and M. F. Lantratov, E.m.f. (648–748°K): Na–Pb system, *Usp. Khim.* 1958 **27** 921.

575. M. P. MOROZOVA and G. EGER, ΔH (combustion): V and V oxides, *Zh. obshchei Khim.* 1960 **30** 3514.
576. M. P. MOROZOVA and T. A. STOLYAROVA, ΔH (combustion): Nb, NbO, NbO_2, *Zh. obshchei Khim.* 1960 **30** 3848.
577. T. E. MORSY, Thermodyn. properties of (SF_6), *Ber. Bunsenges. Phys. Chem.* 1964 **68** 277.
578. C. T. MORTIMER, Keele Univ., Private communication.
579. R. N. R. MULFORD, Thermodyn. properties of Pu, US Atomic En. Comm. LA-2813 (1963).
580. L. E. MURCH and W. F. GIAUQUE, C_p (12–320°K): NaOH, *J. Phys. Chem.* 1962 **66** 2052.
581. M. A. ORANSKAYA, Y. S. LEBEDEV and I. L. PERFILOVA, Equilibrium: $2\langle VCl_3\rangle = (VCl_4) + \langle VCl_2\rangle$, *Zh. neorg. Khim.* 1961 **6** 259.
582. M. A. ORANSKAYA and N. A. MIKHAILOVA, Vapour and dissociation pressures of $PdCl_2$, *Zh. neorg. Khim.* 1960 **5** 12.
583. M. A. ORANSKAYA and I. L. PERFILOVA, Vapour pressure (910–1100°C): VCl_2, *Zh. neorg. Khim.* 1961 **6** 257.
584. A. A. PALKO, A. D. RYON and D. W. KUHN, Vapour pressures of $ZrCl_4$ and $HfCl_4$, *J. Phys. Chem.* 1958 **62** 319.
585. M. B. PANISH and L. REIF, Vapour pressures: Ir (2100–2600°K), Rh (2050–2200°K), Hf (2060–2275°K), *J. Chem. Phys.* 1961 **34** 1915; 1963 **38** 253.
586. L. B. PANKRATZ and K. K. KELLEY, Heat contents of V_3Si (400–1500°K), US Bur. Mines, Rep. Inv. No. 6241, 1963.
587. L. B. PANKRATZ, E. G. KING and K. K. KELLEY, Heat contents (298–1800°K): Eu_2O_3, Gd_2O_3, etc., US Bur. Mines, Rep. Inv. No. 6033, 1962.
588. M. N. PAPADOPOULOS and W. F. GIAUQUE, C_p (15–300°K): Hg_2SO_4, *J. Phys. Chem.* 1962 **66** 2049.
589. R. C. PAULE and J. L. MARGRAVE, Vapour pressures (Langmuir technique): B, Ir, Ru, *J. Phys. Chem.* 1963 **67** 1368, 1896.
590. R. D. PEHLKE and J. F. ELLIOTT, Dissociation pressure (1400–1700°C): αSi_3N_4, *Trans. Met. Soc. AIME* 1959 **215** 781.
591. U. PIESBERGEN, C_p (10–273°K): AlSb, etc., *Z. Naturforsch.* 1963 **18a** 141.
592. G. V. POZHARSKAYA and A. M. EVSEEV, Dissociation pressures (916–963°C): Mn–Ni system, *Zh. fiz. Khim.* 1962 **36** 1358.
593. J. N. PRATT, E.m.f.: $Ag/[Ag^+]/Ag\text{–}Pd$, *Trans. Faraday Soc.* 1960 **56** 975.
594. A. C. P. PUGH and R. F. BARROW, Vapour pressures of alkali metal fluorides, *Trans. Faraday Soc.* 1958 **54** 671.
595. M. H. RAND *et al.*, IAEA monograph *Physicochemical Properties of Plutonium Compounds*, 1966.
596. H. W. RAYSON and W. A. ALEXANDER, Dissociation pressures in the systems Ag–Hg and Au–Hg, *Acta met.* 1960 **8** 833.
597. H. SCHÄFER and K. ETZEL, Equilibria involving (FeCl) and (NiCl) respectively, *Z. anorg. allg. Chem.* 1959 **301** 137.
598. H. SCHÄFER and H. HEINE, ΔH (reaction) of Si halides with AgF and [HF], *Z. anorg. allg. Chem.* 1964 **332** 25.
599. H. SCHÄFER and H. J. HEITLAND, Equilibria involving Ir oxides, *Z. anorg. allg. Chem.* 1960 **304** 249.
600. H. SCHÄFER and W. J. HÖNES, *Z. anorg. alla. Chem.* 1956 **288** 62.
601. H. SCHÄFER and F. KAHLENBERG, Thermochem. properties of Ta chlorides, *Z. anorg. allg. Chem.* 1958 **294** 242; 1960 **305** 178.
602. H. SCHÄFER and F. KAHLENBERG, ΔH (formation): Nb chlorides, *Z. anorg. allg. Chem.* 1960 **305** 291, 327.
603. H. SCHÄFER and F. LIEDMEIER, Reaction of the subchlorides of Nb and Ta with Cl_2 in a bomb, *J. Less-Comm. Met.* 1964 **6** 307.
604. H. SCHÄFER and F. LIEDMEIER, Formation of $NbOCl_2$ from NbO and $NbOCl_3$, *Z. anorg. allg. Chem.* 1964 **329** 225.
605. H. SCHÄFER, G. SCHNEIDEREIT and W. GERHARDT, Dissociation pressure of RuO_2, *Z. anorg. allg. Chem.* 1963 **319** 327.
606. H. SCHÄFER and E. SIBBING, Equilibrium: $Ta_2O_5 + 3(TaCl_5) = 5(TaOCl_3)$, *Z. anorg. allg. Chem.* 1960 **305** 341.
607. H. SCHÄFER, A. TEBBEN and W. GERHARDT, Thermochem. properties of RuO_x, *Z. anorg. allg. Chem.* 1963 **321** 41.
608. H. SCHÄFER, F. WARTENPFUHL and E. WEISE, Thermochem. properties of TiOCl, *Z. anorg. allg. Chem.* 1958 **295** 268.
609. H. SCHÄFER, F. E. WITTIG and H. WILBORN, ΔH (solution): AlOCl, *Z. anorg. allg. Chem.* 1958 **297** 48.
610. N. G. SCHMAHL and P. SIEBEN, Dissociation pressures in the systems Mg–Ni and Mg–Cu, *Metal Chemistry*, NPL Symp. No. 9, 1958, HMSO, 1959, 2K.
611. W. F. SCHOTTKY and M. B. BEVER, ΔH (solution in liquid tin) and heat contents: InSb, GaSb, InAs, *Acta met.* 1958 **6** 320.
612. D. A. SCHULZ and A. W. SEARCY, Vapour pressure (1400–1850°K): CaF_2 (torsion-effusion method), *J. Phys. Chem.* 1963 **67** 103.
613. J. L. SETTLE, H. M. FEDER and W. N. HUBBARD, ΔH (fluorination): Mo, *J. Phys. Chem.* 1961 **65** 1337.

614. L. V. SHCHIPANOVA and P. V. GELD, C_p of Ge, *Chem. Abstr.* 1963 **59** 3371.
615. S. A. SHCHUKAREV and A. R. KURBANOV, Equilibria involving Ta halides, *Vestn. Leningr. Univ.* **17**, Ser. Fiz. Khim. (2), 1962, 144.
616. S. A. SHCHUKAREV and A. N. RYABOV, ΔH (formation): RuO_2, *Zh. neorg. Khim.* (Engl. edition) 1960 **5** 941.
617. S. A. SHCHUKAREV, T. A. TOLMACHEVA and V. M. TSINTSIUS, Equilibrium: $2VBr_3 + (Br_2) = 2(VBr_4)$, *Zh. neorg. Khim.* 1962 **7** 679.
618. E. SHIMAZAKI and K. NIWA, Vapour pressures of rare earth metal halides, *Z. anorg. allg. Chem.* 1962 **314** 21.
619. R. A. J. SHELTON, Equilibrium pressures of solid Cu halides, *Trans. Faraday Soc.* 1961 **57** 2113.
620. D. A. SHIRLEY, C_p (15–325°K): LiCl, *J. Amer. Chem. Soc.* 1960 **82** 3841.
621. R. J. SIME and N. W. GREGORY, Vapour pressures: $FeCl_2$, $FeBr_2$, FeI_2; $CrBr_2$, *J. Phys. Chem.* 1960 **64** 86; *J. Amer. Chem. Soc.* 1960 **82** 800.
622. F. SPANDAU and F. KLANBERG, Vapour pressures: GeS, Ga_2S, ZnS, CdS, *Z. anorg. allg. Chem.* 1958 **295** 291, 300, 309.
623. F. H. SPEDDING, A. H. DANNE, G. WAKEFIELD and D. H. DENNISON, Properties of scandium, *Trans. Met. Soc. AIME* 1960 **218** 608.
624. F. H. SPEDDING, J. J. MCKEOWN and A. H. DAANE, Heat contents of samarium (273–1398°K), *J. Phys. Chem.* 1960 **64** 289.
625. J. W. STOUT and R. C. CHISHOLM, C_p (11–300°K): $CuCl_2$, $CrCl_2$, *J. Chem. Phys.* 1962 **36** 979.
626. J. R. STUBBLES and F. D. RICHARDSON, Equilibria involving molybdenum sulphides, *Trans. Faraday Soc.* 1960 **56** 1460.
627. D. R. STULL and G. C. SINKE, quoted in ref. 99.
628. R. E. TAYLOR and R. A. FINCH, Pulse heating method for measuring C_p of Ta, Mo and Re, *J. Less-Comm. Met.* 1964 **6** 283.
629. J. TERPILOWSKI, E.m.f. of cell: $\langle Mg \rangle/[Mg^{++}]/\{Mg\text{–}Zn\}$ at 680–920°K, *Bull. Acad. Polon. Sci.*, Sér. sci. chim. 1962 **10** 221.
630. J. TERPILOWSKI, E. KUNDYS and H. SLABY, E.m.f. of the cell $Tl/[Tl^+]/Ag\text{–}Tl$, *Arch. Hutnictwa* 1961 **6** 137.
631. T. A. TOLMACHEVA, V. M. TSINTSIUS and L. V. ADRIANOVA, ΔH (solution in KOH·aq): VI_3. Equilibrium: $2\langle VI_3 \rangle = 2\langle VI_2 \rangle + (I_2)$, *Zh. neorg. Khim.* 1963 **8** 553.
632. L. E. TOPOL and L. D. RANSOM, Heat contents: Cd halides and $BiBr_3$, *J. Phys. Chem.* 1960 **64** 1339.
633. A. G. TURNBULL, Reactions of $ZrBr_4$ and ZrI_4 with alkaline solutions, *J. Phys. Chem.* 1961 **65** 1652.
634. A. F. VOROBEV, V. P. KOLESOV and S. M. SKUTAREV, Reaction of SiF_4 and CF_4 with sodium, *Zh. neorg. Khim.* 1960 **5** 1398, 1402.
635. P. N. WALSH, E. W. ART and D. WHITE, C_p (16–296°K): Ag_2S, Ag_2Se, $Ag_{1.88}Te$, *J. Phys. Chem.* 1962 **66** 1546.
636. W. W. WELLER and K. K. KELLEY, C_p (51–298°K): Sc, AsS, GeS, NiS, Ni_2S_3, US Bur. Mines, Rep. Inv. No. 5984, 1962; 6191, 1963; 6511, 1964.
637. W. W. WELLER and E. G. KING, C_p (50–300°K): MgV_2O_6, $Mg_2V_2O_7$, Na_2MoO_4, $CaMoO_4$, $Na_2Mo_2O_7$, Sc_2O_3, $CeO_2\text{–}Ce_2O_3$, US Bur. Mines, Rep. Inv. No. 6130, 1962; 6147, 6245, 1963.
638. E. F. WESTRUM and A. F. BEALE, C_p (5–300°K): CeF_3, CeO_2, *J. Phys. Chem.* 1961 **65** 353.
639. E. F. WESTRUM, CH. CHOU and F. GRONVOLD, C_p (5–350°K): FeTe, $FeTe_2$, *J. Chem. Phys.* 1959 **30** 761.
640. E. F. WESTRUM and G. FEICK, C_p (5–350°K): ZrC, ZrB_2, *J. Chem. Eng. Data* 1963 **8** 176, 193.
641. H. WIEDEMEIER and H. SCHÄFER, ΔG (dissociation): MnS (1401°C). Vapour pressure of Mn (1036–1175°C), *Z. anorg. allg. Chem.* 1964 **326** 225, 230, 235.
642. T. C. WILDER and J. F. ELLIOTT, E.m.f. (700–980°C): $Al[Al^{3+}]/Al\text{–}Ag$, *J. Electrochem. Soc.* 1960 **107** 626.
643. W. S. WILLIAMS, ΔH (formation) of TiB_2, Critical assessment, *J. Phys. Chem.* 1961 **65** 2213.
644. S. S. WISE, J. L. MARGRAVE and R. L. ALTMAN, Heat contents: boron, *J. Phys. Chem.* 1960 **64** 915.
645. S. S. WISE, J. L. MARGRAVE, H. M. FEDER and W. N. HUBBARD, Reactions with F_2 in the bomb: B, BN, Si, quartz, vitr. silica, *J. Phys. Chem.* 1961 **65** 2157; 1962 **66** 381; 1963 **67** 815.
646. W. P. WITT and R. F. BARROW, Equilibrium: $2\langle Al \rangle + \langle AlF_3 \rangle = 3(AlF)$. Vapour pressure of AlF_3, *Trans. Faraday Soc.* 1959 **55** 730.
647. F. E. WITTIG and E. GEHRING, ΔH (mixing): Bi–Sb alloys, *Naturwiss.* 1959 **46** 200.
648. F. E. WITTIG and E. GEHRING, ΔH (mixing) of Ag–Sn alloys at 975°C, *Z. Naturforsch.* 1963 **18a** 351.
649. F. E. WITTIG and G. KEIL, ΔH (mixing) of liquid alloy systems of aluminium at 700°C, *Z. Metallkunde* 1963 **54** 576.
650. F. E. WITTIG and T. KLEINSTEUBER, ΔH (mixing): liquid alloys of sodium and potassium respectively, *XVIIIth IUPAC Congr., Montreal*, 1961.
651. F. E. WITTIG and E. MÜLLER, ΔH (mixing, liquid): In–Bi, Tl–Bi, Tl–In, *Z. physikal. Chem.* (NF) 1959 **21** 47.
652. F. E. WITTIG and P. SCHEIDT, ΔH (solution in mercury) of metals, *Naturwiss.* 1960 **47** 250.
653. F. E. WITTIG and P. SCHEIDT, ΔH (mixing): liquid alloys involving In or Tl, *Z. physikal. Chem.* (NF) 1961 **28** 120.
654. E. WOLF, ΔH (solution in NaOH·aq): $SiCl_4$, $SiHCl_3$, *Z. anorg. allg. Chem.* 1961 **313** 228.

655. W. L. WORELL, Equilibrium (1074–1175°K): $Nb_2O_5 + C = 2NbO_2 + CO$. Estimates of S_{298} for metal carbides; assessment of thermochem. properties for Al_4C_3, *J. Phys. Chem.* 1964 **68** 952, 954; Priv. Comm., 1965.
656. W. L. WORRELL and J. CHIPMAN, Equilibrium (1180–1370°K): $Me_xO_y + (x + y)\ C = x\,MC + y\,CO$, *J. Phys. Chem.* 1964 **68** 860.
657. T. YOKOKAWA and O. J. KLEPPA, ΔH (mixing) of alkali metals, *J. Chem. Phys.* 1964 **40** 46.
658. P. F. YOUNG, Vapour pressure (540–985°C): Rb, *Met. Abstr.* 1963 **30** 426.
659. Y. TAKAHASHI, E. F. WESTRUM and R. A. KENT, C_p (5–350°K): $ThC_{1\cdot 98}$, TlBr, TlI, *J. Chem. Eng. Data,* 1965 **10** 128, 244.
660. E. GREENBERG, C. A. NATKE and W. N. HUBBARD, ΔH (fluorination), Nb, Ta, *J. Phys. Chem.* 1965 **69** 2089.
661. J. L. HOLM and O. J. KLEPPA, Thermodyn. properties of the alsilicates, *Amer. Mineralogist* 1966 **51** 1608.
662. R. L. ALTMAN, Transpiration method, equilibrium: $4\langle MgO\rangle + 2\{Al\} = \langle MgAl_2O_4\rangle + 3(Mg)$, *J. Phys. Chem.* 1964 **68** 3425.
663. I. J. BEAR and A. G. TURNBULL, ΔH (solution in [HF] aq): $BeSO_4$ and its hydrates, *J. Phys. Chem.* 1966 **70** 711.
664. P. BEARDMORE, B. W. HOWLETT, B. D. LICHTER and M. B. BEVER, Calorimetry pertaining to Mg_2Ge, Mg_2Sn, Mg_2Pb, *Trans. Met. Soc. AIME* 1966 **236** 102.
665. R. BENZ and H. SCHMALZRIED, E.M.F. measurements involving the PbO–SiO_2 system, *Z. phys. Chem.* (NF) 1961 **29** 77.
666. R. B. BEVAN, R. A. GILBERT and R. H. BUSEY, C_p (7–310°K): $ReCl_3$, *J. Phys. Chem.* 1966 **70** 147.
667. A. A. BONI, Vapour pressure (275–311°K): $TiBr_4$, $Cr(CO)_6$, *J. Electrochem. Soc.* 1966 **113** 1089.
668. W. BUES and H. VON WARTENBERG, Equilibria: $\frac{1}{2}\langle Ge\rangle + \frac{1}{2}\langle GeO_2\rangle = (GeO)$ and $\{GeO\}_{am.} = (GeO)$, *Z. anorg. allg. Chem.* 1951 **266** 281.
669. P. CHIOTTI, G. J. GARTNER, E. R. STEVENS and Y. SAITO, L_f of some metals and intermetallic compounds, *J. Chem. Eng. Data* 1966 **11** 571.
670. J. F. COUNSELL, R. M. DELL and J. F. MARTIN, C_p (11–320°K): U nitrides, *Trans. Farad. Soc.* 1966 **62** 1736.
671. D. CUBICCIOTTI and H. EDING, Heat contents: BiI_3, SbI_3, *J. Phys. Chem.* 1965 **69** 3621.
672. J. B. DARBY, ΔH (solution in Sn, liq.): Au–Pd, *Acta met.* 1966 **14** 265.
673. J. J. EGAN, EMF (700–970°C): Mo/$\langle Th, ThF_4\rangle$/$\langle ThF_4, ThC_2, C\rangle$/Mo, *J. Phys. Chem.* 1964 **68** 978.
674. E. P. EGAN and B. B. LUFF, ΔH (solution): $PoCl_3$, *J. Chem. Eng. Data* 1966 **11** 92.
675. J. M. ELDRIDGE, E. MILLER and K. L. KOMAREK, p_{Mg} (above 1000°K): Mg–Sn, Mg–Ge, *Trans. Met. Soc. AIME* 1966 **236** 114, 1094.
676. H. A. EICK and R. N. R. MULFORD, p (Knudsen): Np, *J. Chem. Phys.* 1964 **41** 1475.
677. M. FARBER and H. L. PETERSEN, Equilibrium (2200°K): $\langle Al_2O_3\rangle + (AlF_3) = 3(AlOF)$, *Trans. Faraday Soc.* 1963 **59** 836.
678. R. FERRO and R. CAPELLI, ΔH (directly): Pd aluminides, *Atti Accad. naz. Lincei.* 1963 **34** 659.
679. R. FERRO, R. CAPELLI and R. MARAZZA, ΔH (directly): Au aluminides, Private Comm., Sept. 1966.
680. L. G. FASOLINO, ΔH (hydrolysis): BCl_3, *J. Chem. Eng. Data* 1965 **10** 371.
681. M. FISCHER, ΔH (solution): $MnFe_2O_4$ and $ZnFe_2O_4$, and component oxides, *Z. anorg. allg. Chem.* 1966 **345** 134.
682. G. C. FITZGIBBON, C. E. HOLLEY and I. WADSÖ, ΔH (combustion): La, *J. Phys. Chem.* 1965 **69** 2464.
683. R. H. FLOWERS and E. G. RAUH, Dissociation pressures of SrC_2 & BaC_2, *J. Inorg. Nucl. Chem.* 1966 **28** 1355.
684. R. M. GOODMAN and E. F. WESTRUM, C_p (5–350°K): $BaCl_2$, *J. Chem. Eng. Data* 1966 **11** 294.
685. R. L. GRAHAM and L. G. HEPLER, ΔH (solution in [NaOH]aq): MoO_2Cl_2, *J. Phys. Chem.* 1959 **63** 723.
686. G. A. GRAY and R. J. SIME, Vapour pressure (60–130°C): $SbBr_3$, *J. Chem. Eng. Data* 1965 **10** 176.
687. M. A. GREENBAUM, J. WEIHER and M. FARBER, L_f: BeO, *J. Phys. Chem.* 1965 **69** 4035.
688. F. GRONVOLD, and E. F. WESTRUM, C_p (6–347°K): Fe selenides, *Acta chem. scand.* 1959 **13** 241.
689. F. GRONVOLD, T. THURMANN-MOE, E. F. WESTRUM and E. CHANG, C_p (5–346°K): PtS, PtTe, PdTe, *J. Chem. Phys.* 1961 **35** 1665.
690. P. GROSS, C. HAYMAN and J. T. BINGHAM, ΔH (fluorination): Ge, GeO_2, *Trans. Faraday Soc.* 1966 **62** 2388.
691. P. GROSS, C. HAYMAN, P. D. GREENE and J. T. BINGHAM, ΔH (chlorination): Be, Be_2N_3, *Trans. Faraday Soc.* 1966 **62** 2719.
692. P. GROSS, C. HAYMAN, D. L. LEVI and M. C. STUART, US Dept. Comm. Office, Tech. Serv. Rep. 153, 445 (1960).
693. P. GROSS, C. HAYMAN and M. C. STUART, ΔH (fluorination): white phosphorus, *Trans. Faraday Soc.* 1966 **62** 2716.
694. C. E. HABERMANN and A. H. DAANE, Vapour pressure of rare earth metals, *J. Chem. Phys.* 1964 **41** 2818.
695. R. R. HAMMER and J. A. PASK, Vapour pressure (1413–1614°K): MgF_2 (Knudsen and torsion-effusion methods), *J. Amer. Ceram. Soc.* 1964 **47** 264.
696. O. HERSTAD and K. MOTZFELDT, Vapour pressure: Al–Al_2O_3 system, *Rev. Hautes Temp. Réfract.* 1966 **3** 291.

697. D. L. HILDENBRAND, W. F. HALL, F. JU and D. N. POTTER, Torsion–effusion method: LiF, LiCl, MgF_2, $MgCl_2$, *J. Chem. Phys.* 1964 **40** 2882.
698. J. L. HOLM and O. J. KLEPPA, Be_2SiO_4, $BeAl_2O_4$, $BeAl_6O_{10}$, *Acta chem. scand.* 1966 **20** 2568.
699. E. HUBER, ΔH (combustion): NbB_2, 5th Quat. Progr. Rep. RAR–SR–63–183.
700. E. J. HUBER, ΔH (combustion): TiB_2, *J. Chem. Eng. Data* 1966 **11** 430.
701. E. J. HUBER, E. L. HEAD and C. E. HOLLEY, ΔH (combustion): Zr, ZrB_2, *J. Phys. Chem.* 1964 **68** 3040.
702. J. H. E. JEFFES and H. MCKERELL, Assessment of the thermodynamic properties of the metal hydrides, *J. Iron Steel Inst.* 1964 **202** 666.
703. L. D. JENNINGS, R. E. MILLER and F. H. SPEDDING, C_p (15–350°K): Y, *J. Chem. Phys.* 1960 **33** 1849.
704. G. K. JOHNSON, H. M. FEDER and W. N. HUBBARD, ΔH (form.): BF_3, *J. Phys. Chem.* 1966 **70** 1.
705. V. A. KIRILLIN, A. E. SHEINDLIN, V. YA. CHEKHOVSKOI and V. A. PETROV, Heat content measurements combined with published data. C_p (0–3500°K): W, *Zh. fiz. Khim.* 1963 **37** 2249.
706. T. KUROSAWA, R. HASEGAWA and T. YAGIHASHI, C_p, L_f (adiabatic calorimetry): SiI_4, *Trans. Jap. Inst. Metals* 1965 **6** 229.
707. S. M. LETUN, P. V. GEL'D and N. N. SEREBRENNIKOV, C_p (55–2000°K): Mn_3Si, *Izv. Akad. Nauk SSSR, Metally* 1965 (6) **137**.
708. M. LIM and A. W. SEARCY, Torsion–effusion method (1373–1634°K): CeF_3, *J. Phys. Chem.* 1966 **70** 1762.
709. A. C. MACLEOD, H^T_{298} (T = 420–1240°K): sapphire, *Trans. Faraday Soc.* 1965 **63** 300.
710. A. D. MAH, ΔH (combustion): ZrC, HfC, US Bur. Mines, Rep. Inv. No. 6518 (1965).
711. A. D. MAH, S. C., Summary of the thermodynamic properties of vanadium compounds, US Bur. Mines, Rep, Inv. No. 6727 (1966).
712. R. A. MCDONALD, Drop calorimetry: Al, Mg, 21st Ann. Calorimetry Conf., Univ. Colarado, 1966.
713. R. A. MCDONALD and D. R. STULL, H^{1650}_{300}: BN, *J. Phys. Chem.* 1961 **65** 1918.
714. M. P. MOROZOVA and T. A. STOLYAROVA, ΔH (formation): MnSe, MnTe, *Vestn. Leningr. Univ.* **19** (16), Ser. Fiz. Khim. (3) 1964, 150.
715. K. M. MYLES and A. T. ALDRED, Dissociation pressures (1454–1800°K): Fe–V system, *J. Phys. Chem.* 1964 **68** 64.
716. A. B. NIKOLSKI and A. N. RYABOV, Thermodynamic properties of Ru and Os oxides, *Zh. neorg. Khim.* 1965 **10** 3.
717. P. A. G. O'HARE and W. N. HUBBARD, ΔH (fluorination): P modifications, *Trans. Faraday Soc.* 1966 **62** 2709.
718. P. A. G. O'HARE, J. L. SETTLE and W. N. HUBBARD, ΔH (fluorination): S, Se, Te, *Trans. Faraday Soc.* 1966 **62** 558.
719. E. M. OTTO, Equilibrium: $6\langle Mn_2O_3\rangle = 4\langle Mn_3O_4\rangle\alpha + (O_2)$, *J. Electrochem. Soc.* 1964 **111** 88.
720. E. L. PACE and J. S. MOSER, Thermodynamic properties of SiF_4 up to 186°K, *J. Chem. Phys.* 1963 **39** 154.
721. B. I. PANFILOV and N. N. FEODOS'EV, Bomb reactions: $MeCO_3 + TiO_2 = MeTiO_3 + CO_2$; $ZnO + TiO_2$, *Zh. neorg. Khim.* 1964 **9** 2685, 2693; 1965 **10** 3.
722. L. B. PANKRATZ, W. W. WELLER and E. G. KING, C_p (51–300°K): Mo_2C; heat contents (up to 1400 & 1800°K, resp.): Mo_2C, TaC, US Bur. Mines, Rep. Invest. No. 6861 (1966).
723. H. A. PORTE, E. GREENBERG and W. N. HUBBARD, ΔH (fluorination): Ru, *J. Phys. Chem.* 1965 **69** 2308.
724. H. RAU, Thermodyn. prpts. of SnS, *Ber. Bunsenges. Phys. Chem.* 1965 **69** 731.
725. P. M. ROBINSON and M. B. BEVER, ΔH (solution in Bi, liq): CdTe, InTe, SnTe, PbTe, *Trans. Met. Soc. AIME* 1966 **236** 814.
726. P. ROY, p_{Mn} (1240–1510°K): Fe–Mn system, PhD thesis, Univ. Calif., Aug. 1964.
727. E. M. SAVITSKI, V. F. TEREKHOVA and D. P. NAUMKIN, Vapour pressures: Se, Y, Ce, Nd, Eu, Er, Dy, Tu, *Soviet Phys. Uspekhi* 1963 **6** 123.
728. K. SCHWERDTFEGER and H. J. ENGELL, EMF measurements involving {Co–Si} and {Fe–Si} alloys, *Arch. Eisenhüttenw.* 1964 **35** 533.
729. H. E. SCHWIETE and G. ZIEGLER, L_f, L_t (Differnnzcalorimetrie): Li_2SO_4, *Ber. Dtsch. Keram. Ges.* 1958 **35** 193.
730. S. A. SEMENKOVICH, Equilibrium: NaBr + Al = (Na) + (AlBr), *Zh. prikl. Khim.* 1957 **30** 933.
731. R. J. SIME, Vap. press. (435–561°K, 510–629°K): $SbBr_3$, SbI_3, *J. Phys. Chem.* 1963 **67** 501.
732. E. SIRTL, Survey: ΔH_f, ΔS_f of gaseous inorganic compounds, *Z. Naturforsch.* 1966 **21a** 2001.
733. H. SLABY and J. TERPILOWSKI, EMF (673–920°K): $Mg/[Mg^{++}]/Mg$–In, *Bull. Acad. Polon. Sci., Sér. Sci. Chim.* 1965 **13** 319.
734. J. M. STUVE, ΔH (solution in [HCl] aq): $YbCl_3$, $TmCl_3$, US Bur. Mines, Rep. Inv. No. 6902 (1967).
735. R. W. TAYLOR and H. SCHMALZRIED, EMF (800–1200°C): titanates using solid electrolytes, *J. Phys. Chem.* 1964 **68** 2444.
736. J. TERPILOWSKI and H. SLABY, EMF (673–920°K): $Mg/[Mg^{++}]/Mg$–Tl, *Bull. Acad. Polon. Sci., Sér. Sci. Chim.* 1963 **11** 317.
737. J. D. TRETJAKOV and H. SCHMALZRIED, EMF of cells with spinel phases as electrodes and solid electrolytes, *Ber. Bunsenges. Phys. Chem.* 1965 **69** 396.
738. A. A. VECHER, V. A. GEIDERIKH and Y. I. GERASSIMOV, EMF of AlSb formation, *Zh. fiz. Khim.* 1965 **39** 2145.

739. E. F. WESTRUM, H. G. CARLSON, F. GRØNVOLD and A. KJEKSHUS, C_p (5–350°K): PtS_2, $PtTe_2$, $PdTe_2$, *J. Chem. Phys.* 1961 **35** 1670.
740. E. F. WESTRUM and G. A. CLAY, C_p (8–350°K): NbB_2, *J. Phys. Chem.* 1963 **67** 2385.
741. R. E. YATES, M. A. GREENBAUM and M. FARBER, Dissociation pressure: Be_3N_2, *J. Phys. Chem.* 1964 **68** 2682.
742. K. F. ZMBOV and J. L. MARGRAVE, Vapour pressure: NdF_3, etc., *J. Chem. Phys.* 1966 **45** 3167.
743. T. NOZAKI, M. SHIMOJI and K. NIWA, EMF (1000–1200°K): $In/In^+/In$–Ag, *Trans. JIM* 1966 **7** 52.
744. T. NOZAKI, M. SHIMOJI and K. NIWA, Thermodynamic Properties of Ag–Sn, Ag–Sb, *Berichte Bunsenges,* 1966 **70** 207.
745. H. HOSHINO, Y. NAKAMURA, M. SHIMOJI and K. NIWA, EMF of In–Bi, In–Sb, *Berichte Bunsengesell.* 1965 **69** 114.
746. M. NAGAMORI and M. KAMEDA, Thermodynamic properties of liquid Au–Fe, *Nippon Kinzoku Gakkai Shi* 1965 **26**.
747. A. S. ABBASSOV, *Evaluation of the Thermodynamic Properties of Some III–V Compounds,* Teddington, NPL, DCS Report 8.
748. A. S. ABBASSOV *et al., Dokl. Akad. Nauk SSSR* 1964 **156** 1140.
749. G. P. ADAMS, J. L. MARGRAVE and P. W. WILSON, *J. Chem. Thermodyn.* 1970 **2** 741.
750. V. V. AKHACHINSKI *et al., Thermodynamics,* IAEA, Vienna, 1962, p. 309.
751. C. B. ALCOCK, P. GRIEVESON and J. B. CORNISH, *Thermodynamics of Nucl. Mat.*, IAEA, 1965 **I** 211.
752. V. M. AMOSOV, *Izv. vyssh. Zaved, Tsvetnaya Metallurgiya* 1963 **6** 103.
753. R. J. L. ANDON, J. F. MARTIN and K. C. MILLS, *J. Chem. Soc.* (A) 1971, 1788.
754. R. J. L. ANDON and K. C. MILLS, *J. Chem. Thermodyn.* 1971 **3** 583.
755. ATOMIC ENERGY REVIEWS, IAEA, Vienna; No. 1, *Plutonium* (M. H. Rand) 1966; No. 2, *Niobium* (V. I. Lavrentev and Ya. I. Gerassimov) 1968; No. 3, *Tantalum* (Ya. I. Gerassimov and V. I. Lavrentev) 1972; No. 4, *Beryllium* (P. J. Spencer) 1973; No. 5, *Thorium* (M. H. Rand) 1974. Techn. Rep. No. 14, 1963.
756. F. B. BAKER, E. J. HUBER, C. E. HOLLEY and N. H. KRIKORIAN, *J. Chem. Thermodyn.* 1971 **3** 77.
K. A. BECKERS, K. PLIETH and I. N. STRANSKI, *Progress in Inorganic Chemistry* 1962 **4** 1. Interscience Publishers, New York and London.
757. W. E. BELL and M. TAGAMI, *J. Phys. Chem.* 1966 **70** 3736.
758. O. BERNARD and I. ANSARA, Rapport LTCM–1974–TM–O2, ENSEEG—BP44–38401, St. Martin d'Hères, France.
759. Z. BIEGANSKI, *Ber. Bunsenges. Phys. Chem.* 1972 **76** 1183.
760. W. BILTZ and M. HEIMBRECHT, *Z. anorg. allg. Chem.* 1938 **237** 132.
761. R. BLACHNIK, PhD. thesis, TH Clausthal-Zellerfeld, 1967.
762. R. BLACHNIK, R. IGEL and P. WALLBRECHT, *Z. Naturforsch.* 1974 **29** 1198.
763. R. BLACHNIK and A. SCHNEIDER, *J. Chem. Thermodyn.* 1971 **3** 227.
764. B. N. BOKERIYA, D. SH. TSAGAREISHVILI and G. G. GVELESIANI, *Acad. Sci. Georgian SSR, Communications* 1970 **58** 369, 601.
765. L. BREWER, G. R. SOMAYOJULU and E. BRACKETT, *Chem. Rev.* 1963 **63** 111.
766. G. M. CAMPBELL *et al., Pu,* 1970.
D. CAPLAN, M. J. FRASER and A. A. BURR, *Ductile Cr a. Alloys,* Nat. Met. Congr. Philadelphia, Oct. 1955.
767. I. CHADWICK, R. J. IRVING and A. SOUSA-ALONSO, *J. Chem. Thermodyn.* 1970 **2** 611.
768. J. C. CHAN, C. B. ALCOCK and K. T. JACOB, *Can. Met. Quart.* 1973 **12** 439.
769. T. G. CHART, *High Temp.–High Press.* 1973 **5** 241; *Metal Sci.* 1974 **8** 344; 1975 **9** 504.
770. D. CHATTERJI and R. W. VEST, *J. Amer. Ceram. Soc.* 1971 **54** 73.
771. D. CUBICCIOTTI and H. EDING, *J. Phys. Chem.* 1965 **69** 3621.
772. D. CUBICCIOTTI, *Inorg. Chem.* 1968 **7** 208–11.
773. D. CUBICCIOTTI, *J. Electrochem. Soc.* 1968 **115** 1138.
774. L. DENIEL *et al., Compt. rend.* 1971 **272** 1855.
775. H. D. DANNÖHL, PhD thesis, Stuttgart, 1970.
776. D'ANS-LAX, *Taschenbuch für Chemiker und Physiker, Springer-Verlag,* 1967.
777. C. DEENADAS *et al., J. Phys. Chem. Solids* 1971 **32** 1853.
778. E. S. DOMALSKI and G. T. ARMSTRONG, *J. Res. Nat. Bur. Stand.* 1967 **71A** 307.
779. G. P. DUDCHIK, O. G. POLYACHENOK and G. I. NOVIKOV, *Russ. J. Phys. Chem.* 1971 **45** 409.
780. A. S. DWORKIN and M. A. BREDIG, *High Temp. Sci.* 1971 **3** (1) 81; *J. Phys. Chem.* 1968 **72** 1277.
781. J. M. ELDRIDGE, E. MILLER and K. L. KOMAREK, *Trans. Met. Soc.* 1965 **233** 1303; 1967 **239** 570.
782. V. N. EREMENKO *et al., Russ. J. Phys. Chem.* 1971 **45** 1129.
783. —— no copy——
784. A. D. FINOGENOV, *Zh. fiz. Khim.* 1971 **45** (6) 900.
785. G. C. FITZGIBBON and C. E. HOLLEY, *J. Chem. Eng. Data* 1968 **13** 63.
786. A. E. FLOTOW *et al., J. Chem. Phys.* 1969 **51** 583.
787. J. I. FRANCO and H. J. KLEYKAMP, *Ber. Bunsenges. Phys. Chem.* 1971 **75** 934; 1972 **76** 691.

788. D. R. FREDERICKSON and M. C. CHASANOV, *J. Chem. Thermodyn.* 1971 **3** 693.
789. D. R. FREDERICKSON *et al.*, *High Temp. Sci.* 1970 **3** (3) 299.
790. J. FUGER, D. BROWN and J. F. EASY, *J. Chem. Soc.* (A) 1969; *Inorg. Phys. Theor.*, p. 2995.
791. G. GATTOW and B. KREBS, *Z. anorg. allg. Chem.* 1963 **322** 113.
792. L. M. GEDANKSY, G. L. BETRAND and L. G. HEPLER, *J. Chem. Eng. Data* 1967 **12** 135.
793. V. P. GLYBIN and R. B. DOBROTIN, *Russ. J. Phys. Chem.* 1971 **45** 898.
794. N. A. GOKCEN and S. FUJISHIRO, *Trans. AIME* 1963 **227** 542.
795. O. S. GORELKIN *et al.*, *Russ. J. Phys. Chem.* 1972 **46** 431.
796. L. B. PANKRATZ and W. W. WELLER, US Bur. Mines, Rep. Inv. No. 7280, 1969.
797. P. GROSS and C. HAYMAN, Fulmer Res. Inst., Private communications, 1970–1973.
798. P. GROSS, C. HAYMAN and M. C. STUART, *Trans. Faraday Soc.* 1969 **65** 2628, 2856.
799. S. K. GUPTA, *J. Phys. Chem.* 1969 **73** 4086.
800. *Handbook of Chemistry and Physics*, 49th edition, Chemical Rubber Co., 1968–9.
801. H. HARALDSEN, *Z. anorg. allg. Chem.* 1939 **240** 337.
802. A. V. HARIHARAN and H. A. EICK, *High Temp. Sci.* 1971 **3** 123.
803. L. G. HEPLER, *J. Am. Chem. Soc.* 1958 **80** 6181.
804. Reference not used.
805. H. HOLLECK and P. H. KLEYKAMP, *J. Nucl. Mat.* 1970 **35** 158.
806. C. E. HOLLEY JR. and E. K. STORMS, *Thermodynamics of Nuclear Materials, 1967*, pp. 387–426, IAEA, Vienna, 1968.
807. Reference not used.
808. E. J. HUBER, G. C. FITZGIBBON and C. E. HOLLEY, *J. Chem. Thermodyn.* 1971 **3** 643.
809. E. J. HUBER, C. E. HOLLEY and W. G. WHITEMAN, *J. Chem. Thermodyn.* 1969 **1** 579.
810. K. T. JACOB and C. B. ALCOCK, *J. Am. Ceram. Soc.* 1975 **8** 192.
811. K. T. JACOB and C. B. ALCOCK, *Rev. Int. Hautes, Temp. Refract.* 1976 **13**.
812. K. T. JACOB and C. B. ALCOCK, *Met. Trans. B*, 1975 **6B** 215.
813. K. T. JACOB and C. B. ALCOCK, unpublished research, 1975.
814. K. T. JACOB and J. C. CHAN, *J. Electrochem. Soc.* 1974 **121** 534.
815. K. T. JACOB, *Thermochimica Acta* 1976 **15** 76.
816. K. T. JACOB and C. B. ALCOCK, *High Temp.–High Press.* 1975 **7** 43.
817. *JANAF Thermochemical Tables*, Michigan, US Department of Commerce/National Bureau of Standards/Institute for Applied Technology, 1965–68, Juni 1971. Supplements 1974, 1975.
818. G. K. JOHNSON *et al.*, *J. Chem. Phys.* 1969 **1** 89.
819. I. JOHNSON and H. M. FEDER, *Trans. Met. Soc. AIME* 1962 **224** 468.
820. G. P. JONES, private communication, 1973.
821. A. S. JORDAN, *J. Electrochem. Soc.* 1971 **118** 1362.
822. M. KH. KARAPYET'YANTS and M. L. KARAPYET'YANTS, *Osnovye termodynamicheskie konstanty, neorganicheskikh i organicheskikh veshchestv,* Khimiya, Moskau, 1968.
823. I. Y. KASHKOVLI and Z. A. MUNIR, *High Temp. Sci.* 1972 **4** 82.
824. O. V. KEDROVSKI, I. V. KOVTUNENKO, E. V. KISELEVA and A. A. BUNDEL, *Zh. fiz. Khim.* 1967 **41** 414.
825. I. G. KERIMOV *et al.*, *Russ. J. Phys. Chem.* 1972 **45** 1118.
826. E. G. KING, A. D. MAH and L. B. PANKRATZ, *INCRA-Series on Metallurgy of Copper*, Vol. 2, 1973, Copper Res. Assn. and US Bur. Mines.
827. E. P. KIRPICHEV *et al.*, *Russ. J. Phys. Chem.* 1971 **45** 860.
828. H. KLEYKAMP, *Z. phys. Chem. Ffm.* 1969 **67** 277.
829. O. J. KLEPPA and A. NAVROTSKY, *J. Inorg. nucl. Chem.* 1968 **30** 479.
830. B. S. KOGAN, *Russ. J. Phys. Chem.* 1971 **45** (3) 405; (6) 891.
831. N. I. KOLBIN and V. M. SAMOILOV, *Russ. J. Inorg. Chem.* 1969 **14** 330.
832. V. P. KORNYEYEV, V. B. GLUSKOVA and E. K. KELEV, *Izv. Akad. Nauk Neorg. Mat.* 1971 **7** 886.
833. V. N. KOSTRYUKOV and I. N. KALININKINA, *Russ. J. Phys. Chem.* 1964 **38** 422.
834. K. S. KRASNOV and N. V. KAROSEVA, *Opt. Spektroskopiya* 1965 **19** 30; *Opt. Spectry.* 1965 **19** 14.
835. H. J. KRISCHING, K. PLIETH and I. N. STRANSKI, *Z. Kristallographie* 1955 **106** 172.
836. O. L. KRUGER and H. SEWAGE, *J. Chem. Phys.* 1968 **49** 4540.
837. PAN KUAN, *J. Chinese Soc. Taiwan,* Ser. II, 1954 **1** 1.
838. P. KUBASCHEWSKI and H. NÖLTING, *Ber. Bunsenges. Phys. Chem.* 1973 **77** 70.
839. O. KUBASCHEWSKI and G. HEYMER, *Trans. Faraday Soc.* 1960 **56** 473.
840. O. KUBASCHEWSKI, *High Temp.–High Press.* 1972 **4** 1; NPL–DCS Rep. 10, 18, 21 (1971–1973).
841. F. A. KUZNETSOV and T. N. REZUKHINA, *Zh. fiz. Khim.* 1960 **34** 1010; 1961 **35** 956.
F. A. KUZNETSOV, A. N. REZUKHINA and O. GOLUSHENKO, *Zh. fiz. Khim.* 1960 **34** (9); 1962 **36** 1364.
842. V. Y. LEONIDOV, T. N. REZUKHINA and I. A. BEREZNIKOVA, *Zh. fiz. Khim.* 1960 **34** 1562, 1862.
843. A. I. LEONOV *et al.*, *Russ. J. Inorg. Chem.* 1969 **14** 334.
844. A. S. L'VOVA and N. N. FEODOS'EV, *Russ. J. Phys. Chem.* 1965 **39** 1091.
845. G. M. LUKASHENKO and R. I. POLOTSHAYA, *Russ. J. Phys. Chem.* 1971 **45** (9) 1358.

846. W. G. LYON, E. F. WESTRUM and M. CHAVRET, *J. Chem. Thermodyn.* 1971 **3** 571.
847. A. C. MACLEOD, *J. Inorg. Nucl. Chem.* 1971 **33** 2419.
848. A. D. MAH, US Bur. Mines, Rep. Invest. No. 6727, 1966.
849. J. F. MARTIN, F. MÜLLER and O. KUBASCHEWSKI, *Trans. Faraday Soc.* 1970 **66** 1065.
850. S. U. MARTOSUDIRDJO and J. N. PRATT, *Thermochim. Acta* 1974 **10** 23.
851. O. P. MČEDLOV-PETROSJAN, V. I. BABUŠKIN and G. M. MATVEV, *Thermodynamik der Silikate*, VEB-Verlag für Bauwesen, Berlin, 1966.
852. G. M. MEHROTRA, M. G. FROHBERG, P. M. MATHEW and M. L. KAPOOR, *Scripta Metall.* 1973 **7** 1047.
853. B. T. MELEKH, *et al.*, *Russ. J. Phys. Chem.* 1971 **45** (8) 1144.
854. K. C. MILLS, *Thermodynamic Data for Inorganic Sulphides, Selenides and Tellurides*, Butterworths, London, 1974.
855. K. C. MILLS, *J. Chem. Thermodyn.* 1972 **4** 903.
856. K. C. MILLS, *High Temp.–High Press.* 1972 **4** 371. Private communication (1975) $\Delta H^\circ_{298}\langle CrS\rangle$.
857. F. MÜLLER and O. J. KLEPPA, *J. Inorg. Nucl. Chem.* 1973 **35** 2673.
858. V. G. MURADOV, P. V. GELD and P. V. KOCHEROV, *Zh. fiz. Khim.* 1967 **41** 576.
859. M. NAGAMORI and F. HABASHI, *Metall. Trans.* 1974 **4** 523.
860. Nat. Bureau of Standards: NSRDS–Sulfates, 1966, 30 Carbonates, 1969.
861. National Bureau of Standards, *Technical News Bulletin*, January 1971. CODATA Bulletin, No. 10, December 1973. CODATA Special Report, No. 3, September 1975.
862. J. M. NEIL, A. NAVROTSKY and O. J. KLEPPA, *Inorg. Chem.* 1971 **10** 2076.
863. B. G. NEWLAND and R. A. SHELTON, *J. Less-Comm. Met.* 1970 **20** 245.
864. D. B. NOVOTNY and J. F. SMITH, *J. Inst. Met.* 1963 **92** 122.
865. F. L. OETTING, *Chem. Rev.* 1967 **67** 261.
866. I. R. OGREN, N. J. MAGNANI and J. F. SMITH, *Trans. Met. Soc. AIME* 1967 **239** 766.
867. P. A. G. O'HARE and A. C. WAHL, *J. Chem. Phys.* 1972 **56** 4513, 4516.
868. G. ONDRACEK, K. SPLICHAL and H. WEDEMEYER, *Sprechsaal* 1975 **108** 55.
869. A. A. OPALOWSKI *et al.*, *Russ. J. Phys. Chem.* 1971 **45** 1197.
870. H. OPPERMANN, *Z. anorg. allg. Chem.* 1967 **356** 1.
871. L. B. PANKRATZ and W. W. WELLER, US Bur. Mines, Rep. Invest. Nos. 6687, 1965; 7280, 1960.
872. I. E. PAUKOV, L. M. KHRIPLOVICH and A. B. POPOV, *Russ. J. Phys. Chem.* 1970 **44** 309.
873. M. PFEIFFER and H. J. SPANGENBERG, *Z. phys. Chem. (Leipzig)* 1966 **232** 47.
874. B. PREDEL and D. ROTHACKER, *Thermochimica Acta* 1970 **1** 477.
875. B. PREDEL and H. RUGE, *Thermochimica Acta* 1972 **3** 411.
876. See Refs. 23–27.
877. A. V. R. RAO and V. B. TARE, *Scripta Met.* 1971 **5** 807.
878. B. S. RAO and V. V. DADAPE, *High Temp. Sci.* 1971 **3** 1.
879. R. RASCH, *Sprechsaal* 1970 **103** 117.
880. H. RAU, *J. Chem. Thermodynamics* 1975 **7** 27.
881. H. RAU, *J. Less-Comm. Metals* 1971 **24** 291. *Ber. Bunsenges. Phy. Chem.* 1967 **71** 716.
882. L. A. REZNITOKII *et al.*, *Russ. J. Phys. Chem.* 1971 **45** 529.
883. T. N. REZUKHINA, Thesis, Moscow, 1968.
884. T. N. REZUKHINA and B. S. POKAREV, *J. Chem. Thermodyn.* 1971 **3** 369.
885. R. RICHELMI and M. LAFFITTE, *Compt. rend.* 1967 **265** 541.
886. W. A. ROTH, H. BERENDT and G. WIRTHS, *Z. Elektrochem.* 1941 **47** 185.
887. H. SAID, R. CASTANET and H. V. KEHIAIAN, *J. Less-Comm. Metals* (in Press).
888. V. V. SAMAKHVOL *et al.*, *Russ. J. Phys. Chem.* 1971 **45** (8) 1171.
889. M. I. SANDAKOVA *et al.*, *Russ. J. Phys. Chem.* 1971 **45** 901.
890. H. SCHÄFER, H. BRUDERRECK and B. MORCHER, *Z. anorg. allg. Chem.* 1967 **352** 122.
891. H. SCHÄFER, A. TEBBEN and W. GERHARD, *Z. anorg. allg. Chem.* 1963 **32** 41.
892. H. L. SCHICK, *Thermodynamics of Certain Refractory Compounds*, Academic Press, New York, 1966.
893. K. SCHWITZGEBEL, P. S. LOWELL, T. B. PARSONS and K. J. SLADECK, *J. Chem. Eng. Data* 1971 **16** (4) 418.
894. T. M. SEVASTYANOVA and N. V. KARPENKO, *Russ. J. Phys. Chem.* 1971 **45** (7) 1044.
895. A. A. SHCHUKAREV, I. N. BALAKIREVA and A. A. VOSKRESENSKI, *Zh. obzhchei. Khim.* 1961 **31** 1773.
896. A. A. SHIDLOVSKI, I. N. BALAKIREVA and A. A. VOSKRESENSKI, *Russ. J. Phys. Chem.* 1971 **45** 1060.
897. E. SIBBING and H. SCHÄFER, *Z. anorg. allg. Chem.* 1974 **410** 67.
898. N. N. SIROTA and E. M. SMOLYARENKO, *Invest. Akad. Nauk SSSR* 1968 (2) 234.
899. W. H. SMITH, General Electric Res. Lab., Rep. No. 57-RL-1816, Oct. 1957.
900. K. E. SPEAR, H. SCHÄFER and P. W. GILLES, *3rd Intern. Symposium on High Temperature Technology, 1967 (Asiloamov in Pacific Grave, California) USA*, Butterworths, London, 1967.
901. F. H. SPEDDING and D. C. HENDERSON, *J. Chem. Phys.* 1971 **54** 1276, 2476.
902. P. J. SPENCER, *The Thermodynamic Properties of Silicates and Alkali Metal Alloys*, Teddington, NPL, DCS Reports 21 and 10. Private communication, 1976, ΔH°_{298} Fe_3Mo_2.
903. R. D. SRIVASTAVA and M. FARBER, *J. Phys. Chem.* 1971 **75** 1760.

904. L. A. STRETZ and R. G. BATISTA, *Metall. Trans.* 1974 **5** 921.
905. D. R. STULL, D. L. HILDENBRAND, F. L. OETTING and G. C. SINKE, *J. Chem. Eng. Data* 1970 **15** (1) 52.
906. I. V. TANANAEV, V. K. GOROKHOV, N. M. SELINANOVA and N. K. BOLSHAKOVA, *Neorg. Mater.* 1968 **4** 2124.
907. *Termicheskie konstanty veshchestv* (ed. V. P. Glushko and V. A. Medvedev): Part 1 (1965), *O, H, Halogens, Inert Gases*; Part 2 (1966), *S, Se, Te, Po*; Part 3 (1968), *N, P, As, Sb, Bi*; Part 4, Section 1 (1970), Section 2 (1971), *C, Si, Ge, Sn, Pb*; Part 5 (1971), *B, Al, Ga, In, Tl*; Part 6, Section 1 (1972), Section 2 (1973), transition metals, Akademiya Nauk SSSR, Moscow, *Groups VIIIb, Ib, IIb.*
908. N. A. TOVOPOV *et al.*, *Handbook of Phase Diagrams of Silicate Systems,* Israel Program for Scientific Translations, Jerusalem, 1972.
909. D. SH. TSAGAREISHVILI and G. G. GVELESIANI, *Izv. Akad. Nauk, Neorg. Mat.* 1971 **4** 1878.
910. J. O. TVEEKREM and M. S. CHANDRASEKHARAIAH, *J. Electrochem. Soc.* 1968 **115** 1021.
911. V. VASANTASREE, Thesis, Imp. Coll., London, 1971.
912. V. P. VASILEV and H. GRAYSON-SMITH, *Canad. J. Res. Chem.* 1950 **2 BA** 367.
913. V. P. VASILEV *et al.*, *Russ. J. Phys. Chem.* 1971 **45** 1169.
914. Y. VIRMANI and H. A. SKINNER, Manchester Univ., private communication.
915. A. M. VOROB'EV, G. V. EVOSEEVA and L. V. ZENKEVICH, *Zh. fiz. Khim.* 1971 **45** 2650.
916. A. M. VOROB'EV and A. F. BROVIER, *Zh. fiz. Khim.* 1971 **45** (9).
917. G. F. VORONIN, *Zh. fiz. Khim.* 1971 **45** 2100.
918. D. D. WAGMAN, W. H. EVANS, V. B. PARKER, I. HALOW, S. M. BAILEY and R. H. SCHUMM, *Selected Values of Chemical Thermodynamic Properties,* National Bureau of Standards, Technical Note 270–3/7, Washington, 1968–73.
919. D. E. WILCOX, Thesis, quoted by K. J. Sladek, *J. Chem. Eng. Data* 1971 **16** 418.
920. D. E. WORK, H. A. EICK, *J. Phys. Chem.* 1970 **74** 3130.
921. V. M. ZHDANOV, *Russ. J. Phys. Chem.* 1971 **45** 1356.
922. R. BLACHNIK et al., see Ref. 762.
923. P. N. BOKERIJA *et al.*, Drop method: Sr Al_2O_4, *Acad. Sci. Georgian SSR, Communications* 1970 **58** 601.
924. D. CUBICCIOTTI and H. EDING, *J. Phys. Chem.* 1965 **69** 3621.
925. W. A. DENCH and O. KUBASCHEWSKI, Adiabatic calorimetry: iron up to 1450°C, *J. Iron Steel Inst.* 1963 **201** 140.
926. L. DENIELU *et al.*, *Compt. rend.* 1971 **272** 1855.
927. A. S. DWORKIN and M. A. BREDIG, *High Temp. Sci.* 1971 **3** 81.
928. J. L. HOLM, *Acta chem. Scand.* 1968 **22** 361.
929. C. E. KAYLOR *et al.*, *J. Am. Chem. Soc.* 1959 **81** 4172.
930. P. KUBASCHEWSKI and H. NÖLTING, Adiabatic calorimetry (25–900°C): chalcogenides, *Ber. Bunsenges. Physik. Chemie* 1973 **77** 70.
931. V. YA. LEONIDOV, J. N. REZUKHINA and I. A. BEREZNIKOVA, *Zh. fiz. Khim.* 1960 **34** 1562.
932. M. LIM and A. W. SEARCY, *J. Phys. Chem.* 1966 **70** 1762.
933. A. V. R. RAO and V. B. TARE, *Scripta Met.* 1971 **5** 807.
934. H. RAU, *Ber. Bunsenges. Phys. Chem.* 1965 **69** 731.
935. R. RICHELMI and M. LAFFITTE, *Compt. rend.* 1967 **265** 541.
936. T. N. SEVASTYANOVA and N. V. KARPENKO, *Russ. J. Phys. Chem.* 1971 **45** 1044.
937. F. W. SPEDDING and D. C. HENDERSON, *J. Chem. Phys.* 1971 **54** 2476.
938. E. E. SHPILRAIN, D. N. KAGAN and L. S. BARKHATOV, *High Temp.–High Press.* 1972 **4** 605.
939. W. T. THOMPSON and S. N. FLENGAS, *Can. J. Chem.* 1971 **49** 1550.
940. R. A. YAKOVLEVA and T. N. REZUKHINA, *Zh. fiz. Khim.* 1960 **34** 819.
941. B. YA. BEREZIN, V. YA. CHEKHOVSKOI and A. E. SHEINDLIN, *High. Temp. Sci.* 1972 **4** 473.
942. J. S. BINDFORD, J. M. STROHMEYER and T. H. HEBERT, *J. Phys. Chem.* 1967 **71** 2404.
943. C. VASILEFF and H. GRAYSON-SMITH, *Can. J. Res.* 1950 **28A** 367.
944. I. BARIN and O. KNACKE, *Thermochemical Properties of Inorganic Substances*, Springer, Berlin and Verl. Stahleisen, Düsseldorf, 1973 and 1977.
945. J. V. SMITH and D. CHATTERJI, *J. Amer. Ceram. Soc.* 1973 **56** 288.
946. D. CHATTERJI and R. W. VEST, *J. Amer. Ceram. Soc.* 1972 **55** 575.
947. D. CHATTERJI and J. V. SMITH, *J. Electrochem. Soc.* 1973 **120** 889.
948. J. J. BYERLEY and W. K. TEO, *Trans. AIME* 1973 **4** 419.
949. L. ELFORD, F. MÜLLER and O. KUBASCHEWSKI, *Z. Elektrochem.* 1969 **73** 601.
950. J. F. COUNSELL, E. B. LEES and P. J. SPENCER, *Metallurgical Chemistry*, Proc. Symp. Brunel Univ. and NPL, 1971, HMSO, 1972, p. 451.
951. F. MÜLLER and O. KUBASCHEWSKI, *High Temp.–High Press.* 1969 **1** 543.
952. O. KUBASCHEWSKI and T. G. CHART, *J. Inst. Metals* 1964–65 **93** 329.
953. H. B. BELL, J. P. HAJRA, F. H. PUTLAND and P. J. SPENCER, *Met. Sci. J.* 1973 **7** 185.
954. T. G. CHART, NPL Report, Chem. 18, August 1972.
955. N. H. SANTANDER and O. KUBASCHEWSKI, *High Temp.–High Press.* 1975 **7** 573.

956. B. PREDEL and E. ZEHNPFUND, *Z. Metallk.* 1973 **64** 782.
957. B. PREDEL and D. W. STEIN, *Z. Naturforsch* 1971 **26a** 722.
958. O. KUBASCHEWSKI and J. F. COUNSELL, *Monatsh. Chemie.* 1971 **102** 1724.
959. T. G. CHART, *High Temp.–High Press.* 1970 **2** 461.
960. F. H. HAYES, F. MÜLLER and O. KUBASCHEWSKI, *J. Inst. Metals* 1970 **98** 20.
961. B. PREDEL and W. STEIN, *Z. Naturforsch.* 1971 **26a** 722.
962. D. J. FRAY, Private communication, 1972.
963. O. KUBASCHEWSKI, NPL–DCS Rep. 2, November 1968.
964. D. A. R. KAY and A. K. MOHANTY, *Metall. Trans.* 1970 **1** 303.
965. R. CASTANET, Y. CLAIRE and M. LAFFITTE, *J. Chim. Phys.* 1971 **7–8** 1133.
966. R. CASTANET, Int. DATA Series (D), 1975.
967. H. SAKAO and J. F. ELLIOTT, *Met. Trans.* 1974 **5** 2063.
968. K. HACK, Diplomarbeit, Techn. Hochschule, Aachen, 1976.
969. E. ZALESKA, *Rocz. Chem.* 1974 **48** 195.
970. B. PREDEL and D. W. STEIN, *J. Less-Comm. Metals* 1971 **24** 159.
971. P. J. SPENCER, *Thermodynamic Properties of Alkali Metal Alloys*, NPL–DCS Report No. 10, March 1971.
972. K. H. GEIGER, Doktordissertation, Techn. Hochschule, Aachen, 1976.
973. P. J. SPENCER and F. H. PUTLAND, *J. Iron Steel Inst.* 1973 293.
974. F. MÜLLER and F. H. HAYES, *J. Chem. Thermodyn.* 1971 **3** 599.
975. K. H. GEIGER (see ref. 972).
976. C. BERGMAN *et al.*, *Rev. Int. Hautes Tempér. Refract.* 1971 **8** 205.
977. O. KUBASCHEWSKI, NPL–DCS Rep. 7, 1970.
978. L. A. ZHARKOVA and T. N. REZUKHINA, *Zh. fiz. Khim.* 1958 **22** 2233.
979. B. W. HOWLETT, S. MISRA and M. B. BEVER, *Trans. Met. Soc. AIME* 1964 **230** 1367.
980. O. KNACKE, F. MUELLER and E. VAN RENSEN, *Z. phys. Chem.* 1972 **80** 82.
981. A. KUBIK and C. B. ALCOCK, *Metal Sci. J.* 1967 **1** 19.
982. A. A. BONI, Vap. press. (275–311°K): $TiBr_4$, $Cr(CO)_6$; *J. Electrochem. Soc.* 1966 **113** 1089.
983. G. A. GRAY and R. J. SIME, Vap. press. (60–130°C): $SbBr_3$; *J. chem. eng. Data* 1965 **10** 176.
984. H. LIM and A. W. SEARCY, Torsion-effusion: CeF_3; *J. phys. Chem.* 1966 **70** 1762.

Subject Index

Accuracy of reported data 260
Activity, chemical 31 et seq
—calculation from activity diagrams 50 et seq
—coefficient 31
—mutual effect of dissolved elements 62 et seq
Adiabatic calorimetry 67, 103 et seq
Alloys, activity of components 31, 51
—for electrical contacts 252
—entropies of information 56, 188, 229, 386–412
——fusion 186
—oxidation resistance 252, 254
—thermochemical properties 386–412
Alumina
—heat of formation 76, 268
—reduction of 25
Aluminothermic reactions 212–14
Amalgams 39, 84, 87, 172
Ammonia, equilibria with 132
Assessment of standard values 221 et seq
Atomic heat (*see* Molar heat)

Beckmann Thermometer 67, 68
Benzoic acid, calorimetric standard 76
Berthelot bomb 74 et seq, 113
Boiling point method 125
—points of substances 268–322, 359–77
Bond mechanism 27, 45, 56, 187, 324
Bonnier equation 59
Brackets, use of X, 267

Calorie XI
Calorimeters, calibration 70, 76
—descriptions 67–114
Calvet calorimeter 109–11
Carbides 77, 122, 131, 189
Catalytic acceleration of reactions 83
Cells, galvanic 28, 159 et seq
Chemical potential (*see* Gibbs energy, partial molar)
Chill factors 211
Chlorination, calorimetry 79, 96–98
—of oxides 214
Clapeyron's equation 112
Clausius–Clapeyron equation 7, 45
Combustion bomb 74 et seq, 113
Concentration, units 2
Cooling calorimeter 109
—curves, evaluation 109, 112
Coordination number 202–6, 324

Covalent bonds 27, 56, 187, 324
Crucible materials in calorimetry 78, 88

Daniell Cell 18
Darken equation 58
Debye temperature 15
Decarburisation 240, 244–7
Degrees of freedom 179–81
Deoxidation constant 236–42
Dew-point method 119
Differential solution calorimeter 85, 114
Differentiation of heat content equations 12
——thermochemical equations 38
Diffusion
—thermal 133 et seq
Diphenyl ether calorimeter 102
Dissociation of alumina 25
—pressure 7, 18, 30 et seq, 116 et seq
Drop method 88–92
Duhem–Margules equation (*see* Gibbs–Duhem eqn.)
Dulong–Petit's rule 181

Effusion Method 139 et seq, 155
Electrolytes for e.m.f. measurements 157 et seq
Electromotive determination 157 et seq
—force 18, 28–30
Electronegativities 198, 204
Enthalpy (*see* Heat)
Entropy 13–17
—of evaporation 15, 185–7
——fusion 15, 186–8, 258
——reaction 14–17, 20
——solution 29, 33, 55, 265
———transformation 15, 186
———values 386–413
—and order 16, 48, 54 et seq, 185 et seq
—standard 15, 187–92
—values 268–322
Equilibrium box (van't Hoff) 17
—constant 2, 115
—diagrams (*see* Phase diagrams)
—thermodynamics 50 et seq, 232 et seq
Equipartition of energy 179–81
Estimation of entropies 187–92
——heats of formation 192 et seq
———transition 185
——molar heats 179 et seq

Eutectic systems, calculation 232
Evaporation *in vacuo* 139 et seq, 215
Excess functions 34 et seq, 50
—relationship 55

Faraday's constant 8, 18
First Law of Thermodynamics 4 et seq
Fluorides in calorimetry 79, 96
Free energy (*see* Gibbs energy)
Freezing-point depression 45, 46, 112

Gases in metals 47, 123
—molar heats 180
Gibbs energy 17–26
——calculation 262 et seq
——partial molar 28 et seq, 40 et seq, 386–412
——of reaction, values 378–84
——of solution 386–412
——standard 3, 17 et seq
Gibbs–Duhem equation 34
Gibbs–Helmholtz equation 3, 8, 17–19
Glass as electrolyte 172–5
Guggenheim polynomials 38

Haber–Born cycle 197
Halides, combustion 78
—as electrolytes 172, 176, 178
—equilibria with hydrogen 137
—heats of solution 80–83, 202, 233
—vapour pressure 116, 119, 125
Heat, units XI
—capacity (*see* Molar heat)
——deviation function 23
—of combustion 74 et seq, 113
——evaporation 7, 15, 185, 359–77
———estimation 192 et seq
——formation 4–8, 258
———of ions in aqueous solution 201
—content 4, 9–13, 88–92
—of fusion 15, 46, 111, 185, 258, 326–35
——solution 7–8, 41, 80 et seq, 202
———in regular systems 38, 50
———values 386–413
———in water 82, 202
——sublimation 204, 359–77
——transformation 11, 15, 111–13, 326–35
Heat-flow calorimetry 67
Helmholtz function 17
Helmholtz–Gibbs equation (*see* Gibbs–Helmholtz equation)
Henrian constants, values 47
Henry's law 32, 46–47
Hess's law 6, 77, 80, 83
High-speed method 108
High-temperature calorimeters 98 et seq
Homologous series 195
Hydrogen halides, aqueous 82, 201
Hydrogen sulphide, equilibria with 137

Ice calorimeter 84, 101
Initiators of combustion 77
Integration of molar heats 9 et seq, 19 et seq
—of partial molar functions 34 et seq
Interaction coefficient 62 et seq
Ions in water, heat of formation 201
Iron oxygen 33, 40, 44, 133 et seq, 239 et seq
Isoperibol calorimetry 67, 71 et seq, 98 et seq
Isopiestic method 119
Isothermal calorimetry 67, 101

Kirchoff's Law 9–11
Knudsen effusion 139 et seq, 143 et seq
Kohler equation 59

Langmuir effusion 139, 152
Latent heat of (*see* Heat of)
Legendre polynomials 39
Levitation melting 91, 154
Light-absorption cell 117
Liquid-metal calorimeter 100

Magnesium production 217
Manometers 116, 121
Margules equation 59
Mass action law 1–3, 216
Mass spectrometer 144 et seq
Maximum work function (*see* Gibbs energy)
Melting point depression 45, 46, 112
—points of substances 268–322, 326–35
——and heats of formation 193
Mercury, calorimetric solvent 84, 86
Metal-block calorimeter 89
Metallic solutions (*see* Alloys)
Methane, equilibria with 131
Miscibility gaps 53, 234–6
Molar heat 9–12, 20–24, 103–5, 179–85, 260
——estimation 179–85
——measurement 103–5, 114
——values, elements 336–9
———compounds 340–56

Nernst–Einstein equation 160
Neumann–Kopp's rule 184, 262
Newton's law of cooling 73 et seq
Nitrides in calorimetry 79
Non-regular solutions 54 et seq
Non-stoichiometry of compounds 164, 262, 269–323
Notation IX

"Öfchen" calorimeter 94
Optical pyrometer 70
Order-disorder transformation 53, 113, 188
Oxidation, resistance to 252–7
Oxides, calorimetric solvents 87
—chlorination 214
Oxygen probe 167, 169

Packing effect 55, 202–6
Parentheses, use of X, 267
Partial molar functions 29 et seq
Particle size, effect on energy content 83
Phase diagram 27, 45, 52, 193, 227, 232 et seq
——ternary 57, 235
Phosphides, dissociation pressure 121
Phosphorus valve 121
Probability, thermodynamic 49

Quartz-spiral manometer 116, 121

Raoult's Law 31, 45 et seq
Rate of evaporation 139, 215
——reaction 1, 83, 239
Reduction of oxides 25, 42–44, 133, 237 et seq, 212, 217
References to Chapters I, II, III and IV 414–22
——Chapter V and Tables A to F 423–45
Refining of iron 239–42
—lead 242–4
Refractories 134, 247 et seq
Regular solutions 50 et seq, 61, 229
Resistance thermometer 68–69
Reversible process 13
Richard's rule 186

Second Law of Thermodynamics 4, 13, 14
Sievert's law 47
Silicates 52, 87, 190, 217
Silicon monoxides, data assessment 221–6
Slag-metal equilibria 212, 236, 244
Solvents in calorimetry 81 et seq
Specific heat (*see* Molar heat)
Spinoidal decomposition 52
Standard entropies 14
——estimation 187–92
——of gases 191
———substances 187–92
Statistical thermodynamics 48 et seq
Stoichiometry (*see* Non-stoichiometry)
Structural types 324–5
Structure of substances 268–325
Sublimation points 268–323
Sulphides, combustion 78
—dissociation pressures 121–37
—heats of solution 81, 202
—reduction 137
Sulphur valve 121
System, definition 4

Temperature measurement 67–70
Ternary solutions 57, 235
Thermal diffusion 133 et seq
Thermistors 69
Thermocouples 69
Thermometers 67–70
Third Law of Thermodynamics 14–16
Tian–Calvet calorimeter 109–11
Titanium oxides 42
Toop equation 59
Torsion effusion method 155
Tracer method 140
Transformation, magnetic 11
—of substances 268–323
Transpiration technique 153
Transportation method 127 et seq
Transport reactions 218–21
Trouton's rule 185

Van Arkel process 219
Van't Hoff's Isobar 6
Vapour pressure 30 et seq, 259 et seq
——measurement 115 et seq
——values 358–77
Volume changes, heat and entropy of mixing 55, 198–200
——in solids 198–200

Water equivalent 70–71
—vapour, equilibria with 132 et seq

Zero point entropy 16